THE NEW ELEMENTS OF MATHEMATICS

UNITED STATES COAST SURVEY.

CARLILE P. PATTERSON, SUPERINTENDENT.

A QUINCUNCIAL PROJECTION OF THE WORLD

INVENTED BY C.S. PEIRCE, ASSISTANT

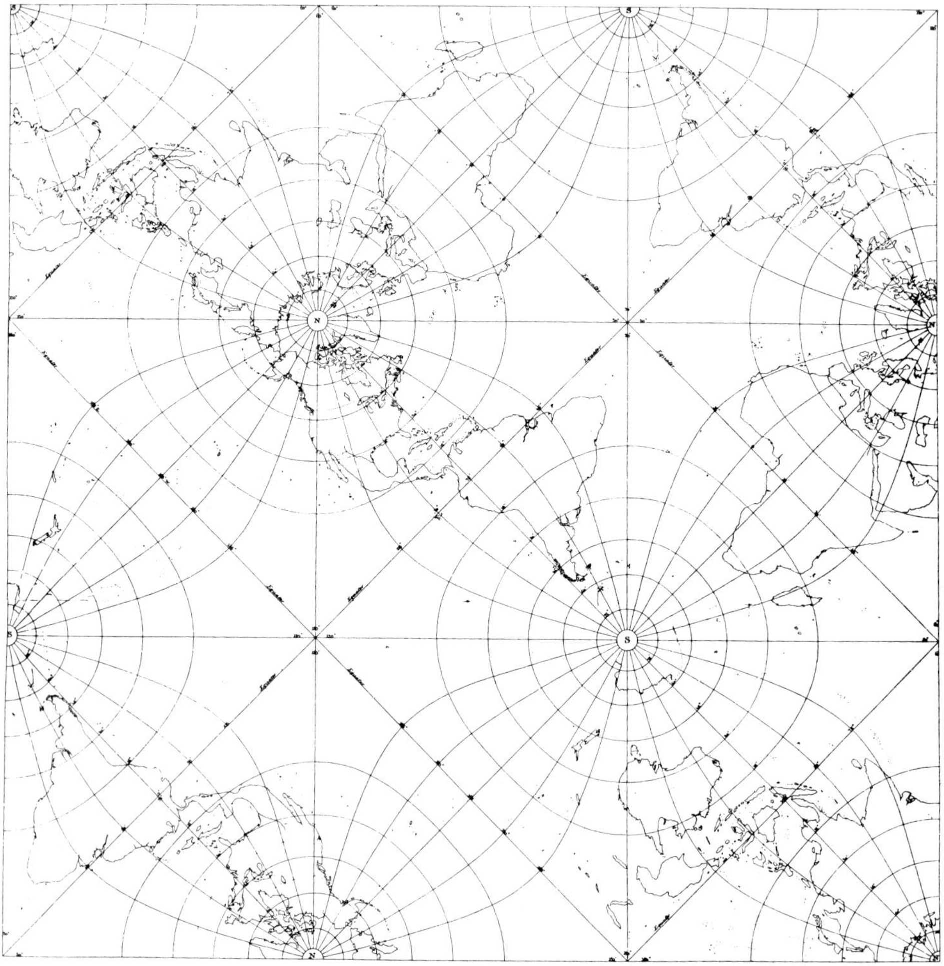

This projection depends on $\int \frac{dt}{\sqrt{1-t^4}}$, where $t = \tan \frac{1}{2} N.P.D. \odot^{\mathrm{Long}\sqrt{-1}}$,

1. The whole sphere is represented on repeating squares.

2. The part where the exaggeration of scale amounts to double that at the centre is only 7 per cent of the area of the sphere.

3. The angles are exactly preserved.

4. The curvature of lines representing great circles is, in every case, very slight, over the greater part of their length.

The Quincuncial Map Projection invented by C. S. Peirce in 1876.

(From the Charles S. Peirce Collection in the Houghton Library, Harvard University.)

THE NEW ELEMENTS OF MATHEMATICS

by

CHARLES S. PEIRCE

Edited by

CAROLYN EISELE

VOLUME III/1

MATHEMATICAL MISCELLANEA

1976

MOUTON PUBLISHERS
THE HAGUE – PARIS

HUMANITIES PRESS
ATLANTIC HIGHLANDS N.J.

ISBN 90 279 3035 X

American edition published by Humanities Press Inc. – Atlantic Highlands, N.J.

Library of Congress Cataloging in Publication Data

Peirce, Charles S.
Mathematical Miscellanea.

(His The new elements of mathematics; v. 3)
Includes index.
1. Mathematical Miscellanea. I. Title.
QA39.2.P42 vol. 3 [QA531] 510'.8s [516'.24] 76-15328
ISBN 0-391-00641-X

Printed in the Netherlands

INTRODUCTION TO VOLUME 3

In an article "On the Definition of an Infinite Number" in *The Monist*, 14 (1903-1904), G. A. Miller wrote that "the developments of mathematics have always been greatly influenced by discoveries of facts which are at variance with what was generally accepted. Of such discoveries in comparatively recent times three are especially noteworthy on account of the fundamental principles involved, namely: (1) There are perfectly consistent geometries in which the sum of the angles of a plane triangle is not equal to two right angles. (2) There are algebras in which the commutative law of multiplication does not hold. (3) There are multitudes or aggregates such that a part is equal to the whole."

Peirce was involved in a major capacity in at least two of the then recent discoveries. One might say that Peirce's general thought was so thoroughly integrated and so much of one piece that new perspectives in any scientific area would then trickle into and permeate his total thought. So it was in mathematics. An upheaval in one field brought his experimental pencil to play on all related materials. These introductory remarks will touch on a number of his experimentations in mathematical fields that were pulsing with life in his time.

THE INFINITE

Peirce believed that what he considered a major mathematical contribution on his part was to be found in his writings on the infinite which began with his paper on the "Logic of Number" (*American Journal of Mathematics*, 1881). In this paper the application of DeMorgan's Syllogism of Transposed Quantity became for him a criterion in the test for finiteness. Dedekind did not publish his *Was sind und was sollen die*

Zahlen until 1888. Therefore a contemporary logician of Peirce's stature found it necessary at an earlier time to seek for himself an analysis of the nature of the infinite since his writings were concerned with the element of continuity in the flow of time, space, thought, and experience in general. Peirce's logic of possibility is grounded in this soil, and leads him in turn to the contemplation of multiple-valued logics that brought him deep insights into what were to be twentieth-century developments.

In his *Grand Logic* of 1893, Peirce clearly stated that "About an enumerable collection certain forms of reasoning hold, which, though they had been used more or less since man began to be a reasoning animal, were first signalized in a logical work by DeMorgan in 1847, and constitute one of his claims to be the greatest of all formal logicians" (4.103).

An infinite collection, on the other hand, admits of the application of the Fermatian inference, as he called it, writing that a "numerable collection is distinguished from an innumerable one by the application to it of a certain mode of reasoning, the Fermatian inference, or as it is sometimes improperly termed, mathematical induction" (6.116). The next step would be to distinguish between the different orders of innumerability, and Peirce's writings included this development of the subject as well. He shows that the Fermatian inference is applicable in turn to a class of "multitude" *aleph-null*, the equivalent of Cantor's "Mächtigkeit" *aleph-null*. Peirce assumed throughout his work that "*to any smaller class some mode of reasoning is applicable which is not applicable to any greater one.* For greater classes," he wrote, "allow greater possibilities." Cantor's mathematical problem of "Mächtigkeit" thus came to be of extreme logical interest to Peirce. The problem of the infinite was so deeply imbedded in Peirce's general thought that notice of his mathematical approach to it was taken in the *Bulletin of the American Mathematical Society* (2nd series, January 1897) where in an account of a National Academy of Sciences meeting it is disclosed that "Mr. Peirce read the only strictly mathematical paper. Its title was 'Mathematical Infinity.'"

In the Houghton Library there is a volume that belonged to Peirce (Math. 205.34*), labelled *Peano and Cantor*. Among the papers bound together therein in January 1904 is a copy of Cantor's *Zur Lehre vom Transfiniten* (Erste Abteilung) with an inscription in the author's hand that reads: "Herrn C. S. Peirce in Milford, Pa., U.S.A., hochachtungsvollst d.V.; Halle, 9ten Juli, 1901." There is but one marginal notation in Peirce's hand while there are several in the hand of Cantor.

A paper by the editor entitled "The Mathematics of Charles S. Peirce," printed in the *Actes du XIe Congrès International D'Histoire des Sciences* (Warsaw, 1966), caught the attention of Dr. I. Grattan-Guinness who tells of a letter in the Philip Jourdain papers in which Cantor speaks of a letter from Peirce "den ich noch nicht beantwortet habe. Ich möchte zuerst wissen ob dies derselbe Peirce ist, welchern von Herrn Schröder in Karlsruhe in dessen Werken oft citirt wird." Grattan-Guinness tells further that Jourdain answered (15 July 1901) that this was the same Peirce. Apparently Cantor acknowledged Peirce's letter with the reprint just cited.

There is also a reprint of a Cantor paper from *Rivista di Matematica* (1892) translated into Italian by G. Vivanti and entitled "Sopra una questione elementare della teoria degli aggregati." A reprint of a paper by the late Cassius J. Keyser of Columbia University entitled "Concerning the Axiom of Infinity and Mathematical Induction" is also there. It had been reprinted in the *Bulletin of the American Mathematical Society* (May 1903), is inscribed to "Mr. C. S. Peirce," and carries Peirce's harsh critical marginal commentary throughout. For example, where Keyser attributes the expression "Fermatian induction" to Peirce, Peirce protests with "I never called it 'induction.'" Where Keyser attempts to answer "why," at a certain point in the thought process, "the judgment imposes itself upon us with irresistible evidence that *P* is a property of all integers," and presupposes an axiom of infinity in Poincaré's statement "C'est qu'il n'est que l'affirmation de la puissance de l'esprit qui sait capable de concevoir la répétition indéfinie d'un même acte dès que cet acte est une fois possible," Peirce scribbles two protests. One reads, "This, I daresay is true, but it is utterly irrelevant to logic; which has nothing to do with 'the mind' considered, as it here is considered, as a human faculty." The other reaction is an emotional "Oh no! Mathematics is simply deduction from general hypotheses. Now the *hypothesis*, — the Postulate, — is that every integer is attainable from 0 by successively proceeding to a *next higher*. And that is equivalent to defining the system of integers as a series for which the Fermatian inference holds." In another place he is moved to write that "A *petitio principii* is not a logical fallacy at all, but is *logically* sound." Finally Peirce uses the blank lower part of the last page to remind whomsoever might peruse this material "that there 'is' (*not* exists) a denumeral collection is shown at once by the fact that the numerals of the secundal notation are such a collection. And once granted that there is a denumeral collection, it follows that there is a collection of every abnumerable multitude." The use of "secundals," the binary notation, serves Peirce well as an instrument in handling the ever-present problem of "betweenness."

Peirce's treatment of the various grades of abnumerality has been analyzed in some detail by Murray Murphey in his *Development of Peirce's Philosophy*.

It is of interest here to note that Peirce's approach to the problem did appeal to some who could put it to practical use. For example a copy of manuscript (229) with the inexplicable "Lefevre" written in Peirce's hand in the upper left hand corner appears in the Appendix of Volume 2 in this collection. Now Arthur Lefevre, when an instructor in mathematics at the University of Texas, wrote a book entitled *Number and its Algebra* (1896) and dedicated it to "the teachers in the common schools," the work being addressed to "inquiring students and teachers." In the fourteenth section, entitled "Mathematics," Lefevre quotes from Peirce's "Law of the Mind" (*Monist*, July 1892) the paragraph that states, "I long ago showed that finite collections are distinguished from infinite ones only by one circumstance and its consequences; namely that to them (*the finite*) is applicable a peculiar and unusual mode of reasoning called by its discoverer DeMorgan, the 'syllogism of transposed quantity.'"

Lefevre then continued with his own observations on the related matter of the infinitesimal and did so in a manner reflecting his agreement with Peirce's stand in the foundations of mathematics. "In regard to infinitesimals (the word is simply the Latin ordinal form of *infinity*), and contending opinions concerning the methods of the Infinitesimal Calculus, it may be remarked that, under the true doctrine of continuity and limits, infinitesimals are presupposed, and that there can be no reason except expediency to shun them in the differential calculus. And since they are indispensable for the integral calculus, Mr. Peirce is probably right in his view of the proper procedure of the whole discipline, when he says, in the paper quoted above, 'as a mathematician I prefer the method of infinitesimals to that of limits, as far easier and less infested with snares.' At all events, any avoidance of infinitesimals as absurdities, or as offering obstacles to sound and lucid reasoning, is unnecessary."

Further evidence that Peirce's stand on this subject was well noted may be found in Volume 22 (1915) of the *American Mathematics Monthly* which carried serially through ten issues a discussion of "The history of Zeno's arguments on motion: phases in the development of the theory of limits" by Florian Cajori. Cajori notes there that "in America C. S. Peirce has adhered to the idea of infinitesimals in the declaration: 'The illumination of the subject by a strict notation for the logic of relatives had shown me clearly and evidently that the idea of an infinitesimal involves no contradiction. Apparently, before he acquired familiarity with the writings of Dedekind and Georg Cantor, C. S. Peirce had firmly recognized that for infinite collections the axiom, that the whole is greater than its part, does not hold.'"

Here, as always, one must exercise great caution to view Peirce's work in the context of the developments of the late nineteenth and early twentieth centuries. In the field of the transfinite that is especially true. The matter of inventing notation that is especially useful and productive of fruitful results always engaged Peirce's attention. One recalls that Cantor's "Beiträge zur Begründung der transfiniten Mengenlehre" was published as late as 1895-1897, and the correspondence between Cantor and Dedekind as late as 1899 on the subject of "multitudes" shows how fresh and new the conceptions of the infinite were even at that time to those most deeply concerned with its proper definition and representation. Consequently, in his paper on "Multitude and Number" (4.170) written in 1897, Peirce used the symbol  for the denumerable multitude. By 1903, however, he had adopted the aleph notation of Cantor as one can see in his lectures of that year at the Lowell Institute.

INFINITESIMALS

Charles Peirce's adamant stand for the retention of the infinitesimal concept in the foundations of the calculus reflected the early influence of Benjamin and Charles's later need of a cohesive element in his treatment of the concept of continuity. It provided that ultimate bit of "welding" material that caused individual points on a line to lose their individual identity. In the Preface to his *Plane and Solid Geometry* published first in 1837, Benjamin argues that "With all its boasted rigor, the ancient geometry can indeed lead to no result more accurate, none more to be dependend upon, than those of the infinitesimal theory; and I doubt if any well constituted mind, well constituted at least for mathematical investigation, ever reposes with any more confidence upon the one than upon the other." It was a traditional stand out of the seventeenth century, out of the calculus of Fermat and Newton and Leibniz. And it served Charles well in the groundwork of his calculus of logical possibilities. In a letter to the Editor of *Science* (Vol. 2, 16 March 1900), Peirce makes clear his stand that collections of discrete point elements on a line would never lead to the continuity he had in mind, where the elements cannot be put into one-to-one correspondence with any collection whatsoever. For him they do not constitute a collection at all. He so often speaks of the elements being "welded" together. "The fact that there is room for any multitude at every part of the line makes it *continuous.*"

When Peirce wrote the mathematical definitions for the *Century Dictionary* (1889) (as well as the terms in logic, metaphysics, astronomy, weights and measures, mechanics), his approach to this subject matter again reflected the logical satisfaction he could find in the infinitesimal base for his general theory. Simon Newcomb was most critical and a correspondence between the two men reveals Peirce's tenacious philosophy regarding the propriety of his own stand. The inability of these two minds to meet on common ground may be seen in "The Charles S. Peirce-Simon Newcomb Correspondence" by the editor in the *Proceedings of the American Philosophical Society*, 101:5 (October 1957). Peirce had explained in a letter to Professor Baldwin at Princeton that Newcomb's views were "very narrow both on the philosophical and on the mathematical side" and that he intended to try to make Newcomb's articles for that *Dictionary* (i.e. Baldwin's) "serviceable to students of philosophy."

In his letter to the Editor of *Science*, Peirce called attention to Royce's statement in *The World and the Individual* that "Mr. Charles Peirce, as I understand his statements in *The Monist*, appears to stand almost alone amongst recent mathematical logicians outside of Italy, in still regarding the calculus as properly to be founded upon the conception of the actually infinite and infinitesimal." Peirce used this observation as a springboard for the retort: "As a logician, I am more comforted by corroboration in the clear mental atmosphere of Italy than I could be by any seconding from a tobacco-clouded and bemused land (if such there be) where no philosophical eccentricity misses its champion, but where sane logic has not found favor." Recent developments in "non-standard analysis" make Peirce a modern precursor in the realm of the infinite and infinitesimal, even as he was in many other fields.

POSTULATIONAL SYSTEMS

It was, of course, Peirce's logical interest in methodology that impelled him to examine critically all aspects of the foundations in mathematics. Reference has already been made to Huntington's "Sets of Independent, Postulates for the Algebra of Logic" in the *Transactions of the American Mathematical Society*. In offering the proof of the theorem on the distributive principle, he explains in a footnote that "This demonstration is borrowed, almost verbatim, from a letter of Mr. C. S. Peirce's, dated December 24, 1903. Mr. Peirce uses the symbol $\prec$ where I have used $\oslash$, and in a slightly different sense; so that he is enabled to state that the principle here called postulate 9 follows from the definition of $P_i \prec C_i$ on p. 18 of his article of 1880. The demonstration was originally worked out for that article [*American Journal of Mathematics*, 3 (1880), p. 33], but is now published for the first time ...." Huntington continues with another letter from Peirce dated 14 February 1904, which runs as follows:

Dear Mr. Huntington:

Should you decide to print the proof of the distributive principle (and this would not only relieve me from a long procrastinated duty, but would have a certain value for exact logic, as removing the eclipse under which the method of developing the subject followed in my paper in vol. 3 of the *American Journal of Mathematics* has been obscured) I should feel that it was incumbent upon me, in decency, to explain its having been so long withheld. The truth is that the paper aforesaid was written during leisure hours gained to me by my being shut up with a severe influenza. In writing it, I omitted the proof, as there said, because it was "too tedious" and because it seemed to me very obvious. Nevertheless, when Dr. Schröder questioned its possibility, I found myself unable to reproduce it, and so concluded that it was to be added to the list of blunders, due to the grippe, with which that paper abounds, — a conclusion that was strengthened when Schröder thought he demonstrated the indemonstrability of the law of distributiveness. (I must confess that I have never carefully examined his proof, having my table loaded with logical books for the perusal of which life was not long enough.) It was not until many years afterwards that, looking over my papers of 1880 for a different purpose, I stumbled upon this proof written out in full for the press, though it was eventually cut out and, at first, I was inclined to think that it employed the principle that *all* existence is individual, which my method, in the paper in question, did not permit me to employ at that stage. I venture to opine that it fully vindicates my characterization of it as "tedious." But this is how I have a new apology to make to exact logicians.

EXISTENTIAL GRAPHS

Peirce's profound interest in topology at a time when that subject was still a fledgling probably accounts for the extensions of his inventiveness into fields closely associated with it. He invented by 1897 a system of logical diagramming that is essentially a topological conception. Peirce realized that by means of the revelation of otherwise unsuspected relationships in diagrammatic representation, hidden truths could suddenly become obvious. He felt that such graphs would teach much about the nature of mathematical reasoning and thus lead to improvements in mathematics. As an example he notes that "geometrical topics stands idle with problems to all appearance very simple staring it unsolved in the face, merely because mathematicians have not found out how to reason about it. Now a thorough understanding of mathematical reasoning must be a long stride toward enabling us to find a method of reasoning about this subject as well, very likely, as about other subjects that are not even recognized to be mathematical" (4.429).

Restrictions on the size of the present collection of Peirce's mathematical writings have severely limited the materials that have been included under the heading of existential graphs. Two doctoral theses have been produced on the subject. One is entitled "The Existential Graphs of Charles S. Peirce" by Don Davis Roberts at the University of Illinois (1963) (a revised version of which has been published by Mouton); the other, "The Graphical Logic of C. S. Peirce" by J. Jay Zeman at the University of Chicago (1964). Moreover, the eight volumes of *The Collected Papers of Charles Sanders Peirce*, edited by Charles Hartshorne and Paul Weiss, contain a good sample of the graphic representation. So much of Peirce's writing is infiltrated with the diagrammatic analysis of existential graphs, however, that no collection could be completely free of the mechanism that he considered his most important contribution to logic. Enough appears in the pages of the present collection to warrant attention to Martin Gardner's tribute to Peirce in his *Logic Machines and Diagrams*: "We must remember, however, that Peirce undertook his Gargantuan project at a time when symbolic logic was in its infancy. ... His logic graphs are still the most ambitious yet attempted, and they are filled with suggestive hints of what can be done along such lines. Peirce himself expected successors to take up where he left off and bring his system to perfection .... In the meantime it stands as a characteristic monument to one man's extraordinary industry, brilliance, and eccentricity."

Similarly, much of Peirce's mathematical logic has already been published in the eight volumes. The present mathematical collection consequently contains only that which naturally permeates Peirce's work in a general way. One must note, however, Peirce's eminent position in the field as reflected in Couturat's statement in the introduction to his *Principles of Mathematics*: "Cette fusion progressive de la Logique et de la Mathématique pure, qui a été implicitement et presque inconsciemment réalisée par les travaux de Boole, de Schröder et de Peirce, d'une part, de Weierstrass, de Georg Cantor et de Peano, d'autre part, constitue évidemment une révolution dans la philosophie des mathématiques, et par suite dans la théorie de la connaissance."

THE FOUR-COLOR PROBLEM

Peirce's belief that the existential graph held the secret to the solution of problems in topology, "to all appearance very simply staring it unsolved in the face," may be associated with his preoccupation with that unconfirmed conjecture treated recently in Oystein Ore's *The Four Color Problem* and defined by Kenneth May in his "Origin of the Four-Color Conjecture" (*Isis*, 1965) as "Any map on a plane or the surface of the sphere can be colored with only four colors so that no two adjacent countries have the same color." Each country must consist of a single connected region, and adjacent countries are those having a boundary line (not merely a single point) in common. It was a problem that came to plague Peirce over many years of trial and error.

The Johns Hopkins *University Circulars* dated January 1880 report that at a meeting of the Scientific Association on 5 November 1879, Dr. Story presented a communication by Mr. A. B. Kempe of London, "On the Geographical Problem of the Four Colors." He gave a resume of Mr. Kempe's proof that every map on a singly-connected surface can be colored with four colors, with an extension of the proof to certain cases not considered by the author. Mr. Peirce is recorded as then discussing a new point in respect to the problem "showing by methods of logical argumentation that a better demonstration of the problem than the one offered by Mr. Kempe is possible." Kempe's so-called proof was published in the *American Journal of Mathematics* (II:3) and was given a notice by Peirce in *The Nation*, 25 December 1879. Peirce announced there that "a first proof of the proposition had been given by Mr. Kempe who was well-known for his investigation into linkage," a subject of great interest to Peirce, also.

It has not been possible to retrieve Peirce's remarks made at that meeting. He spoke on the same subject some twenty years later at a meeting of the National Academy of Sciences (15 November 1899). Again no one manuscript can be pinpointed as the substance of his remarks at that time. But his writings over the years are interlaced with references to the problem; his notebooks are full of sketches and diagrams of various regional possibilities reflecting his continual experimentation and frustration. His correspondence with W. E. Story during the autumn of 1899 reflects his concern with the problem at that time. The draft of an incomplete letter written on 17 August tells of his having come across an old proof of his that only four colors are required to color a map on a spheroidal surface. Peirce was still discussing the solution in terms of Listing's theorem and the periphraxis of a surface on 29 December 1899. An amusing overtone appears in the following letter from William Story to Peirce.

17 Hammond Street,
Worcester, Mass.
Dec. 1, 1900

My dear Peirce:

Last week I sent you by express (Wells, Fargo) to Port Jervis, N.Y. a volume of *Acta Mathematica* containing French translations of G. Cantor's papers in the *Math. Annalen* and elsewhere on Punktmengen. I thought you would prefer to have these papers all in one volume rather than in half a dozen. If you should prefer the German versions, I will send them to you. There are still three of Cantor's papers on the subject not contained in the volume of the *Acta*; namely, No. 6. of the papers:

"Ueber unendliche, lineare Punkten annichfaltigkeiten," *Math. Ann.*, vol. 23, pp. 453-488; and two papers entitled:

"Beiträge zur Begründung der transfiniten Mengenlehre," *Math. Ann.*, vols. 46 and 49; do you want these? If so, let me know, and I will send them. The two volumes of the *Math. Ann.* last mentioned are not in my library, but I can send you the University copies, as I did that of the volume of *Acta Math.*, and you can keep them as long as you please, unless some unforeseen demand for them is made.

As to my not answering your letter about the four-color problem, I am heartily tired of that subject. I have spent an immense amount of time on it, and all to no purpose. Your *first* method had occurred to me years ago, but I did not succeed in getting anything out of it.

Dec. 6, 1900

My delay in sending this off is largely your own fault. You have again reminded me of that fascinating but delusive problem, and I have spent the time since writing the above in trying to solve it, but alas! I believe that the case of exception to Kempe's method requires that the map shall have at least one triangular or quadrilateral district, in which case the pentagon is not the next district to be colored, i.e. the exception does not occur. But I cannot prove it. Some condition is wanting to define the possible cases. Namely, a map having 12 pentagonal and 1 hexagonal district, 33 edges, and 22 vertices in each of which meet 3 edges, satisfies all of Kempe's conditions, and yet no such map exists.

Please give my regards to Mrs. Peirce.

Sincerely Yours,
William E. Story

If one looks into the logic notebook (MS. 339) in which Peirce recorded his first thoughts over the years as he went along, one finds that in 1869 he was creating symbols for his logic of relations beginning on p. 46r. By the time he reached p. 70r, he had developed a reliable calculus for their use. The heading on this page reads: "Example of an application of the forms l^x." The example stated is: "How many colors are required to paint a map?" This is the evidence that as early as October 1869 Peirce attempted a demonstration of the problem expressed in his early notation of the logic of relatives. Peirce's claim that soon after 1860 he offered to the mathematical society at Harvard University a proof of the proposition, which was not challenged at the time, could therefore well be substantiated.

Peirce's handling of this problem, his experimentations with Cayley's trees, with Kempe's linkages, his occasional bouts with Tait's knots, his own existential graphs are all of one piece — all are reflections of his basic interest in the then dormant subject matter of topology, and the work of Listing. For it was in this area that he sought the key to the ever-present problem of the continuum. But a new motivation is found in the four-color struggle. For in his application to the Carnegie Foundation in 1902 for funds to complete at last his great logic, Peirce reveals that he was testing over the years the progress of his own work in heuretic mathematical thought by using the advancement of his own skill in handling the four-color problem as a "landmark."

N-VALUED LOGIC

The problem of the mathematical continuum, as we have seen, was the center from which much of Peirce's thought radiated and he claimed that his great work consisted in carrying the idea of continuity into all parts of philosophy. He has given us 1884 as the date of his first acquaintance with Cantor's work and thus his "Logic of Number" of 1881 could not have been influenced by Cantor. Unlike Cantor he pursued by means of cardinal numbers only the character of a collection which he called its "Multitude." He noted in 1902 that he so pursued the problem to the very end, while Cantor switched off to ordinal numbers. For Peirce, multitudes corresponded to a linear series of objects and one does not use the forms of pure mathematics in investigating them but uses a branch of logic which is directly dependent on mathematics. This is the *logic of substantive possibilities.* He notes as an illustration that the variety of qualities exceeds not only all number but all multitude, finite or infinite. Qualities are general respects in which existing things might agree or differ. They are mere possibilities. They have themselves general respects in which they agree or differ. Again, the idea of human knowledge is very imperfectly realized as long as it is confined to existent individuals; so that the nature of a science is not radically altered until it becomes a study not of existent collections but of classes of possibilities. "Every understanding of experience," he wrote, "the like of which all our useful knowledge is composed relates to an endless series of possibilities." Peirce was not certain that his infinite series of infinite multitudes were the same as Cantor's *alephs*. He called the first multitude denumerable, and the others abnumerable. For him the aggregate of all abnumeral collections leads at last to so crowded a field of possibility that the units of the aggregate lose their individual identity. The aggregate ceases to be a collection and becomes a continuum. In this investigation the binary or secundal notation became for him the key to the notion of betweenness.

Time offered him an example of a primitive continuity and he usually associated distances of points on a line segment with instants in an interval of time to demonstrate that the multitude of instants between two limits of analytic time is the same as the multitude of all possible collections of whole numbers. The instants come to be *welded* together. "I cannot see," he wrote in 1902, "that Cantor has ever got the conception of a true continuum, such that in any lapse of time there is room for any multitude of instants however great."

In permitting himself to be led to a supermultitudinous collection in this manner, beyond all the *alephs*, he held that everything is in reality "welded" together. For Peirce the aggregate of all singulars becomes what we ordinarily call the *truth*, and carried to its greatest possibility becomes continuous. The continuum according to him becomes the true universal. Hence in science, his inductive methodology provides for the passage from the characters of single things to the study of characters of classes of things.

As an Associate Editor of the *Century Dictionary* Peirce received an interleaved copy of the first printed edition before it reached the public and he took the opportunity to modify on the interleaving many of the definitions already printed therein. On 18 September 1903, he wrote a note that reads in part: "Kant's real definition implies that a continuous line contains no points. Now if we are to accept the common-sense idea of continuity ... we must either say that a continuous line contains no points or we must say that the principle of excluded middle does not hold of these points. The principle of excluded middle only applies to an individual .... But places being mere possibles without actual existence are not individuals. I think that Kant's definition correctly defines the common-sense idea, although there are great difficulties with it."

When Peirce spoke of the great difficulties he was referring to the heart of his mathematical concern at the turn of the century. His Dedekindian belief in the inclusion of limiting points themselves in the line continuum is of special interest because he never relinquished his belief in the need for the infinitesimal concept in a context where points clustered about limits and piled up in density so as to merge at last, making of the limit a general. As generals, the principle of excluded middle does not apply to them. The line continuum is no longer a collection. It no longer contains discrete points.

It follows in the same way that were a proposition to be false up to a certain instant and thereafter to be true, at that instant it would be both true and false. Peirce finds that a proposition once true does not necessarily always remain true. "It only follows," he says, "that it remains true through a denumeral series of instants, which is a lapse of time inexpressibly less than any sensible or assignable time, wanting as it does most of the characteristics of duration."

Peirce fusses with the same boundary problem in his topological studies. Boundary problems bring to light this third kind of existence and possibility where an element on the boundary is neither this nor that, or perhaps a bit of both. Peirce tried to state in non-metrical terms that if a series of points up to the limit is included in a continuum, the limit is included. Perhaps this accounts for his deep non-metrical interest in topology. He had criticized Cantor for making his work depend on "metrical considerations."

When the light of trivalency is permitted to shine on Peirce's writings his total thought takes on new overtones. In his preliminary Lowell lecture of 1903 he speaks of the simplest kind of mathematics with only two different values. "There would be an interesting system with three values, which I have slightly examined." In another draft he speaks of "a Mathematics of a System of Three Values which would not be without utility and which has been in some measure developed. The theory of numbers furnishes partial developments of the mathematics of every system having a finite multitude of values." In the section on *n*-valued logic he will be found to remark that "Of these logics of finite systems, only that of the system of three values seems likely to repay study."

Professor Max Fisch has discovered in Peirce's unpublished Logic Notebook evidence of Peirce's formalization in his own mind by 23 February 1909 of a three-valued logic with a matrix treatment similar to that of the two-valued system. Next to the first appearance of the three-valued matrix Peirce was moved to write: "All this is mighty close to nonsense." But three pages beyond this he writes with confidence: "Triadic Logic is that logic, which, though not rejecting entirely the Principle of Excluded Middle, nevertheless recognizes that every proposition, *S* is *P*, is either true, or false, or else has a lower mode of being such that it can neither be determinately *P*, nor determinately not *P*, but is at the limit between *P* and not *P*." Professor Fisch and Professor Atwell Turquette together published a revealing paper on this subject in the *Transactions of the Charles S. Peirce Society* (II:2 [1966]) and Turquette has published additional papers on the subject in the *Transactions* since then (III:2 [1967] and V:4 [1969]). The editor of this collection of mathematics papers is of the opinion that as a result of his attempted mathematical analysis of the nature of the continuum in terms of logical possibility Peirce developed the need for his three-valued logic which gave him the elements of a firm Triadic Logic by 1909.

PROFESSIONAL CORRESPONDENCE

Peirce suffered much in those later years from lack of regular professional contact with mathematicians such as he had enjoyed earlier at the Johns Hopkins University. He carried on an intermittent correspondence with some of America's best mathematicians. Through H. B. Fine at Princeton he was able to borrow the 46th and 47th volumes of the *Mathematische Annalen* from the Princeton Library for the Cantor items. Fine respected Peirce's opinions and part of their correspondence centered on Peirce's objection to Fine's definition of continuity in *Baldwin's Dictionary of Philosophy and Psychology*.

Although many items in the mathematical correspondence are referred to throughout this collection in footnote and introduction, yet two particular items are particularly pertinent here. One of them is a letter from Fine to Peirce (14 January 1901) at a time when Fine was preparing a textbook in algebra for Ginn and Company through Plimpton and he was eager to send a copy to Peirce. He continued in his letter:

> I should value your criticisms above those of any one whom I should know.
> I am mighty glad you have made such a critical study of Cantor. Why don't

you write up your demonstration of the existence of abnumeral multitudes (or transfinite numbers) — the one you outline in your letter to Baldwin — and at the same time your critique of Cantor and publish it in the *Bulletin* or the *Transactions* of the American Math. Soc. It would have a lot of interested readers, Cantor himself among them. I am ashamed to say that I have so far fogotten Cantor's Menge that I cannot now compare it with yours — which I was greatly impressed by. It gave me a realization of the mathematical powers of the logic of relatives which I never had before ...

I hope sometime to have in extenso your views on the foundations of topology; they would interest me tremendously.

I think the development of the fractions very interesting and I have never seen either of them before. I was particularly interested in the second as a method of systematically inserting other fractions between the successive integers and the successive unit fractions. And I should very much like to know more about the treatment of the thickness with which fractions are spread ...

On 28 January Fine again praised Peirce's efforts with the words, "I have read your unfinished paper on fractions ... with a great deal of interest. It brings a lot of curious relations to light and the detail to which you have carried the investigation is simply astonishing." He suggested that Peirce write a "brief statement of principles of arrangement and its consequences" for the *Transactions* or the *Bulletin of the American Mathematical Society* or for the *American Journal of Mathematics.* E. H. Moore and Thomas Fiske had likewise constantly urged Peirce to formalize his mathematical thought in papers for publication in the mathematical journals. C. J. Keyser was similarly attracted to Peirce's mathematical philosophy and was responsible for Peirce's contributions to the *Hibbert Journal.*

The other item of correspondence that must be mentioned here associates Peirce with the great J. Willard Gibbs. Peirce had written to Gibbs on 28 November 1881 regarding usage of notation in the then new Vector Analysis. For he had been sent a gift copy of the privately printed pamphlet *Elements of Vector Analysis* (1881-1884) by Gibbs which is in the Harvard College Library today, having been given as a gift by Mrs. Peirce on 28 June 1915. Their paths surely had crossed at the Johns Hopkins University where Gibbs had lectured on Mechanics in 1880 and had addressed the Scientific Association "On Notations for Vectors." A modest, courteous reply to Peirce's letter dated 1 December 1881 in Gibbs' hand reposes as L162 in the Collection of Peirce Correspondence today. It opens with an acknowledgment of Gibbs's gratification that Peirce "found sufficient interest in the pamphlet to criticize it" and to communicate his criticism. He continued, "I feel quite in the dark in respect to the question of the best signs, and am glad to get all the light

I can on that point. And there is no one whose opinion I should value more than yours." Indeed in *A History of Vector Analysis* by Michael Crowe, the draft of a letter from Gibbs to Victor Schlegel is quoted in which Gibbs in speaking of the pamphlet said: "This I ultimately printed but never published, although I distributed a good many copies among such persons as I thought might possibly take an interest in it. My delay and hesitation in this respect was principally due to difficulty in making up my mind in respect to details of notation, matters trifling in themselves, but in which it is undesirable to make unnecessary changes."

Despite his early interest in quaternions, Peirce was quick to recognize the superiority of Gibbs's vector methods and in his review for *The Nation* (24 January 1907) of the *Scientific Papers of J. Willard Gibbs* Peirce admits that "We can only say that the ease and mastery with which his scholars have handled some of the most thorny problems of physics, as contrasted with the infertility of the quaternionists, incline us to put our trust in 'vector analysis.'" Rebellion on the part of Peirce in this matter was hinted at in an early undated letter to Jem (December 1877, January 1878, according to Max Fisch) in which Peirce said, "I am glad you are making an introduction to quaternions. I hope it is going to be severely practical because it appears to me that what such people as I for instance want is to be taught the fewest number of principles which will launch us into the practical application of quaternions. Quaternions may be as fine as you please, but if it is necessary to hold a vast deal in the memory in order to work *with* it, for ordinary mathematicians it will be but a tinkling cymbal. I greatly wish that a good practical book on the application of it to Mechanics could be made. But it is surprising how little practical even elementary books are."

FINITE DIFFERENCES

Peirce's respect for the mathematical skills of William Story, as has been noted, is in evidence throughout his writings. Long after he had retired from active service in the Coast Survey where his greatest need for the techniques of finite differences had developed he was moved to write the following letter to Story.

Milford Pa.
1909 Jan 26

Dear Story:

How much longer must we wait for the long expected treatise on Finite Differences, which greatly concerns Logic?

I think it would be interesting to have a chapter comparing the different ways of reasoning on the subject critically.

Just now looking into Boole's book of which I only have the first edition, I see that with the notation $x^{(m)}$ to mean $x(x-1)(x-2)\ldots$ to m factors, he very absurdly writes $x^{(-m)}$ for $\frac{1}{x}\,\frac{1}{x+1}\,\frac{1}{x+2}$ to m factors. According $x^{(+0)}=1$ and $x^{(-0)}=\infty$. He ought to make $x^{(-m)}$ denote $\frac{1}{x+1}\,\frac{1}{x+2}$ etc. to m factors when $x^{(-0)}=x^{(+0)}=1$. For then $x^{(-m)}=(x+m)x^{(1-m)}$. So that since $x^{(-1)}=\frac{1}{x+1}$, $x^0=\frac{x+1}{x+1}=1$. He wanted to have the formula $\Delta_x x^{(-m)}=-m x^{(-m-1)}$.

But that holds just the same in either case

$$x^{(-3)}=\frac{1}{(x+1)(x+2)(x+3)}$$

$$E_x\,x^{(-3)}=\frac{1}{(x+2)(x+3)(x+4)}$$

$$\Delta_x\,x^{(-3)}=\left(\frac{1}{x+4}-\frac{1}{x+1}\right)\frac{1}{(x+2)(x+3)}$$

$$=-3\frac{1}{(x+1)(x+2)(x+3)(x+4)}$$

$$=-3x^{(-4)}$$

Here is definition which pleases me: To say that a certain event has a probability, p, means that in an endless series of independent occasions, on each of which the event either occurs or does not, if a tally be kept and a new value of the proportion of occasions on which the event has occurred since the beginning of the series be calculated at every new occasion, then, if there be one value and one value only of this proportion which never ceases to recur, while every other value will some time occur for the last time, that value is the probability of the event. Of course it is assumed that the general conditions do not alter during the whole series. The probability is the value that *would* never cease to recur if the general conditions were to remain the same. That is the only definition of probability that I have ever seen that is *definite*, not utterly *irrelevant*, and involving no *vicious circle*, by the idea of likelihood occurring in the definition itself.

Kindly convey my remembrances to Mrs. Story. It would do me good to meet her again.

Very faithfully
C. S. Peirce

[P.S.] Have you ever been elected to the Academy [?]
I have lost the run of it for the last few years.

Peirce had had need of the methods of finite differences years earlier in the reductions of his observations in astronomy and geodesy. His notebooks are filled with problems solved by finite difference techniques, showing his continuing interest and need over the years. Peirce's acknowledged competence in this field is evidenced in his 5-page review article in the *American Journal of Mathematics* 1 (1878), entitled "Esposizione del Metodo dei Minimi Quadrati per Annibale Ferrero, Tenente Colonnello di Stato Maggiore, ec. Firenze, 1876." The opening paragraph states that "Recent discussions in this country, of the literature of the method of Least Squares, have passed by without mention of the views of the accomplished chief of the geodetical division of the Italian Survey, as set forth in the work above cited, which was first published, in part, in 1871. The subject is here, for the first time, in my opinion, set upon its true value and simple basis; at all events the view here taken is far more worthy of attention than most of the proposed proofs of the method."

In the years 1859-1891 Charles Peirce was affiliated with the Coast and Geodetic Survey of the United States in a most professional way. He started as a member of a party in the field working on the triangulation of the northeast coast of the United States. He was formally appointed as an aide, 1 July 1861. He worked under his father for a time during which his experience as a computer brought him great proficiency and skill in the method of least squares, as exhibited in numerous published reports on his Coast Survey work. He became an Assistant on 1 July 1867, and worked under Winlock at the Harvard Observatory in the years 1868-1875. His astronomical investigations were the subject of his celebrated book, *Photometric Researches*, published in 1878. His was the first attempt at the Observatory to determine the form of the galactic cluster in which we find ourselves, the Milky Way. His spectroscopic researches were to lead him to determine in 1879 the meter in terms of a wave-length of light. The problem of map-projection led to his invention of a new conformal map-projection of the entire world which he called "quincuncial" because of the figuration of medians within repeated squares. He was placed in charge of the pendulum experiments of the Survey in 1872 and was responsible for the work in the measurement of gravity and the determination of the figure of the earth. He invented pendulums that are on display today at the Smithsonian Institution. As has been previously observed, his twenty-three page report entitled "De l'influence de la flexibilité du trépied sur l'oscillation du pendule à reversion" established his reputation with the International Goedetic Association. He was in charge of the Office of Weights and Measures from 1 October 1884 to 22 February 1885. In carrying out the work entailed by these various responsibilities Peirce exhibited a mathematical inventiveness that could have raised him to a Professorship in Applied Mathematics in his later years, had the circumstances of his personal life been less unfortunate. His contributions to the scientific prestige of the Survey were recognized in 1963 in the naming of a newly commissioned Coast Guard Survey vessel the USC and GSS Peirce.

Further scientific biographical detail may be found in "Charles S. Peirce — Nineteenth Century Man of Science," by the editor in *Scripta Mathematica* 24 (1959), and in "Charles S. Peirce and the Problem of Map-Projection" in *Proceedings of the American Philosophical Society* 107:4 (1963). Victor Lenzen has written on different phases of Peirce's scientific life in "The Contributions of Charles S. Peirce to Metrology," *Proceedings of the American Philosophical Society* 109:1 (1965); "Charles S. Peirce as Astronomer" in *Studies in the Philosophy of Charles Sanders Peirce*, Second Series, edited by E. C. Moore and R. S. Robin; "Development of Gravity Pendulums in the Nineteenth Century," by V.F. Lenzen and R. P. Multhauf, Smithsonian Institution, Washington, D.C. (1965); "Charles S. Peirce as Mathematical Geodesist," *Transactions of the Charles S. Peirce Society* 8:2. For a complete biographical statement and bibliography, see the account written by the editor in *The Dictionary of Scientific Biography, Volume 10*, pp. 482-488.

ECONOMETRICS

In a fragment of a letter to his first wife Zina from the Office of the United States Coast Survey in Washington (17 December 1871) Peirce wrote: "... Simon Newcomb came to see me today ... He asked after you ... I have been quite interested in political economy which I generally spend my evenings in studying. I will try my hand at explaining my views to you. 1. Political Economy I conceive treats of the relations of these three quantities, the *price* of a thing, the yearly sales of it which I term the *demand* and the cost to the ...". The letter is incomplete but serves to date the beginning of his serious concern with political economy and questions of economy on which he was to publish his views later. On that very day (17 December 1871) Peirce wrote a letter from his Coast Survey office to Simon Newcomb at the Observatory in which he discussed the "law of Supply and Demand" from an algebraic standpoint, with a "P.S. This is all in Cournot." The letter has been published in the "Charles S. Peirce-Simon Newcomb Correspondence," *Proceedings of the American Philosophical Society* 101:5 (1957), by the editor of these volumes and has been recently cited in *Precursors in Mathematical Economics: An Anthology* (1968) by William J. Baumol and Stephen M. Goldfeld of Princeton University. In the preface to the excerpt that brought Peirce this great recognition the authors remark that "It is also noteworthy that Peirce seems to have been versed in the work of Cournot at a date when this seems to have been true of very few economists. Indeed, assuming the dating of the letter (1871) to be correct, the remark about Cournot is really extraordinary. This was nearly eight years before Jevons's well-known rediscovery of Cournot's *Recherches*, and a year before Jevons had even obtained a copy of the book!"

A notation in the Coast Survey files made on 28 December 1871 mentions a meeting of the Cambridge Scientific Club at Mr. Peirce's. "Subject: Application of Mathematics to certain questions in Political Economy, as price and amount of sale. Conditions of a Maximum. Diagrams. Books on the subject. Present: Winlock, Peirce, Dixwell, Eliot, Lovering, Walker, Bowen (Lovering, Reporter)." A note from J. E. Hilgard in Washington to Benjamin Peirce (19 December 1871) informed Benjamin that "Mr. Charles Peirce took charge of the diagrams you desired made, and I believe was able to forward them today."

It seems, then, that Charles had a part in the newly developing mathematical design of economics theory, as it began to yield to the power of mathematical and statistical analysis. Since Peirce was familiar with the papers in the *Encyclopedia Metropolitana*, one naturally peruses at this point the article entitled "Political Economy" therein (vol. VI; Mixed Sciences, vol. 4, 1845) by Naussau William Senior, Esq. It yields throughout its 94 pages not one mathematical symbol. However Senior complains about the lack of accurate nomenclature. Adam Smith's *The Wealth of Nations* is said to contain scarcely a definition and Ricardo, so often cited by Peirce, is said to have deformed his *Principles of Political Economy and Taxation* "by a use of words so unexplained, and yet so remote from ordinary usage, and from that of other writers on the same subject, and frequently so inconsistent, as to perplex every reader, and not unfrequently to have misled the eminent writer himself." Fault is found with Mr. Malthus too in his *Definitions in Political Economy* and with Mr. Mill.

Yet Peirce felt that Ricardo "may be said to have rediscovere the reasoning of the differential calculus and applied it to the theory of wealth. All the so-called mathematical economists have done nothing which was not quite obvious after Ricardo's examples of analysis" (MS. 1126), and in the Metternich generation "Ricardo carried the analysis of political economy to its highest pitch, and Augustin Cournot treated the subject mathematically (as Ricardo did substantially, too) in a book whose mathematical blunders do not really affect its principal conclusions" (MS. 1123).

By 1893 Peirce must have become acquainted with Irving Fischer's important "Mathematical Investigations in the Theory of Values and Price" to which Thomas Fiske devoted an article-review in the *Bulletin of the American Mathematical Society* (June 1893). In his historical account, Fiske pointed to the early work of Cournot, the "Recherches sur le principe mathématique de la théorie des richesses" (1838), and to Jevons's "Theory of Political Economy" (1881), as well as to Marshall's "Principles of Economics" (1890). Newcomb taught a course in mathematical economics at the Johns Hopkins University and although his book *Principles of Political Economy* (1885) is, according to Baumol and Goldfeld, "rather consistently classical and shows almost no trace of the author's mathematical training," yet the very fact of the existence of course work at the Johns Hopkins University in political economy would have whetted the appetite of Charles Peirce who participated in all the academic activities while associated with that institution. In the "Outline of Remarks" made by Peirce in his introductory lecture on the study of logic early in his logic course at the University, September 1882 (*University Circulars* 19, November 1882), one finds him saying, "The scientific specialists — pendulum swingers and the like — are doing a great and useful work; each one very little, but altogether something vast. But the higher places in science in the coming years are for those who succeed in adapting the methods of one science to the investigation of another. That is what the greatest progress of the passing generation has consisted in." Among the illustrations given, Peirce speaks of Cournot adapting to political economy the calculus of variations. Peirce himself adapted the calculus of variations in his "Note on the Theory of the Economy of Research" in the *Report of the Superintendent of the United States*

Coast Survey for the Fiscal Year ending with June, 1876. Peirce's paper was reprinted recently in *Operations Research* (July-August 1967) by W. Edward Cushen in an article entitled "C. S. Peirce on Benefit-Cost Analysis of Scientific Activity."

This interest in the sphere of economics came to be closely associated with Peirce's epistemological theory, believing as he did that the truth will be revealed in the course of time and that "discovery" means that the knowledge of that truth has been expedited by some carefully selected means. Consequently, he asserted that "the art of discovery is purely a question of economics," a tenet that led him to the prolonged consideration of the economics of research which he considered "the leading doctrine with reference to the art of discovery." In MS. 326, he specifically states that "every scientific man will agree that it is good economy to expend a considerable percentage of the means one has to spend upon a given inquiry in studying out how best to pursue it." In the abductive process — the "guessing" of possible hypotheses to be tested — Peirce maintained it was always the question of economy, economy of money, time, thought, and energy. Although he disapproved of much in the philosophical stance of Mach, he did believe that "it is Professor Ernst Mach who has done the most to show the importance in logic of the consideration of Economy." One can read the extent to which the question of "economy" had permeated Peirce's thought in the following excerpt from MS. 678.

But those who talk of the passion here called "love of knowledge," and in a measure even those who speak of "love of learning," are for the moment thinking of *knowing* any given proposition as if it were as definite a state as that of having any given sum to one's credit on the books of a bank, supposing these to be perfectly kept, although they must know very well that it is far from being so. For instance every chemist who "knew" his atomic weights said in 1904 that that of tantalum was 183; but in 1909 he called it 181, and toward the end of 1910 he made it about $182\frac{1}{2}$, though nobody dreamed of there having been any change mean time. In the more exact of the physical sciences, it is now usual to affix to every determination of a value the sign $\pm$ followed by an estimate (commonly recognized as too small) of the "probable error" of the determination. It is open to doubt whether there be any single sentence whatever of whose truth any man can be absolutely certain. If there be any such, it is probably something like the following: "We cannot be sure that any exact statement of actual fact is absolutely free from error." Thus, we are not quite sure that a body moving uninfluenced by any force would continue to describe (or pass over) equal distances in equal times, unless we measure time by the distances which that body describes, in which case the statement is not one of fact, but is merely an inverted definition of what we mean by "equal times."

But this whole question will come up for consideration in due time. What is pertinent in the present connection is to call attention to the fact that we are already rapidly approaching a time when most of our knowledge of *physics*, at least, can be stated in the form "such and such a quantity (defining it accurately) seems probably not to be larger than one given value nor to be smaller than another given value"; and we can conceive that the other natural sciences, even those which relate to language, history, etc. should ultimately be brought to somewhat similar conditions so that we should no longer look for unexpected additions to our knowledge so much as to narrowing the limits between which it seems likely that each truth will ultimately be found to lie.

When we once reach that point we can calculate what it will probably cost (in the efforts of men of different varieties) to narrow those limits by any given amount, and also what the practical gain of that limitation will be; and these two quantities being calculated for all our problems, the quotients of the probable gains divided by their probable costs will represent the urgencies of the different inquiries at the time, which of course we shall strive to equalize, and thus put into practice a wise scientific economy. The mere prospect of such a future already animates the wiser of the men of science with an ardent desire to reduce the urgencies of the different problems to equality; and this passion shall here be designated by the name of *the love of scientific economy*. It seems to be the highest form that the love of knowledge has been able, as yet, to assume.

DOCTRINE OF CHANCES

Needless to say one of the tools most useful to Peirce in his various investigations was the calculus of probabilities. In an article written for the *New York Evening Post* on Saturday, 12 January 1901, entitled "The Century's Great Men of Science," he declared that "In pure logic, the doctrine of chances, which had been the logical guide of the exact sciences and is now illuminating the pathway of the theory of evolution, and is destined to still higher uses, received at the hands of Jacob Bernoulli and of Laplace developments of the first importance. ... In the nineteenth century, Boole created a method of miraculous fruitfulness, which aided in the development of the logic of relatives, and threw great light on the doctrine of probability, and thereby upon the theory and rules of inductive reasoning." In 1853 Boole had written that the problem which the Chevalier de Méré proposed to Pascal was the first of a series "destined to call into existence new methods in mathematical analysis, and to render valuable service in the practical concerns of life." According to Enriques, Boole, De Morgan, and Peirce were destined in their time to extend those methods in the development of a symbolic logic "where they made a beautiful application of symbolic analysis which consists" in the establishment of a *logical calculus* of events parallel to the numerical calculus of probability. And in the opinion of C. I. Lewis, Peirce's contributions to symbolic logic were more numerous and varied than those of any other nineteenth-century writer. He mentioned especially Peirce's improvement of Boole's methods of applying symbolic logic to problems in probability. Indeed Peirce had been saturated with the spirit of the physical sciences and had absorbed the spirit of fallibilism in his practical contacts with and contributions to astronomical, optical and gravitational theory, so that probabilism became one of the mainstays in the erection of his general architectonic. In Peirce's system the theory of probabilities is simply the science of logic quantitatively treated and thus the problem of probabilities is simply the general problem of logic.

In a letter to William James dated 17 December 1909, he proposes to write an article in which he will, among other things, treat the concept of probability. He writes as follows:

After all this I shall undertake to show (still somewhat imperfectly) that concepts are capable of such phaneroscopic analysis, or in common parlance "logical analysis"; but there are only a few cases in which I pretend as yet to carry the analysis so far as to resolve the concept into its ultimate *elements*. After a few more such questions have been discussed, I show how to go to work to perform the analysis; and then I proceed to show that a definition constructed according to my method at once clears up various puzzles relating to the concept.

For instance, take *probability*. There are some who define this psychologically by what we tend to think etc. But it is impossible that the solid business of Insurance rests on Opinion. It must rest on solid Fact. Others say it is the frequency of a given sort of event in the *long run*. This is vague. *How* long must the run be? And is it *certain* that a pair of perfect dice would turn up doublets precisely in one sixth of the number of throws? No, it is certain they *won't* unless that number happens to be divisible by six. So these persons are either driven, like LaPlace to base the definition on the idea of events "également possibles," which I say has no definite meaning, or to say that the "long run" is an *infinitely* long run, and that they mean that the value of the probability is the *limit* to which the lengthening run tends toward. But if we turn to the books on the infinitesimal calculus to find what is meant by such a limit, we find the following definition laid down substantially or loosely.

A series of numbers a_1 a_2 a_3 etc. or say a_n (where n is any ordinal number) is said to tend to a limiting value $a_\infty = L$, L being a definite number, if, and only if, any positive quantity ε, having been mentioned as too small to be considered, it is possible to assign a value N, such that for any value n' of n that is greater than N it will be certain that $a_{n'}$ differs from L by a quantity less than ε.

Thus, I find in Jordan's *Cour d'analyse* Tome III p. 557 the following: "Soit x une quantité variable, à laquelle on donne successivement une suite illimitée de valeurs $x_1, \ldots, x_n, \ldots$ On dit que la variable x tend ou converge vers la limite c si, pour toute valeur positive ε, on peut assigner une autre quantité ν, telle que l'on ait

$$\text{mod}\,(x_n - c) < \varepsilon$$

pour toutes les valeurs de n supérieures à ν."

I have consulted 5 treatises of dates between 1890 and 1900 and 4 subsequent to 1900 and find that none of them make $\frac{1}{6}$ the "*limit*" of the quotient of the tally of doublets divided by the tally of throws of a pair of dice, any more than Jordan's.

However, I modify the definition as follows:
A series of quantities a_1 a_2 a_3 etc. may be said to converge to the value L *in the long run*, if, and only if, however small $+\varepsilon$ and $-\eta$ may be, there is *some* ordinal number N (whether we can know what it is or not) such that if $n' > N$ then for whatever such value if $a_{n'} > L$, $a_{n'} - L < e$, and if $a_{n'} < L$, $a_{n'} - L > \eta$. This is quite different from the ordinary concept of a limit but in this sense the quotient of tallies will be $\frac{1}{6}$ in the long run.

I wish you would preserve this as a safeguard against a possible charge of plagiarism by an unknown correspondent who wrote me in a general way about probabilities and to whom I gave a hint of this just to see whether he would see what is wanted or not.

John Dewey wrote in 1937 (*The New Republic*, February 3) that "in his [Peirce's] assertion that every scientific proposition is only probable, no matter how exact in itself and how sound the reasoning involved in reaching it, he anticipated the conclusion that science in its own development ... has since been compelled to reach .... The import of the doctrine of probabilities, in accord with the general pragmatic position, lies in the general attitudes and ways of acting that follow from its acceptance .... These attitudes rule out all dogmatism, all cocksureness, all appeal to authority and ultimate first truths; they keep alive the spirit of doubt as the spring of continually renewed inquiry ... which never claims or permits finality but leads to ever renewed effort to learn ..." thus making science "the hope of mankind." Philip Wiener in his *Values in a Universe of Chance* spoke of Peirce's premonition of the statistical conception of the laws of nature, as sketched out by Boltzmann in 1904, as being obscured by his "metaphysical attempts to link the new idea with the older, scholastic ideas of the teleological order of nature."

Rudolph Carnap in his *Logical Foundations of Probability* speaks of Cournot (1843) as combining the classical definition of probability with an interpretation in terms of relative frequency. He then relates that John Venn (*Logic*, 1866), "more than twenty years after Cournot, was the first to advocate the frequency concept of probability, unambiguously and systematically as explicandum and also the first to propose as explicatum for it the concept of the limit of relative frequency in an infinite series. Although his conception influenced the views of some other writers, among them Charles Sanders Peirce (1878), it was only half a century later that comprehensive systematic theories were constructed which took probability$_2$ (relative frequency definition) as their basis." Peirce is grouped with Venn in the original formulation of the definition of a random sample. As regards the problem of the reliability of a value of degree of confirmation, Carnap cites Peirce (1878) as having been the earliest of those who touched on the problem and quotes from his "Probability of Induction" in the *Popular Science Monthly* 12.

In his academic responsibilities at the Johns Hopkins University, Peirce was listed for course work in probability theory as follows: 1879-1880, Course in "Probabilities," developing the mathematics of probabilities; August 1880, "Mr. C. S. Peirce (of the United States Coast and Geodetic Survey) will lecture during the last half-year on the *Theory of Probability* and on other topics in *logic*." Among other topics programmed for the year 1882 there is "The Theory of Probabilities — The fundamental rules of the calculus will be discussed. Its practice will be illustrated by the solution of select problems, beginning with the simplest and proceeding to some of the most difficult. The theory of linear difference equations will be given. The method of least squares will be theoretically and practically treated. Text: Liagre's *Calcul des probabilités*, Boole's *Calculus of Finite Differences*, Ferrero's *Metodo dei Minimi Quadrati*." It is pertinent to our thesis to note here that Peirce also offered at that time a course on "Mathematical Reasoning. The general nature of mathematical demonstration will be explained, the different varieties will be classified, and the particular use to which each can be put will be shown. The methods of mathematical research will be studied in the history of multiple algebra."

LINEAR ASSOCIATIVE ALGEBRA

In James Byrnie Shaw's Synopsis of *Linear Associative Algebra* (1907) reference is made to seven of C. S. Peirce's papers that have already appeared in the *Collected Papers* and therefore do not appear in this collection. These are "Description of a Notation for the Logic of Relatives," "On the Application of Logical Analysis to Multiple Algebra," "Note on Grassman's Calculus of Extension," "Notes on B. Peirce's Linear Associative Algebra," "On the Relative Forms of Quaternions," "On Nonions," "On a Class of Multiple Algebras." In Shaw's classical work,

C. S. Peirce's name is associated with those of Frobenius and Taber. Shaw speaks of Taber's reexamination of the line of work of Benjamin Peirce and mentions two additional lines of development that were followed. The first is characterized by the use of the continuous group as Poincaré announced the isomorphism; the second by the matrix theory as "C. S. Peirce first noticed this isomorphism, although in embryo it appeared sooner." This is the line that was followed by Shaw and Frobenius.

Now Henry Taber's classic paper "On the Theory of Matrices" appeared in Volume XII (1889-1890) of the *American Journal of Mathematics* and in it he referred to the work of Charles Peirce. He spoke of Hamilton as being "the originator of the theory of matrices, as he was the first to show that the symbol of a linear transformation might be made the subject matter of a calculus." Outlining the development of the idea by Cayley, Clifford, Tait, and Sylvester, Taber treats the two Peirces as being subsequent to Cayley but previous to Sylvester and as being led to the consideration of matrices from the standpoint of "the investigation of linear associative algebras involving any number of linearly independent units. In this aspect, the quantities arranged in a square are looked upon as scalar coefficients of the several units or 'vids' of an algebra belonging to a certain class." In footnoting Peirce's terminology, Taber writes that "The term *vid* was introduced by Mr. C. S. Peirce to denote the *units* or *letters* of an algebra."

Furthermore Taber specifically sketches Charles's remarkable contribution to the development of the theory of matrices as follows:

> Subsequent to Cayley's memoir, the next advance was made in 1870 by Charles S. Peirce who, in his investigations upon the extension of the Boolian calculus to the logic of relatives ("Description of a Notation for the Logic of Relatives," *Memoirs Am. Acad. of Arts and Sciences*, n.s. 9, 1870) came upon a set of forms ... constituting a system virtually identical with the calculus of matrices ... As has been stated, the relation of the theory of matrices, as algebras of a certain class, to linear associative algebra in general, was first made clear through the light thrown on the subject by Peirce's systems of vids. ... Charles Peirce has made the great discovery that the whole theory of linear associative algebra is included in the theory of matrices. He has shown that every linear associative algebra has a relative form, i.e. its units may be expressed linearly in terms of the vids (denoted in his notation by $(A : A)(A : B)$, etc.) of a linear transformation; and consequently, that any expression in the algebra can be represented by a matrix. ... Charles Peirce has, moreover, given the relative or matrix forms of all the algebras considered by his father in his *Linear Associative Algebra.*
>
> To Charles Peirce, in conjunction with his father, the identification of quaternions with the quadrate algebra of order two (i.e. the algebra of dual ma-

trices) is also due. ... In his memoir of 1870 Charles Peirce had given, as an example of the infinite system of quadrate algebras, the multiplication table of the quadrate algebra next in order after quaternions, afterwards named *nonions* by Clifford .... The Peirce discovery of the octonomial form of nonions was not published. The priority of publication of this form belongs to Sylvester, who discovered it subsequently to the Peirces without any knowledge of their investigations upon nonions.

Again in the section on *Algebras Analogous to Quaternions*, Taber writes:

If i, j, k, are any three mutually normal unit vectors, any quaternion may be expressed linearly in terms of the four new units

$$\frac{1+i\sqrt{-1}}{2},\quad \frac{j+k\sqrt{-1}}{2},\quad \frac{-j+k\sqrt{-1}}{2},\quad \frac{1-i\sqrt{-1}}{2}$$

which, having the same multiplication table as the four vids of a dual matrix,

$$(A:A),\quad (A:B),\quad (B:A),\quad (B:B),$$

may, consequently, be regarded as respectively identical with them. This identification gives the following values for the ordinary quaternion units,

$$1 = \begin{pmatrix} 1 & 0 \\ 0 & 1 \end{pmatrix},\ i = \begin{pmatrix} -\sqrt{-1} & 0 \\ 0 & \sqrt{-1} \end{pmatrix},\ j = \begin{pmatrix} 0 & 1 \\ -1 & 0 \end{pmatrix},\ k = \begin{pmatrix} 0 & -\sqrt{-1} \\ -\sqrt{-1} & 0 \end{pmatrix}$$

The discovery of this form of quaternions ... is due to Benjamin Peirce; it received its full significance only after the discovery by his son, Charles Peirce, of the unlimited system of quadrates formed from the system of vids $(A:A)$, $(A:B)$ etc., when it appeared that quaternions was only the first of this system of quadrate algebras, and the identification of quaternions with the theory of dual matrices was virtually accomplished.

Victor Lenzen has written a definitive paper on "The Contributions of Charles S. Peirce to Linear Algebra" which has appeared in *Phenomenology and Natural Existence* (SUNY Press, 1973).

Peirce felt that his mission in the world was to "set forth the true nature of logic, and of scientific methods of thought and discovery," and, as he continued in a letter to Jem on 25 October 1885, he had "a great and momentous thing to say on this subject." He felt that "without it, molecular science must remain at a stand-still. It must continue what it is, idle guess-work. The true theory of the constitution of matter, which can only be based on sound scientific logic, must have the most important consequences in every direction. On psychology too, which is to be the great science of the coming hundred years, logic must exert mighty influence. About logic, I have something to say which other men have not thought of, and probably may not soon think of."

Yet he claimed in 1902 that his logical studies had already enabled him to prove some propositions which had arrested mathematicians of power. At that time he disclaimed "all pretension," as he put it, "to having been remarkably successful in dealing with the heuretic department of mathematics. My attention has been directed to the study of its procedure in demonstration, not its procedure in discovering demonstrations. This must come later; and it may be very well that I am not so near to a thorough understanding of it as I may come. I am quite sure that the value of what I have ascertained will be acknowledged by mathematicians."

CONTENTS

1

[ELEMENTS OF TRIGONOMETRY]

SKETCH OF A PROPOSED TREATISE ON TRIGONOMETRY AND THE ART OF COMPUTATION (196)

Intended to be elementary, yet to be kept and be useful as a full handbook.

Most of the propositions to be given without proof. The student to work them out.

CHAPTER 0

A summary statement of the elementary geometry and algebra needed. Also brief explanation of coördinate geometry [and] the elements of the calculus.

I here set down for the student's convenience some things in his earlier studies he will need in the course of the book. I also give some elementary matter concerning Coördinate Geometry & the Calculus.

Algebra

$$a^2 - b^2 = (a-b)(a+b)$$

$$a^n - b^n = (a-b)(a^{n-1} + a^{n-2}b + \dots b^{n-1})$$

$$(x-a)(x-b)(x-c)\dots$$

$$= x^n - (a+b+c+\text{ etc.})x^{n-1}$$

$$+ (ab+ac+bc+ad+bd+cd+\text{ etc.})x^{n-2}$$

$$- (abc+abd+acd+bcd+\text{ etc.})x^{n-3} + \dots \pm abcd\dots$$

$$(x+y)^2 = x^2 + 2xy + y^2$$

$$(x+y)^3 = x^3 + 3x^2y + 3xy^2 + y^3$$

$$(x+y)^4 = x^4 + 4x^3y + 6x^2y^2 + 4xy^3 + y^4$$

$$(x+y)^n = x^n + \frac{n}{1}x^{n-1}y + \frac{n(n-1)}{1.2}x^{n-2}y^2$$

$$+ \frac{n(n-1)(n-2)}{1.2.3}x^{n-3}y^3 + \dots y^n$$

$$\frac{x^n}{n!} = \sum_o^n p \frac{x^p}{p!} \frac{y^q}{q!} \quad \text{(where } p+q = n\text{)}$$

If $\frac{a}{b} = \frac{c}{d}$ then $\frac{a+xb}{a+yb} = \frac{c+xd}{c+yd}$.

CHAPTER 1. IMAGINARIES

Imaginaries have to be used in trigonometry in a way the student need not have known in algebra.

Imaginary quantity is distributed like points on a plane. In studying it, one must continually think of a plane. But the calculus of imaginaries is not plane geometry, because it has no operation like projection and several others, and because many continua besides a mechanical plane are studied by means of it. Hence, in thinking of the plane, when studying imaginaries, the student must only think of it as a particular case of a spread such as imaginary quantity is suited to measure.

Consider a system of rectangular coördinates. We may take the origin as the zero of quantity. Then the point $x = x_1$ $y = y_1$ is (or has for its affix) the expression

$$y_1 + x_1 i.$$

The letter i is not a number but a peculiar unit; and the expression cannot be further reduced. Thus,

$$2 + 3i$$

cannot be made a monomial. It is (or is the affix of) the point $y = 2$, $x = 3$.

We have, as naturally springing from the notation, and enlargements of it

$$y + xi = xi + y$$

$$3 + 5i + 2 - 7i = 5 - 2i.$$

If $\quad y + xi = b + ai$

then $\quad y = b \quad x = a.$

Hence an equation in imaginaries is two equations.

Historically and logically, the conception of imaginaries first arises in the study of the quadratic equation. If a quadratic can be brought to the form

$$x^2 + 2ax + a^2 - b^2 = 0$$

that is, if the absolute term is less than a fourth of the square of the coefficient of x, its solution is plain

$$(x+a)^2 = b^2$$
$$x+a = \pm b$$
$$x = -a \pm b.$$

But if its form is

$$x^2+2ax+a^2+b^2 = 0$$

its solution is

$$x = -a \pm b\sqrt{-1}$$

The square root of -1 here has no direct interpretation in unimensional quantity. If the equation is given in the form

$$x^2+2ax+b^2 = 0$$

its solution is

$$x = -a \pm \sqrt{a^2-b^2}.$$

Here, we have the important form $\sqrt{a^2-b^2}$, with which is naturally connected the form

$$\sqrt{p^2+q^2}.$$

This occurs in probabilities, in geometry, in dynamics, and must be recognized in the imaginary calculus.

Let us suppose we have the equation

$$z^2-8z+25 = 0.$$

Its solution is

$$z = 4 \pm 3i.$$

Let us rebuild the equation upon the diagram.

Then, the distance of z from the origin is

$$\sqrt{y^2+x^2}$$

and the distance of z^2 from the origin is

$$\sqrt{(y^2-x^2)^2+4x^2y^2}$$
$$= \sqrt{y^4-2x^2y^2+x^4+4x^2y^2}$$
$$= \sqrt{y^4+2x^2y^2+x^4}$$
$$= y^2+x^2.$$

The latter distance is the square of the former. This distance from the origin is called the *Modulus* of the imaginary quantity. If then we divide every quantity by its modulus, the quotients will be (or be affixes of) points on the circumference of the circle of unit radius with its centre at the origin. And such quotient for the square of the quantity will, by algebra, be the square of the quotient for the first quantity.

It will be so geometrically too. For draw this circle 10 times exaggerated on the squared paper, and draw radii through z_1 and z_1^2. Now take a pair of compasses and you will see that the arc cut off between the former radius and the axis of y is just half the arc cut off between the latter radius and the same axis! If carefully drawn on a large scale and measured with a large protractor, the angles will be found 36°52′ and 73°44′. There is a marvel of mathematics!

We easily plot on a piece of squared paper, in succession,

$$\begin{aligned} z_1 &= 4+3i \\ 8z_1 &= 32+24i \\ 8z_1-25 &= 7+24i. \end{aligned}$$

Now the equation is

$$z^2 = 8z_1 - 25.$$

But how shall we recognize $7+24i$ as the square of $4+3i$?

Algebraically, it follows from $i^2 = -1$; for

$$\begin{aligned} (4+3i)^2 &= 16+24i+9i^2 \\ &= 16+24i-9 \\ &= 7+24i. \end{aligned}$$

But there is another way, upon which Trigonometry hangs. The point z_1 has for its rectangular coördinates $y = 4$, $x = 3$. That is, x and y being the legs of a right triangle whose hypotheneuse is the distance of z from the origin, this hypotheneuse is

$$\sqrt{4^2+3^2} = \sqrt{16+9} = \sqrt{25} = 5.$$

While the distance of $z^2 = 7+24i$ from the origin is

$$\sqrt{7^2+24^2} = \sqrt{49+576} = \sqrt{625} = 25 = 5^2.$$

This will be so in every case. Namely, let

$$\begin{aligned} z &= y+xi, \\ z^2 &= y^2-x^2+2xyi. \end{aligned}$$

In like manner if we take the cube

$$
\begin{aligned}
(4+3i)^3 &= 64+3.16.3i+3.4.9i^2+27i^3\\
&= 64+144i-108-27i\\
&= -44+117i.
\end{aligned}
$$

If the angle be measured, it will be found to be 110°37′ which is 3 times the first.

Thus, we see (presuming this to be exact) that when the quotient is raised to the nth power, the angle is multiplied by n. That is to say, these angular quantities are of the nature of exponents! Since the point i (that is, the point for which $y = 0$, $x = 1$) is on the circle at an angular distance of 90° from the axis of Y, it follows that

$$i^{\theta/90}$$

is (or is the affix of) the point on the circle θ degrees from the axis of Y measuring round clockwise.

We can test this for $\theta = 45°$. In that case, the hypotheneuse (or radius) being 1, the legs (or coordinates) are both $\frac{1}{\sqrt{2}}$. Accordingly the quantity, which should be $\sqrt{i}$ is

$$\frac{1}{\sqrt{2}}+\frac{1}{\sqrt{2}}i$$

Squaring

$$
\begin{aligned}
\left(\frac{1}{\sqrt{2}}+\frac{1}{\sqrt{2}}i\right)^2 &= \tfrac{1}{2}(1+i)^2\\
&= \tfrac{1}{2}(1+2i+i^2)\\
&= \tfrac{1}{2}(1+2i-1) = i.
\end{aligned}
$$

Again we know that the sides of a regular hexagon inscribed in a circle are equal to the radius [Fig. 1]. Hence, if one of the vertices is at i, the first vertex to the right of the axis of Y has $x = \frac{1}{2}$, $y = \sqrt{1-(\frac{1}{2})^2} = \frac{\sqrt{3}}{2}$; so that i should be the cube of

$$\tfrac{1}{2}(\sqrt{3}+i)$$

Let us try.

$$
\begin{aligned}
\frac{1}{2^3}(\sqrt{3}+i)^3 &= \tfrac{1}{8}(3\sqrt{3}+9i+3\sqrt{3}i^2+i^3)\\
&= \tfrac{1}{8}(3\sqrt{3}+9i-3\sqrt{3}-i)\\
&= \tfrac{1}{8}(8i) = i.
\end{aligned}
$$

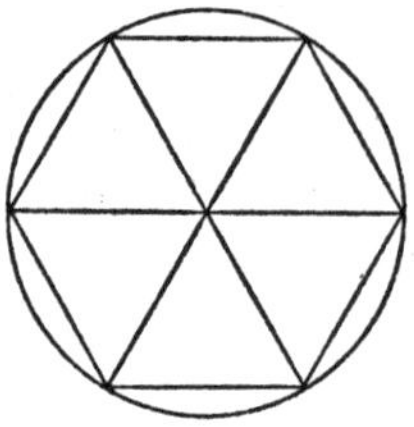

Fig. 1

In order to prove that this is always so, we must get an algebraical expression for an arc of the circle. Suppose we had the value of the arc YP for a known value of x, namely $x = QP$ [Fig. 2]. Of course, the circle being of unit radius, $y = \sqrt{1-(QP)^2}$. Let us find how much longer the arc will be for $x = Q'P' = QP + RP'$, where RP' is a very small quantity d. We shall have

$$(x+d)^2 = x^2 + 2xd + d^2.$$
$$y' = \sqrt{1-(x+d)^2} = \sqrt{1-x^2-2xd-d^2}$$

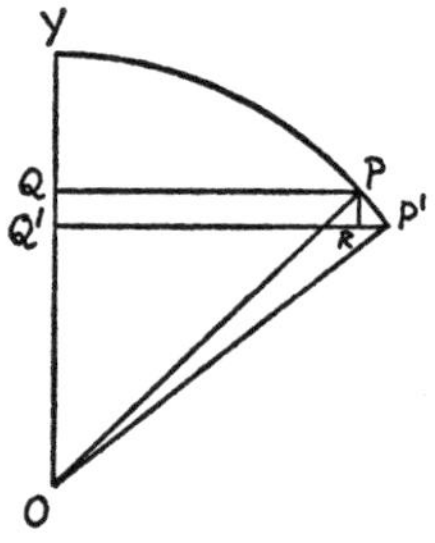

Fig. 2

We neglect the square and higher powers of d. It is shown in the differential calculus, that this is strictly exact. But we cannot go into the niceties here. Suffice it to remark that we can take d as small as we please, say one vigintillionth of a unit. Then d^2 is only one vigintillionth of that; so that when we come to multiply d by a vigintillion or so to make a considerable quantity, the part of the product dependent on d^2 remains excessively small, and as small as we please. (This is not quite logical, but see Klein's recommendation, in his Evanston Lectures.) Then

$$PR = y' - y = -\frac{x}{\sqrt{1-x^2}}\,d$$

The little arc PP' is straight. Hence

$$PP' = \sqrt{d^2 + \left(-\frac{x}{\sqrt{1-x^2}}d\right)^2}$$
$$= d\sqrt{1+\frac{x^2}{1-x^2}}$$
$$= \frac{d}{\sqrt{1-x^2}}$$

Suppose QP to be divided into *a* vigintillion parts each equal to d and this quantity $\dfrac{d}{\sqrt{1-x^2}}$ calculated for each. The result will be the measure of the circular arc YP. Let us write it, as is done in the integral calculus

$$\int_0^x \frac{d}{\sqrt{1-x^2}}$$

Let us now see whether when this arc is doubled, the quantity that is (or is the affix of) its extremity, namely

$$\sqrt{1-x^2} + xi$$

be not squared.

To double

$$\int_0^x \frac{d}{\sqrt{1-x^2}}$$

we add more of the little quantities $\dfrac{d}{\sqrt{1-x^2}}$. Though d has the same value for each, x is growing longer as each one is added, and so $\sqrt{1-x^2}$ grows smaller, and $\dfrac{d}{\sqrt{1-x^2}}$ longer and longer. Consequently, we do not have to double x to double $\int_0^x \dfrac{d}{\sqrt{1-x^2}}$. But each addition of $\dfrac{d}{\sqrt{1-x^2}}$ to the arc changes the quantity

$$\sqrt{1-x^2} + xi$$

to

$$\sqrt{1-x^2} - \frac{x}{\sqrt{1-x^2}}\, d + (x+d)\, i$$

But this is multiplying it by

$$1 + \frac{d}{\sqrt{1-x^2}}\, i; \text{ for } (1 + \frac{d}{\sqrt{1-x^2}}\, i)(\sqrt{1-x^2} + xi)$$

$$= \sqrt{1-x^2} + xi + d\,\frac{\sqrt{1-x^2}}{\sqrt{1-x^2}}\, i + \frac{dx}{\sqrt{1-x^2}}\, i^2$$

$$= \sqrt{1-x^2} - \frac{x}{\sqrt{1-x^2}}\, d + (x+d)i.$$

Thus, when we add to the arc any infinitesimal quantity δ, we multiply the quantity at its extremity by $(1+\delta i)$. Let N be an enormous large quantity, so that $N\delta$ is an ordinary finite arc. If this be added to the arc, the extremity will be multiplied by

$$(1+\delta i)^N = 1 + N\delta i - \frac{N(N-1)}{2}\delta^2 - \frac{N(N-1)(N-2)}{2.3}\delta^3 i + \text{etc.}$$

Since the numbers 1, 2, 3, etc. are very small compared with N, this will be

$$(1+\delta)^N = 1 + N\delta + \frac{1}{2!}(N\delta)^2 + \frac{1}{3!}(N\delta)^3 + \frac{1}{4!}(N\delta)^4 + \text{etc.}$$

Since the numerical coefficients are, after say a vigintillion of them each only a vigintillionth of the last preceding, it follows the terms will at length grow very rapidly smaller; and the sum will not exceed a moderate number. Since the δ enters into this just as the N does it follows we may write the first member $[(1+\delta)^{1/\delta}]^{\delta N}$ and $(1+\delta)^{1/\delta} = 1 + 1 + \frac{1}{2!} + \frac{1}{3!} + \frac{1}{4!} + \text{etc.}$

This being calculated is found to be about 2.7182818. It is called the Neperian base, and may be denoted by e. [Neper is the German form of Napier.]

We find, then, that when the arc is increased by any length, θ, the quantity at its extremity is multiplied by $e^{\theta i}$; and since when the arc is zero that quantity is 1, it follows that if θ be the arc the quantity at its extremity is $e^{\theta i}$.

We now have two ways of writing an imaginary. One by rectangular coordinates

$$y + xi$$

The other by polar coordinates

$$Re^{\theta i}$$

where $R = \sqrt{y^2 + x^2}$.

We may also write it

$$R\left(\frac{y}{\sqrt{x^2 + y^2}} + \frac{xi}{\sqrt{x^2 + y^2}}\right)$$

The values of $\frac{x}{\sqrt{x^2 + y^2}}$ and $\frac{y}{\sqrt{x^2 + y^2}}$ for given values of θ are obtained very simply from the above formula.

$$\frac{x}{\sqrt{x^2 + y^2}} = \theta - \frac{1}{3!}\theta^3 + \frac{1}{5!}\theta^5 - \text{etc.}$$

$$\frac{y}{\sqrt{x^2 + y^2}} = 1 - \frac{1}{2!}\theta^2 + \frac{1}{4!}\theta^4 - \frac{1}{6!}\theta^6 + \text{etc.}$$

These are called sine θ and cosine θ, generally written $\sin\theta$ and $\cos\theta$.

The arc $\theta = \int_0^{\sin\theta} \frac{d}{\sqrt{1 - x^2}}$

Three or more points on one ray are commonly said to be *collir ear*; but in this book the word *coradial* will be used.

An *inscribable* polygon is one which can be inscribed in a circle.

The *sine* of an angle is the ratio of the leg opposite an angle of that value in a right triangle to the hypotheneuse.

The *tangent* of an angle is the ratio of the leg opposite that angle in a right triangle to the leg adjacent.

The *secant* of an angle is the ratio of the hypotheneuse to the leg adjacent an angle of that value in a right triangle.

The *cosine*, *cotangent*, and *cosecant* of an angle are the sine, tangent, and secant, respectively of the complement of that angle.

The *versin*, or *versed sine*, of an angle is unity minus its cosine.

The *haversin* of an angle is the square of the sine of half the angle.

Notation. The following abbreviations are usual

sin for "sine of,"
cos for "cosine of,"
sec for "secant of,"
cosec for "cosecant of,"
tan for "tangent of,"
cot for "cotangent of."

Also $\sin^2 x$, $\tan^2 x$, etc. are written instead of $(\sin x)^2$, $(\tan x)^2$, etc.

Theorem [*The Addition Theorem of Trigonometry*]. If x and y are any angular quantities whatever, then

$$\sin(x+y) = \sin x \cdot \cos y + \cos x \cdot \sin y.$$

Demonstration. Let *aph* be the legs and hypotheneuse of a right triangle of which the angle opposite a equals x; and let *bpk* be the legs and hypotheneuse of a right triangle of which the angle opposite b equals y [Fig. 3].

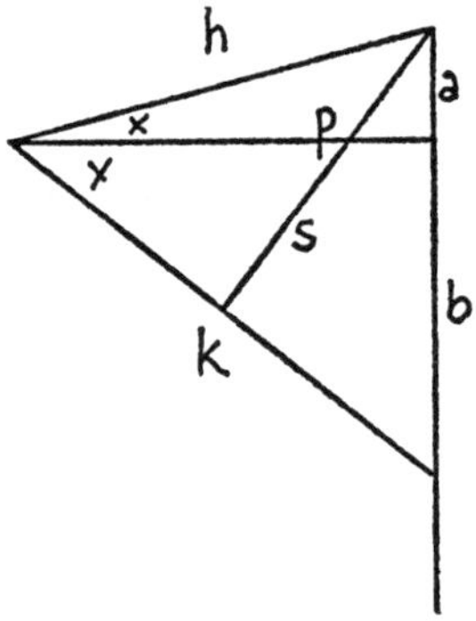

Fig. 3

Let the triangles be so placed that h, k, and $a+b$ form a triangle; and let s be the perpendicular upon k from the opposite vertex of the last triangle. Then, the theorem is that

$$\frac{s}{h} = \frac{a}{h}\frac{p}{k} + \frac{p}{h}\frac{b}{k}.$$

[Twice] the area of the large triangle is sk. It is also $p(a+b)$. Hence,

$$sk = pa + pb.$$

Dividing by hk, we have

$$\frac{s}{h} = \frac{a}{h}\frac{p}{k} + \frac{p}{h}\frac{b}{k}. \qquad \text{Q.E.D.}$$

Corollaries. By turning one triangle over upon the other, we find

$$\sin(x-y) = \sin x \cdot \cos y - \cos x \cdot \sin y.$$

It follows that, no matter what angular values x, y, z may be

$$\sin x \cdot \sin(y-z) + \sin y \cdot \sin(z-x) + \sin z \cdot \sin(x-y) = 0.$$

For, on development, the first member becomes,

$$\begin{aligned} &\sin x \cdot \sin y \cdot \cos z - \sin x \cdot \cos y \cdot \sin z + \cos x \cdot \sin y \cdot \sin z \\ -\ &\sin x \cdot \sin y \cdot \cos z + \sin x \cdot \cos y \cdot \sin z - \cos x \cdot \sin y \cdot \sin z \end{aligned}$$

In the formula just found, put $x = 90°$, $y = 0$. We thus get

$$1 \cdot \sin(-z) + 0 + \sin z \cdot 1 = 0$$

or $\quad \sin(-z) = -\sin z.$

In the same formula, put $y = 90°$, and we recover the formula of [the] Corollary. For we thus get,

$$\sin x \cdot \cos z + \sin(z-x) - \sin z \cdot \cos x = 0$$

or $\quad \sin(z-x) = \sin z \cdot \cos x - \sin x \cos z.$

In the formula of the theorem, put $y = -x$, and we get,

$$0 = \sin x \cdot \cos(-x) + \cos x \cdot \sin(-x).$$

This gives

$$\tan(-x) = \frac{\sin(-x)}{\cos(-x)} = -\frac{\sin x}{\cos x} = -\tan x,$$

or $\quad \tan(-x) = -\tan x.$

The formula

$$0 = \sin x \cdot \cos(-x) + \cos x \cdot \sin(-x)$$

gives $\quad 0 = \sin x \cdot \cos(-x) - \cos x \cdot \sin x,$

or $\quad 0 = \cos(-x) - \cos x,$

or $\quad \cos(-x) = \cos x.$

In the formula

$$\sin x \cdot \sin(y-z) + \sin y \cdot \sin(z-x) + \sin z \cdot \sin(x-y) = 0,$$

put $x = 90°$, $y = 90° - t$. We thus get

$$\cos(t+z) - \cos t \cdot \cos z + \sin z \cdot \sin t = 0,$$

or

$$\cos(t+z) = \cos t \cdot \cos z - \sin t \cdot \sin z.$$

In the last formula, put $z = -s$ and we have

$$\cos(t-s) = \cos t \cdot \cos s + \sin t \cdot \sin s.$$

In the formula of the theorem, put $y = 90°$. We thus get

$$\sin(90° + x) = \cos x.$$

In the formula for $\cos(t+z)$, put $t = 90°$ and we get

$$\cos(90° + z) = -\sin z.$$

In the formula for $\sin(90° + x)$, put $x = 90° + t$ and we get,

$$\sin(180° + t) = \cos(90° + t) = -\sin t.$$

In the formula for $\cos(90° + z)$, put $z = 90° + t$, and we get,

$$\cos(180° + t) = -\sin(90° + t) = -\cos t.$$

$$\tan(90° + t) = \frac{\sin(90° + t)}{\cos(90° + t)} = \frac{\cos t}{-\sin t} = -\cot t.$$

$$\tan(180° + t) = -\cot(90° + t) = -\frac{1}{\tan(90° + t)} = \frac{1}{\cot t} = \tan t.$$

In the formula for $\sin(x+y)$, put $y = x$, and we get,

$$\sin 2x = 2 \sin x \cdot \cos x.$$

In the formula for $\cos(t+z)$, put $z = t$, and we get,

$$\cos 2t = \cos^2 t - \sin^2 t.$$

In the formula for $\sin 2x$, put $x = 45°$, and we get

$$1 = 2 \sin 45° \cdot \cos 45°.$$

But $\cos 45° = \sin(90° - 45°) = \sin 45°.$

Hence $\sin^2 45° = \frac{1}{2}.$

But the definition of a sine shows that the sine of 45° is positive.

Hence $\sin 45° = \frac{1}{\sqrt{2}}.$

The repeated application of the theorem gives

$$\begin{aligned}\sin(x+y+z) &= \sin(x+y)\cos z + \cos(x+y) \cdot \sin z \\ &= \sin x \cdot \cos y \cdot \cos z + \cos x \cdot \sin y \cdot \cos z \\ &+ \cos x \cdot \cos y \cdot \sin z - \sin x \cdot \sin y \cdot \sin z.\end{aligned}$$

So likewise

$$\cos(x+y+z) = \cos(x+y)\cos z - \sin(x+y)\cdot\sin z$$
$$= \cos x\cdot\cos y\cdot\cos z - \sin x\cdot\sin y\cdot\cos z$$
$$- \sin x\cdot\cos y\cdot\sin z - \cos x\cdot\sin y\cdot\sin z.$$

Putting $x = y = z = 30°$

$$1 = 3\sin 30°\cdot\cos^2 30° - \sin^3 30°$$
$$= 3\sin 30° - 4\sin^3 30°.$$
$$0 = \cos^3 30° - 3\sin^2 30°\cdot\cos 30° = 4\cos^3 30° - 3\cos 30°.$$

Since cos 30° and sin 30° are plainly (by the definition) positive, the last equation gives

$$\cos^2 30° = \tfrac{3}{4}$$
$$\cos 30° = \tfrac{\sqrt{3}}{2};$$

and the other equation gives $\dfrac{1}{\sin 30°} = 3 - 4\cdot\tfrac{1}{4} = 2$, or $\sin 30° = \tfrac{1}{2}$.

Art. 2. We write Exp z to denote the exponential of z. It will be remarked that no matter how large a number z may be, the series is *convergent*, that is, it enables us to calculate Exp z accurately to any required number of decimal places. Thus, if $z = 10$, the calculation is

$$1 = 1$$

$$z = 10$$

$$\tfrac{1}{2}z^2 = 50$$

2

$$\tfrac{1}{3!}z^3 = 166.66666666667$$

$$\tfrac{1}{4!}z^4 = 416.66666666667$$

2 5

$$\tfrac{1}{5!}z^5 = 833.33333333333$$

6 5 2 1 4 6

$$\tfrac{1}{6!}z^6 = 1388.88888888889$$

3 6 4 1 5 6 5 4 6 5 6

$$\tfrac{1}{7!}z^7 = 1984.12694126941$$

6 5 5 6 2 1 7 5 2 7 8 3 1

$$\tfrac{1}{8!}z^8 = 2480.15867658677$$

$$\tfrac{1}{9!}z^9 = 2755.7318628742$$

5 2 1 9 4 4 3 1

$$\tfrac{1}{10!}z^{10} = 2755.7318628742$$

1 9 8 9 6 7 6 4 8 4 3

$$\tfrac{1}{11!}z^{11} = 2505.2107844311$$

$$\tfrac{1}{12!}z^{12} = 2087.6756536926$$

$$\tfrac{1}{13!}z^{13} =$$

Each term is now becoming more and more rapidly smaller than the one before it. When $\frac{1}{100!}z^{100}$ is reached, it will be but $\frac{1}{10}$ of the previous term. $\frac{1}{1000000}z^{1000000}$ will begin 5 decimal places to the right of the first figure of the term before it. Thus, eventually the terms become perfectly insignificant, not only singly but collectively.

At the same time, when z is a large number, it is plain that Exp z is *immensely* larger.

Art. 3. We write Exp z to denote

$$\operatorname{Exp} z = 1 + z + \tfrac{1}{2}z^2 + \tfrac{1}{3!}z^3 + \tfrac{1}{4!}z^4 = \sum_{0}^{\infty}{}_p \frac{z^p}{p!}$$

Let us develop into a series Exp tz. Exp z, by simply multiplying the two series by the rules of algebra.

$$\operatorname{Exp} z = 1 + z + \tfrac{1}{2}z^2 + \tfrac{1}{3!}z^3 + \tfrac{1}{4!}z^4 + \text{etc.} = \sum_{0}^{\infty}{}_p \frac{z^p}{p!}$$

$$\operatorname{Exp} tz = 1 + tz + \tfrac{1}{2}t^2z^e + \tfrac{1}{3!}t^3z^3 + \tfrac{1}{4!}t^4z^4 + \text{etc.} = \sum_{0}^{\infty}{}_p \frac{t^pz^p}{p!}$$

$$\begin{aligned}\operatorname{Exp} tz\,.\operatorname{Exp} z = 1 &\\ + (t+1)z &\\ + (\tfrac{1}{2}t^2 + t + \tfrac{1}{2})z^2 &\\ + (\tfrac{1}{3!}t^3 + \tfrac{1}{2}t^2 + \tfrac{1}{2}t + \tfrac{1}{3!})z^3 &\\ + (\tfrac{1}{4!}t^4 + \tfrac{1}{3!}t^3 + \tfrac{1}{2}\cdot\tfrac{1}{2}t^2 + \tfrac{1}{3!}t + \tfrac{1}{4!})z^4 &\\ + (\tfrac{1}{5!}t^5 + \tfrac{1}{4!}t^4 + \tfrac{1}{3!}\tfrac{1}{2}t^3 + \tfrac{1}{2}\tfrac{1}{3!}t^2 + \tfrac{1}{4!}t + \tfrac{1}{5!})z^5 &\\ + (\tfrac{1}{6!}t^6 + \tfrac{1}{5!}t^5 + \tfrac{1}{4!}\tfrac{1}{2}t^4 + \tfrac{1}{3!}\tfrac{1}{3!}t^3 + \tfrac{1}{2}\tfrac{1}{4!}t^2 + \tfrac{1}{5!}t + \tfrac{1}{6!})z^6 &\\ + \text{etc.} &\\ = \sum_{0}^{\infty}{}_p \sum_{0}^{p}{}_q \frac{1}{q!}\frac{1}{(p-q)!}t^qz^p &\end{aligned}$$

But the binomial theorem is

$$\frac{(s+t)^p}{p!} = \sum_{0}^{r}{}_q \frac{1}{q!}\frac{1}{(p-q)!}t^qs^{p-q}$$

Hence

$$\operatorname{Exp} tz \cdot \operatorname{Exp} z = \sum_{0}^{\infty}{}_p \frac{(t+1)^pz^p}{p!} = \operatorname{Exp}\{(t+1)z\}$$

Consequently, if t is an integer and positive

$$(\operatorname{Exp} z)^t = \prod_{1}^{t} \operatorname{Exp} z = \operatorname{Exp} tz$$

Art. 4. The exponential function of 1 as variable is called the Naperian base, and is by most mathematicians denoted by e but by scholars of Benjamin Peirce by $\mathfrak{S}$. That is, if t is a positive integer

$$\mathfrak{S}^t = \operatorname{Exp} t.$$

Take the product Exp $z \cdot$ Exp $(-z)$

$$\operatorname{Exp} z = 1 + z + \tfrac{1}{2}z^2 + \tfrac{1}{3!}z^3 + \tfrac{1}{4!}z^4 + \text{etc.} = \sum_{0}^{\infty}{}_p \frac{x^p}{p!}$$

$$\text{Exp}(-z) = 1 - z + \tfrac{1}{2}z^2 - \tfrac{1}{3!}z^3 + \tfrac{1}{4!}z^4 - \text{etc.} = \sum_{0}^{\infty}{}_p \frac{(-)^p x^p}{p!}$$

$$\begin{aligned}\text{Exp}\, z \cdot \text{Exp}(-z) = {} & 1 \\ & + \tfrac{1}{2}(1-2+1)z^2 \\ & + \tfrac{1}{4!}(1-4+6-4+1)z^4 \\ & + \tfrac{1}{6!}(1-6+15-20+15-6+1)z^6 \\ & + \text{etc.}\end{aligned}$$

But by the binomial theorem

$$\begin{aligned}0 &= (1-1)^2 = 1-2+1 \\ 1 &= (1-1)^4 = 1-4+6-4+1 \\ &= (1-1)^6 = 1-6+15-20+15-6+1 \\ &= \text{etc.}\end{aligned}$$

Hence $\text{Exp}(-t) = \dfrac{1}{\text{Exp}\, t} = \mathfrak{S}^{-t}$

It is evident that $\left(\exp \dfrac{1}{t}\right)^t = \mathfrak{S}$. Hence

$$\text{Exp}\frac{1}{t} = \mathfrak{S}^{\frac{1}{t}}$$

Hence, for any rational fraction

$$\text{Exp}\frac{s}{t} = \mathfrak{S}^{\frac{s}{t}}$$

For irrational values the method of reasoning of Archimedes shows that

$$\text{Exp}\, z = \mathfrak{S}^z$$

Art. 5. In the theory of equations, the student has had some dealings with imaginary quantities. Each of these "quantities" is out of the scheme of ordinary quantity. The aggregate of them form a system of objects having relations of the system of which the relations of real numbers form but a fragment. These quantities follow the formal definitions of algebra, and all theorems strictly deducible from those definitions hold good of them. Other properties of quantity are generally modified. Every imaginary is assumed to be expressible in the form

$$x + yi$$

where x and y are real and $i^2 = -1$. By the principles of algebra

$$\mathfrak{S}^{x+yi} = \mathfrak{S}^x \mathfrak{S}^{yi}.$$

The sole definition of $\mathfrak{S}^{yi}$ is

$$\mathfrak{S}^{yi} = \operatorname{Exp} yi.$$

Now the series gives

$$\operatorname{Exp} yi = 1 - \tfrac{1}{2!}x^2 + \tfrac{1}{4!}x^4 - \tfrac{1}{6!}x^6 + \text{etc.}$$
$$+ [x - \tfrac{1}{3!}x^3 + \tfrac{1}{5!}x^5 - \tfrac{1}{7!}x^7 + \text{etc}] \cdot i.$$

This definition can give rise to no contradiction because it merely generalizes a universal proposition of algebra. Because it does this, it is not only permissible but is forced upon us.

Art. 6. The two real functions of y in the series for $\mathfrak{S}^{yi}$ are termed the *cosine* and *sine* of y. That is

$$\mathfrak{S}^{yi} = \cos y + \sin y \cdot i$$

which is called De Moivre's Principle.

$$\cos y = 1 - \tfrac{1}{2}y^2 + \tfrac{1}{4!}y^4 - \tfrac{1}{6!}y^6 + \text{etc.} = \sum_0^\infty{}_p \frac{(-)^p y^{2p}}{(2p)!}$$

$$\sin y = y - \tfrac{1}{3!}y^3 + \tfrac{1}{5!}y^5 - \tfrac{1}{7!}y^7 + \text{etc.} = \sum_0^\infty{}_p \frac{(-)^p y^{2p+1}}{(2p+1)!}$$

It is evident these series are convergent for all real values of y. We have to study the properties of these functions. These properties all spring from two, as follows:

1st. It is evident from the series that

$$\cos(-y) = \cos y \qquad \sin(-y) = -\sin y$$

and

$$\cos 0 = 1 \qquad \sin 0 = 0.$$

2nd.

$$\begin{aligned}\mathfrak{S}^{(y_1+y_2)i} &= \cos(y_1 + y_2) + \sin(y_1 + y_2) \cdot i \\ = \mathfrak{S}^{y_1 i} \cdot \mathfrak{S}^{y_2 i} &= (\cos y_1 + \sin y_1 \cdot i)(\cos y_2 + \sin y_2 \cdot i) \\ &= \cos y_1 \cdot \cos y_2 - \sin y_1 \cdot \sin y_2 \\ &\quad + (\sin y_1 \cdot \cos y_2 + \sin y_2 \cdot \cos y_1) \cdot i.\end{aligned}$$

That is

$$\cos(y_1 + y_2) = \cos y_1 \cdot \cos y_2 - \sin y_1 \cdot \sin y_2$$
$$\sin(y_1 + y_2) = \sin y_1 \cdot \cos y_2 + \sin y_2 \cdot \cos y_1.$$

This is called the Addition Principle.

From the two principles we have

$$\begin{aligned}1 = \mathfrak{S}^0 = \mathfrak{S}^{0i} = \mathfrak{S}^{(y-y)i} &= \cos y \cdot \cos(-y) - \sin y \cdot \sin(-y) \\ &\quad + [\sin y \cdot \cos(-y) + \sin(-y) \cdot \cos y] i \\ &= \cos^2 y + \sin^2 y.\end{aligned}$$

Art. 7. It is plain that $\sin 0 = 0$ and that $\sin y$ increases as y increases from 0 up until $\sin y$ is nearly (if not quite) 1. For this is true of successive pairs of terms of the development

$$y - \tfrac{1}{3!}y^3 \text{ increases till } y = \sqrt{2}$$
$$\tfrac{1}{5!}y^5 - \tfrac{1}{7!}y^7 \text{ increases till } y = \sqrt{30}$$
$$\tfrac{1}{9!}y^9 - \tfrac{1}{11!}y^{11} \text{ increases till } y = \sqrt{90} \text{ etc.}$$

Now even for $y = \sqrt{2}$

$$\sin y = (1 - \tfrac{2}{3!} + \tfrac{4}{5!} - \tfrac{8}{7!} + \text{etc.})\sqrt{2}$$
$$> \sqrt{\tfrac{8}{9}}$$

We shall therefore get a good idea of the course of the function if, by means of the addition principle, we calculate the sine of successive multiples of H, where $\sin H = \frac{1}{2}$

$$\sin H = \tfrac{1}{2} \qquad \cos H = \sqrt{1-(\tfrac{1}{2})^2} = \tfrac{\sqrt{3}}{2}$$
$$\sin 2H = \tfrac{1}{2}\tfrac{\sqrt{3}}{2} + \tfrac{1}{2}\tfrac{\sqrt{3}}{2} = \tfrac{\sqrt{3}}{2} \qquad \cos 2H = \tfrac{\sqrt{3}}{2}\tfrac{\sqrt{3}}{2} - \tfrac{1}{2}\cdot\tfrac{1}{2} = \tfrac{1}{2}$$
$$\sin 3H = \tfrac{\sqrt{3}}{2}\tfrac{\sqrt{3}}{2} + \tfrac{1}{2}\tfrac{1}{2} = 1 \qquad \cos 3H = \tfrac{1}{2}\tfrac{\sqrt{3}}{2} - \tfrac{\sqrt{3}}{2}\tfrac{1}{2} = 0$$
$$\sin 4H = 1\tfrac{\sqrt{3}}{2} + 0\tfrac{1}{2} = \tfrac{\sqrt{3}}{2} \qquad \cos 4H = 0\tfrac{\sqrt{3}}{2} - 1\tfrac{1}{2} = -\tfrac{1}{2}$$

The student should continue the calculation up to $18H$.

In any imaginary quantity $x + yi$, x is called the real part and yi the imaginary part. It would be quite proper, in imitation of the language of quaternions, to call x the *scalar* of the imaginary, yi the *vector* of the imaginary, and y the *tensor of the vector.*

As z leaves the value $z = 0$, where $ϭ^z = 1$, and becomes a larger and larger vector, yi, the exponential $ϭ^{yi}$ gets a vector part which increases while the scalar of $ϭ^{yi}$ diminishes. When $y = 3H$, the scalar of $ϭ^{yi}$ disappears and the tensor of the vector equals 1. Then, y still increasing, the tensor of the vector begins to diminish while the scalar becomes negative. The sine of y, of $6H - y$, and of $12H + y$ are all equal. Hence the sine of y and $18H - y$ are equal. The cosine of y, of $-y$, and of $12H + y$ are all equal. Hence $\cos(24H - y) = \cos y$ etc.

Art. 8. Let us make a numerical calculation of ϭ

$$1 = 1$$
$$1 = 1$$
$$\tfrac{1}{2} = 0.5$$

22

$$\tfrac{1}{3!} = 0.1666666667$$

$\frac{1}{4!} = 0.0416666667$
255
$\frac{1}{5!} = 0.0083333333$
652 14
$\frac{1}{6!} = 0.0013888889$
36 4
$\frac{1}{7!} = 0.0001984127$
6556
$\frac{1}{8!} = 0.0000248016$

$\frac{1}{9!} = 0.0000027557$
5
$\frac{1}{10!} = 0.0000002756$
1
$\frac{1}{11!} = 0.0000000251$

$\frac{1}{12!} = 0.0000000021$

$\frac{1}{13!} = 0.0000000002$

$\mathfrak{e} = 2.7182818286$
112344445

Some errors may have accumulated owing to neglected decimal places but we shall safely conclude $\mathfrak{e} = 2.71828183 \ldots$

Art. 9. Let us make a numerical calculation of H supposing $\sin H = \frac{1}{2}$. For this purpose we have to solve, for y considered as unknown, the equation

$$y - \tfrac{1}{3!}y^3 + \tfrac{1}{5!}y^5 - \tfrac{1}{7!}y^7 + \text{etc.} = B$$

or at least to get one root of the equation.

If y and thus B also are not too large, y can be developed in series according to increasing powers of B. And since $\sin(-y) = -\sin y$, this will contain only odd powers. Thus if we write a_1, a_3, a_5, etc. for the coefficients whose values have to be determined, begin with the formula

$$y = \alpha_1 B + \alpha_3 B^3 + \alpha_5 B^5 + \alpha_7 B^7 + \alpha_9 B^9 + \text{etc.} = \underset{0}{\textstyle\sum_p}\ a_{2p+1} B^{2p+1}$$

We get successive powers of y by ordinary algebraic multiplication, carefully attending to the general expressions, thus:

$$y^2 = \alpha_1\alpha_1 B^2 + (\alpha_1\alpha_3 + \alpha_3\alpha_1)B^4 + (\alpha_1\alpha_5 + \alpha_3\alpha_3 + \alpha_5\alpha_1)B^6 + (\alpha_1\alpha_7 + \alpha_3\alpha_5 + \alpha_5\alpha_3 + \alpha_7\alpha_1)B^8 + \text{etc.} = \overset{\infty}{\underset{0}{\textstyle\sum_p}}\ \overset{p}{\underset{0}{\textstyle\sum_q}}\ \alpha_{2q+1}\alpha_{2(p-q)+1} B^{2p+2}$$

$$\begin{aligned} y^3 = {} & \alpha_1\alpha_1 \cdot \alpha_1 B^3 + (\alpha_1\alpha_1 \cdot \alpha_3 + \alpha_1\alpha_3 \cdot \alpha_1 + \alpha_3\alpha_1 \cdot \alpha_1)B^5 + \\ & (\alpha_1\alpha_1 \cdot \alpha_5 + \alpha_1\alpha_3 \cdot \alpha_3 + \alpha_3\alpha_1 \cdot \alpha_3 + \alpha_1\alpha_5 \cdot \alpha_1 + \alpha_3\alpha_3 \cdot \alpha_1 + \\ & \alpha_5\alpha_1 \cdot \alpha_1)B^7 + (\alpha_1\alpha_1 \cdot \alpha_7 + \alpha_1\alpha_3 \cdot \alpha_5 + \alpha_3\alpha_1 \cdot \alpha_5 + \alpha_1\alpha_5 \cdot \alpha_3 + \\ & \alpha_3\alpha_3 \cdot \alpha_3 + \alpha_5\alpha_1 \cdot \alpha_3 + \alpha_1\alpha_7 \cdot \alpha_1 + \alpha_3\alpha_5 \cdot \alpha_1 + \alpha_3\alpha_3 \cdot \alpha_1 + \\ & \alpha_7\alpha_1 \cdot \alpha_1)B^8 \\ = {} & \sum_{0}^{\infty}{}_p \sum_{0}^{p}{}_q \sum_{0}^{q}{}_r \alpha_{2r+1}\alpha_{2(q-r)+1}\alpha_{2(p-q)+1}B^{2p+3} \end{aligned}$$

The mode of formation is obvious. We may arrange the αs in the order of their indices multiplying each product of αs by the number of different distinguishable ways of arranging its factors. Thus:

$$\begin{aligned} y^3 = {} & \alpha_1^3 B^3 + 3\alpha_1^2\alpha_3 B^5 + (3\alpha_1^2\alpha_5 + 3\alpha_1\alpha_3^2)B^7 \\ & + (3\alpha_1^2\alpha_7 + 6\alpha_1\alpha_3\alpha_5 + \alpha_3^3)B^9 \\ & + (3\alpha_1^2\alpha_9 + 6\alpha_1\alpha_3\alpha_7 + 3\alpha_1\alpha_5^2 + 3\alpha_3^2\alpha_5)B^{11} + \text{etc.} \\ y^5 = {} & \alpha_1^5 B^5 + 5\alpha_1^4\alpha_3 B^7 + (5\alpha_1^4\alpha_5 + 10\alpha_1^3\alpha_3^2)B^9 \\ & + (5\alpha_1^4\alpha_7 + 20\alpha_1^3\alpha_3\alpha_5 + 10\alpha_1^2\alpha_3^3)B^{11} \\ & + (5\alpha_1^4\alpha_9 + 20\alpha_1^3\alpha_3\alpha_7 + 10\alpha_1^3\alpha_5^2 + 30\alpha_1^2\alpha_3^2\alpha_5 + 5\alpha_1\alpha_3^4)B^{13} + \\ & \text{etc.} \end{aligned}$$

The general formula is

$$\begin{aligned} y^n = {} & \alpha_1^n B^n + n\alpha_1^{n-1}\alpha_3 B^{n+2} + (n\alpha_1^{n-1}\alpha_5 + \frac{n(n-1)}{2}\alpha_1^{n-2}\alpha_3^2)B^{n+4} \\ & + (n\alpha_1^{n-1}\alpha_7 + n(n-1)\alpha_1^{n-2}\alpha_3\alpha_5 \\ & + \frac{n(n-1)(n-2)}{2.3}\alpha_1^{n-3}\alpha_3^3)B^{n+6} \\ & + (n\alpha_1^{n-1}\alpha_9 + n(n-1)\alpha_1^{n-2}\alpha_3\alpha_7 + \frac{n(n-1)}{2}\alpha_1^{n-2}\alpha_5^2 \\ & + \frac{n(n-1)(n-2)}{2}\alpha_1^{n-3}\alpha_3^2\alpha_5 \\ & + \frac{n(n-1)(n-2)(n-3)}{4!}\alpha_1^{n-4}\alpha_3^4)B^{n+8} + \text{etc.} \end{aligned}$$

In forming the coefficient of each power of B we have to remember that there must be in each term of the coefficient as many α factors as the exponent of the power of y. This is obvious from the expression for y. We begin by attaching the index 1 to all but one of these αs. The sum of the indices must equal the power of B, as is obvious from the formula for y. Hence, the index of the last factor is determined. We next increase one index by 2 and diminish one by 2; and we repeat this operation in successive terms of the coefficients upon the same indices until it cannot be performed again without deranging the order of the indices. We then go back to the term resulting from the first such operation and increase one of its smallest indices by 2 while decreasing its largest by 2; and so on. For example, to form the coefficient of B^{n+10}, we first take $\alpha_1^{n-1}\alpha_{11}$. There are n different arrangements of the factors, that is, n arrangements that *show* as different, namely

$$\alpha_{11}\alpha_1^{n-1}$$
$$\alpha_1\alpha_{11}\alpha_1^{n-2}$$
$$\alpha_1^2\alpha_{11}\alpha_1^{n-3}$$
etc.

Hence the first term is $n\alpha_1^{n-1}\alpha_{11}$. The next contains the factors $\alpha_1^{n-2}\alpha_3\alpha_9$. There are $(n-1)$ places in which the α_3 may be inserted among the α_1s and n places in which the α_9 may be inserted among the α_1s and the α_3. Hence this term is $n(n-1)\alpha_1^{n-2}\alpha_3\alpha_9$. The next, on the same principle is $n(n-1)\alpha_1^{n-2}\alpha_5\alpha_7$. We cannot repeat the operation without deranging the order of the indices. We therefore return to $\alpha_1^{n-2}\alpha_3\alpha_9$ and increasing the index of one α_1 and diminishing that of α_9, we get $\alpha_1^{n-3}\alpha_3^2\alpha_9$. There are $(n-2)$ places in which one α_3 may be put among the $n-3$ factors α_1 and $(n-1)$ places in which the other α_3 may be put among the α_1s and the first α_3. But since the two α_3s are indistinguishable each of these $(n-1)(n-2)$ arrangements are indistinguishable; so that there are only $\dfrac{(n-1)(n-2)}{2}$ distinguishable arrangements. Finally there are n places into which α_9 may be placed. Hence the term is $\dfrac{n(n-1)(n-2)}{2}\alpha_1^{n-3}\alpha_3^2\alpha_9$. The next is $n(n-1)(n-2)$ $\alpha_1^{n-3}\alpha_3\alpha_5\alpha_7$. This operation cannot be repeated without deranging the order of the indices. We therefore return to $\alpha_1^{n-3}\alpha_3^2\alpha_9$ and increasing the index of an α_1 and diminishing that of α_9 we get $\alpha_1^{n-4}\alpha_3^3\alpha_7$. The term is $\dfrac{n(n-1)(n-2)(n-3)}{3!}\alpha_1^{n-4}\alpha_3^3\alpha_7$. The next is $\dfrac{n(n-1)(n-2)(n-3)}{2.2}\alpha_1^{n-4}\alpha_3^2\alpha_5^2$. Returning to $\alpha_1^{n-4}\alpha_3^3\alpha_7$, we next get $\alpha_1^{n-5}\alpha_3^4\alpha_5$. The term is $\dfrac{n(n-1)(n-2)(n-3)(n-4)}{4!}\alpha_1^{n-5}\alpha_3^4\alpha_5$. The next combination is $\alpha_1^{n-6}\alpha_3^6$.

The term is $\dfrac{n(n-1)(n-2)(n-3)(n-4)(n-5)}{6!}\alpha_1^{n-6}\alpha_3^6$. No further combination can be made. Hence, the whole term of y^n is $(n\alpha_1^{n-1}\alpha_{11}+$

$$n(n-1)\alpha_1^{n-2}\alpha_3\alpha_9 + n(n-1)\alpha_1^{n-2}\alpha_5\alpha_7 + \frac{n(n-1)(n-2)}{2}\alpha_1^{n-3}\alpha_3^2\alpha_9 +$$
$$n(n-1)(n-2)\alpha_1^{n-3}\alpha_3\alpha_5\alpha_7 + \frac{n(n-1)(n-2)(n-3)}{3!}\alpha_1^{n-4}\alpha_3^3\alpha_7 +$$
$$\frac{n(n-1)(n-2)(n-3)}{4}\alpha_1^{n-4}\alpha_3^2\alpha_5^2 + \frac{n(n-1)(n-2)(n-3)(n-4)}{4!}\alpha_1^{n-5}\alpha_3^4\alpha_5$$
$$+ \frac{n(n-1)(n-2)(n-3)(n-4)(n-5)}{6!}\alpha_1^{n-6}\alpha_3^6)B^{n+10}$$

Having thus obtained expressions for the powers of y in terms of B and the unknown αs, we substitute these in the equation to be solved, keeping terms in the same power of B together. We thus get

$$\begin{aligned} B = {} & \alpha_1 B \\ & + \alpha_3 B^3 - \tfrac{1}{3!}\alpha_1^3 B^3 \\ & + \alpha_5 B^5 - \tfrac{1}{2}\alpha_1^2\alpha_3 B^5 + \tfrac{1}{5!}\alpha_1^5 B^5 \\ & + \alpha_7 B^7 - (\tfrac{1}{2}\alpha_1^2\alpha_5 + \tfrac{1}{2}\alpha_1\alpha_3^2)B^7 + \tfrac{1}{4!}\alpha_1^4\alpha_3 B^7 - \tfrac{1}{7!}\alpha_1^7 B^7 \\ & + \alpha_9 B^9 - (\tfrac{1}{2}\alpha_1^2\alpha_7 + \alpha_1\alpha_3\alpha_5 + \tfrac{1}{6}\alpha_3^3)B^9 + (\tfrac{1}{4!}\alpha_1^4\alpha_5 + \tfrac{1}{12}\alpha_1^3\alpha_3^2)B^9 \\ & - \tfrac{1}{9!}\alpha_1^9 B^9 \\ & + \text{etc.} \end{aligned}$$

This, it must be remembered, is a general expression which holds good for all values of B, so long as it is not so large as to render the series divergent. Hence, it holds for an excessively small value of B. But for an excessively small value of B, all powers of B as high as B^3 are insignificant, and may be neglected. Hence $B = \alpha_1 B$. Subtracting this equation from the above, B^3 is now the lowest value, and by the same reasoning $\alpha_3 B^3 - \frac{1}{3!}\alpha_1^3 B^3 = 0$. In short, an equation which holds good generally is equivalent to as many equations as there are powers of the variable. Indeed, a *general* equation is equivalent to as many equations among the coefficients as we may choose to take of values of the variable. For it holds good of every such value.

Hence, $B = \alpha_1 B$ or $\alpha_1 = 1$

$\alpha_3 - \frac{1}{3!}\alpha_1^3 = 0$ or $\alpha_3 = \frac{1}{3!} = \frac{1}{6}$

$\alpha_5 - \frac{1}{2}\alpha_1^2\alpha_3 + \frac{1}{5!}\alpha_1^5 = 0$ or $\alpha_5 = \frac{1}{2}\cdot\frac{1}{6} - \frac{1}{5!} = \frac{9}{5!} = \frac{3}{2\cdot 4\cdot 5}$ $[= \frac{3^2}{5!}]$

$\alpha_7 - \frac{1}{2}\alpha_1^2\alpha_5 - \frac{1}{2}\alpha_1\alpha_3^2 + \frac{1}{4!}\alpha_1^4\alpha_3 - \frac{1}{7!}\alpha_1^7 = 0$

or $\alpha_7 = \frac{1}{2}\frac{3^2}{5!} + \frac{1}{2}\frac{1}{(3!)^2} - \frac{1}{4!}\frac{1}{3!} + \frac{1}{7!}$

$$= \frac{3^3.7 + 70 - 35 + 1}{7!} = \frac{3^2 5^2}{7!}$$

$\alpha_9 = \frac{1}{2}\alpha_1^2\alpha_7 - \alpha_1\alpha_3\alpha_5 - \frac{1}{6}\alpha_3^3 + \frac{1}{4!}\alpha_1^4\alpha_5 + \frac{1}{12}\alpha_1^3\alpha_3^2 - \frac{1}{9!}\alpha_1^9$

or $\alpha_9 = \frac{1}{2}\frac{3^2\cdot 5^2}{7!} + \frac{1}{3!}\frac{3^2}{5!} + \frac{1}{(3!)^4} - \frac{1}{4!}\frac{3^2}{5!} - \frac{1}{12}\frac{1}{(3!)^2} + \frac{1}{9!}$

$$= \frac{4.9.3^2.5^2 + 56.9.3^2 + 8.7.5 - 9.14.3^2 - 3.8.7.5 + 1}{9!}$$

$$= \frac{3^2.5^2.7^2}{9!}$$

The law is evident. It is true that we have not *demonstrated* it. There might be an appearance of one law in the first terms which was only a special case of a more complicated law if further terms were examined. But without the differential calculus the demonstration, though not impossible, becomes so complicated that it must be dispensed with here. We have in fact as we here *seem* to have

$$y = B + \frac{1^2}{3!}B^3 + \frac{1^2.3^2}{5!}B^5 + \frac{1^2.3^2.5^2}{7!}B^7 + \frac{1^2.3^2.5^2.7^2}{9!}B^9 + \text{etc.}$$

where only the terms are written which have been *proved* to have these values. This may be otherwise written

$$y = B + \frac{1}{2}\frac{B^3}{3} + \frac{1\,3}{2\,4}\frac{B^5}{5} + \frac{1\,3\,5}{2\,4\,6}\frac{B^7}{7} + \frac{1\,3\,5\,7}{2\,4\,6\,8}\frac{B^9}{9} + \text{etc.}$$

We have now only to put $B = \frac{1}{2}$ and perform the numerical calculations to get the value of H. It gives $H = 0.52360$. $6H$ is denoted by π by most mathematicians, but by scholars of Benjamin Peirce by the character ə. It is called *pi*, *circ*, or *circuit*. It has been calculated to 700 places of decimals. Its first figures are 3.14159.

The most instructive feature of this article is that it shows the student that mathematical investigation depends upon observation followed up by demonstration. This demonstration, as will be seen elsewhere, is only observation that certain elements enter into a diagram or array of letters in such ways that they cannot be absent.

Art. 10. It is now time to explain in a general way what the application of the two functions $\cos y$ and $\sin y$ are in geometry. In the right angled triangle ABC [Fig. 1], having its right angle at C, denote the sides thus

$$a = BC \qquad b = AC \qquad h = AB.$$

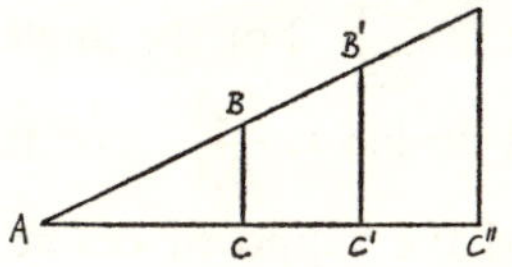

Fig. 1

Then $\frac{a}{h}$ and $\frac{b}{h}$ do not depend on the lengths of the sides, provided the angle A is constant. For by similar triangles, $\frac{a}{h} = \frac{a'}{h'} = \frac{a''}{h''}$, and $\frac{b}{h} = \frac{b'}{h'} = \frac{b''}{h''}$.

Thus $\frac{a}{h}$ and $\frac{b}{h}$ are functions of $\angle A$. They possess the first fundamental property of the cosine and sine inasmuch as if the angle opens in the reverse direction, that is, takes the negative value, $\frac{a}{h}$ becomes negative like the sine, while $\frac{b}{h}$ retains the same value like the cosine [Fig. 2]. Also, when the angle vanishes $\frac{a}{h}$ vanishes like the sine while $\frac{b}{h}$ becomes 1 like the cosine. Further, just as

$$\sin^2 + \cos^2 = 1$$

so
$$\frac{a^2}{h^2} + \frac{b^2}{h^2} = 1$$

for this is the algebraic Pythagorean proposition.

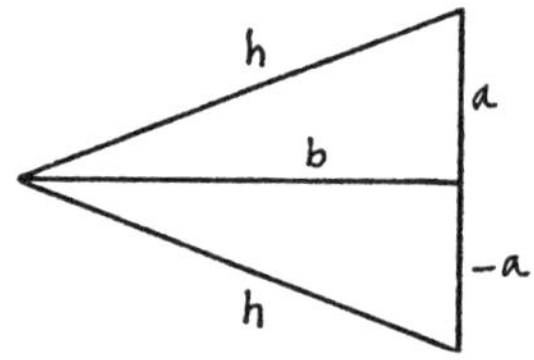

Fig. 2

Let us see whether $\frac{a}{h}$ and $\frac{b}{h}$ have also the addition property of the *sine* and *cosine*; for if they have, they *are* the sine and cosine of some measure of the angle.

Let AC meet BD at right angles at C and let BE meet AD at right angles at E [Fig. 3]. Then for the angle BAC the supposed sine is $\frac{BC}{BA}$ and the supposed cosine is $\frac{CA}{BA}$. For the angle CAD the supposed sine is $\frac{CD}{DA}$ and the supposed cosine is $\frac{CA}{DA}$. For the sum of the angles the supposed sine is $\frac{BE}{BA}$ and the supposed cosine is $\frac{EA}{BA}$. The question is whether or not

$$\frac{EA}{BA} = \frac{CA}{BA}\frac{CA}{DA} - \frac{BC}{BA}\frac{CD}{DA}$$

$$\frac{BE}{BA} = \frac{BC}{BA}\frac{CA}{DA} + \frac{CD}{DA}\frac{CA}{BA}$$

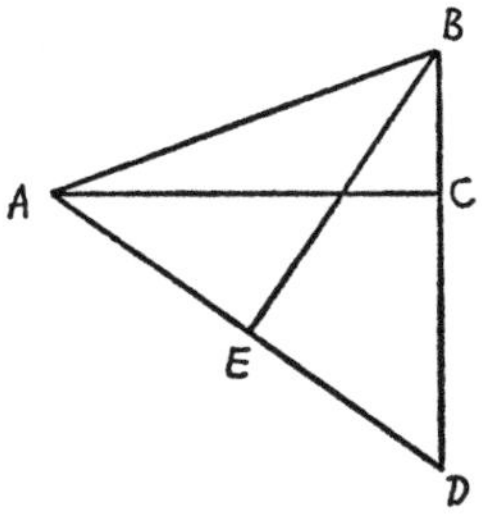

Fig. 3

Clearing the last equation from fractions, it gives

$$BE \cdot DA = (BC + CD)CA$$

which expresses two different ways of measuring twice the area of the two triangles. Clearing the other equation it gives

$$\begin{aligned}(CA)^2 - BC \cdot CD &= EA \cdot DA \\ &= (DA - DE) \cdot DA.\end{aligned}$$

But the two triangles ACD and BED are similar, since they have an angle in common and are both right-angled. Hence

$$DE = CD\frac{BC + CD}{DA}$$

or $\quad DE \cdot DA = CD \cdot BC + (CD^2).$

Hence $\quad (DA - DE)DA = (DA)^2 - (CD)^2 - BC \cdot CD.$

Thus the questionable equation amounts to this

$$(CA)^2 = (DA)^2 - (CD)^2$$

which is again the Algebraic Pythagorean.

Hence $\frac{a}{h}$ and $\frac{b}{h}$ are respectively the sine and cosine of the angle A.

But the angle for which $\frac{b}{h}$ vanishes is a right angle, while the value y for which cos y vanishes is $\frac{1}{2}$ Ͽ. Hence, we must consider 3.14159 as the measure of two right angles. This is called the *analytical* measure of angular quantity [also called *radian* measure].

Now it is proved in the Elements of Mathematics that two right angles at the centre of a circle of unit radius cut off an arc of the circumference whose length is 3.14159. Thus, the analytical measure of an angle is the length of the arc of unit radius which it subtends, if the circle has its centre at the vertex of the angle.

Since two right angles equal 180° the analytical measure of one degree is $\frac{3.14159}{180}$ and the magnitude of the unit of analytical measure is

$$\frac{180^\circ}{3.14159} = 57^\circ \text{ about.}$$

Art. 1. *Definition of* sin z.

Sin z is a function of z having the following properties:

1st, For each value of z, real or imaginary, sin z has a single value.

2nd, The same general formula for $\sin(c+z)$ in terms of c and z holds whether c be real or imaginary.

3rd, There is a constant quantity π such that

$$\sin(z \pm \pi) = \sin(-z).$$

4th, There are no values of z whose sines are equal except those that are deducible from the third property; and there is some value of z which satisfies the equation

$$\sin z = a + bi$$

whatever the value of $a + bi$.

5th, $\sin 0 = 0$.

6th, $\sin \infty = \infty$.

7th, $\sin \frac{\pi}{2} = 1$.

Art. 2. From the third property it follows that the equation

$$\sin z = \sin c$$

is satisfied only by

$$z = 2N\pi + c$$

or $$z = 2N\pi + \pi - c$$

where N is any whole number, positive or negative.

$$\cos x = \left(\sin \frac{\pi}{2} - x\right)$$

$$\sin(x-y) = \sin x \cos y - \cos x \sin y.$$

Put $y = 0$

$$\sin x = \sin x \cdot \cos 0 - \cos x \cdot \sin 0$$

$$\therefore \cos 0 = \sin\frac{\pi}{2} = 1 \qquad \sin 0 = \cos\frac{\pi}{2} = 0$$

Put $x = 0$

$$\sin(-y) = -\sin y$$

Put $x = \frac{\pi}{2} - z$

$$\cos(y+z) = \cos z \cos y - \sin z \sin y$$
$$\cos(y-z) = \cos z \cos y + \sin z \sin y$$

Put $y = z$

$$1 = \cos^2 y + \sin^2 y$$

Put $y = 0$

$$\cos(-z) = \cos z$$
$$\therefore \sin(x+y) = \sin x \cos y + \cos x \sin y$$
$$\sin 2x = 2 \sin x \cos x$$
$$\cos 2x = \cos^2 x - \sin^2 x$$
$$\sin \pi = 0 \qquad \cos \pi = -1$$
$$\sin\frac{3\pi}{2} = -1 \qquad \cos\frac{3\pi}{2} = 0$$
$$\sin 2\pi = 0 \qquad \cos 2\pi = 1$$
$$\sin\left(\frac{\pi}{2}+x\right) = \cos x \qquad \cos\left(\frac{\pi}{2}+x\right) = -\sin x$$
$$\sin(\pi+x) = -\sin x \qquad \cos(\pi+x) = -\cos x$$
$$\sin(2\pi+x) = \sin x \qquad \cos(2\pi+x) = \cos x$$

INTRODUCTION TO TRIGONOMETRY (195)

Don't you think a very small trigonometry for advanced students, treating of errors due to neglected decimals, and such matters of practical computing would go, and pay for the time of preparation.

I should begin something like this:

If $i^2 = -1$, and e is any positive number, then writing

$$z = x + yi$$
$$e^z = e^{x+yi} = e^x \cdot e^{yi}.$$

Now e^x is calculable if x is a rational fraction; and if not, it can be approximated to. The question is what is e^{yi}? Suppose

$$e^{yi} = Fy + \Phi y \cdot i.$$

Then $$e^{(y_1+y_2)i} = F(y_1+y_2) + \Phi(y_1+y_2) \cdot i$$
$$= e^{y_1 i} \cdot e^{y_2 i} = Fy_1 \cdot Fy_2 - \Phi y_1 \cdot \Phi y_2$$
$$+ (Fy_1 \cdot \Phi y_2 + Fy_2 \cdot \Phi y_1)i$$

So that

$$F(y_1+y_2) = Fy_1 \cdot Fy_2 - \Phi y_1 \cdot \Phi y_2; \; F2y = (Fy)^2 - (\Phi y)^2$$
$$\Phi(y_1+y_2) = Fy_1 \cdot \Phi y_2 + Fy_2 \cdot \Phi y_1; \; \Phi 2y = 2Fy \cdot \Phi y$$

Since $$e^{0i} = 1, \quad F0 = 1 \text{ and } \Phi 0 = 0.$$
$$1 = F0 = F(y-y) = Fy \cdot F(-y) - \Phi y \cdot \Phi(-y)$$
$$0 = \Phi 0 = \Phi(y-y) = Fy \cdot \Phi(-y) + F(-y) \cdot \Phi y$$
$$\therefore \frac{Fy}{F(-y)} = (Fy)^2 + (\Phi y)^2 = -\frac{\Phi y}{\Phi(-y)}$$

Now $(Fy)^2 + (\Phi y)^2$, being the sum of the squares of two real quantities, is positive. Hence, when y is infinitesimal. Hence for infinitesimal y

$$\Phi y = -\Phi(-y)$$
$$Fy = F(-y).$$

When y is very small Fy and Φy are developable according to powers of y. Assume

$$Fy = 1 + Ay^M + \text{etc.}$$
$$\Phi y = ay^m + \text{etc.}$$

Then
$$F2y = 1 + 2^M Ay^M + \text{etc.}$$
$$\Phi 2y = 2^m ay^m + \text{etc.}$$

But

$$F2y = (Fy)^2 - (\Phi y)^2$$
$$= 1 + 2Ay^M - a^2y^{2m}$$
$$\Phi 2y = 2Fy \cdot \Phi y$$
$$= 2ay^m + \text{etc.}$$

Hence $m = 1 \qquad M = 2 \qquad 4A = 2A - a^2.$

Hence $\Phi y = ay + \text{etc.}$

But if $e = \varepsilon^t \qquad e^{yi} = \varepsilon^{tyi}.$

Hence if $\Phi_e y = ay + \text{etc.} \qquad \Phi_\varepsilon y = \frac{a}{t} y + \text{etc.}$

Hence by taking a suitable value, $\mathfrak{S}$, of e we can make $\Phi_{\mathfrak{S}} y = y + \text{etc.}$ [Also]

$$\frac{Fy}{F(-y)} = -\frac{\Phi y}{\Phi(-y)} = (\text{say})\ C^{-1}$$

$$\frac{F(2y)}{F(-2y)} = \frac{(Fy)^2 - (\Phi y)^2}{C^2(Fy)^2 - C^2(\Phi y)^2} = C^{-2}$$

Hence by suitably choosing t we can make

$$F(-ty) = Fty$$
$$\Phi(-ty) = -\Phi ty.$$

Write
$$e^t = \mathfrak{S}$$
$$Fty = \cos y$$
$$\Phi ty = \sin y$$

then
$$(\cos y)^2 + (\sin y)^2 = 1.$$

Since $\sin(-y) = -\sin y$

if $\sin y$ is developed according to powers of y only odd powers appear. Let the lowest be the Nth. Then if y is very small

$$\sin y = y^N \qquad \cos y = \sqrt{1 - y^{2N}} = 1 - \text{etc.}$$

Then $\sin 2y = 2^N y^N.$

But $\sin 2y = 2 \sin y \cos y = 2y^N 1.$

$$\therefore 2 = 2^N \text{ or } N = 1.$$

TRIGONOMETRY

§1. Trigonometry is the theory of the exponential function, defined by the series

$$\operatorname{Exp} x = 1 + x + \tfrac{1}{2}x^2 + \tfrac{1}{3!}x^3 + \text{etc.} = \sum_{0}^{\infty}{}_p \frac{x^p}{p!}$$

§2. To find $\operatorname{Exp} tx \times \operatorname{Exp} x$

$$\operatorname{Exp} tx = 1 + tx + \tfrac{1}{2}t^2x^2 + \tfrac{1}{3!}t^3x^3 = \sum_{0}^{\infty}{}_p \frac{t^p x^p}{p!}$$

$$\operatorname{Exp} x = 1 + x + \tfrac{1}{2}x^2 + \tfrac{1}{3!}x^3 = \sum_{0}^{\infty}{}_p \frac{x^p}{p!}$$

$$\begin{aligned}
\text{Product} &= 1 \\
&+ (t+1)x \\
&+ (\tfrac{1}{2}t^2 + t + \tfrac{1}{2})x^2 \\
&+ (\tfrac{1}{3!}t^3 + \tfrac{1}{2}t^2 + \tfrac{1}{2}t + \tfrac{1}{3!})x^3 \\
&+ (\tfrac{1}{4!}t^4 + \tfrac{1}{3!}t^3 + \tfrac{1}{2}\cdot\tfrac{1}{2}t^2 + \tfrac{1}{3!}t + \tfrac{1}{4!})x^4 \\
&+ (\tfrac{1}{5!}t^5 + \tfrac{1}{4!}t^4 + \tfrac{1}{3!}\,\tfrac{1}{2}t^3 + \tfrac{1}{2}\,\tfrac{1}{3!}t^2 + \tfrac{1}{4!}t + \tfrac{1}{5!})x^5 \\
&+ \text{etc.} \\
&= \sum_{1}^{\infty}{}_p \sum_{0}^{p}{}_q \frac{1}{q!}\frac{1}{(p-q)!}t^q x^p
\end{aligned}$$

But the binomial theorem is

$$\frac{(s+t)^p}{p!} = \sum_{0}^{p}{}_q \frac{1}{q!}\frac{1}{(p-q)!}t^q s^{p-q}$$

Hence

$$\operatorname{Exp} tx.\ \operatorname{Exp} x = \sum_{0}^{\infty}{}_p \frac{(t+1)^p x^p}{p!} = \operatorname{Exp}(t+1)p.$$

Hence if t is an integer

$$(\operatorname{Exp} x)^t = \prod_{1}^{t} \operatorname{Exp} x = \operatorname{Exp} tx.$$

§3. The exponential function of §1 as variable is called the Naperian base and is usually denoted by e, but by scholars of Benjamin Peirce by the character Ϭ. That is, if t is a positive integer

$$\text{Ϭ}^t = \operatorname{Exp} t.$$

Take the product $\operatorname{Exp} x \cdot \operatorname{Exp}(-x)$

$$\operatorname{Exp} x = 1 + x + \tfrac{1}{2}x^2 + \tfrac{1}{3!}x^3 + \text{etc.} = \sum_0^\infty{}_p \frac{x^p}{p!}$$

$$\operatorname{Exp}(-x) = 1 - x + \tfrac{1}{2}x^2 - \tfrac{1}{3!}x^3 + \text{etc.} = \sum_0^\infty{}_p \frac{(-)^p x^p}{p!}$$

$$\begin{aligned} \text{Product } = {} & 1 \\ & + (\tfrac{1}{2} - 1 + \tfrac{1}{2})x^2 \\ & + (\tfrac{2}{4!} - \tfrac{2}{3!} + \tfrac{1}{2}\tfrac{1}{2})x^4 \\ & + (\tfrac{2}{6!} - \tfrac{2}{5!} + \tfrac{2}{2\cdot 4!} - \tfrac{1}{3!3!})^6 \\ & + \text{etc.} \end{aligned}$$

But by the binomial theorem all the terms but the first vanish. Hence

$$\operatorname{Exp}(-t) = \text{Ϭ}^{-t}.$$

It is also evident that $(\operatorname{Exp} \frac{1}{t})^t = \text{Ϭ}$. Hence

$$\operatorname{Exp} \tfrac{1}{t} = \text{Ϭ}^{\frac{1}{t}}$$

Hence for any rational fraction $\frac{s}{t}$

$$\operatorname{Exp} \tfrac{s}{t} = \text{Ϭ}^{\frac{s}{t}}$$

The method of limits extends this result to irrational exponents.

§4. The two real functions in the expression for Ϭ^{yi} are called the *cosine* and *sine* of y. That is

$$\text{Ϭ}^{yi} = \cos y + \sin y \cdot i$$

which is called Des Moivre's principle.

$$\cos y = \sum_0^\infty{}_p \frac{(-)^p y^{2p}}{(2p)!} \qquad \sin y = \sum_0^\infty{}_p \frac{(-)^p y^{2p+1}}{(2p+1)!}$$

We have to study the properties of these functions. These properties spring from two that are fundamental.

The first is that, as the series show,

$$\cos(-y) = \cos y \qquad \sin(-y) = -\sin y$$

The second is the addition principle. That is

$$\text{Ϭ}^{(y_1+y_2)i} = \cos(y_1 + y_2) + \sin(y_1 + y_2) \cdot i$$

$$= \mathfrak{G}^{y_1 i} \cdot \mathfrak{G}^{y_2 i} = (\cos y_1 + \sin y_1 \cdot i)(\cos y_2 + \sin y_2 \cdot i)$$

$$= \cos y_1 \cos y_2 - \sin y_1 \sin y_2$$
$$+ (\sin y_1 \cos y_2 + \sin y_2 \cos y_1) i$$

That is

$$\cos(y_1 + y_2) = \cos y_1 \cdot \cos y_2 - \sin y_1 \cdot \sin y_2$$
$$\sin(y_1 + y_2) = \sin y_1 \cdot \cos y_2 + \sin y_2 \cdot \cos y_1.$$

If we put

$$y_2 = -y_1, \text{ since } \mathfrak{G}^{0i} = 1$$
$$1 = \cos y \cdot \cos(-y) - \sin y \cdot \sin(-y) = (\cos y)^2 + (\sin y)^2.$$

2

MULTITUDE AND QUANTITY

A. ON QUANTITY, WITH SPECIAL REFERENCE TO COLLECTIONAL AND MATHEMATICAL INFINITY (14)

§1. PURPOSE OF THIS MEMOIR

Art. 1. The investigation whose chief results are here given had for its aim to apply mathematically exact logic, particularly the logic of relatives, to the solution of the following questions:

1st, What is *mathematics*? That is to say, granted that the theory of numbers, algebra, topology, intersectional and metrical geometry, the differential calculus, etc. are branches of mathematics, it is required to form a distinct and unitary conception of a study embracing those branches (or, at least, what is most [special] and characteristic in the aggregate of them), so limited as to [thus] give it the maximum utility in the natural classification of the sciences, and to be directly connected with a sufficient explanation of the part which mathematics plays in the development of knowledge.

2nd, What is *quantity*? That is to say, granted that the system of the cardinal numbers is a system of quantity, that the system of rational numbers is another, that the system of real numbers is another, that the system of imaginary quantities is another, and that the system of quaternions is another, it is required to form a distinct and unitary conception applicable to those systems, so limited as to give it the maximum utility in the account of cognition, bringing into view the essential figures of such systems, especially those which make them important to science, and explaining why quantity plays the part it does in thought.

3rd, What is *continuity*, such as it is in our natural conceptions about time? What evidence is there that time and space really are continuous? And how far can the mathematician take continuity into account?

4th, Is it possible that there should be two collections not equal in multitude yet of which neither is greater than the other? That is to say, can there be two collections neither of which could in any way be put into a one-to-one correspondence with a part or the whole of the other?

5th, What are the different kinds of *infinity* which concern mathematics or which appear at first sight to do so, and what are their properties? What is the true logic of the different orders of infinity of the calculus? And what are the merits of the method of infinitesimals (as distinguished from the method of limits) in view of the modern doctrines of infinity, imaginaries, etc.? In particular, is [there] any absurdity in the conception of an infinitesimal any more than in the conception of a square root of negative unity?

§2. THE NATURE OF MATHEMATICS

Art. 2. As a general rule, the value of an exact philosophical definition of a term in familiar use lies in its bringing out distinct ideas of the purpose and function of that which is defined. In particular, this is true of the definition of an extensive branch of science. In order to assign the most useful boundaries for such a study, it is necessary to consider what part of the work of science must, from the nature of things, be performed by those men who are to do that part which, it is acknowledged, is to be included in that science. Now the services of the mathematician are called in when the physicist, the engineer, the insurance company, the lawgiver, etc. finds himself confronted with a complicated state of relations of fact and is in doubt whether or not it necessarily involves some other relation, or wishes to know precisely what relation it does involve of a certain kind. The mathematician is not responsible at all for the truth of the premises: that he is to assume. The first task before him usually is to substitute for the excessively complicated facts set before him, often confused in their statement, a comparatively simple system of relations, which while adhering as closely as possible to the given premises shall be within his powers as a mathematician to deal with. This he terms his *hypothesis*. Thereupon, he proceeds to show that the relations stated in that hypothesis involve as a part of them certain other relations, not explicitly taken account of.

Thus the mathematician is not concerned with real truth: he only studies the substance of hypotheses. This distinguishes mathematics from every other science. For even logic deals with some questions of fact. Thus, in reference to each question attacked, the logician must at least *hope*, even if he cannot probably assert, that there is some definite true answer, toward which sincere inquiry will at least converge as to a limit, provided it be pushed far enough, even if it cannot reach it.

Hence, the proper definition of mathematics is that it is the study of the substance of hypotheses, which it first frames and then traces to their consequences.

It would certainly be wrong to limit mathematics to the latter of these tasks; for the former can only be satisfactorily performed by the minds who are to busy themselves with the latter. The framing of the hypothesis of the two-way spread of imaginary quantity with its single point at infinity was certainly a mathematical achievement; and so was the framing of the hypothesis of a Riemann surface. On the other hand, to include all framing of ideal creations under mathematics would be making musicians and poets and the authors of fairy stories mathematicians. In short, while mathematics is the only branch of *science* which has nothing to do with objective truth, it is not the only branch of *intellectual activity* which is purely ideal. Accordingly, a further specification of the characters of mathematical hypotheses is requisite.

There is *pure* mathematics and *applied* mathematics. Pure mathematicians would strenuously object to a definition which should limit their hypotheses to such as are subservient to the discovery of objective truth. A romancer who draws any necessary deductions from the situations he creates (as every romancer does) is beyond doubt doing mathematical work; and the charm of romance is in part due to the natural interest we have in tracing necessary consequences. But this is applied mathematics for the reason that the hypotheses are clothed with accidents which are not relevant to the forms of deduction. Mathematical hypotheses are such as are adapted to the tracing of necessary conclusions; and the hypotheses of *pure* mathematics are stripped of all accidents which do not affect the forms of deduction, that is, the relations of the conclusions to the premises.

We thus finally reach this definition. *Mathematics* is the study of the substance of hypotheses with a view to the drawing of necessary conclusions from them. It is *pure* when the hypotheses contain nothing not relevant to the forms of deduction.

§3. THE NATURE OF QUANTITY

Art. 3. Deductive reasoning consists in first constructing in imagination (or on paper) a diagram the relations of whose parts is completely determined by the premises, secondly in experimenting upon the effects of modifying this diagram, thirdly in observing in the results of those experiments certain relations between its parts over and above those which sufficed to determine its construction, and fourthly in satisfying ourselves by inductive reasoning that these new relations would always subsist where those stated in the premises existed.

Deductive reasoning chiefly consists in compounding relations, or rather, in recognizing relations as compound. But in general relations so far as compounded are dual and any two dual relations if different, when they come to [be] compounded are compounded just as are any other two. There is no difference in form between a lover of a servant and a mother of a benefactor; and in general any relation is compounded with itself just as any other is. Only certain species of relations are either *transitive*, as what is subsequent to something subsequent is itself subsequent, or are *cyclical*, or present some other peculiarity in the character of their combinations with themselves or with other special relations. Hence, pure mathematics will recognize relations only as being the same or different, or, in special cases, as being transitive, cyclical, etc.

It is very useful in tracing effects of compounding relations to be provided with a lettered diagram of all possible combinations of the relations dealt with. Such a diagram is of course merely to have blanks or places for the related objects and is not to take note of the accidental character of the relations, but only of any essential peculiarities of their modes of combinations. It is to show every possible compound of the relations concerned; and therefore every part of it will be precisely like every other part, unless the relation is of such a kind that after certain compoundings, the character of the compoundings changes. An ordered arrangement of dots, everywhere uniform, and therefore unlimited, would, for example, answer the purpose. But the dots must be lettered, or otherwise designated, in order that it may serve any use, for the same reason that indices, or signs of individual objects, have to be made use of in all deductive reasoning, as the logic of relatives clearly shows.

Such a diagram is a schema of the scale of quantity. Thus a *quantity* is nothing but one of the hecceities in the system of possible forms of combination of certain relations, whose characters except in regard to their forms of combination do not concern the scale of quantity.

Art. 4. The simplest of all possible systems of quantity consists of two grades. This system occurs in apodictic logic, where every proposition is either true or false. Negation is the operation by which we pass from asserting the truth of a proposition to asserting its falsity, or the reverse. But when this meaning is given to the two grades, the system belongs to applied mathematics. In pure mathematics there are simply two grades, distinguished by two letters devoid of general signification, as v and f, and there is the operation by which v is changed to f and f to v. Denoting this function by N, we have

$$Nf = v \qquad \text{and} \qquad Nv = f.$$

There will also be functions of two letters, as for instance, a function Φ such that

$$\Phi(f,f) = \Phi(f,v) = \Phi(v,v) = v \ \Phi(v,f) = f.$$

From these will be derived others, such as

$$N\Phi(f,v) = f \qquad\qquad \Phi(Nf,f) = f \quad \text{etc.}$$

Art. 5. Another system of quantity is suggested by the multitudes of finite collections, that is, of collections to which De Morgan's "syllogisms of transposed quantity" apply. A syllogism of transposed quantity is an inference which is similar in principle to the following:

Every Hottentot kills a Hottentot,
But no Hottentot is killed by more than one Hottentot;
∴ Every Hottentot is killed by a Hottentot.

For if two collections, say the *A*s and the *B*s, are such that for every *B* there is not an *A*, but for every *B* but one there is an *A*, then the *B*s are *one more* than the *A*s. Compounding this relation of *one more* than in every possible way, we have an endless succession of grades, which is the scheme of finite multitude. But as long as this signification is retained, this system of quantity is one of applied mathematics. In order to get the scheme of pure mathematics we must evacuate the relation of this special character. We then have a mere endless succession of grades, which must be systematically designated, as for example by

$P, Q, QP, QQ, QPP, QPQ, QQP, QQQ, QPPP$, etc.

or by

$A, AB, B, BA, BAB, BABC, BAC, BC, ABC, AC, C, CA, CAB, CB, CBA$, etc.

But this is not the whole of this system of quantity in pure mathematics, since it is not perfectly homogeneous. It presents a *singularity*, in that the first member has not the same relationships as the others; and this singularity is not inherent, but can be removed by extending the succession endlessly backward. We thus get the system of datary or ordinal number, which extends endlessly forward and backward without a limit, or ∞.

This system may be formally defined as follows: There is a function E, such that a being a known and x an unknown quantity of the system, the equations

$$Ex = a \qquad \text{and} \qquad Ea = x$$

have each, in every case, just one solution, and such that taking any two different quantities a and b of the system one of them may be produced from the other by repeatedly performing the operation E upon it, and no performance of this operation or repetition of it upon a quantity of the system can give that quantity.

Art. 6. In this, as in every system of quantity there is a function, S, of two variables such that

$$S(Ex,y) = S(x,Ey) = ES(x,y).$$

And there is a quantity u such that

$$S(u,u) = u.$$

The operation S is called *addition*, and the quantity, u, *zero*.

Another operation is defined by the general formulae

$$P(Ex,y) = S[P(x,y),y],$$
$$P(x,Ey) = S[x,P(x,y)],$$
$$P(0,0) = 0.$$

This is *multiplication*. From these definitions of the operations all their fundamental properties are readily deducible.

Art. 7. The last operation causes the enlargement of the system of datary quantity, to make a new system. For while the equation

$$S(x,y) = z$$

has just one solution in the datary system, when any two of the three quantities are given datary numbers, the equation

$$P(x,y) = z$$

can in general only be solved for z. In the system of *rational quantity*, there is for any two datary quantities, N and D, a rational quantity, r, such that

$$P(r,D) = N.$$

It is true that some of these are equal; for

$$P[r,P(M,D)] = P[M,P(r,D)]$$

so that the same quantity, r, satisfies both

$$P(r,D) = N$$

and $$P[r,P(M,D)] = P(M,N).$$

It is also true that

$$P(x,0) = 0$$

as a general formula, whatever x may be, so that when $N = D = 0$, the equation is indeterminate. Moreover,

$$P(0,x) = 0$$

so that when $N = 0$, the value of r is *zero* whatever D may be. The case where $D = 0$ while N is not 0 requires special study. Since a change of N to $P(M,N)$ accompanied by a change of D to $P(M,D)$ involves no change of r, it follows that when $D = 0$, it makes no difference what N may be (so long as it is not 0).

Of two datary quantities not equal one can be produced from the other by operations each E, while the latter cannot be so produced from the former. And it is evident that if B can be so produced from A and C can be so produced from B, then C can be so produced from A. In the system of rational quantity, it ceases to be generally true that of two unequal quantities one can be produced from the other by the operation E. But the new quantities which are inserted to make up the system of rational quantity are assumed to be so arranged that the following rule holds:

1st, Any four rational quantities, a, b, c, d, which are equal to datary quantities and which are such that a cannot be produced from c, nor c from a by the operation E or its repetitions without first producing either b or d, are in a certain quadruple relation to one another which may be expressed by writing

$$K_{a,b,c,d}$$

or, what comes to the same thing,

$$K_{b,c,d,a} = K_{c,d,a,b} = K_{d,a,b,c} = K_{d,c,b,a} = K_{c,b,a,d} = K_{b,a,d,c} = K_{a,d,c,b}.$$

This may be expressed by saying that in the arrangement of the quantities a,b,c,d according to their values, a is separated from c by b and d.

2nd, Of any four rational quantities p, q, r, s, all unequal, and such that whatever datary quantity D may be

$$P(p,D) = P(q,ED) = P(r,EED) = P(s,EEED),$$

or such that

$$P(p,D) = E(q,D) = EE(r,D) = EEE(s,D),$$

it is true that

$$K_{p,q,r,s}$$

or p and r are separated in the order of values by q and s.

It follows that $\frac{-1}{1} = -1$, $\frac{1}{0}$, $\frac{1}{1} = 1$, and $\frac{1}{2}$ are such that -1 and 1 are separated from each other by $\frac{1}{0}$ and $\frac{1}{2}$. And $1 = \frac{2}{2}$, $\frac{1}{2}$, $\frac{0}{2} = 0$, $\frac{^-1}{2}$ are such that 1 and 0 are separated from each other by $\frac{1}{2}$ and $\frac{^-1}{2}$. So that 1 and -1 are separated from each other by $\frac{1}{0}$ and 0. Thus, the system of rational quantity forms a ring.

Art. 8. The system of *real quantity* is an enlargement of the system of rational quantity according to the following rules:

1st, The real quantities form a ring in which the rational quantities have the same arrangement as in the system of those quantities.

2nd, Taking any *endless series* of real quantities, that is, a collection of quantities such that of any two of them, the one is *later* than the other in the series (that is, can be produced from that other by repetitions of an operation called *passing to the next* which produces one and only one quantity of the series from each quantity in it), while the latter is not later than the former in the same sense, this *endless series* ought to be called *directly convergent*, if, and only if, taking any four quantities of it, p, q, r, s, such that q is later than p in the series, r than q, and s than r, p is always separated from r in the order of values by q and s. (Note that this is quite different from the ordinary conception of convergency, not only in the trifling particulars of regarding those series which give values alternately on the two sides of the limit as double series, as well as in excluding the first terms of many series from the convergent part, and in allowing one of the terms of the convergent series to be $\frac{1}{0}$, but also in the more important respect of admitting that $\frac{1}{0}$ may be the limit of a directly convergent series.) Now every directly convergent endless series has in the system of real quantities a *limit*; that is to say, there is a quantity L separated in the order of values from the second member of the series, by the first member, and by any other member whatsoever, and L is in the order of values separated from the first member of the series by every other member and by every other quantity which, like L, is separated in the order of values from the second member by the first member and by any other member whatsoever.

This completes all the linear systems of quantity which are employed by those who follow the doctrine of limits. Infinitesimals are to be considered below.

Art. 9. The system of *imaginary quantity* is an enlargement of the system of real quantity formed by multiplying every real quantity by a root of [−] unity. These roots of [−] unity form a ring altogether similar to the ring of real quantity. That is to say, it contains the limit of every endless directly convergent series that it contains. The multiplication of 0 or $\frac{1}{0}$ by a root of [−] unity does not cause any change of value.

Thus, this system of quantity forms a simple artiad, or globular, surface. It is usually said to be *represented* by the surface; but it *is* such a surface in the only sense in which pure mathematics can deal with surfaces, at all.

Art. 10. The system of real *quaternions* is an enlargement of the system of imaginary quantity, formed by adding two linearly independent square roots of negative unity. I have proved in the *American Journal of Mathematics* [Vol. 4, pp. 225-229] that no other system of multiple quantity is possible, so long as the equation $xa = b$ always has a single root (a and b not being both zero).

The quantities of the quaternion system form a simple artiad quadri-dimensional quasi-continuum. It thus appears that a quadri-dimensional space is simpler than a tridimensional space.

Art. 11. The systems of points in the various kinds of space treated by geometry are essentially systems of quantity, although they are not usually so called. The differences between the plane of imaginary quantity and the real plane of plane projective geometry are that the former is an artiad and the latter a perissid surface (that is, on it two topically nonsingular, that is, unlimited and nodeless, and therefore self-returning lines can cut one another in an odd number of points), and that the former is treated by means of one variable and the latter by two. But these are very slight differences as compared with those which distinguish quantities from other objects. In analytical geometry, the algebraical formulae and the spatial image are merely two different modes of representing the same modes of combination of relations which alone are the objects of study of pure mathematics.

Art. 12. Geometry has three great branches *topics* (or topology), *graphics* (or projective geometry, which is the doctrine of rays, or unlimited straight lines), and *metrics* (or metrical geometry). The two last are somewhat special branches suggested by the phenomena of optics and of the displacements of rigid bodies; but the first is a doctrine really embracing both the others as special problems under it, and belonging necessarily to every system of quantity. Mathematics can never be logically presented until the fundamental position and logic of topics is fully brought out.

§4. THE POSSIBLE GRADES OF MULTITUDE

Art. 13. A cat, though she cannot count, has a vague impression of the multitude of her kittens, and can distinguish one multitude from another, much as a man might carry in his head the idea of salmon-color or a standard of musical pitch. But such an idea is not the idea of number and is not mathematical. For the idea of number it is necessary to have a scale of numeral words, the first part of which scale is experimentally brought into one to one correspondence with a given multitude of things in the operation of *counting*, and the last numeral employed is noted. This numeral has various associations in our minds; and a word which is associated with many different ideas having an inward attraction for one another, may perhaps be said to call up in the mind a general concept, however vague. So that it is perhaps not false to say that there is a general conception of seven, or thirteen, or a hundred, or a million. But if there be such a conception, it has nothing to do with the mathematical idea of number. But when a similar experiment of counting is performed upon a second collection, and this being compared with the first, a conclusion is reached say as to which collection is to be preferred for any purpose, that is an application of mathematics. This exact use of number involves no general conception, except the broad one of "more" and "less." The numeral words have no more or other meaning than those corruptions of Romany cardinal numbers which the children use in "counting out."

"Eeny, meeny, mony mi
Bona lena etc."

The cardinal numerals are mere indexical words akin to demonstrative pronouns. They mean nothing. But being remembered in a certain serial order they form a most useful intermediary for the comparison of two multitudes, in order to ascertain which of them is the greater. This is the nature of all quantity. It does not all apply to multitudes; but it is all meaningless, and serving in practical life merely as an intermediary for comparing two things, that comparison conveying an idea which is general, but is quite wanting in the exactitude of mathematics.

It is very often most important to know whether there is a distinct *A* "for" every *B*, that is, whether the *B*s can be put into one-to-one correspondence with a part or the whole of the *A*s.

Art. 14. A *finite* or *enumerable* collection is one of which the inference of transposed quantity holds. This is the best possible definition of it, since, this being a logical characteristic, every other character must follow from this. I call such a collection *enumerable* because it can be counted out, or exhausted in a *one-to-one* correspondence with a part of the numerals, while leaving these unexhausted.

A collection of which De Morgan's syllogism of transposed quantity is not valid, but which can be arranged in serial order so that Fermat's peculiar mode of reasoning holds, namely that a character which belongs if to any member of the series then to the member which comes next after that, and which also belongs to the first member of the series, must belong to every member, — such a collection, I term a *denumerable* one. For such a collection can be put into one-to-one correspondence with the cardinal numbers, but cannot be exhausted before the finite numbers are exhausted.

It is well known that there are still larger collections, and those I term *innumerable*.

The multitude of one collection is defined by G. Cantor to be *less* than another when the members of the former can be put into one-to-one correspondence with a part of the members of the latter, while all the members of the latter cannot be put into one-to-one correspondence with members of the former collection.

A very fundamental question is whether two multitudes can be so vast that the members of neither could by any possibility be in one-to-one correspondence with those of the other.

Art. 15. This question is one of extreme difficulty. It asks whether a state of things is logically possible. Now, that state of things certainly involves no contradiction. Still, that does not settle the question. For in like manner there is no contradiction in supposing that there are only two objects in the universe of ideas. But in the universe of ideas whatever is possible has all the realization that universe admits. Hence if it is possible there are only five ideal objects, it is impossible there should be six. Nothing is easier than to make phrases creative of distinctions apparently explaining this difficulty. But to describe in general terms what is and what is not possible in the sense here pertinent or to determine with certainty whether or not two collections can be such as is supposed above involves the development of a new chapter of logic.

The question is whether two collections can be supposed to have such multitudes that, by reason of that circumstance, neither can have all its individuals in one-to-one correspondence with some (or all) of the individuals of the other.

It seems evident that nothing can prevent the possibility of all the individuals of one collection being in one-to-one correspondence with any others except that the multitude of the former collection is too great. And if two collections were in the relation supposed, it is plain that no addition of further individuals could make it possible for either to be in one-to-one correspondence to a part or all of the other, unless there were some reason why they could be made to correspond with certain individuals but not with others, which supposition is excluded, since it is supposed that it is merely degrees of multitude which prevent the possibility of such correspondence. Hence, if the relation supposed were to exist for any multitudes, it must exist for all equal or greater multitudes.

But it certainly does not exist for all greater multitudes. For considering the aggregate of the two multitudes, either of those multitudes is smaller than that aggregate, since by hypothesis this aggregate could not be placed in one-to-one correspondence with either of the two first multitudes, while each of the latter would by the relation of identity be entirely in one-to-one correspondence with a part of that aggregate.

Hence, the supposition is absurd; and of any two multitudes whatever one can be in one-to-one correspondence with the whole or a part of the other. In other words, of any two multitudes which are not equal, one is greater than the other.

Art. 16. The smallest of all possible multitudes is *none*. There is, by definition, no multitude greater than none and less than *one*; nor is there any multitude greater than a given cardinal number and less than one more than that number. My impression is that Cantor has proved that there is no multitude greater than every finite multitude and less than the multitude of all the cardinal numbers; but I do not remember what his demonstration is. However, it is proved very simply as follows. If there be a multitude to which the syllogism of transposed quantity applies one of its individuals can have the number one attached to it exclusively. But if the cardinal numbers from one up to but not including a given cardinal number attached to part but not to all the individuals of that multitude that cardinal number may be exclusively attached to an individual of the multitude to which no lower number has been attached. Then, it follows, as a syllogism of transposed quantity that there will be a number which being so attached will leave no individuals of the multitude without numbers. Hence, the multitude will be equal to some cardinal number. But if there be a multitude to which the syllogism of transposed quantity does not apply, and if all the cardinal numbers from one up to but not including a given number were attached each exclusively to one of the members of that multitude, then that given number could also be so attached. For if it could not, there would be no individual left without a number, and the count of the multitude [would be] completed; but if the count of the multitude were completed the syllogism of transposed quantity would apply to it. Thus, as said above, if all the numbers up to but not including N could be so attached then all up to but not including $N + 1$ could be so attached. But the number *one* can be exclusively attached to one of the individuals of any multitude. Hence, all the cardinal numbers can be put into one-to-one correspondence with some (or all) of the individuals of any multitude to which the syllogism of transposed quantity does not apply. Hence, the multitude of cardinal numbers is not greater than any such multitude. Hence, the multitude of cardinal numbers is at least as small as any multitude which is not equal to a cardinal number.

Art. 17. We now come to a theorem of prime importance in reference to multitudes. It is that the multitude of partial multitudes composed of individuals of a given multitude is always greater than the multitude itself, it being understood that among these partial multitudes we are to include *none* and also the total multitude. Since we are only inquiring whether the grade of multitude can of itself prevent the formation of partial multitudes whose multitude is greater than the primitive multitude itself, it can make no difference what kind of objects the individuals may be. To fix our ideas, then, let there be a collection of individuals the Ss, which may be numerable or innumerable. Take any predicate, p; and consider all those possible assertions each of which in reference to each *S* either affirms *p* or denies it. Call these assertions the *A*s. Now I say that if there be any relation, *r*, such that every *A* is in that relation to an *S*, — or in briefer phrase, such that every *A* is *r* to an *S*, — then there must be two different *A*s which are *r* to the same *S*. For if this were not the case, an absurdity would result as can readily be shown in two ways, which do not, however, differ substantially.

The first way depends upon the fact that of all possible assertions as to what *S*s are and what are not *p* some one must be true, whatever the facts may be. Suppose, then, that every *S* to which an *A* is *r* had its quality in reference to *p* altered if necessary, so as to make that *A* false of that *S*. If then every *A* were *r* to an *S*, and no two to the same *S*, every *A* would become false. That is, every possible assertion would be false, which is absurd.

The second way of showing the absurdity consists in showing that as long as every *A* is *r* to an *S*, and no two to the same *S*, there is a possible assertion omitted. Namely, form an assertion by taking each *A* finding the *S* to which it is *r*, and contradicting this *A* in reference to this *S*. If there are any *S*s to which no *A* is *r*, it makes no difference whether the new assertion affirms or denies *p* of them. This new assertion is plainly inconsistent with every one of the *A*s, that is with every possible assertion, which is absurd.

It is therefore absurd to suppose that the multitude of classes formed from the individuals of a collection (including 0 and the whole collection) should be no greater than the multitude of the collection itself.

Art. 18. It follows that the possible groups of cardinal numbers (each group denumerable, in general) exceed in multitude all the cardinal numbers. Then, the multitude of such possible groups [is] called the *first abnumeral.* The possible groups of such groups (each of the large groups containing, in general, the first abnumeral collection of the smaller and denumerable groups) will be still greater in multitude. Let this multitude be called the *second abnumeral.* It is evident that there will be a denumerable succession of these abnumerals, numbered according to the finite whole numbers.

There can be no maximum multitude.

Art. 19. It remains to be shown that there can be no multitude intermediate between these multitudes, and none greater than them all.

It seems probable that Cantor or some of his scholars has already proved that there is no multitude larger than that of the cardinal numbers and smaller than that of the surds, which is the first abnumeral multitude. I believe however that I have never seen such proof.

Cantor has shown, however, that the denumerable multitude multiplied into itself any finite number of times is not increased. His proof, slightly modified, is this. If there be *N* variables, each of which can take any integral value, the multitude of variations of the collection of *N* variables is the abnumeral raised to the *N*th power. But write a number of which the units' place and every *N*th place of decimals above it make when the figures are written as units, tens, hundreds, etc. the value of the first variable; of which the tens' place and every *N* place of decimals above it make the second variable, etc. Then, such a number has as many variations as the set of *N* variables; for their values can be read off from the number. For example, if there are three, and their values are 11111111, 22, and 456, the value of the key-number will be 1001001001001401521621, from which the three numbers can at once be read off, when we know the system and the number *N*. Now *N* being a multitude, if it is larger than any enumerable multitude, the smallest it can be is the denumerable multitude. Hence, if a multitude is to be composed of the product of the denumerable multitude into itself a multitude of times, and is to be larger than the denumerable multitude, that is, larger than the denumerable multitude multiplied into itself a finite number of times, it must be equal to the denumerable collection multiplied into itself the denumerable multitude of times. But that will be at least as large as the first abnumeral multitude, which is 2 multiplied into itself the denumerable multitude of times. In point of fact, it is easily shown to be equal to that. But if the multitude is not wholly composed of a continued product of the denumerable into itself, it would be such a product multiplied by some other multitude. These multitudes must not be greater than the denumerable; for if so, the multitude [will] be less than the first abnumeral. They can, therefore, only be enumerable multitudes. But if any enumerable multitude is multiplied into itself the denumerable multitude of times it gives the first abnumeral multitude. Hence, it can only be a finite number multiplied into itself a finite number of times, which gives a finite number. Now if the denumerable multitude multiplied into itself is not increased, *à fortiori* it is not increased when multiplied into a finite number. Hence, the first [abnumeral] multitude is the multitude *next greater* than the denumerable multitude.

Cantor rightly applies his argument showing that the denumerable multiplied into itself a finite number of times is not increased to showing that the first abnumeral multitude multiplied into itself a finite number of times is not increased. For taking any number, N, of surds, expressed to indefinite approximation in decimals, a key-surd can be formed by the same rule, from which these N surds can be read off. But in order to show that there is no multitude intermediate between the first and second abnumeral multitudes, it becomes necessary to show that the first abnumeral multiplied into itself the denumerable multitude of times is not increased. That is to say, it has to be shown that a multitude of possible sets of values is no greater than the multitude of surds, although the multitude of values in each set is equal to that of all the whole numbers and each value can be any surd. In order to prove this, we first remark that it is evident that the multitude of numbers surd or rational between 0 and 1 is equal to the multitude of such numbers between 0 and ∞. For we have only to imagine a key-number, expressed in decimals carried out endlessly, the odd places of decimals representing all the places of decimals in any number from 0 to ∞ while the even places of decimals of the key-number are filled by the figures to the left of the decimal point in the other number, these figures being taken in reverse order. For example

864.97531..

will be represented by

.9476583010...

Then, suppose we have an endless series of numbers each between zero and unity, and each carried out endlessly into the decimal places, the question is whether all possible such series can be represented by the possible variations of a single key-number between zero and unity, this being carried out into the decimal-places without end. The answer is that it can; for we have only to write, in each case, the Mth figure of the Nth number as the Pth figure of the key, where $P = \frac{1}{2}[(M+N)^2 - M - 3N + 2]$. That this will never put two figures in the same place of the key is evident from this that if T is the largest triangular number less than P and t is the smallest triangular number greater than P, then $M = P - T$ and $N = t - P + 1$. Then, by reasoning perfectly analogous to that which proves there is no multitude intermediate between the denumerable and the first abnumeral, it follows that there is no multitude intermediate between the first abnumeral and the second.

For proof that there are no multitudes intermediate between higher abnumerals, I shall endeavor to show that the following reasoning is correct. Let D be the denumerable multitude. Then, the first abnumeral being the multitude of ways in which D things can be distributed into 2 places is 2^D.[1] The second abnumeral, being the multitude of ways in which 2^D things can be distributed into two places, is

$$2^{2^D}.$$

Hence the second abnumeral to the first abnumeral power is

$$(2^{2^D})^{2^D} = 2^{(2^D 2^D)} = 2^{2^{D+D}} = 2^{2^D}.$$

To prove that this reasoning is correct, it will be necessary first to show that for all enumerable, denumerable, or abnumeral multitudes, m and n, however great, we have

$$(2^m)^n = 2^{(mn)}.$$

Now 2^m represents the multitude of ways in which m places can all be filled each by one or other of two figures, say 0 and 1. $(2^m)^n$ represents the multitude of ways in which n places can all be filled, each some one of 2^m figures, or sets of figures.

Here it must be shown that all the individuals of any enumerable, denumerable, or abnumeral collection can be arranged in a linear series. I am not saying that there is room in space to do so. That is beside the question. It is merely meant that such a relation can exist between the individual[s] that there is one of them to which no other is in that relation; that of any two different ones one only is in that relation to the other; that if A is in that relation to B and B is in that relation to C, then A is in that relation to C; and that given any one individual of the collection (except perhaps one) there is another to which it is in that relation without being in that relation to any intermediate one that is in that relation to that second. That this is true of any enumerable or denumerable collection is evident; for *coming before* in the succession of whole numbers is such a relation. That, if it is true of any collection whose multitude is m,

[1] It is apparent that this manuscript was written before Peirce became acquainted with Cantor's notation, probably even before Cantor had invented it. Cantor's "alephs" seem not to have been used by him in his publication before 1895, in his "Beiträge zur Begründung der transfiniten Mengenlehre" in the *Math. Annalen*, Bd. 46. In the Peirce manuscript one is aware of Peirce's ingenuity in the creation of symbols. In the above

D represents Cantor's $\text{aleph}_0 = \aleph_0$
2^D represents Cantor's $\text{aleph}_1 = \aleph_1$
2^{2^D} represents Cantor's $\text{aleph}_2 = \aleph_2$ etc.

A few paragraphs beyond this Peirce's new notation makes

♈ = Exp^0♈ the equivalent of Cantor's $\aleph_0$
$2^{♈}$ = Exp^1♈ the equivalent of Cantor's $\aleph_1$
$2^{2_1^{♈}}$ = Exp^2♈ the equivalent of Cantor's $\aleph_2$ etc.

In 4.196 of the *Collected Papers*, the symbolism in Peirce is *not* $\aleph$ as given there, but ☘ and ☘^{2}. That Peirce manuscript is dated "1897" by Hartshorne and Weiss.

it it also true of any collection whose multitude is 2^m is easily shown. For the latter collection can be put into complete correspondence with the collection of possible ways in which a collection of places of multitude m can all be filled, each either with a 0 or a 1. Now this collection of multitude m being ranged in such a series, which has a *first* member, and a member *next* after each member (except the *last*, if there is a last), the collection of multitude 2^m is so arranged by taking that individual as *first* which corresponds to filling all the m places by 0s and that one as *last* which corresponds to filling all the m places by 1s, and that individual as *next after* a given individual, which corresponds to that distribution of 0s and 1s which differs from the distribution corresponding to the individual it comes next after in changing its first 0 to 1 and changing any 1s which may precede that 0 to 0s. It follows, then, that every abnumeral collection can be arranged in such a linear series.

It may seem at first sight that this proves that all the abnumeral collections are equal; but a moment's reflection will show that this does not follow, since the multitudes m are different for different 2^ms.

Returning, now, to the consideration of $(2^m)^n$, let the m places be arranged in serial order, and the n places be also arranged in independent serial order. Then, the ways representing $(2^m)^n$ will be ways of filling the block of places consisting of m rows and n columns, each with either a 0 or a 1. This may be written $2^{(mn)}$.

Now the block mn, if m is not greater than n and is either enumerable or denumerable, agrees pretty clearly with n in being enumerable, denumerable, or of any given order of abnumerality. However, it is not logically requisite to prove this; but it will be sufficient to consider the case in which both m and n are abnumeral and therefore of the forms 2^μ and 2^γ. In that case, I say that for the block, we have

$$2^\mu \cdot 2^\gamma = 2^{\mu+\gamma}$$

This is true because $2^\mu \cdot 2^\gamma$ is the multitude of pairs of ways in which μ places can be each filled either with 0 or 1 and in which ν other places can be each filled with 0 or 1. But each of these is simply a way in which the aggregate of μ and ν places, which we may represent by $\mu + \nu$, can each be filled with 0 or 1. Now the aggregate of μ and ν places agrees with that multitude of the two which is not smaller than the other in being enumerable, denumerable, or of a given order of abnumerality.

It has been proved that

$$[2^{(2^D)}]^{(2^D)} = 2^{(2^D \cdot 2^D)} = 2^{(2^{D+D})} = 2^{(2^D)}$$

that is that any abnumeral multitude multiplied into itself a multitude of times of less abnumerality is not increased.

Thence it follows, by the same reasoning which was used to show the impossibility of any multitude intermediate between the denumerable and the first abnumeral that there is no multitude intermediate between two successive abnumerals after the second.

Consequently, if ♈ (a form of Aries resembling ∞, which I employ in the algebra of logic) represents the denumerable multitude, the only multitudes possible, are, 1st, the denumerable series of enumerable multitudes,

$$0, 1, 2, 3, 4, 5, \text{etc.}$$

and, 2nd, the denumerable series of inenumerable multitudes,

$$\text{Exp}^{\,0}\,♈ = ♈,\ \text{Exp}^{\,1}\,♈ = 2^{♈},\ \text{Exp}^{\,2}\,♈ = 2^{2^{♈}},\ \text{Exp}^{\,3}\,♈ = 2^{2^{2^{♈}}},\ \text{etc.}$$

Art. 20. What goes before is ample demonstration that there is no difficulty in reasoning mathematically and with exact certainty about infinite multitudes. At the same time, the abnumeral multitudes are not, strictly speaking, *quantities*, in the sense in which quantity was defined above. This does not arise from their being infinite; because anybody can also reason mathematically about $\frac{1}{\infty}$, if he proceeds logically and not by the rule of thumb which in many brains, even of mathematicians, takes the place of logical thought, — and with great advantage, too; for it is a species of instinct, and the operation of instinct is far less liable to err, within its limited sphere, than that of reason under its celestial vault. The abnumeral multitudes are not quantities, because they do not represent *places upon a scale*.

Let us form a scale of quantity adapted to multitudes. Let us call the quantities, or grades of this scale *arithms*. These arithms, then, will all be, like any grades of quantity, in themselves perfectly similar, and will only be distinguishable by their relative places on the scale. The first arithm will be 0, and will mark the beginning, or preparation, for counting. The second arithm will be the number 1: the third will be the number 2: and so on until the numbers are exhausted, the Nth arithm being the number $N-1$. The arithm of the denumerable multitude will be infinity, and will be recognized by having an endless series of arithms before it. An arithm following *next after* this could not be distinguished from it, any more than the one *next before* it could be distinguished from it by its place on the scale (which is the only distinction there is between arithms or any quantities). Accordingly, all the abnumerals must take the arithm of the denumeral; and on the scale of quantity these multitudes can only be distinguished as "infinite." Thus, *quantitative* reasoning, in the narrow sense here given to "quantity," is unable to cope with questions relating to grades of inenumerable multitude, although those questions can perfectly well be mathematically discussed.

§5. OF THE NATURE OF THE CONTINUITY OF TIME AND SPACE

Art. 21. Cantor, in effect, defines the continuity of a line as consisting in that line's containing *all* its points. This is a singular *circulus in definiendo*, since the very problem was to state how those points were related. But I should not have noticed it, were it not that the phrase seems to imply that the line contains as many points as it could contain. Now we have seen in the last section that there is no maximum grade of multitude. If, therefore, a line contains all the points there could be, these points must cease to form a multitude.

Logic, as it was conceived by Aristotle and as it was apprehended even by the subtlest realists of the middle ages and as its ideas are embodied in every development of syllogistic, rests as upon bed-rock upon the principle that a "general" exists only in so far as it inheres in individuals, which are the "first substances," having absolute, independent, and ultimate existence. Many philosophers have denied this in metaphysics; but they have never shown what would be left of formal logic after the havoc that denial would bring into that field. In the English logic of this century, generals appear as "class-names;" and a class is a multitude, or collection, of individual things, each having its distinct, independent, and prior existence. Such a class cannot have as many individuals as there could be; because "as many as there could be" is not a possible grade of multitude; and the result of insisting upon that would inevitably be that the individuals would be sunk to a potential being, and would no longer be unconditionally and *per se* there. The discovery of such a state of things would be an earthquake in logic, leveling its whole fabric; and it would be incumbent upon the philosopher who should accept it to begin at the very beginning and build up the elementary rules of reasoning anew.

We must either hold that there are not as many points upon a line as there might be, or else we must say that points are in some sense fictions which are freely made up when and where they are wanted.

As far as points upon a spatial line go, no doubt a large party would be quite disposed to regard them so. But the continuity of space seems unquestionably to be derived from the continuity of time; and common sense would find very grave difficulty in admitting that the smallest portion of past or future time was immediately present; and if not, the present instant would seem to be the most indubitable and independent reality in all our knowledge.

I consider that it is pertinent to the present investigation, first, to analyze the nature of the continuity of space, and especially of time, as logically involved in the common-sense ideas of those continua, and second, to consider what the evidence is that objective time and space possess such continuity. For although mathematics has nothing to do with positive truth, yet its hypotheses are suggested by experience, and any theory for which there may be even imperfect evidence ought to be erected into a mathematical hypothesis, provided it be of such a nature that a great body of deductions can be drawn from it. In short, though this part of the inquiry can only shed a side-light on the main question, which pertains to the infinitesimals of the calculus, yet that illumination may be strong, and in my opinion will be so.

Art. 22. According to natural common-sense, only the state at a single instant, the present instant, is ever immediately present to consciousness, and yet we are conscious of the flow of time, we imagine events as in time, and we have a real memory, not merely of states, but also of motions. Our present task is, not to criticise this idea of time, but to endeavour to gain a distinct comprehension of its elements and of how they are related to one another. Imagine a series of instantaneous photographs to be taken. Then, no matter how closely they follow one another, there is no more motion visible in any one of them than if they were taken at intervals of centuries. This is the common-sense idea of that which is immediately present to the mind.

Opinions will differ as to whether common-sense holds that the flow of time is directly perceived or not. Let us first suppose that it does not. Then, according to common-sense, we can come to believe that events happen in time only by a sort of vaticination, the idea springing up in the mind without any reason. For reasoning must be conscious that it is reasoning, or it ceases to be reasoning. Now reasoning is essentially a process. Consequently, we cannot reason without having already the idea of time. A greater difficulty is that an instantaneous photograph, though it may contain a symbol of time, or even an indication that time exists, can certainly not contain a true likeness of time. To imagine time, time is required. Hence, if we do not directly perceive the flow of time, we cannot imagine time. Yet the sense of time is something forced upon common-sense. So that, if common-sense denies that the flow [of] time is directly perceived, it is hopelessly entangled in contradictions and cannot be identified with any distinct and intelligible conception.

But to me it seems clear that our natural common-sense belief is that the flow of time *is* directly perceived. In that case, common-sense must hold that something more than an instant is immediately present to consciousness. As to what this "something more" is, several hypotheses might be made. But there are two propositions about time which, if they are acknowledged to be involved in the common-sense idea, determine the character of its continuity. One is, that there is in a sensible time room for any multitude, however great, of distinct instants. The other is, that the instants are so close together as to merge into one another, so that they are not distinct from one another. I do not think that anybody fairly considering the matter can doubt that the natural idea of time common to all men supports both those judgments. The former seems to express what there is that is true in Cantor's statement that a line includes *all* the points possible. Its consonance with the common idea is further shown by the circumstance that it has occurred to nobody to object to the orders of infinitesimals of the calculus that there would not be room in time or space for so many distinct points as they create, although the vagueness of the multitude needed has been strongly felt. Nor has there ever been any doubt that surds and transcendentals of all real kinds could be conceived as measured off in space (whose continuity is recognized as precisely like that of time), although there have been doubts as to whether the variations of those quantities were adequate to representing all the points of space. On the whole, then, this proposition may be accepted as a dictum of common-sense. The other proposition, that the instants of time are so crowded as to merge into one another and lose their distinct existence, seems to be involved in the conception of the "flow" of time. For this phrase likens time to a homogeneous fluid in which the "particles" are mere creations of the mind, made for convenience of calculators. Again, nobody has, as far as I know, ever suggested that two lines might cross one another without having a common point, although if their points were distinct from one another, — two multitudes in series order, — there is no reason why the points of one line might not slip through between those of the other. The very word *continuity* implies that the instants of time or the points of a line are everywhere welded together. This proposition may then, likewise, be accepted as true to the common idea.

According to that idea, then, the instants of a time are not a multitude. Each of the two propositions proves that. For, first, since any multitude whatever of instants exists among the instants of any given time, and since there is no maximum multitude, it follows that the instants of time do not in their totality form a multitude. In this sense, they may be said to be "more" than any multitude; that is, there is among them a multitude greater than any multitude which may be proposed. Second, [they] are not in themselves distinct from one another, as the units of any multitude are, even if they happen to be joined together.

Moreover, because the instants do not preserve their distinct identities, it follows that taking any proposition whatsoever, if that proposition is true of any instant then a later instant can be found such that it remains true to that instant, and in like manner an earlier instant can be found such that it has been true since that instant. Were a proposition to be false up to a certain instant and thereafter to be true, at that instant it would be both true and false. It does not follow that a proposition once true remains always true; it only follows that it remains true through a denumerable series of instants, which is a lapse of time inexpressibly less than any sensible or assignable time, if it can properly be called a lapse of time, at all, wanting as it does most of the characteristics of duration. A denumerable succession of instants may be called a *moment*.

Even infinitely longer than the moment will a proposition of the slightest latitude in respect to the change which is taking place, though this latitude be far too small to be detected, remain true.

In particular, if an instant be *immediately present*, — since this is a proposition concerning that instant, a denumerable succession of instants before and after it are fully present; and even infinitely longer will the proximity of instants be so close, in the case of the past with respect to the action of facts upon the mind, through sense, and in the case of the future with respect to the action of the mind upon facts, through volition, that they are practically present. Thus, temporal succession is immediately present to consciousness, according to the logical explication of the common-sense idea of time; and even temporal continuity is practically present; and so there is no difficulty in accounting for the suggestion of the idea of time, nor for our being persuaded of its truth.

We are now prepared to define a *continuum* after the exemplar of the common-sense idea of time. Namely, a *continuum* is whatever has the following properties:

1st, it is a whole composed of parts. We must define this relation. The parts are a logical aggregate of mutually exclusive subjects having a common predicate; and that aggregate regarded as a single object is the whole.

2nd, these parts form a series. That is, there is a relation, l, such that, taking any two of the parts, if these are not identical one of them is in the relation, l, to everything to which the other is in that relation and to something else besides.

3rd, taking any multitude whatever, a collection of these parts can be found whose multitude is greater than the given multitude. Consequently, the indivisible parts, that is, parts such that none is a collective aggregate of objects one of which is in the relation, l, to everything to which another is in that relation and to more besides, — are not distinct. That is to say, the relation l cannot be fully defined, so that in any attempted specification of it, l', any part which appears indivisible, becomes divisible into others, by means of a further specification, l''.

Art. 23. I will now briefly note the evidence known to me that continuity really exists. That space is an objective reality and not merely a form of intuition (the Kantian argument for which I consider illogical) seems to be shown by three characters, first, that it is tridimensional, or perhaps quadridimensional, while there is no necessity for its having any particular number of dimensions, second, that calculations founded upon proper motions of stars show distinct indications that it is hyperbolic,[2] and thus has a constant the value of which cannot be necessary; and third, as Newton pointed out, velocity of rotation is absolute and not merely relative. That time and space are of the same nature seems most probable. That time and space are innate ideas, so far from proving that they have merely a mental existence, as Kant thought, ought to be regarded as evidence of their reality. For the constitution of the mind is the result of evolution under the influence of experience.

In the same way, the fact that the continuity of space and time is a natural belief is perhaps evidence that it is true. Better evidence is that it explains the personal identity of consciousness in time, which is almost

[2] Peirce later changed his mind. It was in the late fall of 1891 that calculations led him to the conclusion in the text. See pp. 421-423 of "The Charles S. Peirce–Simon Newcomb Correspondence" by the editor in *Proceedings of the American Philosophical Society* 101:5 (October 1957).

if not quite incomprehensible otherwise.

Moreover, if the whole universe and its laws are the result of evolution, as a logical method requires us to assume unt itil is disproved, space and time must, I think, have been continuous when they were first formed; for, as Spencer says, the course of evolution is from homogeneity to heterogeneity. And it seems impossible that there should have been any action making time and space discontinuous after they were once continuous.

B. ON MULTITUDE (26)[1]

Art. 1. Mathematics is a study of exact hypotheses, in so far as consequences can be deduced from them. To limit mathematics to the deduction of those consequences would be to separate from it some of the greatest of the achievements of modern mathematicians, — achievements which nobody but mathematicians could have performed, — such as the formation of the idea of the system of imaginaries, and of the idea of Riemann surfaces. It must be allowed, therefore, that the formation of its hypotheses is a part of the business of mathematics.

Art. 2. The purpose of the present paper is to inquire what grades of multitude of collections are mathematically possible, that is are possible as purely ideal, but exact hypotheses.

I can hardly contend that this inquiry comes within the scope of mathematics, although it must be conducted chiefly by means of mathematical reasoning. But it is not the usage of mathematicians to investigate the possibility of fundamental hypotheses; they content themselves with adopting such hypotheses as are manifestly possible. If they inquire into the possibility of a secondary hypothesis, their problem is to determine whether or not it is compatible with a fundamental hypothesis. An inquiry like the present, not merely requires like all mathematics a strict *logica utens*, but it must be based on the principles, or must develop new principles of *logica docens*. It must, therefore, be classed as a study in logic and not as merely mathematical; although it will, undoubtedly, interest mathematicians.

Art. 3. It is requisite, in the first place, to define a collection. Now, it is impossible to give a genetic definition without taking for granted, at the

[1] There are several drafts of MS. 26. In one of them Peirce associates footnotes (17)-(23) with "Rules of π and ε" and "Practical Rules." Pages 1-17 are continuous in the manuscript. Pages 18-24 belong to another draft not included in this edition.

outset, all the main results of the investigation. We must, therefore, content ourselves with the following:

A *collection* is whatever stands to a general predicate of single subjects in a certain relation *sui generis*, such that for every such predicate there is a single collection and for every collection there is such a predicate.

Of predicates some have single subjects and are expressible by intransitive verbs, while others have plural sets of subjects and are expressed by verbs requiring in our languages a direct object and perhaps a certain fixed set of indirect objects. These objects are essentially subjects; and there is no one of the great families of human speech which does not embrace languages which habitually put the subject of a sentence into an oblique case, thus contrasting it no more with the other objects than these are contrasted with one another. Among the living European languages, the Gaelic is an example in point.

To say that a collection is in a one-to-one relation *sui generis* to a predicate of single subjects amounts to saying that it is such a predicate spoken of with a peculiar phraseology. For a relation *sui generis* is indefinable, and an indefinable relation has no meaning. When we speak of a predicate as a collection, we call it, roughly speaking, larger the more subjects it has; but when we call it a predicate we consider it as less the more subjects it has. But, then, the two relations of "being more than" and of "being less than" are relations of precisely the same form, and do not differ from one another in any intelligible way. The maximum collection is that which embraces all possible units, while this is the zero of predication. The minimum collection is that which embraces no unit; but the predicate which is applicable to no possible subjects without falsity is the compound of all predicates. But these statements, far from establishing any intelligible contrast between a predicate of single objects and a collection, simply go to identify them, by showing that their absolute limits of measurement are the same. They are, it is true, interchanged; but there is no intelligible difference between them.

Art. 4. By a *postulate* of a given branch of mathematics, I mean a propositional formula embodying a part or the whole of the fundamental hypothesis which is adhered to throughout that branch. By an *axiom* of a given branch of mathematics, I mean a propositional formula embodying a principle which, being established in logic or in some other branch of mathematics, is accepted in the former branch without further inquiry.

There are three axioms relating to collections. I shall express these in the General Algebra of Logic, which I have explained in *The Monist*, VII, pp. 189 *et seqq.* §§8 and 9, and elsewhere, and which is also explained in the third volume of Schröder's *Logic*, though with a slightly different notation.

The first axiom relates to a single collection, and is that every possible object belongs to a collection from which all other objects are excluded. That is, if we write

$$q_{\alpha b}$$

to mean, "*b* is a unit of the collection α," the first axiom is

(1) $\Pi_x\Sigma_\alpha\Pi_y\, q_{\alpha x}\cdot(\mathrm{I}_{xy}\,\Psi\,\bar{q}_{\alpha y})$. [See p. 776 and p. 907 for Ψ.]

The second axiom relates to two collections, and is, that any collection whatever has a *complementary* collection such that every possible object is included in one collection or the other and is excluded from one or the other. That is,

(2) $\Pi_\alpha\Sigma_\beta\Pi_x\,(q_{\alpha x}\,\Psi\,q_{\beta x})\cdot(\bar{q}_{\alpha x}\,\Psi\,\bar{q}_{\beta x})$.

This may occasion some doubt, which can, however, only be due to a misunderstanding. If any object is possible, by virtue of that possibility it belongs to every collection whose characteristic predicate belongs to it.

The third axiom relates to three collections, and is that, given any two collections whatever, there is a third collection which embraces every unit of either and nothing else. That is,

(3) $\Pi_\alpha\Pi_\beta\Sigma_\gamma\Pi_x\,(q_{\alpha x}\,\Psi\,q_{\beta x}\,\Psi\,\bar{q}_{\gamma x})\cdot(\bar{q}_{\alpha x}\,\Psi\,q_{\gamma x})\cdot(\bar{q}_{\beta x}\,\Psi\,q_{\gamma x})$.

Art. 5. A few of the consequences of these axioms may advantageously be noticed, at once. Those deductions may be called *corollaries* which follow from single propositions already established or assumed; while consequences of two or more propositions, however simple, are to rank as *theorems.* What I mean by the *order* of a deduction is explained in *The Monist*, Vol. VII, p. 191.

Theorem 1. Every possible object is excluded from some collection to which all other possible objects belong. That is,

(4) $\Pi_x\Sigma_\alpha\Pi_y\,\bar{q}_{\alpha x}\cdot(\mathrm{I}_{xy}\,\Psi\,q_{\alpha y})$.

This is a consequence of the first order in axiom (1) and of the second order in axiom (2); so that we may write $(1)\cdot(2)^2 \prec (4)$.

For the first factor of (4), or $\Pi_x\Sigma_\alpha\bar{q}_{\alpha x}$, will evidently follow from the first factor of (1); while the last factor of (2) will follow from the last factor of (1) and the first factor of (2). But there is only one index in (2) referring to a possible object, while there are two in (4). Hence (2) must be multiplied into itself. This paragraph explains the reasoning required to *invent* the demonstration. Being irrelevant to the demonstration itself, the usual style of writing mathematics omits such explanations. In other words, the whole current of thought except the mechanical portions is left unexpressed and concealed.

In multiplying any proposition by itself we may begin at any point of the quantifying part to diversify the indices, provided we carry the diversification to the end. Thus, in squaring (2) we may change x to y in one factor and so get

$$\Pi_\alpha \Sigma_\beta \Pi_x \Pi_y (q_{\alpha x} \Psi q_{\beta x}) \cdot (\bar{q}_{\alpha x} \Psi \bar{q}_{\beta x}) \cdot (q_{\alpha y} \Psi q_{\beta y}) \cdot (\bar{q}_{\alpha y} \Psi \bar{q}_{\beta y}).$$

Dropping useless factors and multiplying by (1), and distinguishing the indices of $(2)^2$ by accents we get

$$\Pi_x \Sigma_\alpha \Pi_{\alpha'} \Sigma_\beta \Pi_y \Pi_{x'} \Pi_{y'} q_{\alpha x} (1_{xy} \Psi \bar{q}_{\alpha y}) \cdot (\bar{q}_{\alpha' x'} \Psi \bar{q}_{\beta x'}) \cdot (q_{\alpha' y'} \Psi q_{\beta y'}).$$

Identifying α' with α, x' with x and y' with y, the general formulae

$$p \cdot (q \Psi r) \prec p \cdot q \Psi r$$
$$s \cdot \bar{s} \prec 0 \qquad 0 \Psi t \prec t$$

give $\Pi_x \Sigma_\alpha \Sigma_\beta \Pi_y \bar{q}_{\beta x} \cdot (1_{xy} \Psi q_{\beta y})$.

Dropping from the quantifying part the index α, which does not occur in the Boolian, and then changing β to α, we get (4). Q.E.D.

Theorem 2. There is a collection which includes all possible objects. That is

(5) $\Sigma_\alpha \Pi_x q_{\alpha x}$.

This is a consequence of the first order from axioms (2) and (3); so that $(2) \cdot (3) \prec (5)$. For the product of the last two factors of (3) is

$$\Pi_\alpha \Pi_\beta \Sigma_\gamma \Pi_x \bar{q}_{\alpha x} \cdot \bar{q}_{\beta x} \Psi q_{\gamma x}.$$

Multiplying this by the first factor of (2), whose indices we accent, we get

$$\Pi_\alpha \Pi_{\alpha'} \Sigma_{\beta'} \Pi_\beta \Sigma_\gamma \Pi_x \Pi_{x'} (q_{\alpha' x'} \Psi q_{\beta' x'}) \cdot (\bar{q}_{\alpha x} \cdot \bar{q}_{\beta x} \Psi q_{\gamma x}).$$

Identifying α' with α, β with β', and x' with x, we have by the application of the same general formulae that were used in Theorem 1, and are almost always applied,

$$\Pi_\alpha \Sigma_{\beta'} \Sigma_\gamma \Pi_x q_{\gamma x}.$$

Dropping from the quantifying part the indices α and β', which do not occur in the Boolian, and changing γ to α, we get (5). Q.E.D.

Corollary 1. From (5), by carrying the Σ to the right we get immediately

(5*bis*) $\Pi_x \Sigma_\alpha q_{\alpha x}$.

That is, every object is a unit of some collection.

Scholium 1. This corollary can also be deduced immediately from (1); so that there is an inelegant pleonasm in the statement of the axioms.

Theorem 3. For every two collections, there is a third which includes every unit of either, and nothing else. That is,

(6) $\Pi_\alpha \Pi_\beta \Sigma_\gamma \Pi_x (q_{\alpha x} \Psi \bar{q}_{\gamma x}) \cdot (q_{\beta x} \Psi \bar{q}_{\gamma x}) \cdot (\bar{q}_{\alpha x} \Psi \bar{q}_{\beta x} \Psi q_{\gamma x})$.

This is a consequence of the third order in (2) and of the first order in (3); or $(2)^3 \cdot (3) \prec (6)$.

For multiplying (2) three times into (3) substituting in the three expressions of (2), for α, β, x, respectively, δ, ε, u; ζ, η, v; θ, ι, ω; we get

$$\Pi_\delta \Sigma_\varepsilon \Pi_\zeta \Sigma_\eta \Pi_\alpha \Pi_\beta \Sigma_\gamma \Pi_\theta \Sigma_\iota \Pi_x \Pi_u \Pi_v \Pi_w$$
$$(\bar{q}_{\delta u} \Psi \bar{q}_{\varepsilon u}) \cdot (\bar{q}_{\zeta v} \Psi \bar{q}_{\eta v}) \cdot (q_{\alpha x} \Psi q_{\beta x} \Psi \bar{q}_{\gamma x}) \cdot (q_{\beta w} \Psi q_{\iota w}) \cdot$$
$$(q_{\delta u} \Psi q_{\varepsilon u}) \cdot (\bar{q}_{\alpha x} \Psi q_{\gamma x}) \cdot (q_{\zeta v} \Psi q_{\eta v}) \cdot (\bar{q}_{\beta x} \Psi q_{\gamma x}) \cdot (\bar{q}_{\theta w} \Psi \bar{q}_{\iota w}).$$

Identifying α with ε, β with η, θ with γ, u, v, and w with x, we get by the usual formulae of reduction,

$$\Pi_\delta \Sigma_\varepsilon \Pi_\zeta \Sigma_\eta \Sigma_\gamma \Sigma_\iota \Pi_x (\bar{q}_{\delta x} \Psi \bar{q}_{\zeta x} \Psi q_{\iota x}) \cdot (q_{\delta x} \Psi \bar{q}_{\iota x}) \cdot (q_{\zeta x} \Psi \bar{q}_{\iota x})$$

and dropping from the quantifying part the indices not found in the Boolian, we get

$$\Pi_\delta \Pi_\zeta \Sigma_\iota \Pi_x (\bar{q}_{\delta x} \Psi \bar{q}_{\zeta x} \Psi q_{\iota x}) \cdot (q_{\delta x} \Psi \bar{q}_{\iota x}) \cdot (q_{\zeta x} \Psi \bar{q}_{\iota x}).$$

Q.E.D.

Theorem 4. There is a collection which includes no unit at all. That is

(7) $\Sigma_\alpha \Pi_x \bar{q}_{\alpha x}$.

This is a consequence of the second order in (2) and of the first order in (3); so that $(2)^2 \cdot (3) \prec (7)$.

For multiplying (5) by the second factor of (2) we get

$$\Sigma_\alpha \Pi_{\alpha'} \Sigma_\beta \Pi_x \Pi_{x'} q_{\alpha x} \cdot (\bar{q}_{\alpha' x'} \Psi \bar{q}_{\beta x}).$$

Identifying α' with α and x' with x, we get by the usual formulae of reduction

$$\Sigma_\alpha \Sigma_\beta \Pi_x \, q_{\beta x}.$$

Dropping the index α which does not occur in the Boolian, we get

$$\Sigma_\beta \Pi_x \, \bar{q}_{\beta x}.$$

Q.E.D.

Art. 6. A *dyadic relation* is a predicate of an ordered pair of subjects, of which one, the grammatical subject, is called the *relate*, and the other, the grammatical object is called the *correlate.*

We may write

$$r_{\alpha bc}$$

to mean, "the individual b is in the relation α to the individual c," or, to use more convenient locutions, "b is α of c," or "b is α to c," or "c is α'd by b."

Two axioms must be added to our list to formulate the properties of relations.

Axiom 4. Given any two collections, whether different from or identical with each other, there is a relation such that every unit of either collection is in that relation to every unit of the other; and if anything is in that relation to anything, the relate is a unit of the former collection and the correlate of the latter. That is,

$$(8) \qquad \Pi_\alpha \Pi_\beta \Sigma_\gamma \Pi_x \Pi_y \, (\bar{q}_{\alpha x} \,\Psi\, \bar{q}_{\beta y} \,\Psi\, r_{\gamma xy}) \cdot (q_{\alpha x} \,\Psi\, \bar{r}_{\gamma xy}) \cdot (q_{\beta y} \,\Psi\, \bar{r}_{\gamma xy}).$$

Axiom 5. Given any relation whatever and any collection whatever, there is a collection such that no unit of the former collection is in that relation to anything but a unit of the latter collection, and every unit of the former collection is in that relation to every unit of the latter. That is,

$$(9) \qquad \Pi_\alpha \Pi_\beta \Sigma_\gamma \Pi_x \Pi_y \, (\bar{r}_{\alpha xy} \,\Psi\, \bar{q}_{\beta x} \,\Psi\, q_{\gamma y}) \cdot (r_{\alpha xy} \,\Psi\, \bar{q}_{\beta x} \,\Psi\, \bar{q}_{\gamma y}).$$

Art. 7. There are sundry propositions deducible from the last two axioms in conjunction with those previously stated, which it will be advantageous to notice, at once.

Theorem 5. Given any object whatever and any collection whatever, there is a relation which that object bears to every unit of that collection but to nothing else, and which no other object bears to anything. That is,

(10) $\Pi_x \Pi_\alpha \Sigma_\beta \Pi_y \Pi_x (\bar{q}_{\alpha y} \Psi r_{\beta xy}) \cdot (1_{xx} \Psi \bar{r}_{\beta xy}) \cdot (q_{yy} \Psi \bar{r}_{\beta xy})$.

This is a conclusion of the first order in (1) and the second order in (8); so that (1) · (8)² ⤙ (10).

For first squaring (8) after transposing x and y in the quantifying part, and then diversifying x in the two factors, and dropping certain factors, we get

(11) $\Pi_\alpha \Pi_\beta \Sigma_\gamma \Pi_y \Pi_x \Pi_x (\bar{q}_{\alpha x} \Psi \bar{q}_{\beta y} \Psi r_{\gamma xy}) \cdot (q_{\alpha x} \Psi \bar{r}_{\gamma xy}) \cdot (q_{\beta y} \Psi \bar{r}_{\gamma xy})$.

Multiplying this by (1) in which for x, a, y, we write u, δ, υ, we get

$$\Pi_u \Sigma_\delta \Pi_\alpha \Pi_\beta \Sigma_\gamma \Pi_y \Pi_x \Pi_x \Pi_\upsilon \, q_{\delta u} \cdot (\bar{q}_{\alpha x} \Psi \bar{q}_{\beta y} \Psi r_{\gamma xy}) \cdot (1_{u\upsilon} \Psi \bar{q}_{\delta \upsilon}) \cdot (q_{\alpha x} \Psi \bar{r}_{\gamma xy}) \cdot (q_{\beta y} \Psi \bar{r}_{\gamma xy}).$$

Identifying α with δ, x with u, and υ with z, and employing the usual reductions, we get,

$$\Pi_u \Sigma_\delta \Pi_\beta \Sigma_\gamma \Pi_y \Pi_x (\bar{q}_{\beta y} \Psi \bar{r}_{\gamma uy}) \cdot (1_{ux} \Psi \bar{r}_{\gamma zy}) \cdot (q_{\beta y} \Psi \bar{r}_{\gamma uy}).$$

Dropping from the quantifying part the index δ which does not appear in the Boolian, we get (10) with u written for x, β for α, and γ for β. Q.E.D.

Theorem 6. Given any two objects whatever, there is a relation which either of those objects bears to the latter and to nothing else, and which no other object bears to anything. That is,

(12) $\Pi_x \Pi_y \Sigma_\alpha \Pi_u \Pi_\upsilon \, r_{\alpha xy} \cdot (\bar{r}_{\alpha u\upsilon} \Psi 1_{ux}) \cdot (\bar{r}_{\alpha u\upsilon} \Psi 1_{\upsilon y})$.

This is a conclusion of the second order both in (1) and in (8); so that (1)² · (8)² ⤙ (12). For squaring (8), and diversifying both x and y, we get, on dropping factors,

(13) $\Pi_\alpha \Pi_\beta \Sigma_\gamma \Pi_x \Pi_u \Pi_y \Pi_\upsilon (\bar{q}_{\alpha x} \Psi \bar{q}_{\beta y} \Psi r_{\gamma xy}) \cdot (q_{\alpha u} \Psi \bar{r}_{\gamma u\upsilon}) \cdot (q_{\beta \upsilon} \Psi \bar{r}_{\gamma u\upsilon})$.

Now multiplying [by] the square of (1) where for x, a, y, we substitute in one factor s, δ, i, and in the other t, ε, j, we get,

$$\Pi_s \Sigma_\delta \Pi_t \Sigma_\varepsilon \Pi_\alpha \Pi_\beta \Sigma_\gamma \Pi_x \Pi_u \Pi_y \Pi_\upsilon \Pi_i \Pi_j$$
$$q_{\delta s} \cdot q_{\varepsilon t} \cdot (\bar{q}_{\alpha x} \Psi \bar{q}_{\beta y} \Psi r_{\gamma xy}) \cdot (1_{si} \Psi \bar{q}_{\delta i}) \cdot (q_{\alpha u} \Psi \bar{r}_{\gamma u\upsilon}) \cdot (1_{tj} \Psi \bar{q}_{\varepsilon j}) \cdot (q_{\beta \upsilon} \Psi \bar{r}_{\gamma u\upsilon}).$$

Identifying α with δ, β with ε, x with s, y with t, i with u, j with v, we get after the usual reductions

$$\Pi_s \Sigma_\delta \Pi_t \Sigma_\varepsilon \Sigma_\gamma \Pi_u \Pi_\upsilon \, r_{\gamma st} \cdot (1_{su} \Psi \bar{r}_{\gamma u\upsilon}) \cdot (1_{t\upsilon} \Psi \bar{r}_{\gamma u\upsilon}).$$

Dropping δ and ε, which do not occur in the Boolian, changing s to x, t to y, and γ to α we get (12). Q.E.D.

Theorem 7. Taking any collection whatever, there is a relation which every unit of that collection bears to everything, and which nothing else bears to anything. That is,

(14) $$\Pi_\alpha \Sigma_\beta \Pi_x \Pi_y (\bar{q}_{\alpha x} \Psi r_{\beta xy}) \cdot (q_{\alpha x} \Psi \bar{r}_{\beta xy}).$$

This is a consequence of the first order in (2), (3), and (8); so that $(2) \cdot (3) \cdot (8) \prec (14)$.

For multiplying (8) by (5) in which for a and x we substitute δ and z, we get

$$\Sigma_\delta \Pi_\alpha \Pi_\beta \Sigma_\gamma \Pi_x \Pi_y \Pi_z \, q_{\delta z} \cdot (\bar{q}_{\alpha x} \Psi \bar{q}_{\beta y} \Psi r_{\gamma xy}) \cdot (q_{\alpha x} \Psi \bar{r}_{\gamma xy}) \cdot (q_{\beta y} \Psi \bar{r}_{\gamma xy}).$$

Identifying β with δ and z with y, this becomes

$$\Sigma_\delta \Pi_\alpha \Sigma_\gamma \Pi_x \Pi_y \, q_{\delta y} \cdot (\bar{q}_{\alpha x} \Psi r_{\gamma xy}) \cdot (q_{\alpha x} \Psi \bar{r}_{\gamma xy}).$$

Omitting the first factor, dropping δ, which then no longer appears in the Boolian, and substituting β for γ, we get (14). Q.E.D.

Theorem 8. Taking any relation whatsoever, there is a collection, to every unit of which, and to nothing else, everything is in that relation. That is,

(15) $$\Pi_\alpha \Sigma_\beta \Pi_x \Pi_y (\bar{q}_{\beta y} \Psi r_{\alpha xy}) \cdot (\bar{r}_{\alpha xy} \Psi q_{\beta y}).$$

This is a consequence of the first order from (2), (3) and (9); so that $(2) \cdot (3) \cdot (9) \prec (15)$. For multiplying (9) by (5) in which we change a to δ and x to z, we get

$$\Sigma_\delta \Pi_\alpha \Pi_\beta \Sigma_\gamma \Pi_x \Pi_y \Pi_z \, q_{\delta z} \cdot [\bar{q}_{\beta x} \Psi (\bar{r}_{\alpha xy} \Psi q_{\gamma y}) \cdot (r_{\alpha xy} \Psi \bar{q}_{\gamma y})].$$

Identifying β with δ and z with x, and employing the usual reductions, we get

$$\Sigma_\delta \Pi_\alpha \Sigma_\gamma \Pi_x \Pi_y (\bar{r}_{\alpha xy} \Psi q_{\gamma y}) \cdot (r_{\alpha xy} \Psi \bar{q}_{\gamma y}).$$

Dropping δ, which does not occur in the Boolian and changing γ to β, we get (15). Q.E.D.

Corollary 2. In (15) if we carry the Σ to the right, we get

(15bis) $$\Pi_\alpha \Pi_x \Sigma_\beta \Pi_y (\bar{q}_{\beta y} \Psi r_{\alpha xy}) \cdot (\bar{r}_{\alpha xy} \Psi q_{\beta y}).$$

That is, taking any relation whatsoever and any object whatsoever, there is a collection to every unit of which, and to nothing else, that object is in that relation.

Theorem 9. For every relation there is a *negative* relation, such that every object stands to every object in the one relation or the other, and is either not in the one relation or not in the other to that object. That is,

(16) $\Pi_\alpha \Sigma_\beta \Pi_x \Pi_y (r_{\alpha xy} \,\Psi\, r_{\beta xy}) \cdot (\bar{r}_{\alpha xy} \,\psi\, \bar{r}_{\beta xy}).$

This is a consequence of the second order in (2) and of the first in (3), (8), and (9); so that $(2)^2 \cdot (3) \cdot (8) \cdot (9) \prec (16)$.

Thus, let s mean "—slanders—to—" and let p mean "—praises—to—." Then, we have

$$\Sigma_i \Pi_j \Sigma_k \, s_{ijk} \cdot p_{ijk} \prec \Sigma_i \Pi_j \Sigma_k \Pi_u \Sigma_v \, s_{ijk} \cdot p_{iuv}.$$

But it evidently would not be true that

$$\Sigma_i \Pi_j \Sigma_k \, s_{ijk} \cdot p_{ijk} \prec \Sigma_i \Pi_j \Pi_u \Sigma_k \, s_{ijk} \cdot p_{iuk}.$$

For there is no reason to assume that two different persons one slandered and the other praised would be slandered and praised to the same person.

It follows, by contraposition, that when the Boolian consists of two aggregants, under certain circumstances identification may take place. But the conditions are rarely fulfilled, except when the aggregants of the Boolian are of the same form, so that the whole is the aggregate of a proposition with itself.

Any index can be changed to a new letter throughout.

Any quantifier of the Percian whose index does not occur in the Boolian can be omitted.

An even number of obeli over the same proposition are equivalent to none, and therefore an odd number to one.

An obelus extending over more than one letter of the Boolian can be broken up into obeli over each single one of those letters and over any copula of inclusion covered by the long obelus, by changing every sign of aggregation which had been covered by the long obelus to a sign of compredication and conversely.

If the obelus extends also over a part or the whole of the Percian, each Π that is uncovered must be changed to a Σ, and conversely.

If from one proposition another necessarily follows, then *by contraposition* from the negative of the latter would necessarily follow the negative of the former. That is

(24) If $x \prec y$, then $\bar{y} \prec \bar{x}$.

Arithmetical Relations. A number with indices attached to it not fewer than that number, means that the object denoted by those indices form a collection of that number of different things existing in the universe of discourse. But the only relations required are the following:

0 means the proposition essentially (i.e. by definition) false, of which one expression is "something false is true."
0_x means "x does not exist."
1_x means "x is an existing thing," but it will be best to write this 1_{xx};
1_{xy} means "x is identical with y."
2_{xy} means "x is not y."

Instead of $\bar{0}$, the sign ♉ Taurus, or any of the signs of the zodiac may be used. For for certain purposes it is advantageous to have a numberof equivalent signs for $\bar{0}$.

It will be observed that $\bar{0}_x = 1_x$, $\bar{1}_{xy} = 2_{xy}$, $\bar{2}_{xyz} = 3_{xyz} \Psi 1_{xyz}$, etc.

For all numbers higher than 1, we have equations like the following:

$$3_{abcd} = (3_{abc} \Psi 3_{abd} \Psi 3_{acd} \Psi 3_{bcd}) \cdot (\bar{3}_{abc} \Psi \bar{3}_{abd}) \cdot$$
$$\cdot (\bar{3}_{abc} \Psi \bar{3}_{acd}) \cdot (\bar{3}_{abc} \Psi \bar{3}_{bcd}) \cdot (\bar{3}_{abd} \Psi \bar{3}_{acd}), \text{ etc.}$$

But it is not true that $1_{xy} = (1_x \Psi 1_y) \cdot (\bar{1}_x \Psi \bar{1}_y)$; because $\bar{1}_x$ is absurd, and nothing involves an absurdity if there is any possible interpretation not absurd, while for all higher numbers than 1, there is no relation implied in the counting that does not consist of relations of pairs. Those relations are subject to the following conditions.

(25)	$x \Psi 0 \prec x$	$x \prec x \cdot$ ♉
(26)	$0 \prec x$	$x \prec$ ♉
(27)	$\Sigma_x 0_x \prec 0$	♉ $\prec \Pi_x 1_{xx}$
(28)	$\Sigma_y l_{xy} \cdot 0_y \prec 0_x$	$1_{xx} \prec \Pi_y l_{xy} \Psi 1_{yy}$
(29)	$\Pi_y l_{xy} \Psi 0_y = \Pi_y l_{xy}$	$\Sigma_y l_{xy} \cdot 1_{yy} = \Sigma_y l_{xy}$
(30)	$\Pi_x 0_x = \Pi_x \Pi_y 2_{xy}$	$\Sigma_x 1_{xx} = \Sigma_x \Sigma_y 1_{xy}$

(31) Usually
$\Pi_x 0_x \prec 0$ and ♉ $\prec \Sigma_x 1_{xx}$,

because of a universe utterly devoid of objects there can hardly be discourse. Still, it cannot be said that these propositions are absolutely necessary.

(32)	$l_{xy} \cdot 1_{yx} \prec l_{xx}$	$l_{xx} \prec l_{xy} \Psi 2_{yz}$.

The general method of drawing conclusions by means of this algebra. The process of reasoning consists, first, in compredicating the premises, and secondly, in transforming the compredicate so resulting. Each premise may be introduced as a compredicant any number of times; and the conclusions will always differ as this number differs. If a premise is introduced *N* times, the conclusion is said to be of the *N*th *order* in that premise. The interest of a conclusion increases with *N*, when *N* is small, until a maximum interest is reached when it diminishes with the increase of *N*.

When the object in view is to explore in a general way the consequences of given premises, the above process may advantageously be employed without modification. But when the object is to find whether a given proposition can be demonstrated it is better to proceed *apagogically*, that is, by the *reductio ad absurdum*.* (*The common expression *reductio ad absurdum* in strict logical parlance ought to be restricted to that particular *deductio ad absurdum* which establishes the validity of such a mood of syllogism as *Bocardo* or *Baroco*.) That is, we commence with the precise denial of the proposition to be proved, and having performed upon it suitable illative transformations we resolve it into aggregants each of which is the precise denial of an accepted truth. We may, then, start with these truths as premises, and having compredicated them, may go through the inverse of all the previous processes in inverse order and so reach the conclusion by direct reasoning. But it is far better to state the demonstration in its apagogical form; because the usual style of writing of mathematicians is to omit all explanation of their reasons for performing their different operations of inference. Now, in the apagogical demonstration the reader can more readily perceive what those reasons are.

If the object is to prove a philosopheme, we may naturally employ dilemmatic reasoning. That is, we show that the hypothesis, or premise, leads to an aggregate of propositions. We, then, show separately that each of these by illative transformation leads to the possibility of that which we are endeavoring to show is possible.

The apagogical transformation of the dilemma starts with denying the possibility of that whose possibility is to be established and by different kinds of transformation show that each one of several propositions is deducible from that denial. We finally show that from the compredicate of these follows something incompatible with the fundamental hypothesis.

All these processes have to be performed intelligently with an adaptation of means to ends. The algebra does not relieve the mind from any important portion of the process of thought. Whether or not, in any case, it facilitates reasoning depends upon whether it is more easy to reason about the algebraical forms or about the ideas for which these forms stand. The only advantage which the algebra affords in all cases is that it insures absolute logical precision in the demonstrations.

I will now describe in detail the method of direct demonstration of an ordinary theorem. The premises are first to be multiplied together. In this process all the indices of different premises should usually be different letters. But when a premise is multiplied into itself those of the indices of the different factors are to be the same whose quantifiers in the Percian do not come to the right of a given Σ, while all the rest of the indices are to be diversified. The same thing may be true of different premises; but that is rarely the case. It is best to begin by compredicating those propositions in which such partial identification of indices is to be made. When one of the arithmetical relatives 0, ȣ, 1, 2 occurs in the expression of a premise, — as 1 and 2 frequently do, — it is often necessary to use one of the formulae (25)-(32), or a consequence of it, as a premise. It is generally best to express each premise as an aggregate of aggregants none of which is a compredicate of two compredicants or as a compredicate of such aggregants. In the arrangement of the Percian of the compredicate of the premises, our general effort is to put Σs as far to the left as possible. We generally have our choice between having one Σ well to the left and another quite far to the right or the reverse. In such a case, it is necessary to scrutinize the formula, so as to see which alternative will best answer the purpose in view. One arrangement may lead to one conclusion and the other to another. We next identify certain indices attached to Πs each with another index which in the Percian is to the left of it. This we do with a view of being able afterward, by transformations of the Boolian, to produce compredicants of one of the forms $(a \Psi \beta) \cdot \bar{b}$ and $(a \Psi b) \cdot (\bar{b} \Psi c)$, where a, b, c, stand each for some letter on the line together with its indices, or even for a more complex proposition relating to a set of individuals. One of these forms having been produced, we transform the Boolian according to one of the following formulae

$$(a \Psi b) \cdot \bar{b} \prec a \Psi b \cdot \bar{b} \prec a \Psi 0 \prec a$$
$$(a \Psi b) \cdot (\bar{b} \Psi c) \prec a \Psi b \cdot (\bar{b} \Psi c) \prec a \Psi b \cdot \bar{b} \Psi c \prec a \Psi 0 \Psi c \prec a \Psi c$$

We, thus, get rid of b and shall very likely have destroyed every occurrence in the Boolian of some index, which can, therefore, now be dropped from the Percian.

[ON MULTITUDES] (29)

Generality is a potential infinity: that is, there may be infinitely many things that possess a given quality. Infinity, on the other hand, as we shall soon see reposes upon generality.

Infinity is a word often misunderstood by philosophers; and the word itself is calculated to mislead. According to its etymology, it would naturally mean the absence of a limit. It is imitated from the Greek ἀπειρία, in the sense of endlessness.* (*The Greek also means inexperience or ignorance. In both meanings the root is PAR, to go through, found in our *fare*.) But that is not at all what it means. On the contrary, the being confined within limits is favorable to infinity, and *vice versa*. A circle is unlimited; and this is almost reason enough for its being finite, that is for a finite number of steps completing its circuit. But if measurement is such as to be necessarily confined between two limits, unless it is abruptly interrupted it must be infinite. Infinity may be of measurement; and it then means that the multitude of units is infinite. Originally, and properly it is a term of multitude; and in that sense, it requires examination. When we say God is infinite, this may be understood as meaning that His measurable qualities are infinite, or it may be understood as an unclear and mystical comparison with space and time, implying incomprehensibility, etc.

Some multitudes are finite. That is, when you go to count any of them, or put it into one to one correspondence with the cardinal numbers, the operation is interrupted by the exhaustion of the multitude. This is an exception and a complication. It makes reasoning about such multitudes, as we shall find in due time, to be more intricate than about others which do not present this peculiar feature. For the present we pass such multitudes by.

There are, besides those, many other multitudes which can be arranged in a row, each one having on each side another *next* to it. If a multitude when so arranged has no first or no last, or if the row can be severed into two parts one of which has no first and the other no last, it is said to be *inenumerable*, or *indefinite*. The cardinal numbers are an example. Another is all the numbers between 0 and 1, *exclusive** (*Pronounce the final *e*. This is the Latin adverb, not the English adjective.), which are exactly expressible in decimals. Arranged in the order of their values, these last numbers are in a row, but each has not another nearest to it on each side. But if we reverse the order of the figures with which each is written, and then remove the decimal point, — so, for example, as to convert .461 into 146, and .0012 into 2100, — we turn them into whole numbers. If now, we arrange the original numbers in the order of values of these whole numbers, we have them in a row where each has next neighbors, and the row has no last. It will begin, .1, .2, .3, .4, .5, .6, .7, .8, .9, .01, .11, .21, .31, .41, .51, .61, .71, .81, .91, .02, .12, .13, etc.

But such collections, though *inenumerable*, are *numerable*, that is, a whole number can be assigned to each one, though the whole numbers are none too many for the purpose. All collections of this description must be reckoned as equally numerous; the disarrangement of them cannot alter their number. Consequently, the numbers divisible by ten are just as numerous as all the numbers. For putting a zero after every number, we get a number divisible by ten. Thus, it is not true of inenumerable parts that the whole is greater than its part.

There are greater multitudes. The square of such a multitude is, indeed, no greater than the multitude itself. For the square would be the multitude of pairs of all whole numbers. Now if we write all the figures of one number in the even decimal places and those of the other number in the odd decimal places, we get one number from which those two numbers can be read off. Thus if one is 93715 and the other 24, the number resulting from writing the figures together is 903072145. Now the multitude of all such numbers, each of which will give a distinct pair of numbers, is of course not greater than the multitude of all whole numbers. But if, instead of raising that multitude to a power whose exponent is a finite number, we raise a finite number to a power indicated by such a multitude, no such method of expressing the multitude of such multitudes [is] applicable; and in point of fact we have a greater multitude. Thus, if we raise ten to such a power, we have the multitude of inenumerable multitudes resulting from putting in place of each of the dots of an inenumerable row of dots, in every possible way all the ten Arabic figures 0, 1, 2, 3, 4, 5, 6, 7, 8, 9. If before each such endless row of numbers we put a decimal point, the entire multitude is the multitude of all the numbers to whose values the decimals from 0 to 1, *exclusive*, will indefinitely approximate and which differ from one another by finite fractions. If we reverse those endless decimals, as we did the others, after omitting for the sake of simplicity those which contain 0s, we get whole numbers in which *all* places of decimals, however high, are filled with significant figures. These may be called indefinitely great numbers. We will not call them infinite, because there is not an infinite but only an indefinitely great number of decimal points. Accordingly, the number of finite whole numbers is indefinitely great, and the new multitude is the number of indefinitely great numbers. It was first proved by Dr. Georg Cantor that this multitude is greater than the multitude of finite whole numbers.

The proof is easiest stated in the form of a *reductio ad absurdum*. Suppose that the multitude of numbers between 0 and 1 using all the decimal places at a finite number of places from the decimal-point were each to have a distinct cardinal number attached to it. It can be shown that such a supposition involves us in absurdity. For take any portion of the interval between 0 and 1, and let *A* be the number which in that portion has the lowest cardinal number attached to it. Let *B* be that number between *A* and the last of the portion which receives the lowest cardinal number. Let *C* be that number between *A* and *B* which receives the lowest cardinal number; let *D* be that number between *B* and *C* that receives the lowest cardinal number; and so on, endlessly. There will always be a finite interval between numbers successively lettered; for they are different numbers, and there is a finite interval between any two numbers. Now every one of the numbers *A*, *C*, *E*, *G*, etc. is less than every one of the numbers *B*, *D*, *F*, *H*, etc. If we compare any two successive numbers as *E* and *F*, *E* from the increasing series, *F* from the diminishing series, they will agree down to some place of decimals, after which they will not agree. So far as they agree, all the numbers following them, *G*, *H*, *I*, *J*, etc. will agree with them. But there is always some place of decimals which in subsequent numbers will get altered (since all the numbers differ from one another); and it will ultimately get changed to a number which will be altered no more. If, then, we consider the number which has in every place of decimals the figure which ultimately takes that place, and is never after changed, that number will be larger than any of the numbers *A*, *C*, *E*, *G*, etc. and smaller than any of the numbers *B*, *D*, *F*, *H*, etc. It will, therefore, not be any of the numbers which have cardinal numbers attached to them; and therefore it is absurd to suppose that all the numbers have cardinal numbers attached to them. Thus, the multitude of numbers in which all the finite decimal places are filled is too great to be put into one-to-one correspondence with the cardinal numbers.

Such multitudes may be called, not inenumerable, merely, but *innumerable*. These collections may also properly be termed *infinite*.

Reasoning about these matters is exceedingly puzzling. Almost all metaphysicians, — and even a mathematician, here and there, — have fallen into some of the pitfalls with which the ground is spongy. For example, one might be tempted to reason as follows: Between any two numbers exactly expressible in decimals there are evidently innumerable surds (that is, numbers not so expressible nor expressible in circulating decimals). But no matter how small a power of $\frac{1}{10}$ we may take, two numbers exactly expressible in decimals may differ by less than that. Hence, there must be surds whose difference is less than any power of $\frac{1}{10}$ whatever. Now this does not follow. The thread which will guide us safely through this labyrinth is a perfectly tangible one. We have only to remember that to say that some of the *A*s love every one of the *B*s is saying much more than that Everyone of the *B*s is loved by some of the *A*s. For the first means: I can find an *A* such that, after I have chosen it, you cannot find a *B* such that my *A* does not love your *B*. But the second only means: You cannot find a *B* but such that, after you have chosen it, I can find an *A* such that my *A* shall love your *B*. If I have to find an *A* that loves your *B*, of course it is an immense advantage to know what *B* you take. If then when *some*, *all*, *every*, are spoken of, before coming to the substance of the statement we begin by setting out how the objects to be selected are supposed to be selected, and notice that making a *suitable* selection before a random selection produces possibilities of falsity which do not exist when the suitable selection is not made till after the random selection, then almost all risk of erroneous reasoning will be avoided. Let us apply this rule to correct the false reasoning just offered as an example. Numbers exactly expressible in decimals are numbers such that, having chosen any one of them, it is possible to find a place of decimals such that, after that is taken, taking any place of decimals to the right of it, that place of decimals will have a zero in it. Taking any two such numbers, we can find innumerable surds between them. Taking [a] place of decimals, we can find two numbers exactly expressible in decimals, whose figures agree down to that place. Hence, taking any place of decimals no matter what, we can find two surds that differ by less than that. That is obvious enough. But it is a very different thing to conclude, as we pretended to do, that we can find two surds such that, after they are chosen, no matter what place of decimals we take, those numbers will agree in that place of decimals. Thus, the reasoning which, without our maxim, seemed so conclusive, is seen, by the microscope of that maxim, to have its conclusion quite cut off from all logical connection with its premises.

The square, or any finite power, of an innumerable multitude is plainly equal to that multitude itself, by the same reasoning that was applied to inenumerable collections. But there are multitudes greater than the multitude of all surds.

We will first show how to construct such a collection, and then show that it exceeds the multitude of surds. But by way of preparation, it may be as well to mention one or two little things which conflict with prejudices derived from familiarity with finite collections.

A row of objects without beginning or end is not necessarily larger than one with beginning and end. Thus, take all numbers exactly expressible in decimals between 0 and 1, *exclusive*. That has no beginning nor end and, as we have seen, is simply inenumerable. Take all real numbers (that is all numbers that are expressible in decimals, or in circulating decimals, or that are surds) from 0 to 1, *inclusive*. That has both beginning and end, but is innumerable, and so innumerably greater than the other.

A row of objects of which each has one on each side next to it, and which has both beginning and end, is not necessarily finite. Thus reverse the order of the even numbers and place the whole series after the odd numbers taken in their proper order. The whole makes one row; for of any two numbers one comes after the other. Every one has a next neighbor before and after it. Yet the whole is inenumerable. But this row can be severed into two the one coming before having no end and the one coming after having no beginning. A row which cannot be so severed may be called a *simple* row. A row in which every object has a next neighbor on each side of it may be called a *sparse* row. A simple, sparse row having both beginning and end is finite.

The series of numbers expressible by decimals is not a simple series. For it may be severed into those less than $\frac{1}{9}$ and those greater than $\frac{1}{9}$; and the former part will not have an end, nor the latter a beginning. But the series of real numbers is a simple series. Wherever it is severed, either the preceding part has an end or the following part has a beginning. This is a fact which has an essential bearing upon the question we are presently to consider. Let the series of real numbers between 0 and 1 be severed. Of the numbers in the preceding part, take the highest figure that occurs in the first decimal place, then the highest figure which in that preceding part next follows the figure just taken, and so on without end. A certain real number will result. This number will not necessarily belong to the preceding part; for although, no matter what place of decimals may be taken, there is some number in the preceding part which agrees with that limiting number down to that point, it does not follow that a number can be found in the preceding part such that no matter what place of decimals be taken, it agrees with the limiting number in that place of decimals. Suppose, for example, that the limiting number is $\frac{1}{9}$ = .1111111 etc. Then, it may be that this number belongs in the second part; so that, taking any place of decimals, a number of the preceding part can be found which has nothing but 1s down to that point. It will have a zero, however, somewhere further on. Yet taking the place of the first zero, another number can be found that has all 1s down to that point. Still, it will have a zero somewhere further on, again. However, there always is such a limiting number. Now every real number less than that limiting number, is less than it in some finite decimal place to the left of which it agrees with it. Such number will, therefore, from the way in which the limiting number is formed, belong to the preceding part. If the limiting number, then, does not belong to the preceding part, it is the least number of the following part; and thus the following part has a beginning. If on the other hand, the limiting number belongs to the preceding part, then since every greater number is greater than it in some decimal place down to which it agrees with it, it follows from the rule of the formation of the limiting number that such number belongs to the second part. In this case, therefore, the limiting number is the end of the preceding part. Thus in every case, either the preceding part has an end or the following part has a beginning; and by the definition of a simple series the series of real numbers is such a series ....

C. [MULTITUDE AND CONTINUITY] (28)

A collection formed by aggregating an enumerable multitude of denumerable collections is a denumerable collection. For, first, two denumerable collections aggregated make a denumerable collection. For let one of them consist of M_0 and an individual which is *r* of each individual of the collection, nothing being *r* of two different things, and the *r* of anything being other than that thing, while the other consists of N_0 and an individual which is *s* of each individual of the collection, nothing being *s* of two different things, and the *s* of anything being other than that thing. The aggregate of the two consists of M_0 and an individual which is ϱ to each individual of the aggregate, where no two things are ϱ of the same thing and nothing is ϱ of itself. Namely, the ϱ of M_0 is N_0 and the ϱ of any *N* is the *r* of that *M* of which that *N* is the ϱ, while the ϱ of any other *M* than M_0 is the *s* of that *N* of which the *M* is ϱ. Thus, the aggregate conforms to the definition of a denumerable collection. Hence, obviously, by the Fermatian inference, it follows, secondly, that the aggregate of any enumerable collection of denumerable collections is a denumerable collection.

A collection formed by aggregating a denumerable multitude of denumerable collections is a denumerable collection. For if we take two denumerable collections, the *M*s and the *N*s, defined as before, the collection of possible pairs each consisting of an *M* and an *N* is the aggregate of a denumerable multitude of denumerable collections. For there is a denumerable collection of such pairs containing each single *N* and there is a denumerable collection of *N*s. But the aggregate is a denumerable collection. For it consists of the pair M_0N_0 and a pair which is σ of each pair it contains, where no pair is σ of two different pairs, and no pair is σ of itself. For the rule is that the *M*s and the *N*s having first been put into one to one correspondence so that N_0 is *c* of M_0 and the *N* which is *s* of any given *N* say N_i is *c* of the *M* that is *r* of that *M* of which N_i is *c*, then the σ of the pair consisting of any *M* say M_j and N_0

is the pair consisting of M_0 and of the N which is s of the N which is c of M_j, while the σ of any pair consisting of any M, say M_m, and of any N, say N_n, other than N_0 is the pair which consists of the M which is r of M_m and the N of which N_n is the s. For example, if the Ms are M_0, M_1, M_2, M_3, etc. and the Ns are N_0, N_1, N_2, N_3, etc., then the succession of pairs are

$$M_0N_0,\ M_0N_1, M_1N_0, M_0N_2, M_1N_1, M_2N_0, M_0N_3, M_1N_2, M_2N_1, M_3N_0, \text{ etc.}$$

Thus, the collection of pairs is shown to conform to the definition of a denumerable collection.

The multitude of possible sets of objects, each set formed by taking one object out of each of an enumerable multitude of denumerable collections, is denumerable. This is obviously true by the Fermatian inference, proceeding from the last result. We shall soon see, however, that this would cease to be true, if the multitude of denumerable collections contributing to the sets were denumerable.

Dr. Georg Cantor has shown that the individuals of a denumerable collection may be so arranged in a linear sequence, that between any two of them there is a denumerable collection of intermediate individuals. For this is true of the series of rational quantities arranged in the order of their values between $\frac{0}{1}$ and $\frac{1}{1}$. Namely, the whole collection is denumerable, for it consists of $\frac{0}{1}$ and a quantity which is τ to each such quantity, where no quantity is τ to two quantities and no quantity is τ to itself. Namely, the τ of any quantity, is obtained by expressing that quantity in its lowest terms, and then increasing the numerator by the smallest amount which will leave the fraction irreducible and not allow its value to exceed the limit, or if no such fraction exists, the τ is the fraction whose denominator is greater by 1 and its numerator is 1. Thus, the first terms of the series are

$$\tfrac{0}{1}, \tfrac{1}{1}, \tfrac{1}{2}, \tfrac{1}{3}, \tfrac{2}{3}, \tfrac{1}{4}, \tfrac{3}{4}, \tfrac{1}{5}, \tfrac{2}{5}, \tfrac{3}{5}, \tfrac{4}{5}, \tfrac{1}{6}, \tfrac{5}{6}, \tfrac{1}{7}, \tfrac{2}{7}, \tfrac{3}{7}, \tfrac{4}{7}, \tfrac{5}{7}, \tfrac{6}{7}, \tfrac{1}{8}, \tfrac{3}{8}, \text{ etc.}$$

It is plain that there is such a denumerable series between any given limits whatever.

So much for the denumerable multitude. Let us now ask, what is the smallest multitude which exceeds a denumerable multitude? That there is a multitude which exceeds a denumerable multitude has already been shown. Namely, the possible combinations of whole numbers is such a multitude. The possible combinations of whole numbers each of which consists of an enumerable collection of whole numbers is merely denumerable. But the possible combinations of denumerable whole numbers form a collection which exceeds a denumerable collection. I call this multitude the *first* [*abnumeral*] *multitude*.

That there is no multitude greater than the denumerable multitude and less than the first abnumeral multitude, I argue as follows. The demonstration does not seem to be perfect; but I think it is only because I do not succeed in stating it quite right. The addition, or aggregation, of denumerable collections is entirely without effect upon the multitude, even if the multitude of collections aggregated is denumerable. The multiplication of denumerable collections, that is the formation of all possible compound objects each being a set composed of an individual out of each collection, is also without effect, if the multitude of denumerable factors is enumerable. But if the multitude of factors is denumerable, the result is the first abnumeral. Now since there is no multitude greater than every enumerable multitude and less than the denumerable multitude, it follows that there is no multitude greater than the product of every enumerable collection of denumerable factors and less than the product of a denumerable collection of denumerable factors which is the first abnumeral multitude.

I have already substantially proved that the constituent individuals of a first abnumeral collection are distinct from one another.

As an example of a first abnumeral collection, we may take all the real quantities between any two limits, that is all the quantities which are capable of expression to any desired degree of approximation by means of decimals carried out to an indefinite number of places. It is evident that the number of quantities between zero and one capable of being distinguished from one another in N decimals place is 10^N. If N is denumerable, this is the product [of] a denumerable multitude of enumerable factors. Whether each factor is an enumerable or a denumerable multitude makes no difference, so long as it exceeds unity. If we use the binary system of numerical notation, it is 2^N instead of 10^N. Now 2^N is the multitude of combinations of N objects, and if N is the denumerable multitude, 2^N is the first abnumeral multitude.

The product of an enumerable multitude of factors each of the first abnumeral multitude is itself of the first abnumeral multitude. This was remarked by Dr. Georg Cantor. Thus, the multitude of a collection of objects which are in one to one correspondence with all the possible values of $x+y\sqrt{-1}$ where the possible values of x are of the first abnumeral multitude and those of y equally many is no greater than the multitude of possible values of x. For suppose each of the variables,

x and y, could be any real quantity between zero and one. Then, a single quantity z can be found, of the same description from the value of which the values of both x and y can be determined to any desired degree of approximation. For example, the rule for so determining x and y might be after writing z as a decimal carrying it out to an indefinite number of places, to pick out all the figures in odd places and write them together for x and to pick out all the figures in even places and write them together for y. Thus if

$$z = .237114812589128625834559 9\ldots$$

we should have

$$x = .2718281828459\ldots$$
$$y = .314159265359\ldots$$

This would be applicable to any enumerable multitude of factors.

But the product of a denumerable multitude of factors each of the first abnumeral multitude, is the same as the multitude of ways of distributing the individuals of a first abnumeral collection into two abodes, which has already been proved to be greater than the first abnumeral multitude. I call this the *second abnumeral multitude.*

It has already been substantially demonstrated that the constituent individuals of such a collection are distinct from one another.

Mathematics affords no example of such a multitude. Mathematics has no occasion to consider multitudes as great as this. The multitude of possible combinations of real quantities is this multitude. Were the combinations of real quantities limited to enumerable or even to denumerable collections, their possible multitude would be no greater than that of the real quantities themselves. But it [is] the multitude of first abnumeral collections of real quantities which amounts to the second abnumeral multitude. It is 2^{2^N} where N is denumerable.

The product of an enumerable multitude of factors each of the second abnumeral multitude is itself of the second abnumeral multitude. But if the multitude of factors is denumerable, it is the same as the multitude of combinations of objects of the second abnumeral multitude. That is, it is the *third abnumeral multitude.*

And so the multitudes succeed one another indefinitely; and the constituent individuals of collections of those multitudes are distinct from one another.

Thus, the whole series of multitude, so far as yet made out, begins with the multitude of a non-existent collection, or zero, and then comes the multitude of a single object, and then the multitude of 2, and so on increasing by one without end. After these multitudes comes the denumerable multitude which may be called the zero abnumeral multitude, then the first abnumeral multitude, then the second abnumeral, and so on increasing in order by one without end. All these multitudes thus form two denumerable series, and consequently there is only a denumerable multitude of different possible multitudes, so far as yet made out.

Let us now suppose that there is a collection of distinct objects of each of those multitudes. Then, taking any one of those collections, no matter what, there is, among the whole collection of those collections, a denumerable multitude of collections each of which is greater than the collection chosen. Let us then throw together all the distinct individuals of those collections so as to form an aggregate collection. This aggregate collection is greater than any of the single collections; for it has, as we have just seen, a denumerable collection of parts greater than any one of those collections. I shall call it a *supermultitudinous* collection.

It seems to be sufficiently evident that there is no collection at once greater than every abnumeral collection and less than such a supermultitudinous collection.

The collection of possible ways of distributing the individuals of a supermultitudinous collection, S, into two abodes is no greater than that supermultitudinous collection, S, itself. For denoting by ♈ the denumerable multitude, the abnumeral multitudes are $2^{♈}$ $2^{2^{♈}}$ $2^{2^{2^{♈}}}$ etc., or Exp ♈, $(\text{Exp})^2$♈, $(\text{Exp}$♈$)^3$♈, etc.; and the magnitude of the supermultitudinous collection is the limit of this series. It is, in short, the result of a denumerable succession of exponential operations upon the denumerable multitude. But the magnitude of the collection of possible ways of distributing the individuals of a collection into two abodes is simply the result of an exponential operation upon the magnitude of the collection itself. Hence the magnitude of the ways of distributing the individuals of a supermultitudinous collection into two abodes is obtained by adding one more to the collection of exponential operations successively performed upon the denumerable multitude. But this collection of operations being denumerable, the addition of one operation to it does not increase its magnitude. Hence, the collection of possible ways of distributing the individuals of a supermultitudinous collection into two abodes equals that collection itself.

But we have already seen that the multitude of ways of distributing the individuals of a collection of distinct individuals into two abodes exceeds the collection itself. Hence, it follows that a supermultitudinous collection is so great that its individuals are no longer distinct from one another.

This is not in conflict with the fact that a supermultitudinous collection is a denumerable collection of distinct collections in each of which the individuals are all distinct from one another.

A supermultitudinous collection, then, is no longer *discrete*; but is *continuous*. As such the term "multitude" ceases to be applicable to it; and I shall speak of its *order of magnitude*, meaning that character by which, considered as a collection, it is greater than one collection and less than another.

We have to consider whether or not all supermultitudinous collections are of the same order of magnitude. For that purpose, we need to develop a distinct notion of the relationship of such a collection to its individuals.

Let us begin with two individuals, which we may mark 0 and 1. We take a third individual; and [begin] to construct a relation, *r*, which is to have the general property that if anything *A* is in the relation, *r*, to anything *B*, then *B* is not in the relation *r* to *A*. (Whence nothing is *r* to itself.) We mark the third individual .1 and say that .1 is in the relation *r* to 0, while 1 is in the relation *r* to .1. We next add two more individuals, which we mark .01 and .11 we say that .01 is *r* to 0, that .1 is *r* to .01, that .11 is *r* to .1 and that 1 is *r* to .11. We next add 4 new individuals, which we mark as .001, .011, .101, and .111. We say that .001 is *r* to 0, that .01 is *r* to .001, that .011 is *r* to .01, that .10 is *r* to .011, that .101 is *r* to .10, that .11 is *r* to .101, that .111 is *r* to .11, and that 1 is *r* to .111. Our next addition will be of eight new individuals, .0001, .0011, .0101, .0111, .1001, .1011, .1101, .1111. Our next addition will be of sixteen new individuals. We go on until we have carried the additions as far as they can be carried without using an inenumerable number of characters to mark one individual.

Then all the individuals which are marked each by an enumerable collection of characters form a denumerable collection of individuals.

But now I call your attention to a very remarkable circumstance. All those individuals stand distinct and independent. Any one of them or any collection of them may be taken away without affecting the remainder; but yet there are already symptoms of incipient cohesiveness in them, a premonition of continuity. Remember that the multitude of grades of enumerable multitude is denumerable; and the multitude of all the grades of multitude is no greater. In dealing with finite numbers, where each multitude differs from the next but by one, we become accustomed to think that the multitude *numerals*, or marks of grades of multitude, below a given numeral distinguishes that multitude from every other. But this ceases to be true when we leave the enumerable collections. If we use the term *arithm* to mean the multitude of grades of multitude below the multitude to which the arithm is attached, then the arithm of zero is zero, the arithm of 1 is 1, the arithm of 2 is 2, and in short the arithm of any enumerable multitude is that multitude. But the arithm of all higher multitudes is the same. It is the denumerable multitude, which may be called *infinity*, ∞. We shall, therefore, lose ourselves in a labyrinth [of] hopeless confusion if we allow ourselves for an instant to judge of a higher multitude by its arithm.

But I repeat that this unity of the arithms of all the higher multitudes is the first embryon of continuity. I proceed to explain this remark. In interpolating those individuals, I was just speaking of, the first interpolation being of 1 individual, the second interpolation being of 2 individuals, the third interpolation being of 4, the fourth interpolation being of 8 individuals, the fifth interpolation being of 16 individuals, and so on (I repeat this so many times in order to impress upon you what I here mean by the word interpolation), so long as there has been only an enumerable collection of interpolations, it is plain that not all of the individuals which are designated by enumerable collections of characters have as yet been inserted. The entire collection of those individuals that are marked by enumerable series of characters, is *denumerable*. But as soon as the denumerable multitude of interpolations has been made, the collections of characters attached to the last inserted individuals are denumerable, and the collection of individuals there is of the first abnumeral multitude. Stop it at any point you please, however early, if all the individuals marked by enumerable series of characters are there, there is also along with them a first abnumeral multitude of individuals marked by denumerable series of characters. *They stick together*. This is what I meant by saying there was an incipient cohesiveness, a germinal [element] of continuity.

The cause of the phenomenon is so easily traced that hasty thinkers will say it is nothing but a fallacy. I do not agree with them. It is not true continuity, but only an appearance of cohesion; but in my opinion it is genuinely the first stage in the development of continuity.

The explanation which I allude to ought to be pretty obvious after what I have just said about the arithm. If the succession of insertions of individuals stops in such a way that there is a last inserted individual, this is either one of those individuals which are marked by enumerable collections of characters, in which case not all of them have been inserted, or else it is one of those which are marked by denumerable collections of characters, in which case, in addition to the individuals marked by enumerable collections of characters, others amounting to the first abnumeral multitude have been inserted besides. But it is not necessary that the succession of insertions should be so broken off that there is any last individual inserted, and if this is the case, it may be that all the individuals marked by enumerable collections of characters have been inserted and no others. If all those individuals *are* inserted, the multitude of interpolations is precisely the same, namely denumerable, whether those interpolations do or do not include interpolations of the individuals marked by denumerable collections of characters.

Thus far, all the symptoms of cohesion which manifest themselves depend upon the order of succession. An enumerable collection does not cling together at all in whatever order it be taken. A denumerable succession does not cling together if taken in the order of its generation, any further than this, that no part can be struck off the latter end without so much is struck off as to reduce it to an enumerable collection. A denumerable collection can be so arranged that a denumerable collection of denumerable sequences can be struck out from it without reducing its multitude. Suppose for example we arrange the whole numbers in the following order.

1st, powers of the lowest prime in their order

1, 2, 4, 8, 16, 32, 64, 128, 256, 512, 1024, etc.

2nd, each of these successively multiplied by powers of the next lowest prime in their order,

3, 9, 27, 81, etc.
6, 18, 54, 162, etc.
12, 36, 128, 324, etc.

3rd, each of these successively multiplied by powers of the next lowest prime in their order

5, 25, 125, 625, etc.
10, 50, 250, 1250, etc.
20, 100, 500, 2500, etc.
etc.
15, 75, 375, 1875, etc.
45, 225, 1125, 5625, etc.
135, 675, 3375, 16875, etc.
etc., etc.

and so on *ad infinitum*. Another arrangement would be

1st, all numbers in their order whose highest prime factor is the lowest prime

1, 2, 4, 8, 16, 32, 64, 128, 256, 512, 1024, etc.

2nd, all numbers in their order whose highest prime factor is the next lowest prime

3, 6, 9, 12, 18, 24, 27, 36, 48, 54, etc.

3rd, all numbers in their order whose highest prime factor is the next lowest prime

5, 10, 15, 20, 25, 30, 40, 45, 50, 60, 75, 80, etc.

and so on *ad infinitum*.

There is also an abnumeral collection of arrangements of any denumerable collection such that it is impossible to reduce it to an enumerable collection without striking out more distinct parts than there are individuals remaining. Such an arrangement of the whole numbers is obtained for example by arranging them in what would be their natural order if the succession of figures with which they are written were reversed. Thus, the first sixteen numbers in the binary notation would occur in the order

```
                         △
0000 = 0
          ....8
1000 = 8
          .......  −4
 100 = 4
          ....8
1100 = 12
          ...........  −10
  10 = 2
          ....8
1010 = 10
          .......  −4
 110 = 6
          ....8
1110 = 14
          ................  −13
   1 = 1
          ....8
1001 = 9
          .......  −4
 101 = 5
          ....8
1101 = 13
          ...........  −10
  11 = 3
          ....8
1011 = 11
          .......  −4
 111 = 7
          ....8
1111 = 15
```

The numbers from −13 to +13 in the ternary system [would occur in the following order where each T is $\bar{1}$ or −1].

△

TTT = −13
.... +9
TT = −4
.... +9
1TT = 5
.... −15
T0T = −10
.... +9
T = −1
.... +9
10T = 8
.... −15
T1T = −7
.... +9
1T = 2
.... +9
11T = 11
.... −23
TT0 = −12
.... +9
T0 = −3
.... +9
1T0 = 6
.... −15
T00 = −9
.... +9
0 = 0
.... +9
100 = 9
.... −15
T10 = −6
.... +9
10 = 3
.... +9
110 = 12
.... −23
TT1 = −11
.... +9
T1 = −2
.... +9
1T1 = 7
.... −15
T01 = −8
.... +9
1 = 1
.... +9
101 = 10
.... −15
T11 = −5
.... +9
11 = 4
.... +9
111 = 13

They may also be arranged in a ring thus like a compass card with 3 cardinal points thus [Fig. 1].

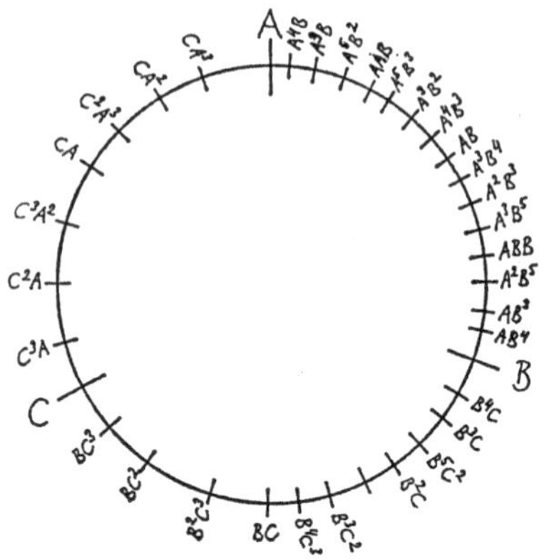

Fig. 1

For any such arrangement of a denumerable collection there must be some general rule connecting the place of an individual in the arrangement with its designation or symbol. *This rule necessarily assigns places not only to the denumerable collection but also to a collection of first abnumeral multitude of other possible individuals.* Thus a denumerable collection cannot be even in idea without the potential accompaniment of a first abnumeral collection.

And yet the constituent individuals of the abnumeral collection are distinct and discrete. Any one of them may be annihilated without affecting the others.

Observe that the precise manner in which the denumerable collection involves the first abnumeral collection is this, that in order to be able to say, here is complete the entire denumerable collection, it is necessary to have stopped somewhere and to have put down a last, — for by the word "complete," we mean to a last; — and before there can be any last after the denumerable collection is all there, a first abnumeral collection must already be there.

It is equally true that it is impossible to have completed the entire first abnumeral collection of sets of individuals of a denumerable collection without also having a second abnumeral collection; but this does not strike us so forcibly, because we never trouble ourselves to imagine that the entire first abnumeral collection of sets of denumerable individuals *is* complete. Another point of difference in the two cases is this, a denumerable collection seldom presents itself with a last. It *may* do so, as when the numbers are arranged

1, 3, 5, 7, 9, 11.....12, 10, 8, 6, 4, 2, 0

And there are arrangements in which there is no break at any particular point as

$$\frac{0}{1}\cdots\frac{1}{10}\cdots\frac{1}{9}\cdots\frac{1}{8}\cdots\frac{1}{7}\cdots\frac{1}{6}\cdots\frac{1}{5}\cdots\frac{2}{9}\cdots\frac{1}{4}\cdot\frac{2}{7}\ \frac{3}{10}\cdots\frac{1}{3}\cdots\frac{3}{8}\cdots\frac{2}{5}\cdots\frac{3}{7}\cdots\frac{4}{9}\cdots\frac{1}{2}$$

But in order that all the gaps may be *completely* filled it is necessary that infinitely high denominators should be used, and thus a first abnumeral collection is there. In order that the entire system of this abnumeral collection should be there it is necessary that a second abnumeral collection should be there.

In order to show this, we must imagine the individuals of the first abnumeral collection so expressed as to give the imagination something to lay hold upon in order to conceive of their being all there. Let us imagine for example we have lists of all possible sets of rational quantities. The multitude of those lists is the first abnumeral. But in order to be able to say that the collection of all possible such lists is *complete* they must be carried so far that irrational quantities are included; and as soon as that is done, the multitude of lists is the second abnumeral.

We become habituated to think that numbers are capable in themselves of expressing magnitude or at least proportional magnitude. If by magnitude be meant *multitude*, of course this is so; but taking magnitude in the sense of continuous magnitude, numbers in themselves can express neither magnitudes, nor the ratios of magnitudes. Numbers express nothing whatsoever except order, *discrete order*. The fraction $\frac{7}{12}$ expresses nothing whatever except something greater than $\frac{6}{11}$ and less than $\frac{8}{13}$, greater than $\frac{8}{14}$ and less than [$\frac{6}{10}$] etc. In other words, the rules of arithmetic prescribe that the values [of] fractions shall follow a certain sequence, but in regard to the equality of the different parts into which the unit is cut, it can take no further cognizance than to *reckon* them as all units on a par. The logic of number can never be mastered until this idea is fully grasped.

Number cannot possibly express continuity. We can perfectly well mark a point to express π, or 3.14159 and, drawing two horizontal lines [one] to the right and [one to the] left of it, say that the perimeters of inscribed polygons shall be measured off on the left hand line and those of circumscribed polygons on the right hand line.[1] And thus we bring

[1] Peirce illustrates this by the following diagram: ——— . ———

before our eyes what ought to be clear enough to the eye of reason, that there is nothing in the nature of numbers to forbid the interpolation of any multitude of quantities between all the approximations of a convergent series and its limit. There is nothing about numbers which can possibly forbid there being between the points representing all the values of a convergent series and its limit, any abnumeral collection of other points whatsoever.

On the contrary, just as much and just in the same way, as the supposition that the denumerable collection of rational points are *completely* present on a line involves the existence of a primipostnumeral collection of irrational points, so the supposition that the system of irrational points on a line is *complete* involves the existence upon it of a secundipostnumeral collection of other points intermediate between the series and their limits.

No doubt that the ordinary conception of a *limit*, namely that the limit of an increasing convergent series is the *least* quantity which is greater than all the approximations of the series, so far begs the question that it arbitrarily limits the system of possible quantities to a primipostnumeral multitude; and when it is then assumed that there is just one point for every assignable quantity, this completes the *petitio principii.*

I may here remark that the ordinary argument used by writers on the doctrine of limits about "assignable" quantities equally begs the question. For by an *assignable* quantity is meant a quantity to which numerical notation can indefinitely approximate. It is very easy by a careful analysis of the argument to convince oneself that nothing more is meant. Hence, that argument begins by assuming that the system of quantities is a primipostnumeral collection. But the only thing the argument is designed to do is to exclude a larger multitude of quantities. It is, therefore, completely illogical. However, there are forms of presentation of the doctrine of limits which are perfectly unobjectionable; and the doctrine has its value if rightly presented. I only say that the cruder forms of it, such as will be found in Newcomb's treatise, are illogical and out of agreement with modern conceptions of quantity.

While I am on the subject of fallacies, I may as well notice a point which might possibly puzzle you. It is substantially proved by Euclid that there is but one assignable quantity which is the limit of a convergent series. That is, if there is an increasing convergent series, A, and a decreasing convergent series, B, of which every approximation exceeds every approximation of A, and if there is no rational quantity which is at once greater than every approximation of A and less than every approximation of B, then there is but one surd quantity so intermediate. Now it might seem to you as if it followed that there was but one surd quantity intermediate between every pair of rational quantities so that the multitude of surds could not be greater than the multitude of rational quantities. But there is no end to a denumerable series and there are, therefore, no two adjacent rational quantities. There is one surd quantity and only one for each convergent series, calling two series the same if their approximations all agree after a sufficient number of terms, or if their difference approximates toward zero. But this is only to say that the multitude of surds equals the multitude of denumerable sets of rational quantities, which is, as we have seen, the *primipostnumeral* multitude.

Going back to our representation of π [on a line] we remark that there is plenty of room to insert a *secundipostnumeral* multitude of quantities between the convergent series and its limit. Any one of those quantities may likewise be separated from its neighbors, and we thus see that between it and its nearest neighbors there is ample room for a *tertiopostnumeral* multitude of other quantities, and so on through the whole denumerable series of postnumeral quantities.

But if we suppose that *all* such orders of systems of quantities have been inserted, there is no longer any room for so inserting any more. For to do so we must select some quantity to be thus isolated in our representation. Now whatever one we take, there will always be quantities of higher orders filling up the spaces on the two sides.

We therefore see that such a supermultitudinous collection sticks together by logical necessity. Its constituent individuals are no longer distinct and independent subjects. They have no existence, — no hypothetical existence, — except in their relations to one another. They are not subjects, but phrases expressive of the properties of the continuum.

From a line as it is usually conceived in analysis, that is, as a *primipostnumeral* succession of points, its extremity, which is its last point, may in logical possibility be taken away; and when that is done the line is left without an extremity at that part. So whenever a line is severed in the middle, one of the parts will necessarily have an extreme point while the other will necessarily be left without any extremity.

But supposing a line to be a supermultitudinous collection of points, nothing of the sort is logically possible. To sever a line in the middle is to disrupt the logical identity of the point there, and make it two points. It is impossible to sever a continuum by separating the connections of the points, for the points only exist by virtue of those connections. The only way to sever a continuum is to burst it, that is, to convert that which was one into two.

It has hitherto been the opinion of mathematicians, — I speak only of those who are thoroughly acquainted with the most modern achievements [in] this particular branch of mathematical philosophy, — that the points upon a line [are] of that multitude which I call *primipostnumeral*. But I hold this class of thinkers in such extraordinary esteem that I believe that when that opinion is refuted they will hold to it no longer. As for the swarm of pedagogues who infest this land, where pedagogy is so terribly overdone that instruction is generally supposed to be the chief purpose of a university, they have never heard of the opinion itself and never will hear of its refutation.

But that the collection of points upon a line is really supermultitudinous, is, I am confident, made evident by the following considerations. Across a line a collection of blades may come down simultaneously, and so long as the collection of blades is not so great that they merge into one another, owing to their supermultitude, they will cut the line up into as great a collection of pieces each of which will be a line, — just as completely a line as was the whole. This I say is the intuitional idea of a line with which the synthetic geometer really works, — his virtual hypothesis, whether he recognizes it or not; and I appeal to the scholars of this institution where geometry flourishes as all the world knows,[2] to cast aside all analytical theories about lines, and looking at the matter from a synthetical point of view to make the mental experiment and say whether it is not true that the line refuses to be cut up into points by any discrete multitude of knives, however great. If this be the case the *lines* into which any line can be cut exceed any discrete multitude whatever. A line consists wholly of points, in one sense; for it is generated by a moving particle. But in order to chop a line up into its constituent points the blades of the chopper would have to be in incipient mergency into one another. They would have to be supermultitudinous; and so the points are supermultitudinous. Here then are two proofs. One is this:

The possible lines into which any line may be cut at one chop exceed any discrete multitude. Now the points on a line form a collection at least as great as the collection of the possible lines into which it can be chopped. Hence, the points of a line are supermultitudinous.

The other proof is this:

A line consists wholly of points; but in order to chop a line into points, the two ends of each piece must unite; and to do that without shrinkage they must merge into one another. Hence the collection of blades of the

[2] This manuscript is evidently part of a lecture, probably given or planned in the period 1895-1900.

chopper must be so great that its constituent individual blades are no longer distinct. In other words, they must be supermultitudinous; and the points into which this chopper severs the line must form an equal collection.

This I declare to be the synthetic geometer's hypothesis of the relation of a line to its points. But it does not affect my argument if it be not so. It is sufficient for my main purpose that it is a perfectly consistent hypothesis. For all I am trying to do is to elucidate the conception of a supermultitudinous collection and show that it involves no contradiction. In order to clinch my argument I am going presently to restate the matter in exact logical terms.

But before doing so, there is one possible objection which I wish to answer. It may be objected: you say that a supermultitudinous collection is one which is so great that the constituent individuals are no longer distinct. Now, even if we grant all that you have said, you have only shown that these individuals cohere when they are in a certain order of arrangement. In other cases, cohesion is a matter of arrangement and not of multitude; and it is evident on general principles that it must be so; for there is nothing in the nature of the mere extension of the boundaries of what individuals we will include in the collection we consider which can have the slightest effect on the mode of existence of those individuals, while on the other hand their arrangement includes their relations to one another and may easily condition their inseparability. Hence, your supermultitudinous collection appears to derive its essential character from its arrangement, not its multitude, and you have said nothing to obviate this objection. It appears, therefore, not to be a grade of multitude but only a mode of arrangement.

This is an exceedingly plausible objection, and is worthy of a careful reply. My hearers must not understand me as admitting that the objection is substantially sound, when I begin my reply by calling attention to the fact that I have myself carefully abstained from using the word multitude in connection with supermultitudinous collections. Multitude implies an independence in the individuals of one another which is not found in the supermultitudinous. But I deny that we can perceive *a priori* that the mere extension of the boundaries of what we will take for the collection to be considered cannot condition any existential or real relation between the individuals. The word "boundary" here must not be permitted to lead us into confusion. There is no question of a geometrical boundary. It is only a logical boundary, a ὁρισμός, or definition, which is in question. Now I say that if the individuals in the hypothetical universe of our discourse are really cemented together, then in case our logical boundaries are so narrow that the cementing individuals can be omitted, the result will be that those individuals that are included will exhibit no necessary cementation, and consequently we shall say that generally speaking in such a collection they are not cemented, and if they happen to be so, it is but an accidental circumstance extraneous to the essential multitude of the collection; while if our logical boundaries are so wide that the cementing individuals cannot be omitted, then it is perfectly true that a collection as large as that must by logical necessity exhibit the phenomenon of cementation, and hence the cementation is logically involved in the greatness of the collection. This argument does not suppose, either, that the cementing individuals are any peculiar class of individuals. They may be merely any intermediate ones so related that without them the remainder are not in direct connection. Neither can I admit that in the case of other multitudes we have found the cohesion of the individuals to be a matter of arrangement if by arrangement be meant an adventitious system of relationship. On the contrary, one of the most obtrusive phenomena in regard to all inenumerable collections is that the arrangement is essential to the multitude. What for example is the definition of a denumerable collection? It is one which contains first one individual and in the primal arrangement, next after each individual it contains, contains one other, and which contains no individuals except what this description requires it to contain. You see that this primal arrangement is essential. True, the denumerable collection may be arranged otherwise. But what of that? Such derangements of it introduce adventitious systems of relationship; but they never can abrogate that system of relationship between its individuals in which its primal arrangement consists. That system of relationship continues undisturbed through all derangements, as is seen by its effect upon the cohesion. For though in a certain sense the cohesiveness of a denumerable collection is connected with its arrangement, it will be found on examination that it is determined solely by that primal or normal arrangement which is never really destroyed. I will not again discourse at large on the phenomena of cohesion presented by denumerable collections. But I will recall to you that the first and most striking of these phenomena is that each denumerable collection in connection with that part of it which comes last in the primal arrangement, — its endless part, — the potential accompaniment of an exterior primipostnumeral collection. Thus, the succession of integer numbers has never been completed until infinitely large numbers (or rather indications of infinite collections standing for numbers) have been added to it; and of such infinitely large numbers if they are to be there at all, since the series of them has no first, there must be a primipostnumeral multitude. If the series of integer numbers is deranged in any way, it is still in connection with the indefinitely large numbers that the cohesion with the potential primipostnumeral collection takes place. It is so if the odd numbers come first in ascending order followed by the even numbers in descending order. It is so in any of the arrangements which break the whole series into a denumerable succession of denumerable sequences. Finally if the whole numbers are arranged like the series [of] rational fractions taken in the order of their values, so that the indefinitely large numbers are disseminated through every part of the series, then the primipostnumeral appendix is equally scattered through the whole. And so it is in every case. Wherever phenomena of cohesion manifest themselves it is not in immediate connection with any adventitious arrangement but in connection with an intrinsic arrangement which is essentially inseparable from the particular grade of multitude in which those phenomena of cohesion are found.

In a supermultitudinous collection the intrinsic connection is so strong that it forbids, to a great extent, adventitious derangements. I will here explain a point of geometrical terminology. I think it tends to perspicuity to avoid, so far as we can, speaking of loci as moving. Of course, there is no absolute objection to the idea; nor can it be dispensed with altogether. But where it is just as easy to conceive of the loci as fixed and of that which moves as a visible thing occupying a place, I think the locution has an advantage in facility of apprehension. Accordingly, I speak of a *particle* meaning an indivisible visible thing. A *point* is the place of a particle at an instant. A *line* is a place which a particle may occupy in the instants of a lapse of time. By a *filament* I mean a visible thing which at any one instant occupies a line. A *surface* is a place which a filament constantly leaving its place entirely vacant can occupy in the course of a time. By a *film* I mean a visible thing which at any one instant occupies a surface. A *space* is a place which a film constantly leaving its place totally vacant can occupy in the course of a time. By a *body* I mean a visible thing which at any one instant occupies a space. And so on to higher dimensions. Now a filament may have its parts deranged, so far as its continuity permits. A particle of it may burst into two. But still the one fragment preserves its continuity with one part of the filament and the other. It is impossible that all the particles should burst without their being annihilated for the most part at least. For what would result from such bursting would be a discrete multitude of particles. Now the collection of particles in a filament exceeds any given discrete multitude.

There is, I confess, a paradoxical aspect in the proposition that a collection may be so great that its individuals lose their separate identities. But the key of that paradox will probably ultimately be discovered to lie in some unnoticed condition in the general hypothesis of a collection which requires this mergency of individuals.

Meantime, it would be a logical blunder to surrender the proposition merely because it seems paradoxical in the face of the clear argument in its favor which is as follows: There can be no maximum discrete multitude, for the multitude of sets of individuals in any discrete collection is greater than the multitude of individuals. But an aggregate collection formed out of one collection of each possible discrete multitude is a collection, at least as great as any possible discrete collection. Since therefore there is no maximum discrete multitude, this aggregate collection must be too great to be discrete.

I promised that I would carefully restate the relation of a supermultitudinous collection to its individuals. That promise I proceed to fulfill. [...]

D. THE LOGIC OF CONTINUITY (from 948)

Of all conceptions Continuity is by far the most difficult for Philosophy to handle. You naturally cannot do much with a conception until you can define it. Now every man at all competent to express an opinion must admit as it seems to me that no definition of continuity up to quite recent times was nearly right, and I maintain that the only thoroughly satisfactory definition is that which I have been gradually working out, and of which I presented a first *ébauche* when I had the honor of reading a paper here in Cambridge in 1892, and of the final form of which I have given you sufficient hints in these lectures. But even supposing that my definition, which as yet has not received that sanction which can only come from the critical examination of the most powerful and exact intellects, is all wrong, still no man not in leading strings as to this matter can possibly think that there was anything like a satisfactory definition before the labors of Dr. Georg Cantor, which only began to attract the attention of the whole world [in recent years].

But after a satisfactory definition of continuity has been obtained the philosophical difficulties connected with this conception only begin to [be] felt in all their strength. Those difficulties are of two kinds. First there is the logical difficulty, how are we to establish a method of reasoning about continuity in philosophy? And second there is the metaphysical difficulty; what are we to say about the being, and the existence, and the genesis of continuity? As to the proper method of reasoning about continuity, the dictate of good sense would seem to be that philosophy should in this matter follow the lead of geometry, the business of which it is to study continua.

But alas! The history of geometry forces upon us some sad lessons about the minds of men. That which had already been called the Elements of geometry long before the day of Euclid is a collection of convenient propositions concerning the relations between the lengths of lines, the areas of surfaces, the volumes of solids, and the measures of angles.

It concerns itself only incidentally with the intrinsic properties of space, primarily only with the ideal properties of perfectly rigid bodies, of which we avail ourselves to construct a convenient system of measuring space. The measurement of a thing was clearly shown by Klein, twenty-five years ago, to be always extrinsic to the nature of the thing itself. Elementary geometry is nothing but the introduction to *geometrical metric*, or the mathematical part of the physics of rigid bodies. The very early Greek geometers, I mean for example [Hippocrates of Chios], who is said to have written the first *Elements*, I have no doubt, considered metric as the philosophical basis and foundation, not only of geometry, but of mathematics in general. For it is to be remarked that considerably the larger part of Euclid's Elements is occupied with algebra, not with geometry; and since he, and all the Greeks, had a much stronger impulse to get to the logical foundation of any object of study than we have, and since it is only the first book of Euclid in which the logic has been a matter of deep cogitation, it is plain that it was originally, at least, conceived that those geometrical truths in the first book of the Elements lay at the foundation even of algebra itself. But Euclid certainly, and in my opinion much earlier Greeks, had become acquainted with that branch of geometry which studies the conditions under which different rays of light indefinitely prolonged will intersect in common points or lie in common planes. There is no accepted name for this branch. It is sometimes called descriptive geometry; but that is in violent conflict with the principles of nomenclature, since descriptive geometry is the accepted name of a branch of geometry invented by Monge and so named by him, — a branch closely allied to this other doctrine but not the same. Clifford called the branch of which we are speaking, Graphics (which conveys no implication); other writers call it synthetic geometry (though it may be treated analytically), geometry of position (which is the name of something else), modern geometry (when in fact it is ancient), intersectional geometry (though projection plays as great a *role* as section in it), *projective* geometry (though section is as important as projection), perspective geometry, etc. I would propose the name *geometrical optic.* Euclid, I say, and earlier Greeks were acquainted [with] this geometrical optic. Now to any person of discernment in regard to intellectual qualities and who knows what the Greeks were, and especially what Greek geometers were, and most particularly what Euclid was, it seems to me incredible that Euclid should have been acquainted with geometrical optic and not have perceived that it was more fundamental, — more intimately concerned with the intrinsic nature of space, — than metric is. And indeed *a pos-*

teriori evidences that he actually did so are not wanting. Why, then, did Euclid not say a single word about this optic in his elements? Why did he altogether omit it even in cases where he must have seen that its propositions were indispensable conditions of the cogency of his demonstrations? Two possible explanations have occurred to me. It may be that he did not know how to prove the propositions of *optic* otherwise than by means of *metric*; and therefore, seeing that he could not make a thorough job, preferred rather ostentatiously and emphatically (quite in his style in other matters) omitting all mention of optical propositions. Or it may be that, being a university professor, he did not wish to repel students by teaching propositions that had an appearance of being useless. Remember that even the stupendous Descartes abandoned the study of geometry. And why? Because he said it was *useless*. And this he said *a propos* of conic sections! That he should have thought conic sections useless, is comparatively pardonable. But that he the Moses of modern thinkers should have thought that a philosopher ought not to study useless things, is it not a stain of dishonor upon the human mind itself?

In modern times the Greek science of geometrical optic was utterly forgotten, all the books written about it were lost, and mathematicians became entirely ignorant that there was any such branch of geometry. There was a certain contemporary of Descartes, one Desargues, who rediscovered that optic and carried his researches into it very far indeed. He showed clearly and in detail the great utility of the doctrine in perspective drawing and in architecture, and the great economy that it would effect in the cutting of stones for building. On the theoretical side he pushed discovery to an advance of a good deal more than two centuries. He was a secular man. But he worked alone, with hardly the slightest recognition. Insignificant men treated him with vitriolic scorn. His works, though printed, were utterly lost and forgotten. The most voluminous historians of mathematics though compatriots did not know that such a man had ever lived, until one day Michel Chasles walking along the *Quai des Grands Augustins* probably after a meeting of the *Institut*, came across and bought for a franc a MS copy of one of those printed books. He took it home and studied it. He learned from it the important theory of the Involution of Six Points [Fig. 1];[1] and from him the mathematical

[1] The figure given here was found in fragments and interpolated here. The explanation is that "Six collinear points are said to be *in involution*, provided that four points can be found such that every pair of them is in one straight line with *one* of the six but not with *all* the six. Thus *AA'BB'CC'* are in involution because of the four points *PQRS*."

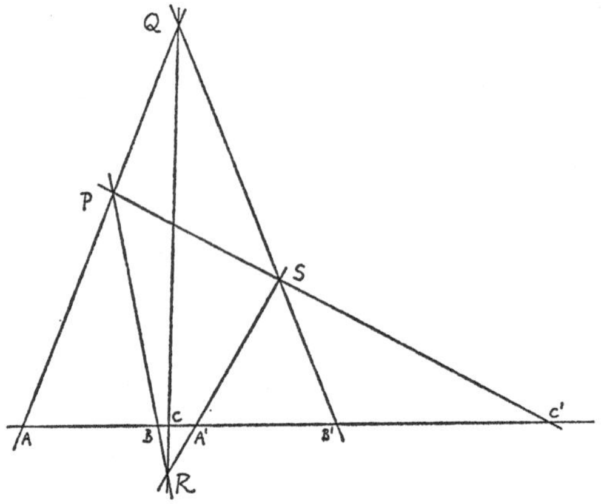

Fig. 1

world learned it; and it has been a great factor in the development of modern geometry. There can be no possible doubt that this knowledge actually came from the book of Desargues, because the relation has always borne the strange name Involution which Desargues had bestowed upon it, — and the whole theory is in his book although it had been totally unmentioned in any known treatise, memoir, or programme previous to the lucky find of Chasles. When will mankind learn the lessons such facts teach? That had that doctrine not been lost to all those generations of geometers, philosophy would have been further advanced today, and that the nations would have attained a higher intellectual level, is undoubtedly true, — but that may be passed by as a bagatelle. But why will men not reflect that but for the stupidity with which Desargues was met, — many a man might have eaten a better dinner and have had a better bottle or wine with it? It needs not much computation of causes and effects to see that that must be *so*.

In 1859, Arthur Cayley showed that the whole of geometrical metric is but a special problem in geometrical optic. Namely, Cayley showed that there is a locus in space, — not a *kind* of locus, but an individual place, — whose optical properties and relations to rigid bodies constitute those facts that are expressed by space-measurement.

It attracted the admiration and assent of the whole mathematical world, which has never since ceased to comment upon it, and develop the doctrine. Yet a few years ago I was talking with a man who had written two elementary geometries and who perhaps was, for aught I know still is, more influential than any other individual in determining how Geometry shall be taught in American schools at large, and this gentleman never heard of Projective Geometry neither the name nor the thing and his politeness never shone more than for his not treating what I said about Cayley with silent contempt.

But many years before Cayley made that discovery; a geometer in Göttingen, Listing by name, — a name which I will venture to say that Cayley, learned as he was in all departments of mathematics heard for the first time many years later, probably from Tait, who knew of him because he and Listing were both physicists, — this Listing had in 1847 four years before Riemann's first paper discovered the existence of quite another branch of geometry, and had written two very long and rich memoirs about it. But the mathematical world paid no heed to them till half a century had passed. This branch, which he called Topology, but which I shall call Topic, to rhyme with *metric* and *optic*, bears substantially the same relation to optic that optic bears to metric. Namely, *topic* shows that the entire collection of all possible *rays*, or unlimited straight lines, in space, has no general geometrical characters whatever that distinguish it at all from countless other families of lines. Its only distinction lies in its physical relations. Light moves along rays; so do particles unacted on by any forces; and maximum–minimum measurements are along rays. But the whole doctrine of geometrical optic is merely a special case of a topical doctrine.

That which *topic* treats of is the modes of connection of the parts of continua. *Geometrical topic* is what the philosopher must study who seeks to learn anything about continuity from geometry. I will give you a slight sketch of the doctrine. We have seen in a previous lecture what continuity consists in. There is an endless series of abnumeral multitudes, each related to the next following as M is related to 2^M, where we might put any other quantity in place of 2. The least of these abnumeral multitudes is 2^N where N is the multitude of all whole numbers. It is impossible that there should be a collection of distinct individuals of greater multitude than all these abnumeral multitudes. Yet every one of these multitudes is possible and the existence of a collection of any one of these multitudes will not in the least militate against the existence of a collection of any other of these multitudes. Why then, may we not suppose a collection of distinct individuals which is an aggregate of one collection of each [of] those multitudes? The answer is, that to suppose an aggregate of *all* is to suppose the process of aggregation *completed*, and that is supposing the series of abnumeral multitudes brought to *an end*, while it can be proved that there is no last nor limit to the series.[2]

Let me remind you that by the *limit* of an endless series of successive objects we mean an object which comes after all the objects of that series, but so that every *other* object which comes after all those objects comes after the limit also. When I say that the series of abnumeral multitudes has no limit, I mean that it has no limit among multitudes of distinct individuals. It will have a limit if there is properly speaking, any meaning in saying that something that is *not* a multitude of distinct individuals is *more* than every multitude of distinct individuals. But, you will ask, can there be any sense in that? I answer, yes, there can, in this way. That which is possible is in so far *general* and, as general, it ceases to be individual. Hence, remembering that the word "potential" means *indeterminate yet capable of determination in any special case*, there may be a *potential* aggregate of all the possibilities that are consistent with certain general conditions; and this may be such that given any collection of distinct individuals whatsoever, out of that potential aggregate there may be actualized a more multitudinous collection than the given collection. Thus the potential aggregate is, with the strictest exactitude, greater in multitude than any possible multitude of individuals. But being a potential aggregate only, it does not contain any individuals at all. It only contains general conditions which *permit* the determination of individuals.

The logic of this may be illustrated by considering an analogous case. You know very well that $\frac{2}{3}$ is not a whole number. In the whole collection of whole numbers you will not find $\frac{2}{3}$. That you know. Therefore, you know something about the entire collection of whole numbers. But what is the nature of your conception of this collection? It is general. It is potential. It is vague, but yet with such a vagueness as permits of its accurate determination in regard to any particular object proposed for examination. Very well, that being granted, I proceed to the analogy with what we have been saying. Every whole number considered as a multitude is capable of being completely counted. Nor does its being aggregated with or added to any other whole number in the least degree interfere with the completion of the count. Yet the aggregate of *all* whole numbers cannot be completely counted. For the completion would suppose the *last* whole number was included, whereas there is no last whole number. But though the aggregate of all whole numbers cannot

[2] The following four paragraphs appear also in the *Collected Papers* (6.185-6.188). The other materials from MS 948 appearing in the present collection were omitted from the earlier presentation in the *C.P.*

be completely counted, that does not prevent our having a distinct idea of the multitude of all whole numbers. We have a conception of the entire collection of whole numbers. It is a *potential* collection, indeterminate yet determinable. And we see that the entire collection of whole numbers is more multitudinous than any whole number.

In like manner the potential aggregate of all the abnumeral multitudes is more multitudinous than any multitude. This potential aggregate cannot be a multitude of distinct individuals any more than the aggregate of all the whole numbers can be completely counted. But it is a distinct general conception for all that — a conception of a potentiality.

A potential collection, more multitudinous than any collection of distinct individuals can be, cannot be entirely vague. For the potentiality supposes that the individuals are determinable in every multitude. That is, they are determinable as distinct. But there cannot be a distinctive quality for each individual; for these qualities would form a collection too multitudinous for them to remain distinct. It must therefore be by means of relations that the individuals are distinguishable from one another.

Suppose, in the first place, that there is but one such distinguishing relation, *r*. Then since one individual is to be distinguished from another simply by this that one is *r* of the other, it is plain that nothing is *r* to itself. Let us first try making this *r* a simple dyadic relation. If, then, of three individuals *A*, *B*, *C*, *A* is *r* to *B* and *B* is *r* to *C*, it must be that *A* is *r* to *C* or else that *C* is *r* to *A*. We do not see, at first, that there it matters which. Only there must be a general rule about it, because the whole idea of the system is the potential determination of individuals by means of entirely general characters. Suppose, first, that if *A* is *r* to *B* and *B* is *r* to *C* then in every case *C* is *r* to *A*, and consequently *A* is not *r* to *C*. Take then any fourth individual, *D*. Either

A is *r* to *D* or *D* is *r* to *A*

<table>
<tr>
<td>If A is r to D, since D is r to A
D is r to C
Then either B is r to D or D is r to B
<table><tr><td>C is r to B
absurd</td><td>Since B is r to C
C is r to D
absurd</td></tr></table>
</td>
<td>Either C is r to D or D is r to C
<table><tr><td>A is r to C
absurd</td><td>Either B is r to D or D is r to B
<table><tr><td>C is r to B
absurd</td><td>Since B is r to C
C is r to D
absurd</td></tr></table>
</td></tr></table>
</td>
</tr>
</table>

That rule, then, when you come to look into it will not work. The other rule that if A is r to B and B is r to C then A is r to C leads to no contradiction, but it does lead to this, that there are two possible exceptional individuals one that is r to everything else and another to which everything else is r. This is like a limited line, where every point is r that is, is to the *right* of every other or else that other is to the right of it. The generality of the case is destroyed by those two points of discontinuity, — the extremities. Thus, we see that no perfect continuum can be defined by a dyadic relation. But if we take instead a triadic relation, and say A is r to B for C, say to fix our ideas that proceeding from A in a particular way, say to the right, you reach B before C, it is quite evident, that a continuum will result like a self-returning line with no discontinuity whatever [Fig. 2]. All lines are simple rings and are

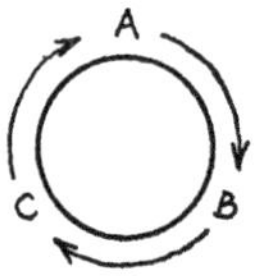

Fig. 2

topically precisely alike except that a line [may] have *topical singularities.* A *topical singularity* of a place is a place within that place from which the modes of departure are fewer or more than from the main collection of such places within the place. The topical singularities of *lines* are singular points. From an *ordinary* point on a line a particle can move two ways. Singular points are points from which a particle can move either no way, or in *one* way, or else in *three* ways or more. That is they are either, first, *isolated points* from which a particle cannot move in the line at all, or secondly, *extremities*, from which a particle can move but one way, or thirdly, *furcations*, from which a particle can move in three or more ways [Fig. 3]. Those are the only topical distinctions there are

Fig. 3

among lines. Surfaces, or two dimensional continua, can also have singularities. These are either *singular points* or *singular lines.* The singular lines are either isolated lines, which may have singular points at which they are not isolated, or they are *bounding edges*, or they are lines at which the surface splits into different sheets. These singular lines may themselves have singular points, which are subject [to] interesting laws. A student would find the singular lines of surfaces a good subject for a thesis. Isolated singular points of surfaces are either entirely detached from the surface or they are points at which different sheets or parts of the same sheet are tacked together. But aside from their singularities surfaces are of different kinds. In the first place, they are either *perissid* or *artiad.* A perissid surface is one which, although unbounded, does not enclose any space, that is, does not necessarily cut space into two regions, or what comes to the same thing, it has only one side. Such is the plane surface of geometrical optics, and in fact, such is every surface of odd order. The perissid surfaces are mathematically the simpler; but the artiad surfaces are the more familiar. A half twisted ribbon pasted together so that one side becomes continuous with the other side is an example of a bounded perissid surface. If you pass along a plane in geometrical optic, you finally come back to the same point, only you are on the other side of the plane [Fig. 4]. An artiad surface, on the other

Fig. 4

hand, is for example the bounding surface between air and the stone of any finite stone, however curiously it may be cut. Moreover, a surface may have a *fornix* or any number of fornices. A fornix is a part of the surface like a railway-tunnel which at once bridges over the interval between two parts of the surface, and so connects them, and at the same time, tunnels under that bridge so that a particle may move on the surface from one side of the bridge to the other without touching the bridge. A flat-iron handle, or any handle with two attachments has a surface which is a *fornix* of the whole surface of which it forms a part. Both perissid and artiad surfaces can equally have any number of fornices, without disturbing their artiad or perissid character.

If I were to attempt to tell you much about the different shapes which unbounded three dimensional spaces could take, I fear I might seem to talk gibberish to you, so different is your state of mental training and mine. Yet I must endeavor to make some things plain, or at least not leave them quite dark. Suppose that you were acquainted with no surface except the surface of the earth, and I were to endeavor to make the shape of the surface of a double ring clear to you. I should say, you can

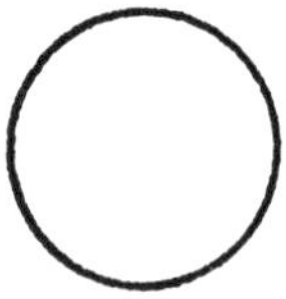

Fig. 5

Fig. 6

imagine in the first place a disk with an outer boundary [Fig. 5]. Then you can imagine that this has a hole or holes cut through it [Fig. 6]. Then you can imagine a second disk just like this and imagine the two to be pasted together at all their edges, so that there are no longer any edges. Thus I should give you some glimmer of an idea of a double ring. Now I am going in a similar way to describe an unbounded three-dimensional space, having a different shape from the space we know. Begin if you please by imagining a closed cave bounded on all sides. In order not to complicate the subject with optical ideas which are not necessary, I will suppose that this cave is pitch dark. I will also suppose that you can swim about in the air regardless of gravity. I will suppose that you have learned this cave thoroughly; that you know it is pretty cool, but warmer in some places, you know just where, than others, and that the different parts have different odors by which they are known. I will suppose that these odors are those of neroli, portugal, limette, lemon, bergamot, and lemongrass, — all of them generically alike. I will further suppose that you feel floating in this cave two great balloons entirely separated from the walls and from each other, yet perfectly stationary. With the feeling of each of them and with its precise locality I suppose you to be familiarly acquainted. I will further suppose that you formerly inhabited a cave exactly like this one, except it was rather warm, that the distribution of temperature was entirely different, and that the odors in different localities in it with which you are equally familiar, were those of frankincense, benzoin, camphor, sandal-wood, cinnamon, and coffee, thus contrasting strongly with those of the other cave. I will further suppose the texture-feeling of the walls and of the two balloons to be widely different in the two caves. Now, let us suppose that you, being as familiar with both caves as with your pocket, learn that works are in progress to open them into one another. At length, you are informed that the wall of one of the balloons has been reduced to a mere film which you can feel with your hand but through which you can pass. You being all this time in the cool cave swim up to that balloon and try it. You pass through it readily; only, in doing so, you feel a strange twist, such as you never have felt, and you find by feeling with your hand that you are just passing out through one of the corresponding balloons of the warm cave. You recognize the warmth of that cave, its perfume, and the texture of the walls. After you have passed backward and forward often enough to become familiar with the fact that the passage may be made through every part of the surface of the balloon, you are told that the other balloon is now in the same state. You try it and find it to be so, passing round and round in every way. Finally, you are told that the outer walls have been removed. You swim to where they were. You feel the queer twist and you find yourself in the other cave. You ascertain by trial that it is so with every part of the walls, the floor, and the roof — they do not exist any longer — there is no outer boundary at all.

Now all this is quite contrary to the geometry of our actual space. Yet it is not altogether inconceivable even sensuously. A man would accustom himself to it. On the mathematical side, the conception presents no particular difficulty. In fact mathematically our own shaped space is by no means the easiest to comprehend. That will give you an idea of what is meant by a space shaped differently from our space. The shape may be further complicated by supposing the two balloons to have the shape of anchor-rings and to be interlinked with one another.

After what I have said, you cannot have much difficulty in imagining that in passing through one of the balloons you have a choice of twisting yourself in either of two opposite ways, one way carrying you into the second cave and the other way into a third cave. That balloon surface is then a *singular surface.*

I will not attempt to carry you further into geometrical topic. You can readily understand that nothing but a rigidly exact logic of relations can be your guide in such a field. I will only mention that the real complications of the subject only begin to appear when continua of higher dimensionality than 3 are considered. For then first we begin to have systems of relations between the different dimensions.

A continuum may have any discrete multitude of dimensions whatsoever. If the multitude of dimensions surpasses all discrete multitudes there cease to be any distinct dimensions. I have not as yet obtained a logically distinct conception of such a continuum. Provisionally, I identify it with the *uralt* vague generality of the most abstract potentiality.

Listing, that somewhat obscure Göttingen professor of physics, whose name must forever be illustrious as that of the father of geometrical topic, — the only intrinsic science of space, — invented a highly artificial method of thinking about continua, which the great Riemann independently fell into in considering the connectivity of surfaces. But Riemann never studied it sufficiently to master it thoroughly. This method is not all that could be desired for continua of more than 3 dimensions, which Listing never studied. Even for our space, the method fails to throw much light on the theory of knots; but it is highly useful in all cases, and is almost all that could be desired for tridimensional space. This method consists in the employment of a series of numbers which I proceed to define. By a *figure*, modern geometers do not mean what *Euclid* meant at all. We simply mean any place or places considered together. An indivisible place is a *point*. A movable thing which at any one instant occupies a point is a *particle*. The place which a particle can occupy during a lapse of time, one point at one instant and another at another instant, is a *line*. A movable thing which at any one instant occupies a line is a *filament*. The place which a filament can occupy in a lapse of time is a *surface*. A movable thing which at any one instant occupies a surface is a *film*. The place which a film can occupy in a lapse of time is a *space*, or as I would call it a *tripon*. A movable body which at any one instant occupies a *tripon* is a *solid*, or as I would call it a *trion*. Thus for higher dimensional places we have the *tetrapon*, the *pentapon*, etc. And for the movable things in them the *tetron*, *penton*, etc.

Now then for Listing's numbers.

I call the first of them the *Chorisis*. He calls it simply the number of separate pieces. I give it a name to rhyme with his Cyclons and Periphraxis; and with the analogous name I give it an analogous definition. Namely, the *chorisis* of a *D*-dimensional figure, is the number of simplest possible *D*-dimensional places that must be removed from it, in order to leave room for no particle at all. That is but a roundabout way of expressing the number of separate pieces. At first, I wished to define it as the number of simple *D*-dimensional places that must be removed in order to leave no room for a pair of particles which cannot move within the figure so as to coalesce. This would be one less than the number of pieces. But I subsequently surrendered to Listing's own view.

The second Listing number he calls the *Cyclosis*. The cyclosis of a *D*-dimensional figure is the number of simplest possible places of *D*-1 dimensions which have to be cut away from the figure, to preclude the existence in the remaining place of a filament without topical singularities which cannot gradually move within that place so as to collapse to a particle. For example, this line has a cyclosis equal to 1 [Fig. 7]. For you must cut it through to preclude a ring shaped filament which is in-

Fig. 7

capable of collapsing by any gradual motion within the figure. On the other hand the surface of the blackboard has its cyclosis equal to *zero*, because any ringshaped filament in it has room gradually to shrink to a particle. It is true that when the ringshaped filament shrinks to a particle there is a breach of continuity at the last instant; but when we define cyclosis we *except* that final breach of continuity. A similar remark applies to all the other numbers. An annular surface bounded by two rings has a cyclosis equal to 1. For it has to be cut through on one line to prevent the existence of a ring-shaped filament which cannot gradually shrink within the annular surface to a particle. The surface of an anchor ring has a cyclosis equal to 2. For to preclude the existence of the noncollapsable filament it is necessary first to cut round the bar of the ring and after that to slit the bar along all its length. The space which the solid iron of the anchor ring fills has a cyclosis equal to 1. For simply sawing it through in any plane is sufficient to preclude a noncollapsable filament. A spiral line having one end and winding in toward the centre at such a rate as to be infinitely long, has a cyclosis *zero*. For although it be infinitely long in measure, measure does not concern topical geometry. The line has an end at the centre, and its infinite windings will not prevent any filament in it from shrinking in it to a particle. The plane of imaginary quantity which the theory of functions studies has a cyclosis equal to zero. For take a straight line extending through the zero point and the point of infinity. Though the modulus of its point of greatest modulus is infinity, yet measure does not concern topical geometry and it may continuously contract to a circle and finally to the origin [Fig. 8]. But the plane of perspective geometry is of an entirely different shape. Its cyclosis is 1. For consider a ray or unlimited straight line. That ray

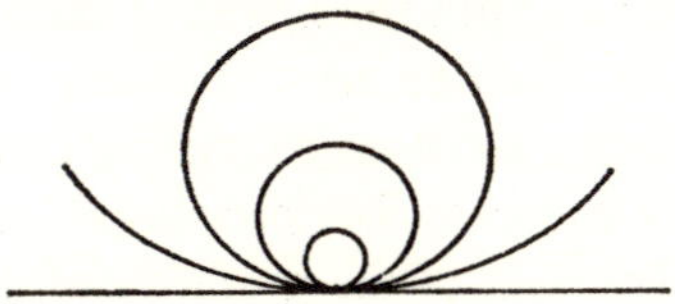

Fig. 8

Fig. 9

Fig. 10

cuts the ray at infinity. In the plane of imaginary quantity, there *is* no line at infinity, but only a point. Here, there is a ray at infinity. Here it is [Fig. 9]. Here is the movable filament cutting it [Fig. 10]. That filament cuts that ray *once* and once only. But the ray at infinity returns into itself as anybody would see who stood upon a boundless desert plane and viewed his horizon round and round. That horizon would be that ray. Now, distort and move it as you will, the filament will always cut that fixed ray in an odd number of points. Thus it can never cease to cut it. For if it did it would cut it in *zero* points and zero is an even number. Thus we see that any ray, or any curve of odd order, is the place of an unshrinkable filament. But every such line cuts every other. Hence if the surface were separated along a single line of that kind, no unshrinkable filament could any longer exist in it. Those of you who are acquainted with non-Euclidean geometry might ask me how it would be in hyperbolic space. I reply it is the same thing precisely. Perspective geometry no more concerns itself with measure than does Topical, or Intrinsic, geometry. It is true that according to the mode of measurement used in hyperbolic geometry, all the parts of the plane at finite distance are enclosed within one circle. The parts of the plane outside the circle have then *no existence*, that is to say we are utterly cut off from volitional reactions with them. But they are none the less real on that account. You must recognize them, or fill your mind with senseless exceptions to all the laws of optics. Two rays continue to intersect in one point, although that point may be outside our universe. Only it now becomes a matter of indifference to us what the shape of the plane may be, since whether it have a cyclosis zero or a cyclosis 1, the part of the plane in our universe is a simple disk.

Listing's third number he calls the Periphraxis. The periphraxis of a figure of *D* dimensions is the number of simple places of *D*-2 dimensions which must be taken away to prevent a non-singular film from gradually collapsing to a filament within the figure. Thus a cave with 2 balloons in it has a periphraxis equal to zero. For it is necessary to build out two barriers to preclude a sac from containing one balloon or the other, so that the sac cannot collapse. The periphraxis of the space of perspective geometry is 1. But that of the geometry of quaternions is zero.

Listing's fourth number [he] calls the *Immensity*. It might be called the "fourth Listing." It is the number of simple places of D-3 dimensions which must be taken away to preclude a non-singular noncollapsable solid, or trion. As Listing remarks, for all figures in our space this number is equal to *zero* excepting only for the entirety of space itself, for which it is 1.

For anybody who wishes to study this subject I will say that one of Listing's two papers is in the *Göttinger Studien*, page 771, and my impression is that it is in the excellent library of Columbia University. The other memoir which relates to Listing's Census Theorem, which is mainly if not wholly an artificial theorem, — true indeed but yet a mere formality, or affair of book-keeping (although this memoir is most important and it is in it that the author develops his numbers) is in the *Göttinger Abhandlungen*. I think Vol. VII. Both memoirs are full of interest and excessively easy to read. Listing himself makes the Cyclosis and Periphraxis of space equal to zero, showing how little he knew of mathematics.

I have occupied far too large a part of my hour with this matter and must now leave entirely untouched two methods of my own invention for treating problems upon which Listing's numbers fail to throw much light. [...][3]

[3] The remainder of this lecture may be found in *The Collected Papers of Charles Sanders Peirce*, beginning in the middle of 6.188.

E. ACHILLES, OR ACHILLES AND THE TORTOISE
(interleaf from *THE CENTURY DICTIONARY*)

"The Achilles," or "the paradox of Achilles and the tortoise" is an argument proposed by Zeno of Velia (or Elea) to show that motion is impossible. The argument (restated) is this. Suppose Achilles, a rapidly moving body, or rather a mathematical point marked on such a body, runs after the Tortoise, a mathematical point marked on a slowly moving body, which at the start is in advance of Achilles. We may suppose they move on parallel lines toward the North and when Achilles is as far North as the tortoise, then if he ever is so far and then only he shall be considered to have overtaken the tortoise. The tortoise being in advance of Achilles, Achilles cannot overtake the tortoise without going as far North as the tortoise now is. But when he gets so far, the tortoise will have gone further North and the same condition will have been reproduced. Thus, before Achilles overtakes the tortoise an endless series of times *all finite* must have elapsed and therefore he will never overtake the tortoise. Probably in our time nobody, sane or insane, doubts the reality of space and time; but numberless persons are influenced by this argument to suppose that continuous space and time, though actually existent, are nevertheless essentially violations of logic. If this were so, the simple remedy would be to change the rules of logic. It is not continuity only, but also the succession of rational fractions which this argument, were it sound, would refute. It is an instructive problem to a person capable of reasoning with accuracy (of which few of those who are much struck with it are capable) as leading one to understand what can be endless and what cannot. There is either an endless succession of possible whole numbers or else there is some finite number to which unity could not logically be added. This shows, at once, that some endless series is possible. If the series of places are every one possible all are possible; and points and instants are no more existent than are numbers. The supposed difficulty about an endless series being ended is too puerile for notice. That an infinite series of finite quantities need not be infinite is not serious. In any system of exact logic the difficulty disappears, but those who are deceived by the sophism are incapable of exact thought and cling to their difficulties, insisting on [adhering to a] logic which leads from true premisses to false conclusions.

ACHILLES AND THE TORTOISE (814)

If a fallacious argument is not stated pretty fully, the logician may be at his wit's end to know what sort of a fallacy to call it, because he cannot quite see from what point of view anybody could regard it as an argument at all. The "Achilles and the Tortoise" is an illustration in point. It begins by insisting that in order to overtake the tortoise Achilles must first go as far as where the tortoise starts, next to where the tortoise then is, next again to where the tortoise then is, "and so on *ad infinitum.*" Granted, of course; but what of it? Why, then it is said to be impossible for Achilles to overtake the tortoise, at all. Good gracious, what a jump! No semblance of a proof is given for it, and how people think there is ryhme or reason in it is difficult to imagine.

Is it that the performance of an endless series of actions is deemed too fatiguing?

Is it that the "so on *ad infinitum*" means that it is supposed Achilles has to do something analogous to an infinite number of sums in addition? Naturally, the distinction between addition proper and the evaluation of the limit of an infinite series whose general term is given would be overlooked.

Can it be, one might reasonably ask, that Zeno had the strange idea that the sum of an endless series of finite quantities is necessarily infinite? If we had asked him why he thought so, would he have said with Full-Private James,

> "A glimmering thought occurs to me,
> Its source I can't unearth,"

Or would he have responded that the sum of a series of unequal positive numbers exceeds the product of the number of terms into the value of the least term? If so, the obvious answer would be that in a series of quantities each smaller than the preceding, the least is the last; but an *endless* series has *ex vi termini* no last, and consequently no least.

Zeno lived 500 years before Christ, and is called by Aristotle the earliest logician. If he really conducted his attack on motion so very feebly as he is represented to have done, he is to be forgiven. But that the world should continue to this day to admire this wretched little catch, which does not even turn upon any peculiarity of continuity, but is only a faint rudimentary likeness to an argument directed against an endless series, is less pardonable.

I was once in a yacht skirting along the shores of the Principato Citeriore and admiring the two mountains, Monte Sacro and Monte Cervato are their names, if my memory does not fail me, which remind one of two sweet-corns upon the instep of Italy, when I fancied I caught sight of some ruins. We managed to make a landing and climb the cliff, and there I found myself, sure enough, among the few remains of that most metaphysical of towns, Velia, which gave its name to the school of Parmenides and Zeno. Desiring to drink to their august memories, I sent a peasant to the neighboring village of Vallo, the same name modernized, for a bottle of wine. But it was so long coming that I fell asleep; and in my sleep I had the honor of a visit not from the noble Parmenides himself, but from the great logician, Zeno. Of course, the reader will understand me to mean that I dreamed this; and not wishing to be thought out of my wits, I will let him think so. The sage expounded to me those four arguments; he showed me what they really had been, and why just four were needed. Very, very different from the stuff which figures for them in Simplicius. He recognized now that they were wrong, though not shallowly wrong; and he was not a little proud of having rejected the testimony of sense in his loyalty to reason. He called them the "Four Seals upon the Secret of the Universe." I am sworn not to tell what they really are; but I will permit myself to put just two questions to the reader.

Suppose that Achilles and the Tortoise ran a race; and suppose the tortoise was allowed one stadium of start, and crawled just one stadium per hour. Suppose that he and the hero were mathematical points moving along a straight line. Suppose that the son of Peleus, making fun of the affair, had determined to regulate his speed by his distance from the tortoise, moving always faster than that self-contained Eleatic by a number of stadia per hour equal to the cube root of the square of the distance between them in stadia. Then, I ask (mind, *not* what the answer to the mathematical problem is) but whether the two following arguments contain any fallacy, or fallacies, and if so precisely in what it or they consist:

1. Achilles, so long as he is behind the tortoise, always moves (1) to-

ward, and (2) faster than the tortoise. *Ergo*, he must overtake him.

2. Achilles can only move beyond the tortoise by a motion relative to the tortoise. But his rule implies that when he coincides with the tortoise in place, he moves at precisely the same speed as the tortoise. Consequently, after he has caught up with the tortoise he has no motion relative to the tortoise. *Ergo*, he never can get ahead of him; and the race must finish, if not before Achilles reaches his competitor, then with the two neck and neck.

A person really skilled in modern logic should have no difficulty in answering those questions in the regular limit of a telegram for each. If anybody begins his reply "Oh! that is not the way to proceed: you should reason, so and so," set him down at once, as a person unable so much as to see the point. I should be very happy to receive answers to the above questions, and to comment upon any that may be remarkable.

The doctrine of limits is a pretty thing. In its best form, it rests upon the hypothesis of a system of numbers from which infinitesimals are excluded. All mathematical reasoning consists simply in tracing out the consequences of hypotheses. It is a familiar result to find the hypothesis self-contradictory. But attempts have been made to apply the method of limits to show the absurdity of the idea of infinitesimals. Is it not a strange logical procedure to deduce as a consequence of one hypothesis that a conflicting hypothesis is untrue? Of course, it is untrue if the first one remains true; but how is a mere hypothesis to decide that it cannot itself be set aside?

F. THE QUESTION OF INFINITESIMALS (s-14)

I

Let points be placed upon a line so that between any two points there are points either way round. Then those points are related as the system of rational numbers.

Calling one way of going round the line forward, the other backward, let points be so placed that if a series of points be taken so that, without passing a certain fixed point, each has others of the series forward of it, and if another series of points be taken so that without passing that fixed point each has other points of the series back of it, and if the latter series be altogether forward of the former, without passing the same fixed point, then between those two series there always will be a point. In that case, the points are related to one another as the system of real numbers.

Since the last property involves no contradiction,[1] it will involve none

[1] Peirce writes in MS. S-15:

"Accordingly Cantor, finding no contradiction involved, assumes that just as the denumeral multitude follows after the enumerable multitudes as the multitude of an aggregate of collections of enumerable multitudes of which there is no greatest, so following after all the alephs, or abnumerable multitudes, there is a multitude which is that of an aggregate of abnumerable collections among which there is no greatest; and that there is thereupon another endless series of multitudes and so on, unceasingly. But the present writer is in possession of what seems to be a clear demonstration that a contradiction is involved of a kind against which Cantor was not upon the watch. Namely, the contradiction lies in supposing the members of such an aggregate to be independent of one another, so that any one is present whether the others are present or not, and any one might be removed without affecting the others. Such independence is essential to the theory of multitude which reposes upon the idea that each member of one collection may have a relation to a single one of another collection. This is precisely what would not be possible for the aggregate considered. It is, however, of secondary importance at what particular place in the series of multitudes this phenomenon occurs. It must occur somewhere; otherwise our ordinary conception of a terminated line would not be possible. Cantor, indeed, conceives a line as an ordered collection of independent points. He talks, for example, of removing the terminal point, and so leaving the line without a terminal point. But according to our ordinary conception the single points of a line have no if in place of *points* we substitute oranges, cats, or anything else. Let us substitute in place of points collections of points similar to the whole collection. The series are now series of ranges of innumerable points, and the property consists in there always being a range of innumerable points between them[2]. In this case, the whole system of real numbers is inadequate to distinguishing between the points.

Professor W. F. Osgood (*Bulletin Am. Math. Soc.*, 2nd Series., Vol. I., p. 147) says this is demonstrably impossible. It seems to me manifest that it involves no contradiction. But it is easier to make sure that a hypothesis is contradictory than that it is not so. Therefore, he may be right. Let him produce his demonstration. I believe it will prove fallacious; but perhaps not.[3]

> separate identity. If a single point flies off from the end of a line, the line continues to have a terminal point ..."

A recent addition to MS. 961 reveals further that Peirce believed that

> "the doctrine of limits has been invented to evade the difficulty, or according to some as an exposition of the signification of the word infinitesimal; that this doctrine, in one form or another, is taught in all the text-books; it is satisfactory enough for the purposes of the calculus.
>
> I was myself of the opinion that the conception of an infinitesimal involved contradiction, until I had applied to the subject a notation for the logic of relations which seemed to me against all danger of fallacy, when I found that opinion was erroneous. Subsequently, I became acquainted with the memoirs of Dr. Georg Cantor; and I wish to acknowledge that while I do not altogether accept all of that writer's views, yet my ideas about infinity and ..."

2 This conception makes Peirce a precursor in the currently developing *nonstandard analysis* in which the infinitesimal is being resurrected with honor. See A. Robinson.

3 On page 6 of an unidentified letter which was removed from MS. 482 Peirce writes: "I have devoted sufficient time to the examination of the question to be quite sure, but my impression is that excepting with a few very profound thinkers the doctrine of literal infinitesimals never has been in much favor. Whether or not it ought to be so I will not decide until I have examined Professor Osgood's supposed demonstration of the impossibility of infinitesimals."

Peirce apparently felt challenged by Osgood's stand, in a review of Forsyth's *Theory of Functions of a Complex Variable* in the *Bulletin of the American Mathematical Society*, March 1895, where Osgood said at one point:

> "Here is the doctrine of infinitesimals so popular in earlier times. That doctrine regards an infinitesimal not as a variable, but as a constant, an 'infinitely small constant,' having geometric existence (e.g., an 'element' of arc, area, etc.); and hence two fixed points on a line that do not coincide need not be situated a definite distance apart from each other. But this does not correspond to our intuition of a line, and geometric intuition was formerly the foundation on which arithmetic rested. In recent times, arithmetic has been developed independently of geometry, and it now stands on the firmest foundation known to mathematicians. In its turn, it has thrown light on the nature of our geometric axioms; and not the least important result that it has achieved is the demonstration of the impossibility of such a doctrine of infinitesimals as the one the author holds."

II

Prof. Osgood also makes a historical remark the truth of which I doubt. Namely, he says the literal doctrine of infinitesimals was "in earlier times," "popular." At what time and in what country?

III

It is singular that nobody objects to $\sqrt{-1}$ as involving any contradiction, nor, since Cantor, are infinitely great quantities much objected to, but still the antique prejudice against infinitesimally small quantities remains.

IV

As an example of the convenience of thinking about infinitesimals, — supposing the idea does *not* involve any contradiction, — I may mention the conception of a translation as an infinitesimal rotation. In the non-Euclidean geometry, if a rigid body filling all space (which I term a *metron*) rotates about an axis, the particles on the ray, which is the polar conjugate of that axis with respect to the absolute, move along that ray. When the Absolute becomes a double plane, the polar conjugate of any ray not lying in that plane is a ray of the plane, and thus our generalization forces us to regard a translation as a rotation about an axis in the plane at infinity. It is an infinitesimal rotation, yet the angular quantity of it is proportional to the linear displacement. There is an infinitesimal perfectly definite and measurable. The rotation through 180° round the axis at infinity is also perfectly definite. It brings the particles on the line back to their positions, but turned upside down, and so the turn is distinguishable from a complete revolution.

V

In case the idea of an infinitesimal does not involve contradiction, there is some evidence to show that the supposed system of points really has its counterpart in nature.

Among the offprints in the Peirce papers is a copy of Osgood's "The Law of the Mean and the Limits $\frac{0}{0}$, $\frac{\infty}{\infty}$," from the *Annals of Mathematics* (University of Virginia) 12 (June 1898), 65-78. It carries a long note of protest and two short ones in Peirce's hand. In one case Peirce writes, "So he thinks that logic is here abandoned"; in another, "Not so!"

It is difficult to explain the fact of memory and our apparently perceiving the flow of time, unless we suppose immediate consciousness to extend beyond a single instant. Yet if we make such a supposition we fall into grave difficulties, unless we suppose the time of which we are immediately conscious to be strictly infinitesimal.

There are other reasons, — but such questions belong to physics, not mathematics. I have mentioned the instance of consciousness merely because the very proposition of Professor Forsyth upon which Professor Osgood pours out so much scorn finds an application in that case, — supposing the idea of infinitesimals not to involve contradiction.

VI

Nominalistic analyses, of which the doctrine of limits is an example, are certainly exceedingly illuminative. Nevertheless, in the most signal cases they have turned out inadequate, after all.

VII

Nominalists have always been given to framing most extraordinary men of straw, remarkable at once for their clearness and their perverseness of thought. They do not resemble anything in historical fact; but neither they nor still less the historical characters have taken a sufficiently bold stand against nominalism.

Everybody knows by this time that the Scotists and other scholastics never did occupy the position the modern nominalists have represented them as occupying. But their fault, as it is now more and more appearing, was that they were not realistic enough.

Near a hundred years ago, it became the fashion to declaim against some imaginary early psychologists who held to the existence of "faculties" of the soul, as having a real existence, and not springing out of a logical [...]

G. [ON CONTINUOUS SERIES AND THE INFINITESIMAL] (718)

... It is now to be shown that the whole series of numbers, rational and irrational, — that is, expressible by decimals, expressible by circulating decimals, and not expressible either way, but differing from one another by more than one or another power of $\frac{1}{10}$, — do not constitute a continuous series. The reason cannot be fully appreciated in advance of the argumentation; but it will be usefully stated. If an infinite series of numbers each comes after (that is, if each is larger than or each is smaller than) the number next before it in the series, and if there be numbers that come after all the numbers of the series, then by the *limit* of the series mathematicians understand that number which comes *next after* all the numbers of the series. That is, the limit comes after all the numbers of the series, but does not come after any other number that comes after all the numbers of the series. A series of the above description always has such a *limit* in the series of rational and irrational numbers. The reason, briefly sketched, is that the endless series can only be described by a general rule; and when that rule is supposed to be carried out without exception and breaking off, one of the rational or irrational numbers results, and there is no number before this by a finite difference among those numbers which has not members of the infinite series after it by finite differences. Thus, there is a certain kind of *next-ness* in the series of rational and irrational numbers; and this constitutes a breach of continuity. To see how it does so, let us call to mind a certain maxim of which some mathematicians have made use. It runs, "What is true up to the limit is true at the limit." If a man is broad awake at every instant previous to midnight, he cannot but be broad awake at midnight. What is meant is not that it is true, as a matter of fact, that he cannot pass from being wide awake to being asleep between the infinite series of instants and the limit to that series, but that it would be absurd to suppose such a thing. The series is continuous in the sense that it cannot be broken abruptly even in thought. All the instants before one instant, *exclusive*, is in the continuous series a self-contradictory description. Applied to the series of numbers, rational and irrational, it is not self-contradictory. If, then, we can show how to construct a perfectly clear idea of a series for which such description involves an absurdity, we show that the series of numbers is not continuous in the strictest sense; and if we can further show that the instants of time form a series for which such a description is absurd, all will agree that the numbers ought not to [be] called continuous.

Let it be granted that we have a clear idea of the series of numbers, rational and irrational. Then, we have a clear idea of their order of arrangement. We may, then, perfectly well suppose any things whatever capable of arrangement, and sufficiently multitudinous, to be arranged in that same order. They may be points, or instants, or peacocks, or anything. Let us suppose each one to be a beginningless and endless series of points, related to one another like the rational and irrational numbers between 0 and 1, *exclusive*. We may use the numbers to designate, each, some one point of a distinct one of those series of points. The number will not show precisely what point is meant; but it will show to which series the point meant belongs.

We have, thus far, supposed that each number applies to a series of points. We will now change that feature of the hypothesis, and make them no longer series of points, but series of series; each of the last series being a series of points analogous to the whole series of rational and irrational numbers. We will make another change, so that the objects of every series shall be minuter series. We may now make a further change in no longer forbidding two numbers to be attached to the same series of the first order. It will be sufficient that they are attached to different series of *some* order. The result is, that we have altogether eliminated points. We have a series of series of series, *ad infinitum*. Every part, however closely designated, is still a series and divisible into further series. There are no points in such a line; there is no exact boundary between any parts.

We are conscious only of the present time, which is an instant, if there be any such thing as an instant. But in the present we are conscious of the flow of time. There is no flow in an instant. Hence, the present is not an instant.

Let any mode of measure be carried to its limit of precision. Still, each number will designate not an indivisible part; but a series of series, *ad infinitum*. Hence, there are parts immeasurably smaller than any given part.

When the scale of numbers, rational and irrational, is applied to a line, the numbers are insufficient for exactitude; and it [is] intrinsecally doubtful precisely where each number is placed. But the environs of each number is called a point. Thus, a point is the hazily outlined part of the line whereon is placed a single number. When we say *is* placed, we mean *would be* placed, could the placing of the numbers be made as precise as the nature of numbers permits.

When we say that lengths on the line are equal, we mean that the numbers which measure those lengths are equal. Lengths immeasurably shorter than measurable lengths are equal to zero. Yet they are lengths, just the same. Numbers are equally applicable to these also; and then they are algebraically treated as infinitesimals. Again, there will be lengths not measurable by such numbers, nor by limits of series of them. These, when numbers are applied to them, become infinitesimals of the second order.

H. MULTITUDE AND CONTINUITY

A LECTURE TO STUDENTS OF PHILOSOPHY TO BE DELIVERED IN HARVARD UNIVERSITY 1903 MAY 15. BY CHARLES S(ANTIAGO?) PEIRCE (316a)

Gentlemen (and Ladies?):

I address you as students of philosophy. You are no doubt aware that the last generation has seen the most wonderful development of pure mathematics. It was in 1862 that I took my degree in Chemistry; and chemistry was at that time in such a state that one could carry in his mind pretty much the whole of it, and as we students used to say, could keep up with the progress of it provided one only devoted twenty-five hours a day to reading, and had a good chemical memory. The Jahrbuch which covered the whole subject consisted of one thick volume. Today the Berichte of the German society of chemists consists of several unwieldy volumes a year, and is almost confined to one branch of the subject. But the world of ideas is no less rich and teeming than the world of outward perception and I do not think that the progress of chemistry has been a bit greater than that of mathematics, or that its rate of advance is today undergoing greater acceleration. Nobody can think of keeping up with the pure mathematical discoveries in anything more than some very special department. And the quality of the work, its profundity, generality, originality, improves almost as rapidly as its quantity. In particular in precision of reasoning mathematics has undergone a radical revolution during the time that I have known something of it. But do not understand me as meaning that this revolution has been accomplished once for all, so that a tranquil state of logical security has been attained. Not at all: it is easy to see that very great changes cannot be long delayed, although we cannot say precisely what they will consist in. There are great lessons in logic here. Mathematicians always have been the very best reasoners in the world; while metaphysicians always have been the very worst. Therein is reason enough why students of philosophy should not neglect mathematics.[1] But during the last thirty years, there has been an extraordinary mathematical development of the general doctrine of multitude (including, of course, infinity) and of continuity. Philosophers would fall short of their well earned reputation as dunces if they paid much attention to this until it begins to ring in their ears from all quarters. But you would not have come here tonight unless you were philosophers rather in the etymological sense of the word than according to the wont of those who profess logic and metaphysics — unless you were philosophers of the school of Royce rather than of the school of Hamilton.

The leader in this investigation, with whom no other ought to be put into comparison, is Dr. Georg Cantor.

My own first contribution to this branch was made in the autumn of 1867. It contained an idea to which I intend to return, and I received some warm letters about it. But I myself became doubtful of its value.

I may remark that this has invariably been my experience. I have no itch at all to go into print and never will do so until I have gone over the subject so many times and have elaborated so clear an idea of it that it seems a great pity that it should not be recorded. It is always anywhere from five to twenty years from my first having written the matter out. Of course I don't speak of newspaper contributions, though I sometimes leave these standing in type for months before I will consent to their going to press. This is because I know of the castigation that is coming to me from myself. For no sooner is the paper out that I set to work to raise all the objections I possibly can both of the trifling kind and of

[1] In MS. 823 Peirce carries this thought further.

"The writers of logic-books, with rare exceptions, are themselves but shambling reasoners. How wilt thou say to thy brother, Let me cast out the mote out of thine eye; and lo, the beam is in thine own eye? In fact we may say of philosophers at large, both small and great, that their reasoning is mostly so loose and fallacious that it would be derided in mathematics, in political-economy, and in physical science. I felicitate myself that I was made to see this at an early age; for even in my teens, when I was reading Kant, Spinoza, and Hegel, my father, who was the celebrated mathematician Benjamin Peirce, not a powerful analyst of thought, so that his demonstrations were sometimes faulty, but a mind who never once failed, as well as I can remember, to draw the correct conclusion from given premises, unless by a mere slip, my father, I say, would induce me to tell him about the proofs offered by the philosophers, and in a very few words would almost invariably rip them up and show them empty. He had even less mercy for such philosophers as Hobbes, Hume, and James Mill. In that way, the bad habits of thinking which would have been impressed upon me by those mighty powers were in great measure, though I confess not entirely, overcome. Would that every young student of philosophy could enjoy a similar companionship with a stalwart practical reasoner."

the dynamitic kind. And the result has always been that I have found that there were other men who were far better satisfied with them than I myself have been.

The truth is that I am far too well acquainted with the depths of my own stupidity to know what it is to be satisfied with any product of my mind ....

My second contribution was 1881 ...

My third ...

That is all I have printed. My work has been, I believe, completely independent of Cantor. I never knew anything definite about him until 1884.

I have seen it stated in some book that I modified the statements of Dedekind. But the truth is that Dedekind's *Was sind und was sollen die Zahlen* first appeared in [1888].

It contains not a single idea which was not in my paper of [1881?], of which an extra copy was sent to him and I do not doubt influenced his work. It is true that certain of Dedekind's appendices to Le Jeane Dirichlet's Theory of Numbers developed the application of the concept of the *Abbild* to numbers. But I had never read those appendices and of course the *Abbild* idea had been perfectly familiar to everybody for decades before either of us wrote.

Understand me. I am simply defending my own position which is that of Goethe's "*Quidam,*" except that my results coincide in the main with those which any competent analyst much reach in the nature of things on these subjects.

As for Dedekind's book, it is worked out with great ability, although it might have been done better.

I will first show you my way of analyzing these ideas. Namely I require they should be expressed in a certain form of diagram.

The system I call that of *existential graphs*. I do not believe it the best possible; but it is the best I have been able to invent for the purpose ....

I proceed to explain my system of existential graphs.

According to the doctrine of pragmatism, this must connect them with familiar everyday ideas.

The first principle explains what I shall mean by writing a proposition on the board. Namely, I suppose that the board on which I write, whether anything is already written on it or not, represents *a universe* perfectly familiar and well-recognized by the *writer* and the *writee*, the person who writes and the person who interprets the graph.

And if I write any proposition on the board *quite outside of anything else that is written there*, I simply say that this proposition is true as well as what may be written.

2nd Principle That a line of a certain color (or texture) called a line of identity will mean that there is an individual existing i.e. occurring in this universe whom every part of this line denotes, and anything is true of it which is written abutting on this line. ——lives——loves—— ——loves——loves—— ——loves——hates——

3rd Principle That a slightly drawn oval denies whatever is written within it and denies the identity of parts of a line crossing it.

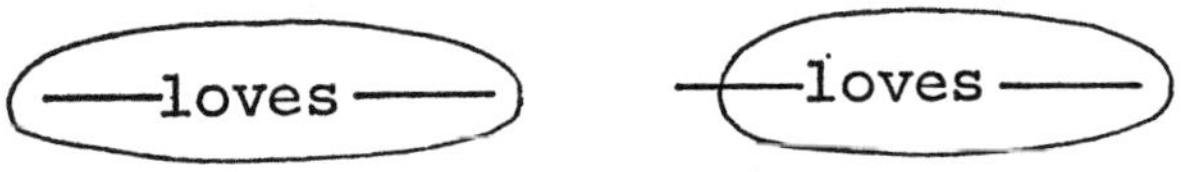

No transformations except insertions and omissions.

5 Rules of *insertions* and *omissions*

Rule I. What is written can be erased.

Rule II. Whatever transformation of all that is written is permissible of any part under even enclosures (i.e. ovals) and the reverse under odd enclosures.

Rule III. Double ovals with nothing between are of no effect and can be erased or inserted.

Rule IV. Any part can be *iterated* preserving its connections, under the same enclosures or any additional enclosures *already written.*

Rule V. A line of identity may be broken where unenclosed.

3

PROBABILITY

A. PROBABILITY (748)

PART I

DESCRIPTIONS

CHAPTER I

THE QUESTION IN PROBABILITY[1]

Into Probability enter three factors:

1st. Descriptions.

2nd. The Order of applying the Descriptions to Things.

3rd. The Number of Things believed to be covered by the Descriptions.

Probability may be considered the Theory of Descriptions applied to Things, or the Theory of Things determined under Descriptions.

It is only under Descriptions that I know Things. Descriptions cover Qualities and Relations. True, I believe something real underlies these Qualities and Relations. But what that real thing is apart from its Qualities and Relations I do not know and cannot conceive.

A description is to me unmeaning unless I believe it can apply to things.

A thing to which no description applies is to me a void nonentity.

Descriptions and Things to me imply each other. Neither can be com-

[1] The following outline is found in MS. 748.

PLAN AND OBJECT OF THIS WORK

"In this Work we seek to present only so much Psychology, so much Logic and so much Mathematics as may be necessary to develop Probability as a complete system.

The Work is elementary and simple in its nature.

We have endeavored to make each chapter complete in itself. However, the chapters upon Development, Elimination, Interpretation and Consolidation constitute the foundation of the subject. Probability as a whole cannot be understood without these four principles.

The first part of the book is devoted to Descriptions. This constitutes the logical and psychological side of Probability.

The second part of the book is devoted to counting the Things determined by the descriptions. This naturally falls under the mathematics of Probability.

pletely conceived apart from the other.

One is the general element in Knowledge; the other is the individual element.

One is the Intellectual side of Knowledge; the other is given on the Sensuous side.

Descriptions are never complete for it is impossible to enumerate all the Qualities and Relations even of the simplest thing.

Descriptions indicate classes or possible classes. No matter how far I carry a description the description can never completely identify a thing. Identity is a question of imagination and belief, not of absolute knowledge, and cannot be proved by descriptions. If the description of two things is the same, to me the things are perceived as the same, though I may imagine and believe them to be different in fact.

TABLE OF CONTENTS

PART I. DESCRIPTIONS

PART. II. THINGS

Defective descriptions are supplemented by the Imagination.

Descriptions are connected with things by Belief.

The increase in my knowledge concerning any individual thing consists in increasing the description under which I know the thing.

I may unite in my mind as many descriptions as I please and consider them one description. I may decompose a description and consider each part a separate description.

Multiplication is the process of applying different descriptions in the mind to the same thing. Multiplication is the establishment of a condition, determination, qualification, limitation, restriction.

Division is abstraction; the removing of a mental determination, qualification, limitation, restriction or condition.

Addition is the process of introducing in the mind different individual cases under the same general description.

Subtraction is the mental process of taking an individual case out from under a general description.

Addition and multiplication imply each other.

For, to add two things I must first discriminate the two things as different from each other and this can be done only by multiplication.

There are thus two distinct methods of uniting descriptions in thought.

One is the process of uniting in thought different descriptions in the same individual thing. This is multiplication.

The other is the process of uniting in thought different individual cases under one general notion. This is addition.

The union of Things in thought follows the same laws as the union of Descriptions in thought, for Descriptions are known to me only through the Things to which they apply, and Things are known to me only through the Descriptions that apply to them.

Suppose I know an individual thing under the generic imperfect description $\mathscr{G}$.

Suppose I now discover that the thing also falls under the more specific description $\mathscr{G}.\mathscr{S}$. I will express this increase of knowledge in the following form:

That which until now I have imperfectly known under the description $\mathscr{G}$, I now more completely know under the more specific description $\mathscr{G}\mathscr{S}$.

I may put this proposition succinctly in the following form:

$$\mathscr{G} = \mathscr{G}\mathscr{S}. \text{ (Man is Man Mortal)}$$

But these descriptions being imperfect the proposition may apply to a good many individuals besides this particular one.

The proposition may thus be interpreted either as a general proposition or as a singular proposition.

Uninterpreted the equation is unmeaning. The descriptions $\mathscr{G}$ and $\mathscr{G}\mathscr{S}$ upon their faces are not equal. The equality consists not in the descriptions themselves but in the application of the descriptions to things.

In the broadest interpretation the proposition means that all $\mathscr{G}$s are $\mathscr{G}\mathscr{S}$s. But through imagination and belief I may narrow this interpretation down to this individual case.

The descriptive equation:

$$\mathscr{G} = \mathscr{G}\mathscr{S}$$

is thus a general synthetic real proposition. It asserts that the things under contemplation covered by the generic description $\mathscr{G}$ are also covered by the more specific description $\mathscr{G}\mathscr{S}$. This will mean all $\mathscr{G}$s are $\mathscr{G}\mathscr{S}$s unless in imagination I restrict the meaning.

A synthetic real proposition is in contrast to any analytical proposition or a definition.

I may assume:

$$\mathscr{A} = \mathscr{B}\mathscr{C}\mathscr{D}$$

where the description $\mathscr{A}$ is merely a short name for the description $\mathscr{B}\mathscr{C}\mathscr{D}$ or where the description $\mathscr{B}\mathscr{C}\mathscr{D}$ is a definition of the description $\mathscr{A}$. This equality is assumed as a definition, an agreement, a convention. The equality is not proved by application to things as in real synthetic propositions.

Every real synthetic proposition is of the form:

$$\mathscr{G} = \mathscr{G}\mathscr{S}.$$

The proposition unrestricted asserts that all the things that fall under the description $\mathscr{G}$ fall also under the more specific description $\mathscr{G}\mathscr{S}$.

The above proposition may be wholly true or wholly false or partially true and partially false.

Suppose that I seek to apply the description $\mathscr{G}$ exhaustively to the Universe of Things and in this attempt discover one thousand individual cases to which I think the description $\mathscr{G}$ applies, everything else being not $\mathscr{G}$. Suppose that among the $\mathscr{G}$s I find only 250 that are $\mathscr{G}\mathscr{S}$s. That is to say, in the Universe, to the best of my knowledge, information and belief, there are one thousand $\mathscr{G}$s and only 250 $\mathscr{G}\mathscr{S}$s. The proposition that all $\mathscr{G}$s are $\mathscr{G}\mathscr{S}$s or that the descriptions $\mathscr{G}$ and $\mathscr{G}\mathscr{S}$ cover the same things is to me only 25% true, under this evidence. The probability that $\mathscr{G}$ is $\mathscr{G}\mathscr{S}$ is 25% under this evidence.

Probability deals with and is to me the percentage of truth in general synthetic real propositions under the evidence before me at the present time.

True, as the question in Probability is sometimes put, Probability seems to relate to an individual thing, but what do I know of the individual thing? I know it only under an imperfect description. I must generalize the question before I answer it. The answer that I give belongs to the class covered by the imperfect description and not to the individual thing alone.

Attempts have been made to develop theories of purely "Objective Probability" based upon purely (?) "Objective Series."

True, my belief in an Objective Series constitutes to me the basis of Probability, but all I know about the Objective Series is through a subjective description.

Probability involves the order of applying the descriptions to things. This order indicates the path of the mind in its investigations.

Given a thing is a $\mathscr{G}$ Is it also a $\mathscr{G}\mathscr{S}$

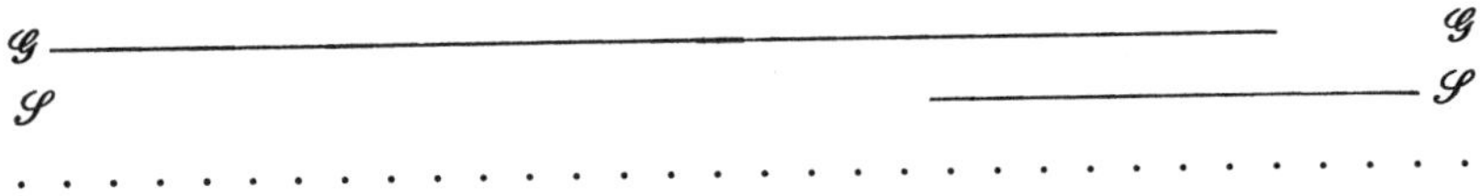

In the above diagram let the line $\mathscr{G}$ represent the description $\mathscr{G}$ and the line $\mathscr{S}$ the description $\mathscr{S}$. Let the dotted line represent the different things in the Universe. Gather under the description $\mathscr{G}$ all the things that are $\mathscr{G}$s and under the description $\mathscr{S}$ all the things that are $\mathscr{S}$s. From the diagram it will be seen that very few of the $\mathscr{G}$s are $\mathscr{S}$s and, therefore, the probability that a $\mathscr{G}$ is an $\mathscr{S}$ is very small.

But from the diagram it will be seen that nearly all of the $\mathscr{S}$s are $\mathscr{G}$s. Thus the probability that an $\mathscr{S}$ is a $\mathscr{G}$ is almost certainty.

Probability differs from an ordinary investigation in the fact that in Probability I seek to ascertain the number of cases under the generic description and the number of cases under the specific description and thus to obtain in figures the percentage of truth in the proposition.

What is the probability of rain tomorrow? If I had a complete description of tomorrow the description would determine the fact and there would be no Probability about it. But all I know about tomorrow is its yesterday; to wit, to-day. My imperfect description of to-day is the only description I have of tomorrow. I count the tomorrows that had yesterdays like to-day. I count how many such tomorrows were rainy. These numerical figures determine the Probability based upon this evidence.

The subject or predicate or both of any proposition may split up into compounds.

We may have, for instance, the proposition:

$$\mathscr{G} = \mathscr{G}(\mathscr{S}_1 + \mathscr{S}_2 + \mathscr{S}_3 + \text{etc.})$$

which may be interpreted either conjunctively or disjunctively or singly as follows:

The genus determined by the generic description $\mathscr{G}$ includes the species $\mathscr{S}_1$, $\mathscr{S}_2$, and $\mathscr{S}_3$, etc.

Or any $\mathscr{G}$ that we may select is either an $\mathscr{S}_1$ or an $\mathscr{S}_2$ or an $\mathscr{S}_3$, etc.

Or this particular $\mathscr{G}$ under discussion is either an $\mathscr{S}_1$ or an $\mathscr{S}_2$ or an $\mathscr{S}_3$, etc.

If the species $\mathscr{S}_1$, $\mathscr{S}_2$ and $\mathscr{S}_3$, etc., make up the whole genus then the probability that $\mathscr{G}$ is $\mathscr{S}_1$ or $\mathscr{S}_2$ or $\mathscr{S}_3$, etc., is unity.

We may in fact interpret any equation between descriptions hypothetically. Thus:

$$\mathscr{A}\mathscr{B} = \mathscr{A}\mathscr{B}\cdot\mathscr{C}\mathscr{D}$$

may be interpreted, if $\mathscr{A}$ is $\mathscr{B}$, $\mathscr{C}$ is $\mathscr{D}$.

Any two descriptions $\mathscr{X}$ and $\mathscr{Y}$ may be put together thus ($\mathscr{X}\mathscr{Y}$) and interpreted in either of three ways:

1st. As the description.

2nd. As the proposition that $\mathscr{X}$ is $\mathscr{Y}$ using in this notation the first description as the subject.

3rd. As the amount of truth in the proposition that $\mathscr{X}$ is $\mathscr{Y}$. That is to say, the probability that $\mathscr{X}$ is $\mathscr{Y}$.

The general problem in Probability consists in ascertaining the probabilities of certain general real synthetic propositions and from these probabilities as known to pass through a system of reasoning to other unknown probabilities.

Just as descriptions may be exclusive of each other or consistent, and may cover the same things or different things, and just as descriptions may thus have boundless relations as between themselves, so propositions, that are made up of descriptions put together in order and considered as applicable to things, may have boundless relations as between themselves, and the resultant probabilities may have as many relations.

The language and symbols of ordinary life are short, defective and figurative. As little as possible is spoken, as much as possible is left to implication, imagination and belief.

But scientific symbols and methods should be complete. As little as possible should be left to implication, imagination and belief.

Probability rests upon general real synthetic propositions and propositions rest upon descriptions.

The general method in Probability should, therefore, be as follows:

The short, defective, figurative language of ordinary life should be translated into exact scientific symbols. These scientific symbols represent descriptions and descriptions represent things and the combination of these descriptions into propositions represent statements concerning things and probabilities relating to things.

We observe certain probabilities. From these observed probabilities through our exact methods we pass to other unknown and required probabilities.

The theory of Probabilities is thus the theory of the percentage of truth in general real synthetic Propositions and the theory of Propositions is the theory of Descriptions.

Suppose we have given a certain set of propositions expressed in the short, defective, figurative language of ordinary life. When these propositions are handled by exact scientific methods the results will fall into four classes:

1st. A set of propositions expressly enumerating in detail all the facts expressly or impliedly given in the data.

2nd. A set of propositions disclosing the ambiguities in the data.

3rd. A set of propositions concerning which the data discloses nothing.

4th. A set of propositions enumerating the conditions expressed or implied to which the procedure is subject.

Whatever relations exist between these propositions exist between the probabilities of the propositions, for the probabilities are the percentage of truth in the propositions.

The procedure falls into two classes:

Part I. The descriptive elements covering the logical relations.

Part II. The numerical elements covering the number of cases included under the descriptions, which is the mathematical side.

B. [THE CONCEPT OF PROBABILITY] (706)

[CHAPTER I]

The Greeks, although the evidences are strong, that of all breeds of men they were the most intellectually gifted that have yet trod our planet, never had the least idea of such a thing. The ponderous lexicon of their language contains no word that expresses or enwraps the principal ingredient of the modern scientific concept of probability; and though gambling was a common enough vice among them, — though not so virulent as among the Italians, — it is strange to note to what awkward devices of speech they are put to express the idea of one of the parties to a wager giving odds to the other.

Carry our minds back to Pyrrho, the famous philosopher of Elis. Elis was a town near the westernmost point of the Morea 22 miles North-west by North of that Olympia where the quadrennial games were held. It was then as now a place where ideas were in a relatively primitive stage of development; — and when I speak of harking "back" to Pyrrho, I do not mean to imply that he was so very early *in time*; for he was only about forty years old when Socrates died; but I mean that in the order of rational development, he was antique, — antically antique; — and so one would think must the mass of his fellow citizens have been to have entertained the exorbitant respect they did for the intelligence of such a goose as he. So one *would* think, I say, if one were to forget the caliber of those teachers who in these United States have attracted enthusiastic disciples by the regiment and the brigades.

The earliest state of opinion is that which makes no other appeal than that "people do so and so." The next is that which divides men into the sheep and the goats and divides everything else into the absolutely one thing and the absolutely other thing. Pyrrho had gone so far into philosophy as to conceive of knowledge and of ignorance. Knowledge for him meant that which could not possibly be wrong, absolute knowledge. Ignorance meant absolute ignorance. Now since any man might possibly be mistaken about any given thing, it followed to his mind that we are and always must be in a state of absolute ignorance about everything. But he would not stand aside from the front of a mad bull. The Greeks had at all times, a singular distrust of the senses; and his fellow citizens regarded this fellow who conducted himself more like a hen or a goose than like a rational being, as a man endowed with high wisdom.

Arcesilaus, whose very name, being one of the two that were alternately carried by the kings of Cyrene, bespeaks his Spartan connections, of Pitana, or Pitane, a small town on the North shore of the Aeolian Bay in Lat. 38°54′, where education had a military direction, its λόχος or cohort being noted in the Peloponnesian war, and where pure intellect was presumably at a discount, held a position so nearly like that of Pyrrho, that but for a single sentence in Sextus Empiricus I should have been unable to distinguish them. By the sentence, we are assured that Arcesilaus did recognize a state of mind intermediate between infallible knowledge and utter ignorance; and his successor Carneades, of Cyrene itself, which was the very essence of Sparta, distilled over, to the shores of Africa about Barca, who instituted the Third Academy, made very much of the *probabile*, as Cicero translates it, though the Greek was merely τὸ εὔλογον the dictate of common sense; but we find no vestige of a quantitative view of modern science in the somewhat large fragments of his philosophy that we possess.

Let us now turn our eyes to the very cradle of the modern conception. The seven most wonderful mathematical minds of the modern period previous to Leibniz and Newton appear to me to have been in the order of their powers: Pierre Fermat (of Toulousain or Quercy), Gaspard Desargues (of Lyonnois), François Viète (of Poitou), René Descartes or Des Quartes (of Touraine), Blaise Pascal (of the Auvergne), Christiaan Huygens (of the Hague). All of them except perhaps Viète were of distinguished families. Were I to name a seventh it might be Thomas Harriot's of Oxford.

Although of these six gentlemen I rate Pascal as only the fifth in mathematical power, yet there is no doubt that he labored more than any of them, or perhaps than any man with whose life we are acquainted under that affliction which we name "genius." From 1650 or 1651 to the autumn of 1654, that is to say from the age of seventeen or eighteen to that of twenty-one and a quarter, being a wealthy young gentleman, he mingled much in society; and among his friends was a famous gamester, the Chevalier de Meré, who wrote him two letters from which long extracts are given by Bayle in his Dictionnaire under the head of Zenon de Sidon. In one of these he put the two following questions: The first was whether it would be advantageous to bet that one would throw double sixes with two dice in a given number of throws; and the question shows a great advance over the state of mind of the ancients in respect to the comprehension of the nature of probability. The other question was supposing two players had staked a certain sum upon which of them would first throw three double sixes, and supposing the game were unavoidably interrupted after one of them had thrown one double six while the other had thrown none, how ought the stakes to be divided between them? To this Pascal gave a solution which, although it involves a minute error, is so ingenious and at the same time so exhibits so precisely how far he had mastered the conception of probability that I desire to bring it to my reader's notice. Designating the two players as A and B, let us, he says, begin by supposing that A has at the time of the interruption already thrown 2 double sixes and B one. Then, had they thrown more it would be equally likely that the next double six would be thrown by A and that it would be thrown by B. In the one case A would have won all the stakes: in the other they would be reduced to an equality, in which case each would be entitled to half the stakes. A then should in any case receive one half the stakes and while the other half should be divided between them equally. A then should receive $\frac{1}{2}$ the stakes *plus* one half of the other half, or $\frac{3}{4}$ of the whole while B should receive only one half of a half or $\frac{1}{4}$ of the whole.

Next he says suppose at the time of the interruption A had thrown 2 double sixes and B none. Then if A after the interruption would have thrown the next double six he would have received the whole of the stakes, while if B would have thrown the next double six, we have the last case; and A ought then to receive $\frac{3}{4}$ of the stakes. So much then belongs to him in any case and of the remaining $\frac{1}{4}$ the stakes he would have had as good a chance of getting it as B, and therefore should receive a half of it. A should therefore receive $\frac{3}{4}+\frac{1}{2}\cdot\frac{1}{4} = \frac{7}{8}$ of the stakes and B the other $\frac{1}{8}$.

Lastly let us consider the case proposed in which A has at the time of the interruption thrown 1 double six and B none.

Now even if the next double six were to be thrown by B the two players would be even and A would then be entitled to $\frac{1}{2}$ the stakes.

But if the next sixes should be thrown by him he would by the last result be entitled to $\frac{7}{8}$ of the stakes. These two results being equally likely he should receive the mean between them or $\frac{1}{2}(\frac{1}{2}+\frac{7}{8}) = \frac{11}{16}$ and B should receive $\frac{5}{16}$.

The method is perfectly sound; but there is a slight error in the application of it due to the fact that the two players having each thrown sixes twice might throw their third sixes *at the same time*, a consideration which would make a difference of about $\frac{1}{4}$ of one *per cent.*

For sixes will be thrown 35 times by each party when the other party does not throw sixes to once that both throw sixes together which will only happen in one throw out of every 1296 in the long run.

In the case then of *A* having 2 sixes and *B* one at the time of the interruption

there are 35 chances of *A* throwing sixes again before *B* and so winning the whole stake;

35 chances of *B* throwing sixes before *A* and so equalizing matters when *A* will have $\frac{1}{2}$ the stake;

and 1 chance of both throwing sixes together then *A* will have the whole stake.

Consequently *A* should receive $\frac{36}{71} + \frac{1}{2}\frac{35}{71} = \frac{214}{284}$ of the stakes instead of $\frac{213}{284}$ as Pascal found.

In the case of *A* having 2 sixes and *B* none at the interruption

there are 35 chances of *A* having the next sixes and getting the whole stake;

and 35 chances of *B* throwing the next sixes while *A* does not in which case *A* will be entitled to $\frac{214}{284}$.

$\therefore$ *A* should have $\frac{36}{71} + \frac{35}{71}\frac{214}{284}$ or $\frac{35428}{40328}$ instead of $\frac{35287}{40328}$ that Pascal gave him.

In the case of *A* having 1 sixes and *B* none

In 35 cases *A* will have 2 and *B* none $\frac{35}{71} \cdot \frac{8857}{10081} = \frac{35}{71} \cdot \frac{17714}{20164}$;

35 cases *A* will have 1 and *B* 1 $\frac{35}{71} \cdot \frac{1}{2} = \frac{35}{71}\cdot\frac{10082}{20164}$;

1 case *A* will have 2 and *B* 1 $\frac{1}{71}\frac{107}{142} = \frac{142 \cdot 107}{71 \cdot 20164}$.

Pascal's method is very ingenious and quite correct. But on account of his overlooking the circumstance that after both play[er]s had thrown sixes twice, they might both throw sixes the third time simultaneously when of course they would divide the stakes, he gives *A*, the player who had thrown sixes already, before the interruption a share of the stakes which is too little by about one seventy-first part of itself. However, it is very seldom worth while to consider such slight probabilities as that, inasmuch as the Doctrine of Chances is essentially a rough approximation unless the number of trials is stupendously large. For unless there are at least 25 trials or unless the *pros* and *cons* are very [unequal], it cannot be expected that the very first figure of the number that will come out in any predicted way will be verified in fact; and there must be at least 2500 trials to warrant reliance on the first two figures and two and a half million of actual trials will be requisite to justify the third figure; and indeed because it is impossible to take account of all the possibilities, all these numbers of trials are usually too small. Pascal's error, therefore, is wholly insignificant, especially to a person so regardless of small sums as any gamester must be.

But what is a trifle better worth notice is that Pascal does not show that he at all knew how to take account of the possibility of both players throwing sixes at the same throw. For his whole reasoning is based upon a single principle which, as happens, is clear and simple enough in any concrete case, but cannot easily be stated in general terms without considerable grammatical complication. The best general statement of it I can make is this:

Suppose that it is not evident how the stakes ought justly to be divided at the time of the interruption, but suppose how the division of the stakes would justly be made were the liquidation postponed until an additional doublet of sixes had been thrown, whichever of the two equally likely events should occur, the one being that this additional doublet were thrown by *A*, the other that it were thrown by *B*. Then if in either case *A* would become entitled to something, it would be just that in advance the lesser of the two sums to which he would become entitled should be set aside for him; and as to the remainder, in one of the events he would not become entitled to any of it, while in the other equally likely event he would become entitled to the whole or to some known fraction of it. Then that part to which it would be equally likely that justice would allot to *A* or to *B*, should be equally divided between the two; while the remainder to which *A* would have no just claim even after the additional throw should go to *B*. This principle could readily have been generalized so as to take account of the possibility of both players throwing double six simultaneously. For this purpose it will be convenient to suppose that one of each of the two player's dice is stained yellow on all six faces while the other remains white. Then when the first double of sixes is thrown there will be 71 equally likely events

that both players throw double sixes	1 event
that *A* throws double sixes, that *B*'s yellow die comes up six, and his white die either ace, deuce, tray, catter or sank	5 events
that *A* throws double sixes, that *B*'s yellow die comes up sank and his white any side	6 events

that A throws double sixes, that B's yellow die comes up catter and his white any side — 6 events
that A throws double sixes, that B's yellow die come up tray — 6 events
that A throws double sixes, that B's yellow die comes up deuce — 6 events
that A throws double sixes, that B's yellow die comes up Ace — 6 events

In all that A throws double sixes and B not — 35 events
that B throws double sixes and A not — 35 events

35 cases in which A will be entitled to nothing more than this minimum
1 case in which he will be entitled to half the doubtful remainder
35 cases in which he will be entitled to all the doubtful remainder.

First: Suppose then that when the interruption occurs A has thrown sixes twice and B once.

Then were liquidation to be postponed until double six were thrown there would be 35 cases in which B would throw sixes and A not when the stakes would be equally divided and A would be entitled to one half

and there would be 36 equally likely cases when A would be entitled besides to the other half.

He will therefore be entitled to $\frac{1}{2} + \frac{36}{71}\,\frac{1}{2} = \frac{1}{2}\,\frac{142}{142} + \frac{72}{142}\,\frac{1}{2} = \frac{71}{142} + \frac{36}{142} = \frac{107}{142}$ of the stakes.

Second, Suppose that at the time of the interruption A has thrown sixes twice and B not at all.

Then were liquidation postponed until sixes had been thrown again there would be of the 71 equally likely cases

35 in which B alone would throw sixes and A would be entitled to $\frac{107}{142}$ of the stakes.

36 in which A would be entitled to the remaining $\frac{35}{142}$ of the stakes.

He will therefore be entitled to $\frac{107}{142} + \frac{36}{71} \cdot \frac{35}{142} = \frac{107}{142} \cdot \frac{71}{71} + \frac{36 \cdot 35}{142 \cdot 71} = \frac{7597}{10082} + \frac{1260}{10082} = \frac{8851}{10082}$ of the stakes leaving $\frac{1231}{10082}$ for B.

Third, Suppose A has thrown sixes once and B not at all.

Then were liquidation postponed until sixes were thrown once more there would be of the 71 equally likely cases

35 in which B alone would throw sixes and A would receive $\frac{1}{2}$ the stakes [or] $\frac{71}{142}\,\frac{71}{71} = \frac{5041}{10082} = \frac{10082}{20164}$

1 in which both would throw sixes and A would be entitled to $\frac{107}{142}$ of the other half [of] the stakes or $\frac{107}{284} \cdot \frac{71}{71} = \frac{7597}{20164}$

35 in which A would be entitled to $\frac{17702}{20164}$ of the stakes.

He would therefore be entitled to $\frac{35}{71}\,\frac{1}{2}+\frac{1}{71}\,\frac{107}{142}+\frac{35}{71}\,\frac{8851}{10082}=\frac{35}{71}\,\frac{5041}{10082}+\frac{107}{10082}\cdot\frac{71}{71}+\frac{35\cdot 8851}{10082}=\frac{176435}{715822}+\frac{7597}{715822}+\frac{309785}{715822}=\frac{493817}{715822}$

That is to say

In any case he gets $\frac{1}{2}$ the stakes

And in 36 of the 71 cases he gets $\frac{36}{142}=\frac{18}{71}$ of the other half making $\frac{107}{142}$ of the whole. [1]

It will be observed that although all Pascal's reasoning consists in a quantitative discussion of uncertainties, he never uses the word "probability" at all. For to do so would have involved a violation of the fundamental principles of French composition, which does not permit the use of any expression whose exact signification is not clear to both writer and reader. If he had been writing Latin, Cicero, the supreme arbiter of expression, would have authorized the use of *probabile*: but in French you must not utter a word whose meaning is not clear; and Pascal always wrote French. All his talk is about the share of the stakes to which the player A would have been justly entitled in case the progress of play were interrupted and an immediate liquidation had to be made. If a modern logician had been on hand to ask him what he meant by "*justly entitled*," a question that, as any clear-headed modern logician would have put, his answer would be that if the liquidation were to be postponed, so that in the interval of a certain number, n, of equally likely events, some one should occur, and no other could occur and it was known what would either be the just or the inevitable mode of partition of the stakes after whichever one might occur, then the just mode of partition at the time of the actual interruption would consist in giving to each player one nth part of the sum of the different equally likely shares which he would either justly or inevitably receive in the case of such postponement. Now what is meant by the more modern expression "the present probability of a future event" is precisely that share of the stakes to which according to Pascal's principle a person who had laid an even wager upon the occurrence of that event would be justly entitled now to receive in case of a liquidation at this moment.

It is to be observed that Pascal's principle may be separated into two Rules. One of these Rules is that if the event upon whose occurrence a player has laid a wager may come about in several entirely inconsistent ways, then the share of the stakes to which he would be justly entitled in

[1] More arithmetic computation follows — leading to no particular conclusion.

case of immediate liquidation is the sum of the shares to which he would be entitled if he had bet upon the event coming about in those different ways singly. The other Rule is that if the event, *E*, upon whose occurrence a player has laid a wager consists of two parts, say the event *C* and the event *D*, so that if *E* occurs *C* and *D* must occur, and if *C* and *D* both occur, *E* must occur, then the share of the stakes to which that player would be justly entitled in case of an immediate liquidation will be the product of two fractions, one being the share, or proportion of the whole stakes to which he would have been entitled if he had merely laid his wager upon the occurrence of *C*, and the other being the share to which he would have been entitled, if after *C* had happened or was absolutely known to be about to happen, he had laid his wager upon the occurrence of *D*. To these two rules are to be added two axioms; first, that each of the two parties to a wager bets upon the occurrence of an event which consists in the non-occurrence of the event upon which the other party bets; and secondly, that as soon as it is certainly known that the event upon which a player has bet will occur he becomes entitled to receive at once the whole of the stakes; and these two Rules and two Axioms constitute the entire basis of the Doctrine of Chances, so far as it is confined to games and the like, where the issues are certain soon to become positively known. But if the two Rules and two Axioms were proposed as a Definition of Probability, this definition would be open to the logical objection that it involved a *Vicious Circle*. By a *Vicious Circle*, or *circulus in definiendo*, is meant that fault in the definition of a concept which consists in that very concept entering into the definition of itself. This would be the case here inasmuch as the first Rule speaks of different events being *equally likely*.

The history of the Science of Probabilities, during the quarter of a millennium since Pascal inaugurated it, has involved a tale of practically serious blunders into which mathematicians have fallen in consequence of not rightly apprehending the nature of Probability; and yet examination will leave no doubt that, even as it has been, [it] has raised every other science to which it has been applied to a distinctly higher plane; and it has already been applied to the majority of the large divisions of science.

In the next chapter, I shall inquire why this should be. In the chapter following the next, I shall show that every definition of probability hitherto given either is vague, or defines something that is not probability, or falls into a vicious circle; whereupon I shall give for the first time a correct definition, and shall answer possible objections to it.

CHAPTER II

Why should the Doctrine of Chances raise Science to a higher Plane?

In the first place, let us make sure that it does do so.

In Mathematics there is little room for its application; for as to the so-called "problems in geometrical probability," or "local probability" (such as the question what is the form of the plane contour which gives for the probability a minimum that four points taken within it at random shall be the vertices of a reëntrant quadrilateral?), I doubt whether they are, properly speaking, questions concerning probability, at all. I will allow that there is a very small element of probability in such a problem; but in the main it is a question of the integral calculus, or calculus of variations. I know of but one purely mathematical problem that must be allowed to be fairly a problem in the doctrine of chances. I mean the question how many prime numbers we should expect to find greater than one number and less than another, — a very abstruse sort of question which Chebichef, Glaisher, and others have studied. When we consider that the prime numbers are those that remain over after all the composites have been sifted out by the so-called *cribrum arithmeticum*, it must be allowed that this problem is much more closely allied to ordinary problems of probability than it is to any ordinary problem of mathematics.

Passing over the philosophical sciences for the present, we find that the kinetical theory of gases, now extended to liquids and solids, has veritably transformed pure physics already, and, being nearly coëxtensive, as it is, with the physics of nonreversible actions, such as the diffusion of matter, the conduction of heat, and the action of viscosity and other varieties of friction, it seems not unlikely in the future still more fundamentally to revolutionize physics, until instead of such actions being regarded as exceptional, it may be that it will be the reversible processes such as motion under gravity, the action of the dynamo, etc., which will in future appear so. Now this all important theory is the direct offspring of the calculus of probabilities. In chemistry, the doctrine of chances, has hitherto accomplished very little, — directly and overtly, at least; but I am persuaded that there will be great triumphs for it when the science has reached the stage at which it can be applied more thoroughly. For to mention but one field for its influence, the proportions of the different products of organic reactions, which are commonly so very complicated, must, it would seem, depend upon probabilities. In biology, that tremendous upheaval caused in 1860 by Darwin's theory of fortuitous variations was but the consequence of a theorem in probabilities, namely, the theorem that very many similar things constituting one class are subject to very many slight fortuitous variations, as much in one direction as in the opposite direction, which when they aggregate a sufficient effect upon any one of those things in one direction must eliminate it from nature, while there is no corresponding effect of an aggregation of variations in the other direction, the result must, in the long run, be to produce a change of the average characters of that class of things in the latter direction. For example, if each member of the class contains a large number of certain minute parts, and the variations consist in any one [of] them from time to time gaining or losing, one or more of these parts; and if the loss of all of them will destroy the thing or otherwise eliminate it from the class while there can be no such effect from any gain in these parts, then, in the long run the average number of such parts in members of the class must gradually increase, according to a quantitative law which is determined by the principles of the doctrine of chances. Anybody who is old enough, as I am, to have been well-acquainted with the spirit and habits of science before 1860, must admit that in this case, at any rate, the work of elevating the character of science that has been achieved by a simple principle of probability has been truly stupendous. In all those sciences in which measurements made in exactly the same way, though perhaps under widely different circumstances, are very many times repeated, as is the case especially in astronomy, the mere habit of applying the method of least squares, which is a method of drawing the proper conclusion from the measures according to the principles of probability, — even the mere requirement that every result should be accompanied by a statement of its probable error, has had a great effect in stimulating precision of observation. That, however, is by no means the only influence that the doctrine of chances has had upon astronomy. There are a number of the highest branches of the science, which simply could not be carried on without this doctrine. The distances of no fixed star, for example, could be ascertained without combining hundreds of observations according to principles of probability; and such facts as the direction of motion of the solar system in space could not otherwise be ascertained at all. We may say that so far as astronomy has taught us anything about the sidereal universe as a whole, it has only done so by means of a constant attention to probabilities. All the statistical branches of science, which on the whole touch human life the most, exist at all only by virtue of the doctrine of probability. The psychical sciences, under which head I include not merely psychology but the whole of that congeries of sciences called collectively anthropology, together with linguistics, textual and historical criticism, archeology, history, comparative biography, stand in need of the same support even more than do the physical sciences; but it is one of the greatest misfortunes of our species, comparable to the prevalence of tuberculosis, that, owing chiefly to bad teaching in childhood, the majority of students of those sciences are unable to concentrate their minds sufficiently to understand any mathematical reasoning; and consequently fail to make use of the principles of probability as it is their duty to do.[2]

There is the fact; and the question now is, What is the reason of it? Why should it be so?

My first answer shall be that there is nothing strange about it: every man of sense must see why it is. Only, there are slight differences between the reasons of the great stimulus which the doctrine of chances has brought to studies, 1st, of the sciences of exact measurement; 2ndly, to sidereal and physical science; and 3rdly, to biology.

That which these three slightly different reasons have in common is that the general mass, all but the whole of the science of probabilities consists in the tracing out by mathematical deduction of the phenomena that must necessarily result when a vast multitude of precisely similar objects of any description that under the same general influences are subjected to a great number of small causes of diversification. Everybody who is acquainted with the theory of probabilities will admit that that truthfully characterizes it. Now the reason why it has been so beneficial to sciences of exact measurement is that these measurements have been the result of many men each many times daily for many years having made these measurements by taxing the control of their nerves just as far as it could be taxed without manifestly beginning to break down; and that under these circumstances every measurement made has been influenced by many slight and obscure causes of perturbation, no one of which was singly perceptible and some of these causes have worked one

[2] There is evidence in the MS that Peirce intended to eliminate the following but failed to do so. "Letters between Pascal and Fermat on probability were circulated as early as 1656. In 1657 Huygens published a memoir on methods of reasoning about chances. In 1708 Pierre Rémond, usually called Monmort from an estate he had purchased, published an *Essai d'analyse sur les jeux de hasard*, and in 1711 appeared an important tract of the Englishman (?) Abraham De Moivre on the measurement of chance. All this material was used by Nicolas Bernoulli the first, in editing and rewriting the posthumous work of his uncle James (Jakob) Bernoulli, the first's treatise *Ars Conjectandi* which was published in quarto at Bâle in 1713. This work contains the whole of the three general principles of doctrine that are really sound. I will briefly state them." Instead Peirce experiments with formulas and arithmetic computations.

way and some in the opposite way to about the same extent. How, under these circumstances these innumerable slight and obscure causes of error would affect the statistics of measurement is just the sort of problem in which the doctrine of chances is at home. The consequence was that as soon as the theory of the subject, which is contained in the method of least squares, was developed, every man could calculate his own liability to error, or at least to that part of his error that depends upon the variability of his nervous condition, and thus self-criticism was stimulated. Now everybody knows that self-criticism is the head teacher of self-control. Moreover, public opinion soon required every observer to publish the exact result of his self-criticism, together with full data whereby anybody else could check its accuracy; and thus emulation was excited which is a powerful assistant and awakener to self-criticism. It is no wonder, therefore, that sciences of measurement made mighty progress after the doctrine of chances was called to its aid. It is true that no modern observers could manage the instruments of Tycho Brahe and Helvetius as they themselves managed them, or could at all vie with the early Egyptian rope-stretchers. But that is only because instruments have been so vastly perfected, new methods have been devised, and greater attention is now paid to all the circumstances of observation, under the influence of this same relentless criticism of observational results.

The influence of the study of probabilities upon physics has been of a somewhat different kind. For physicists are seldom obliged to work in the open air under the nervous influences of darkness, cold, distorted postures, and a trembling atmosphere; so that their results [are] much less subject to the aggregate effects of a hundred small perturbations of their frames each by itself inappreciable.

But here every observation made is affected by something like a trillion, — I mean, a million millions of millions, — of molecules, while no effect that depends upon the concurrence of fewer than a thousand billions of those trillion molecules can come to their cognizance at all. I am speaking, for example, of observations of weight, pressure, temperature, light, electricity, and other physical forces. A few of Crookes's observations first recognized effects of smaller collections of molecules. Now imagine that every time gamblers threw their dice, instead of throwing one, two, or three they had to throw a trillion of independent dice and could perceive no fewer than a thousand billion of spots turned up. Had this been the case, what would have been the result? By the so-called "law of high numbers" or of averages, which is in reality no law at all, but is only the mathematically necessary effect of the throw of each individual die being unaffected by any of the others, there would have been, in the whole history of dicing, no instance of a throw in which more or fewer pips turned up than in any other; and any report of any other throw, though chance *might* bring it about at any time would be so unprecedented, that to believe that it had actually occurred would be universally looked upon as a belief in a miracle. But because that is the way in which the bodies that physicists observe are constituted, and because the purely mathematical doctrine of chances is able to predict all the statistical consequences of such a state of things. Everybody knows that an atomic theory of the constitution of matter was held by a very important school of ancient philosophers and found its advocates all through the middle ages. Boyle made it the foundation of his chemical doctrine. Daniel Bernoulli, who was only thirteen years old when his cousin Nicolas (the first, as he is called, his father who bore the same name not being counted by scientific men) put forth his enlarged edition of his uncle James's work on probabilities, the *Ars Conjectandi*, which sets forth all the principles of the subject that are at once general and sound, this Daniel, I say, a quarter of a century later showed that Boyle's Law of the relation between the volume and the pressure of gases could be explained through the doctrine of chances on the assumption that gases consist of atoms moving along straight lines until they either strike one other, or impinge upon the walls of the vessel that holds them, or are drawn down by gravity. But it was not until the mechanical theory of heat had been thoroughly proved by Clausius and Joule (Kelvin, Rankine, and others, — even Sadi Carnot, — taking the same step independently, though with less probative force), that Clausius demonstrated the kinetical theory of gases from the fact that the free expansion of a gas does not cool it by anything approaching the amount that it would do if the tension of gases were due to repulsion between their molecules. But it was Maxwell who first clinched the proof by not only showing that the theory accounted satisfactorily for the known laws of the diffusion, conduction, and viscosity of gases, but also by predicting, quite contrary to the belief of physicists and to the analogy of liquids that the viscosity of gases would be found to increase with the temperature and then showing by new experiments that the prediction was correct. Boltzmann, also, elaborated the application of mathematical theory of probability to gases, by important and difficult mathematical reasonings. Then, the wonderful experiments of Crookes with vacua so high that molecules in large number could travel for inches before colliding with others removed every vestige of doubt; and last of all came, what scientists are in the habit of expecting to complete the proof of any theory, namely, its leading to new discoveries of great importance, came when Stoney put forward the Electron Theory and so rounded out Maxwell's work. All of this vast improvement in our comprehension of the physical universe was due to mathematical speculations upon chances. It was owing to its turning out that that concept which man had evolved from games of his own invention, — the concept of probability, — was already embodied in God's material creation.

But magnificent as was this contribution to the teachings of science, no judicious physicist can hesitate to admit that it is far outshone by the heliac effulgence that has been emitted from Darwin's immortal volume. It has always been a matter of personal regret to me that I missed the first impression the work made in Cambridge, Massachusetts, where I lived, owing to my having set out to go under instructions from the Superintendent of the Coast Survey more than a month before its appearance on Nov. 24, 1859 to take part in triangulation on the east coast of Louisiana. Once arrived in the field of work, I spoke absolutely with no soul except two Coast Survey Officers not particularly interested in biology, a densely benighted sailing master, half a dozen men before the mast, a negro cook, and a cabin boy until April. However, in the course of the winter, a letter from my mother told me what a sensation the book had made; and thereupon I wrote to my friend Mr. Chauncey Wright that I felt confident that Darwin had received a hint of his idea from Malthus *On Population*. Long after, Wright inquired of one of Darwin's sons whether this was the case; and reported the answer to be that it was nearly certain the father had never looked into Malthus's celebrated book. But I believe it has since transpired that he really did derive just such a hint as I surmised from that very quarter. I mention this circumstance as evidence that that element of the theory that at the time of its broaching struck a young chemist as the most novel idea in it was the proposition that the very existence of species is continually at stake in a desperate struggle. For the cardinal principle of the hypothesis, that given what might be called a state of incessant tendency to instability and (so to speak) of shakings up of the species in the fortuitous variations at every birth of an individual of it, and granted the proposition that a small modification of the characters of the species would suffice for its extirpation, it must be all the time adjusting itself more and more perfectly to its environment appeared to the chemist as familiar. But though the general mode of reasoning was familiar, the application of it to biology, fraught as it was with the highest hopes for that science, was by no means familiar. It was an example of the advantage to each science of having among its devotees, some man of great general ability who may thoroughly comprehend and be skillful in the ways of thinking of each other science. But still more strikingly did it exemplify the advantage for a science of understanding what it is that is the supreme need of that science at a given time, and of considering that need in its broadest form, and with the power of determining that broad conception so as to bring it down to the case in hand without losing any element of its importance.

[CHAPTER] III

The reasons why the application of the conception of probability to three branches of science was as important as it was have thus been, I believe, correctly stated. Yet the explanations do not fully satisfy either myself or, I am persuaded, my thoughtful reader. In the first place, I have given three different reasons for the advantages of applying the conception of probability to as many different branches of science, whereas I cannot help fancying there should be some single reason applicable to the three cases; and in the second place though my reasons show why the introduction of the conception of probability should be advantageous in the three cases, they do not seem to necessitate its being so supremely important as it has been. I earnestly wish that some more powerful intellect than mine would attack the problem. In the original papers in the *Popular Science Monthly* of March and April, 1878, of which the present essay is a revision, I contented myself with applying the "common observation" that "a science first begins to be exact when it gets quantitatively treated." Now this is, as I called it, "an observation." But observation only supplies us with individual facts: it does not, by itself alone, supply a reason for the necessity of the facts observed. It may be a fact that any science takes a higher science as soon as it becomes quantitative, or, what comes to the same thing, so far as observation can show, that any science as it matures is led to pay more and more heed to quantities; for naturally we first remark the rough discriminations of quality and later the finer differences of quantity. Each of these two forms of interpreting the observable facts represents, I do not doubt, a side of the truth that the other leaves unexpressed. When we come to slight degrees of difference we resort to quantity, simply as a convenient means of expressing and of finding slight dissimilarities; but moreover quantitative expressions always tend to stimulate our attention to small distinctions. But it certainly is not a universal truth that progress in a science will render it more attentive to quantity; for it has not been so with nineteenth-century mathematics, [and] the early modern mathematicians. The geometers of the XVIIth century were quite content to abide by the old definition, "Mathematics is the science of quantity," a saying all three of whose nouns had had in the original enunciation of it borne quite other significations than those that the lips and ears of three centuries ago attached to them. For "quantity" had originally been understood "quantities" in the concrete, things numerable or measurable. For it is expressly stated by ancient writers that music must be reckoned as a branch of mathematics inasmuch as the pitch of sounds is measurable. "Science," I need hardly say, conveyed to the ancients the idea of infallible certainty, which forbade any inductive study's receiving the name of a science; and this accounts for the origin and frequent application by the ancients of the word *philosophy*, — wisdom-loving, — to *studies*; whereas only necessary theorems could go to the composition of sciences. And as for *mathematics*, — which had been the collective title of whatever discipline had *quanta* for its object, and but four such disciplines were recognized: what they called arithmetic, with geometry, music, and astronomy. Of their "arithmetic," the art of computation formed no part. It was called *logistic* by the Greeks, who were not very expert at it; and later when our so-called "Arabic notation" for numbers came in, it was called in English "augrim," a corruption of the Latin *algorism* or *algorithm*, from the Arabic *Al-Khwârizmî*, — "the man of Chorasmia, or Khiva," the ordinary designation of the author of that treatise which, so far as rules of computation could, fitted the sons of the desert for taking account of the gold that their new religion was showering upon them, as it subsequently and indirectly instructed the inhabitants of Europe, too. But everything like this ignoble art was excluded altogether from the liberal education of the Romans and the Greeks, who probably regarded ciphering as toil fit for slaves, and as such gave it no place in liberal science. These facts were more open to the generality of XVIIth century mathematicians than to ours; and yet they were content to accept the words of the ancient definition "mathematics is the science of quantity" as expressing under their modern acceptions of the words the nature of the different science that had come to be called mathematics. How entirely their geometry was engrossed with ideas of measurement is shown by Fermat's writing to Pascal that he cannot imagine how his *Hexagramma Mirificum* which had nothing to do with measurement could be proved. In the last half century, however, this nonquantitativeness has been characteristic of the main part of geometrical research and all the metrical relations seem so relatively unimportant that I would wager that not half the professors of mathematics living today have a definite recollection of the relation between Tarry's point and Pascal's hexagram, although everything connected with the latter is of deep interest to them. But the relation [of] Tarry's point to it being metrical is likely, I think, to be dimmed in their memory as being an unimportant detail, though it is far from being unimportant metrically.

[CHAPTER] IV

Just as there are doubtless flitting through the physical universe trains of vibrations of all wave-lengths, while there are in the whole range of possible wave-lengths but a few scattered intervals within which the waves can produce any impression upon our senses, and to all the vast majority of them we remain as utterly insensible as if we were made of sand; so I thoroughly believe (and possibly I may be right) it is with our intellects. For it seems to me that of the two branches of science in which man has had any success, namely, the physical and the psychical sciences, the former is a development of instincts which always were indispensable to him for getting his food, for which business some comprehension of mechanics was necessary — and all animals possess such ideas; namely those [of] time, space, and force — while the latter is a development of the instincts of reproduction, for which some comprehension of what is passing in his neighbour's mind is necessary — and all animals possess the ideas that fit them for understanding their own kind. But there is not the slightest reason that I can discern which the laws of happenings should be limited to such as are related to the relations of animals whether with inanimate objects or with their fellows. I naturally surmise that there will be others. But supposing there were such, how would they betray themselves? Evidently by puzzling phenomena manifestly inexplicable whether on the principles of mechanics or on those of the association of ideas. How would such phenomena be recognized? By their, *now and then*, being such as no purely physical conditions could explain, nor yet any more psychical conditions. Everybody who does not insist upon shutting his eyes to the evidence must admit that the "psychical researchers" have met with such phenomena, which although they persist in thinking them explicable by psychical conditions, are nevertheless not to be paralleled in the ordinary life of sane men, and which therefore do not come within the domain of psychology.[3]

[3] This MS was written 23-31 January 1909.

C. [PROBABILITY AND INDUCTION] (L231)

Milford, Pa.
1911 June 22

Dear Mr. Kehler:

Yours of the 20th inst has reached me this afternoon. It informs me that the Hon. Lady Welby has written you many letters and has so often been kind enough to mention me that you would like to know who I am and what I have done. I am a person 72 years old. I have accomplished nothing in particular. But since Lady Welby seems, from what you say, to desire that you and I should be acquainted, and I think a gentleman to whom she has written many letters must be a very desirable acquaintance, I dare say that my 72 years have equipped me with sufficient garrulity to develope a voluminous epistle by way of conveying to you a notion of the particular variety of nonentity that I can boast myself.

When I was thirteen years old, being one day in the room of my elder brother, I picked up from his table a copy of Whately's Logic, and asked him what "logic" was. Being answered, I stretched myself out on the carpet with the book, and in a few days had mastered all the Archbishop had to say. From that week until I had reached my three score years and ten the central passion of my being was to find out, — not by any means what passed in my organism and my consciousness when I thought, but something anterior to all such knowledge, namely, what are the fundamentally different ways of reasoning, what kind and degree of assurance each could supply, and under precisely what conditions, and by what methods to proceed in order to gain such knowledge as is possible for human beings. The more I studied this subject the more and more deeply I felt the *shocking levity and looseness of thought with which these basic questions had been treated*; and although in modern times men have reasoned much better than ever before concerning whatever subjects they really have seriously investigated, and although if a man works soberly and zealously enough over any subject, and has the means of detecting the errors into which he will get seduced, he will become very expert in reasoning about that particular subject, or any other sufficiently like that, yet outside the particular corner of science that each one has learned how to make a particular kind of investigation, the stupendous errors into which they have fallen about reasoning and the slap-dash fashion in which they have taken up opinions which if they had only questioned them and tried to doubt them they would soon have seen to be utterly false, — all these errors seem to be more and more glaring the more triumphantly truth is pursued in those branches in which it has been seriously, ardently pursued.

Almost the first book which is seriously studied that offered any difficulty was Kant's Critik der reinen Vernunft. Somebody is likely to tell you that I betray my ignorance, in writing "Critik" instead of "Kritik." I shall not defend myself. I spell it as Kant himself spelled it in his first edition, — which in some respects I thought was the better of the two, — and I studied both until I very nearly knew both by heart. But I particularly like the old-fashioned spelling *Critik*, because it calls to mind the source from which Kant got the word, namely from the English of Locke, who used "critick" to mean the science of criticism, and Kant in his long preface is most careful to explain that *that* is what he means by "Critik," and that he does *not* mean by it a critique, which has not prevented the title of the work being translated "Critique of Pure Reason," where there are two errors; for another is representing Kant as criticizing *reason*. That is downright ridiculous. It is not *reason* that he thinks is unreasonable, but only *the* reason, i.e. the human faculty in its effort to attain reason. It was a stupendous work, though naturally in great part erroneous. But the strange illustration it affords of the truth of what I was saying in my three first pages of this letter is that although he makes the great abstract ideas which he calls "categories" to be derived from the logical forms of assertions, which he calls in "Prolegomena zur Metaphysik" §21, and elsewhere "die verschiedenen Momente des Verstandes in Urtheilen," it is evident that, according to him, the correctness of his results depends entirely on his getting his list of these "moments "or "functions," or logical forms of assertion rightly constituted. Well, I am tolerably confident that I cannot be mistaken in saying that Kant never read a book of logic in his life. For it is well known that he read nothing of that sort in his later life; and when — I *think* it was when he was first appointed Privat-Docent, — in order (quite evidently to my mind) to attract students to lectures on an uninviting subject, he got out a pamphlet called "Ueber die falsche Fitzspindigkeit der vier syllogistischen Figuren," and it is truly comical to see how that — *skit*, I should call it, — has been lauded to the skies by philosophical Germans. Kant evidently fancied it was quite original. In fact, it is not only entirely given in Aristotle's Analytics, but it happens to relate to a branch of the subject that throughout the middle ages from the middle of the XIIIth century was dilated upon by every Doctor of the schools who touched upon logic; and the whole doctrine is detailed in the "Summulae Logicales" which was put into every boy's hands as soon as he was through with grammar. So that to suppose Kant was not a conscious "fakir," it is necessary to conclude that he was utterly ignorant of the best known parts of Aristotelian logic.

Now I was saying how essential it was to Kant's purpose to get the forms of judgments right. But he dismisses the subject in half a dozen very short pages, and in arguments so weak that it needs all that one knows of the extraordinary power of Kant's thought not to speak of [them] with unqualified contempt. He doesn't really come within sight of the real difficulties of the subject. I spent two years, absolutely solid, on nothing but the study of Kant, — chiefly the *C.d.r.V.* I read every medieval scholastic work that I could procure, after I had read everything of a logical or philosophical nature that has been preserved of the Greeks. These people seem to have been the only ones who ever constructed an original logical doctrine; and they constructed several, none of which, however, are thoroughly sound. I also read all I could get hold of of a logical nature of modern origin. But all of this, except what originated in mathematical thought, is so loosely thought as to tempt contempt. The mathematical genius George Boole, who, like the rest, only reasons accurately when he has mathematical machinery to hold him straight, invented a method of very limited applicability that has real merit. I will make a statement here of the essence of it freed from the dross. Boole would not have approved of the statement. Yet it brings out his one just logical idea. Suppose we take any two numbers and call one of them the *value* of any assertion that is true, and the other the *value* of any assertion that is not altogether true. Let v (for *verum*) be the former and f (for *falsum*) be the latter. These two numbers being different, of course, no assertion, x, can be equal to v and also equal to f. But it must have one or the other value; and this principle can be represented most simply by writing

$$(v-x)(x-f) = 0.$$

For the product of the two factors $v-x$ and $x-f$ being zero, one or other of them must be zero. That is either $v-x = 0$ and $x = v$, or else $x-f = 0$ and $x = f$, so that this equation may be adopted as the general expression of the principle that every assertion is either true or false. Here, then, is a machinery by which a certain kind of reasoning may be tested. For instance, if x and y represent two states of things, then $(y-f)(v-x) = 0$ will express that if y is true x must be true; or, in other words, if y is not false, x must be true. In like manner, $(z-f)(v-y) = 0$ will express that if z is true y must be so. Now from the two equations let us eliminate y. For that purpose, multiply the first by $(z-f)$ and the second by $(v-x)$. We so obtain

$$(z-f)(y-f)(v-x) = 0$$
$$(z-f)(v-y)(v-x) = 0$$

Adding these two equations, we get

$$(z-f)(v-f)(v-x) = 0.$$

But the middle factor, $v-f$, cannot be zero, since v and f are different numbers. Consequently, we have a right to divide by this middle factor; and doing so we get

$$(z-f)(v-x) = 0$$

which means, if z is true, x is true, the correct conclusion from the premisses.

Boole's book *Laws of Thought*, in which his system is developed and is applied to probabilities, is a work of genius, in which much is true and much false or confused to the point of meaninglessness.

One of my earliest works was an enlargement of Boole's idea so as to take into account ideas of relation, — or at least of all ideas of existential relation. By an existential relation I mean any relation, R, such that anything that is R to x (where x is some particular kind of object) is non-existent in case x is non-existent. Thus lovers of women of bright green complexions are non-existent in case there are no such women.

I invented several different systems of signs to deal with relations. One of them is called the general algebra of relations, and another the algebra of dyadic relations. *I was finally led to prefer what I call a diagrammatic syntax*. It is a way of setting down on paper any assertion, however intricate, and one so sets down any premisses, and then (guided by 3 simple rules) makes *erasures* and *insertions*, he will read before his eyes a necessary conclusion from those premisses. This syntax is so simple that I will describe it. Every word makes an assertion. Thus — man means "there is a man" in whatever universe the whole sheet refers to. The dash before "man" is the "line of identity" (man / eats—man; this means "Some man eats a man." To deny that there is any phoenix, we shade that assertion which we deny *as a whole* [Fig. 1]. Thus what I have just scribed means "It is false that there is a phoenix." But the following [Fig. 2] only means "there is something that is not identical with any

Fig. 1

Fig. 2

phoenix." Fig. 3 denies Fig. 4, which asserts that it thunders without lightening. For a denial shades the unshaded and unshades the shaded. Consequently Fig. 3 means "If it thunders, it lightens." In order to make

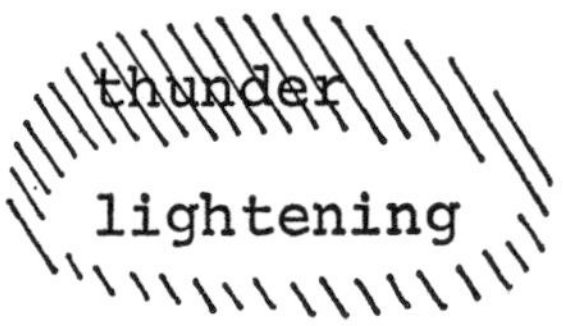

Fig. 3

Fig. 4

the lines of identity in their connection with shading and its absence perfectly perspicuous, I must provide you with a bit or two of nomenclature. By an "*area*," then, I mean the whole of any continuous part of the surface on which graphs are scribed that is alike in all parts of it either shaded or unshaded. By a "*graph*" I mean *the way* in which a given assertion is scribed. It is the general kind, not a single instance. For example there is in English but one single "word" that serves as definite article. It is the word "the." It will occur some twenty or more times on an average page; and when an editor asks for an article of so many thousand "words" he means to count each of those instances as a distinct word. He speaks loosely of *instances* of words as words, which they are not. Now in like manner a *graph* is one thing and a "*graph instance*" is another thing. Any expression of an assertion in this particular diagrammatic syntax is an *Existential Graph*, of which I use the single word "*graph*" as a commodious abbreviation as long as I have nothing to do with another kind of graph. A graph then may be complex or indivisible. Thus

male
– human
African

is a graph-instance composed of instances of

three indivisible graphs which assert "there is a male," "there is something human" and "there is an African." The syntactic junction or point of teridentity asserts the identity of something denoted by all three. Indivisible graphs usually carry "*pegs*" which are places on their periphery appropriated to denote, each of them, one of the subjects of the graph. A graph like "thunders" is called a "*medad*" as having no peg (though one might have made it mean "*some time* it thunders" when it would require a peg).

A graph or graph instance having 0 peg is a *Medad*
A graph or graph instance having 1 peg is a *Monad*
A graph or graph instance having 2 pegs is a *dyad*
A graph or graph instance having 3 pegs is a *triad.*

Every indivisible graph instance must be wholly contained in a single area. The line of identity can be regarded as a graph composed of any number of dyads "—is—." or as a single dyad. But it must be wholly in one area. Yet it may abut upon another line of identity in another area. Thus Fig. 5 *denies* that there is a man that will not die, that is, it asserts that Every man (if there be such an animal) will die. It contains two lines of identity. It denies Fig. 6 which asserts, "there is a man that is something that is something that is not anything that is anything unless it be something that will not die." I state the meaning in this way to show how the identity is continuous regardless of shading; and this is *necessarily* the case. It is the nature of identity. That is its entire meaning. For the shading denies the *whole* of what is in its area but not each part except disjunctively. Fig. 6 may be read, "There is a man that

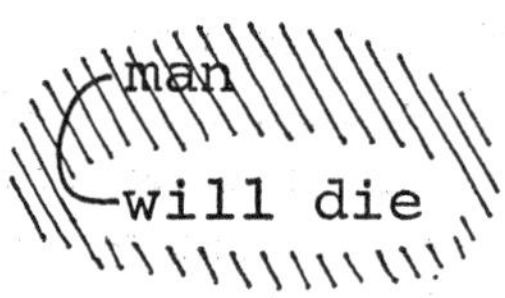

Fig. 5

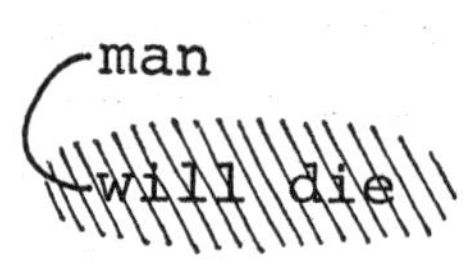

Fig. 6

is identical with something that is not identical with anything or only with something that is not identical with anything unless it will not die."

I dwell on these details which from our ordinary point of view appear unspeakably trifling, — not to say idiotic, — because they go to show that this syntax is truly *diagrammatic*, that is to say that its parts are really related to one another in forms of relation analogous to those of the assertions they represent, and that consequently in studying this syntax we may be assured that we are studying the real relations of the parts of the assertions and reasonings; which is by no means the case with the syntax of speech.

A line which is composed of two or more lines of identity abutting on one another is called a "ligature." Of course it is not a graph, of itself. Or it may be regarded as a graph meaning either "nothing is anything that is anything that is" (in case the shaded end is exterior to the unshaded end) or "something is identical with something that is not identical with anything but what" (in case the shaded end lies in an area enclosed by the unshaded area where the other end is). The rule of interpretation which necessarily follows from the diagrammatization is that the interpretation is "endoporeutic" (or proceeds inwardly). That is to say a ligature denotes "something" or "anything not" according as its *outermost part* lies on an unshaded or a shaded area respectively.

There are three simple rules for modifying premisses when they have once been scribed in order to get any sound necessary conclusion from them. Of course I do not count among these rules two recommendations which are nevertheless of the highest importance. One is to be sure to scribe every premiss that is really pertinent to the conclusion one aims at. The other is to scribe them with sufficient analysis of their meaning, and not by any means to neglect *abstractions* which modern philosophers think most foolishly are of little or no importance or are even *unreal* because they are of the nature of *signs*. They tell us that it is we who create the laws of nature! That is *Real* which is true just the same whether you or I or any collection of persons *opine* or otherwise *think* it true or not. The planets were always accelerated toward the sun for millions of years before any finite mind was in being to have any opinion on the subject. Therefore the law of gravitation is a *Reality*. I do not say that Newton's formulation of the law is quite right, because when Newcomb was at work on the inferior planets, Mercury and Venus, I wrote to him and called his attention to the fact that certain motions of Mercury go to show that the attraction is not precisely inversely as the 2nd power of the distance but is rather proportional to the -2.01 power or thereabouts; and I see that in his tables not only of Mercury but also of Venus he has introduced such a correction. He says he introduces it to make his tables accord with observation. He does not say that the cause of the discrepancy of observation [lies] with Newton's law. But that is the only way I can see of accounting for it. I had not supposed it would be perceptible in so circular an orbit as that of Venus. No doubt all our other formulations of laws are merely approximate; but the laws, as they really are, are Real.

I will now state what modifications are permissible in any graph we may have scribed.

1st Permission. Any graph-instance on an unshaded area may be erased; and on a shaded area that already exists, any graph-instance may be inserted. This includes the right to cut any line of identity on an unshaded area, and to prolong one or join two on a shaded area. (The shading itself must not be erased, of course; because it is not a graph-instance.)

2nd Permission. Any graph-instance may be *iterated* (i.e. duplicated) in the same area or in any area enclosed within that, provided the new lines of identity so introduced have identically the same connexions they had before the iteration. And if any graph-instance is already duplicated in the same area or in two areas one of which is included (whether immediately or not) within the other, their connexions being identical, then the inner of the instances (or either of them if they are in the same area) may be erased. This is called the Rule of Iteration and Deiteration.

3rd Permission. Any ring-shaped area which is entirely vacant may be suppressed by extending the areas within and without it so that they form one. And a vacant ring-shaped area may be created in any area by shading or by obliterating shading so as to separate two parts of any area by the new ring-shaped area.

It is evident that neither of these three principles will ever permit one to assert more than he has already asserted. I will give examples the consideration of which will suffice to convince you of this. Fig. 7 asserts

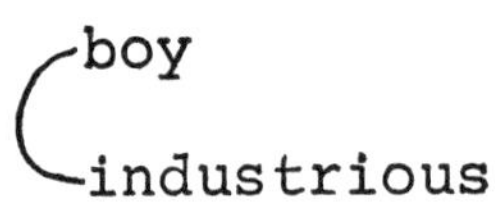

Fig. 7

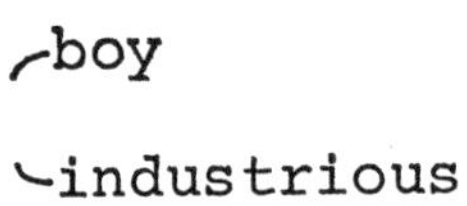

Fig. 8

Fig. 9

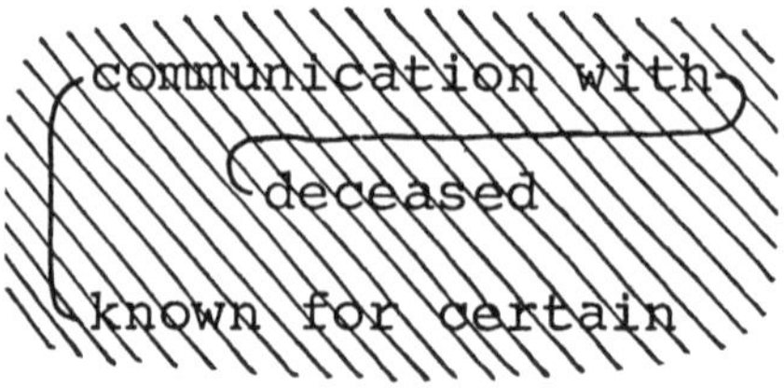

Fig. 10

that some boy is industrious. By the 1st permission it can be changed to Fig. 8, which asserts that there is a boy and that there is an industrious person. This was asserted in Fig. 7, together with the identity of some case. Fig. 9 asserts either there is nothing known for certain or else there is no communication with anybody. By the same permission this can be changed to Fig. 10 which asserts that no communication with anybody deceased is known for certain. But this is fully included in the state of things asserted in Fig. 9.

In illustrating the application of the Second Permission, I am obliged to notice one of the faults of the system of logic which has been taught to every generation of young men for some sixty odd generations. One of the syllogisms that they have all been taught is a sound apodictic argument called *Darapti*, and whose validity nobody has questioned furnishes a fair sample of the quality of intellect of the Doctors and Regents of the world's most famous and proudest universities. Here is a sample of it.

Any Phoenix would be a bird
Any Phoenix rises from her own ashes
∴ Some bird rises from its own ashes.

They might try to crawl out of this absurdity by saying that they do not state the premisses as

Any Phoenix there may be is a bird
Any Phoenix there may be rises from its ashes

but

Every Phoenix *there is* is a bird, etc.

But the reply to that (passing over the fact that Sir Wm. Hamilton, lauded as the highest of authorities, insists that *Any* and not Every is the right word) is that by "Contradictories" they mean two propositions which, by their very meaning, can neither both be true nor both false, and they all agree that every simple proposition has a simple contradictory, and that the contradictory of "Some *S* is not *P*" is "Any, all, or every (Greek πᾶς) *S* is *P*." Now if this latter implied the existence of some *S*, Every *S* is *P* and Some *S* is not *P* could both be false by there not existing any *S*. That would be a much graver fault with their logic than that which I charge against it. For I only charge that two "moods" or species of syllogism are false (i.e. not necessary, as they profess to be). And curiously Aristotle never mentions these with examples, as he does in all other cases; but merely says ——— But this letter will be long enough without discussing Aristotle and his Greek commentators, a subject on which I should soon tire you, interesting as it is to me.

So I will break off that and just give an illustration or two of how this Syntax of Existential Graphs works. But before doing that I wish to draw your attention, in the most emphatic way possible, to the purpose this Syntax is intended to subserve; since anybody who did not pay attention to that statement would be all but sure, not merely to mistake the intention of this syntax, but to think that intention as CONTRARY to what it really is as well he could. Namely he would suppose the object was to reach the conclusion from given premisses with the utmost facility and speed, while the real purpose is to dissect the reasoning into the greatest possible number of distinct steps and so to force attention to every requisite of the reasoning. The *supposed* purpose would be of little consequence, and it is the business of the mathematicians to furnish inventions to attain it; but the real purpose is to supply a real and crying need, although logicians are so stupid as not to recognize it and to put obstacles in the way of meeting it.

I will now, by way of an example of the way of working with this syntax, show how by successive steps of inference to pass from the premisses of a simple syllogism to its conclusion. Fig. 11 shows the two premisses "Any *M* is *P*" and "Any *S* is *M*."

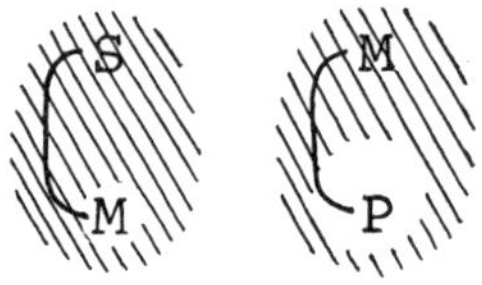

Fig. 11

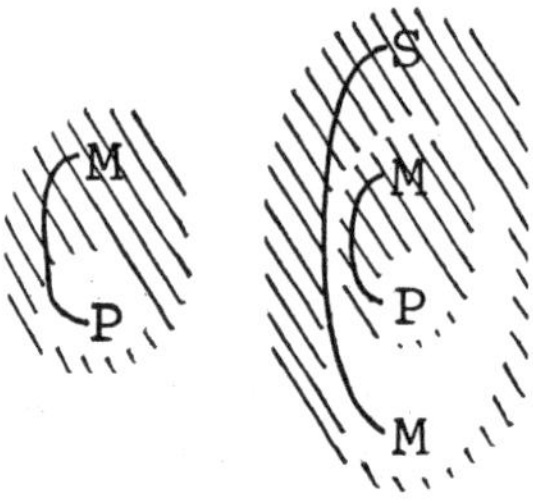

Fig. 12

The first step consists in passing to Fig. 12 by the 2nd Permission.

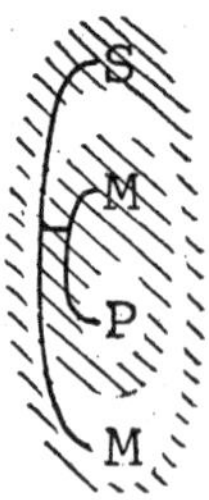

Fig. 13

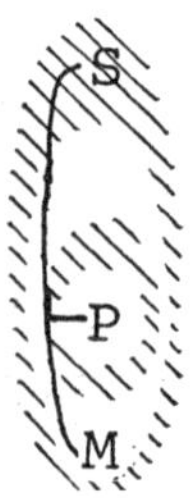

Fig. 14

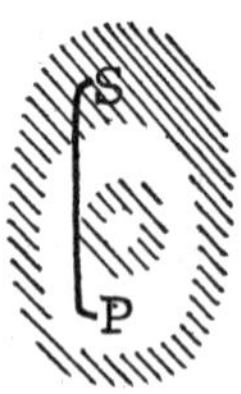

Fig. 15

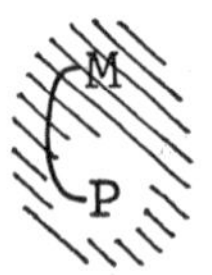

Fig. 16

The second step is simply to erase by the 1st Permission [as in Fig. 16]. The third step is to join the two ligatures by the 1st Permission as shown in Fig. 13. It will be observed that in iterating the major premiss, I had a right to put the new graph instance at any part of the area into which I put it; and I took good care to have the ligature of the minor premiss *touch* the shaded area of the iterated graph-instance. Now by the 1st Permission I have a right to insert what I please into a shaded area, and without making the new line of junction leave the shaded area, I make it touch the unshaded line of identity of the minor premiss. This gives me a right in the fourth step to deiterate *M* so as to get Fig. 14 by the second permission. The fifth step is to delete the *M* on an unshaded field giving Fig. 15 while the sixth step, authorized by permission the third, consists in getting rid of the empty ring-shaped shaded area round the *P*, giving Fig. 17.

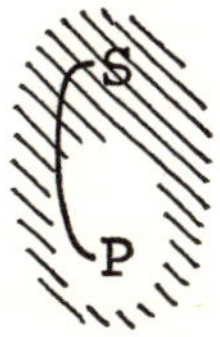

Fig. 17

By the space that I have occupied in explaining this syntax, you will surely think it is my chief work. On the contrary, it is one of the smallest; but it is the only one of which I could put you into a position to gain some understanding without writing a book about it.

You will ask perhaps "If all one has to do is to avail oneself of those 3 permissions, how is it that mathematics (which is nothing but deductive reasoning) is so difficult and demands high genius? There are several circumstances which go to clearing up this. The first of these is that the mathematician is not supplied in advance with a definite list of premisses; nor is he asked whether or not a definite conclusion can or cannot be drawn. His usual first approach to a problem is something like the following entirely fanciful situation which serves to illustrate what one of his difficulties is like. An astronomer comes to a mathematician and says, " I want to consult you about something." — But hold! I can perfectly well substitute a historical case about which I am fully informed. Toward the end of October 1604 the astronomer Tycho Brahe died and left a mass of observations including continual measurements of the apparent places of the planet Mars extending over 15 years. Keppler, who was a remarkable mathematician and who had had the advantage of training in observations under Tycho, had possession of the MSS and had continued the observations of Mars some years longer so as to make the series extend over 20 years; and it devolved upon him to take these measurements of the Latitude and Longitude of Mars (remarkably fine observations considering they were made with the naked eye) and by means of them to construct tables by which the Lat. and Long. of ♂ could be calculated for any future time. Of course it was assumed that Mars would continue to move as it had been moving and therefore one could calculate just what its Lat. and Long. were at any instant during those 20 years, except at those times of each year when it had been too close to the sun to be observed. I note that John Stuart Mill in his Logic says that Keppler only had to make a *general description* of facts known in the shape of observation. But Mill was a constant writer of reviews who had at the same time almost the responsibility of governing India on his shoulders and it would have been beyond human powers for a man every three months to turn out an article of high literary excellence in the Westminster Review, and conduct the business of India House and add to that any profound study of logic. He had evidently no conception of what Keppler had to do. He had before him the latitudes and longitudes. But since Tycho was a single observer at a field station he could make no observations of *parallax* that is of the *third* coördinate of ♂'s position, its *distance* from the observer. For observations of its position when on or near the horizon. Anyway the smallest angle visible, — the *minimum visibile* is about one minute of arc and the greatest parallax of ♂ is only about $\frac{1}{3}$ of that. Keppler it is true found an ingenious method of measuring the distance of Mars or any planet from the earth and from the sun. But it requires the theory of the motion of the planet to be complete, or nearly so, first. In short, Keppler's reasoning was not, and could not have been, purely mathematical. It was, on the contrary, the greatest piece of *Inductive* reasoning ever yet conducted. Had the parallaxes, or distances of Mars from the earth, been known to Keppler with the requisite degree of accuracy, it must have saved Keppler that marvellous piece of reasoning by which he ran down the truth like an indefatigable detective, with hardly a wasted day. It would have been a great loss to students of reasoning. The Conic Sections were not understood in 1610 as they are today, for Pascal was only born in 1623, or thereabout; and his theorem, that if six points are taken on a conic section and straight lines are drawn through the 1st and 2nd, the 2nd and 3rd and so round to a line through the 6th and 1st, then the intersection of the 1st and 4th of these lines, that of the 2nd and 5th, and that of the

3rd and 6th will all lie on one straight line, no matter how the original six points are chosen (whether they are consecutively passed in going round the curve or not). This proposition, which is the foundation, one may say, of the modern theory of Conics, would have been unknown to him. But Keppler was enough of a mathematician to be himself the first discoverer of the "foci" of such curves, and he would undoubtedly have made great advances in the subject, had he been in search of the law of the motion of Mars with knowledge which rendered it a purely "deductive" problem that is a matter of necessary reasoning. However I have after all been rather sidetracked by choosing this example; for I wished to show that mathematical problems *of a new kind* are generally first presented in a form which is simply a bewildering mass of facts into which the mathematician has to dive and fetch up first the *question* to be asked and then with great subtlety pick out the appropriate premisses. Then the second difficulty is that mathematical problems are apt to be so fearfully complicated (for example there are over 80 equations in the moon's longitude) that it requires a most capacious brain to embrace the question. Finally there comes a difficulty in many problems which has led me to divide mathematical reasonings into the *corollarial deductions* and the *theorematic deductions*. The terms "corollary" and "theorem" have no definite meanings and never did have. The original "*theorems*" of geometry were those propositions that Euclid proved, while the *corollaries* were simple deductions from the theorems inserted by Euclid's commentators and editors. They are said to have been marked by the figure of a little garland (or *corolla*) in the margin. But I use the adjectives which I form from the words "theorem" and "corollary" with exact meanings. The ultimate premisses of geometry are called by present day geometers "hypotheses," because the mathematicians, *as such*, do not accept any responsibility for their truth. They are of three kinds, *definitions*, *axioms*, and *postulates*. The axioms are, in *my* opinion, all false, if one insists on their rigid accuracy, in all cases. The "postulates" were originally understood to be premisses expressing that certain lines could be drawn, though everybody knew they could not, *exactly*. But in my opinion it is far better to consider them as statements that space contains certain kinds of *places*. For instance, the two old postulates that a straight line *can* be drawn from any point to any other and that a straight line can be "produced" (that is lengthened) at either end, I would supersede by the one postulate that, considering a line as a place, or "locus," as mathematicians have universally considered it since Descartes, "an unlimited straight line *is* through every pair of points, or places without parts." (Euclid's definition of a straight line is that it is a line that lies "evenly" between its extremities; by which I suppose he means, perhaps a little vaguely, that there are points from which such a line would appear as a point, or from the modern standpoint, it is a line whose shadow, if the source of light were a point on the line, would be a point. It is a question whether this is the better definition (as I decidedly think) or whether we ought to say that a straight line might be the path of a particle not acted on, during its motion, by any force. The definition that a straight line is the shortest distance between two points ought I think to be regarded as an *axiom* presumably only approximately truer.) Now one of the great difficulties of geometry is that no proposition of the kind I should call a "theorem" can be proved without introducing subsidiary lines of surfaces, that are not mentioned either in the proposition to be proved nor in the previously proved propositions. The right to assume these subsidiary *loci* is derived from the postulates.

I pass over crowds of points deeply interesting to anybody who cares to explore these fields, and come to another division of deductive reasonings, — that into what I call *necessary deductions* and *probable deductions*. All deductions are necessary reasonings in the sense that the conclusion must be true so long as the premisses are so. But I use the expression "*probable deduction*" as a convenient abbreviation of "deduction of a probability." Probable deductions include all the logically sound parts of the doctrine of chances, otherwise called, the calculus of probabilities. This includes so much of that doctrine as could safely be made the basis of the business of insurance. There is a lot more in the books, — particularly in LaPlace's book, which is the base of all nineteenth century works on the subject, — LaPlace being the idol of the French mathematicians, — there is I say a lot of it that is utter rot. He says a probability expresses in part knowledge and in part ignorance. This statement is a fair specimen of the loose thought of the book. LaPlace's *mathematics* is sometimes clumsy, but it is correct as long as his premisses mean anything. But when he attempts to define anything at all difficult he writes utter nonsense. In the sense in which he means it, that which expresses ignorance is utterly worthless and is no part of true science. If two possibilities, he says, are "*également possibles*" their probabilities are equal, and if two events that are mutually exclusive have equal probabilities, the probability is double that of either, that one or other will occur. "What is the probability that the inhabitants of Saturn have red hair?" asked Mill in the first edition of his logic. That it is *red*, that it *is not* red are "également possibles" since we are absolutely ignorant about it, is true for *possibility* that a thing *may be* admits of no more or less. If it is possible, that is, if we do not know that it isn't so, which is certainly the case if we are utterly ignorant, then the two are "equally possible" in the only sense the phrase can have, that we don't know anything against the truth of either, but how could an insurance company fare who should try to do business on such a basis? A basis for business has got to be knowledge and not ignorance. As a specimen of LaPlace's results I will mention something he deduces from his principle of the "également possibles" and which is copied into all the textbooks of the subject, — all the usual ones, — to this day. Namely LaPlace says that if a man on occasions entirely new to him sees a phenomenon equally new on every one of those occasions up to N occasions (N being any whole number) then the probability is $\frac{N+1}{N+2}$ that the same phenomenon will occur on the next such occasion. I say this is nonsense, that it is trying to conclude by mathematical reasoning that which requires a radically different kind of reasoning. And what proves that it is nonsense is that if $N = 0$ the probability is $\frac{1}{2}$. That is to say that on a wholly new occasion it would be a reasonable thing to make an even bet that an unheard of event would take place. That is the nonsense that results from trying to reason mathematically on matters of fact on the basis of pure ignorance. LaPlace was renowned for lack of sound good sense, and his doctrine about these inverse probabilities shows it. A probability, if it is correct, is a basis for business. But there can be no such basis except experience and the idea of deducing any quarter of fact from anything but knowledge is absurd.

Now you will ask me "How do you define probability?" I will define it in a concrete example. Suppose I say "I have a die and owing to its being somewhat ill made, instead of the probability of its turning up six at any one throw being $\frac{1}{6}$, or $0.16\frac{2}{3}$, as it should be, the probability of that event is only 0.16." Now you ask what I mean by that. I mean that (the result of any one throw not having any effect or consequence as to the result of any other throw) the throws in which six is turned up will be 0.16 of all the throws "*in the long run*." If you ask me what I mean by the "long run," — (you will see that in defining what I mean by probability, I must *not* introduce that same concept) — I reply that "I mean an endless succession of throws in the order in which they are thrown." Thereupon you will ask what I mean "by 0.16 of an infinite number of throws. It is certainly itself infinite and equal to the whole. (For there are certainly as many even numbers as there are of odd and even numbers taken together, since every number has a double, which is the double of no other number

1 2 3 4 5 6 7 8 9 10 11 12 etc.

doubled give

2 4 6 8 10 12 14 16 18 20 22 24 etc.

and thus there is no distinct even number for each and every whole number.) That is there is the same multitude of even numbers as there is of odd and even numbers together; and on the same principle 0.16 of an endless series is equal to the whole endless series." This I must admit; and therefore I proceed to define what I mean by 0.16 of an endless series of throws. But first let me observe that beside "probability" there are other conceptions which express the same facts precisely. It depends upon the kind of relation which the facts one knows bear to the facts one wants to know which of the different conceptions is most convenient in a given case. The *probability* that if an antecedent condition is satisfied, a consequent kind of event will take place is the quotient of the number of occasions, "in the long run," in which both the antecedent will be satisfied and the consequent kind of event will take place, divided by the total number of occasions on which the antecedent conditions will be satisfied. The *odds* in favor of a kind of event occurring on any occasion on which an antecedent condition is fulfilled, is the quotient of the number of those occasions on which the condition is fulfilled and the kind of event in question occurs divided by the number of occasions on which the condition is fulfilled and the kind of event in question does not occur. The *weight of reasons* for believing that if the condition is fulfilled a given kind of result will occur is the remainder after subtracting from the logarithm of the number of occasions on which the reasons apply and the given kind of result occurs, the number of occasions (in the long run, of course) to which the reasons do apply but the result in question will not occur.

I now proceed to define what I mean by the number of occasions in the long run etc. I mean that if the die were thrown over and over again *without cessation* and if one person, *A*, were to keep tally of the throws which brought the ⚅ side of the die uppermost, and a second person, *B*, were to keep tally of the throws which did *not* turn up the ⚅ but some other side, and a third person *C* were after each throw to divide the number of throws *A*'s tally showed by the number of throws *B*'s tally showed while a fourth person *D* were to divide the number *A*'s tally showed by the sum of the numbers that *A*'s and *B*'s tallies showed, and a fifth person, E set down the logarithm of the result of C's division, [then] after the first throw one of the two tallies will show 0 and the other 1; so that C's result will either be 0 or ∞ (infinity), D's either 0 or 1, and E's either $+\infty$ or $-\infty$ (the logarithm of 0 being $-\infty$). After the first throw that one of the two tallies that shows *zero* may continue to do so for an indefinite number of throws. But as soon as both tallies have left the zero point, *every throw must change* the result of C, that of D, and that of E. For suppose C's had been $\frac{y}{n}$. Then after the next throw it will either be $\frac{y+1}{n}$ or $\frac{y}{n+1}$ and $\frac{y+1}{n} > \frac{y}{n}$ while $\frac{y}{n+1} < \frac{y}{n}$ unless $n = 0$. So D's will change from $\frac{y}{y+n}$ either to $\frac{y+1}{y+n+1}$ or to $\frac{y}{y+n+1}$; that is its *reciprocal* will change from $1 + \frac{n}{y}$ either to $1 + \frac{n}{y+1}$ or to $1 + \frac{n+1}{y}$. Now as long as neither n nor y is zero $\frac{n}{y}$ can neither be equal to $\frac{n}{y+1}$ nor to $\frac{n+1}{y}$.

Accordingly as soon as the results of both A and B have left the zero, each of the results of C, D, and E must change with every throw. Now when we say that some definite fraction will result or come to pass "in the long run," we mean in the first place that it will be so, whether we begin with the first occasion or the millionth or any other. And therefore, since we use this expression in regard to the results equally of C and of D (we may omit those of E since this is merely the logarithm of C), we mean that for one thing, that those values, the results of C and of D, oscillate, however irregularly, from one side to the other, C, of $\frac{4}{21}$, or $0.\dot{1}9047\dot{6}$, and D, of $\frac{16}{100} = \frac{4}{25}$ (For since $\frac{x}{x+y} = \frac{4}{25}$, $1 + \frac{y}{x} = \frac{25}{4}$ and $\frac{y}{x} = \frac{21}{4}$ and $\frac{x}{y} = \frac{4}{21} = 0.\dot{1}9047\dot{6}$

$$1 = 0.\dot{9}9999\dot{9}$$
$$\tfrac{1}{3} = 0.\dot{3}3333\dot{3}$$
$$\tfrac{1}{7}\,\tfrac{1}{3} = 0.\dot{0}4761\dot{9}$$
$$4 \times \tfrac{1}{7} \times \tfrac{1}{3} = 0.\dot{1}9047\dot{6})$$

One thing we mean then is that D's quotient will never cease to oscillate about the value 0.16 and C's about the value $\frac{4}{21}$. Moreover, we mean that 0.16 is *the only* value of D's quotient that it will not sooner or later become larger than or smaller than *for the last* time, although it is impossible to know *when* that last time will be. This *sounds* self-contradictory; but it is not so; and when one has attained full command of the concept of endlessly diminishing oscillations and of the endless generally, it ceases even to *sound* self-contradictory in his ears. This definition of the "long run" (which can, of course, be made easier to use by being expressed in Existential Graphs) removes absolutely all the errors and paradoxes of the calculus of probabilities, and leaves it all its utility without its pitfalls. It thus restores all its great utility, so patent in the great business of insurance, and at the same time brings out all the high moral lessons of the uncertainty of anything, or rather the *fate* of everything that is not anchored to eternal verity and to universality. Every insurance company as well as every other particular human creation is bound to ultimate disaster. But of course in this letter I can only set forth fragments. In my effort to avoid extreme prolixity, I see that I have omitted considerations that must not be omitted in the briefest statement. Therefore here please insert the next sheet *A*.[1]

[1] There is a "next sheet A" in the manuscript today which reads:

A

MORE ABOUT PROBABILITY

"The illustration of probability by means of throws of dice conceals one of the most essential points on almost all kinds of reasoning about probability, conceals them, I may go so far as to say, more effectually than almost any other sort of illustration does. For instance drawing cards from a pack or a package from a 'grab-bag,' though these sin in the same way, do not near so much so as the dice-throwing.

For example, all whist-players have remarked the frequency of short suits after a misdeal. The fact is often doubted because no reason for it is apparent. Well, I will not waste time in calculating the probability of such a thing. But suppose the first two cards of a suit to be dealt to one person, as the first two cards dealt to him, what is the probability that no one of the other eleven will be dealt to him? It is

$$\frac{39 \times 38 \times 37 \times 36 \times 35 \times 34 \times 33 \times 32 \times 31 \times 30 \times 29}{50 \times 49 \times 48 \times 47 \times 46 \times 45 \times 44 \times 43 \times 42 \times 41 \times 40}$$

which I judge to be about $\frac{1}{22}$. Now when one thinks that there are 77 other ways in which the same two cards might be dealt to him and 77 other couples of cards in the same suit, one sees that it must be a good deal more likely than not that some one of the 4 players *will* have a short suit, IF THE CARDS WERE THOROUGHLY SHUFFLED. But owing to their being picked up in tricks and very little shuffled, *unless there is a misdeal*, which shuffles that pack tolerably, the cards seldom get any shuffling to speak of. The calculus of probabilities is a very precious aid to anybody who is trained in using it; but it cannot be made to answer half the questions men try to make it answer, and the amount of fallacious reasoning about probabilities is ten times as great here as it is in any other field of discussion.

In order to practice or understand Probable Deduction, it is necessary to study Induction, by which alone the fact that there is about a stated value to the prob-

Deduction, or necessary reasoning, is only one, and certainly not the highest one, of three absolutely disparate ways of reasoning. I believe I was the first to prove this, perhaps the first even to assert it. I am unable yet quite to *prove* that the three kinds of reasoning I mean are the *only* kinds of sound reasoning; though I can show reason to think that it can be proved, and *very strong* probable reasons for thinking that

---

ability that an object taken at random from a given class or collection will have a specified character, can be inferred from experiments. The first requisite is to understand just what is meant by taking an object 'at random.' If instead of a 'class,' which consists of *would-bes* and not of actually existing individuals (I mean, for example, that if we say '*any* man is mortal' we are not speaking *merely* of existing men, but of any man there may be) if, I say, we are speaking of a finite collection, then we may say that an individual instance is taken 'at random' if it be chosen by a method which, used successively, would 'in the long run' result in any one individual of the collection being taken as often as any other. Mechanical devices such as pitching coins, throwing dice, spinning teetotems etc. to govern the selection are useful. When we are unable to devise such a method, it is necessary to make a strong effort of will not to be influenced in one's selection by any motive which in the long run would lead to one choosing objects more from one part of the class than from another, unless one has good reason to think that the part more favored is on the whole just like the part less favored in respect to the character to ascertain the probability of which is the purpose of forming the sample. A man should not undertake to do any sampling in order to base any opinion on it until he has undergone such a course of discipline of his will that he can, under a full sense of the responsibility involved in urging any opinion, or in embracing one, be reasonably sure that he not only does not lean one way or the other in making his sample, but further that he has taken all the precautions against self-deceit that a deep sense of his responsibility would make him regard as a duty.

Of calculations, whether deductive nor inductive, relating to mathematically expressed probability, excluding those concerned with insurance and those relating to games, and excluding (for another reason) applications of the method of least squares (which is but fair since a large minority *at least* of the men who make these do not regard the conclusions as meaning what they say, using such calculations in order to escape more serious dangers) — with these exceptions the proportion out of those one meets with of such as are seriously fallacious is very large; and these fallacies are largely, if not mostly, due to unskillful or unfair sampling. They might be avoided without much additional labor by making at the same time control calculations of probabilities whose true values could otherwise be ascertained. Such false calculations deceive those who have not mastered this branch of logic, and for the most part deceive nobody else. The worst I have seen of late years related to 'Bacon cryptograms in Shakespeare' etc. One deception common in such pretended science consists (to express its nature by an example) in making a very complicated kind of cipher and showing that it gives sense in one or two cases, and then calculating the probability that it would do so by chance just as if that rule were the one that Bacon would necessarily adopt, and as if it would make sense of *any* passage in Shakespeare, instead of being a very complicated one among many tried and rejected and fitting only one or two passages in Shakespeare's work. This is like calculating the probable error from an equation involving many unknowns as if there were but one."

there is no fourth kind. I call the three, Deduction, Induction, and Retroduction; though the last only is a word invented by me.

A scientific inquiry must usually, if not always, begin with retroduction. An Induction can hardly be sound or at least is to be suspected usually, unless it has been preceded by a Retroductive reasoning to the same general effect. Induction chiefly serves to render more certain ideas that have already been otherwise suggested. I use "Induction" in a wider sense than usual. It is usually regarded as a reasoning by which one passes from asserting something of a number of single things to asserting the same of the whole class to which those things belong. I give the definition a somewhat different turn, at least, and throw much light upon Induction by defining it as any reasoning from a *sample* to the whole sampled. For example, if I sample a cargo of coffee, and find throughout different samples that about the same proportion of the coffee-beans are "male" beans, and thence conclude that the same is true of the unexamined parts of the cargo, — while it may be said, in one sense, that this is a reasoning from what is true of parts to what is true of the whole, — yet it sheds a flood of light upon the rationale of the inference to say that it is an inference not so much to the whole as to a certain fraction of the whole. The graph can equally be read as "If A is true, C is true," and "If C is false, A is false" [Fig. 18].

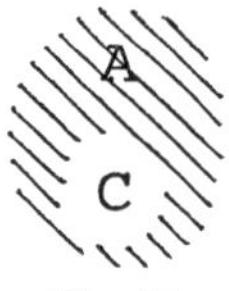

Fig. 18

In most cases, accordingly, to reason from having found that A is true to the conclusion that C is true is just as sound as, and no sounder than, if one had not found anything except that C is false, and thence reasoned that A is false. In probabilities, however, we are surprised to find that these two inferences may be of a radically different kind. Suppose, for instance (if so wild a fancy may be permitted) that in digging a well (that in fact I mean to dig when I put up some cottages) I should come across a deposit of a reddish brown color, opaque, not to be scratched with my nail but with not much difficulty with a knife, almost as heavy as so much copper, almost dull looking but almost metallic, capable of being cut without breaking every time yet apt to do so; with some tendency to uneven cleavage, and finally when scratched showing a bright red streak. I should say, "This is pure cinnabar and more than $\frac{6}{7}$ of it is Mercury. If I were to weigh out 7 ounces of it and heat it in a retort with the neck under water I should distill over $6\frac{1}{29}$ ounces of Mercury." I take the liberty of imagining you to say thereupon, "I wish you would try the experiment," whereupon I should reply, "I have neither the retort nor the heater nor the balance handy. Besides, what's the use? You see, here is a list of all the atomic weights. Mercury is set down as 200.0 and sulphur as 32.06, and this is therefore $\frac{200.0}{232.06}$ Mercury or $\frac{1}{1.1603}$, which is 86.18461% and $\frac{6}{7}$ is 85.714. This is cinnabar, you must admit; and the atomic weights are as accurate as the figures are given." Thereupon you say, "Yes, I admit all that." I say "Well the conclusion is a necessary one, is it not?" "Yes, you say; but still I should like better a direct experiment." (I am attributing to myself faults quite contrary to my real tendencies.) The direct experiment was tried and gave a little *more* mercury than was expected. However, metallic mercury is frequently found in minute quantities in deposits of cinnabar; and we decided to make a careful determination of the percentage of mercury in the mineral, in the wet way. Three such were made (of course, it is all pure fiction designed to show the kind of reasoning under typical circumstances). They were in excellent agreement and gave, in the mean, for the percentage of mercury in the mineral 86.90. In order to ascertain whether the excess of this result over the theoretical 86.185 was due to the presence of metallic mercury (as I felt almost sure was the case), or to mercurous sulphide, or to mercuric or mercurous fluoride or oxide (the red oxide), we took four specimens of vermillion, one Chinese, one prepared in the dry way, and two prepared in different wet ways, and having brought the content of mercury in separate portions of each up to that of our mineral in the six ways mentioned, we compared the effect upon each of various solvents with the corresponding effect upon our mineral, and so reached the final conclusion that the latter was composed of $\frac{6}{7}$ by weight of cinnabar and $\frac{1}{7}$ of the red oxide, which is called "montroydite" by the mineralogists.

The first step toward a comprehension of Logical Critics must be the recognition of the gulf which separates Deduction, — exemplified on the one hand in the reasoning from the recognition of a character of a chemical species to the recognition that that character must belong to any specimen known to belong to that species, and on the other hand the reasoning from the fact that the character belongs to the specimen either to the conclusion that the same character belongs to the whole species or to the conclusion that the specimen belongs to the species. The last reasoning is the kind of reasoning that in the fiction we have been imagining we were represented as performing. The first of these three kinds of reasoning would be a deductive or necessary reasoning. If instead of *knowing* that the specimen belongs to the species, we only knew that it had been taken "at random" from a set of three specimens two of which were known to belong to the species while the third was known *not* to have the character in question, we could conclude that the probability was $\frac{2}{3}$ or the odds 2:1 that the specimen drawn had the character in question, or if we knew nothing about the third specimen, we could conclude that the probability was *at least* $\frac{2}{3}$ that the specimen drawn had the character in question. We should be unable to say, more exactly, what that probability was. Each of these three reasonings as to the character of the specimen would be a "deduction," and as such, a necessary reasoning, although, in the last two cases it would be what I call a "probable deduction," that is, a necessary reasoning about a probability. If you will pardon the repetition, by a "necessary reasoning" is meant (by all correct writers) one whose conclusion everybody is *forced* to admit (supposing him candid), if he fully admits the truth of the "copulate premiss" (i.e. the premisses) and fully comprehends the reasoning. But he would not be absolutely *forced* to admit the conclusion, if he could deny it without falling into a contradiction "*in adjecto*," that is, in denying it, though he might not *directly* deny anything stated *explicitly* in the copulate premiss (to do which would be a contradiction "*in terms*"), yet he would deny the reality of a state of things the reality of which had been affirmed, though perhaps not in one sentence, yet in the copulate premiss taken in its entirety. If you still ask for a further explanation of what I mean by saying that the reality of a given state of things has been affirmed in a given copulate premiss, I do not just now see my way to an exacter definition of my meaning than by saying that if that premiss be expressed in the syntax of existential graphs then the three "permissions" attached to my account of that syntax will enable one to modify the total graph that has been scribed so that it shall assert any fact, or state of things whose reality was in any way asserted in the total graph before those modifications. The first permission amounts to saying that if one has asserted that some fact *A* and some fact *B* are true, one may modify that assertion by saying that *A* is true without saying anything about *B*. Thus if one has asserted that James loves Eliza, one may [make the] assertion by saying that James loves something and that somebody is Eliza, without saying who is Eliza or whom James loves. On the other hand, if one has denied that any assertion *A* is true, that is, has denied that it implies nothing false, one may deny that *A* and *B* are both true. And so if one has asserted that if *A* is true then *B* is true and *C* is not so, one may modify that so as to say that if *A* and *Z* are both true then *C* and *Y* are not both true. The second permission amounts to saying that if one has asserted that *A* is true and that further if *A* and *B* are both true, *C* is true, then one may modify this so as to assert that if *A* is true and that if *B* is true *C* is so, or so as to assert that if *B* is true *C* and *A* are both so. And if one has asserted that *X* is true and that if *Y* is true, then so is *Z*, one may modify it so as to say that if *Y* is true *X* and *Z* are both true, etc. The third permission amounts to declaring that to say that if anything whatever is true, *P* is true is precisely as much as to say that *P* is true, and *vice versa.*

Probable reasoning involves numbers, and I have not stated how I would deal with the logic of numbers, for the reason that that field of thought is so distasteful to many persons, and for the further reason that during the time that I have been studying logic, mathematicians have corrected most of their fallacious reasonings except in certain special departments, such as in reasoning about continuity and sometimes about infinite multitudes and infinite ordinal numbers, upon which special fields I have demonstrated some current errors. But although I have passed over the logic of number for fear it might be too distasteful to you, yet I will say that my way of dealing with it compares favorably with any part of my system that I have described whether in the rigour of its proofs, its elegance, or its surprising simplicity; and I have the pleasure of confidently counting upon what I have contributed to mathematical journals helping in the near future to enable mathematicians to correct the more fundamental of their still remaining causes of error. If you have any curiosity to know about this part of my work, I should take pleasure in gratifying it. I ought to acknowledge that some fundamental points in the logic of probabilities might never have been so completely cleared up in my mind, if I had not had the advantage of standing on [the] shoulders of George Boole. For although Boole was not himself a very strict logician, yet he did see that LaPlace's reasoning on parts of the doctrine of chances was all wrong, and by a stroke of genius was able to put his objections to LaPlace's procedure into such a form that the overwhelming truth of them must appear plain to everybody who followed the somewhat difficult, but perfectly evident, footsteps of his thought. This was the more admirable in that Boole did not himself reach a perfectly clear notion of probability, being somewhat in the

Laplacian fogs. But without the suggestions which I derived from his reasonings I doubt whether I ever should have been able to attain the crystal clearness of conception in regard to probabilities that I have attained.

Mathematicians may and do employ all kinds of reasonings in getting their *suggestions*; but "mathematical reasoning," that is to say, all the reasoning with which mathematicians appeal to their readers for their assent, is *deductive.*

Induction is that kind of reasoning which from what is true of a part, concludes what is true of the whole. It is evident, therefore, that it is not deductive; and as soon as it is shown that a supposedly inductive conclusion can be proved, either certainly or probably, according to my definition of probability, from a copulative premiss, so soon is it proved that the reasoning was not (or need not have been) Inductive.

Yet this is a position in which I [am] almost in a minority of one, though I am perfectly confident that the world must and will come round to it.

As far as *necessary* deduction goes, I think most people would assent to what I say, as long as the question was put baldly, in spite of its really conflicting, though they don't see that it does, with other opinions of theirs that they will continue to hold for a good while yet.

But when it comes to *probable* deductions, though these are really as completely necessary reasonings as are "necessary deductions," all the treatises on the doctrine of chances that I have ever come across are flatly against me, as well as are the great majority of those who have undertaken to explain why an Induction should be as good a reasoning as it has often seemed to be, and as they think (and I with them) that it often really is. For most of the attempts to explain Induction really are (though often not perceived by their sectators to be) attempts to reduce it to probable induction; and the astonishing feebleness of these attempts, considering the general intelligence of their authors, renders the fact all the more significant.

I will pass in as rapid a review as I possibly can all those attempts to explain the validity of Induction that are really worth examination. Of course they are not numerous. It has only been at exceptional moments that it has been possible for a man here and there, say one in a lac (100000) or so, to have this attention at all focussed on Logical Critics. It has been a sufficiently rare historical node that permitted a proportion of grown men to engage, with any earnestness of sincerity, in any inquiry *whatever.* (I admit that children have a much greater spirit of science; and here and there a child in growing into manhood would preserve his interests in some particular line of thought. I am rather inclined to surmise that this accounts perhaps for the majority of cases of scientific genius. In the immense majority of individuals, puberty occasions a supervention of stupidity, along with its other effects, good and bad.) Of the few who do become inquirers, what a minute proportion are led to interest themselves in the question of precisely what it is, in general, that justifies an inference! And I can testify from perhaps the widest reading of anybody of my time on this subject, that the literature is scant, and very little of it goes to the point.

The Greek word for induction, of which our word is simply the latinization, — ἐπαγωγή — was apparently introduced by Socrates. It expresses the reasoning by the metaphor of a body of soldiers led up to make an attack on a position. Socrates was the first to employ induction systematically. He offered no theory on the subject, but simply relied on common sense.

Aristotle shows that an induction is simply a syllogism; only that which is the conclusion of the syllogism becomes in induction one of the premisses, while the major premiss of the syllogism becomes the conclusion of the induction. Further, this syllogism of which induction is a transposition has for its minor term, i.e. the subject of the syllogistic conclusion, an enumeration of instances.

Syllogism	*Induction*
Nothing long lived has bile	Man, the horse, the mule are long lived
Man, the horse, the mule are long lived	Man, the horse, the mule have no bile
Ergo, man, the horse, the mule have no bile	*Ergo*, nothing long lived has bile.

This doctrine, as far as it goes, is quite correct. But to it Aristotle adds two assertions that are mistaken and which I will state presently, as soon as I have made some improvements upon this correct but not sufficiently definite statement. Namely, it is not from a simple syllogism such as Aristotle instances that induction results by turning its conclusion into a premiss and its major premiss into a conclusion. But the syllogism from which induction does result from such transpositions of its premisses is what I call a "Statistical Syllogism." Even so, it is not *every* kind of induction which so results, but only one of three classes of inductions; — the strongest of the three, however. I call it "Quantitative Induction."

There is a second kind of induction, which results from interchanging the conclusion of a syllogism (a peculiar kind of syllogism different from that from which induction results) with the *minor*, instead of the major, premiss; and this I now call Qualitative Induction. Finally, we are sometimes forced to resort to a third, the weakest of all kinds of Inductions, which I call "Crude Induction" which seems to me to have no relation to a syllogism. The syllogism from which Quantitative Induction results is like this in most cases:

Premiss: Among all individual objects of a given class (call it the *M*s) that are likely to have any interest for me, a percentage, say 10% will have a certain special interest, consisting, we will say in each of them having the character μ, and these will be distributed among all occurrences of *M*s in an entirely irregular way, so that whether or not those of the character μ have been occurring at any time more frequently or less frequently than usual is no reason for thinking either that the next ones to occur will be more likely to have the character μ, or not to have it, than if recent occurrences has been of any other character.

Conclusion: It follows that the next occurrences of *M*s up to any number *specified in advance* (so that we cannot by examining them as they occur and allowing the occurrences to run on until they take a particular average character and then stopping the experiment, exercize any measure of control over this average character) will probably and approximately (i.e. with more probability the less close the specified approximation, according to a certain law made known in the doctrine of chances) *resemble the whole class* in having *P* % of its *M*s of the character μ.

In this statement, owing to my habit of thinking in the syntax of existential graphs, I have lumped the major and minor premisses into one copulate premiss. In order to study Aristotle's account of induction (while taking induction in the general sense of reasoning from the character of a sample or random part of the class to that of the whole class sampled) I must distinguish the major and minor premisses. The major premiss is so much of the copulate premiss as affirms or denies the character of the class concerning a part of which the conclusion partly affirms and partly denies the same characters. The minor premiss is that which affirms the character that defines the general class (i.e. the character of consisting of *M*s) of that part of this class to which the conclusion asserts something (namely that probably and approximately it will agree with the whole class in that respect which the major premiss partly affirms or partly denies of the class).

Now any of Aristotle's syllogisms remains and necessary reasoning when the conclusion and either premiss are interchanged, *provided that at the same* [time] *the interchanged propositions be transmuted into their precise denials.* This is of course true since the very same graph of Fig. 19 that expresses that if both premisses be true the conclusion is true, equally expresses that if one premiss be true while the conclusion is false, then the other premiss is false. Only it might be rather more natural to scribe this consequence (not consequentem, but consequentiam) in the shape of Fig. 20, though there is really no difference of inter-

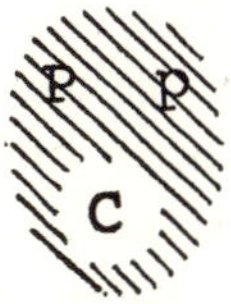

Fig. 19

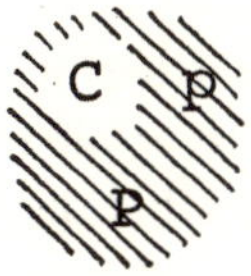

Fig. 20

pretation between the two. (*P* and *p* are the premisses, *C* the conclusion.)

Instead of making such an indeterminate and algebraic supposition, I shall make myself more readily understood if I make the following supposition. Let there be an urn with three balls in it, one Red, one Green, and one Violet. Let a person put his hand in the urn and draw out a ball, and whatever its color may be let it be immediately replaced by another ball of the same color, so that the urn shall still contain 3 balls, one Red, one Green, and one Violet. Let the same person who drew the first ball now draw another and let a new one of the same color be put in the urn to replace it so that once more there shall be one ball in the urn of each of the three colors. Let him go on in this way until he has drawn 6 balls. Then there are $3^6 = 729$ possibilities, and I will give at once the values of various probabilities, each multiplied by 729.

That all the balls drawn will be of 1 color...	3
That all the balls drawn will be of 2 colors	186
That all the balls drawn will be of 3 colors	540
That all the balls drawn will be of one or other or both of 2 given colors	64
That 5 will be alike in color and the 6th different	36
That 4 will be alike and the other two alike	90
That 4 will be alike and the other two unlike	90

That 3 will be alike and the other three alike	60
That 3 will be alike and the other three not	60
That there will be 2 of each color	90
That 6 will be of a given color	1
That 5 will be of a given color	12
That 4 will be of a given color	60
That 3 will be of a given color	160
That 2 will be of a given color	240 [;] $\frac{8 \cdot 9}{27}$
That 1 will be of a given color	192
That 0 will be of a given color	64

If he had only drawn 3 balls the probability that just 1 would be of a color named in advance would have been $\frac{12}{27}$. This illustrates the fact that the departure of a sample from the *ratios* of the whole class is for large numbers inversely proportional to the square root of the size of the sample so that the probable departure in the actual number of occurrences between theory and fact is *directly* proportional to the square root of the size of the sample. (This day is the hottest I have experienced since I came to live here nearly thirty years ago; and I hardly know what I am writing.)

There are writers of such inaptitude for the subjects on which they write, that they think that because it is probable that a large sample will be distributed into varieties nearly proportional to those of the corresponding varieties of the class of things sampled, that therefore the finding that the sample is divided into varieties in certain proportions renders it probable that the class sampled has such varieties in approximately the same proportions.

But, of course, it is only in fictitious cases that we can ever know the *exact* value of a probability, and we rarely know one nearer than to the nearest *per cent*. One of my colleagues of the U.S. National Academy of Sciences says that by throwing a stick on the floor and keeping account of the number of cases in which it fell across the line between two floor boards, he has calculated the value of π accurately to four figures! When we consider how many times it would so fall that one might count the case as one of the stick's falling across a crack or otherwise according as one was inclined in one's heart, I think it may be questionable whether what he has really added to our knowledge be not something about his scientific probity. For to calculate a *real* probability to one ten-thousandth of itself would be a matter of such difficulty that I do not think a serious man would undertake without some serious purpose, such as is entirely wanting for that way of calculating the value of π. The *theory* of the method is all right, but the whole difficulty is that without a high power microscope there would be too many cases of doubt; and it might be a long job to resolve some of them.

I have a great mind before leaving the subject of probable induction to run over Laplace's book and pick out the cases in which he pretends to have solved problems *by mathematical reasoning* when it is plain that the state of things concluded is in no form asserted in the copulate premiss, so that it cannot be reached by such reasoning. It would be too much of a job for me as well as an intolerable bore for you if I were to go through the book and extract all such cases. But I will take a few. (All the old editions of the book are made up from the sheets printed for the 1st edition (with some additions) so that if I give the page, the problem referred to can always be found.)

The first I notice is on p. 207. "Une urne étant supposée renfermer le nombre x de boules, on en tire une partie ou la totalité, et l'on demande la probabilité que le nombre des boules extraites sera pair." I should not venture to quote that absurd question without giving the page number so that you can verify it, it is so contrary to common sense. He says that the probability that the number of balls taken out will be even is $\frac{2^{x-1}-1}{2^x-1}$ and this conclusion he obtains by supposing that each possible consitution of a draw is equal to every other. For instance, if there were five balls, *A B C D E*, there are five equally likely ways of taking out one, which I admit; and that there are ten equally likely ways of taking out two, namely, *AB AC AD AE BC BD BE CD CE DE* which I equally admit. But he further assumes that each of these ten ways of taking out two is equally probable with each of the five ways of taking out one, which I deny to be implied in any way in the supposition. He concludes that if anybody makes it a rule to bet that an odd number of balls have been taken he will be morally certain to win in the long run! I don't believe a word of it, as a matter of fact. It is not, however, a question of *fact* but the question is whether the conclusion follows *mathematically* from the premiss. I say it obviously does not.

Chapitre VI of his Livre Deux is entitled: "De la probabilité des causes et des événements futurs, tirée des événements observés." Now to begin with, I object that the expression "la probabilité d'un événement" is utterly meaningless. Probability is the ratio of frequency of a specific *kind* of event to that of a generic kind of event to which the specific kind belongs. Being of the nature of a *ratio*, it is necessary to specify the *two* terms between which the ratio of frequency is desired. Furthermore, Laplace is no authority except *in mathematics*; and mathematics is the science by which one deduces necessary consequences from assumed premisses, generally of a quantitative nature (not always, since in projective geometry there is little question of numbers and none at all of *measures*). Consequently it is impossible to deduce the probability of a *cause* from any observed event. That requires *induction*, a kind of reasoning radically unlike mathematical reasoning or any other mode of deduction.

I have mentioned I believe Laplace's absurd result given in all the textbooks on Probabilities that if any event has been observed on N successive occasions, the probability will be $\frac{(N+1)}{(N+2)}$ that it will occur on the next occasion.

One of his problems professes to calculate from the fact that all the balls in an urn are numbered 1, 2, 3, etc. and the fact that a ball has been drawn and found to bear a number N, what the probable number of balls in the urn is. But no deductive conclusion on the subject can be drawn from those premisses correctly.

Since you want to know what I have done, I ought to mention that my studies of deduction have led me to various mathematical truths that are of more or less interest, and some of them not without importance. I will only briefly mention one or two.

One is that no infinite multitude is the largest possible (as had been assumed by several mathematicians). They form a series analogous to that of the whole numbers; and the theorematic proof of this is interesting, too.

Quite an important proposition of which I gave a very simple proof was that every multiple associative algebra can be represented by a square matrix.

I also proved that there are but three possible associative algebras in which the quotient of one quantity divided by another that is not zero has a definite single value. These three are ordinary algebra of real quantities, ordinary algebra of imaginaries, and real quaternions.

I also gave a rule for the numerical solution of any algebraic equation, that is the most expeditious possible; and it has the curious property that if a numerical error in the calculation is made one need only keep right on according to the rule, and one will come out all right.

I also showed that fractions can preserve all their properties and yet have only an *ordinal* signification; and I gave two simple rules, one by which all numerical fractional values can be set down in their lowest terms in the order of their values reducing them to a common denominator; and the other a rule by which without any arithmetical operation whatever except counting all fractional expressions in numbers can be set down with their relations of being equal or the one larger than the other. This is of interest as showing that measurement and the equality of parts is not necessary for the employment of fractions.

There are a lot more of such things.

I will now proceed to *Induction*. My doctrine here is the only correct one, and will have to be received. By induction I mean any reasoning from a part to a whole. Of course, it cannot be a necessary reasoning. Nor is it a probable deduction. Though its conclusion may be more or less false, yet it is justified by the circumstance that though the conclusion may be more or less erroneous, yet one has only to persist in following the same method, and the conclusion will get corrected. It is the only way there is of assuring oneself of *real* truth beyond what direct perception furnishes.

I recognize *three* important classes of Inductions, or perhaps they should rather be called Orders, since they differ chiefly in the complexity of their precautions against error. I know no way of measuring the probative force of Inductions. If there be any such way I suppose I shall have to leave the discovery of it for my successors. But the three Orders of Induction I speak of I call respectively Quantitative, Qualitative, and Crude Induction.

I cannot find, though my reading in Logic has been pretty near exhaustive I imagine (miserable trash it is, too, on the average, though here and there a Leibniz or a Boole rears his head and though very liable to error does advance the science), that anybody before me had any *distinct* conception of the nature of Induction and I will now run through history and make remarks on the different theories of this kind of reasoning, — *all* of which, except Gratry's (one of the least credible) I can absolutely disprove.

Before doing that I will remark upon the name "induction." Before me, this term was always taken by logicians to mean a reasoning by which one was led from the knowledge of instances of a class and from finding that all known instances (or all those examined) had a certain character to conclude that this character would be found to belong to every member of the class. But I understand by "induction" something so different from this in several particulars that I now begin to think that I had better adopt a different word to express my conception.

The word "induction" is, of course, taken from the Latin inductio, and this first occurs, in its logical sense, in two passages of Cicero (unless I have forgotten. I will consult the Lexicon. — I have done so; and while I am confirmed as to the two passages of Cicero, I have learned something that quite surprises me; and from it you will see how imperfect is my mastery of Latin.) In both those passages, which I remembered very well, Cicero speaks as if the Latin word were formed in imitation of the Greek ἐπαγωγή, and one naturally gathers that he himself had first applied the Latin word (or perhaps invented it) because it is formed from *in* and *duco.* "I lead," just as the Greek from ἐπί "up against" and ἄγω "I lead." But I had always been prevented from believing that it was Cicero who had invented the logical term because I said to myself "Cicero, with his fine sense for the meanings of words would never have turned ἐπαγωγή, which means, outside of logic, the *leading of a body of individuals up against* a military position especially a fortification. Cicero," I said to myself, "would have seen that, not *inducere,* but *adducere,* was the appropriate equivalent of ἐπάγειν and would have adopted for the logical word, not *inductio,* but *adductio.*" But now in reward for my energy in jumping up and fishing out the Lexicon, when there seemed no great need of it, I learn that *adductio* is not in the Lexicon at all, so that it cannot have occurred in ancient Latin. It may have been avoided because of its suggesting the "procuring" of a woman. In Medieval Latin, in which I am more at home, it is common enough and in English, since logical remarks have been common, we have spoken of the "adduction" of passages or examples in support of any statement exactly as the word ἐπαγωγή is used in Greek. If one spoke of the "induction" of such passages, one would hardly be understood. Yet the bringing forward of instances is just the characteristic of the kind of reasoning or argumentation called "induction." In view of this, I intend to take seriously into consideration, in view of my conception of the essential nature of such reasoning being as different from that of all who preceded me as it is, whether I ought not to have a different word for what I mean and call it *adduction.* If it were a mere scientific term, I should not hesitate. In science the only proper way is to invent new terms for new significations. But the wider logical words are or ought to be like the names of other great classes, in being fit for the use of the generality of those who read and write; and what would do well enough for mere terms of science may not suit the taste of those who know much about matters of taste but little about matters of science. It was about 1870 — I don't think it could have been as late as 1872 — that I invented the word "pragmatism" to mean that way of thinking that regards thinking as consisting not necessarily in talking to oneself because an algebraist like Boole plainly thought in algebraic symbols; and so did I, until, at great pains, I learned to think in diagrams, which is a much superior method. I am convinced there is a far better one, capable of wonders; but the great cost of the apparatus forbids my learning it. It consists in thinking in stereoscopic moving pictures. Of course one might substitute the real objects moving in solid space; and that might not be so very unreasonably costly. — Well I was saying that by pragmatism I meant a philosophy which should regard thinking as manipulating signs so as to consider questions. But attention, whether voluntary or not, is always an *act*; and a general conception is a *habit*, and *believing*, *real* genuine belief, consists in a habit with which one is contented and which one usually recognizes (though not always) this habit consisting in the general fact that under certain circumstances one would act in a definite way, and would be content to do so. And so real a thing is imagination that not only will a habit be created by acting over and over again in a particular way, but it is almost as effectually created by repeatedly *imagining* the same performance in an intense way, this intensity consisting in a certain stress that one puts upon a mental self-command or self-imperative at the same time calling up all the essential and all the characteristic circumstances that belong to the kind of performance that one wishes to inculcate in oneself. I don't mean matters of morals; because in *that* field everybody knows the truth of what I am saying, but I have rather in mind things that [are] hard to do. For instance, if you cannot with one foot describe horizontal circles in one direction while simultaneously describing with the fist of the same side (right or left) circles in the opposite direction and then at a word of command given by another person, reverse the direction of motion of hand or foot (whichever be commanded) or both if commanded, and that in a perfectly smooth and facile manner; I say if you cannot do this, then you can acquire the habit of doing it, meaning that you can acquire the power of doing it easily at will, without any actual practice at all; although you may need to put hand or foot or both into certain positions and note how it feels to move them one way or the other, just enough to acquaint yourself with the kinds of efforts you have to learn to make, but not practising the motions at all. I will venture to guess that you will be surprised to learn by such an experience how true it is that a habit can be acquired by imaginary practice. Out of such considerations which turn, as if upon a pivot, about the idea that a thought is nothing but a habit connected with a sign, one can build up quite a little philosophy which is what I meant by "pragmatism." I think the idea was suggested to me by Berkeley's two little books about vision; and while the idea was a fresh one for me in the early seventies, I used to talk my friends to death about "pragmatism." I never, however, saw fit to use the word in print, and even in 1889, when I had entire charge of the philosophical, logical, metrological, and various other departments of the *Century Dictionary*, I decided against inserting the word PRAGMATISM.

William James who had heard me talk about it so much, but who had no head for logic at all, took it up and made the man in [the] street get some notions of what pragmatism was. I still think I was right in not inserting it in the *Century Dictionary*. It is a grave responsibility to introduce a new word even if one has the power to do so. Therefore I will not rashly decide to call what in my logic corresponds to induction by the name Adduction. But in order to try it on, as it were, I will so call it in this letter.

The first man to make a systematic use of this great method of reasoning, Adduction, the only way in which [one] can get anything like certainty beyond what the senses afford us of real things, — since Deduction only traces the consequences of premisses that may be, for ought Deduction can teach us, mere arbitrary suppositions, — was as living a man as ever did live. He naturally would be so. It was Socrates. It is curious that there is not the least indication that he called the method ἐπαγωγή, though we know he was a good soldier. Though I haven't Ast's Lexicon and cannot be absolutely sure, yet if that word had occurred in any of the writings attributed to Plato it could not possibly, I think, have failed to have attracted my attention. Xenophon does use the word, but not in the *Memorabilia*, I feel pretty sure. (I am sorry to say I haven't a copy of that book as I ought to have. But there is nothing for me to do but regret it.)

The Adductions of Socrates are all of the "crude" order, and one may say the crudest of the crude; but all classical Greek is entirely anterior to all science, naturally. When men get so advanced as seriously to have reached the scientific way of looking at matters, that delicious naïveté is necessarily gone. Here is one of Socrates's genuine Adductions given by Xenophon himself (for though I haven't the book I pick up scraps of it here and there in footnotes to German books that I have etc. They only serve to make me feel that I ought not to publish anything until I can get books and refresh my memory about the books I read in my early years.) — Well I am unable to find the passage I wanted. But examples of Socrates's way of reasoning are given in Plato's earliest dialogues, Euthyphron (written, say, 400 B.C.), Crito (398 B.C.), Charmides (395 B.C.), Hippias Minor? (397 B.C.), Lysis (395 B.C.), Laches (385 B.C.), Euthydemus (394 B.C.). (The dates are the results of an investigation of my own and cannot be far wrong.)[2]

It has suddenly occurred to me where the little example of Socrates's style of Adduction was. It is in *Cicero's* De Inventione Rhetorica, Book 1 cap 31. But as long as it is not from Xenophon himself, though Xenophon and his wife are in it, it loses its interest.

Aristotle was the first, so far as I know, to use the Greek word ἐπαγωγή in its logical sense, and there is a passage in the last chapter of the first book of his Topics which seems to me convincing proof that the proper English translation of the noun (supposing our dear Lady Welby's principle be true that what was originally unwisely done ought to be corrected now at once no matter whether a few millennia of usage may seem to some to have consecrated it) ought to be "adduction". (My diagrammatic syntax has utterly spoiled my English.) However I shall call it Induction as long as it is not just what I call Adduction. Aristotle's theory of Induction is that it is the inference of the major premiss of an ordinary syllogism (he does not say this must be in the first figure, but it would be odd if it were in the third) from the minor premiss and conclusion. He gives this example

The Syllogism

No long lived things have gall
Man, the horse, the mule are long lived
∴ Man, the horse, the mule [have no gall]

The Induction

Man, the horse, the mule are [a sample of] things long lived
Man, the horse, the mule have no bile
So Whatever is long lived is without bile

This is Adduction, though only Crude Adduction. There is also the germ of a penetrating truth in the relation which Aristotle finds between the

[2] Peirce made a study of the chronology of the Dialogues of Plato (MSS. 974-982).

Syllogism and the Induction, although in the bald way in which he states it, there seems to be little or no reason for asserting any particular relation between the two. If however instead of its being an ordinary Deductive Syllogism with which he starts, it had been a Probable Deductive Syllogism, it would be seen that there really was just that relation which he asserts between the Syllogism and the Adduction which would result. He follows an illustrative example of this.

Probable Deduction

Here is a large chest full of little balls the size of peas. Just one third of them are marked each with the letter *E*; and for a year, well considered machinery, moving with great rapidity day and night, has been mixing the balls up. It will be seen that most of the balls carry over twenty characters, letters of the alphabets, Roman, Greek, Anglo-Saxon, English, Russian, numeral characters, astronomical signs, etc. in great variety. It is, however, only the letter *E* that is to form the subject of this experiment.

Now I propose to draw out 125 balls at random. That is, the chest being nearly cubical, I shall endeavour, as nearly as I conveniently can, to imagine each dimension of the chest to be equally divided into five parts, and I shall endeavour to draw one pea from each of the 125 cubes (or rectangular parallelopipeda) so imagined. And the probability will be (calculating *not* according to the exact rule, which would be ridiculous, since almost all probabilities are somewhat doubtful, but as is usual, by means of the Θ function of probabilities, which must not be confounded with those of the same name in elliptic functions. Tables of Θ are given in all works on probability.) 0.1505 that the number of balls marked *E* among the 125 drawn will be either 41 or 42. That the number will be 43, 42, 41, or 40 is 0.2957. That it will be from 39 to 44 inclusive will be 0.4025; that it will be from 38 to 45 inclusive will be 0.5521; and that it will be from 32 to 51 inclusive will be 0.9422.

There are two conditions of this reasoning being legitimate that are apt to be overlooked; and they need mention here because corresponding conditions apply to Quantitative Adduction.

The first is that the fact that it was the probability of the balls carrying a letter *E* and not any other distinguishing mark ought to be *decided upon before the drawings*. This is the reason it was mentioned in the premisses that there were many other letters on balls. Among so many, it is very likely that some will occur oftener on the 125 drawn than the rules of probability indicate and others less often. If then it were not decided what the conclusion was to relate to, one might be tempted either to use it for some character which had been remarked to be frequent on the balls drawn, or the reverse, according as one either trusted or distrusted calculations of probability more than one ought to do.

The other condition is that in drawing the balls they shall really be drawn at random. That is why I supposed that special machinery should have been acting incessantly for a year in mixing up the balls; and instead of this being an exaggerated precaution, it would on the contrary be altogether inadequate if one proposed to state the probabilities, as I have done, to 4 figures of decimals.

These reasonings about the balls that have so far been considered have all been deductive. That is to say, nothing has been asserted in the conclusions that had not been asserted in the premisses; — at least, such was my aim. I don't know as I quite succeeded. In saying that nothing is asserted in the conclusions that had not been asserted in the premisses, I mean that when I conclude that a *probability has a certain value*, I mean that in the long run the event would happen so often. But when I say that the drawings are made at random (and independently of each other) from a collection of balls as described, I *mean precisely* that they are so made that in the long run they will turn out in the same way. I see that in one respect my statement was faulty. For the method of drawing out the balls amounts nearly to drawing them from 125 distinct parts of the chest instead of drawing each one at random from the whole sheet. What ought to have been done was to have subjected the whole lot to a new mixing operation. I didn't because I thought you might get tired waiting 125 years for the experiment to be concluded. But I might have had 125 chests prepared beforehand. My method was too favorable to the doctrine of chances. It made the drawings conform to theory more closely than they were entitled to do. If I had calculated the probabilities of my probabilities, experiment would not have confirmed them.

But what I mean is that inferences as to mathematical probabilities, *if they are correctly drawn*, — for which purpose it must be known in advance that events *will occur just so often in the long run*; from which supposing further that the different events are absolutely independent of one another, so that one's turning out one way has not the slightest influence on the frequency of the other, then the reasoning being confined strictly to concluding the probabilities of combinations of those events is strictly deductive.

Of course, it is purely mathematical; and the Arabian Nights' Entertainments is not more purely fictional than mathematics in its exactitude.

If anything is stated to be exactly so and so, and it is veraciously so stated, then what one is talking about is pure, — the most exquisitely pure of fictions. One and one make EXACTLY two. That is if one ever *should* put down just one thing and nothing else and then another and nothing more one *would* have put down just two. But then one never *did* do that. And even this is probably not *exact* for one may once or twice have done so. But when one speaks of doing something twice there is no exact idea there. I once carried a platinum kilo, in a velvet-lined étui into which it fitted as accurately as anything can fit into velvet, in my hand-bag, from the Office of Weights and Measures in Washington to Dr. B... (his name escapes me) at the International Office of W. and M. in the Pavillon de Breteuil in the Park of St. Cloud. He remarked that the weight would probably have suffered loss from rubbing in the transport. "True," I said, "but so it was sent to us; and at the time it was made they did not know how to cast platinum, nor did they know that platinum includes hydrogen in great quantities."

> But that I am forbid
> To speak the secrets of my prison house,
> I could a tale unfold.[3]

And not one only, but one or more from my personal experience of *each* of the most renowned of the rational prototype standards that would cause you to admit that the porcupine would not evince any great fretfulness if he should let fly a quill or two at the kind of keeping those standards have had. But if such be the exactitude of weights and measures, what shall we say to expressing probabilities to several places of decimals, which is done every day?

Not only are the above reasonings deductive, but the following would be equally so.

The probability would be just one third that if a sample of 125 balls had been taken at random out of a certain chest of balls, and that if one ball were taken from that sample at random it would prove to be marked with an *E*.

Therefore the probability is just one third that a ball taken at random from the whole chest would be marked with an *E* and exactly one third of all the balls in that chest were marked with an *E*.

[3] Peirce revealed his knowledge of these matters many times over, in reviews, in articles, in his editing of McCormack's English translation of a section on metrology in Ernst Mach's *Die Mechanik in ihrer Entwickelung*, published by the *Open Court* in 1902.

This is deductive as long as the premisses do not speak of what *has happened* merely but of what would happen in the long run.

But now I will turn the leaf and write down a reasoning related to the first just as Aristotle says that Induction is related to *ordinary necessary* deduction; and you will see that this will be an *Adduction*, and utterly different from a deduction. Here is the

Quantitative Adduction.

These 125 balls have been carefully drawn strictly at random from this chest of balls, for the purpose stated in advance of trying [to find] how many of them are marked with an *E*.

On examination, it is found that 42 of them are so marked. Hence presumably and provisionally, until further evidence is obtained, we ought to hold that *about* $\frac{42}{125}$, or very nearly $\frac{1}{3}$, of all the balls in the chest are so marked.

All scientific reasoning, outside of mathematics and the Arabian Nights, *is* provisional. Every scientific man knows it. It was only the other day that the second law of motion was exploded. The same force that would accelerate a slowly moving body very much, will have hardly any effect if the body affected is moving nearly as fast as light.

But many reflexions force themselves upon our attention at this stage of our studies. In the first place what a gulf there is between the logical character of the last reasoning and that of the antepenultimate one, — the Adduction and the Deduction. The deduction adds nothing to our knowledge, whatever: it only calls our attention to our having already admitted something that we may have noticed. The adduction does teach us something new. It furnishes us with information, albeit only approximate and provisional concerning the whole lot sampled, on the evidence of an insignificant sample. For it shows that about $\frac{1}{3}$ of all the balls carry an *E* and that the other $\frac{2}{3}$ do not.

In the next place, we remark that the connexion between the probable deduction and the adduction is evident, undeniable. The one depends on the same two conditions as the other. One may say that we accept the conclusion of the adduction as showing us how it comes about that $\frac{1}{3}$ of our sample carry *E*'s and the rest do not. The conclusion is accepted as accounting for this.

Next, we note that this is just Aristotle's theory of Crude Induction, and we are disposed to admit his theory because in the case of Quantitative Adduction it is evidently correct. In fact, nobody could miss that theory in the Quantitative case. Aristotle's own instance by no means allows the close relationship to be perceived so easily. For my part, it seems to me so surprising that Aristotle should have discovered the connexion, or that having discovered it, he should see that that was the secret of the whole matter, that the more I think of it, the more astounding it seems. There is some mystery about this. A mystery? Greece was a great place for mysteries, — especially Thessaly and probably Macedonia. And Aristotle, the boss of Athens, enormously wealthy, is believed to have spent money in prying into mysteries. He seems to have had no head for mathematics. Trade secrets are the secrets that are kept the best. I know something of one or two, and of their histories. Gambling is the kind of trade that might have had secrets undisclosed for many centuries. I think the doctrine of chances is an instance in point. For in Todhunter's History of the subject, which is the standard work, it is told how this doctrine took its rise in a correspondence between Pascal and a certain Chevalier de Méré. This would be in or about 1654. You can find some account of it in Bayle's Dictionary under the article Zenon. As far as I remember (the History of Probability is not the only work of Todhunter's that I have to lament that I have forgotten parts that I sadly need — fit to tear my hair, which I could ill spare!), as far as I remember Todhunter mentions nothing earlier on the subject. At any rate, most works date the beginning of the Doctrine of Chances from de Méré. But in so well-known and entertaining a book as Libri's *Histoire des Mathématiques en Italie*, Vol. II, p. 188 footnote, there is copied from a MS a passage from a commentary on Dante, which Libri says was *printed* in 1477 so that it is earlier than that; and this passage undertakes to derive the frequencies of the different totals of the throws of 3 dice from the number of ways in which they can come about. It only gives specimen throws, and one of these is wrong, for he says that 4 can be thrown in only one way. But I think that may have been in order not to let the reader know too much. Anyway, as soon as the idea was hit upon any clear-headed gambler, — and gamblers are made clear-headed by the survival of the fittest, — could easily see for himself that it can be thrown in 3 ways: 112, 121, 211. Now the same game with three dice is so frequently referred to in Greek tragedies, comedies, and philosophical works that it is pretty evident that the Greeks were a good deal addicted to it. And who can believe that none of the gamblers of Athens had found out how to calculate the chances of the game? Of course they would keep it secret, but Aristotle would know it, probably under an obligation of secrecy, which he was not the man to be unfaithful to. I really think this is antecedently more likely than not. And when I consider that there seems to be no other explanation of his having discovered the true nature of the relation of Induction (or Quantitative Adduction) to Deduction, I am all but ready to embrace the opinion that he did know something of the doctrine of chances. I will mention one other circumstance that squints, as it seems to me, most singularly toward Aristotle's knowing something of the doctrine of chances.

Namely, if Aristotle knew nothing of the doctrine of chances, how can he have been so stupid, so triply dull, as not to say to himself, "If so fine and precious a method of reasoning results from transposing a syllogistic conclusion of the first figure with the *major* premiss, could we not gain another way of reasoning by transposing a syllogistic conclusion and its minor premiss?" *Must* he not have asked himself that question? Yet if he had thought of trying the experiment, he would at once have found that there *are* such reasonings in great numbers. For instance, I see an animal somewhat like a dog; yet entirely unlike any breed I ever saw. I watch him. I note that he is decided by a home with human beings. I say the syllogism is

> The dog is fond of human beings
> Suppose this animal is a dog
> Then he should be fond of humans.

But that is just what he is. I note that I say

> The dog takes several naps during the day and sleeps lightly at night
> Supposing then this animal is a dog
> He ought to take several naps in the daytime and sleep lightly at night.

I watch him and find he does, and the syllogism would *explain* his doing so, if he is a dog. But I should say

> When a dog lies down he goes round and round several times before finally lying down
> Supposing, then, this animal to be a dog
> He will turn round three or more times before he lies down.

I watch him, and find he does that. All these syllogisms *explain* his behavior if he is a dog and are inexplicable if he is not a dog.

Surely Aristotle would have made a chapter on this kind of reasoning without fail. But he doesn't! He never mentions it. Explain that most singular omission if you can!

Well, it needs no explanation if we suppose that Aristotle was looking at Induction from the point of view of the doctrine of chances. Because then he would find that he must keep the minor premiss as a premiss, since otherwise he could not make any application of the doctrine of chances not having two premisses both speaking of [an] accountable number of objects.

In that way the omission is explicable and I do not see how else it is. This is the kind of reasoning that I call Qualitative Adduction. It enumerates qualities and circumstances though they are things not capable of being counted, or rather, they have no sharp unmistakable boundaries so that there can be no doubt how they ought to be *counted.* Indeed we don't want to count them but we need to *weigh* them. But there is no simple unmistakable way of measuring them.

As it is, Aristotle says in the chapter I have mentioned and in several other places that there are but two ways of reasoning, By Syllogism and By Induction.

And now I will give the socdolager[4] that conclusively shows that either Aristotle never dreamed of the doctrine of chances or if he did was determined not to let all his students who we know were allowed to refer to the MSS of his works from the numerous cross-references they contain (i.e. One treatise referring the reader to a second, and then this second referring the reader to the first. And many such references plainly show that the persons who made them did not understand what they read.) And that is that referring to his example of an induction where he speaks of "man, the horse, the mule" and he notes that though he names merely the three, yet it is to be understood that *all* are meant and then adds "ἡ γὰρ ἐπαγωγὴ διὰ πάντων," — "for Induction draws its conclusion from all instances." He doubtless means from all of which one has sufficient knowledge although lots of logicians, I suppose the majority, understand him to mean all there are and call an argument which concludes something to be true of all of a class by making the premisses enumerate every instance of the class, this I say they call a "perfect induction." It isn't an Adduction at all. It is a necessary inference and therefore deductive, i.e. supposing a premiss to state, as Sir W. Hamilton for example, *requires* that the instances are a complete list of the members of the class.

But understanding Aristotle to mean as I say all the instances of which the reasoner has sufficient knowledge to use them at all, or all he remembers, then the reasoning is what I should call "crude adduction."

[4] "Socdolager" is a variant of sockdologer.

The original stoics were of a very different order of minds from those philistine moralists that the word "stoic" is apt to bring up to our fancy now. They were strict logicians. I do not say they were sound logicians, for, on the contrary, they were nominalists; and nominalism is deadly poison to any living reasoning. As to anything like adduction, they, very consistently with their nominalism, altogether condemned. They thought its *rôle* in the development of science could be filled by the *reductio ad absurdum*, a futile idea. They were already decidedly philistine, — by which I mean that they stuck to certain opinions as if they were known to be revealed to them by God; and would not criticize them as they did every suggestion opposed to them.

It is a striking illustration of how slight and fragmentary our knowledge of the world of ideas of the Greeks is, that were it not for a single roll of papyrus found in Herculaneum of which (according to my recollection) no one single sentence remains intact and complete, and most of it is mere despair-bringing words showing that a deeply interesting and most intelligent logical discussion had originally been written there, — I say, if it were not for this single fragmentary roll, we should be unable to deny, what many historians of philosophy writing since the discovery of this roll have continued to assert, namely, that the Epicureans had no system of logic. But fragmentary as this roll is, it enables us to assert with certainty that they did have a logic, and that that logic was as highly developed as that of John Stuart Mill, which it greatly resembles, while curiously contrasting with that. I note this as showing that the fact that we hear nothing of a knowledge of probabilities in Ancient Greece has hardly any force as an argument against such knowledge having existed, and in my opinion it does not affect the antecedent likelihood that there would be such knowledge. As to the special nature of the Epicurean doctrine of induction, it will be time enough to consider that when I come to Mill's theory.

As for Francis Bacon (for there was nothing worth mention in the way of studies of the theory of Induction between Cicero's youth and the Novum Organum) I am pretty well satisfied that he never had any definite theory on the subject. He simply thought that some system of registry of observations and of planning and making experiments might be devised by him that would render it possible to discover what he called the real "forms" of natural phenomena. But he never did and it was quite visionary. However, it is most remarkable that he should have lit upon the true nature of heat and should have rendered the true doctrine decidedly plausible and even likely, although he did not succeed in impressing the minds of physicists, whom even Count Rumford could not convince by experiments that ought to have convinced them. So that it was only when I was old enough to be keenly interested in physical science and its reasonings that the physicists ultimately came round to the "mechanical theory of heat" and to the doctrine of "energy" which they now treat as if it were much more logically established than it really is. For in this world of uncertainties nothing whatever is more certain than that absolute certainty is unattainable, especially concerning real things; and if my system of logic is not greatly at fault, about the next nearest to certainty is that the quantitative exactitude of any statement about real things can never receive so much as the smallest presumption for sound logic.

Whately, whose logic was first published in 1826, and the great majority of English logicians who have written since, base the trustworthiness of inductions on *uniformities*. There is an endless variety of uniformities or partial uniformities; but the simplest kinds are these four.

1st The members of a class may be known to be alike in certain respects. Thus all specimens of a substance of any definite chemical consitution will agree in a great many respects well known to chemists, such as specific gravity, solubility, etc.; so that a chemist will need only to examine a single specimen in order to pronounce with confidence on all specimens as far as those characters are concerned.

2nd Certain sets of characters are intimately concerned. Zoology is full of such connections so that if two or three of them are found in the same fossil a paleontologist will be pretty sure the animal possessed all the others.

3rd Some characters belong merely to individuals, others are uniform through any variety, others are specific, others generic, etc.

4th An object, especially a particular man, may be known usually to possess or to want the whole of some groups of characters if he possesses or wants any of them.

The knowledge of any such *uniformity* enables us to make certain quantitative or qualitative inductions with great confidence from only a few instances.

People have generalized all these so as to declare that "Nature is uniform; under the same circumstances you meet the same phenomenon." But this is only a vague expression of the fact that there are a great many special uniformities. Some will say "Oh it is more than that, I know that under precisely the same circumstances one finds *precisely* the same state of things *every time*." I ask him, "How many times have you ever been in precisely the same circumstances?" One of the circumstances that must have been the same was the year, the day, the hour, and the minute. I dare say that that single circumstance being the same everything else was so. No it means nothing. If you modify it so that it will mean anything it is easy to prove that it is false. The character of nature, its true characteristic is not uniformity but variety. The uniformities are to the varieties as one in a million. But owing to the fact that uniformities are useful to us and that varieties are not, we take note of the former and dismiss the latter from our minds.

The uniformities do *help* inductions. It is absurd to say they are the sole basis of Adductions.

But how were the uniformities one and all found out? Only by Induction or Adduction. Now it is strange doctrine that uniformities are the sole ground of inductions while inductions are the sole ground of uniformities.

No[w] uniformities *help* some Inductions and render them more certain. But they have this power only because they are supported by Inductions. The *application* of a uniformity to an induction takes place by Deductive reasoning. But the only reason we have for believing in Adduction is that in the long run it must lead to the truth.

Skipping a great deal, I now take up the third great class of Reasonings, which I call Retroductions. By *Deduction* one infers that if certain premisses are exactly true, then a certain conclusion must be true, either always or once in a certain proportion of cases in the long run. By Adduction one infers that a certain state of things *is* true, at least approximately. By the third class of reasonings one only infers that a certain state of things *may be* true and that the indications of its being so are sufficient to warrant further examination. We must accept the Deductive conclusion because we have already assented to it, and consistency requires it. We accept the Adductive conclusion because though it may not be accurate, yet persistence in the same method of reasoning will diminish the error, if there be one, until it indefinitely approaches exactitude. The reason for accepting the Retroductive conclusion, is that man must trust to his power of getting at the truth simply because it is all he has to guide him; and moreover when we look at the instincts of various animals, we are struck with wonder at how they lead those creatures toward rational behaviour. For instance how wasps having no experience of the destiny of an egg will lay its eggs so that the young will find a kind of food that would not be food at all to the mother, even if she knew that any young were to come forth. Now man has equally powerful instincts though we do not recognize them any more than the wasp does hers, and for the same reason, because in following our instincts we, like her, seem to ourselves to be acting according to manifest good sense. So it is to us, though it would not be so but for our instincts. The reason we go right in the main in following our instincts, is the same as the reason that the wasp does so. Namely, it is because these instincts have been formed under the influence of those very laws of nature that they lead us to conform to. For instance, our instinct that a straight line and the nearest to a straight line that can be on a rounded [surface] is the shortest distance between two points is to be trusted because it has been developed in us under the influence of the fact that it really is so; or under the influence of facts which cause our instincts to conform to them. At the same time, there is danger in following what appear to be our natural instincts too closely; and partly because we cannot distinguish between true instincts and mere prejudices; as we see when in the book of Genesis, the Creator is represented as having become so disgusted with the creatures that He had created with entire understanding of how they would act, that He resolves to exterminate them, until reflexion shows Him that He is acting like a dam fool, and He decides to save one family. There is a striking example of how we may be grossly misled by trusting too much to what seem to us "natural" beliefs and which, in a sense, really are so. Therefore all conclusions of retroduction ought to be submitted to Adductive criticism, when there is an opportunity to do so. At the same time, one may often be misled by a modest willingness to surrender one's own Retroduction to the presumed wisdom of one who without any such divine light, brings apparent Adductive objections against one's Retroductive conclusions.

I do not, at present, feel quite convinced that any logical form can be assigned that will cover all "Retroductions." For what I mean by a Retroduction is simply a *conjecture* which arises in the mind. Now does not this often happen before we can formulate any judgment about the state of things that we are experiencing and which leads us to the conjecture? In my youth, however, I was accustomed to think that that very state of being unable to characterize what we experience may be regarded as predicating a confused jumble of characters of it. Thus, a child out of doors alone on a dark night and far from home has a feeling of timidity and looking at a dark object and trying to make out what it is, does not definitely make any description to himself of what he sees, but sees a dark object which frightens it and says to himself

That thing has a peculiar kind of dreadfulness
A bear would have that very kind of dreadfulness
I guess it must be a bear.

This would be the same form of reasoning as a Qualitative Adduction. Thus, the question arises whether there is any difference in form between a Retroduction and a Qualitative Adduction. But the distinction that is undoubtedly of highest importance between Reasonings (and I call anything a "Reasoning" where one belief or tendency to believe causes another) is that which consists in the *nature of the assurance* being different. Now Adduction is the very coolest of all classes of Reasonings; for its conclusion is accepted provisionally *because* if it deceives us we have only to persist in the same kind of reasoning in order to be undeceived. Retroduction on the other hand is the most impulsive of reasonings. There is really *no* reason to accept the conclusion except that *we cannot help it* or in its least impulsive form feel that it is the natural, the reasonable, the Human way of thinking. If the child after this first fright begins to look closer and compare the looks of the object before him with the way a bear would look, he is no longer merely *conjecturing*; he is adducing new qualities.

But I am confident that not all Retroductions take on the [form] of Qualitative Adductions. Walking along one of those interminable French roads bordered by two interminable rows of poplars I notice that on one side each tree has a white-washed stone of about ten kilos at the foot of it. I am seized with an impulse to lift one of the stones and see what is under it. I yield to the impulse and dam if I don't find a *louis d'or* under it! I say to myself "There certainly can't be a gold piece under every one; and yet it's very strange that I should have been impelled to lift the very stone that had that strange deposit. That is a manifest case of Retroduction. But should I go on and lift two or three successively and I should find a goldpiece under each it would be getting to be a Quantitative Adduction. Here seems to be a case where Retroduction takes the form of Quantitative Adduction. Though it might be said to be a case of reasoning from consequent to antecedent.

"If every stone covers a gold-piece" being the antecedent, a "This stone covers a gold piece" being the consequent.

However one of the most marvellous and unmistakable cases of what I call (quite unsuitably according to my present views) Retroduction seems to me clearly to have had neither the Quantitative nor the Qualitative form of Adduction. Namely, when Dalton discovered (as he supposed for the first time) that the chemical elements combine in the simplest of multiple proportions, he was seized with such an intense conviction that they were composed of atoms that he never thereafter for a moment doubted it; and stranger still, to my mind, other chemists when they heard of Dalton's theory at once accepted it with hardly more of doubt than Dalton himself entertained. And the most marvellous circumstance of all is that there is now *other* evidence, absolutely proving (in the sense in which such things can be "absolutely proved") that bodies *are* composed of chemical *molecules* and single *electrons* or corpuscles have also been isolated; so that there seems to be no reason to doubt the reality of atoms though I doubt if one ought to think it anything approaching a certainty; though I have not thoroughly considered the question. The first scientific paper I ever published was an attack on the atomic theory, and Sterry Hunt said I was right and continued to do so after I had quite given up my contention. But all this is aside from the point; which is that the existence of atoms no more account for the simple ratios of multiple proportions than the fact that a bag of coffee consists of separate "beans" goes toward proving that if two kinds of coffee are mixed in a bag it must be in some simple proportion. It was no reason at all; but what it was perhaps, was one of those cases in which men have made guesses apparently utterly unfounded and yet correct. No doubt many of these are cases of instinct. Twice in my life I have had extraordinary experiences of that sort.

I consider Retroduction (a poor name) to be the most important kind of reasoning, notwithstanding its very unreliable nature, because it is the only kind of reasoning that opens up new ground. Deduction doesn't teach anything but only draws attention to knowledge we may have overlooked. Adduction only increases our knowledge in various respects without producing any new knowledge. At least, it is not at all likely to teach us anything quite novel and also important unless where Retroduction furnishes a hint. But Retroduction gives hints that come straight from our dear and adorable Creator. We ought to labour to cultivate this Divine privilege. It is the side of human intellect that is exposed to influence from on high. With this investigation starts. Having once formed a conjecture, the first thing to be done is to draw Deductions from it and compare them with observation. So we correct the errors of our Retroductions by processes of Adduction.

So Retroduction comes first and is the least certain and least complex kind of Reasoning.

Deduction follows. It is as certain as are its premisses and no more so.

Finally Adduction ranges in complexity from a simple Crude Adduction up to elaborate Quantitative Adductions which offer material for new Retroductions.

I have now sketched my doctrine of Logical Critic, skipping a good deal. I recognize two other parts of Logic. One which may be called *Analytic* examines the nature of thought, not psychologically but simply to define what it is to doubt, to believe, to learn, etc., and then to base critic on these definitions is my real method, though in this letter I have taken the third branch of logic, *Methodeutic*, which shows how to conduct an inquiry. This is what the greater part of my life has been devoted to, though I base it upon Critic.

Of course in order to study methodeutic it is necessary to make researches in as great a variety of sciences as possible, — *real* researches, not the two penny half penny "research work" that students of colleges do. Now I have always been poor; and my father when I was a boy was always on the fringe of penury. In France as distinguished a man of science as he was who rescued his country from contempt in regard to mathematics and astronomy would not have been so poor. But even today the United States is governed or led by men actually *below* the average. Witness the miserable tyranny that is exercized over the great businessmen, the barbarous "Sherman act," the bug-bear of monopolies. Monopolies, such as *can* be created in our days, are most beneficent for the public.

Well, as soon as I had taken my academic degree, I was appointed an aid in the Coast Survey at $15 a month. That was an opportunity to learn my lesson in one science. I managed three years later to take a degree in chemistry, and I was the first in Harvard to take a degree in that science summa cum laude. In the Coast Survey I particularly made myself master of the subject of weights and measures. Later I was appointed to the charge of all the investigations of the Survey into gravity. I got leave to go abroad to study European methods of investigating gravity. While I was in Paris, there happened to be a conference of all the European Surveys. It was held in the *Palais des affaires étrangères*; and I received an invitation to attend the meetings. At the first I attended, the subject of gravity was discussed; and I was taken completely by surprise when the president, Gen. Ibañez, called upon me for my opinion of the work they had been doing. Of course, I was obliged to express my real opinion. They thought they were measuring gravity with errors not exceeding 1 or at most 2 millionths of itself. But the pendulum was swung from a brass tripod and I expressed the opinion very decidedly from an examination I had made of that tripod in Geneva that it swayed under the pendulum to an extent which though not directly observable, I had been able to get a notion of the amount of, by measuring how much the part where the pendulum rested would be moored by a horizontal pull of 1 kilo's weight. Whence I concluded that all the values of gravity which they had been publishing during the past ten years were too small by about $\frac{1}{10000}$ of themselves, or a hundred times the error they thought they were excluding. Now when you reflect that it takes usually 100 times as much labour of all kinds to reduce an error to $\frac{1}{10}$ of its previous amount. That is, if they had in *one month* succeeded in measuring anything to a millimetre, then it must be expected that a *hundred* months labor mental and physical would be required to measure the same thing accurately to a tenth of a millimetre; when you remember this you will see how my statement that this error was a *hundred times* as great as they supposed sounded! Well, the idea was evidently completely new to them and they didn't venture to say much about it. However, the next year they had another meeting in Brussels when three of the members who were supposed to be the most competent reported certain experiments (of the most ridiculous nature in my opinion. For instance, one of them had put a delicate spirit level on the stand of a half-second pendulum and it didn't budge. This really seemed too much like an ostrich sticking his head under the sand. Those delicate levels require nearly a minute to oscillate!). Anyway they reported that "our American colleague," — in brief had found a mare's nest. I did not receive the report of that meeting until nearly a year later within about a fortnight to three weeks of a third meeting. Meantime with the apparatus I had procured abroad, a duplicate of their most approved pattern made for me with the utmost care by the first mechanician of Europe, I had done a lot of work and was by that time perfectly sure that the amount of the error that I had stated in Paris was as nearly as I stated it exactly right. I had had a good safe stand made too and had proved the pendulum swing quicker on that by the calculated amount. So I instantly applied for permission of absence. Well, the people in Washington were awed by the fame of the gentlemen who had reported against me in Brussels. Besides that, there was that stomach-turning *jealousy* — fough! — to a scientific man who looks upon scientific work as an humble act of worship of his God and his Creator, — it is unspeakable! So the leave of absence was refused me. I was in New York, stationed there by official order. There happened to be at the time over at the Brooklyn Navy Yard or somewhere (I am sure I forget where but near at hand) another Coast

Survey fellow who thought he was under some obligation to me, and who was a perfect master of Newspaper English. I sent to him to meet me the next morning at the Century Club (the old 15th St. house). And there he and I concocted half a dozen paragraphs — which my friend copied in his exquisite handwriting, — after we had put our heads together in the composition of them. Then that night I took a coupé so as to be down in Park Row as near midnight — just a bit later — and I would rush into each of the half dozen leading morning papers — and tell the fellow at the desk with an air of great authority "Send that up to the Night Editor, and tell him to put it right in, without fail!" and then I would leave in equal haste. Well the next morning, sure enough, three of the best papers had our paragraph, and among them the *Tribune*. When I saw that, for the paragraph had almost the tone of an imperative order to be executed at once, I laughed heartily and felt the thing was done. And early — I think it was about noon — I had not a leave of absence but an order to go and represent the Survey. You may believe that I got on board the first steamer. I was landed at Plymouth and travelled right through night and day to Stuttgart where was the meeting. I got to the hotel in the evening during dinner. I knew there were 2 men who believed in me, — or rather $1\frac{1}{3}$. The one was Genl. Baeyer the leader of European geodesy. The $\frac{1}{3}$ was a fraction of Mr. Emile Plantamour, who had seen me at work in Geneva. I met Genl. Baeyer and his daughter in the corridor of the hotel as I was being shown to my room and the old General who had been fighting for me all day but really did not *know* much about the subject was so delighted to see me that he threw his arms round me and kissed me on both cheeks! The next morning I went into the meeting which was a particularly distinguished gathering, several men who were not regular geodesists being among them, as Henri St. Claire Deville — M. Faye — etc. I began with the mathematical theory which I had, in coming across, succeeded in putting into a form in which every man of them could see the correctness of it. Then I described the instrument by which I had automatically registered the instants of the passage of the pendulum over the vertical, while it was swinging on the brass tripod and when it was on a properly stiff support. I had the chronograph sheets with me, and the whole demonstration was complete, and when I sat down each of my three antagonists at Brussels got up one after another and very handsomely admitted that I was entirely right. And from that time I was acknowledged as the head of that small branch or twig of science.

Another time when I was in a minority of about 2 with the entire brunt of the argument to sustain was when I saved the country from having the metric system made compulsory. Of course the only unit that is of any particular importance is the *inch*. *That* certainly was so at that time and probably is so now. I won't tell the story in writing because it was in a secret session. There were very near a hundred men red-hot against me, but I carried the point, [...]

D. ON THE FOUNDATION OF AMPLIATIVE REASONING (660)

Reasoning from express premisses is either *Explicative* or *Ampliative.* Explicative Reasoning, which embraces all syllogistic and sound mathematical reasoning, concludes nothing that has not been asserted, though it be in an involved way, — or, as we phrase it, "implicitly" in its premisses.[1] All other reasoning, as going beyond what is asserted in the premisses, is called Ampliative. It includes almost all the reasoning upon which any stress is laid except by mathematicians and by formalists and pedants. To explain how such reasoning can lead to the truth, and under what conditions it is valid, is the most obligatory task of the logician, as well as the most important. I distinctly claim that in my writings, and so far as I know, in them alone, is to be found the proved solution of the main part of this problem, and from my doctrine alone the rules that need to be followed in such reasoning follow as corollaries. I further claim that in regard to the secondary parts of the problem, by which I mean the question of how an Ampliative Reasoning can be strengthened, I have brought to light and proved as much more as had been known before.

Several of the unsuccessful attempts to explain ampliative induction merit study because there is some truth in them. In particular, there are two theories that correctly explain, at least in great part the validity of certain inferences which are easily confounded with pure induction.... These two theories, too, are so plausible that many minds justly wielding great authority, — such is their manifest vigor, though only in other departments than logic, — have accepted, some the one and some the other. Some eminent authorities seem, indeed, to have accepted both these two theories at once, in spite of their inconsistency with each other. But the truth may have been that on some days one of the two recommended

[1] Peirce directs that an insertion be made here. But that material is no longer in the manuscript.

itself to them, and on other days the other. One of these two, generally referred to Laplace, certainly its most glorious champion, makes inductive conclusions to follow from their premisses as probable, by the doctrine of chances, otherwise named the calculus of probabilities, that is to say, by mathematical reasoning, which is merely explicative. I shall devote the present section to an examination of this theory. The other of the two theories, which shall form the subject of the next section is that of John Stuart Mill, as set forth in his *System of Logic*, according to which the inductive conclusion follows from its premisses because of the uniformities of Nature, especially because of that great uniformity, that *under precisely the same circumstances, precisely the same phenomena take place.* The clear distinction between a "uniformity" and a law was first made by students of the doctrine of chances. If a person pitches a coin, and it turns up heads a dozen or fifteen times in succession, as has doubtless often happened, this is a mere uniformity, not a law: it is merely what does happen, if it does happen, and not what was *compelled*, nor in the slightest degree *influenced*, to happen. Now Mill does not believe there are any *laws* of nature, though he does not pretend to have any positive reason to think there are none; but he holds that as long as the uniformities last, their existence would be sufficient to warrant decided inclinations to believe in inductive conclusions. For my part, I hold, and shall endeavour to show the reader, in the next section, 1st, that neither uniformities nor laws would, in the least, afford the slightest warrant for an *ampliative* influence, such as is required to discover any law; 2nd, that until we probably know a law we cannot be warranted in making any prediction, without some ampliative reasoning. But, 3rd, by means of ampliative reasoning, namely, first, by Retroduction, we are warranted in conjecturing, and then by Induction, we are warranted in believing that certain Laws there really are, that is to say, *general habits* of Nature's behaviour, though we are not warranted in believing any that we know to be eternal, nor everlasting; 4th, that though Mill was as far as possible from being warranted in believing that there are *no* laws in nature, it is true that we are not warranted in believing that any *given* law is absolute; and on the contrary, it is presumable that any given law is continually violated in an excessively small measure, and with excessive infrequency in perceptible measure; and this presumption is confirmed by the extreme *variety* of nature, which cannot otherwise be so satisfactorily accounted for; and there are several other reasons, some of them weightier still, which go toward proving this important conclusion. Since it now appears necessary to the representation of the motions of Mercury and Venus to introduce a rotation of the line of apsides that signifies that gravitation is not precisely proportional to the inverse square of the radius vector, since Lorenz's extraordinary conclusions concerning time and space, and since Newton's laws of motion are recognized as not exact when the moving body has nearly the velocity of light, scientific men must be ready to examine the evidence that the laws of nature are subject to irregular violations. I seem to be almost prepared to take another step in my investigation of this question: I mention this as an encouragement to any other man of science who may find himself on the same track.

I now turn to Laplace's Theory of Probability. I venture to opine that Laplace introduced no signal improvement of this calculus except the now almost obsolete theory of generating functions, and that the erroneous procedures that he encouraged (I do not know how far he originated them) due to his vague conception of probability did great harm in several ways, but especially in preventing investigation into the true warrant and rules of induction.

Because my principal objection to his treatment is that his conception of probability is wrong, — an unusual sort of objection since it unquestionably is a conception, and the reader may well be at a loss to comprehend why Laplace should be blamed for developing any conception, and it may be surmised that I am merely disputing about a word, the clearest arrangement of my criticisms will be to begin by defining what I mean by "*probability*," for which it may at first seem that the word "probability" is not well chosen. It is, however, substantially, what Mr. Venn called "objective probability" in the first edition of his *Logic of Chance* (1866). It was already the conception I had long entertained myself; and doubtless others had done so; especially J.S. Mill in the first edition of his *System of Logic* (1843). I will then show that all the ordinary problems about probabilities in games and about life insurance will receive precisely the same treatment, whether one adopts my view of probability or that of Laplace. The treatment of observations, too, ought to remain as it is whatever be the rationale of the subject. But most of Laplace's treatment of problems on causes, especially his assumption that in the absence of any knowledge about a kind of event all values of the probability are "equally *possible*," as Laplace phrases it, meaning equally *probable* (which seems to me a juggle with words), are entirely unwarranted, from the point of view of my definition, and seem to me to be arbitrary even according to Laplace's own definition, and they imply self-contradictions. Now self-contradiction is the characteristic of an indefinite position. Problems concerning judicial testimony and the credibility of witnesses also lead to reflections against adopting Laplace's definition of probability as the basis of that useful "doctrine of chances" for which Bernouilli and others gave the rules; and I shall show that Laplace's own formula for the probability of crude induction contradicts his definition of probability. By "crude" induction, I mean that inartificial, irreflective kind usually but very inappropriately termed (I suppose in imitation of Francis Bacon) *inductio per simplicen enumerationem.* Arrived at this stage of the discussion, I anticipate no longer any difficulty in convincing the open-minded reader firstly, that mathematical reasoning is *explicative* reasoning, that is to say, that, unless it be unsound and quite worthless, it concludes no more than is asserted in its premisses; secondly, that the reasoning of the doctrine of chances is altogether mathematical, and in particular that the process by which Laplace professes to reach any inductive conclusion from its premisses is either utterly worthless or is necessary reasoning and mathematical reasoning; thirdly, that ampliative induction is not necessary reasoning nor explicative reasoning, but is of an altogether different nature, and therefore cannot be represented by any kind of mathematical reasoning; and that the reasoning by which an assurance company's actuary constructs a mortality-table and the reasoning by which the company is led to undertake a given risk are so utterly different that if the conclusion of the latter is to be called *probable* (though it is plainly necessary), then some other word must be sought to describe the conclusion of the former. I propose to speak of it as a *likely*, or *verisimilar* conclusion. It is the very best kind of reasoning in the world, — better than the demonstrative reasoning of mathematics, because it assures us of what we had not previously known even in dismembered fragments; and although this assurance is not absolute and may be erroneous, yet perseverance in the same kind of reasoning will gradually correct every error, and will enable us to approach indefinitely near to certainty. On the other hand, it is a far higher kind of reasoning to that instinctive operation by which on considering a surprising experience we are enabled to make a good guess at its cause or reason. Galileo called this power *il lume naturale.* I liken it to instincts of the lower animals, such for example as that whereby a wasp is led to lay her eggs where the young will best find food, although she has never seen such a thing as an egg; even if she remembers that former state of her existence when she used to eat the kind of food that will, at a later time, grow in the place where she is going to lay her first eggs. Induction is a higher kind of reasoning than that for the very reason that it understands itself, and consequently able to regulate itself and to attain a higher degree of certainty without end; while in most human beings the instinctive power of dividing truth is almost atrophied by their habit of self-control, which they come to regard as essential to rational thought and to reasonable conduct. And in this they are mainly right, though inspiration is requisite if one is to be a veritable discoverer. In order that one may *start* an induction, two things are requisite: 1st, that one should be in some small degree inspired with a power of divination; and 2nd, that one should be profoundly sceptical of any such power.

I now proceed to set forth my own conception of probability, which word I strictly limit to the meaning it ought to have in the doctrine of chances. I came substantially to my present views on this subject in the year 1864, by long and deep reflexion on the chapters about it in Boole's *Laws of Thought*, although I regarded Boole's own views as insufficiently considered. I have found mathematicians to be, by far, the best reasoners of any social class that excludes them; and yet it seems that an exclusive absorption in their studies, which more than any others demand exclusive devotion, tends to blind them to other kinds of reasoning. This was more true under the Pre Weierstrassian dispensation but has not ceased to be true yet. It is not many months since a professor of mathematics who has taken to writing on topics of ancient history wrote to me to set me down in the regular old magisterial fashion, informing me that a certain argument he had put forth concluding as to whether a certain event had or had not occurred centuries ago, was demonstrative. It was not a loose expression. He mentioned his own argument merely as an example of a doctrine of logic, namely, that ancient history is capable of rigorous demonstration; and this doctrine I was to accept on his authority. For such a man, you will say, Weierstrass might as well not have lived. ...

We must begin by constructing a definition of probability; and to that end we shall do wisely to begin by defining it in the sense that the word bears in those problems of the doctrine of chances concerning which there is no dispute, everybody acknowledging that the results of the calculus are correct and useful. When we once have a perfectly distinct apprehension of what probability is in those cases, we can consider the problems in dispute. So far as we find that our definition holds perfectly in those cases, the definition must remain unchanged. But we shall find that is not the case; and in each such case the question will be whether we ought to broaden our definition or whether we ought to give another name to what is called probability by those who accept the dubious solutions; and this question must be answered on the principle of making the meaning of the word broad enough to include everything which it can include without prejudice to its utility, but making that meaning strict enough to exclude every case whose admission would prejudice that utility.

In order to form the preliminary definition which this method requires us to begin by forming, there is a little catalogue of questions which we must first consider. The first of these is, What sort of objects are they which are said to have probabilities of such and such a value in the undisputed problems? By writers on the calculus of probabilities, that which has a probability is called an "*event*," which seems about as inappropriate [a] designation as could have been selected, inasmuch as the word *event* was originally a passive participle, applied to that which has turned out, or has resulted, while the probability of anything that might result ceases to apply to it as soon as it is certain of what sort it is. We may regard the term, "event," as an abbreviation for "mode of possible event." It came into our language, however, as a translation of the French *événement*, which not being a participle could, with no great violence, receive the secondary meaning of a possible mode of resulting, just as a *raisonnement* usually means a possible mode of reasoning, an *échappement*, a mode of release (in watchmaking), a *dénouement*, sometimes a mode of untying, or, as we rather oddly say in English, of winding up, etc. At any rate, it means the as yet unknown way in which some known occurrence will result. Thus if an urn is known to contain a number of balls all just alike except that all are of different colors, and somebody puts his hand into the urn and draws out a ball, then before the color of the ball has been seen, the "event" of its being *red* will be the subject of a probability. For the balls being just alike in every respect except color, which cannot influence the person who draws since he is to draw without seeing the ball he draws, it seems certain that if ball after ball were drawn out and thrown back again, without cessation, the whole being well stirred up before every drawing, any one ball would in the *long run* be drawn as often as another. The expression "long run" must here be understood to mean a run *endlessly* long, since we are aiming to construct a definition of the conception of probability. For a definition can obviously not be allowed to contain that which it professes to define. If that were permissible, definition would be child's play. We should only need to say, for example, "Definition is definition." In the present case, we are defining, not the *word* probability, but the *conception*, we must exclude that idea, however expressed. We must analyze that concept. Now if we were to speak of a long but finite succession of drawings of balls, it would not be absolutely certain that any one of them would be drawn just as often as any other. On the contrary, it would be an amazing event, if when one of them had been drawn just 10000 times, it were found that every other had also been drawn precisely 10000 times, too. ...[2]

[2] This manuscript was begun 24 October 1910, 2:45 P.M. and worked on at different times, last page marked 28 October 1910, 10:00 P.M.

E. [LOWELL LECTURE OF 1866] (354)

LECTURE III

I have here a bag of balls. I shake them up well; and draw out one. It is red. I draw out another. It is red. I draw out another. It is red. I shake them up again and draw out another. It is red, again. Another: red. Another: red. One more: still red.

Now I suppose you have doubt that almost all the balls in that bag are red. Why? Upon the correct answer to that question much depends. Our fate hangs upon it. All difficult questions require an understanding of the reason of our faith in experience as a witness to the future and unexperienced. There are several common answers to the question, and the first which I shall consider is that it is mathematically more probable that most of the balls in the bag are red. This is the answer given by Laplace and the mathematicians. In order to test its accuracy it is necessary that we should inquire a little into the theory of probabilities.

By a probability in Mathematics we must not understand a likelihood but only a numerical ratio; namely, the ratio of the number of occurrences of a specific event to the whole number of occurrences of a supposed *certainly* known generic event of which the first event is a special kind. For example: since we know that when a *die* is cast very often, the side numbered *six* will turn up about $\frac{1}{6}$ of the whole number of times, we say that the probability of throwing a six is $\frac{1}{6}$.

If I put into this bag an equal number of black and white balls, then since I know that in the long run white balls will be drawn half the time, we say that the probability of drawing a white ball is one half.

Thus probability is the ratio of the frequency of the occurrence of the event in question to the frequency of the occurrence of the event we know happens.

Suppose I take three bags; in the first, I place only blue balls, and mark it with chalk below. In the others I put three yellow balls for every blue ball. And I put the three bags in a basket. Now if I draw out balls at random from that basket; experience will show that I shall draw out blue balls half the time. I take one of the bags at random then and am going to take out one of the balls at random, the probability of my drawing a blue ball is $\frac{1}{2}$. But now I look to see whether it is the bag which had the chalk mark upon it. I find that it is not that bag. And as I know that in the others, there were three yellow balls for every blue one, and that consequently in the long run blue balls would be drawn only a *fourth* of the time, the probability of drawing a blue ball is *reduced* from $\frac{1}{2}$ to $\frac{1}{4}$.

This shows that the generic event which frequency we take as the denominator of our fraction, must be as specific and determinate as any relatively to which we know the frequency of the event whose probability we are considering.

It cannot escape your attention that mathematical probability expresses an *outward fact* — the relative frequency of an event — and therefore *not* the degree of expectation with which the event ought to be looked for. It is, indeed, supposed to *correspond to* the intensity of a proper feeling of expectation, and it is on that account that it is called probability, but that correspondence is not a fact which it belonged to the mathematician to discover; and all that he has a right to decide is the relative frequency of an event.

Suppose I take two bags and fill them with blue balls and fill one bag with three blue balls for every white one, and then put these bags together into a basket. What is the probability of the first ball drawn being white? I should in the long run draw out a ball from the right bag $\frac{1}{3}$ of the time, as there are three bags and only one contains white balls. And I should draw out a white ball from that bag a fourth of the time that I drew from that bag, because only a fourth of the balls in this bag are white. Since therefore, I get a white ball a fourth of the time that I draw from the right bag, and draw from the right bag a third of the time that I draw; it follows that I draw a white ball a *third of a fourth* of the time that I draw. And, therefore, a third of a fourth is the probability of my drawing a white ball.

Let us take another somewhat similar case. What is the probability of throwing *sixes* with two dice, at the first throw. One die turns up *six* one sixth of the times because it has six equal sides and is of the same specific gravity throughout, and the 2nd die turns up *six* one sixth of those times that the first turns up six, for it turns up *six* one sixth of all the times and no oftener or less often when the other turns up six, and therefore turns up six one sixth of the times that the first comes six. The two sixes come up therefore $\frac{1}{6}$ of $\frac{1}{6}$ of the time. Hence $\frac{1}{6}$ of $\frac{1}{6}$ is the probability of throwing sixes at any one throw.

In general then the probability of the concurrence of two events is equal to the probability of one of them, multiplied by the probability of the *other, if the first one occurs.*

Suppose I take the bag of balls which contains three blue ones for every white and draw one out and put it back, shake up the bag, and draw out another. What is the probability that one of these will be white and the other blue? There are two ways in which this can happen: either the 1st may be blue and the 2nd white or the 1st may be white and the 2nd blue. The probability that the first will be blue, since $\frac{3}{4}$ of the balls are blue, is $\frac{3}{4}$. The probability that if the first is blue the second will be white, since in that case $\frac{1}{4}$ of the balls will be white is $\frac{1}{4}$ and therefore $\frac{1}{4}$ of $\frac{3}{4}$ is the probability that the 1st will be blue and the 2nd white. The probability that the 1st will be white is $\frac{1}{4}$ and the probability that if the 1st is white the second will be blue is $\frac{3}{4}$. Hence the probability that the 1st will be white and the 2nd blue is $\frac{3}{4}$ of $\frac{1}{4}$, or $\frac{3}{16}$. In the long run then one ball will be blue and the other white, in the first way, $\frac{3}{16}$ of the time, and in the second way $\frac{3}{16}$ of the time, and as it cannot happen in both ways at the same time, it will happen in one way or the other $\frac{3}{16}$ and $\frac{3}{16}$ or $\frac{6}{16}$ of the time. Therefore the probability of drawing first a ball of one color and then one of the other in one trial is $\frac{6}{16}$ or $\frac{3}{8}$. Again let us put into a bag nine yellow balls, six white ones, and twelve black ones. What is the probability of drawing a light colored ball. A light colored ball is either yellow or white. There are nine yellow balls and 27 balls in all. Hence, the probability of drawing a yellow ball is $\frac{9}{27}$ or $\frac{1}{3}$. There are six out of twenty-seven white balls; hence the probability of drawing a white ball is $\frac{6}{27}$ or $\frac{2}{9}$. Hence the probability of drawing either a yellow or white ball is $\frac{1}{3}$ added to $\frac{2}{9}$.

In general, then, the probability of an event which can happen in two ways, but not in both at once is the sum of the probabilities of its happening in each way.

We now know what the signification of the *sum* of two probabilities is, and what the signification of the *product* of two probabilities is.

The sum of the probabilities of two events is the probability of one or other happening supposing both cannot happen, at once. The product of the probabilities of two events is the probability of both happening at once, supposing that they are quite independent, that is, that one happens no oftener or less often when the other happens than when it does not.

As we now understand the addition and multiplication of probabilities, we can easily find what the meaning of subtraction and division is.

If we know that the probability of drawing out a light ball from this bag is $\frac{3}{4}$ and the chance of drawing out a *white* ball is $\frac{1}{4}$ we know that the probability of drawing out any light ball except a white one is $\frac{2}{4}$. Thus the remainder of one probability subtracted from another is the probability of the first happening in any other way than by the other, supposing that whenever the second happens the first happens.

Suppose that in the case of my having 3 bags in a basket, I know that the probability of drawing out a white ball is $\frac{1}{4}$ and that the probability of drawing from the only bag which contains white ones is $\frac{1}{3}$. Then I know that the probability if I do draw from the right bag of getting a white one is $\frac{1}{4}$ divided by $\frac{1}{3}$ or $\frac{3}{4}$. So that the quotient of one probability divided by another is the probability of the first happening if the second does supposing that the first cannot happen unless the second does.

We have now the fundamental principles of probabilities.

Let us now turn our attention to the connection between probability and weight of evidence. If there are ten balls in a bag and all are black. Then the probability of the first ball drawn being black is by the definition of probability the ratio between the number of times a black ball will be drawn to the total number of balls drawn. This ratio is $\frac{10}{10}$ or 1. But in this case we are certain to draw a black ball the first time. And you can easily see that the probability of anything else which is certain is 1. On the other hand the probability of drawing a white ball is nothing; because a white ball would never be drawn. Now it is impossible that a white ball should be drawn the first time, and therefore the probability *zero* corresponds to impossibility.

Moreover the oftener an event happens, the more apt it is to occur in the long run, the more apt it is to occur in the special case, the more evidence we have that it will occur in the special case.

Let us ask ourselves why is this: — Suppose there are 2 blue balls for every yellow one in this bag. Then we say,

All the yellow balls are in $\frac{1}{3}$ of the balls
The balls of the color drawn are not in $\frac{1}{3}$ of the balls
∴ The balls of the color drawn [are] not yellow.

The two last propositions are only *likely*, not *certain*.

But what does this word *likely* mean? That is *likely*, I should say, which is got at by a method which will yield truth oftener than falsehood. Is not this the meaning of likely? I say it is *likely* a yellow ball will not be drawn; that is, I acknowledge a yellow ball *might* be drawn, but the opinion that the first ball is not yellow, is obtained by a sort of inference which will not *often* deceive me.

Some, I know, will deny that this *is* the meaning of the term likely; they will say that *likely* is a simple and unanalyzable conception. [Note A follows in parentheses.]

(This is a favorite mode of getting over a difficulty with some philosophers. They present some hypothesis of things, of the antecedent probability of which we know nothing. The only possible objection, then, which can be made to such a theory is that there are facts which it will not explain. But the moment you urge one of these facts in the way of objection, the philosopher replies, "Oh *that fact* is inexplicable." You know what would be said to such a plea in a court of law. If a party were to say, this fact is inexplicable, the judge would say you *must* have your theory, or you will lose your case. The facts which these philosophers call inexplicable, are so upon their theory, but they are not so upon another theory. Inexpicable, with them, means inexplicable to me. But suppose, neither party could explain the facts; the question in a court of law would in that case be decided by a legal presumption, that is to say, by a rule by which unproved cases are decided, a rule which would conform to general experience if there be any such experience, but any rate is framed so as to dispose of the case somehow. These rules are necessary in the law, because it is for the well-being and peace of the community that every quarrel should be decided, and it is better to have it decided wrongly than not at all; but it is entirely different in speculative philosophy; here it is our interest to get at the truth if centuries are required to reach it, and it is better not to decide than to decide without reason. If, therefore, neither party can explain the facts, they cannot continue to hold their respective opinions, they must both give up and wait till the explanation comes. But, it is said, there must be something ultimate, something inexplicable. Undoubtedly; — in one sense. That is to say, there must be a point where each system fails; man never can attain to absolute insight. But this is not to admit that there is a point which no system can explain. Man can advance indefinitely; he cannot go to infinity, but he may be able to pass beyond any assignable point. Every system must fail; but when it has failed, it must be given up.)

Let me point out with reference to this mode of doing away with a difficulty that it is so very *easy* and so generally applicable instead of a logical analysis, that we should be on our guard against it and should be careful that we do not admit any conception to be simple without a positive test. Everything must be presumed to be explicable till proved inexplicable, otherwise we fall back everywhere into mental stagnation. But allow that the term *likely* is itself simple; then I ask what is the immediate justification of its employment. Do we know by intuition that something is likely? If so; it implies we know not what that is objective, if it has no definition, it is a mere peculiar feeling, and there is no use and no need of any search for a justification of it.

Others perhaps will admit that likelihood may be defined, but will not admit that I have given the true definition. They may say that it is an approximation to certainty. All will admit this; but wherein does this approximation to certainty consist if not in what I have said? In a preponderance of evidence for, over evidence against? Certainly not; because the preponderance of evidence supposes that each piece of evidence by itself yields some approximation to certainty which is far from complete. "Such evidence," says Dr. Reid, "may be compared to a rope made up of many slender filaments twisted together. The rope has strength more than sufficient to bear the stress laid upon it, though no one of the filaments of which it is composed would be sufficient for that purpose." But as each filament of the rope must have some strength, so each piece of evidence must give some likelihood to the conclusion. Probability cannot therefore arise from a preponderance of evidence, since there can be no such preponderance unless there is some probability without opposition of evidence. What, then, is meant by the approximation to certainty afforded by each portion of evidence?

A piece of evidence which yields a likelihood always yields that likelihood by a process which would more often yield truth than the reverse; and every process which is known to yield truth more often than the reverse gives likelihood. This is the only property of likelihood we know of, but the meaning of the word must be some constant and peculiar character of that to which it is applied and must be one which we know; hence this must be what we mean by likely, if we mean anything.

To this argument, I can add the authority of a great analyst, Locke in his Essay on the Human Understanding says: — (see iii, p. 149).

This then being what we mean by likely; it is plain that that which is mathematically the more *probable* of two events is also the more likely. For example, when we conclude that a yellow ball will not be drawn the first time from this bag because only a third of the balls it contains are yellow, the frequency with which a conclusion drawn from this sort of argument is true is just equal to the frequency with which non-yellow balls will be drawn from the bag. So that the mathematical probability is just equal to the probability in the sense of likelihood. And so it will evidently always be.

If the probability of an event is equal to $\frac{1}{2}$, if for example the number of yellow balls in a bag be just equal to the number of black and white ones together, then there is no more reason to think that this event will occur than that it will not; for instance, the reasons for believing a yellow ball will be drawn are just balanced by the reasons for thinking that another one will be drawn, inasmuch as it is just as likely that the ball drawn will be yellow as that it will be of another color.

Let us compare this with a state of complete ignorance upon the contents of the bag. Here is a bag whose contents we do not know, except that it is balls. We have no reason to think that the first ball will be blue, and no reason to think that it will not be blue except our general knowledge that blue balls are not so common as balls of all other colors put together; but suppose that we did not know even this. Then as we should have no reason to think either one way or the other, our reasons would be balanced because *zero* balances *zero.* Thus our state of mind would be in some respects the same as though we knew that half the balls were blue.

If a bet were to be made as to the ball being blue, in both cases it would be just that the bets upon both sides should be equal. It would not be imprudent for a person who was in the habit of betting to bet a small sum; if he knew that half the balls were blue, because he would know that in the long run he would not lose much by such bets, the probability of $\frac{1}{2}$ implies that. It might not be imprudent for him to bet in the other case, because he may know that in the long run he would not lose much by such bets either; but this is supposing that he has some further knowledge upon the subject. In a state of complete ignorance it would be imprudent for him to bet.

But whether the proper state of mind in the two cases would be exactly the same or somewhat different; there would be a great difference in the fraction which expresses the mathematical probability. In one case, that namely in which $\frac{1}{2}$ of the balls are known to be blue, the fraction is $\frac{1}{2}$; in the other case where nothing is known upon the subject, it would be totally unknown.

Some mathematicians have sought to represent absolute ignorance by the probability of $\frac{1}{2}$. But the principle by which alone such a proceeding could be justified is self-contradictory. Since we do not know what proportion of blue balls there are in this bag, the probability of the first ball being blue is $\frac{1}{2}$ according to that principle; and since we do not know how many non-blue balls there are, the probability of the first ball not being blue is $\frac{1}{2}$. But blue balls are either light-blue or dark blue balls and hence the probability of the blue ball if drawn being dark blue is $\frac{1}{2}$, and of its being light blue is $\frac{1}{2}$. But since all the balls are either light blue, dark blue or not blue and we do not know any reason for thinking either will predominate the probability of each is $\frac{1}{3}$ and consequently the probability of the first ball not being blue is $\frac{1}{3}$ only and yet is $\frac{1}{2}$ which is self-contradictory.

I will put this argument in another form. The doctrine that, absolute ignorance is represented by $\frac{1}{2}$, can only be justified by the principle that possible events of which nothing more is known are equally probable. But suppose *A* and *Z* are the only possible events and suppose *A* may be of two kinds, *b* and *c*. [Then] $A\begin{cases} b \\ c \end{cases}$ and *Z* [are the events]. Then since *A* and *Z* are equally probable upon this principle and yet *b*, *c*, and *Z* are also equally probable, which is impossible. There are only two modes of escaping this contradiction, 1st to forbid the division of any event into its kinds and second to insist upon the division of every event into all its possible kinds. But 1st if no event is to be divided into its kinds, what is the probability of the event *b* upon this principle? Answer: $\frac{1}{2}$. What is the probability of the event *A*? Answer: $\frac{1}{2}$. Thus we are landed in a new contradiction. If on the other hand each event is to be divided into all its possible kinds, since every event has an infinite number of kinds, the probability of any event is infinity divided by infinity — the value of which we do not know. For instance there are any infinite number of shades of colors; so the probability of drawing a ball which is blue is equal by that principle, to the number of shades of blue, divided by the number of all shades and as both these numbers are infinite, the probability is infinity divided by infinity which is we know not what.

The notion, therefore, that absolute ignorance can be represented by a probability of one half — or an even chance must be entirely given up. Is it an even chance, asks Mr. Mill in the first edition of his Logic, that the inhabitants of Saturn have red hair? In subsequent editions he seems to answer that it is; but we have seen that this is wrong.

Let us now return to our original question. We draw a few balls from this bag as samples of its contents; and as we find that there are all red we infer that all the balls in the bags are red. *Why?* Almost all the mathematicians who treat of probabilities tell us that their science can afford the solution of this question; although the principles upon which different mathematicians solve it are widely different. Let us then undertake the examination of the competency of mathematics to answer this logical question.

The question, then, is: how frequently when seven balls successively have been drawn from a bag and all are found to be red, will it prove that all the balls in the bag are red, or that any large proportion are red. Now this plainly depends upon the frequency with which a large proportion of the balls in a bag are red, because if it were very very rarely that this happened, it might be a more common occurrence to draw out the same ball seven times successively, extraordinary as that is, than to find a bag of which all the balls were red.

Suppose I were to open a book and look at seven pages, and were to find that the first letter upon each of these pages was a vowel. I should not conclude that every page of the book began with a vowel; because that would be even more extraordinary than that the first pages that I lit upon began with vowels.

It is the same with the red balls; unless we know how frequently any proportion of balls in bags in general are red we cannot say how frequently that proportion will be found to be red after we have found the first seven drawn were red; unless the frequency of the latter has been made a matter of observation.

If the mathematicians are allowed to assume, as they usually do, that any one proportion of red balls is as probable as any other, they certainly can solve the problem; but they cannot do this upon the basis of absolute ignorance, and it is supposed no observations have been made as to frequency of the occurrence of red balls.

I will yield to no one in my admiration for the genius of great mathematicians; in this very matter, it is wonderful how well they have made up for a want of the logic of induction; but I am bound as a student of logic to say that the premisses upon which they have proceeded, are logically invalid and that consequently their results do not apply to those problems to which they have applied them.

Mathematical transformations, it is admitted, are deductive; that is, they can only infer what was implicitly given in the premisses. But a probability is a matter of fact; namely, a frequency of a species of events relatively to its genus. Hence, no correct ciphering can begin with ignorance and land us in a knowledge of a probability.

And, hence, mathematics cannot explain why we are justified in believing that all or nearly all of the balls in this bag are red, when we have drawn out only *seven*.

The argument to prove this which I have presented to you in a general way will be found given with mathematical precision in the great work of Professor George Boole — a man who united a genius for mathematics with a high originality as a logician. His book is Logic put into Algebra. Every proposition makes an equation and the process of inference is an algebraical process performed upon equations. The system is a marvellous piece of ingenuity, and as sound as it is ingenious. And the result which I have presented to you, so imperfectly, will be found capable of the exactest proof by his method. I would recommend anyone who desires to know any easy and certain method of solving questions of probabilities to study Boole's *Laws of Thought.*

The inference which we have made with reference to all the balls in this bag from only seeing a sample of them, is what we call an Inductive inference. It is very similar in its nature to hypothesis and may be taken as the type of scientific reasoning in general. Those capitalists who layed out so many millions in laying the Atlantic telegraph — not because they were improvident and overweening madcaps for they were solid Englishmen intrepid, unflinching, and cool; upon what did they rest their hopes of success? Why, upon just as an argument as this about the balls. If a telegraph can work across the channel, they said, it can work across the Atlantic. Every argument by which we get to any new truth is also of such a kind as this. The faculty for this sort of reasoning makes up shrewdness, and is the essence of genius. The alliance of man with the divinity is more plainly seen here than anywhere. He observes the regularities of the animal kingdom now, and he knows from that how it was in some geological era — a million ages ago. He observes that a thousand, or a million or a billion men have died, and he leaps to the fact that all men *will* die; — he has not *observed* it of those who now live but he *knows* it of them and all other men who ever shall be, though they be so numerous that a billion will be to them but as the number of grains of sand in ten thousand cartloads to all that lie upon the sea-shore. In short, he observes the finite and he seems to know the possible infinite.

How is it possible that this should be; how can we comprehend such a proceeding? We call such inference, resting upon observation; but when we have drawn seven balls from this bag, we have not observed a single one of the remainder which we infer are all red. So astounding does such a faculty seem; that some logicians have said — it is inspiration and nothing less. This you may be sure is said not by any sneerer by trade who is bound to see nothing wonderful in the totality of things, and who means that Inspiration is only Shrewdness; but by pious and spiritual minds who can see that even Shrewdness is essentially Inspiration. For my part, I could not imagine a more sublime manifestation of the Deity than that which thus appears in the nature of inference itself. Nor can

I conceive of a function so appropriate to Infinite Power as that of regulating not the universe merely but the very consistencies upon which the possibility of a universe depends. And it seems to me that the power of judging the unseen by this seen, — even if it be applied to the most sordid and wicked objects — affiliates man to the Creator of all things.[1]

Nevertheless I must say that these theological conclusions do not explain Induction; they are no part of *Logic*. Man requires to comprehend his own arguments; and unless he can comprehend them he is dashed from this lofty pinnacle to the level of an irrational machine. If he is impelled, he knows not upon what principle, without any conscious principle from one belief to another, he has no more reason than the pen with which he writes. But the very faculty of judging the whole from the parts, of classifying objects presented to him, enables him to classify arguments, to state what they all have in common and thus to enounce the principle upon which they proceed. The lofty speculations of the theological logicians are therefore not needed in logic, for the mere faculty of colligating facts and drawing general conclusions from them, a faculty however which appears to be as divine as any other — is quite sufficient to show the rational nature of induction.

The same logicians who take this view of scientific presumption, also sometimes tell us that induction rests upon the goodness of God. There is one sense in which this would be true, but irrelevant; namely if it were intended to say that the faculty of making inductions is evidently given by the goodness of God and so testifies to that goodness. That would be true; but would not explain upon what principle our knowledge of the truth of induction rests. But that is not what is meant by the logicians I refer to. They mean that our knowledge of the goodness of God is

[1] A marginal reference to *B* seems to indicate that a note *B* found with this manuscript was intended to be inserted in the lecture somewhere in this paragraph. The note reads in part as follows:

> "Yet, already, I am put upon my defence, for some persons think there is no sense in this question; no need of it and no answer to it. They also decry the question as an unnatural one. Without stopping to reflect upon the odd objection that the question is unnatural, I shall go on to consider whether it means anything, whether it is of any use, and whether it admits of any answer.
>
> It is urged that the question *why* the future generally resembles the past, and what we come to know what we have hitherto known, means nothing, because the question *why* asks only for a rational account, an account in accordance with the principles of inference, but the fact we now consider is a principle of inference and therefore it is itself an answer to *why*, and this question cannot be applied to it.
>
> It is also urged that the question is meaningless because the question *why*, seeks only for a general statement of the facts, but the fact here presented is a general one and therefore *why* cannot be asked of it ...."

our Evidence of the truth of induction. I think they forget that if the logic rests on Theology, theology cannot in its turn rest upon logic. One or the other must be known otherwise than on the testimony of the other. Theologians drive their arguments more recklessly than any other class of thinkers in the world. They have such confidence in their ultimate conclusion, that they care not what risks they make their premisses run. Some Divines of every stripe, from him who thinks he could not go the heaven if it were not for his total depravity, to him who thinks that the infallible accuracy of the account of creation in Genesis is indispensable to our faith, will tell you that their whole religion hangs on this or that extremely doubtful proposition. But of all logical improvidence, it seems to me that the greatest is that of those men who would deprive themselves of the advantage of using Induction and Scientific reasoning generally, as an evidence of the goodness of God. If this most goodly frame the earth, and this most excellent canopy the air, look you, this brave o'erhanging firmament, this majestical roof fretted with golden fire, will not prove that God is good, think you that any syllogism which is at best but a barren turning about of what we know already is going to do it? Fie! There is no sense in such a thought. And yet this is the logical consequence of resting Induction upon Theology.

At the next lecture I shall consider another common answer to our question.

F. [LETTERS TO COUSIN JO] (L366)

Milford, Pa.
26 June 1909

My dear Cousin Jo:

I seize the first time that has been at my disposal since I had the pleasure of receiving your letter to answer it; and the only difficulties I have in doing so are 1st that I am not quite sure I understand your way of using those troublesome words, the selective pronouns, *some* and *any*; and 2nd that I do not find in your letter any indication, one way or the other, of how much you are given to mathematical ways of thinking. By "mathematical" ways of thinking, I mean making a diagram on paper or in imagination, whether it contain lines or is merely an array of points, and experimenting on the possible variations of it, and generalizing the results. The only safe way will be to assume that you are little given to the mathematical kind of thought, though your interest in the problem seems to belie that, since from any *other* point of view, the notion that Bacon wrote Shakespeare's plays seems to me to have precious little to recommend it. The limitations of Bacon's style are so obtrusive, — especially when he writes English, — that to me the notion in question seems utterly out of the question.

As to the treatment of it by the Doctrine of Chances, I am sorry I have not a spare copy of the little book entitled "Studies in Logic, by Members of the Johns Hopkins University. Boston: Little, Brown, & Co. 1883." But you can probably find it in one of the libraries. In that I discuss the whole subject of probable inference. I have learned a great deal about it since then, and have learned that there are various ways of reasoning of which I then took no account, and which I then may have pronounced illogical, but which I have found are not absolutely so, and though usually of little or no value, sometimes rising to high importance. In fact, my whole notion of reasoning and of the mind has been quite remodelled since then. Nevertheless, the general contents of that essay have only been strengthened by my subsequent criticism of it, as far as it goes. In particular what I there say of the utter nonsense of one great division of the modern Theory of Probabilities, as taught in all the text-books, is entirely sound. My natural disposition makes me, as soon as I have once published an essay, pooh-pooh it, try to pick flaws in it, and search for considerations by which I may escape its conclusions; and therefore when this has gone on for several years, and I can find only confirmations of it, and at length yield to my former reasoning, I am all the more confident that it was correct, either just as it stood or with such subtractions as I have been able to make. In the present case, I have been attacked, besides, both publicly and in letters from some of the most distinguished thinkers in the world, and my view has only been made more and more evident. I hold that when we *know* in the beginning, either by statistics, as in insurance business, or by the nature of things, as in reasoning about throws of dice, that there are certain probabilities that on certain kinds of occasions certain kinds of results will occur; or in other words, know that were certain general states of things to remain unchanged, those kinds of results would in the "long run" occur in certain proportions of such occasions; which again means that, the general conditions remaining unchanged, if we were to keep tally of all the occasions of a given kind which turn out in a proposed way, and at the same time keep another tally of all the occasions of that same kind that do not turn out in that way, and after each occasion divide the number of the first tally by the sum of the numbers of the two tallies, and were to keep up this practice endlessly, then if the ratio which thus changes its value at each new occasion, its denominator being increased by one, while its numerator either remains the same or is increased by one (so that two successive values *cannot* be equal unless all the values from the beginning have been equal, and either all equal to *zero* or to *unity*), if that ratio I say be one that will have any definite value at all in "the long run," — that is to say in an infinitely long succession of occasions, — this state of things will consist in an irregular narrowing of the limits of its fluctuations of values; so that taking any one possible value of the varying ratio that you please, there will either be a time (though we can never know when it comes) after which the varying values will never transcend that value, but will all be greater than it or all less, with the exception of one single value which the ratio will never cease to cross over; and that single value is what is meant by the value of the ratio "in the long run." And the value of the ratio in the long run is the *real* (or "objective," — a word I avoid) probability of the kind of event whose every occurrence adds one to the numerator, or "antecedent," of the ratio, on that kind of occasion, whose every occurrence adds one to the denominator, or "consequent" of the ratio. This is the only accurate definition of "probability," in the sense of the Doctrine of Chances, that I have ever seen; for the best of those I have seen fall into a vicious *circulus in definiendo* by virtually, if not expressly, using in the definition the very concept that was to be defined, and the other definitions either make probability to be something unreal, and therefore no safe basis for such a vast business as Insurance, or assume some highly dubious theory (as when they say it is the measure of the strength of belief that we ought to have in the event, which modern Metrics quite refutes), or they introduce phrases of no meaning at all (like Laplace's "également possibles"). There are *other* quite proper senses in which the word "probable" can be used; but the sense I have defined is the only one that can properly be treated by the calculus of probabilities. "The measure of the proper belief," say that Mars is inhabited by rational beings, must be a ratio to some arbitrary standard adopted as a unit or standard of measurement, and the *ratio*, or numerical part of it will consist of the logarithm of any good reckoning of the strength of the experience in our possession that favors that opinion *minus* the logarithm of the strength of experience which would render it absolutely certain that Mars was inhabited *minus* the logarithm of the actual strength of the experience that is averse to its being inhabited, *plus* the logarithm of the strength of the experience that would render it absolutely certain that Mars was not inhabited; but I am not yet prepared to say how these "strengths of experience" ought to be reckoned. I only say that when that is ascertained, this measure being added to the similar measure of any new argument bearing on the same conclusion will give the whole weight of evidence in its favor. That is as near as we can come at present to saying what "the measure of the just belief in Mars being inhabited" will be; and it shows just what is needed to enable us to measure such arguments.

Since 1543, the epoch of the publication of the *De Revolutionibus* of Copernicus, there has been, one may say, *no* scientific study of Logic, meaning by "scientific" study, investigation by a considerable group [of] men who devote their lives to pursuing it according to the best established methods of their times and working in coöperation. I hold it to be self-evident that Logic, if it were such a science, would be the most important of all sciences; for I use the word "reasoning" to include the whole operation passing from one state of knowledge to another; and by Logic I mean all that investigation which aims to discover the nature and workings of reasoning. In ancient times the Greeks alone opened up a deliberate theory of reasoning; and they originated no less than 5 distinct systems of logic. For in the first place there were those disjointed and fragmentary attempts of the sophists, the philosophers of Megara and others to teach reasoning. Then there was Aristotle. Then there was the method of the mathematicians, with their diagrams. They recognized, what Aristotle had completely overlooked, that in order to reason to any effect (they might have said, *at all*), it is necessary to have a concrete representation of the premised state of things, that representamen being capable of alteration, and that it is by experimenting upon the effect of alterations that new truths are discovered. They also began the study of the different kinds of Signs. I define a sign as anything, — be it an existent thing or actual fact, or be it, like what we call a "word," a mere possible form to which an audible sound, visible shape, or other sensible object may conform to, or be it a property or habit of behaviour of something either experienced or imagined, — which is on the one hand so determined (i.e. affected either by causation or through the medium of a mind) by an object other than itself and on the other hand, in its turn so affects some mind, or is capable of doing so, that this mind is thereby itself mediately determined by the same object. This is a very broad conception, but the whole breadth of it is pertinent to logic. I will remark, by the way, as to my statement that the object of the sign must be other than the sign itself, that we can imagine a map of the United States, or better, of the District of Columbia to be laid down on Lafayette Square which should be so microscopically accurate that it should show Lafayette Square and the map lying upon it and a point on that map which should represent that very point itself. Indeed, if a map represents a region where itself lies on the ground, it must be a very peculiar kind of projection if there is *not* some point of it that represents itself. But though a part of a sign may thus have itself for its object, and although a whole map might represent nothing but itself, yet each part of it cannot represent nothing but itself without losing all the essential character of a sign. The Greek mathematicians recognized as essential to reasoning not only a concrete diagram of the subject of the reasoning, but also arbitrary signs, such as the lettering of the point of a geometrical figure, as well as signs which denote whatever there may be that conforms to certain conditions, such as their words σημεῖον, ἐπιφάνεια, γωνία, κύκλος, etc. And these three kinds were recognized as generally requisite for reasoning. Besides that, they had terms for propositions fulfilling different logical functions such as ὅρος, αἴτημα, πρότασις, ἔκθεσις, κατασκευή, ἀπόδειξις, with the certain set phrases, such as ὅπερ ἔδει δεῖξαι, etc. Here was therefore an independent and very valuable development of logic. One might almost say there were two, since the terminology of the mathematicians as reported by Plato, Aristotle and other nearly classical writers, is entirely different from that of Euclid, Apollonius, and Archimedes.

In the *4th Place*, we have the logic of the Epicureans, which is entirely different from any other of ancient times, but most singularly close to some modern logic. We are pretty well informed upon this subject because in addition to Sextus Empiricus and others we have a treatise on inductive logic by Philoponus, found in Herculaneum; and although this was horribly mutilated in unrolling it, enough fragments remain to enable us, by the aid of the reports about him in Cicero, who attended his lectures, to enable us to reconstruct what was probably the main substance of the tractate. This substance is easily comprehended; for it is, excepting in minutiae, the very same as the Logic of John Stuart Mill. He says that all our knowledge comes from induction, and the validity of induction is due to uniformities in nature. Only while Mill attends almost exclusively to the kind of uniformities which consist in certain characters generally belonging to the whole of a large class of objects, if they belong to any of them, as we note when he says (Logic, Bk. III, ch. iii, § 3) "When a chemist announces the existence and properties of a newly discovered substance, ... we feel assured that the conclusions he has arrived at will hold universally, though the induction be founded but on a single instance.... Now mark another case, and contrast it with this. Not all the instances which have been observed since the beginning of the world in support of the general proposition that all crows are black would be deemed a sufficient presumption of the truth of the proposition to outweigh the testimony of one unexceptionable witness who should affirm that in some region ... not fully explored, he had caught ... a crow, and had found it to be grey. Why is a single instance, in some cases, sufficient while in others myriads [are not so]. Whoever can answer this question... has solved the problem of induction." One would naturally expect to find that his answer would be that we know in general that different samples of the same chemical consitution have almost identical properties except in the rarest cases, while different species of the same family of birds are known to exhibit endless variety. And this is, in fact, the kind of uniformity that Philodemus emphasizes. Mill, on the other hand, as we find as we go on further means by a uniformity the tendency of certain characters to extend to the whole of any large class in which they appear at all ("generic character" and "family character" etc.) while others, called "specific" or "individual" are usually confined to small classes. Of course besides this Mill fancies (as for that matter most people do) that it is a characteristic of nature that under similar circumstances similar results occur. It is not so. This is purely subjective, due to our only paying attention to the characters of which this is true, for the obvious reason that they alone are *significant*. Suppose you phrase the doctrine in another way and say, Almost all characters are significant. Is that true. I sneeze at the moment when a star is occulted by the moon. Is it not evident that there are millions and millions of such insignificant facts for every one that is significant? The Epicureans were, moreover, like Mill in being extreme Nominalists. I do not wonder, therefore, that, as far as I remember none of the eminent physicists of ancient civilization were Epicureans. In the XIXth century nothing prevented a disciple of Mill from being an eminent physicist, because the methods of physics were so well-settled that it was no longer dependent upon logic, in those many directions in which nominalism was not *obviously* hostile to scientific research. I think it probable that the deeper questions of physics, in which it has made no progress except what was due to Atomism, which has no necessary connexion with Nominalism, would have been better understood in our time had Nominalism not been prevalent.

The fifth system of Greek logic was that of Chrysippus and the Stoics. This was worst of all the Greek systems. It was eminently a Philistine philosophy, — I mean one attractive to persons who had never thought deeply upon philosophy. To any worthy ignoramus, its Ethics, which was the heart of it, would seem to be just what was needed to correct the degeneracy of later antiquity; and such excellent persons as Epictetus and Marcus Aurelius probably knew nothing at all about Logic. Zeno himself knew very little about it. There was, nevertheless, a very marked stoic system of Logic. Most of all the ancient systems it was opposed to induction which it pronounced altogether fallacious. No other did so. It was also thoroughly Nominalistic. Notwithstanding this, several of the Greek mathematicians and physicists are set down as stoics. (*Stoic* ought not to be capitalized, since it is not derived from a proper name. I capitalize Logic, Ethics, Nominalism on account of their supreme importance.) Probably they were not life-devoted at all to philosophy, but had merely heard some Ethical lectures in the stoa, and approved of them.

Contemplating the five ancient systems collectively, they appear to me to be fairly adequate to the range of ancient reasoning. This was quite unquestionably the case with the Middle Ages, when logic (including the philosophy of grammar) was the only true science there was. It is a shame that there is no good history of it. Hauréau's *Philosophie du Moyen Âge* (as if there was any philosophy at that time except logic, the philosophy of grammar, and nonsensical theology crammed down men's throats by terror) is utterly superficial, and Prantl's much admired *Geschichte der Logic im Aben[d]lande* is the most ridiculous book I know of. It happens that Prantl's account of Aristotle's logic is distinctly good; and I suppose that appreciating that, and knowing absolutely nothing about any other system of Logic, people have taken the rest of the book on trust. But there is no kind of fault except timidity that does not mark this work above all others. It is true that, living in Munich, whose library, I fancy, is the richest in the world in incunabula, he has read the colophons of an immense number of books, and of a good proportion of them he has read a few pages. But he has never consulted a manuscript except of one William Shyreswood. He has quoted in full, as remarkable, passages that contain only words the medieval author expected every reader to recognize as familiar quotations, has passed over in silence some of the matter that is most remarkable, he pronounces *ex cathedra*, but in the accents of an angry woman, the most ill-founded judgments, and he is very much addicted to baseless hypotheses, which he might sometimes have readily refuted by looking into MSS, but which in all cases he simply states as indubitable truths. He has one axiom by which he solves all the historical problems relating to the logic of the middle ages; namely, that everything written about Logic in Western Europe during the middle ages must have been *taken* bodily from some previous writing. Now this is one of his hypotheses as baseless as any could be. Had he said the same of *modern* logic, there would have been a good deal more ground for it. Except those two works, so far as I know, there is nothing else about the Logic of the Middle Ages excepting detached dissertations on special points; and so far as I am aware, and I am pretty well read in that direction (to state the matter modestly), people are so profoundly ignorant of medieval logic, except some catholic priests who have read that great work the *Summa Theologica* of St. Thomas Aquinas, the *doctor angelicus*, and a good many of them the same doctor's *Summa contra Gentiles*, and a good many have dipped into the *Doctor Subtilissimus*, John Duns, the Scot. (Probably, his family belonged to the Town of Duns, in the SE corner of Scotland, although there is ground for surmising that he may have been born 36 miles from there, in the purlieus of the castle of Dunstanburgh on the shore of Northumberland.) Their reading, however, has mostly been perfunctory, and extremely few or none of them are qualified to form a valuable judgment of the substance of what they have read, because their studies of logic were too superficial. I think that I was, when I began my readings of scholasticism, as well prepared for it as anybody, not only by the extent of my other logical reading and the depth of my thinking, but likewise by my comparative freedom from prejudice; and yet it was only after many years of reading and reflexion before I did justice to scholastic logic.

In the first place, I consider the style, say of Duns Scotus, to be remarkably adapted to the subject. He asks a definite question, and then states briefly all the considerations which are likely to occur to his students in favor of the side of the question which he opposes and then all the reasons which are likely to occur to them for holding his own opinion. Next he points out the capital distinction or other decisive point upon which the right view hinges and shows how that leads to the right opinion. Then, if there are apparent objections to this, he takes them up, one by one, and refutes each. Sometimes, this part of the discussion is very intricate. Finally, he returns to reasons originally given against his opinion, and shows how, in the light of what has been said each one loses all its force. This manner seems to me superior to that of Aristotle in treating, say, Metaphysics; and Aristotle held his own as a rhetorician at Athens against Isocrates, for several years, and is praised by Cicero to the skies for his "eloquentia." The truth is that those who complain of the style of the scholastics are only echoing the impressions of the Greeks who came over to Italy about 1453, men whose own Greek is neither classical nor in good taste, and who judge the scholastics from the point of view of those who regard reading merely as an amusement, and find fault with all Latin but the language of Cicero.

The scholastics, — the really good ones, — are as far as possible from being dull. They are subtile thinkers; above all, Scotus. It is true that they value only necessary reasoning, which is a fundamental error. But it was practically the error of Aristotle, who though he admitted the indispensableness of induction, always reasons deductively. It is, however, quite true that scholasticism could not discover much because its reasoning aims to be syllogistic. Fortunately, there is an undercurrent of thought of another character ....

Milford, Pa.
29 June 1909

My dear Cousin Jo:

I feel that it would be an act of treachery merely to answer your questions without pointing out how little they have to do with your ultimate purpose. I have also a difficulty in giving helpful answers owing to my not knowing how much of your mathematics you remember.

The Doctrine of Chances is a branch of Logic; and therefore it has necessarily been an object of close study with me. As a branch of logic, it must rest, first of all on *logical analysis*, and secondly, as considering complicated states of things, on *mathematical analysis*. Since the epoch of Copernicus, A.D. 1543, the science of logic has been totally neglected. So far as I know, I am absolutely the voice crying in the wilderness, *vox clamantis in deserto*, on the subject, — with the exception of (of course I don't count imbeciles) a few who have studied certain *special* points. But it is not special studies that are needed *as yet* in logic, but a general conception of the science and its possible efficient methods. The result has been that the Doctrine of Chances has been left to the Mathematicians, with the result that on the mathematical side, it is finely developed. But mathematicians have been the merest babes in logical analysis, and have fallen into gross absurdities in consequence. They have a superstitious awe of all that Laplace lays down in his great work on the subject, although Laplace never, as far as I remember, attempts a definition of probability, and makes the foundations of the subject depend on the phrase "*également possibles*," which means nothing at all unless it means *equally probable*, in which case he falls into a vicious circle in using it as he does.

We cannot reason about probabilities until we know what *probability* is. Now according to *pragmatism*, which word I invented to denote my method of getting at the meanings of words and other signs, external or in the mind, the first step toward ascertaining what probability means is to consider what is proposed to be done with it. The most important thing we do with probability is to found the business of insurance upon it. In order that it should afford a secure basis for insurance, — secure for the underwriters and safe for the insured, it must be a *fact* to say that such and such as probability exists that in a given year or other occasion a given sort of event will happen. What objective fact can that be? The fact that once in so often that sort of event will happen *in the long run*. But what is a "long run." I suppose the underwriters would say "Oh, if it works as long as we live it will do for us." But I ask them, well do you mean to say that you are absolutely certain that so and so often an event will turn out in a way favorable to you during your life-time. He may reply "Oh, not precisely; but there is no doubt it will with *about* the frequency calculated." To which I reply, You may have no doubt of it, but it is not absolutely certain, according to your own calculations, that the fact will be anywhere near what you calculate. Therefore, I want to know what you mean by the "*long run*." He will be forced to say that he means an endlessly long run. But what does he mean by saying that the ratio of favorable events to all the events has a given value when all the events are innumerable? I suppose he might thoughtlessly say that he meant just what we usually mean by the ratio of two infinite numbers in the infinitesimal Calculus. But I shall at once show him that that reply will not do at all. If one looks into any modern introduction to the calculus, he will find that by saying, in *that* branch of mathematics, that a ratio y/x has a certain value, r, when x is infinite, means that if ε be any error, no matter how small, then there is a value of x such that for *all* greater values $\pm(y/x - r)$ is $< \varepsilon$, where the $\pm$ merely expresses that the absolute value unaffected by the sign $+$ or $-$, the "modulus" of $(y/x - r)$ is less than ε. Now no such thing can be said in the case of probabilities. It cannot be said, for example, that when a coin is pitched up a great number of times, that the quotient of the times it comes up heads divided by the total number of times it is pitched up is *certain* to differ from $\frac{1}{2}$ by a quantity less than a given quantity. There is no certainty about any value whatever in the future. So we certainly do *not* use the language of the calculus when we say that an event will happen once in so often in the "long run." What *do* we mean then? It is absolutely indispensable to know what one means if one desires to avoid error. Well, the *only* way to express in terms of *certainties* what we mean by a *probability*, and even then the *certainty* is only of the kind termed strictly *moral certainty*, or inductive certainty, and not *apodictic certainty* is some way like the following (of which there are obviously innumerable variants). Namely, what we mean when we say that the probability is $\frac{1}{2}$ that a perfectly formed cent, if tossed up will come down heads, is that if we keep two tallies, the one of the heads that come up and the other of the tails that come up, and if we were to toss it up unceasingly, and if after each trial we were to divide the number of heads so far, by the sum of the numbers of heads and tails so far, there would certainly come a time after which the quotient so obtained would never again amount to 1 or be so small as 0; and there would certainly come a time (though we never could be sure that it had come) after which the quotient would never again be so small as $\frac{1}{3}$ nor so great as $\frac{2}{3}$; and there would be a time after which the quotient would never again be so small again as $\frac{2}{5}$ nor so great as $\frac{3}{5}$; and whatever whole number N may be, there would be a time after which the quotient would never again be so small as $\frac{N}{2N+1}$ nor so great as $\frac{N+1}{2N+1}$. And in short, while the value of the quotient would necessarily change after any pitching-up, whatever value less than $\frac{1}{2}$ would some time or other cease forever, with all smaller values to reappear, and so given any ratio greater than $\frac{1}{2}$ there would be a time after which no quotient so great as that would reappear; but there never would be a time after which there would not be values of the quotient that were less than $\frac{1}{2}$ as well as values greater than $\frac{1}{2}$; all these certainties being *moral certainties*. But of course we never can know with absolute certainty the value of any probability of an *experiential* event. We know that a perfect cent tossed up in such a manner that the result of one pitching had no influence on the result of another, nor any tendency to result in heads rather than tails or the reverse, would have a probability of precisely $\frac{1}{2}$ of coming up heads. But we never could be certain that those conditions were fulfilled; nor could we without enormous labor with any ordinary cent and with ordinary fairness in pitching have any serious grounds for supposing that there was a decided tendency for it to come up one way rather than the other in the long run.

If anybody will procure ten half-eagles at each of a hundred ordinary places, — banks and reputable money changers, — and keeping each decad separately will toss up each half-eagle 250000 times and give me his thousand numbers of heads, I will tell him what the probability is that a half eagle, obtained in any prudent way will have a probability differing not more than a given amount of coming up heads. For I shall know what he means by that question.

But if anybody presents me with a lot of data of any kind and asks me to deduce from them the mathematical probability that Francis Bacon wrote the plays of Shakespeare or had any hand in them, I shall not have the slightest notion of what he means, but shall shrewdly suspect that he does not mean anything that ought to be called by the same name as the Insurance companies' "probabilities"; and nothing numerical, ...[1]

... convergent series which gives the value with accuracy amply sufficient for the calculation of probabilities. A semi-convergent series is one which begins by converging toward the truth, that is, the addition of each new term gives a value nearer and nearer the truth, but after a while (depending on the particular value you take), the values begin to diverge from the truth, and faster and faster. But in practice such formulae are often all that could practically be desired.

I will first write $\frac{(mn-a)(mn-a-1)(mn-a-2)}{mn \quad (mn-1) \quad (mn-2)}$ so as to introduce two usual and useful notations. The first is $\frac{(mn-a)^{\underline{|3}}}{mn^{\underline{|3}}}$, as the English write it, or $\frac{(mn-a)^{(3)}}{mn^{(3)}}$ as Continentals often write it, the 3 being a sort of exponent, an *exponens facultatum*, as the phrase is in those works that are written in Latin, or a *factorial index* in English. This notation is important because just as we write D_x for $\frac{d}{dx}$ and $D_x x^m = mx^{m-1}$, so Δ i.e. $(x+1)^{\underline{|m}} - x^{\underline{|m}} = mx^{\underline{|m-1}}$. Moreover by $x^{\underline{|-m}}$ is meant $\frac{1}{x(x+1)\ldots(x+m-1)}$ where there are m factors, each decreasing owing to a change of 1 in the value of x; and $\Delta x^{\underline{|-m}} = -m^{\underline{|-m-1}}$. Another notation, still more frequently used, where N is any positive whole number is $N!$ meaning the continued product of all whole numbers from 1 up to and including N. Hence, if x is a whole number, $x^{\underline{|m}} = \frac{x!}{(x-m)!}$ and this is important as showing that in order to calculate a "factorial" (or *facultas*) it is only necessary to be provided with a convenient formula for calculating $N!$ Now there is such a formula, called "Sterling's formula," which involves the semiconvergent series I alluded to above. I will show how it is derived. For that purpose, I must first explain several notations that are used in the Calculus of Finite Differences. In the first place, the *idea* of a function is different in this calculus from the idea in the infinitesimal calculus. In the latter by fx we mean some quantity

[1] A page is missing here. It undoubtedly served to highlight the usefulness of a knowledge of the Calculus of Finite Differences in probability contexts.

produced by mathematical operations upon x, but the function of finite differences, which is preferably written u_x or v_x or any letter with an x subscript, means any quantity which may change when x is changed. Thus x may be the ordinal number of a man's name in the City Directory and u_x may mean the number of that man's house in the street he lives in. Then for every value of x not exceeding the number of names in the Directory there is a value of u_x, though in the Differential Calculus we should not ordinarily say that the street-number of a man's house was a function of the ordinal number of his name in the directory; because in that calculus we conceive variables to vary *continuously*. We may define $\Delta_x u_x$ as $u_{x+1} - u_x$; so that, as a *notation* merely, we might write

$$\Delta_x = \frac{u_{x+1} - u_x}{u_x}.$$

We have

$$u_{x+1} = u_x + \Delta_x u_x$$
$$u_{x+2} = u_x + 2\Delta_x u_x + \Delta_x^2 u_x$$

where $\Delta_x^2 = \Delta_x \Delta_x, \Delta_x u_x = u_{x+1} - u_x$ *and* $\Delta_x^2 u_x = \Delta_x(u_{x+1} - u_x) = u_{x+2} - u_{x+1} - u_{x+1} + u_x = u_{x+2} - 2u_{x+1} + u_x$. And transposing, and substituting $2\Delta_x u_x$ for $2(u_{x+1} - u_x)$ we get $u_{x+2} = u_x + 2\Delta_x u_x + \Delta_x^2 u_x$; and so we find $u_{x+3} = u_x + 3\Delta_x u_x + 3\Delta_x^2 u_x + \Delta_x^3 u_x$. Now we write also $E_x u_x$ for u_{x+1} and call this the "enlargement according to x" *of* u_x. Therefore $E_x^2 u_x = u_{x+2}$ and $E_x^n u_x = u_{x+n}$. Then it is easy to make out that $u_{x+n} = u_x + n\Delta_x u_x + \frac{n!}{2!(n-2)!}\Delta_x^2 u_x$ + etc. or $E_x^n = \Delta_x^0 + n\Delta_x + \frac{n(n-1)}{2}\Delta_x^2$ + etc. $= (1 + \Delta_x)^n$ where $(1 + \Delta_x)^n$ means the expansion by the binomial theorem *in form*.

I do not know whether you ever studied the calculus; although a general acquaintance with the differential and integral calculus and theory of functions is indispensable to understanding modern logic or modern reasoning about such subjects as economics. That being the case, in my opinion those easy branches of "higher" mathematics are really essential to a liberal education; and no man without a liberal education ought to be assigned to the duties of casting a vote in the government. Had that been the rule with the ignorant fathers of the republic these labor and capital difficulties would not have existed. For any mechanic can learn the differential calculus, and would have done so if his having a vote had depended on it. However, since the colleges do not reckon any acquaintance whatever with the differential calculus indispensable, you very likely never knew what MacLaurin's theorem is. I shall, therefore, have to tell you. $D_x fx$, the *derivative*, or more explicitly, the *differential coefficient* of any function, relatively to x, is the *rate* at which, when x begins to increase, fx consequently increases relatively to x. That is to say, if we give a very small increment i to x, fx becoming $f(x+i)$, we shall have an equation $f(x+i) = fx + if'x +$ etc. where the *etc.* stands for further terms which will be smaller the smaller the value of i, and where $if'x$ is that part of the increment of fx which is proportional to i when fx is increased (or diminished) by the very small increment i. Suppose we make a diagram of fx, laying off the values of x horizontally, and those of fx vertically, and get such a diagram as this [Fig. 1]. Then

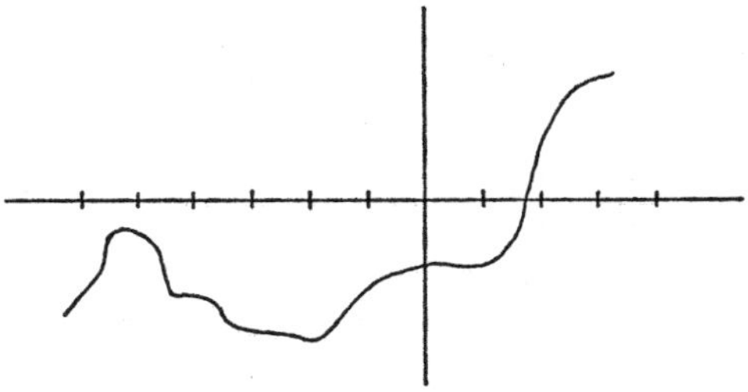

Fig. 1

at any point on this curve we draw a tangent to it. We may say that for an infinitesimal (or vanishing) distance the curve coincides with the tangent.

I have posted on this sheet a bit of squared paper on which the little crosses are simply mistakes to be altogether disregarded [Fig. 2].[2] The

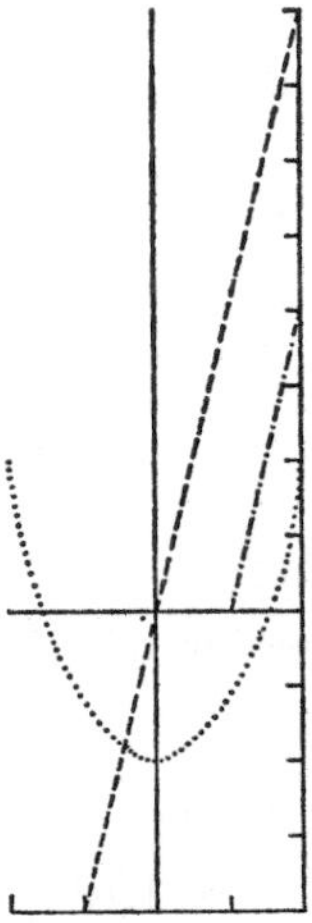

Fig. 2

[2] This is the dot-dash line in Fig. 2.

two [perpendicular] lines I have drawn are the axes of x and y. The fine black dots lie on the curve $y = fx = x^2 - 2$. The [dashed] dots are at heights above the axis of x equal to the increase of y (for the black dots) per unit of x at the same value of x, i.e. on the same vertical. That is to say, the [dashed] dots have $y = D_x fx$, fx being always $x^2 - 2$.

When a symbol for an *operation*, such as D_x has an *exponent*, that exponent expresses how many times over the operation is performed. For example, $D_x^2 fx$ means $D_x(D_x fx)$ and $D_x^3 fx$ means $D(D^2 fx)$, etc.

We have the habit of writing $D_x^n f0$ to mean that $D_x^n fx$ is to be algebraically expressed in terms of x and then x is to be put equal to zero. For example when $fx = x^2 - 2$, $D_x fx = 2x$, and $D_x^2 fx = 2$. Consequently f having this meaning $D_x f0 = 0$, but $D_x^2 f0 = 2$. But it seems to me more consistent to write $D_0^n f0$ instead of $D_x^n f0$, and I shall do so· Now MacLaurin's theorem is, that for ordinary functions, which are expressible in a series according to powers of x,

$$fa = f0 + aD_o f0 + a^2 \frac{D_o^2 f0}{2!} + a^3 \frac{D_o^3 f0}{3!} + a^4 \frac{D_o^4 f0}{4!} + \text{etc. ad infinitum.}$$

Now I must mention that if F and Φ and Ψ are [three] symbols of complex operations composed only of additions, multiplications, and raising to positive integral powers of nothing but such compounds of D and constant numbers, then $(F\Phi)D = F(\Phi D)$ and $(F + \Phi)\Psi = F\Psi + \Phi\Psi$ and in short the rules of algebra generally hold good.

This is the favorite method of English mathematicians, and with proper caution, it is a very powerful method.

If, then, e is the number that is called the Napierian base, being $e = 1 + \frac{1}{1!} + \frac{1}{2!} + \frac{1}{3!} +$ etc. ad infinitum which is nearly $e = 2.718281828$ then

$$e^x = 1 + \frac{x}{1!} + \frac{x^2}{2!} + \frac{x^3}{3!} + \text{etc. ad inf.}$$

Therefore the long formula on the last page may be written

$$fa = e^{aD_o} f0,$$

which is the symbolical expression of MacLaurin's Theorem; or

$$E_x = e^{D_x}$$

and we may treat the symbols E_x, D_x, Δ_x exactly like symbols of numbers, when no inverse operations are involved. You now see what $\Delta_o^n 0^m$ means. It means that you are to calculate $\Delta_x^n x^m$ and then put $x = 0$. Here is a table of its values

	0	0^2	0^3	0^4	0^5	0^6	0^7	0^8	0^9	0^{10}
Δ_o^1 –	1	1	1	1	1	1	1	1	1	1
Δ_o^2	0	2	6	14	30	62	126	254	510	1022
Δ_o^3	0	0	6	36	150	540	1806	5796	18150	55980
Δ_o^4	0	0	0	24	240	1560	8400	40824	186480	818520
Δ_o^5	0	0	0	0	120	1800	16800	126000	834120	5103000
Δ_o^6	0	0	0	0	0	720	15120	191520	1905120	16435440
Δ_o^7	0	0	0	0	0	0	5040	141120	2328480	29635200
Δ_o^8	0	0	0	0	0	0	0	40320	1451520	30240000
Δ_o^9	0	0	0	0	0	0	0	0	362880	16329600
Δ_o^{10}	0	0	0	0	0	0	0	0	0	3628800

You will remark that if $n > m$, then $\Delta_o^n 0^m = 0$ while $\Delta_o^m 0^m = m!$

We now come to inverse operations, for which we can no longer absolutely rely upon symbolical algebra. Yet its suggestions are always valuable.

The first inverse operation is Σ, which is the inverse of Δ. That is $\Sigma = \Delta^{-1}$. It is that operation upon which Δ being performed, gives 1. Now $\Delta = E - 1$

$$\frac{1}{E-1} = \frac{E^{-1}}{1-E^{-1}}$$

Perform the long division

$$\begin{array}{rl} 1-E^{-1})\,E^{-1} & \quad (E^{-1}+E^{-2}+E^{-3} \\ \underline{E^{-1}-E^{-2}} & \\ E^{-2} & \\ \underline{E^{-2}-E^{-3}} & \\ E^{-3} & \\ E^{-3}-E^{-4} & \end{array}$$

It is thus suggested that $\Sigma = E^{-1} + E^{-2} + E^{-3} +$ etc. and this is correct; for

$$\begin{aligned} &(E^{-1}+E^{-2}+E^{-3}+\text{etc.})(E-1) = \\ &1+E^{-1}+E^{-2}+E^{-3}+\text{etc.} \\ &-E^{-2}-E^{-2}-E^{-3}+\text{etc.} \\ &= 1. \end{aligned}$$

but where is the succession of Es to stop? There is nothing in the problem to show any cessation. The last multiplication will not come out right if it stops at all; and since we cannot carry the summation on unceasingly, we must stop where we must and add an "arbitrary constant" whose value in any case must be discovered by some other condition of the problem.

I must first give you the "conjugate form of Maclaurin's Theorem."

Let $\quad \Phi x = a + bx + cx^2 + dx^3 + ex^4 + \text{etc.}$

where a, b, c, d, etc. are no matter what

$$\Psi x = A + Bx + Cx^2 + Dx^3 + Ex^4 + \text{etc.}$$

no matter what A, B, C, D, etc. are.

Now let us compare $(\Phi D_x) \cdot (\Psi x)$ with $(\Psi D_x) \cdot (\Phi x)$

$$\begin{aligned}
(\Phi D_x)\cdot(\Psi x) &= \\
&= (a+bD_x+cD_x^2+dD_x^3+eD_x^4+\text{etc.})(A+Bx+Cx^2+Dx^3+Ex^4+\text{etc.}) \\
&= aA+aBx+aCx^2+aDx^3+aEx^4+aFx^5+aGx^6+\text{etc.} \\
&\quad +bB+2bCx+3bDx^2+4bEx^3+5bFx^4+6bGx^5+7bHx^6+\text{etc.} \\
&\quad +2cC+6cDx+4{\cdot}3cEx^2+5{\cdot}4cFx^3+6{\cdot}5cGx^4+7{\cdot}6cHx^5+8{\cdot}7cIx^6+\text{etc.} \\
&\quad +3!dD+\frac{4!}{1!}dEx+\frac{5!}{2!}dFx^2+\frac{6!}{3!}dGx^3+\frac{7!}{4!}dHx^4+\frac{8!}{5!}dIx^5+\frac{9!}{6!}dJx^6+\text{etc.} \\
&\quad +4!eE+\text{etc.}
\end{aligned}$$

Now we need go no further. What I wanted to show was that if we now put $x = 0$ we have

$$(\Phi D_0)(\Psi 0) = aA + bB + 2!cC + 3!dD + 4!eE + \text{etc.}$$

and evidently

$$(\Psi D_0)(\Phi 0) = Aa + 1!Bb + 2!Cc + 3!Dd + 4!Ee + \text{etc.}$$

In other words, $(\Phi D_0)(\Psi 0) = (\Psi D_0)(\Phi 0)$, no matter what functions Φ and Ψ may be, as long as they can both be developed according to ascending powers of the variable. Hence Maclaurin's theorem, which is

$$fa = f0 + D_o f0 \cdot a + D_o^2 f0 \frac{a^2}{2!} + D_o^3 f0 \cdot \frac{a^3}{3!} + \text{etc.}$$

can be put in the form

$$fa = f0 + f(D_o 0)a + f(D_o 0^2)\frac{a^2}{2!} + f(D_o 0^3)\frac{a^3}{3!} + \text{etc.}$$

Now please have patience, for we are slowly but surely coming to the point. There would be a somewhat shorter cut by the use of integrals of imaginary quantities *round* infinite values; but I think you would find that harder to understand than this, which only requires patient attention. It is the work of the great mathematical genius, and very ingenious logician, George Boole.

$$e^{D_x} = 1 + D_x + \frac{D_x^2}{2!} + \frac{D_x^3}{3!} + \text{etc.}$$

$$e^{D_x} - 1 = D_x + \frac{D_x^2}{2!} + \frac{D_x^3}{3!} + \text{etc.}$$

Divide 1 by this by long division

$$D_x + \frac{D_x^2}{2!} + \frac{D_x^3}{3!} + \text{etc.}) \; 1 \qquad\qquad (D_x^{-1} \frac{1}{2!} + \frac{D_x}{12} + \text{etc.}$$

$$1 + \frac{D_x}{2!} + \frac{D_x^2}{3!} + \text{etc.}$$

$$- \frac{D_x}{2!} - \frac{D_x^2}{3!} - \text{etc.}$$

$$- \frac{D_x}{2!} \quad \frac{D_x^2}{2!2!}$$

$$\frac{D_x^2}{12} + \text{etc.}$$

Therefore, if C is the arbitrary constant, $\Sigma u_x = D_x^{-1} u_x - \frac{1}{2} u_x + \frac{1}{12} D_x u_x +$ etc. where $D_x^{-1} u_x$ involves the arbitrary constant.

We now want to find out what the law of the coëfficients is. The quotient of our long division shows that

$\dfrac{D_x}{e^{D_x} - 1} = 1 - \dfrac{D_x}{2} + R$ where R is the sum of all the remaining terms.

Transposing $\dfrac{D_x}{2}$ to the other side $\dfrac{D_x}{e^{D_x} - 1} + \dfrac{D_x}{2} = 1 + R$

Reducing the terms of the first member to a common denominator

$$1 + R = \tfrac{1}{2} D_x \frac{2 + e^{D_x} - 1}{e^{D_x} - 1} = \tfrac{1}{2} D_x \frac{e^{D_x} + 1}{e^{D_x} - 1} = \frac{D_x}{2} \cdot \frac{e^{D_x/2} + e^{-D_x/2}}{e^{D_x/2} - e^{-D_x/2}}$$

Now when we change the sign D_x in this expression it becomes $-\dfrac{D_x}{2} \cdot \dfrac{e^{-D_x/2} + e^{+D_x/2}}{e^{-D_x/2} - e^{D_x/2}}$. There is a change in the sign of the denominator and a change of the sign of the whole expression, without any other change at all. This proves that all the terms, each and every of them, involve only even powers of D_x. Since this is thus seen to be true of $\dfrac{D_x}{e^{D_x} - 1}$ it follows that $\dfrac{1}{e^{D_x} - 1}$ involves no other even power of D_x except the term $-\frac{1}{2}$. At this point Boole makes one of his flying leaps quite over the bars of mathematical reasoning which by some secret logic that he certainly did not himself understand invariably landed him safely.

He asks us to denote $e^{D_x} - 1$ by θ, – a most harmless request. With that substitution $\dfrac{D_x}{e^{D_x} - 1}$ will take the form $\dfrac{\text{nat log}\,(1 + \theta)}{\theta}$. Now if θ, instead of standing for an operation, stood for a number less than 1, we know that we should have

$$\log(1 + \theta) = \theta - \tfrac{1}{2}\theta^2 + \tfrac{1}{3}\theta^3 - \tfrac{1}{4}\theta^4 + \text{etc.}$$

$$\frac{\text{nat log}\,(1 + \theta)}{\theta} = 1 - \tfrac{1}{2}\theta + \tfrac{1}{3}\theta^2 - \tfrac{1}{4}\theta^3 + \text{etc.}$$

Accordingly he writes

$$\frac{D_x}{e^{D_x} - 1} = 1 - \frac{e^{D_x} - 1}{2} + \frac{(e^{D_x} - 1)^2}{3} - \frac{(e^{D_x} - 1)^3}{4} + \text{etc.} \quad [...]$$

4

FINITE DIFFERENCES

A. A TREATISE ON THE CALCULUS OF DIFFERENCES (91)

INTRODUCTION

Did the ethics of language not forbid the disturbance of established scientific terminology, until it positively interferes with the advance of science, it were better that the branch of mathematics that goes by the name of the Calculus of Differences, or of Finite Differences, should be christened the theory, or calculus, of successions. For its whole study is of successions of quantities, successions of successions of quantities, etc.

Its aim is to discover how to take advantage of numerical relations subsisting between the members of a succession, be those members quantities or themselves successions, so as to establish rules and methods for the algebraic expression and numerical computation of quantities arranged in successions, successions of successions, etc. It seeks to make its rules and methods as general as possible, and therefore disregards the value and relations of value of the quantities that are arranged in successions, except so far as these values and relations are involved in conditions of the applicability of the rule or method set forth.

The general way in which this calculus pursues its aim is by introducing into algebra single letters to signify operations, which letters are subject either to the ordinary rules of algebra or to some modification of them. Thus, we write u_n, where n stands for any ordinal number, to denote the number or other quantity which stands in the nth place in a certain succession of numbers, the u's. For example, n might be the ordinal number of the place of any man's name in the New York City Directory for 1903; and u_n might be the street-number of his residence. Then u_{n+1} would be the street-number of the residence of the man whose name followed next after. But instead of writing u_{n+1}, we often write $E_n u_n$, where E_n, called by Mr. McClintock the *enlargement* relatively to n, is the operation of taking the next following ordinal number in place of n. Now if anybody were to discover that there was a well-known succession of numbers, which might be denoted by A_1, A_2, A_3, etc. and another well-known succession of numbers, B_1, B_2, B_3, etc. such that the equation

$$A_n u_n + A_{n-1} u_{n-1} + A_{n-2} u_{n-2} = B_n$$

should hold good whatever value might be taken for n, from 3 up (for $n-1$ would have no meaning if n were less than 3) to $n = N$, N being the total number of names in the directory, then that equation, called an "equation of differences" (owing to an exploded way of regarding it as analogous to a differential equation) would be simply a compact way of writing the $N-2$ different equations that that man would have discovered to hold good. The single equations which it embodies would be

$$\begin{aligned} A_3 u_3 + A_2 u_2 + A_1 u_1 &= B_3 \\ A_4 u_4 + A_3 u_3 + A_2 u_2 &= B_4 \\ A_5 u_5 + A_4 u_4 + A_3 u_3 &= B_5 \\ A_6 u_6 + A_5 u_5 + A_4 u_4 &= B_6 \\ \text{etc.} \end{aligned}$$

That equation of differences could be written

$$A_n E_n E_n u_{n-1} + A_{n-1} E_n u_{n-1} + A_{n-2} u_{n-1} = B_n$$

or $$A_n E_n^2 u_{n-1} + A_{n-1} E_n u_{n-1} + A_{n-2} u_{n-1} = B_n$$

or again $$(A_n E_n^2 + A_{n-1} E_n + A_{n-2}) u_{n-1} = B_n$$

But $E_n A_n = A_{n+1}$ and $E_n A_{n+i} u_{n+j} = A_{n+1+i} E u_{n+j}$. Consequently, the equation might equally be written

$$(E_n^2 + E_n + 1) A_{n-1} u_{n-1} = B_n.$$

We shall find that in such a case there would be a simple algebraic expression for any u_n in terms of the A_ns, the B_ns, u_1, and u_2. As for the numerical computation of the street-number of a man's residence from the number of his place in the directory, its facility or complexity would depend on what the two successions of known numbers, the As and the Bs, were. The calculus would render it as easy as the nature of the case would permit.

The whole subject may be divided into four parts. The first will treat the subject in the most general way without regard to the nature of the known quantities, except so far as some of them or some difference between pairs of them may vanish. This part will mainly be occupied with equations of differences.

The second part will consider successions of whole numbers and the relations of other successions to whole numbers. In this part, we shall approach the theory of numbers. It is natural to expect that the doctrine of discrete successions should have special affinity with the theory of integer numbers.

The third part will consider the bearing of this calculus upon the mathematics of continuous quantity. Before the logic of the differential calculus was as well understood as it is today (and it is even now far from satisfactory) the analogy between that calculus and this was much dwelt upon. No doubt a differential is a sort of difference, or very similar to a difference. The custom is universal of writing $\Delta_n u_n$ for $u_{n+1} - u_n = (E-1)u_n$. If, then, u_n is a function of x_n, say for example its square, $\frac{\Delta_n u_n}{\Delta_n x_n}$ has a certain superficial analogy with $\frac{du}{dx}$. There is, accordingly, a certain general parallelism between the theory of equations of difference and that of differential equations. But two circumstances prevent the likeness from extending far. In the first place, the differential of a function has the differential of the independent variable as a factor; so that

$$du = D_x u \cdot dx,$$

where $D_x u$, the differential coefficient, remains absolutely unchanged, however dx may be altered. Substituting the difference for the differential, this proposition fails, unless u is a linear function of x. It is true that for any integral function, there is an analogous equation. Thus,

$$\Delta_n x_n^3 = x_{n+1}^3 - x_n^3 = (x_{n+1}^2 + x_{n+1}x_n + x_n^2)\Delta_n x_n,$$

where the coefficient of $\Delta_n x_n$ is algebraically independent of $\Delta_n x_n$. But that does not suffice to verify the proposition, for which purpose that coefficient would have to remain unchanged however $\Delta_n x_n$ might be changed, x_n remaining unchanged meantime. Since the coefficient in question equals $(\Delta_n x_n)^2 + 3\Delta_n x_n + 3x_n^2$ that is evidently not true.

This divergency of the two theories would be sufficient of itself to preclude any close parallelism between the two kinds of equations. It will appear, for instance, that it is the cause of a great dissimilarity in respect to what are called "singular solutions." But there is another difference far more essential yet. It is that owing to the differential calculus dealing with a system of quantity having such a quasicontinuity that all the values of a function are bound together by a law which does not admit of any discontinuity other than a passage through infinity or from real to imaginary values.* (* It is commonly said that the infinite series generally called Fourier's is capable of representing any discontinuous function. But if that means that as the number of values exactly represented is increased by taking into account more terms of the series, there is a closer approximation to intermediate values, it is not true. The contrary is true: the greater the number of terms included, the worse is representation of the discontinuous function.) Consequently, the only functions contemplated in the differential calculus are such that if a sufficient multitude (perhaps infinite) of values are given, many others may be indefinitely approximated to (as the multitude of given values increases) without any further knowledge of the particular function than that it is one of the class considered in the differential calculus. Nothing in the least like this is true in the calculus of differences, where, if one uses the inappropriate expression that u_n "is a function of" x_n, it merely means that the us are a succession of numbers and the xs another, and that if there is a u in any place of the succession of us, designated by an ordinal number, there is an x in a place in the succession of xs that is designated by the same ordinal number. Thus the equation of differences

$$(\Delta_n u_n - 1)(\Delta_n u_n - 2) = 0$$

is satisfied by any succession of numbers, u_n, such that each number exceeds the one next preceding it either by 1 or by 2, and these values of the successive differences may succeed one another in any manner, however irregular. But the corresponding differential equation,

$$\left(\frac{du}{dn} - 1\right)\left(\frac{du}{dn} - 2\right) = 0$$

can only be regarded as having the two solutions. It cannot break off from one law to follow the other. We cannot but admit that the contrary opinion concerning this equation is quite tenable, in the sense that the calculus itself cannot absolutely refute it. But the very reason of this is the utter inapplicability of the differential calculus to cases of discontinuity; as a hundred examples show. Discontinuity is demonstrably beyond its jurisdiction; and therefore a discontinuous solution of a differential equation cannot be admitted to be correct.

This third part of the calculus of differences must however consider equations in which differentials occur in equations of differences, and has further to study the effects of its operations upon continuous functions.

The fourth part will be devoted to applications of the calculus of differences. It certainly will not yield in interest to any of the others. For it must be admitted that the pure calculus of differences is not one of the richer branches of mathematics. It is somewhat poor in ideas. Its mathematical methods are simple and somewhat monotonous. It does not, like the differential calculus, analytical, projective, and topical geometry, the theory of functions, mathematical logic, the theory of multitudes, etc. aid us in forming distinct conceptions applicable to other than mathematical reasoning. Perhaps the reason is that those subjects are not so purely mathematical as they seem to be. At any rate, when we come to the applications to interpolation, to the solution of numerical algebraic equations and other approximation, to combinations, permutations, arrangements, and the partitions of numbers, to statistics in general, to probabilities in particular, to the evaluation of series, to the calculus of functions, etc. all are of more or less intellectual value.

PART I. EQUATIONS OF DIFFERENCE

CHAPTER 1. THE PRINCIPAL SYMBOLS

We shall have to do with a mode of dependence of one quantity upon another constituting the former what may be called a *logical function* of the latter. In order to get a distinct notion of the difference between a *logical* and a *mathematical* function, it may be well to note, first, some features of mathematical functionality.

If there is any one general and invariable arithmetical rule by which, from any given numerical values whatever, assigned to the quantities x, y, z, etc. a finite collection of quantities, all the precise corresponding values of another quantity, u, may be computed, or indefinitely approximated to, then, undoubtedly, u is a function of x, y, z, etc. This is so, no matter what sort of complex quantities, of a finite number of terms, x, y, z, etc. may be. Under some limitation, it may even be so when x, y, z, etc. form an endless succession. But unless there be, in the true sense, one general and invariable *rule* by which u may be arithmetically computed, either directly or indirectly, at least to indefinite approximation, from the values of x, y, z, etc. there is no mathematical functionality. No mere *description* must be passed off for such a *rule*. Let a die be thrown time after time in endless succession and the number of points shown at any throw is not a mathematical function of the ordinal number of that throw. No doubt, an endless series may be conceived to be constructed, such as that whereby Fourier pretends to represent discontinuous functions, which should so connect those two quantities. But it would be a mere description, not a rule, because it would have received a distinct modification in order to represent each value that it does represent. Or rather, since some values may be satisfied accidentally, without special provision, we must insist that a genuine rule must be capable of making an infinite multitude of predictions, and that if it be *true*, these predictions either are or would be, if occasion offered, fulfilled; or at least that what we admit to be a genuine rule must be of such a nature to convince us that if there were room for an infinite multitude of predictions based on this rule they would all be fulfilled. We say that an octogon is *regular* if its eight sides are equal to one another and its eight angles are equal to one another. This means that we are so constituted that equality has for us a certain esthetic simplicity, and especially a concomitant equality of all parts whose measurement depends on long measure and of all parts whose measurement depends on angular measure. *This* is a genuine rule applicable to and verifiable in an infinite multitude of cases. But it must be denied that there is any strictly objective rule in any relation or formula holding good of sixteen objects regardless of the possibility of others. The experience of a regular octogon, or even of one approximately regular, may suffice to convince us that it is due to a *cause*, when it acted, always would produce regularity; and if that were true there would be a *rule* in the nature of the possible action of that cause. But this would not be *constituted*, though it might be deemed sufficiently *proved*, merely by the equalities of the eight sides and of the eight angles of the octogon.

But one quantity is a *logical function* of another if the former stands in a given general relation to another, whether by virtue of any arithmetical rule or not, at least by virtue of some matter of fact. Thus if the 25th of a certain succession of throws of a die turns up the number 4, then 4 is a logical function of 25; namely it is the number shown by a throw of the succession whose ordinal number is 25. Mathematical functions form a special class of logical functions. The subject of the present part of the calculus of differences is logical functions in general.

A single quantity is termed a *value*. All the values which, at the outset of any problem, are contemplated as possible are commonly related to one another in some regular way, and form what is termed the *system of values*. But at present it is not necessary to restrict ourselves to cases in which there is any regular relation among all the possible values, and therefore it will be better to term the entire collection or continuum of them the *universe of values*. A general description of quantity which may be applicable to many values of the universe, or even to a continuum of them, or to a single value only, or to none at all, is termed a *general quantity*. In the last case, it is usually called, not a quantity, but an absurd expression; but it is more consistent to term it a non-existent quantity.

Mathematicians seek to express all relations of quantity. But there will be no harm in providing ourselves with the means for expressing every logical relation between two general quantities in the directest and simplest manner. Let the sheet upon which we write be supposed to have a place for every value of the universe of values; and let two ovals be supposed to enclose respectively all the values of two general quantities *u* and *v* and no others; thus [as in Fig. 1]:

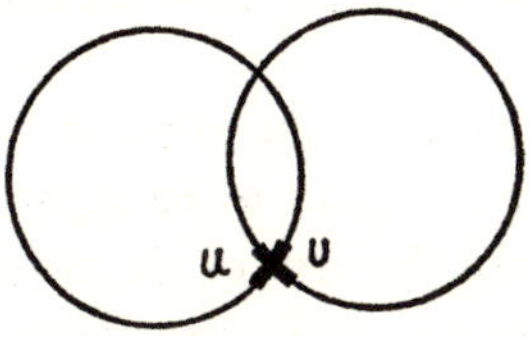

Fig. 1

It will be unnecessary to draw the entire ovals: the cross which is more heavily drawn in the figure will suffice. Let us then adopt the following:

Rule of Notation 1. The four angles of a decussate cross (especially if its two intersecting lines are concave upwards), which cross is placed between two expressions descriptive of quantities (or of ordered sets of quantities) to the right and left of it, and called its *members*, shall refer

the upper angle to the entire collection of values common to the two members,

the two angles to the right and left to the two collections of values each of which is composed of all those values of the nearer member that are not values of the further member,

the lower angle to the entire collection of values of the value-universe that belong to neither member.

If any angle has a bar drawn across it, then a sign of multitude in the angle, inside the bar, shall express a *maximum* which the collection of values to which that angle refers does not exceed in multitude.

A sign of multitude in an angle that has no bar or outside the bar, if it has one, shall express a *minimum* that does not exceed the multitude of the collection of values to which the angle refers.

As signs of multitude,

a *blank* shall stand for *none*,

one, two, or three dots for as many values (or sets),

an Arabic numeral for that number,

∞ for the [denumerable] multitude (that of all integers),

⟨1⟩ , ⟨2⟩ , ⟨3⟩ , etc. for the first, second, third, etc. [abnumeral] multitude,

① , ② , ③ , etc. for a continuum of one, two, or three dimensions,

a general expression of quantity (or of sets) enclosed in square brackets for the multitude of values (or sets of values) of that expression,

a letter or other conventional sign for such multitude as may be agreed upon,

a short line crossing one of the arms of the cross may stand for a value (or for a multitude of values to be separately stated) whose place on one side or the other of the arm of the cross is left uncertain.

When the members are complex expressions, it will be convenient to leave *m*-spaces (as the printers say) between them and the cross; but it will be advisable to avoid enclosing them in parentheses, etc.

For the convenience of the printer the bars which should be heavy, and the numerals may be placed outside the cross.

Instead of (X the cursive form $\propto$ may be used, instead of X the cursive form Ƴ, and instead of (X) the cursive form V. Also instead of (X), Recorde's sign = will naturally be [used].

I append some examples.

u $\propto$ *v* signifies that whatever value *u* may have will be a value of *v*. Thus, 2 $\propto \sqrt{4}$ or $\sqrt{4}$ X) 2.

u X̊ *v* signifies that *u* and *v* have no common value.

u Ẍ *v* signifies that *u* and *v* have at least two common values.

u Ẋ *v* signifies that *u* and *v* have just one common value.

u $\overset{\frac{1}{2}[u]}{X}$ *v* signifies that at least half the values of *u* are common to *v*. The same thing is signified by $u \mid \frac{1}{2}[u]$ X v.

u (X *v* signifies that either *u* has no value at all but is an "absurd expression," in the recognized universe of values, but *v* has at least one value, or else the sole value of *u* is a value of *v*.

I need not stop to work out in detail the various modes of inference from these formulae, which we may term *decussations*. It will suffice to remark the two keys to all such inference. If $u \mid m$ X v, then $u \overset{[u]-m}{X} v$, and u X $[v]-[u]+m \mid v$. If $u \mid m$ X v and $v \mid n$ X w then $u \mid m+n$ X w. There are, besides, under special conditions, other inferences, the consideration of which does not belong to this part of the calculus.

We are thus far confined to expressing arithmetical relations between *two* multitudes of values (or sets). To remedy this let us adopt the following:

Rule of Notation 2. In a decussation of which one member is a decussation, the values of the latter are those which it would make the value-universe to consist in if it stood alone.

For example, $v \mid n \times w$ signifies that the value-universe consists of whatever values of w there may be, of whatever values there may be that belong neither to w nor to v and besides of at most n values of v that are not values of w. Then $u \mid m \times (v \mid n \times w)$ that the values so described are not, perhaps, all that there are. The others, which are necessarily values of v that are not values of w and are not included among the n of this description mentioned consist of whatever there may be that are not values of u together with at most m others that are values of u. This comes to saying that all the values of the value-universe consist of those which are values of w, of those which are not at once values of u and of v, of at most m values at once of u and of v but not of w, and of at most n values which are values of v and not of w, and of course there is nothing to show that they are not values of u. Or, to put the matter still more compactly, there may be at most $m+n$ values at once of u and v while they are not values of w. This is precisely the same thing that is signified by

$$(u \otimes v) \mid m+n \,(\!\times\, w.$$

For $u \otimes v$ describes the value-universe as consisting exclusively of values at once of u and v. The complex decussation asserts that the values so described, instead of being the only ones, form the class of which there are at most $m+n$ that are not values of w.

It will be remarked that by the effect of the last rule of notation

u 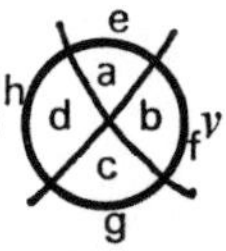 v not only makes an assertion, but is also a name for a certain

general description of quantity. Namely, it describes a quantity as being either any one of certain unspecified common values of u and v from [e] to a in number (without saying how many such values there may be in all), or one of from f to b values of v that are not values of u, or one of from g to c values that belong neither to u nor to v, or of one of from h to d values which belong to u but not to v. This is the meaning of the decussation when it is a member of a decussation. But when it is written alone, it is the assertion that that description comprises all the values of the value-universe.

In order to set forth systematically the principal transformations of decussations it will be convenient, just as we speak of the different cube roots of a number, to speak of a decussation in the plural, meaning the different values which it can take considered as a general description of a quantity (or set of quantities). It will also be convenient to borrow from the theory of probabilities (a theory itself an application of the calculus of differences) the following notation:

Rule of Notation 3. Any denotative sign with a heavy bar over it, called an *obelus*, shall denote all the objects of the universe to which such sign in the same place would refer, barring only those denoted by that sign.

Thus $\bar{u}$ shall have every value of the value-universe that is not a value of u.

NOTE ON THE NOTATION OF THE CALCULUS OF FINITE DIFFERENCES (92)

The calculus of finite differences is not a rich theory. It consists, pretty much, in a notation, without performing any such service of facilitating thought as the great mathematical theories perform. All the more attention should be paid to making its notation consistent and convenient.

The analogy of this calculus to the differential calculus is quite superficial. In the latter, when we speak of a "function," we mean a function with a differential coefficient, and therefore subject to a rational law. But in the former, the calculus of differences, to say that y_n is a "function" of x_n merely means that these two numbers occupy corresponding places in two series of numbers to which they respectively belong. For example, the street-number of a New Yorker's house may be x_n, and the number of the street, y_n, n being the ordinal place of his name in the directory. True, the calculus may conveniently consider the successions of values in regular functions; but this is not its principal business. In any case, there must be a reference to ordinal series; and therefore it is a useless perplexity to make n, the subscript number, increase by any other increment than 1. At most, it may occasionally be convenient to regard its values as descending, instead of ascending.

The calculus has no essential reference to the differences of successive values, as x_{n+1} and x_n, rather than to their quotients. It is merely the fact that subtraction is a simpler operation than division which gives a prominence to the difference. But as we write $\Delta_n u_n = u_{n+1} - u_n$, so we ought to provide ourselves with a symbol, say Λ, such that $\Lambda_n u_n = \frac{u_{n+1}}{u_n}$. And as we have a symbol, $\Sigma = \Delta^{-1}$, so that $\Delta_n \Sigma_n = 1$, $\Sigma_n \Delta_n = [\infty - \infty +]1$, where the square brackets enclose an application to the *result* of the operation 1, not to this operation itself, and $\Sigma_n u_n = u_{n-1} + u_{n-2} +$ etc. $+ C_1 = C_2 - u_n - u_{n+1} - u_{n+2} -$ etc., where $C_2 - C_1 = \sum_{-\infty}^{+\infty}{}_n u_n$, so likewise we may generalize the ordinary use of Γ (which use is foreign to this calculus) so that $\Gamma = \Lambda^{-1}$, and $\Lambda_n\Gamma_n = 1$, $\Gamma_n\Lambda_n = \left[\frac{\infty}{\infty}\times\right]1$, *and* $\Gamma_n u_n = C_3 u_{n-1}\cdot u_{n-2}\cdot u_{n-3}\cdot\text{etc.} = C_4\frac{1}{u_n\cdot u_{n+1}\cdot u_{n+2}\cdot\text{etc.}}$ where $C_4/C_3 = \Gamma_{n\,-\infty}^{\;+\infty}\, u_n$. The objection to the employment of Π instead of Γ is that many mathematicians assume $\Pi u_n = u_n\cdot u_{n-1}\cdot u_{n-2}\cdot$ etc., so that they have the awkward formula, $\Pi e^{B_n} = e^{\Sigma B_{n+1}}$. Boole uses the curiously inconsistent notation, $u_n^{(m)} = u_n\cdot u_{n-1}\cdot u_{n-2}\,\ldots\, u_{n-m+1}$ and $u_n^{(-m)} = \frac{1}{u_n\cdot u_{n+1}\cdot u_{n+2}\ldots u_{n+m-1}}$. But there is considerable gain in simplicity of expression, both in this case and in others, if, while we use the note of admiration in its usual mathematical sense, we write

$$\langle m\circ n\rangle = \langle n\circ m\rangle = \frac{(m+n)!}{m!n!}$$

Then, although $\langle m\circ -n\rangle = 0$, yet $(-n)!\langle m\circ - n\rangle = \frac{(m-n)!}{m!}$. Now $\Delta_m\langle m\circ n\rangle = \langle m+1\circ n-1\rangle = \Sigma_n\langle m\circ n\rangle$. Accordingly, Boole's $m^{(n)} = n!\langle m-n\circ n\rangle$ and $\Delta_m m^{(n)} = n!\langle m-n+1\circ n-1\rangle = n\cdot m^{(n-1)}$; while his $m^{(-n)} = (-n)!\langle m+n-1\circ -n\rangle$ and $\Delta_m m^{(-n)} = (-n)!\langle m+n\circ - n+1\rangle = -nm^{(-n-1)}$. This, it appears to me, exhibits the true relation between the two cases. That is, not only has n changed its sign, but the value of m is altered. With the proposed notation the binomial theorem is

$$(p+q)^l = \sum_{m+n=l}\langle m\circ n\rangle p^m q^n.$$

and a common formula of interpolation is

$$f(x+m) = \sum_{n\,0}^{\infty}\langle n\circ\infty\rangle m^n\Delta_x^n fx\ldots$$

B. A TRADE SECRET (212)

Among the many little dodges that are generally known among professional computers, the following is curious because there is no general mathematical necessity for its working as well as it usually does. I trust my brethren may not inflict the death-penalty upon me for making it public.

Suppose one has to compute values of a function from an infinite series. Then, if the last six terms of the series of which one proposes to take account are $u_m + u_{m+1} + u_{m+2} + u_{m+3} + u_{m+4} + u_{m+5}$, one begins by calculating the five quantities v_m, v_{m+1}, v_{m+2}, v_{m+3}, v_{m+4}, by the formula

$$v_{m+n} = \Sigma u_{m+n} - \frac{u^2_{m+n}}{\Delta u_{m+n}},$$

that is,

$$v_m = \frac{u^2_m}{u_m - u_{m+1}}, \qquad v_{m+1} = u_m + \frac{u^2_{m+1}}{u_{m+1} - u_{m+2}},$$

$$v_{m+2} = u_m + u_{m+1} + \frac{u^2_{m+2}}{u_{m+2} - u_{m+3}}$$

One next calculates three quantities w_m, w_{m+1}, w_{m+2} by the formula

$$w_{m+n} = u_{m+n} - \frac{(\Delta u_{m+n})^2}{\Delta^2 u_{m+n}}$$

that is,

$$w_m = v_m - \frac{(v_m - v_{m+1})^2}{v_m - 2v_{m+1} + v_{m+2}},$$

$$w_{m+1} = v_{m+1} - \frac{(v_{m+1} - v_{m+2})^2}{v_{m+1} - 2v_{m+2} + v_{m+3}}, \text{ etc.}$$

will furnish three triads of successive approximations giving three successive approximations of the second series; and from these three the same rule will show how to obtain a still better approximation of a third series. It will be still better to begin with seven, or even nine, successive approximations of the first series, when one can make series after series of approximations. From ordinally *corresponding* approximations of three successive series, a still better approximation can often be obtained; but the advantage of this procedure is dubious and at best very slight.

As an example, let us calculate the natural logarithm of 2.1 from the first six terms of the divergent series.

$$\log(1+x) = x - \tfrac{1}{2}x^2 + \tfrac{1}{3}x^3 - \tfrac{1}{4}x^4 + \tfrac{1}{5}x^5 - \tfrac{1}{6}x^6$$

This gives seven approximations as follows:

a	Δa	$\Delta^2 a$
0.0	+1.1	−1.705
1.1	−0.605	+1.04866667
0.495	+0.44366667	−0.80969167
0.93866667	−0.36602500	+0.68812700
0.57264167	+0.32210200	−0.61736217
0.89474367	−0.29526017	+0.57364832
0.59948350	+0.27838816	−0.54633676
	−0.26794860	

The true value is 0.741937344729, so that the best of the approximations of the a series has an error of 0.14245384, or about one fifth of the quantity. The second series of approximations will be

b	Δb	$\Delta^2 b$
0.70968	+0.04128	−0.05432
0.75096	−0.01304	+0.01909
0.73792	+0.00605	−0.00933
0.74397	−0.00328	+0.00536
0.74069	+0.00208	−0.00351
0.74277	−0.00143	
0.74134		

The error of the fifth of this series is only −0.00124246. The third series of approximations will be

c	Δc	$\Delta^2 c$
0.74105	+0.00100	−0.00122
0.74205	−0.00022	
0.74184		

The error of the last is only −0.000110. Finally the fourth series of approximations is represented by the single value

$d = 0.74188$, the error of which is only −0.000061.

Finally, if we take the first approximations of the four sets, we have

	Δ	Δ^2
$a_1 = 0.0$	$+0.70968$	-0.67830
$b_1 = 0.70968$	$+0.03137$	
$c_1 = 0.74105$	$+0.00082$	0.03055
$d_1 = 0.74188$		

Treating these in the same way, we get these new approximations

a'
0.74250
0.74190

The error of the last is -0.00003888. Comparing this with the result the ordinary arithmetician would get, we find that the latter would reduce from the six terms of the formula a result more than 3664 times as erroneous as that which the expert computer will derive from the same data. Of course, there must be cases where the method breaks down badly, though these must, I think, be very rare when one sets out from a series of nine approximations.

C.S. Peirce

5

BOOLIAN STUDIES

A. BOOLIAN ALGEBRA (s-37 and s-38)

FIRST LECTION

§1. *Introductory.*

The algebra of logic (which must be reckoned among man's precious possessions for that it illuminates the tangled paths of thought) was given to the world in 1842; and George Boole is the name, an honored one upon other accounts in the mathematical world, of the mortal upon whom this inspiration descended. Although there had been some previous attempts in the same direction, Boole's idea by no means grew from what other men had conceived, but, as truly as any mental product may, sprang from the brain of genius, motherless. You shall be told, before we leave the subject, precisely what Boole's original algebra was; it has, however, been improved and extended by the labors of other logicians, not in England alone, but also in France, in Germany, and in our own borders; and it is to one of the modified systems which have so been produced that I shall first introduce you, and shall for the most part adhere. The whole apparatus of this algebra is somewhat extensive. You must not suppose that you are getting it all in the first, the second, or the third lection. But the subject matter shall be so arranged that you may from the outset make some use of the notation described, and even apply it to the solution of problems.

A deficiency of pronouns makes itself felt in English, as in every tongue, whenever there is occasion to discourse concerning relations between more than two objects; so that, to supply the place of the wanting words, the designations, *A*, *B*, and *C*, are resorted to, not only by geometricians for points, but also by lawyers and economists for persons and other parties. This device is already a long stride toward an algebraical notation; and in any mode of expression whose only elegance is to consist in absolute clearness and in the aid it affords to the mind in reasoning, the use of letters in place of words ought to be further extended.

Another serious imperfection of ordinary language, in its written form at least, belongs to our feeble marks of punctuation. The illustration of how a phrase may be ambiguous when written, from which the pauses of speech would remove all uncertainty, is now too stale a joke for the padding of a news-paper. But in algebra we find a method of punctuation which answers its purpose to perfection and is at the same time of the utmost simplicity. The plan is simply to enclose a phrase in parenthesis to show that it is to be treated as a unit in its combination with other phrases or single words. When one such parenthesis is included within another, the appearance of the ordinary curvilinear marks ( ) is varied, either by use of square brackets [ ] or braces { }, or making the lines heavier or longer ( ). Sometimes, a vinculum or straight line drawn over the phrase or compound expression is used instead of the parenthesis. By this simple means, we readily distinguish between the black (lady's veil) and the (black lady)'s veil; or between the following:

The {(church of England)'s [($\overline{\text{gunpowder}}$ plot) services]},
[The (church of England)'s] [($\overline{\text{gunpowder}}$ plot) services],
{(The church) of [England's ($\overline{\text{gunpowder}}$ plot)]} services,
The {[(church of England)'s gun] [(powder plot) services]},
etc. etc. etc.

Another fault of ordinary language as an instrument of reasoning is that it is more pictorial than diagrammatic. It serves the purposes of literature well, but not those of logic. The thought of the writer is encumbered with sensuous accessories. In striving to convey a clear conception of a complicated system of relations, the writer is driven to circumlocutions which distract the attention or to polysyllabic and unfamiliar words which are not very much better. Besides, almost every word signifies the most disparate and even contrary things in different connections (for example, the "number of millimetres in an inch" is the same as "an inch in millimetres"), so that if the reader seizes the idea at all, he only does it by substituting for the signs in which it is expressed some mental diagram which embodies the same relations in a clearer form. Games of chess are described in old books after this fashion: "The white king's pawn is advanced two squares. The black king's pawn is advanced two squares. The white king's knight's [pawn] is placed on the square in front of the king's bishop's pawn," etc. In ancient writings, arithmetical processes are performed in words with the same intolerable prolixity. To remedy this vice of language, what is required is a system of abbreviations of invariable significations and so chosen that the different relations upon which reasoning turns may find their analogues in the relations between the different parts of the expression. (Please to reflect on this last condition.) Among such abbreviations of quasi diagrammatical power, we shall find the algebraical signs + and × of the greatest utility, owing to their being familiarly associated with the rules for using them.

§2. *The Copula.*[1]

In the special modification of the Boolian calculus now to be described, which I shall designate as Propositional Algebra, the letters of the alpha-

[1] Two pages of a draft of the letter to W. S. Jevons (L227) written 16 May 1875 and mentioned in the introductory remarks read as follows:

> "Algebraic notation is a system of arbitrary signs fit to represent the relations of objects in such a manner as to facilitate necessary reasoning about them.
>
> The first class of algebraic signs consists of signs of inference. These must indicate the probability of the inference. As yet only one such sign that of demonstrative illation ∴ has been used.
>
> The second class of algebraic signs are copulas. The first of these is $\prec$. It signifies that whatever corresponds to that which precedes it has all the common characters of all the objects which correspond to what follows it. According to the manner of the correspondence the signification of the sign may be varied. Thus 'man $\prec$ sinner' may be used to denote that every man is a sinner or that every quality of all men is a quality of all sinners or that the number of men is as great or as small as the number of sinners, etc. Accents may be affixed to the general sign when it is necessary to distinguish the different meanings. The definition just given of this sign is precisely equivalent to three algebraical conditions: — viz
>
> 1st If $x \prec y$ and $y \prec z$ then $x \prec z$ whatever x, y, and z may be.
>
> 2nd $x \prec x$.
>
> 3rd If $x \prec y$ and $y \prec x$ then x and y may everywhere be substituted for one another.
>
> Every signification of the copula of inclusion $\prec$ has a converse signification $\prec'$ such that $x \prec y$ is the same as $y \prec' x$.
>
> Two logical term-signs are suggested and defined by the copula. They are zero 0, and ∞ infinity. Their definitions are as follows: —
>
> Zero is such a term that $0 \prec x$ whatever x may be.
>
> Infinity is such that $x \prec \infty$ whatever x may be.
>
> Of course the zero of one progression may not be the zero of another progression; and therefore I have greatly enlarged the usual meanings of the words in these definitions.
>
> The copula of inclusion is subject to rather complicated rules of algebraic manipulation. To simplify the process we may use the copula of identity =. The definition of this is that $x \prec y$ is the same as $x \prec y$ and $y \prec x$.
>
> The third class of algebraic signs consists of signs of operation. Signs of operation are of three orders according as they are applicable to all terms or only to relative and conjugative terms or to conjugative terms alone. The sign of conjunction + is the first sign of the first order. It is defined by three conditions, as follows: —
>
> 1st $x \prec x +, y$
>
> 2nd $y \prec x +, y$
>
> 3rd If $x \prec z$ and $y \prec z$ then $x +, y \prec z$."

bet are used to signify statements, the special statement signified by each letter depending on the convenience of the moment. The statement signified by a letter may be one that we believe or one that we disbelieve: it may be very simple or it may be indefinitely complex. We may, if we

---

The following statement is of special interest here. It comes from MS. 431a and may be read after 4.261 of the *Collected Papers.*

"One assertion will admit every possible set of values: it is the empty form of assertion $x \times y$. One will violate the hypothesis, by excluding every set of values: it is the absurd assertion $x \otimes y$. Six assertions will admit two sets of values and exclude two. Of these, four may be called *degenerate* propositions. They determine the value of one of the two quantities, and leave the other entirely indeterminate. They are x ⊗ y, x ⊗ y, x ⊗ y, x ⊗ y; which may, perhaps, be more cursively written x ◿ y, x ◸ y, x ◺ y, x ◹ y. The remaining two of the six are correspondential and distinctive. They are $x \infty y$, and x 8 y. It happens, however, that an excellent and well-established character is already in use for the former of these; and there is no sufficient reason for breaking with the usage, by which $x = y$ is written to express that x and y have the same value. Still, for the sake of the system, we may sometimes give the sign the form $x \infty y$. The sign = was introduced by Robert Recorde, the author of the first arithmetic in the English language, called the *Ground of Artes.* But this sign first appeared, in 1556, in his treatise on algebra, entitled the *Whetstone of Witte* (socalled because he mistook the name *cos,* usual at that time for the unknown quantity, for the Latin word for 'whetstone,' it being in fact, from the vulgar Latin and Italian *cosa,* for *caussa,* 'the matter in hand'). It was the first time a regular sign for equality was introduced into algebra. Recorde says that he adopts it because the two manifestly equal lines form an icon of equality ('because no 2 thynges can be more equalle'). It had, however, been used in MSS. to mean '*est.*' Inequality is, by algebraists, expressed in various ways; as $x \lessgtr y$, $x \neq y$, none of which is old or exclusively favored. I shall, therefore, conform to the general method of the notation here used, by turning ∞ into a vertical position to express inequality. Thus, x 8 y, instead of x 8 y. It ought to be mentioned that it was Mrs. Franklin who first proposed to put the same character into four positions in order to represent the relationship between logical copulas, and that it was a part of her proposal that when the relation signified was symmetrical, the sign should have a right and left symmetry. It will be seen that the notation here used simply carries out that proposal in a particular way.

Of the remaining eight possible assertions, four admit only one set of values, and four exclude only one set. The latter are shown, according to our plan, by

x ɣ y

x ≻∘ y   x ∘≺ y

x ʎ y

Of these the upper and lower are symmetrical with respect to x and y; or, as the mathematicians say, are commutative; or, as the logicians say, are *convertible* or *equiparant.* Of the two placed to the right and left, each is the converse of the other; that is x ≻∘ y and y ≻∘ x are equivalent assertions.

In practice, I think we shall do well slightly to modify these characters, for the following reasons. We shall find that the forms of assertion which they denote are even more important logically than mathematically. Now, from a logical point of view, the relation between these forms expressed by giving them signs choose, employ a single letter to designate the whole contents of a book, or the sum of omniscience, or a falsehood as such. To use the consecrated term of logic which Appuleius, in the second century of our era, already speaks of as familiar, the letters of the alphabet are to be PROPOSITIONS.

---

which result from putting one character into four positions, is not the predominant, or even principal, truth about them. For that reason, I think it is better that the characters should be so modified that each should have a certain individuality, which will also be an aid to the memory in using them. Besides, the introduction of these forms into dichotomic algebra is due to the labors of men, whose work, I think, ought to be commemorated in the signs we use.

In place, therefore, of *x* ɣ *y*, to express that either *x* or *y* is *v*, I shall write *x* ♈ *y*, using a character which can be made with equal facility at one stroke of the pen. The reason is, that it was William Stanley Jevons who, in 1864, introduced the operation corresponding, as we shall see, to this sign, into dichotomic algebra, thus taking an important step. (The whole significance of it was first shown by me in 1867, in ignorance of Mr. Jevons's work.) Now Jevons recommended the sign ·|·. This, however, is inadmissible, because it is liable to be mistaken for three signs. Besides, we wish to adopt Mrs. Franklin's idea, not merely for its intrinsic elegance, which is incontestable, but also that we may have the pleasure of using a piece of woman's work. Moreover, we wish to use characters upon which logicians and mathematicians will agree to accept; and this is not possible without combining different considerations. As long as the character commemorates Jevons's step, as the two dots and upright part of ♈ do, justice is done to him; while the loop below sufficiently recalls the diagram, and thus satisfies the requirements of Mrs. Franklin's idea.

In place of *x* ♎ *y*, to express that either *x* or *y* is *f*, I would write *x* ♎ *y*. The reduction of the loop to a small size gives the sign a wedge shape, and commemorrates the use by Mrs. Franklin of a wedge to the same effect in 1882.

In place of *x* ≻o*y*, to signify that *x* is *v* if *y* is *v*, I propose *x* ꝏ *y*. Descartes and other early algebraists employed this sign for equality. The only justification there can have been for their rejection of Recorde's =, would be that they did not mean equality precisely. They wrote $\sqrt{4}$ ꝏ 2 meaning that *one* of the values of $\sqrt{4}$ was 2. So, as I propose to use it, *x* ꝏ *y* means that the value of *x* is either that of *y* or is *v*. It must be admitted that this is a rather small reason, hardly worth the fetching from so far. So I shall be very ready to adopt a different modification of ≻o whenever a better suggestion shall be presented.

In place of *x* o≺ *y*, to signify that if *x* is *v*, so is *y*, I ask leave to write *x*⤙*y*. In 1870, I introduced into algebra the sign $-\!\!<$, to which I gave the cursive form, ⤙, with the following remark: 'I use the sign $-\!\!<$ in place of $\leqq$. My reasons for not liking the latter sign are that it cannot be written rapidly enough, and that it seems to represent the relation it expresses as being compounded of two others which in reality are complications of this. It is universally admitted that a higher conception is logically more simple than a lower one under it. Whence it follows from the relations of extension and comprehension, that, in any state of information, a broader concept is more simple than a narrower one included under it. Now all equality is inclusion in, but the converse is not true; hence inclusion in is a wider concept than equality, and therefore logically a simpler one. On the same principle, inclusion is also simpler than being less than. The sign $\leqq$ seems to involve a definition by enumeration; and such a definition offends against the laws of definition.' Since it was I who in the same memoir first took the important

The final letters x, y, z, will be specially appropriated to the expression of formulae which hold good whatever statements these letters may represent; so that in such a formula each of these letters may be replaced throughout by any proposition whatever.

---

step of giving to this form of assertion its due place in dichotomic logic, I am rightfully entitled to prescribe the notation for it. For only under this plain rule of ethics can we come to any settled usage as to notation. Of course, if the progress of science should imperatively demand the employment of another sign, that would be a different matter; but thus far no premonition of such an event has appeared. I, therefore, request those who apply dichotomic, or logical, algebra to respect my recommendation, and use my sign in one of its shapes −<, ≺, ⊸, or in such other slight modification as may seem advantageous. It is in the third form made at one facile stroke of the pen. It is only a simplification of the diagram, than which nothing could be more expressive. I am confident that reasonableness and fairness will in the long run prevail over lower passions in this matter, and will bring a settlement to usage.

The four forms of assertion which admit only one of the four states of things are shown in the following diagrams:

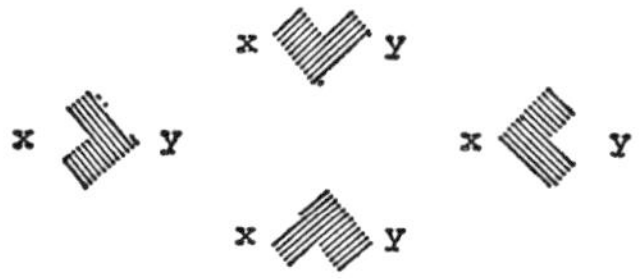

These diagrams naturally suggest the signs,

$$x \vee y$$

$$x > y \qquad\qquad\qquad x < y$$

$$x \wedge y$$

I shall modify the two symmetrical ones, as follows:

In place of $x \vee y$, to signify that both x and y have the value v, I propose to use, sometimes a heavy dot above the line, $x \cdot y$, which may be regarded as an extreme simplification, sometimes as the last degree of simplification to leave a blank, or write the letters together, while sometimes using the form $\vee$. The successive steps of simplification will be

x  y, $x \vee y$, $x^{\vee} y$, $x^{\scriptscriptstyle\vee} y$, $x \cdot y$, $x\;y$, $\dot{x}y$.

Mrs. Franklin, in 1882, used the sign $\vee$ for substantially the same purpose. But Boole, in 1847, had written the letters together. The relation signified, being the commonest in logical algebra, needs a very simple shape, and a variety of shapes will be convenient.

In place of $x \wedge y$, to signify that x and y both have the value f, I propose to write $x \curlywedge y$. There is not much reason for this. But DeMorgan, in a system of logical symbols, *not* in what one would ordinarily understand by a dichotomic algebra, wrote $x)(y$ to signify this relation, substantially. I preserve the lower halves of his spiculae and bring them together, in order to preserve a reasonable conformity to Mrs. Franklin's proposal.

The signs $x > y$ and $x < y$, to signify that x is greater than, and is smaller than, y, were the earliest used in algebra for this purpose, having been introduced

The idea is to express the degree of truth of propositions upon a quantitative scale, as temperatures are expressed by degrees of the thermometer scale. Only, since every proposition is either true or false, the scale of truth has but two points upon it, the true point and the false point. We shall conceive truth to be higher in the scale than falsity. In that branch of the art of reasoning which this algebra immediately subserves, we are to study the modes of necessary inference. A proposition or propositions, called *premises*, being taken for granted, the question is what other propositions, called CONCLUSIONS, these premises entitle us to affirm. The truth of the premises is not now to be examined; for that is assumed to have been satisfactorily determined, already; and any process of inference (i.e. formation of a conclusion from premises) will be satisfactory, provided it be such that the conclusion is certainly true unless the premises are false. That is to say, if *P* signifies the premises and *C* the conclusion, the condition of the validity of the inference is that either *P* is false or *C* is true. For human reason cannot undertake to guarantee that the conclusion shall be true if the premises on which it depends are false. It is true that we can imagine inferences which satisfy this condition and yet are illogical. Such are the following:

P true, *C* true. The world is round; therefore, the sun is hot.
P false, *C* true. The world is square: therefore, the sun is hot.
P false, *C* false. The world is square; therefore, the sun is cold.

The reason why such inferences would be bad is that nobody could, in such cases, know that either *P* is false or *C* true, unless he knew already that *P* was false (when it would not properly be a premise), or else knew independently that *C* was true (when it would not properly be a conclusion drawn from *P*). But if the proverbial Angel Gabriel, who has been imagined as making so many extraordinary utterances, were to descend and tell me "Either the earth is not round, or the further side of the moon is blue," it would be perfectly logical for me, from the known fact that

---

by Thomas Harriotts in the sixteenth century. The date assigned to this algebraist in the histories is far too late, owing to the fact that his devotion to practical duties, as Henry Stevens showed,* prevented the publication of his mathematical works before his death in 1621; and after that, the conceit and obstinacy of another individual delayed the publication of the algebra until 1631 and prevented that of other works. We shall find reason for regarding *v* as greater than *f*. (*Thomas Hariot. London: 1900. Privately printed. I use that spelling of the name which Harriotts himself used in signing his will. The final s looks like a modern e. The printing of Stevens's book was begun in January 1878, but was not finished till February, 1900.)"

the earth is round, to conclude that the other side of the moon is blue. It is true that the inference would not be what is called a *complete* or *logical* one; that is to say, the principle that either P is false or C true could not be known from the study of reasonings in general; but it would be a perfectly sound or *valid* inference.

From what has been said it is plain that that relation between two propositions which consists in our knowing that either the one is true or the other false is of prime importance as warranting an inference from the former to the latter. It is therefore, desirable to have an abbreviation to express this relation. The sign $\prec$ is to be used in such a sense that $x \prec y$ means that x is at least as low on the scale of truth as y. The sign $\prec$ is to be called the COPULA, and for the sake of brevity it may be read "gives," that is, warrants the inference of. A proposition like $x \prec y$ will be called a HYPOTHETICAL, the proposition x preceding the copula will be called the ANTECEDENT, and the proposition y following the copula will be called the CONSEQUENT. The meaning of $\prec$ may be more explicitly stated in the following propositions, which, for convenience of reference, I mark A, B, C.

A. If x is false, $x \prec y$.

B. If y is true, $x \prec y$.

C. If $x \prec y$, either x is false or y is true.* (* In using the conjunctions "either ... or," I always intend to leave open the possibility that both alternatives may hold good. By "either x or y," I mean "either x or y or both.")

Rules of the Copula.

The sign $\prec$ is subject to three algebraical rules* (* A "rule" in algebra differs from most other rules, in that it *requires* nothing to be done, but only *permits* us to make certain transformations.) as follows:

Rule I. If $x \prec y$ and $y \prec z$, then $x \prec z$. This is called the *principle of the transitiveness of the copula.*

Rule II. Either $x \prec y$ or $y \prec z$.

Rule III. There are two propositions, u and v, such that $v \prec u$, is false.

To these is to be added the following:

RULE OF INTERPRETATION. If y is true, $v \prec y$; and if y is false $y \prec u$.

Rule I can be proved from propositions A, B, C. For by C, if $x \prec y$ either x [is] false or y is true. In the statement of A, substitute z for y. It then reads that if x is false, $x \prec z$. Hence, if $x \prec y$, either $x \prec z$ or y is true. Call this proposition P. In the statement of C, substitute y for x, and z for y. It then reads that if $y \prec z$, either y is false or z is true. Combining this with P, we see that if $x \prec y$ and $y \prec z$ either $x \prec z$ or z is true. Call this proposition Q. In the statement of B, substitute z for y. It then reads that if z is true $x \prec z$. Combining this with Q, we conclude that if $x \prec y$ and $y \prec z$, $x \prec z$. Q.E.D.

Rule II can be proved from propositions A and B alone. For in the statement of A, substitute y for x and z for y. It then reads that if y is false, $y \prec z$. But by B, if y is true, $x \prec y$. Hence, either $x \prec y$ or $y \prec z$. Q.E.D.

Rule III can be proved from proposition C alone. For in the statement of C, substitute v for x and u for y; and it reads that if $v \prec u$, either v is false or u is true. If, therefore, v is any true proposition, and u any false one, it is not true that $v \prec u$. Thus, it is possible so to take u and v that $v \prec u$ shall be false. Q.E.D.

The rule of interpretation evidently follows from propositions A and B.

That these four rules fully represent propositions A, B, C, can be shown by deducing the latter from the former. It is left to the student to construct these proofs.

Let us consider the three algebraical rules by themselves, independently of the rule of interpretation. Rule I shows that $x \prec y$ expresses a relation between x and y analogous to that of numbers on a scale, the number x being at least as low on the scale as y. For if x is at least as low as y, and y at least as low as z, then x is at least as low as z. Rule II shows that this scale has not more than two places upon it. For if one number y could be lower on the scale than a second x, and at the same time higher than a third, z, neither $x \prec y$ nor $y \prec z$ would be true. Rule III shows that the scale has at least two places. For if it had but one any one number would be as low as any other and we should have $x \prec y$ for all values of x and y. Finally the rule of interpretation shows that the higher point on the scale represents the truth and the lower falsity.

(Note A. In the ordinary logic, the fact that there are not more than two varieties of propositions in respect to truth is expressed by the so-called *Principle of Excluded Middle*, which is that every proposition is either true or false (or A is either B or not B); while the fact that there are at least two different varieties is expressed by the so-called *Principle of Contradiction*, which is that nothing is both true and false (or A is not A).)

I now proceed to deduce a few useful formulae from Rules I, II, III. In Rule II, substitute x for z, and we have

(1). Either $x \prec y$ or $y \prec x$.

In (1), substitute x for y, and we have

(2). $x \prec x$.

By Rules II and III,

(3). $x \prec v$.

(4). $u \prec x$.

By Rule I,

(5). If $x \prec y$, while it is false that $x \prec z$, then it is false that $y \prec z$.

(6). If $y \prec z$, while it is false that $x \prec z$, then it is false that $x \prec y$.

(7). If it is false that $x \prec z$, it is either false that $x \prec y$ or that $y \prec z$.

By Rules I and III,

(8). It is either false that $v \prec x$ or false that $x \prec u$.

(9). Either $v \prec x$ or $x \prec u$.

(Note B. An entire calculus of logic might be made with the sign $\prec$ alone. Some of the most important formulae would be, as follows: if $x \prec x(y \prec z)$, then $y \prec (x \prec z)\,(v \prec x) \prec x$. But such a calculus would be useless on account of its complexity.)

(Note C. The equation $x = y$ means, of course, that x and y are at the same point of the scale. That is to say, the definition of $x = y$ is contained in the following propositions:

A. If $x = y$, $x \prec y$.

B. If $x = y$, $y \prec x$.

C. If $x \prec y$ and $y \prec x$, then $x = y$.

From these propositions, it follows that logical equality is subject to the following rules:

I. $x = x$.

II. If $x = y$, then $y = x$.

III. If $x = y$ and $y = z$, then $x = z$.

IV. Either $x = y$ or $y = z$ or $z = x$.

The Proof of these from A, B, C, by means of Rules I, II, III, is left to the student.)

§3 *Logical Addition and Multiplication*

The sign $\prec$ as explained above, is, we may trust, free from every trace of ambiguity. But while it does not hesitate between two meanings, it does carry two meanings at one and the same time. The expression $x \prec y$

means that either x is false or y is true; but it also means that x is at least as low as y upon a scale. In short, $x \prec y$ not only states something, but it states it under a particular *aspect*; and though it is anything but a poetical or rhetorical expression, it conveys its purport by means of an arithmetical simile. Now, elegance requires that this simile, once adopted, should be adhered to; and elegance, as we shall find, is every whit as important a consideration in the art of reasoning as it is in the more sensuous modes to which the name of Art is commonly appropriated. Following out this analogy, then, we proceed to inquire what are to be the logical significations of addition, subtraction, multiplication and division.

Any two numbers whatever (say 5 and 2) might be chosen for u and v, the representatives of the false and the true; though there is some convenience in making v the larger. Then, the principle of contradiction is satisfied by these being different numbers; for a number x cannot at once be equal to 5 and 2, and therefore the proposition represented by x cannot be at once true and false. But in order to satisfy the principle of excluded middle, that every proposition is true or false, every letter x, signifying a proposition must, considered as a number, be supposed subject to a quadratic equation whose roots are u and v. In short, we must have

$$(x-u)(v-x) = 0$$

Since the product forming the left hand member of this equation vanishes, one of the factors must vanish. So that either

$$x-u = 0 \text{ and } x = u, \text{ or } v-x = 0 \text{ and } x = v.$$

Another way of expressing the principle of excluded middle would be:

$$\frac{1}{x-u} + \frac{1}{x-v} = \infty$$

It will be found, however, that occasion seldom arises for taking explicit account of the principle of excluded middle.

The propositions

Either x is false or y is true, and
either y is false or z is true,

are expressed by the equations

$$(x-u)(v-y) = 0$$
$$(y-u)(v-z) = 0$$

For, as before, to say that the product forming the first member of each equation vanishes, is equivalent to saying that one or other factor vanishes.

Let us now eliminate y from the above two equations. For this purpose, we multiply the first by $(v-z)$ and the second by $(x-u)$. We thus get

$$(x-u)(v-y)(v-z)=0$$
$$(x-u)(y-u)(v-z)=0$$

We now add these two equations and get

$$(x-u)(v-u)(v-z)=0$$

But the factor $(v-u)$ does not vanish. We, therefore, divide by it, and so find

$$(x-u)(v-z)=0$$

The signification of this is,

Either x is false or z is true;

and this is the legitimate conclusion from the two propositions

Either x is false or y is true, and
Either y is false or z is true.

Suppose, now, that we seek to find the expression of the precise denial of x (which in logical terminology is called the *contradictory* of x). Call this X. Then it is necessary and sufficient that X should be true when x is false and false when x is true. We may therefore put

$$X=u+v-x$$
or
$$X=uv/x$$

These two expressions are equal, by the equation of excluded middle.

[The simplest expression] of that proposition which is true if x,y,z are all true and is false if any of them are false is

$$u+\frac{(x-w)(y-u)(z-u)}{(v-u)^2}$$

The simplest expression for the proposition which is true if any of the propositions x,y,z, is true, but is false if all are false is

$$v-\frac{(v-x)(v-y)(v-z)}{(v-u)^2}$$

It is now easy to see that some values of u and v are much more convenient than others. For example, the proposition which asserts that some two at least of the three propositions, x,y,z, are true, is, if $u = 2$, $v = 5$,

$$7\tfrac{7}{9} - 2\tfrac{2}{9}(x + y + z) + \tfrac{7}{9}(xy + xz + yz) - \tfrac{2}{9}(xyz)$$

but if $u = -1$, $v = +1$, the same statement is simply

$$x + y + z - xyz$$

Perhaps the system which would most readily occur to a mathematician would be to take the true, v, as an odd number, and the false, u, as an even one* (* The Pythagorean notion was that odd was good, even bad) and not to discriminate between numbers except as odd or even. Thus, we should have

$$v = 1 = 3 = 5 = 7 = \text{etc.}$$
$$u = 0 = 2 = 4 = 6 = \text{etc.}$$

In other words, we should measure round a circle, having its circumference equal to 2; so that 2 would fall on 0, 3 on 1, etc. On this system, every possible algebraical expression formed by means of the addition and multiplication of propositions would have a meaning. Thus, $x + y + z +$ etc. would mean that some odd number of the propositions x,y,z, were true. While xyz etc. would mean that the proposition x,y,z, were all true. For these would be the conditions of the expressions representing odd numbers. Subtraction would have the same meaning as addition, for we should have $-x = x$. A quotient, as $\frac{x}{y}$ would not properly signify a proposition, since it would not necessarily represent any possible whole number. Namely, if x were odd and y even, $\frac{x}{y}$ would be a fraction.

Logical equality is, besides, subject to a fourth rule, namely, that, either $x = y$ or $y = z$ or $x = z$. I prove this. By Rule 2, as stated above either $x \prec y$ or $y \prec z$, substituting z for x, x for y, and y for x, either $x \prec y$ or $z \prec x$. Hence, either $x \prec y$ or both $y \prec z$ and $z \prec x$. In like manner, either $y \prec x$ or both $z \prec y$ and $x \prec z$.

In like manner, either $y \prec x$ or both $z \prec y$ and $x \prec z$.

Hence, either both $x \prec y$ and $y \prec x$ (when $x = y$), or both $y \prec z$ and $z \prec x$ or both $z \prec y$ and $x \prec z$. Call this proposition P. Again by Rule 2, substituting y for x, z for y, and x for z, either $y \prec z$ or $z \prec x$. Hence, by P, either $x = y$ or both $y \prec z$ and $z \prec x$ or both $z \prec y$ and $y \prec z$ (when $y = z$) or both $x \prec z$ and $z \prec x$ (when $x = z$). Call this proposition Q. Again by Rule 2, substituting z for y and y for z, either $x \prec z$ or $z \prec y$. Then, by Q, either $x = y$ or $y = z$ or $x = z$ or both $y \prec z$ and

$z \prec y$ (when again $y = z$) or both $z \prec x$ and $x \prec z$ (when again $z = x$). So that, finally, either $x = y$ or $y = z$ or $z = x$. Q.E.D.

Let us now call the true point on our scale t, and the false point f. To fix the ideas, we may take any two numbers arbitrarily for t and f, though it will be convenient to make t the longer. You might assume, for example, $t = 5, f = 2$. These letters are subject to the following algebraical rule.

Rule 3. It is *not* true that $t \prec f$.

A. If $t = 5, f = 2$, $x \prec y$ means that x is at least as small as y.

Rule 1 will then obviously be true.

Rule 2 that either $x \prec y$ or $y \prec z$, is also true remembering that each letter must have one or other of the two values 2 and 5. For $x \prec y$ unless $x = 5$, $y = 2$; but if $y = 2$, $y \prec z$ whether $z = 2$, or $z = 5$. Rule 3 simply is that 5 is not so small as 2.

From Rules 2 and 3, we draw these two corollaries, which I number (11) and (12) for convenience.

$$(11) \quad f \prec x \qquad\qquad (12) \quad x \prec t.$$

For, by Rule 2, either $t \prec f$ or $f \prec x$. But by Rule 3, $t \prec f$ is not true. Hence, $f \prec x$ must be true. (12) is proved in a similar way.

To express that any proposition, x, is true, we write $x = t$. Or, since this is only to say that $t \prec x$ and $x \prec t$, and the last is true in any case we may simply write $t \prec x$. In like manner, to express that x is false we write $x = f$ or $x \prec f$.

From Rules 1, 2, 3, we may recover propositions A, B, C. If x is false, that is, if $x \prec f$, it cannot be that $f \prec x$; for if both these were true, we should, by Rule 1, have $t \prec f$, contrary to Rule 3. Call the proposition "if x is false, $t \prec x$ is false," proposition P. By Rule 2, either $t \prec x$ or $x \prec y$. Hence, by P, if x is false, $x \prec y$; which is proposition A. The deduction of B and C is left to the student.

Taking any two numbers arbitrarily for f and t, every proportion, x, will be subject to the quadratic equation $(x-f)(t-x) = 0$. This expresses that x is either true or false. For to say that the product $(x-f)(t-x)$ vanishes is the same as to say that one of its factors vanishes, that is that $x-f = 0$ or $f-x = 0$, i.e. $x = f$ or $x = t$. If we write $(x-f)(t-y) = 0$ this will express that if y is true, z is true. Multiply the last equation but one by $t-z$ and the last by $x-f$. They then become

$$(x-f)(t-y)(t-z) = 0$$
$$(x-f)(y-f)(t-z) = 0.$$

Adding these we have $(x-f)(t-f)(t-z) = 0$. Striking out the finite factor $t-f$ we have $(x-f)(t-z)=0$, which is the syllogistic conclusion, if x is true, z is true. In this system, the expression

$$f+\frac{(x-f)(y-f)(z-f)}{(v-f)^2}$$

will signify that proposition which is true if x, y and z are all true, but is false if any of them are false; while the expression

$$t-\frac{(t-x)(t-y)(t-z)}{(t-f)^2}$$

will signify that proposition that is true if any of the propositions x, y, z, is true, but is false if all are false.

Boole chose to make $t = 1, f = 0$. If that choice is made, we need only distinguish numbers as odd and even, calling 0 even; so that the true is odd and the false even.* (* According to the Pythagoreans, odd and even correspond respectively to good and evil.) The principle of excluded middle will then be expressed, $x(x+1) =$ Even, which is, in fact, true of every number. In place of $x=y$, we may write $x+y=$ Even, for to make this true, x and y must be both odd or both even. The expression of the proposition which is true only if x, y and z are all true, being false if any of them are false, becomes xyz. The proposition which is true if either x, y, or z is true, and is false only if all are false is $xyz+xy+yz+zx+x+y+z$, an intolerable complicated expression.

The best way is to exclude all negative numbers and not to distinguish between positive numbers except as to whether they are nothing or something, putting $f =$ nothing and $t =$ something. In that case, xyz is the proposition which is true if and only if x, y, and z are all true; while $x+y+z$ is the proposition which is true if either x, y or z is true. I shall adopt these results, without, however, considering what the particular values of f and t may be.

Addition and multiplication, together with their cognate words and logical signs, are used in such senses, that $x+y$ signifies that either x or y is true, and xy that both x and y are true. More explicitly, these signs are defined by the following propositions.

D. Either x is false or $x+y$ is true.
E. Either y is false or $x+y$ is true.
F. Either $x+y$ is false or x is true or y is true.
G. Either xy is false or x is true.
H. Either xy is false or y is true.
I. Either x is false or y is false or xy is true.

BOOLIAN ALGEBRA — ELEMENTARY EXPLANATIONS

There is a very convenient system of signs by which very intricate problems of reasoning can be solved. I shall now introduce you to one part of this system only, and after you are well exercised in that, we will study some additional signs which give the method increased range and power. We use letters in this system to signify statements or facts, real or fictitious. We change their signification to suit the different problems. Two statements a and b are said to be equivalent when equal, provided that in every conceivable state of things in which either is true, the other is true, so that they are true and false together, and we then use a sign of equality between them, and write $a = b$. We use the words addition, sum, etc., and the symbol $+$ in such a sense, that if a is one fact, say that the moon is made of green cheese, and b is another fact, say that some nursery tales are false, that is $a + b$, or a added to b, or the sum of a and b signifies that one or the other (perhaps both) of the facts added are true, so that $a + b$ is a statement; true if one or both of the statements a and b are true and false if both are false. Giving to a and b the above significations, it would mean that the moon is made of green cheese, or some nursery tales are false, or both. In translating it into ordinary language, you generally omit the words, or both, as unnecessary.

We use the words multiplication, product, factor, etc., and the signs of multiplication, or we write the two factors one after the other with no sign between them to mean that both of the two statements multiplied are true, so that ab is a statement which is true only if both the statements a and b are true, and [is] false if either a or b [is] false, with the above significations it would mean that the moon is made of green cheese, and that some nursery tales are false. When we wish to signify the multiplication of a whole sum by any factor, we write that sum in parenthesis, thus $(a + b)c$ would mean the product of $a + b$ into c while $a + bc$ would mean the sum of a and of the product of b and c; giving the above significations to a and b, and letting c mean some proverbs were false, $(a + b)$, there we signify the combined statements of, some proverbs are false, and that either the moon is made of green cheese, or some nursery tales are false, while $a + bc$ would mean that either the moon is made of green cheese, or else some proverbs and some nursery tales are false. There are certain rules which facilitate the application of these symbols to reasoning, thus, $a + a$ will mean neither more nor less than a written alone, so that we may write $a + a = a$, for $a + a$ according to what has been said, is that statement which is true if a is true, and is false only if a is false.

The statement aa is also the same as a standing alone, for it merely asserts the fact a twice over so that we may write $aa = a$. We also say that $a + b$ is the same as $b + a$ and that ab is the same as ba. This is usually expressed by saying that addition and multiplication are commutative operations. Also that $(a + b) + c$ is the same as $a + (b + c)$, and $(ab)c$ is the same as $a(bc)$. This is usually expressed by saying that addition and multiplication are associative operations. We also have $(a + b)c = ac + bc$, for if we say that c is true and also that either a or b is true, we state neither more nor less than if we say that either both a and c are true, or both b and c are true. In like manner we have $a + bc = (a + b)(a + c)$ for if we say that either a, or else both b and c are true we state neither more nor less than if we say that either a or b is true, and also that either a or c is true. As this is perhaps not quite evident, I will give a proof of it. We have seen already that $(a + b)c = ac + bc$. Now this has nothing to do with the particular letters used, but will be as true for any other three letters. We will therefore write $(a + b)x = ax + bx$. Now x may be any statement whatever. Let it then be the statement $a + c$ and substitute this in the place of x in the conclusion, then we get $(a + b)(a + c) = a(a + c) + b(a + c)$. Now, on the same principle the first term of the second member of this conclusion $a(a + c)$ is equal to $aa + ac$ and aa we have just seen to be equal to a, so that the first term is $a + ac$; the second term $b(a + c)$ is equal to $ba + bc$, so that the whole expression $(a + b)(a + c)$ equals $a + ac + ab + bc$. Now it is plain that $a + ac$ equals a, for $a + ac$ is only false if both a and either a or c are false. Now if a is false plainly, either a or c is false, that is one of those two, a and c, is false, so that $a + ac$ is false whenever a is false and only then.

And on the same principle $a + ab$ is equal to a, and thus the second member of the last conclusion reduces to $a + bc$, and the whole conclusion is $(a + b)(a + c)$ equals $a + bc$, which is the very conclusion we had to prove. The two principles that $(a + b)c = ab + ac$, and $a + bc$ equals $(a + b)(a + c)$ are commonly referred to by saying that multiplication is distributive with reference to addition, and that addition is distributive with reference to multiplication. I shall now introduce two statements which have special symbols in the system; the first is $\$$, and means any fact necessarily true; all facts that are necessarily true are equal, because we agreed that we should say that two statements are equal provided they are true and false together in all conceivable states of things. The other special symbol for a statement is 0. This signifies any statement that is false. All false statements are equal for the same reason that all true statements are equal. There are a number of rules facilitating the use of these symbols $\$$ and 0. The first is $a+0 = a$. This means that to say that either a is true, or else a false statement is true, is the same as to say at once that a is true. In like manner $\$a = a$; for this means that to say a is true and also that any undesignated true statement is true, is no more than to say that a is true. Second: $\$ + a = \$$, for this means that to say that either a is true or something true is true, is no more than to say that something true is true, which is not saying anything at all. And in like manner $0a = 0$; for this means that to say that a is true and that something false is true is to say something false. Third: Since every statement is either true or false, if we replace any letter, say a, by $\$$ throughout any formula and find the formula is then necessarily true, and if, on afterwards replacing the same letter by 0, we find that the formula so resulting is true also, then the original formula must be true any way. This affords quite a valuable means of proving any doubtful formula. For instance, let us apply it to proving the formula demonstrated above, $a+bc = (a+b)(a+c)$. First, replace a by $\$$ and the formula becomes, $\$+bc = (\$+b)(\$+c)$. Now, $\$$ added to anything gives $\$$; so that $\$+bc = \$$, $\$+b = \$$, and $\$+c = \$$. The whole formula thus reduces to $\$ = \$\$5$ which is true. Now replace a by 0 and the formula becomes $0+bc = (0+b)(0+c)$. Now 0 added to anything does not alter it, so that we may drop these added 0s, and the formula reduces to $bc = bc$, which is true. Thus, it has been shown that the formula is true when a equals $\$$ or a equals 0; and as a must equal one or the other, it is true any way. We must now introduce a new sign. $a = \$$ is the same as a written alone; it means that the statement a is true. But we have as yet, no simple expression for $a = 0$, meaning that the statement a is false. Let us denote this by making a line over the a; thus, $\bar{a}$; this we call the negative or denial of a. There are several rules facilitating the use of denials. First, $a\bar{a}$ equals 0, or nothing can be true and false at the same time, this is called the principle of contradiction. Second $a+\bar{a}$ equals $\$$, or everything is either true or false; this is called the principle of excluded middle.

We will now proceed to show this system of signs is to be used for the purpose of drawing conclusions from premises. The simplest possible kind of reasoning is the immediate application of a rule, thus, a little girl says that whatever mamma forbids is wrong, but mamma forbids this, therefore, this is wrong. Let a mean that anything is forbidden by mamma, b that it is wrong. Then, to say that anything is forbidden by mamma is wrong is the same as to say that either it is not forbidden by mamma, or else it is wrong. This proposition is therefore written $a + b$. The other proposition is a. These propositions are asserted to be both true and therefore, they must be multiplied together, and we have, $a(\bar{a} + b)$. On performing this multiplication, that is, on applying the distributive principle, we get $a\bar{a} + ab$, but $a\bar{a}$ is 0 by the principle of contradiction, and may therefore, be dropped. We therefore have ab. ab is therefore asserted of both the propositions a and b. It therefore asserts B, and therefore the act in question is wrong. Now it would of course, be perfectly ridiculous to use this cumbrous system of signs for the purpose of bringing out the conclusion of such a simple mode of argument as this, but it will be found that the system is well adapted to complicated cases, but this very feature makes it cumbrous for simple ones.

It will be observed in the above example after we get the conclusion ab, we drop the factor a, leaving only b. We obviously have the right to do this at any time. We are always entitled to drop a factor from any additive term, and we are also at liberty to add a term to any factor. In consequence of this, whenever we have given an expression in the form $a(b + c)$ we are at liberty to drop the parenthesis and write $ab + c$. For the distributive principle gives us $ab + ac$, and on dropping the factor a from the last term, we get $ab + c$.

In my different publications I have used a sign like Y turned over on its side, $\prec$ to signify the relation between the antecedent and consequent of an hypothetical proposition. It is a very convenient sign, but as I have no such sign on this type-writer, I shall use a colon for the same purpose, according to the practice of Mr. Hugh McColl. Then, we may write $a : b = \bar{a} + b$. But although the use of this sign does simplify some cases, and I have been one of its principal advocates, and it certainly is useful for a learner, yet it makes a good deal of difficulty in complicated problems. The best way for a beginner to do is to use this sign first, to write down the relations of antecedent and consequent, and afterwards to replace it by $+$, at the same time negativing the antecedent.

Let us now take a slightly more complicated kind of reasoning, the direct syllogism. If you tell one lie you will tell a hundred, and if you tell a hundred lies you will corrupt your integrity. Let a mean that you tell a lie, b that you tell a hundred, c that you corrupt your integrity. Then, the premises are $a : b$ and $b : c$. Multiplying them together we have $(a{:}b)(b{:}c)$. This is equivalent to $(\bar{a} + b)(\bar{b} + c)$. Breaking down the first parenthesis, according to the rule just given, we have $\bar{a} + b(\bar{b} + c)$. Now breaking down the second parenthesis, according to the same rule, we have $\bar{a} + b\bar{b} + c$. But $b\bar{b}$ is 0 and may be dropped. Thus, we reach the conclusion $\bar{a} + c$ or $a{:}c$, if you tell a lie you will corrupt your integrity.

I will now show how to treat a little more complicated arguments called indirect syllogisms. Take these premises: if Enoch and Elijah are mortal the Bible errs; but all men are mortal. Let a mean that any given person is Enoch or Elijah, b that he is a man, c that he is mortal, and d that the Bible errs. Then $a : c$ means that if any person is Enoch or Elijah he is mortal, or what is the same thing, that Enoch and Elijah are mortal. Then $(a : c) : d$ means that if Enoch and Elijah are mortal the Bible errs, which is the first premise. The second premise is $b : c$, or if any person is a man he is mortal. Multiplying the two premises together, we have $[(a{:}c)d](b{:}c)$. Now $a{:}c = \bar{a} + c$ and $(\bar{a} + c){:}d$ is to be converted into the regular form by negativing the antecedent and putting $a +$ instead of the colon. We have, therefore, to find the negative of $\bar{a} + c$. The rule for finding the negative of any expression is this: put a line over every letter that has no line over it, and take a line off every letter that has a line over it, and everywhere substitute multiplication for addition and addition for multiplication. Applying this rule, the negative of $\bar{a} + c$ is $a\bar{c}$. That is to say, to deny that anything is either not Enoch or Elijah, or else is mortal, is equivalent to asserting that something is Enoch or Elijah and at the same time is not mortal. To prove that this is so it will be sufficient to show that these two expressions satisfy the formulae of contradiction and excluded middle. By the principle of contradiction their product ought to vanish. Now their product is $(\bar{a} + c)a\bar{c}$. By the distributive principle this is the same as $\bar{a}a\bar{c} + ca\bar{c}$, but $a\bar{a} = 0$ and $c\bar{c} = 0$, so that the whole is $0\bar{c} + 0a$. Now $0\bar{c} = 0$ and $0a = 0$, so that it comes to $0 + 0$ which is 0.

Thus, the principle of contradiction is satisfied. According to the principle of excluded middle, the sum of any expression and its negative gives \$. Adding the two expressions we have $\bar{a} + c + a\bar{c}$. By the distributive principle of addition with respect to multiplication this is the same as $(\bar{a} + c + a)(\bar{a} + c + \bar{c})$. Now $\bar{a} + a = \$$ and $c + \bar{c} = \$$, so that the whole becomes $(\$ + c)(\bar{a} + \$)$.

But $\$ + c + \$$ and $\$ + \bar{a} = \$$, so that it reduces to $\$ + \$$ which is \$.; and thus, the principle of excluded middle is also satisfied. Our first premise then, is $a\bar{c} + d$, and the product of the two premises is $(a\bar{c} + d)(\bar{b} + c)$. We may arrange this by the associative principle in the following order, $(\bar{b} + c)\bar{c}a + d$. We now put in a new parenthesis, which we are entitled to do by the associative principle, so as to write $[(\bar{b} + c)\bar{c}]a + d$. We now break down the inner parenthesis and thus have $[\bar{b} + c\bar{c}]a + d$ and since $\bar{c} = 0$ this is $\bar{b}a + d$, or $(a:b):d$ which means that if Enoch or Elijah is a man the Bible errs.

I will now give an example of another variety of indirect syllogism. Take the premises, translated persons are not mortal and all men are mortal. Let *a* mean that any person is translated, *b* that he is man, *c* that he is mortal. The first premise, no translated persons are mortal might be written $ac = 0$. But if we take the negative of both members of this conclusion we have $\bar{a} + \bar{c} = \$$, or simply $\bar{a} + \bar{c}$. The other premise is $\bar{b} + c$. The product of the two premises is this, $(\bar{a} + \bar{c})(\bar{b} + c)$. Treating this precisely as in the case of a direct syllogism we get the conclusion $\bar{a} + \bar{b}$, or no translated persons are men.

We now go on to a slightly more complicated kind of reasoning, the dilemma. The dilemma is a reasoning in which you show that there are two (or more) possible alternatives, and then show that in either case a consequence follows. This kind of reasoning, although treated in books on Rhetoric, was first introduced into the treatises on logic about the year 1500 by Laurentius Valla. In point of fact the dilemma was very little used during the middle ages and it forms the most elementary example of the falsity of the traditional and Aristotelian notion that all reasoning is syllogistic. In the present system of signs, however, we fail to see anything peculiar about the dilemma, for the reason that we have in this system arbitrarily twisted every syllogism into a dilemmatic form, by writing all *a* is *b* in the form either non-*a* or *b*. The truth is that this system of signs is altogether framed to meet the case of the dilemma. Syllogistic reasoning is so easy that it is got rid of by a little artifice. ...

OUTLINE OF LOGIC (s-41)

CHAPTER I. BOOLIAN ALGEBRA

The deficiency of pronouns in English, as in every other tongue, is felt as soon as there is occasion to discourse of the relations of three or more objects, and forces the lawyers of today, as it did Euclid of old, to distinguish them by the letters *A*, *B*, *C*, etc. This is already a great stride toward an algebraical notation. Two other kinds of signs are, however, very useful. The first embraces the parentheses and brackets which are the punctuation-marks of algebra. Everybody knows how imperfect the ordinary system of punctuation is, how inferior it is even to the pauses of speech. There ought to be no objection, then, to our using, in logical studies, the algebraic system, which answers its purpose sufficiently well, and which simply consists in enclosing within parenthesis-marks any expression which is to be treated as a whole. Square brackets are employed when one such whole is to be included within another; and braces and other forms of enclosing marks are adopted when distinctness requires it. Finally, there are such signs as $+$, $-$, etc., which are mere abbreviations, though they are, at the same time, great aids to thought.

The algebra described in this chapter is called Boolian because it [is] but a modification of the system of the English mathematician, George Boole.

Each letter of the alphabet is used, as occasion requires, to express any statement, true or false, simple or complex. A single letter may signify the entire substance of a book, or the sum of human knowledge, or the content of omniscience. To use the consecrated term of logic, which Appuleius, in the second century, speaks of as familiar, the letters are propositions. The letters x, y, z, will be appropriated to the expression of formulae which are true, whatever statements these letters may signify.

Equality and cognate words, together with the sign $=$, are so used that $x = y$ (no matter what statements x and y may signify) means that x and y are equally true, that is, are either both true or both false. More explicitly, the following propositions, marked L, M, N, for convenience of future reference, sum up the meaning of equality in this system of algebra.

L. If $x = y$, either x is false or y is true.
M. If $x = y$, either x is true or y is false.
N. If x and y are either both true or both false, then $x = y$.

It follows that the sign of equality is subject in this algebra to precisely the same rules as in ordinary algebra. These rules are as follows:

Rule 1. $x = x$.
Rule 2. If $x = y$, then $y = x$.
Rule 3. If $x = y$ and $y = z$, then $x = z$.

These rules are proved as follows: Rule 1. The proposition x is either true or false. Hence, by N, $x = x$. Rule 2. By L and M, if $x = y$, since neither x nor y can be both true and false, either x is false and y is false, or x is true and y is true. But in either of these cases, by N, $y = x$. Rule 3. Suppose y to be true. Then, by M, if $x = y$, x is true; and by L, if $y = z$, z is true; so that, if $x = y$ and $y = z$, both x and z are true, and by N, $x = z$. Next, suppose y to be false. Then, by L, x is false; and by M, if $y = z$, z is false; so that if $x = y$ and $y = z$, both x and z are false, and again by N, $x = z$. But y is either true or false; hence, in any case, if $x = y$ and $y = z$, then $x = z$.

Addition and multiplication, with their cognate words and algebraical signs, are used in such senses that $x + y$ means that either x or y is true, and xy that both x and y are true. More explicitly, the meanings of the sum and product are summed up in the following propositions, lettered for reference, A, B, C, X, Y, Z.

A. Either x is false or $x + y$ is true.
B. Either y is false or $x + y$ is true.
C. Either $x + y$ is false or x is true or y is true.
X. Either xy is false or x is true.
Y. Either xy is false or y is true.
Z. Either x is false or y is false or xy is true.

From these definitions it follows that in this logical algebra all the ordinary rules of addition and multiplication [hold]; that is to say, we are permitted to perform everything that we are permitted in ordinary algebra to perform, as far as addition and multiplication are concerned. But there are besides other rules which permit us to [do] what would not be legitimate in ordinary algebra. The rules common to ordinary and logical algebra are the following:

Rule 4. The associative principle of addition. $(x+y)+z = x+(y+z)$.
Rule 5. The associative principle of multiplication. $(xy)z = x(yz)$.
Rule 6. The commutative principle of addition. $x+y = y+x$.
Rule 7. The commutative principle of multiplication. $xy = yx$.
Rule 8. The distributive principle of multiplication in reference to addition. $x(y+z) = xy+xz$.

All the other general formulae are deducible from the above together with the following two.

Rule 9. $x+x = xx$.
Rule 10. If $x+y = xz$, then $x+y = xz = x$.

All these rules are very easily proved as follows: Rule 4. If $(x+y)+z$ is true, by C, either $(x+y)$ is true or z is true, and if $x+y$ is true, either x is true or y is true; so that if $(x+y)+z$ is true either x or y or z is true. If x is true, by A, $x+(y+z)$ is true. If y is true, by A, $y+z$ is true, and by B, $x+(y+z)$ is true. If z is true, by B, $y+z$ is true, and again by B, $x+(y+z)$ is true. Thus, if either x or y or z is true, $x+(y+z)$ is true; or if $(x+y)+z$ is true $x+(y+z)$ is true. If $(x+y)+z$ is false, by B, z is false, and by A, $x+y$ is false. But if $x+y$ is false, by [B], y is false, and by A, x is false. This, if $(+x+y)+z$ is false, x, y, and z are all false. If y and z are both false, by C, $y+z$ is false, and if x and $y+z$ are both false by C, $x+(y+z)$ is false. Thus, if $(x+y)+z$ is false, $x+(y+z)$ is false. But $(x+y)+z$ is either true or false, so that either $(x+y)+z$ and $x+(y+z)$ are both true or both false, and consequently by N, $(x+y)+z = x+(y+z)$. The proof of Rule 5 is exactly similar to that of Rule 4, only {interchanging / transposing} everywhere addition and multiplication, true and false, A, B, C, and X, Y, Z. Rule 6. By c, if $x+y$ is true, either x or y is true. By A and B, if either x or y is true, $y+x$ is true. By A and B, if $x+y$ is false, both x and y are false. By C, if both x and y are false, $y+x$ is false. Hence, $x+y$ and $y+x$ are either both true or both false, and hence, by N, $x+y = y+x$. Rule 7 is proved similarly to Rule 6. Rule 8. Suppose $x(y+z)$ to be false. Then by Z, either x is false or $y+z$ is false, and in the latter case, by A and B, both y and z are false. If x is false, by X, xy is false, and for the same reason xz is false; and by C, if xy and xz are both false, $xy+xz$ is false. If y is false, by B, xy is false; and for the same reason, if z is false xz is false;

so that if y and z are both false both xy and xz are false. But in this case, by C, $xy + xz$ is false. We have thus shown that if $x(y + z)$ is false, either x or else both y and z are false.

B. BOOLIAN ALGEBRA (s-39)

In this system of signs, each letter of the alphabet is the abbreviated statement of a fact, simple or complex. Thus, x might be taken to signify that twice two is four, and that either all men are mortal or else Napoleon Bonaparte was born in Salem, Mass. Every letter is an abbreviated statement. It may be of all that is stated a book, it may be of something very simple.

The sign = is used in ordinary algebra to signify equality. Thus, $x = y$, read "x equals y," can be written when the quantity named x is known to be equal to the quantity named y. In the Boolian Algebra the same sign is used, and is read in the same words, but a different meaning is attached to the sign and to the words "equal," etc. with which it is associated. Two facts are considered to be equal when every possible state of things in which either is true is or would be a state of things in which the other is or would be true, and *vice versa.* Thus, "this polygon has three sides," and "this polygon has three angles" are equal or equivalent facts, although the statements refer the one to the sides, the other to the angles. If you ask what is meant by a "possible" state of things, I reply that the possible is that which is not known to be non-existent, and — that either in the state of knowledge in which we find ourselves, or else in some feigned condition of ignorance. The "possible" of this Algebra may be any variety of the possible, — as the logical possible, the physically possible, etc. Only the same meaning must be attached to the possible throughout any one discussion. Or if different varieties are used, these must be carefully distinguished. It must be understood that when $x = y$ is written the facts x and y are not stated; it is only stated that they are either both actual or neither.

The sign + is used in ordinary algebra to denote addition. Thus $x + y$, read "x plus y," denotes the sum of the quantities named x and y. This sign, with the words "plus," "sum," "add," etc. associated with it, is also used in the Boolian Algebra, but by most writers (myself included) in a sense not strictly analogous to the mathematical sense. The expression $x+y$ is a *statement*, and it states that either x or y (perhaps both) is a fact, without saying which. The expression $x+y$ is that statement which is true if either of the statements x and y is true, and is false if both x and y are false. It follows that $x+x=x$ contrary to the analogy of arithmetic.

The signs $\times$ and $\cdot$, and the writing one after another of two expressions, are used in ordinary algebra to denote multiplication. Thus, $x \times y$ or $x \cdot y$, or xy, read "xy,", or "x into y," or "x multiplied by y," denote the product of x and y. The same signs and words are used in the Boolian Algebra in a different sense. The expression xy is a *statement*, and it is the statement of both facts x and y. More precisely, it is the statement which is true if both the statements x and y are true, and is false if either of them is false. It follows that $xx = x$, which in ordinary algebra would be an equation satisfied by $x = 0$, $x = 1$, and $x = \infty$.

Exercise. Ascertain whether the following formulae necessarily hold good whatever statement x, y, and z may be:

$$x+y=y+x \qquad xy=yx \qquad x(yz)=(xy)z$$

(N.B. Parentheses are used to signify that the statement in parenthesis is to be combined as one statement by addition or multiplication with the one outside of the parenthesis.)

$$x+(y+z)=(x+y)+z; \quad x(y+z)=xy+yz;$$
$$x+yz=(x+y)(x+z); \quad (x+y)(y+z)(z+x)=xy+yz+zx$$

SECOND LESSON

N.B. The exercises on the other side [of the page] should be done, and the formulae verified by the definitions of logical addition and multiplication in the way shown in the following example. $x+yz$ is true if either x or yz is true; otherwise it is false. But yz is only true if both y and z are true. Hence $x+yz$ is true first if x is true and second if both y and z are true. The statement $(x+y)(x+z)$ is true only if $(x+y)$ and $(x+z)$ are true. $x+y$ is true first if x is true, and second if y is true. $x+z$ is true, first if x is true, and second if z is true. Hence both $(x+y)$ and $(x+z)$ are true, first if x is true, and second if both y and z are true. But these are the same circumstances under which $x+yz$ is true. Hence, the two statements $x+yz$ and $(x+y)(x+z)$ are equivalent, and we can write $x+yz=(x+y)(x+z)$. Do all those exercises before going further.

The equations $x+(y+z)=(x+y)+z$ and $x(yz)=(xy)z$ are called respectively the Associative Principle of Addition and the Associative Principle of Multiplication. The formulae $x+y = y+x$ and $xy = yx$ are called the Commutative Principle of addition and multiplication. The formula $x(y+z) = xy+xz$ is called the Distributive Principle of Multiplication with respect to addition. The formula $x+yz=(x+y)(x+z)$ is called the Distributive Principle of Addition with respect to multiplication.

From those formulae with $x+x = x$ and $xx = x$ all others can be deduced without recourse to the meanings of the operations. Thus, to prove $(a+b)(c+d) = ac+ad+bc+bd$, we proceed as follows. By the distributive principle $(a+b)(c+d) = (a+b)c+(a+b)d$. By the commutative principle this equals $c(a+b) + d(a+b)$ and, again applying the distributive principle, this equals $ca+cb+da+db$ or by the commutative principle $ac+bc+ad+bd$. The associative principle is assumed in leaving off the parentheses.

Exercises. Prove in this way the following:

$$ab+cd = (a+c)(a+d)(b+c)(b+d)$$
$$a+ab = a; \quad a(a+b) = a; \quad (a+b)(b+c)(c+a) = ab+bc+ca;$$
$$(a+b)(b+c)(c+d)(d+a) = ac+bd$$
$$(a+b+c)(a+b+d)(a+c+d)(b+c+d) = ab+ac+ad+bc+bd+cd$$
$$(a+b+c)(a+b+d)(b+c+e)(c+a+f)(a+d+f)(b+d+e)(c+e+f)(d+e+f) = ae+bf+cd$$
$$(a+e)(b+f)(c+d) = abc+abd+acf+bce+adf+bde+cef+def$$
$$(b+f)(c+g)(d+h)(a+e)(b+d)(b+g)(d+g)(c+f)(c+h)(f+h) = abcd+abgh+adfg+bche+cdef+efgh$$

THIRD LESSON

We now introduce a new sign. A line drawn over a statement denies it. Thus, $\bar{x}$ is a statement that x is not true. Then, show that $x = \bar{\bar{x}}$. Also that $\overline{x+y} = \bar{x}\bar{y}$. Also that $\overline{xy} = \bar{x}+\bar{y}$. Also that $\overline{\bar{x}\bar{y}} = x+y$. Also that $\overline{\bar{x}+\bar{y}} = xy$. Also, that $\overline{x+\bar{y}} = \bar{x}y$. Also that $\overline{x\bar{y}} = \bar{x}+y$.

Let the sign of Dollars \$ be used to signify a statement that is true in every possible case, that is, which is known to be true. Then, to write $x = \$$ is the same as to write x simply; or $x = (x = \$)$. Let naught, 0, be used to signify a statement known to be false. Then, to write $x = 0$ is the same as to write $\bar{x}$ simply; or $\bar{x} = (x=0)$.

Exercises. Show that $x+0 = x$; $0x = 0$; $\$+x = \$$; $x\bar{x} = 0$. (This is the principle of contradiction.)[1]

[1] An additional sheet, apparently part of a sequence on propositions, tells that "the rules for using the sign $\prec$ are so difficult, that we discard the use of it entirely. In place of it we use the sign for logical addition and multiplication. The logical sum of two propositions is that proposition which is true if either of those two are true and false only if both are false. If A and B are the two propositions, their logical sum is written $A+B$. It is equivalent to $\bar{A} \prec B$.

If A is false and B is false, $A+B$ is false.
If A is false and B is true, $A+B$ is true.
If A is true and B is false, $A+B$ is true.
If A is true and B is true, $A+B$ is true.

Suppose A signifies 'the earth moves round the sun,' and B 'the sun moves round the earth'; then $A+B$ signifies 'either the earth moves round the sun, or the sun moves round the earth.'

The logical product of two propositions is that proposition which is false if either be false and is true only if both be true. The logical product of A and B is written AB. It is equivalent to $A \overline{\prec} \bar{B}$.

If A is false and B is false, AB is false.
If A is false and B is true, AB is false.
If A is true and B is false, AB is false.
If A is true and B is true, AB is true.

If A and B have the significations given above AB means 'the earth moves round the sun, and the sun round the earth.' The sign $=$ means that the proposition to the right and left of it are either both true or both false."

C. LOGIC OF SCIENCE (342)

LECTURE III

Closely allied to the subject of logic, if not actually a part of it, is the theory of probabilities, — or doctrine of chances. It is a matter with which it is of considerable advantage to have a practical acquaintance. But there are two obstacles to acquiring a thorough knowledge of it; the first is that the more complicated problems tax the resources of the mathematician; and the second and more serious is that the mere putting a problem of chances into equations or bringing it to the point where the mathematical work first begins requires the subtlety of a metaphysician. Fortunately, the late Professor Boole has entirely overcome this second difficulty by means of a distinct calculus; which is very readily comprehended and may be applied with the utmost facility. Everyone can master this curious branch of mathematics in a short time; and as the knowledge of it enables us to solve readily all simple questions of probability and to understand the general principles of solution of the most difficult ones; I propose to devote this lecture and the next to the exposition of it.

The first point is that instead of writing words, we write letters. Thus instead of horses we write *h*, instead of black *b*, and so forth. *h* stands not for one horse but for all horses, — the whole class of horses collectively. *b* stands for all black things.

Then, if we wish to write *all black horses*, it will be natural to put *bh* — black horses. This is analogous to multiplication in arithmetic for as three times two means *three twos* and six times seven, six sevens, so we may say that *black horses* are black into horses. Three times two implies a *three* ⋮ each of whose units is a two ⋮⋮. In the same way *black horses* implies all black things each of which is a horse. Perhaps you will think this analogy fanciful; I do not say that it is not. But it will serve our purpose very well as you will see.

Suppose *c* represents cows. Then how shall we represent all horses and all cows together. Clearly it is horses added to cows or $h + c$.

Suppose w stands for Washington City and u for the Capitol of the United States. How shall we say that Washington City and the capitol of the United States are the same? By writing $w = u$, w equals u. By equality then we mean, in this case, not [an] identity in respect to number but complete identity.

In arithmetic any number added to nought remains the same

$$3+0 = 3$$
$$4+0 = 4$$

Now in this new calculus which we are studying we have as yet not introduced any *naught*. Let us introduce it, with such a meaning that such equations as these shall hold. Let

$$h+0 = h$$
$$c+0 = c$$

Or all horses together with *naught* constitute all horses. And all cows together with *naught* constitute all cows. What, then, does *naught* mean? It plainly means nothing; not nothing in respect to one measure merely as it does in arithmetic but absolutely nothing.

You can now see the real appropriateness of making *black* multiplied by *all horses* mean all black horses. For in Arithmetic, any number multiplied by naught is naught.

$$3 \times 0 = 0$$
$$4 \times 0 = 0, \text{ etc.}$$

Now just in the same way in the Calculus of Logic. Horses which are nothing are nothing; horses which do not exist are nothing.

$$h \times 0 = 0$$

and so for any other letter

$$c \times 0 = 0 \qquad b \times 0 = 0.$$

In arithmetic, any number multiplied by one gives that number

$$3 \times 1 = 3$$
$$4 \times 1 = 4, \text{ etc.}$$

Now thus far we have not adopted any meaning at all for *one* in our Calculus of Logic. We may therefore assign it any convenient meaning. Let us, then, take it in such a sense that these equations shall hold. So that for instance

$$h \times 1 = h \qquad c \times 1 = c, \text{ etc.}$$

Now what meaning must one have? The product of two letters represents those objects which are common to both the two classes represented by the letters multiplied together. Now, *one* represents that class which multiplied by any other gives that other; that is it is that class which has the whole of the object of every class under it. In other words it is *everything* or *whatever is*.

$$h \times 1 = h$$

means that whatever horses *are* the same as all horses. $c \times 1 = c$ means that whatever cows *are* are the same as all cows. *One* then means all that *is*. Just as h means all horses, c all cows, etc.

I must now draw your attention to one of those peculiar and original procedures which characterize Professor Boole's work. $h+c$ means horses and cows, *besides*. Now you cannot take anything beside itself; you cannot say horses and horses besides; although you can say nothing and nothing besides. And, therefore, if you meet with such an expression as

$$a + a$$

or a and a besides, you may be sure that a is nothing. Thus the mere addition of a thing to itself, in this system, quite contrary to the whole analogy of mathematics, determines what the value of that thing is.

As $a+a$ makes $a = 0$ so does $a+a+a$, $a+a+a+a$ and so forth or in other terms $2a$, $3a$, $4a$ make $a = 0$, etc. Now this determines the meaning of all the other numbers besides *one* and *zero*. These numbers mean that that which they are multiplied by is nothing.

We thus see that in this Boolian calculus no letter can have any numerical value except $\overset{\text{unity}}{1}$ and $\overset{\text{zero}}{0}$. But unity means *all things* and zero means nothing. It is plain then that h which stands for all horses, since it is neither all things nor nothing, has a value which is *not numerical*. This, again, is a very peculiar and interesting point. Every letter, unless it denote everything or nothing, has a value which is not numerical. At the same time we may have a general expression, in number, for a class. For since

$$h \times 0 = 0, \qquad c \times 0 = 0, \qquad b \times 0 = 0, \quad \text{etc.}$$

If we divide these quantities by *zero*, we have

$$h = \tfrac{0}{0} \qquad c = \tfrac{0}{0} \qquad b = \tfrac{0}{0}$$

Zero divided by zero then denotes any class — without discriminating and determining what class. This meaning of zero divided by zero is similar to its meaning in arithmetic. In arithmetic one number divided by another is the number of times a pail which holds the number of quarts denoted by the divisor can be emptied into a pail which holds the number of quarts denoted by the dividend. Now suppose neither pail holds anything. Suppose the sides of both have been cut down till nothing but the bottom is left, then the number of times that one can be emptied into the other is *zero* divided by *zero*. But how many times is that? We cannot say; it is indeterminate. Just so in the calculus of logic zero divided by zero is an indeterminate class, we cannot say what one.

If we have $a+b = c$; that is, if the classes a and b together make up the class c then $c-a = b$ that is b is c except a or the class c after the class a is taken away. Let b denote everything black and $\bar{b}$ denote everything not black; then, $b + \bar{b} = 1$ that is everything black and everything not black make up everything then $\bar{b} = 1-b$ or everything not black is denoted by b subtracted from 1 or everything except what is black.

In the same way $1-h$ stands for all that is not a horse and so with every letter.

The peculiarities of the system may all be summed up in one form of equation $b(1-b) = 0$ or that which is black and is not black is nothing. $h(1-h) = 0$ that which is a horse and is not a horse is nothing. Observe that there are only two numbers *zero* and *one* which will satisfy such an equation for since the product of b and $(1-b)$ is *zero* either b is *zero* or $(1-b)$ is zero $\begin{matrix} b=0 \\ 1-b=0 \end{matrix}$. If $1-b = 0$, b $= 1$ so that this equation itself implies that b has only two numerical values, *one* and *zero*.

I have now given explanations of the signification of the more simple combinations of letters; but for the benefit of those who may have lost a word here and there I will briefly repeat what I have said.

In this logical mathematics, then, the single letters denote the whole collection of individuals under a class. Thus, m may be taken to stand for all men, w for all women, a for all animals, o for all that is old, d for all that is dead. *Secondly* 1 minus a letter, or the difference between one and a letter, stands for the whole collection of things which do not belong to the class denoted by the letter. Thus $1-m$ will stand for all things not men; $1-w$ for all things not women, and so forth. *Thirdly*, one class multiplied by another stands for those things which belong to both of the two classes. Thus dm stands for dead men, $(1-d)m$ for men not dead, $d(1-m)$ for everything not men that are dead, ow for all old women, $a(1-w)$ for all animals not women, and so forth. *Fourthly*, one letter added to another, stands [for] all things denoted by the first together with all the things denoted by the second besides; and implies that there are no things which belong to both classes at once. Thus $m+w$ stands for all men together with all women. $d+(1-d)$ stands for all things dead and all things not dead. $dm+(1-d)w$ stands for all dead men and all women not dead. *Fifthly*, 1 or as we often write it $\frac{1}{1}$ stands for all that *is*; and consequently $\frac{1}{1}m$ stands for all the men there are, that is for all men; the same as m. *Sixthly*, 0 or as we often write it $\frac{0}{1}$ stands for what is *not*, or nothing, hence $\frac{0}{1}m$ stands for no men, and is the same as $\frac{0}{1}$ alone. *Seventhly*, $\frac{0}{0}$ stands for some class we know not what, perhaps for all things, perhaps for nothing perhaps for some things. Accordingly $\frac{0}{0}m$ stands for *some all* or *no* men; $\frac{0}{0}w$ for *some all* or *no* women and so forth. *Eighthly*, any number except *one* and *zero* when multiplied by any letter implies that there is no such class as is denoted by that letter. Thus $2d(1-d)$ implies that the class of dead things which are not dead, does not exist. Instead of writing a number we generally write $\frac{1}{0}$ so that

$$\frac{1}{1} \quad \frac{0}{1} \quad \frac{0}{0} \quad \frac{1}{0}$$

are the four numerical forms which have a use in this system. *Finally* every letter and every combination of letters is equal to *one* or *zero* or else has no numerical value. This is expressed by the equations $m(1-m) = 0$, $w(1-w) = 0$, $d(1-d) = 0$, $a(1-a) = 0$, $0(1-0) = 0$, etc.

Supposing, then, that so much is firmly fixed in the mind, I proceed to explain how we can interpret the mean of complicated expressions such as

$$\frac{a}{m} \quad \frac{w}{W+m} \quad \frac{1-d}{a+Wd} \quad \text{and so forth.}$$

We will begin with expressions which contain only one letter, such as $\frac{m}{m}$, $\frac{1+m}{2-m}$. Now if we can reduce such an expression to the form $\frac{1}{1}m + \frac{1}{1}(1-m)$ or $\frac{1}{0}m + \frac{0}{0}(1-m)$ or $\frac{0}{1}m + \frac{1}{1}(1-m)$ or any form where we have m and $(1-m)$ separately multiplied by $\frac{1}{1}$, $\frac{0}{1}$, $\frac{0}{0}$, or $\frac{1}{0}$ and then added; I say if we can show that any expression has the same meaning as such an expression as this, we can tell what that meaning is. For we know what $\frac{1}{1}m$, $\frac{0}{1}m$, $\frac{1}{0}m$ and $\frac{0}{0}m$ stand for and what $\frac{1}{1}(1-m)$, $\frac{0}{1}(1-m)$, $\frac{1}{0}(1-m)$, and $\frac{0}{0}(1-m)$ stand for

	$\frac{1}{1}m$ stands for all men	$\frac{1}{1}(1-m)$ for all things not men
	$\frac{0}{1}m$ for no men	$\frac{0}{1}(1-m)$ for no things not men
	$\frac{0}{0}m$ for *some all* or *no* men	$\frac{0}{0}(1-m)$ for some all or no things not men
and	$\frac{1}{0}m$ implies that there are no men	$\frac{1}{0}(1-m)$ implies that there are no things not men

We know what each of these means and we know what addition means, and therefore we know what any such expression as $\frac{1}{1}m + \frac{0}{0}(1-m)$, $\frac{0}{1}m + \frac{1}{1}(1-m)$ means. The first means all men together with *some all no* things not men. The second means all things not men.

We may denote such an expression in a general way by $Am + B(1-m)$. That is, m is multiplied either by $\frac{1}{1}$ or $\frac{0}{1}$ or $\frac{0}{0}$ or $\frac{1}{0}$ but as we do not know which, we put A merely to show that it is one or the other of these, and we put $B(1-m)$ for the same purpose.

Now I can give you a very simple rule by which you can reduce any expression to this form $Am + B(1-m)$ for if I write down this form twice —

$$Am + B(1-m) \qquad\qquad Am + B(1-m)$$

and on one side put 1 instead of m and on the other put 0 instead of m

I have $\quad A \times 1 + B(1-1) \quad$ and $\quad A \times 0 + B(1-0)$

now A times one is A and 1 minus 1 is nothing so that on that side we have $A + B \times 0$ or as B times nothing is nothing, it is simply A. On the other side $A \times 0$ is nothing; and 1 minus zero is 1 so that we have $B \times 1$ or simply B. So we have A on the side where 1 was put instead of m and B on the side where zero was put instead of m. Now we have only to write in the ms and

$$\begin{array}{lcl} A & & B \\ \text{becomes } Am & + & B(1-m) \end{array}$$

which is the original expression. Now suppose we perform this same process on an expression of a different form; say $\frac{m}{m}$

$$\begin{array}{lcl} \frac{m}{m} & & \frac{m}{m} \\ \frac{1}{1} & & \frac{0}{0} \\ \frac{1}{1}m & + & \frac{0}{0}(1-m) \end{array}$$

and this shows that $\frac{m}{m}$ stands for *all* men and some all or none of the things not men. Now I can show you in a different way that this is what $\frac{m}{m}$ stands for.

$$am = m$$

that is all men who are animals are the same as all men. Now divide by m and we have

$$a = \frac{m}{m}$$

but all animals are $a = m + \tfrac{0}{0}(1 - m)$ and therefore $\dfrac{m}{m} = m + \tfrac{0}{0}(1 - m)$. The general rule then for reducing an expression to a form we can understand is to put it down twice side by side; and put 1 for the letter on the left and zero for the letter on the right and then write the letter after the first and 1 minus the letter after the second and connect them with the sign +.

What does $\dfrac{1-m}{3-m}$ stand for

$$\frac{1-m}{3-m} \qquad\qquad \frac{1-m}{3-m}$$

$$\frac{1-1}{3-1} \qquad\qquad \frac{1-0}{3-0}$$

$$\tfrac{0}{2} \qquad\qquad \tfrac{1}{3}$$

for $\tfrac{1}{2}$ put $\tfrac{0}{1}$ and for $\tfrac{1}{3}$ put $\tfrac{1}{0}$ which mean the same

$$\tfrac{0}{1}m \qquad + \qquad \tfrac{1}{0}(1 - m)$$

Now let us consider expressions which contain two letters. Take for instance $\dfrac{mw}{m+w}$. We may treat it first as though m were the only letter

$$\frac{mw}{m+w} \qquad\qquad \frac{mw}{m+w}$$

$$\frac{w}{1+w} \qquad\qquad \frac{0}{w}$$

$$\frac{w}{1+w}m \qquad + \qquad \frac{0}{w}(1 - m)$$

Now what does this mean? We can answer this if we know first what $\dfrac{w}{1+w}m$ means and then what $\dfrac{0}{w}(1 - m)$ means. 1st what does $\dfrac{w}{1+w}m$ mean? Let us treat it as if w were the only letter

$$\frac{w}{1+w}m \qquad\qquad \frac{w}{1+w}m$$

$$\frac{1}{1+1}m \qquad\qquad \tfrac{0}{1}m$$

$$\tfrac{1}{0}mw \qquad + \qquad \tfrac{0}{1}m(1-w).$$ This is what $\frac{w}{1+w}m$ means that there are no men who are women and it includes none of the men not women. Next what does $\frac{0}{w}(1-m)$ mean

$$\frac{0}{w}(1-m) \qquad\qquad \frac{0}{w}(1-m)$$

$$\tfrac{0}{1}(1-m) \qquad\qquad \tfrac{0}{0}(1-m)$$

$$\tfrac{0}{1}(1-m)w \qquad + \qquad \tfrac{0}{0}(1-m)(1-w).$$ This is the equivalent of $\frac{0}{w}(1-m)$. It includes no women not men and some all or none of the things which are neither women nor men.

Now if we add together the expressions for $\frac{w}{1+w}m$ and $\frac{0}{w}(1-m)$ we have the equivalent of $\frac{mw}{m+w}$ because this is the same as $\frac{w}{1+w}m + \frac{0}{w}(1-m)$. $\frac{mw}{m+w}$ then is

$$\tfrac{1}{0}mw + \tfrac{0}{1}m(1-w) + \tfrac{0}{1}(1-m)w + \tfrac{0}{0}(1-m)(1-w)$$

That is it is the class which includes some, all, or none of the things which are neither men nor women and none of the women not men and none of the men not women and it implies further that there are no men who are women.

Now I can give you a shorter rule for getting the same result. We start with $\frac{mw}{m+w}$. Write it down four times

$$\frac{mw}{m+w} \qquad \frac{mw}{m+w} \qquad \frac{mw}{m+w} \qquad \frac{mw}{m+w}$$

First we put

$$\begin{matrix} m=1 \\ w=1 \end{matrix} \text{ then } \begin{matrix} m=1 \\ w=0 \end{matrix} \text{ then } \begin{matrix} m=0 \\ w=1 \end{matrix} \text{ then } \begin{matrix} m=0 \\ w=0 \end{matrix}$$

This gives $\frac{1}{1+1}$ or

$$\tfrac{1}{0} \qquad \tfrac{0}{1} \qquad \tfrac{0}{1} \qquad \tfrac{0}{0}$$

then wherever we put one for m write an m and wherever we put zero for m write $1 - m$, and so with w, and we have

$$\tfrac{1}{0}mw \quad + \quad \tfrac{0}{1}m(1-w) + \quad \tfrac{0}{1}(1-m)w + \quad \tfrac{0}{0}(1-m)(1-w)$$

which is the same result we had before.

I will give one more example of this process.

What does $\dfrac{m}{a}$ stand for?

$\dfrac{m}{a}$	$\dfrac{m}{a}$	$\dfrac{m}{a}$	$\dfrac{m}{a}$
$m1$	$m1$	$m0$	$m0$
$a1$	$a0$	$a1$	$a0$
$\frac{1}{1}$	$\frac{1}{0}$	$\frac{0}{1}$	$\frac{0}{0}$

$$\tfrac{1}{1}ma \quad + \quad \tfrac{1}{0}m(1-a) + \quad \tfrac{0}{1}(1-m)a + \quad \tfrac{0}{0}(1-m)(1-a)$$

Suppose we have the equation

$$1 = \tfrac{1}{1}xy + \tfrac{1}{1}x(1-y) + \tfrac{0}{0}(1-x)(1-y)$$

This reads that all things consist of all those things which are both x and y, of all which are x and not y, and perhaps of some that are neither x nor y. Observe that it is not implied that there are any things which are both x and y; or that there are any that are x and not y. But as something certainly exists, there must be something of one or other of these two classes.

When then we have 1 made equal to anything; it is asserted 1st that something among all the classes multiplied by $\frac{1}{1}$ *exists*, 2nd that nothing among all the classes multiplied by $\frac{0}{1}$ exists; and 3rd it is left in doubt whether anything among all the classes multiplied by $\frac{0}{0}$ exists. In the same way it might be shown that if we have such an expression as the following

$$\frac{xy + x(1-y) + (1-x)y + (1-x)(1-y)}{xy + x(1-y) + (1-x)y + (1-x)(1-y)} = 0$$

or any other expression derived from this by striking out any term or terms from numerator or denominator or both that

1st, No individual among all the classes appearing in the numerator exists;

2nd, Some individual among all the classes not in the numerator but in the denominator exists.

Hence if we wish to say

All men are animals we write $\dfrac{m(1-a)}{ma + m(1-a)} = 0$

No men are women $mw = 0$

Some men are Kings $\frac{0}{mK} = 0$

I will not explain how reasoning is to be conducted upon this system. To do this we wish to be able to strike any letter out of an equation. Suppose for instance we have $ab + c(1 - b) = 0$ and wish to get rid of b

$$\begin{array}{cccc} ab + c(1-b) & ab + c(1-b) & ab + c(1-b) & ab + c(1-b) \\ b + (1-b) & b & (1-b) & 0 \\ 1 & & & 0 \end{array}$$

$$ac \quad + \quad ba(1-c) \quad + \quad (1-b)(1-a)\,c \quad = 0$$

$$ac = 0$$

Now this result may be got also by writing

$$ab + c(1-b)$$

Put b

$= 1$ and $\qquad a$

$=$ zero $\qquad c$

$$ac = 0$$

Now let us put an argument into syllogisms. All men are Animals:

$$\frac{m(1-a)}{ma + m(1-a)} = 0, \text{ Socrates is a Man: } \frac{s(1-m)}{sm + s(1-m)} = 0$$

$$\frac{m(1-a)}{ma + m(1-a)} + \frac{s(1-m)}{sm + s(1-m)} = 0$$

As I know your minds must be wearied with this mathematics, I will [now] postpone the further consideration of it for another lecture and will take up now a lighter subject.

The art of Logic began with the conflict of pantheistic and sensational philosophies. This conflict led to acute arguments on the one side to show the contradictions involved in the doctrine that many things exist. And to curious instances on the other side to show that there is no absolute truth. Those who took this latter ground were called Sophists; and the purpose of their sophistry was to show that valid argument contradicts itself and that therefore truth is nothing more than what any man believes. Aristotle was brought up in a school that for two generations had rejected the conclusions of the sophists. Their sophisms, therefore, were to him merely matters of curiosity. They are useful to philosophy, he says, for two reasons; 1st because they generally arise from ambiguities of speech and, therefore, are instructive in reference to the nature of propositions and arguments and 2nd because practice in detecting them makes us less liable to deceive ourselves with false arguments. I believe that it was in fact from the study of these bits of sophistry, which smaller minds pass over as unworthy of serious attention, that Aristotle was led to the knowledge of the true laws of reasoning. To illustrate: An ancient philosopher Melissus had advanced this argument to prove that the world had no beginning:

The universe is not born	*E*	} tr.
Everything born has a beginning	*A*	
∴ The universe has no beginning.	*E*	

Now Aristotle quotes this argument and remarks that is it not conclusive because it does not follow that everything which has a beginning is born; from the fact that everything born has a beginning. In fact, it is easy to see that if we could say

Everything which has a beginning is born, *A*. Therefrom the fact
that The universe is not born, *E*, it would follow that
The universe has not a beginning, *E*

Thus Aristotle would be led by the consideration of this deceptive argument or paralogism to make a distinction between two figures of reasoning, in the first of which the mood *AEE* is bad; though in the second it is good. It seems to me, therefore, that Aristotle's remarks upon the advantages of picking to pieces even the most obvious fallacies and detecting precisely where the fallacy lies, are [exceedingly] just; and that the contempt with which such things [are] commonly received is only another instance of what will be found a general rule, that the unintellectual man considers the objects to which the intellectual man applies himself as too trifling to bestow any attention upon. But, in fact, as Aristotle says, the analysis of sophisms not only sharpens the wits and makes us wary in inferring, but is also the great means for the discovery of logical forms.

On that account, I propose to consider some of the ancient sophisms and also some which I have invented myself.

I will begin with one which I find in a commentary upon Aristotle but whose origin I cannot trace. It runs thus: — It is impossible for a person who is silent to speak; but this man is silent; therefore it is impossible for this man to speak. Yet this man has spoken and will speak. This is to be explained in this way. When we say it is impossible that a person who is silent should speak, we mean that without considering whether in fact any one exists who is silent, if in any state of things whatever such a person should exist he would not speak. Or in other words whenever any person is silent he is not speaking. This is the expression of a general rule. Now when we say that this person is silent, we subsume a case under that rule, and the result follows that this person is not speaking. We ought not to infer that this person would never speak, for we have not said that this person would always be silent. Thus we see that if for our rule we have

I would always be *C*	and for our case
D is *I*	the result is not
D would always be *C*	but
D is *C*.	To infer *D* would always be *C* we

must have *D* would always be *I* for our case. The proposition

S would be *P*	is termed a *necessary* one
S is *P*	a contingent one.

This is the established meaning of the word necessary. When we say that a straight line is necessarily the shortest distance between two points we mean that it is so not merely in this or that state of things but in every state of things; it always has been so, it is so; and it always will be so. When we say that all men are sinful; we make a merely contingent assertion because we generally allow that there was a state of innocence in Eden and that there will be another in the Millennium. In this case, therefore, we speak only of existing men; and the proposition is contingent. This is what philosophers have generally meant by necessary. But, of late, an entirely different meaning has been given to the word. It has been said that that is necessary whose contradictory we cannot conceive. Thus we cannot conceive of a straight line which should not be the shortest distance between two points, and there every straight line is necessarily the shortest distance between two points. Now this is to confound what we necessarily think with what we think to be necessary. A great philosopher has laid down the principle that nothing is observed to be necessary. For instance, we have observed that trees have roots; but we have not observed that they have roots under all circumstances, for the simple reason that all circumstances are not within the range of our experience. Mr. John Stuart Mill thinks he has triumphantly refuted this principle by adducing the following instance. The ancient could not understand how people could be upon the other side of the earth and therefore thought it impossible. You will wonder how he supposes that this is inconsistent with the principle laid down, since he must admit that they had not observed that there were no antipodes. He thinks it bears on the matter in hand first because he understands that [philosophy] to hold that whatever we have tried in vain to imagine is false; and second because he thinks the ancients tried in vain to imagine antipodes. But neither one nor the other of these propositions is true. No one ever meant to say that we can be sure that further discipline will not enable us to imagine what we cannot now imagine; nor were the ancients unable to imagine antipodes. They thought antipodes were impossible, but a race who had produced the science of geometry were certainly able to imagine a man in what they thought an impossible position, as the very existence of the word antipodes in their tongue proves. Such are the reflections which are suggested by this first sophism.

The next argument which I shall take up is one which was not intended to be fallacious but was urged seriously. It runs as follows.

Whatever moves in an instant must move in the place in which it is at that instant or in a place in which it is not in that instant. But nothing can move in the place in which it is at an instant nor in any place in which it is not during the motion. Hence nothing moves in an instant. [What] has been moving at no instant has not been moving at all. Hence nothing has ever moved at all. ...

A much more difficult case to resolve is this. Every proposition is either true or not true and is not both. The proposition

What is here written is not true.

if it be true — is not true for it says it is not true but if it be not true — is true for it says it is not true. Now ask yourselves *Is* this true? It either is or it is not; and it can't be both. If it is true then *what is here written is true* but the proposition contradicts this and therefore contradicts what is true. Therefore if it is true it is not true. But suppose it is not true then what is true what is here written is not true; but this is just what is said; therefore, if it is not true, it is true. Again, suppose it is true, then what it *says* is true, but it says it isn't true; therefore, it isn't true. But if it isn't true what it says isn't true; but it says it isn't true; therefore it is true. Do you say then that it is true or not? That it is not true? What is not true, what is here written or what is here written about what is here written? Because if what is here written is not true, what is here written about what is here written is true; and if what is here written is true what is here written about what is here written is not true. But after all what is here written is identical with what is here written about what is here written, for nothing is here written about anything except what is here written.

We have then the statement what is here written

↖

and we have the statement that that is false

↖

[and we have] the statement that that is false

↖

[and we have] the statement that that is false
and so on to infinity.

And moreover what is here written is a statement that something is false and thus we go to infinity in the other direction also. Now what all the statements are about if they are about anything is the last one of this infinite series. But there is no last one and therefore they are about nothing and have no meaning at all. The statement here written then is not true because it has no meaning; and therefore it is true, after all because it says so. We can say; this proposition so far as it is spoken about by itself is false but so far as it speaks about itself is true. But this is a distinction without a difference. The question is whether this proposition is in all respects true. If it is not in all respects true, then it is in all respects true; for two reasons, 1st because what it says cannot in that case be altogether true, and what it says is that it isn't in all respects true, and 2nd because it is seen to accord precisely with what is in all respects true; namely, that it isn't in all respects true. The fact is that in this proposition truth and not truth — affirmative and negative — this and other — coincide. It stands upon the boundary of the true and the false; and is therefore in both.

Here is a sheet of paper of which one part is red and the other blue. Every point is either red or blue. The boundary between them forms a line. Now is that line red or blue? You cannot say it [is] red on one side and blue on the other unless like Hudibras you can distinguish and divide a hair twixt south and southwest side. A line is not double and has not different strips of color. It is all red or all green or all both at once or all neither. It is plainly as much either one as it is the other. We must therefore say that it is *both* or neither. Most persons will say that it is neither; because color can only reside in surfaces not in lines, but that is to fall into an ambiguity similar to that between motion in an instant and motion at an instant. It is true that a surface is required to constitute color; but we apply the term to every point of that surface. In the same way a community is requisite to make a city but every individual of that community is called a citizen. If that line is not red it lies without the red part of the sheet therefore if I simply draw away the red portion, I cannot affect the color of that line which lies without it. Accordingly that line and whatever the moving boundary passes over is neither red nor green; but it may pass over the whole sheet and therefore the whole sheet is neither red nor green. But it clearly is green. It seems to me, therefore, that the proper answer is that the boundary is both red and green; — the distinction between them vanishing at this point. And this is the answer which was made by Hegel and which mathematicians give to similar questions. It seems to me that this is a parallel instance to the proposition we have on the board. We cannot say that this proposition is neither *true* nor *not true*. That is self-contradictory. It is not self-contradictory to say that it is both true and not true because this [is] a special case when the difference between affirmation and negation vanishes. Now these questions may seem trifling and puerile; but I have no hesitation in saying that I know of none upon the correct solution to which man's happiness depends more; for the paradoxes which beset our highest practical interest — our religion — the puzzles of free will, of divinity, of immortality are precisely of such a character as these.

LOGIC OF SCIENCE (344)

LECTURE VI

BOOLE'S CALCULUS OF LOGIC

Perhaps the most extraordinary view of logic which has ever been developed with success is that of the late Professor Boole of Dublin. His book is entitled "An Investigation of the Laws of Thought on which are founded the mathematical theorics of Logic and Mathematics." It is destined to mark a great epoch in logic; for it contains a conception which in point of fruitfulness will rival that of Aristotle's Organon.

For two centuries many schemes have been proposed for representing logical processes in some other *set of symbols* than those *two* which are commonly in use; — I mean *words* and *thoughts*. These different systems are divisible into two classes. The first class embraces such symbols as have in their original nature something corresponding to logical laws. An example of this class is Euler's circles. Upon this system — the subordination of one term to another is represented by  one circle within another. Thus the larger term may represent animals and the smaller one vertebrates. Logical intersection is represented by intersecting circles . Thus, one might be *marine animals* and the other *mammals*. Exclusion — the relation between such terms as *men* and *ships* is represented by circles side by side. It is clear that these circles may be made to represent all the principal kinds of demonstration — which depend entirely upon relations containing and contained. These circles are merely concrete representations of the action of the very same laws. For this reason, they are most useful in enabling us to think rapidly and easily about logic; but for the same reason they cannot explain the laws of logic, at all.

The other class of logical symbols is more algebraical. It consists of arbitrary signs upon which the laws of logic, so far as they are understood, are arbitrarily imposed; but since the symbols are not naturally subject to any such laws the application of the laws scarcely reaches any further than has been explicitly supposed. This is the case with Hamilton's system of notation. He has one mark to signify the subject, another to signify the predicate, another for affirmation, another for negation, another for distribution or universality of a term, another for non distribution. But as these marks have no connection with each other except such as is explicitly laid down; the system is of no practical value whatever. A very similar instance is the Notation of Ploucquet, of which I will give an example. Take the propositions

Every man is a creature
Some man is not an Ethiopian.

Ploucquet would write the first sentence thus *Mc*. *M* stands for man and *c* for creature. Writing them together without anything between them, shows that they are affirmatively connected. Writing the *M* first shows that it is the subject. Writing the *M* in capital shows that it is distributed and writing the *c* in small letter shows that it is undistributed. In every respect, therefore, except the mere outward marks, this is like Hamilton's notation *M*: ▶ , *C*. 'Some man is not an Ethiopian' would be written by Ploucquet thus *m* > *E*. The crooked line makes negation. Hamilton would write it *M*, ▶̸ :*E*. Now Ploucquet can write his two propositions together taking care to make every letter small which occurring in both propositions is not a capital in both. Thus *Mc* and *m* > *E* become *mc* > *E*. In the same way Hamilton may write *E*: ◀̸ , *M*: ▶ , *C*. Next Ploucquet can drop the *m* and the conclusion is *c* > *E* or some creature is not an Ethiopian. So Hamilton can drop the *M* and have either *E*: ▶̸ , *C* or *E*: ◀̸ , *C* which amount to the same thing.

Such a notation as this has one point of superiority over the geometrical systems; in that it involves some analysis of the laws of logic. But as it will do nothing with these laws except what we have decided before hand that it ought to do it is utterly useless both in practice and as the basis of a conception of the science.

The notation which Boole invented combines the excellencies of both these classes of symbols. For like the literal notations it is abstract and deals with the laws of logic themselves and like the geometrical notations it brings out a harmony between logic and mathematics, so as to render the former easier to think about. In addition to this, it has a peculiar excellence of its own for it reflects upon mathematics a new light from logic and immensely facilitates the solution of difficult questions of probabilities. I am very far from saying that the system is a perfect representation of logic, on the contrary, I shall point out immense gulfs in its notation which were entirely overlooked by its author and shall show that but a very small fraction of all judgments can be expressed in this way. But then it must be remembered that the method is in its infancy yet, while even now it throws a light upon many points which is invaluable.

I will briefly describe it. The sign of equality, which in algebra signifies agreement in number, in other branches of mathematics denotes some other kind of agreement. In logic, then, it should signify logical agreement, or *identity*. Accordingly,

$$a+b+c = d$$

will mean not merely that a and c and d taken together are as many as d but that taken together they make up d. Thus a might be animals, b vegetables, c minerals, and d natural objects. Then the equation would signify that animals, vegetables, and minerals constitute natural objects, or *what is the same thing*, that every natural object is either an animal or a vegetable, or a mineral. The sign *plus*, then, makes an extensive sum of two classes. $x+y$ denotes a class which has the extension of x and that of y also or which has the comprehension which is common to x and y. The rule of transposition is nothing but the definition of the sign *minus*. It must therefore hold good in every system. We may therefore assume that it holds and then inquire what the sign *minus* must mean in order that it should hold. Transposing c then in the last equation, we get,

$$a+b = d-c$$

Now what are animals and vegetables in terms of natural objects and minerals? They are all natural objects except minerals. The sign minus then signifies extensive subtraction; that is, $x-y$ denotes a class which has the extension of x not shared by y and the comprehension of x and that of non-y together. Observe, however, that in addition to this something is implied in the expression $x-y$ in regard to the relation of y to x, namely that y is subordinate to x in extension for otherwise $x-y$ would be an absurdity and incapable of interpretation. It is important to remember what is implied in this inverse process.

The Rule of Transposition not only defines the sign minus, it also defines zero for by transposition $a-a = 0$. Zero then denotes that class whose extension is nothing and whose comprehension is non-existence.

The idea of multiplication in all branches of mathematics is combination into one. It is supposed that the two factors are two measures of the quantity completely independent and that their product is that which is measured by both. Now the measures of a symbol is its comprehension. It is the comprehension which measures the extension, not *vice versa.* Accordingly if R means Roman and C means christian rc would mean that which has the comprehension Roman and that of Christian at once. Or in other words it would be a Roman Christian. xy, then, has the comprehension of x and of y both and the extension which is common to x and y. Just as $x+y$ has the extension of both, and the comprehension which is common to the two.

The rule for clearing from fractions is the definition of division. Let $x = \frac{b}{a}$. Then the question we have now to ask is what is the meaning of x in terms of b and a. The answer is $ax = b$. That is x is a class which when determined by the comprehension of a gives b. In other words $\frac{b}{a}$ is a class which includes b and nothing but b that is at the same time a; that is, it comprises all b and some all or none of what is not a beside. At the same time, just as the inverse process of subtraction implies in itself that the extension of the subtrahend includes the extension of the minuend; so the inverse process of division implies that the comprehension of the dividend includes the comprehension of the divisor. Thus take $\frac{x}{y}$. Now unless x contains in itself y as a factor the division cannot be performed and the expression is incapable of interpretation.

Just as the rule of transposition serves to define not only subtraction but also the *zero*, so the rule of clearing from fractions defines not only division but also the *unity*. For by this rule $6 = 6 \times 1$. Now what is that class whose comprehension makes a part of that of every other class and whose extension includes that of every other class. It is the *existent.* All existent includes everything. Existence is implied in every class. Hence if a class = *zero* or is non-existent, there is thereby an absurdity in it.

Exponentials, Logarithms, Signs and other functions can be interpreted upon Boole's system only by developing them into algebraic forms.

Let us now endeavor to express upon Boole's system the three fundamental laws of logic — those, namely, of Identity, Contradiction, and Excluded Third.

In order to express the law of Identity or all A is A, let us first express All X is Y and then substitute A for both X and Y. To say All X is Y is the same as to say that the class X is identical with the class which has the comprehension of X and Y together or $X = XY$. Now substituting A for x and y we have $A = A^2$. Now does this express the law of identity? Unless the formula A is A means that A is B that is that A *is* something or in other words that every logical term is capable of an affirmative predicate or that it is *real*, I say unless the law of identity means this then *John is John* does not represent the law at all. For *John is John* means absolutely nothing more than that *John is X*, or *John is*. What the law may be taken to mean is this: — that all the individuals composing a class have the class-character.

Let A denote a certain class
Let a denote the individuals in it
Let α denote whatever has the class character.

Then the proposition that a class is composed of individuals having a certain character will be

$$A = a\alpha$$

but A a α are all identical; hence we may write

$$a = a^2$$

as the symbol of the law.

But if the law be considered to mean merely that every logical term is capable of an affirmative predicate insofar as it obeys logical laws, then Boole's system will not express this law at all. The reason that it will not, I will explain shortly.

The law of excluded third is that A is either B or not B. Here again if this law be taken as it usually is to mean A vel est B vel est non-B then the law cannot be expressed upon Boole's system, because the existence of A is here implied. I mean that kind of existence which is implied in an affirmative copula, and in Logical Identity in general. But if the law mean[s] A vel est B vel non est B, then the law may be expressed. For this will mean that A is something or is not or in other words that A either is or is not. In other words that a has two roots *unity* or the existent and *zero* or the non existent. In algebra if x has two roots x_0 and x_1 this fact is expressed in the following equation

$$(x - x_0)(x - x_1) = 0$$

for this equation can only be true when $x - x_0 = 0$ and when consequently $x = x_0$ or else when $x - x_1 = 0$ and when consequently $x = x_1$. Expressing then in this way that a equals either 1 or 0 we have

$$(a - 1)(a - 0) = 0.$$

The law of contradiction, the third of the triad of fundamental laws may undoubtedly be expressed upon Boole's system. For it is merely that *A* is not not-*A* or what comes to the same thing, there is *A* which is not-*A* is nonexistent. The class not-*A* is that one which with *A* makes up the existent or in other words it is all the existent except *a*. That is $1 - a$. *A* which is not-*A* is then $a(1 - a)$ and this equals *zero* or $a(1 - a) = 0$.

These three fundamental laws as thus expressed are algebraically one and the same. And they amount to this, that in the calculus of logic every letter has one of two values *unity* or *zero*.

It is now incumbent upon me to exhibit the capacities of the system for expressing different kinds of propositions. I regret that we must come to the examination of this point now, because it will be sure to leave you with much too low an opinion of the method. I warn you therefore not to infer from the enormous deficiencies that the utility of the system is small nor need we give up the hope that they may be hereafter supplied.

Ordinary language is capable of distinguishing judgments in Modality, Relation, Quality, and Quantity. In modality judgments are either Problematic, Assertory, or Apodeictic. Apodeictic judgments can certainly not be distinguished from the Assertory upon Boole's system. It may be doubted whether problematic judgments can be expressed or not. *xy* may be taken to mean '*x* may be *y*'; and perhaps it does. In Relation, judgments are Categorical, Hypothetical and Disjunctive. Hypotheticals and Disjunctives cannot be expressed upon Boole's system. The author endeavored to express them in the following way. Take the judgment "If there is an east wind, the barometer will rise." He would say Let $a =$ express There is an east wind; and let *b* express the barometer will rise. Then $a = ab$ will mean, If there is an east wind the barometer will rise. But in the first place, this alters the meaning of the sign of equality; and belongs therefore to a system inconsistent with that upon which Boole expresses Categoricals. In the second place it entirely destroys the possibility of expressing problematic propositions, except by single letters. In quality, ordinary language expresses the difference between affirmatives and negatives. An affirmative I define to be a proposition which implies the reality of its terms. Logicians might call this statement in question. The word *reality*, which is the correct one here, is certainly liable to be misunderstood. In an analytical proposition, in which an identity of concepts merely is asserted, it is merely implied that the concepts are real. But that this is implied is evident from the fact that there is no contradiction between any two propositions if affirmatives admit the case of the nothingness of both terms. Thus if though griffins do not exist it is true to affirm anything whatever of griffins. If anything can be affirmed of a foursided-triangle it is that is is foursided, but this may certainly be denied of it, for otherwise we could not show it an absurdity. Of course, in the case of a problematic proposition this reality is only problematic. In the case of an analytic proposition it merely concerns concepts. But in all synthetic assertory affirmatives, there is a reality asserted in the ordinary sense.* (* As a further illustration of the implication of Entity in Affirmatives, take the following reasoning which follows strictly unless either affirmatives or particulars imply the Entity of their subject.

	[Scratch work]
No black is white	$bw = 0$
No non-smooth black is white	$vbw + (1 - v)bw = 0$
If any black is not smooth, it is not white	
If any black is white it is smooth	$vbw = 0 \quad bw = bw(1 - v)$
All white black is smooth	
Some smooth is white black	$x(1 - v) = (1 - v)bw$
Some smooth black is white black	$xb(1 - v) = (1 - v)bw$
Some smooth black is white	$xb(1 - v) = (1 - v)WM$
Some black is white	$LB = wM$

The fallacy here consists I believe in the inference from No black white is non-smooth. To All black-white is smooth.) Taking then this definition of affirmatives, we find that affirmatives cannot be expressed upon Boole's system.* (* In this respect, the system of Leibniz from which Boole may have derived the hint for his own, is superior to the latter one.) If he wishes to say all men are mortal letting h be men and m mortal, he writes $h = hm$. The men are identical with the mortal men; but yet there may be *none* men. In fact, *man* may be an absurdity. $h = hm$ therefore means No man is immortal but not All men are mortal. In Quantity, ordinary language expresses the difference of Universal and Particular. Professor Boole gave as the expression for the particular negative Some X is not Y

$$vx = v(1 - y)$$

where $v =$ denotes that indefinite class some. But the absurdity of this is evident from the fact that by transposing we get

$$vy = v(1 - x)$$

or Some Y is not X. But it does not follow from Some X is not Y that Some Y is not X. This expression is therefore wrong. The cause of the defect of it is evident. He has represented *some* as being merely an indefinite class. As though we were to say instead of Some animals are not men

Four legged animals are not men.

Now it does follow that if Four legged animals are not men, Four legged men are not animals. Boole's system then as he has left it can only express assertory, categorical, negative, universals. But it is possible that all its defects might be remedied. In fact, the particular quantity can be easily supplied. When we say some animals are not men some is not a wholly indefinite class for it is understood to be a class of animals; in other words there are none of the whole class *some* who are non-animals. Now this is expressed by the equation

$$vxy + v(1-x)y + v(1-x)(1-y) = 0$$

Or by $v = vx(1-y)$ Or $v(1-x(1-y)) = 0$

While Boole's system is insufficient to express most of the kinds of judgment expressed in speech, it is capable of noting some points which common forms of language cannot. Thus the difference between an analytic and synthetic judgment is easily noted.

$\frac{x}{y}$ shows that x in itself implies that it is not y while $xy = 0$ leaves it doubtful.

Language has no way of distinguishing disjunctives from divisives. Boole's system can express the latter unequivocally. Definitions can hardly be expressed precisely [...] elegantly in language; we can say All A is all B but it is an awkward and unnatural phrase. Boole simply writes $a = b$ that is there is no A which is not B and no B which is not A.

It is curious that extensive combination should be represented by addition; and comprehensive combination by multiplication when the extensive and intensive quantities stand in reciprocal relation. It is not obvious at first glance how multiplication can undo the work of addition and *vice versa*. An example, however, will make this plain.

Let a and b be two distinct classes and to represent the fact that they are distinct we may write them $a(1-b)$, $b(1-a)$. Now let us add them $a(1-b) + b(1-a)$. Next let us multiply by a. $a \times a(1-b)$ is $a(1-b)$ for $a^2 = a$, $a \times b(1-a) = 0$ for $a(1-a) = 0$. The multiplication therefore separates this term $a(1-b)$ which the addition combined with another. We might have begun with the multiplication. Let x and y be two terms. Multiply them we have xy. Add x, $xy + x$. This $x = xy +$

$x(1-y)$. We have then $2xy + x(1-y)$ but the coefficient means nothing. It may be struck off. We have then $xy + x(1-y)$ or x.

The application of Boole's calculus to ordinary reasoning depends upon two very simple theorems which I proceed to give.

The use of the first is to enable us to interpret complex expressions. For example $\frac{(a-b)^2}{1-(a-b)^2}$ could hardly be interpreted as it stands. In general, it is required then to simplify fx. Now x has one of two values. Calling these x_0 and x_1, we have the algebraical law

$$(x_1 - x_0)fx = (x - x_0)fx_1 + (x_1 - x)fx_0$$

This equation is always true for x = either x_1 or x_0. When it $= x_0$ the equation becomes

$$(x_1 - x_0)fx_0 = (x_0 - x_0)fx_1 + (x_1 - x_0)fx_0$$

where the first term of the second member disappears and when $x = x_1$ it becomes

$$(x_1 - x_0)fx_1 = (x_1 - x_0)fx_1.$$

This equation then being true let us substitute for x_1 and x_0 the values which they have in logic, namely unity and zero and we have

$$fx = xf1 + (1-x)f0$$

Now I will show how to use this formula by developing the expression $\frac{1-(a-b)^2}{a}$. This has to be developed in terms of a and b. We have to complicate the formula a little $f(a,b) = a(f1,b) + (1-a)f(0,b)$ developing this again according to b we have

$$f(a,b) = abf(1,1) + a(1-b)f(1,0) + (1-a)bf(0,1) + (1-a)(1-b)f(0,0)$$

Now to apply it. When $a = 1$ and $b = 1$

$$\frac{1-(a-b)^2}{a} = 1 \text{ The first term then is } ab$$

When $a = 1$ $b = 0$ $\frac{1-(a-b)^2}{a} = 0$ The second term disappears

When $a = 0$ $b = 1$ $\frac{1-(a-b)^2}{a} = \frac{0}{0}$ The third term then is $\frac{0}{0}(1-a)b$

When $a = 0$ $b = 0$ $\frac{1-(a-b)^2}{a} = \frac{1}{0}$ The fourth term then is $\frac{(1-a)(1-b)}{0}$

We have then $$\frac{1-(a-b)^2}{a} = ab + \tfrac{0}{0}(1-a)b + \frac{(1-a)(1-b)}{0}$$

We have here two coefficients $\frac{0}{0}$ and $\frac{1}{0}$ which we have not yet interpreted. To find what $\frac{0}{0}$ means put it equal to x, $\frac{0}{0} = x$, multiply by zero and we have $0xx = 0$. Now this is true whatever be the value of x. $\frac{0}{0}$ therefore is a wholly indeterminate class and means *all*, *some* or *none*. $\frac{0}{0}(1-a)b$ means *some all* or *none* of b which is not a.

To find what $\frac{(1-a)(1-b)}{0}$ means we must remember that when one letter is divided by another as $\frac{x}{y}$ it must be that the dividend contains the divisor as a factor. Now to say that $(1-a)(1-b)$ contains zero as a factor is to say, not merely that $(1-a)(1-b)$ does not enter into the meaning of $\frac{1-(a-b)^2}{0}$ but that it enters into the meaning of nothing and does not exist. This same result may also be obtained thus. Let $\frac{(1-a)(1-b)}{0} = y$ then multiplying by zero $(1-a)(1-b) = 0$ while

the value of y is wholly indeterminate.

Such being the meaning of $\frac{0}{0}$ and $\frac{1}{0}$ I pass to the second fundamental theorem which enables us to suppress a letter from an equation. Let $fx = 0$ be an equation from which we wish to eliminate x. As x is either $= 1$ or $= 0$ we have

Either $f(1) = 0$ or $f(0) = 0$. In any case $f(1)f(0) = 0$

Take for example the equation $ab + (1-b)c = 0$

Required to eliminate b. When $b = 1$ this becomes $a = 0$. When $b = 0$ it becomes $c = 0$. In any case then $ac = 0$.

I will now show you how to perform a reasoning process in this calculus. We shall always have two equations which are to be combined.

Each term therefore is separately equal to zero. The two equations may then be added together and the superfluous expressions eliminated. Take this example.

No animals are vegetables
All men are animals

The first is $av = 0$
The second is $m(1-a) = 0$

$$m(1-a) + av = 0$$

Eliminating a we have $am = 0$. No men are vegetables. I would give a more complex instance but have hardly time.

Since x is equal either to *one* or *zero*

Either $f(1) = 0$ or $f(0) = 0$
in either case

$$f(1)f(0) = 0$$

which is the equation with x left out.

We now come to the property of these symbols which enables us to draw logical conclusions. This property is that in this calculus any number of different equations can be combined into one equation which shall express all the facts contained in the several equations. I will show you how this can be done. Let us take two equations $a = b$ and $c = d$ to combine into one.

We have first $a - b = 0$ and $c - d = 0$
Square these equations $a^2 - 2ab + b^2 = 0$, $c^2 - 2cd + d^2 = 0$ but as $a = a^2$ for either *one* or *zero* this is the same as $a - 2ab + b = 0$, $c - 2cd + d = 0$ which may be written $a(1-b) + b(1-a) = 0$, $c(1-d) + d(1-c) = 0$
Now as the value of a letter never exceeds *unity* all these terms are positive and consequently if they are all added together into one equation $a(1-b) + b(1-a) + c(1-d) + d(1-c) = 0$, this equation itself implies that the terms are separately equal to zero and consequently that $a(1-b) = b(1-a)$. That is $a - ab = b - ab$ or $a = b$ and in the same way it is implied that $c = d$; so that this one equation contains now all that was contained in the original two. The practical value of this method can only be illustrated by very complicated cases; but in order to show how it is used I will take a very simple case. From these two premisses what follows?

> The ancestors of [humans] had no tails
> Monkeys have tails

Let m be monkeys, n ancestors of [humans], t whatever has a tail. Then the premises are $0 = nt$ or there are no tailed ancestors of [humans] and $m = tm$ or monkeys are all tailed monkeys. The second equation gives $m - mt = 0$ or $m(1-t) = 0$. Add the two equations and we have $nt + m(1-t) = 0$. We wish now to get rid of t. Now t must equal either *one* or *zero*. In the former case we have $n = 0$, in the latter case $m = 0$, in either case $mn = 0$ or none of the ancestors of [humans] were monkeys.

The application of this calculus to questions of probabilities is very beautiful. I wish to call particular attention to this application because it shows conclusively that we can draw no argument for the validity of induction from the doctrine of chances. It was for the sake of showing this that I brought Boole's method to your notice. The theory of its application to probabilities is exceedingly simple and is best exhibited in an example.

Given the probability that one or both of two events happen and let it equal p. Given also the probability that one or both of them fail to happen, let it equal q. What is the probability that one only will happen.

Let the two events be x and y

Let the case that one or both happen be s; that is

$s = xy$ (that they both happen) $+ x(1-y)$ (or x happens without y) $+ (1-x)y$ (or y happens without x). Let the case that one or both fail be t; that is $t = x(1-y)$ (that x happens without y) $+ (1-x)y$ (that y happens without x) $+ (1-x)(1-y)$ (or neither happens). Let the case that only one happens be w; that is $w = x(1-y)$ (that x happens without y) $+ (1-x)y$ (that y happens without x).

We must first reduce these three equations to one. Taking the first one and transposing the second member we have

$$s - xy - x(1-y) - (1-x)y = 0$$

Applying to this the formula $fs = sf(1) + (1-s)f(0)$. When $s = 1$ the equation becomes $(1-x)(1-y) = 0$. When $s = 0$ it becomes $xy + x(1-y) + (1-x)y = 0$. In general then it is $(1-s)xy + (1-s)x(1-y) + (1-s)(1-x)y + s(1-x)(1-y) = 0$. By the same development the other two equations become

$$txy + (1-t)x(1-y) + (1-t)(1-x)y + (1-t)(1-x)(1-y) = 0$$
$$wxy + (1-w)x(1-y) + (1-w)(1-x)y + w(1-x)(1-y) = 0$$

We may now take the sum of these three equations *equal to zero*. The next thing to be done is to get rid of x and y because there is nothing said about the simple events in the problem. For this purpose if we find what our equation becomes

1st	2nd	3rd	4th
when	when	when	when
$x = 1$	$x = 1$	$x = 0$	$x = 0$
$y = 1$	$y = 0$	$y = 1$	$y = 0$

We shall have four expressions all equated to *zero*, one or other of which is really equal to *zero*, and therefore the product of them is certainly equal to *zero*.

When $x = 1$ $y = 1$ everything except the coefficients of xy will disappear and we shall have

$$(1-s) + t + w = 0$$

When $x = 1$ $y = 0$ everything except the coefficients of $x(1-y)$ will disappear and we shall have

$$(1-s) + (1-t) + (1-w) = 0$$

When $x = 0$ $y = 1$ everything except the coefficients of $(1-x)y$ will disappear and we shall have

$$(1-s) + (1-t) + (1-w) = 0$$

When $x = 0$ $y = 0$, everything except the coefficients of $(1-x)(1-y)$ will disappear and we shall have

$$s + (1-t) + w = 0$$

We must now multiply these four equations together. I will not stop to perform this process but will merely give the result which is

$$st(1-w) + s(1-t)w + (1-s)t(1-w) + (1-s)(1-t)w + \\ + (1-s)(1-t)(1-w) = 0$$

And this by a very simple reduction gives

$$w = \frac{st + (1-s)t + (1-s)(1-t)}{st + (1-s)t - s(1-t)}$$

Let us now develop this fraction by the formula made on the same principle as the one I have just used; namely

$$f(s,t) = stf(1,1) + s(1-t)f(1,0) + (1-s)tf(0,1) + \\ + (1-s)(1-t)f(0,0)$$

When $s = 1$ $t = 1$ $w = 1$
When $s = 1$ $t = 0$ $w = 0$
When $s = 0$ $t = 1$ $w = 0$
When $s = 0$ $t = 0$ $w = \frac{1}{0}$

Hence in general $w = st + 0s(1-t) + 0(1-s)t + \frac{1}{0}(1-s)(1-t)$.

We have now completed the logical solution of the problem. We must next apply the theory of probabilities. For this purpose instead of insisting any longer upon allowing s and t only the two values *zero* and *unity*, we must allow them values proportionate to the number of cases in which they will occur. Then the probability of s will be represented by a fraction whose numerator is s and whose denominator is the sum of all the possible cases. Now the cases which are impossible are those represented by $(1-s)(1-t)$ as is shown by the coefficient $\frac{1}{0}$. Hence, the probability of s or

$$p = \frac{s}{st + s(1-t) + (1-s)t}$$

In the same way

$$q = \frac{t}{st + s(1-t) + (1-s)t}$$

And in the same way the probability of w which is sought, is, since by the equation st represents all the cases in its favor,

$$X = \frac{st}{st + s(1-t) + (1-s)t}$$

From these three equations we must now eliminate s and t. For this purpose we have $\frac{p}{s} = \frac{q}{t} = \frac{X}{st}$. $s = \frac{X}{q}$ $t = \frac{X}{p}$. Substituting these values in that of X we have

$$X = \frac{\frac{X^2}{pq}}{\frac{X}{p} + \frac{X}{q} - \frac{X^2}{pq}} = \frac{X}{q+p-X}$$

$$1 = \frac{1}{q+p-X}$$

$$X = q+p-1$$

which is the probability required.

I will now take a case where an attempt is made to work inductively.

Let the probability that it *thunders* on a given day be p

the probability that it *thunders and hails* be q

Required the probability that it *hails* or X

Let x be It hails

y It thunders

u It thunders and hails

Then $u = xy$

$x = \frac{u}{y}$. Developing this we have $x = uy + \frac{1}{0}u(1-y) + 0(1-u)y + \frac{0}{0}(1-u)(1-y)$.

This solves the logical question; proceeding with probabilities we have

$$p = \frac{y}{y+(1-u)(1-y)},\quad q = \frac{uy}{y+(1-u)(1-y)},\quad X = \frac{uy+\frac{0}{0}(1-u)(1-y)}{y+(1-u)(1-y)}.$$

Elimination of u and y results in the following value of X

$$X = q + \tfrac{0}{0}(1-p)$$

which is as you perceive an indeterminate result; showing merely that the data were insufficient.

Let z be some complicated system of phenomena; let y be the concomitant circumstances under which these phenomena are observed. Let x be a physical hypothesis which perfectly accounts for them. Required X large or the unknown probability of this hypothesis.

In this case we have the equation $xy = xyz$ which expresses that if the hypothesis holds, then under these circumstances the actual phenomena are sure to appear or are accounted for. Observe that there cannot be a stronger case for the hypothesis, than that. The value of x from this

$$x = \tfrac{0}{0}yz + 0y(1-z) + \tfrac{0}{0}(1-y)z + \tfrac{0}{0}(1-y)(1-z)$$

These indeterminate coefficients will necessarily appear in the solution; which is in fact completely indeterminate.

I shall conclude by reading some remarks of Boole's upon this subject [p. 365].

6

LOWELL LECTURES, 1903

LOWELL INSTITUTE.

EIGHT LECTURES

— ON —

Some Topics of Logic Bearing on Questions Now Vexed,

— BY —

CHARLES S. PEIRCE, Esq.,

MONDAY AND THURSDAY EVENINGS,

BEGINNING MONDAY, NOVEMBER 23, 1903.

Hall doors opened at 7.30, **CLOSED AT 8.00** o'clock.

This Ticket entitles the bearer to reserved seat

NO. 501 CENTRE,

in Huntington Hall, Rogers Building, 491 Boylston Street, during Mr. Peirce's Course.

RESERVED till 7.57, Hall Time.

Some Topics of Logic Bearing on Questions Now Vexed.

Nov. 23. What Makes a Reasoning Sound?

Nov. 27. (**Friday**) A System of Diagrams for Studying Logical Relations. Exposition of it begun.

Nov. 30. The Three Universal Categories and their Utility.

Dec. 3. Exposition of the System of Diagrams Completed.

Dec. 7. The Doctrine of Multitude, Infinity and Continuity.

Dec. 10. What is Chance?

Dec. 14. Induction as Doing, not mere Cogitation.

Dec. 17. How to Theorize.

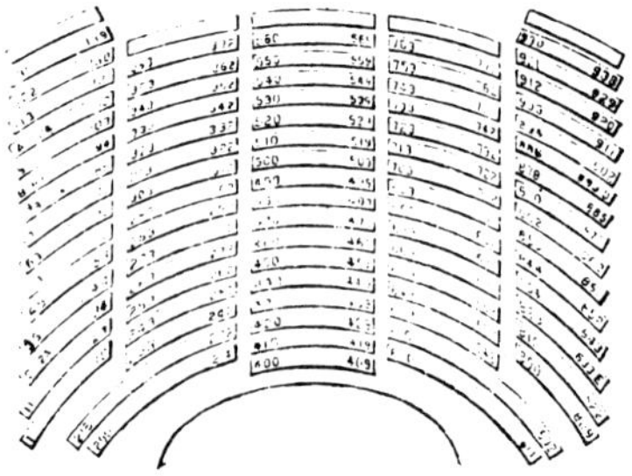

A ticket of admission to C. S. Peirce's Lowell Institute Lectures of 1903.

(From the Charles S. Peirce Collection in the Houghton Library, Harvard University.)

A. LECTURE III. 1ST DRAUGHT (458)

I am going to devote this evening's curtate hour in endeavoring to convey to you some notion of the conceptions of mathematics.

What is mathematics? Traditionally it is defined as the science of quantity; but at the time when that definition was first adopted, three words bore entirely different acceptations from those they bear today. Those three words are *mathematics*, *science*, and *quantity*. Under mathematics were included the four sciences of Geometry, Arithmetic, Astronomy, and Music. Buτ by Arithmetic was not meant what is now so called, but was then called *Logistic*. The arithmetic of those days was a wordy system of naming *ratios* mainly. It is now deservedly forgotten. That was not considered as any part of mathematics. The word *science* in those days did not mean what it now means, single-minded research. Neither did it mean what Coleridge called science, i.e. *systematized knowledge*. It meant *comprehension*, understanding the metaphysical principles of things; nearly what we call philosophy. Finally the four branches of philosophy meant were said to relate to quantity because they all at that time were confined to the study of *measures*. Thus the definition of Mathematics as the Science of Quantity is simply a case of the frequent phenomenon of a phrase or formula being stoutly adhered to long after its original meaning has been forgotten. The truth is that *quantity* plays a very considerable part in mathematics; but there is much mathematics that has nothing at all to do with it.

If you take a paper ribbon, give it a half twist, paste the ends together to make a ring, and slit it along the middle till you come round, the result will be a single ring. That is a proposition in mathematics. But it has nothing to do with quantity.

The truth is that it is only within the last generation that mathematicians have come to understand the nature of their own science. My father was the first, in 1870, to define Mathematics as the science which draws necessary conclusions. This is now universally acknowledged by competent men to be substantially correct. Some self conceited fool may assert that there are necessary conclusions not mathematical; but that is simply due to superficial study or no study at all. There is no room for a difference of intelligent opinion. Dedekind, one of the leaders of modern mathematics, whose little book published by the Open Court Company I strongly recommend to you, goes so far as to call mathematics a branch of logic. But that overlooks the circumstance that the mathematician and the logician though standing on the same gound face in diametrically opposite directions. The mathematician is trying to draw the necessary conclusions, the logician is trying to find out how inferences necessary and probable are composed. One study is synthetic, the other analytic.

Mathematics is the science of hypotheses, — the science of what is supposable. Supposable does not mean directly imaginable; it means what makes sense. For instance, we cannot *imagine* an endless succession of instants. We can, however, imagine *a graph* which asserts that between midnight and any instant before midnight, there is an instance; and we can suppose that that graph is true. If so, there is an endless succession of instants before midnight. It makes sense, although we cannot actually call up the idea of so many distinct instants. It would obviously require an endless series of exertions for which an endless life would be requisite.

Mathematics may be divided according to the complexity of its hypotheses. The simplest kind of mathematics would be the mathematics of existential graphs. Let us look at this subject mathematically. Any graph has one or other of two grades of value. The lower grade is that of a *false* graph, which must not be scribed on the sheet of assertion. The higher grade is that of a true graph, which may be scribed on the sheet of assertion.

Now the pure mathematical way of looking at this matter would be something like this. The mathematician always thinks in diagrams. He would put two spots for truth and falsity marking them with letters to distinguish them in writing about them. Below these he would imagine a whole collection of spots, to represent the graphs, and would actually draw some of them. Each graph spot would be joined either to the *f* spot or to the *v* spot, — None to both. He might now draw arrowhead lines from one to another of the graph-points, attaching a new graph-point to each. These would denote preconditionals *de inesse*. Then if it were possible to pass from the *v* points to the *f* point along the line of such a graph-point in the direction of its arrow, he would connect that point with *f*. Otherwise with *v* [as in Fig. 1].

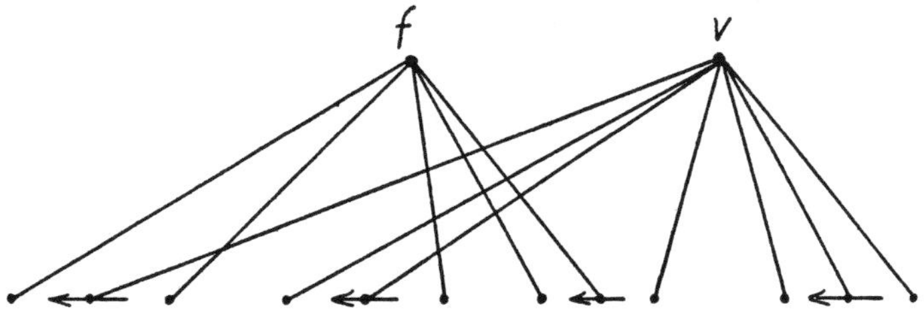

Fig. 1

By some such means as that he would study out the laws of these things. I have not time to go further. What I have said will give you an inkling into the pure mathematics of logic.

This is the very simplest kind of mathematics, — this system with only two different values. There would be an interesting system with three values, which I have slightly examined. But the simplest mathematics with sufficient complexity to be of any great importance is the mathematics of whole numbers, which goes by the name of the *Theory of Numbers*, but is better called Higher Arithmetic. For it does not deal with *all* numbers but only with finite numbers. I shall use the term Theory of All Numbers to mean the branch that deals with all numbers.

I will endeavor to give you an idea of what numbers are, as regarded from the point of view of pure mathematics. Before entering upon that subject, however, I will tell you something about *multitude*, or maniness. The doctrine of multitude is not a theory of pure mathematics. It is, rather, an *application* of the general theory of all numbers to the logical subject of maniness. Nevertheless, I shall begin by this. The recognized leader of this branch is Dr. Georg Cantor, but I began my studies of it and pushed them to considerable results before Cantor took up the subject, and I had made out the main outlines of the doctrine before I knew anything about his work; and have developed it in my own way quite independently, reaching some results not made out clearly by him, and not at all without the aid of the Pure Mathematical theory, which I, as a logician treating a subject properly logical, do not do.

I shall, therefore, give you my own exposition of the matter. My results agree with Cantor's but are reached differently. I do not in the least dispute his priority of publication and I desire to pay the tribute of my warmest admiration for the enormous value of his work.

The definition of *multitude*, what precisely it means to say that one collection has greater multitude than another, was first made clear long before Cantor's time and mine by the Austrian logician *Bernard Bolzano* whose logic, in four volumes, appeared before my birth. But this definition only occurs in a posthumous work. For Bolzano was a catholic theologian; and so clear-headed a logician as he was did not escape persecution, by which the publication of his chief work was delayed until after his death in 1848.

What had caused the subject to remain so long a mystery was that Euclid had laid it down as an axiom that "A whole is greater than its part." The reason Euclid said this was that the Greek conception of a whole was that of something which had received its last element. Consequently, they did not think of the *infinite* or *endless* as a whole. Whether it is so or not depends on what one means by a whole. If you simply mean that which is conceived as making up a *one*, the Greeks themselves spoke of *the infinite* meaning *all space*. But if by a whole, you mean something which has a *last element*, then it is a mere question of arrangement whether an *infinite* is a *whole* or not.

Using *whole* in the sense of that which is regarded as *one* thing made of many, let us see whether it is true that a whole is always greater than every part.

Consider all possible ordinal numbers; 1st, 2nd, 3rd, etc. without end. From that series of numbers, take away the first million and consider the part of the series that remains, namely one million and first, one million and second, one million and third and so on without end. Now surely this partial series has its 1st, its 2nd, its 3rd, and there is not a single possible one of all the ordinal numbers which is not occupied in designating the place of some member of this partial series. The partial series in no way differs from the whole series except that each ordinal number is increased by one million. Surely, then, the partial series is quite as multitudinous as the whole series.

Perhaps you will say "Ah! No doubt all infinite collections have the same multitude." But there you are mistaken. For I will demonstrate to you presently that if we compare all the values to which by carrying out decimals to indefinitely many places we can indefinitely approximate, which values we may call for convenience the *analytical positives*, — if, I say, we compare the collection of all such analytical positives with the collection of all values that can be exactly represented by *proper fractions*, which are called the *rational positives* — I will demonstrate, I say, in a few moments, that it is impossible from the nature of the two collections that every analytical positive should have a distinct rational positive exclusively assigned to it, or placed opposite to it, or put in place of it. The fractional values are infinitely multitudinous, since the denominator of a fraction can be any positive integer. But nevertheless, they do not suffice to go round among the *analytical positives* and furnish a separate tag for each.

You will now understand the following definition. One collection, say that of the *A*s is said to have a *greater multitude* than another collection, that of the *B*s, if, and only if, there is no relation whatsoever in which any *A* stands to a *B* to which no other *A* stands in the same relation. In other words, there is no way of pairing every *A* with a distinct and separate *B*.

That is substantially the definition of Bolzano which is the basis of the whole doctrine of multitude.

Accepting that definition we are confronted, at the very threshold of the subject with this question: Supposing that it is impossible for every one of the *A*s to be paired with a separate *B*, may it not be that the collection of *B*s is so great, that it is equally impossible for every *B* to be paired with a separate *A*? In that case, according to our definition, the two collections would be each greater than the other.

This question has not received from Cantor any answer generally deemed conclusive; and it is impossible that it should, since Cantor looks at the subject from a purely mathematical point of view, while this question is a logical one. Perhaps I ought not to omit to say that Cantor has also multitude and number from a metaphysical point of view. But that is, if possible, still less to meet this question properly.

In order to answer it, the first step requisite is to produce a definition of a *collection*. By a *definition* I do not mean a statement conveying the exact sense of a *word*, although that will naturally be involved in a definition. But what I mean is the explanation of the relations of an intellectual conception to other conceptions. That is what is wanted.

When I speak of a *collection* I do not mean that its members are existentially brought together: they may be or they may not be. Nor do I even mean that they are actually brought together in thought. For example, I should say that all the five-legged calves that had ever existed on this planet [were] a collection, scattered though they have been.

I should define a collection as a single object such that the truth of whatever predicate is true of it consists in a corresponding predicate being true of what exists.

For example the truth of whatever is true of the collection of five-legged calves consists in some proposition being true concerning such single five-legged calves as exist. It may be of all of them it may be of most of them it may be of some of them. For example, were anybody to say that the collection of five-legged calves is strictly confined to the north temperate zone, he would mean that any five-legged calf you may select will be found to be in the temperate zone. If he says that the collection of five-legged calves has been an object of close study he means that almost any five-legged calf you can find a record of has been examined. If he says that the collection of five-legged calves embraces calves of all breeds he means that some are of this breed some of that. If he says that the collection of five-legged calves does not amount to a thousand he means that he could place numbers so that any five-legged calf you might choose would bear some number of less [than] four figures and that you could not find two calves with the same number.

In short the truth of anything true about the collection consists in the truth of what exists that can be truly called a five-legged calf.

The definition implies that whatever collection may be taken there will always be some general predicate that is really true of every member of that collection and is true of nothing else in existence.

Without going into the full proof of this, I will remark that no matter what collection be taken, there is one thing that is true of all the members of that collection and of nothing else, namely, that whether anybody ever has his attention drawn to this collection or not, these things are capable in their own nature of having some sign attached to each of them which is attached to nothing else.

Now let us consider what a one-to-one relation is, in which every member of one collection stands to a single distinct member of another collection, like the relation of the collection of husbands to the collection of wives. I use the word *dyad* to mean an ordered pair, the word *duette* to mean an unordered pair. Thus

$$AB \qquad BA$$

are the same *duette* but are different *dyads*. Now imagine a collection of entirely distinct *dyads*. This will be like any other collection, and there will be some general predicate true of all the dyads of that collection and true of nothing else. To say that one collection is in one-to-one correspondence with another is nothing more than to say that there is some general predicate applicable to mutually exclusive dyads of the one and the other collection and so that every member of either collection is a member of such a dyad.

You will observe that such a pairing off of the members of two collections is not something that we bring about. It is a fact which is there whether we recognize it or not.

Any description of a state of things which does not involve a contradiction is logically possible; and if a state of things logically possible is not realized, it is because existence cannot exhaust all possibilities. Grasp this point well, I beg you. If any general description of a logically possible state of things is not realized, it is simply because the single things do not exist that would realize it.

Now that two collections should be in one-to-one correspondence with one another throughout, every member of each being paired with a member of the other, is a general description of a logically possible state of things. If therefore in the case of two existing collections it is not realized, it can only be that in one or other of those collections, the single things do not exist that would realize that state of things. That is, if not every member of one collection is paired to a member of the other, it is because that other collection *lacks members* with which all might be paired.

Suppose then that we have two collections A and B

$$\begin{matrix} A \\ B \end{matrix}$$

of which A is greater than B, so that there is no relation in which every A stands to a B to which no other A is in this relation. Then it must be from lack of members in B. Add to B another suitable collection, C and now the collection

$$\begin{matrix} A \\ B+C \end{matrix}$$

A and the collection $B+C$ will be in one-to-one correspondence with one another. That is there will be a relation in which every member of A stands to a member of $B+C$ to which no other A is in that relation and there will be a relation in which every member of $B+C$ stands to A to which no other member of $B+C$ stands in the same relation. But then this relation is one in which every member of B stands to a member of A to which no other B stands in the same relation. This however is the same as to say that the collection B is not greater than the collection A.

Thus I have given you an outline sketch of the perfect proof that if one collection is greater than another that other is not greater than it. It is not a mathematical demonstration; it is drawn from the principles of logic.

To this proposition there is no exception, so long as it is understood to apply to two *different* collections of absolutely distinct individuals.

There is, however, one collection, and only one, which is, according to the definition, larger than itself.

Take any other collection but that and there is a relation in which every member of it stands to a member to which no other member stands in the same relation. Namely, one such relation is the relation of *identity*. The exceptional collection of which this is not true is the collection called *Nothing*. *Nothing* falls short of every collection whatever; even of itself.

What makes the doctrine of multitude of extreme logical interest is that for every multitude there is a valid mode of reasoning which does not hold good of any higher multitude.

The modes of inference peculiar to the different finite collections would, no doubt, appear to you quite trivial, although they are really important.

The peculiarity of the collection "nothing" is that of its members any predicate whatever is universally true. For there is no falsity in any assertion if it is not applied to anything and every proposition is either true or false.

The peculiarity of a collection having a single member is that every predicate is either universally true of it or universally false of it.

The peculiarities of the collection of two members are those which give rise to logic which refers to the *values* of propositions, there being but two values. Passing over all the modes of inference peculiar to universes of individual finite multitudes, although the properties of a system of three is really very important we come to the mode of inference which holds good of every finite collection but of no infinite collection.

This form of inference was discovered by De Morgan in 1860* (* He says there are instances of it in his book published in 1847), although it had no doubt been *used* since time immemorial. He called it the *syllogism of transposed quantity*. The appellation "*De Morgan inference*" might be better. But it was not until my paper of 1881, on the Logic of Number, that the limitation of its validity to finite collections was first noticed. The example of it given by me was

Every Hottentot kills a Hottentot
But no Hottentot is killed by more than one Hottentot
∴ Every Hottentot is killed by a Hottentot.

As long as the Hottentots compose a finite collection this follows: Every Hottentot cannot kill a distinct Hottentot unless some one kills the first killer [Fig. 2].

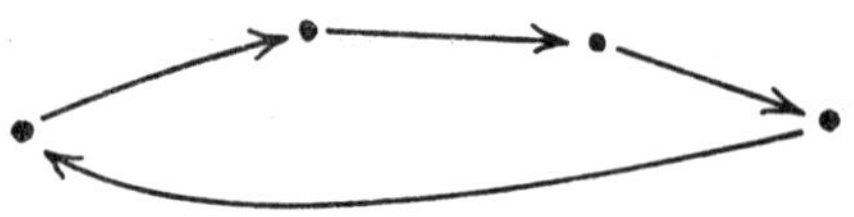

Fig. 2

But if you suppose the race of Hottentots to continue forever every Hottentot might kill some one of the following generation and still leave many unkilled.

The question of whether an infinite multitude of distinct things exist or not has vexed metaphysicians. The logician need not trouble himself to answer that question. But it is sufficiently obvious that in a finite lifetime a man cannot make an endless series of distinct efforts, or receive an endless series of distinct experiences.

Now in every case, our universe of discourse will refer to the individual subjects of experiencive force which we have hitherto distinctly recognized, or to some part of them; and therefore if we use the word *exist* to mean occurring as a member of the universe of discourse, it is certainly true that only a finite collection can *exist* in the sense of being composed of existing members.

But it is also indispensable that the logician should recognize that all understanding of experience, whether it be a correct understanding or not, consists in the assertion of some general rule which informs us what sort of events *will be* experienced in the future, supposing our experience extends far enough. For example, to say that water is composed of hydrogen 2 parts and oxygen 16 parts means that taking any drop or other portion of water whatsoever if your experience extends so far as to note the result of decomposing that water and noting the result then you will note that it is composed of hydrogen 2 parts to every 16 of oxygen. If your experience is cut off without noting the result of the decomposition, then the assertion amounts to nothing. Thus every understanding of experience, of the like of which all our useful knowledge is composed, relates to an *endless series of possibilities.* Whatever is endless is composed, then, of what partakes more or less of the nature of *possibility*, of *idea.*

But we not only apply our understanding to what is to happen, but also with success to what has happened. Whenever we do that, we assert that the existential conforms to the ideal and careful logical analysis will show that we assert that an endless series has been gone through and is done. There is not the slightest contradiction in that. The fact that the series has no distinct terminal member does not prevent it from being accomplished. It would prevent our having accomplished it by a distinct effort for each member of the series; but that is an impossibility attaching to the nature of past experience and not to the nature of the series in itself. It is a curious fact that there are persons of the most excellent judgment in practical matters, and even in some kinds of theoretical matters, who are nevertheless so destitute of intellectual subtlety as to be puzzled by the silly paradox of Achilles and the Tortoise.

Achilles and the tortoise are supposed to be particles without size. The tortoise is at a distance from Achilles. Achilles runs after him. The tortoise runs away but Achilles comes up to him. Everybody admits that that is what would happen; but some people say it is contrary to *logic*. It is not contrary to *my* logic; and if it were I should think the remedy for that state of things would be to adopt a logic that did not lead me into manifest falsity. But the greatest curiosity of their reasoning is that although they themselves admit that their logic has this inconvenience they nevertheless try to induce *me* to give up mine that has no such defect and adopt theirs and that for no better reason than that their discredited logic requires it. What singular notion of reasoning is that!

If one of those people were to come to me and ask me how I would have him think (which I am quite sure he will not), I should say, "State the course of your logic." He would say "When Achilles starts the tortoise is at a certain point." Undoubtedly. "But they run on the same path, Achilles must move to that place." No doubt. "And when he gets there, the tortoise will be at another place." Of course. "And Achilles must move to that second place." Yes. "And when he gets there the tortoise will be in a third place." Quite so. "But there will be an endless succession of such places through all of which Achilles must pass in order to catch up with the tortoise." Well, what of it? "Why, he could not pass through all these places unless he passed through the last of them." I reply "He could not pass through *all* without passing through whatever one of them may be the last. But you have just said the series is endless, that is, has no last. He is not obliged to pass through a last one when there is no last one to be passed through." "Then how can he pass through them all?" I reply "That question supposes there is some difficulty, which my logic does not recognize. What difficulty does your logic find?" "The difficulty that Achilles would have to accomplish an endless series of performances." I reply, "If your conception of the problem is that Achilles at the start sees where the tortoise is and says 'I will go there' and having got there finds the tortoise is not there but at the second place and says I will go there" and so on, then I grant that because it is metaphysically, not logically, impossible that a man should have performed an endless series of distinct volitions or distinct acts of attention, it would not only be logic that would forbid Achilles to catch up with the tortoise but the fact would bear out the conclusion. He never would catch up. That, however, is not the case supposed which is merely that Achilles has a velocity presumably by one act of volition and by virtue of that greater speed he catches up with the tortoise. The places which you say Achilles must pass through have nothing corresponding to them in the action of Achilles. They are what are usually called *convenient fictions*, invented by you to describe the space over which Achilles passes. But in truth they are neither *convenient* nor are they *exactly fictions*. They represent possibilities, what might be measured off. Now every possibility whatever consists of an endless series of possibilities. For it may be logically subdivided without end and the subdivisions have the same mode of being as that which is divided. But that does not prevent the possibility from being realized; and hence my logic says that an endless series of possibilities can be completed as a whole although the sequence of its members is not broken off by a last. For example, a decimal fraction expressed by a row of 7 ones .1111111 equals $\frac{1111111}{10000000}$ or $\frac{9999999}{90000000}$ while $\frac{1}{9}$ is $\frac{10000000}{90000000}$ which is more by $\frac{1}{90000000}$. It would be the same if instead of 7 ones we had taken any other finite number of ones. In every case, $\frac{1}{9}$ would be more. So then there is an endless series of decimal fractions

.1
.11
.111
.1111
.11111
.111111
.1111111
.11111111

and so on for all finite numbers of ones and that series is endless because there is no greatest whole number. Yet $\frac{1}{9}$ is evidently greater than any one of these. Is there anything at all extraordinary about that?

Very well. Suppose the tortoise has in the beginning a start of .1 kilometre ahead of Achilles and runs .1 of a kilometre every hour while Achilles runs 1 kilometre every hour. Then Achilles will reach the spot where the tortoise first was in .1 of an hour. The tortoise in that .1 hour will have run .01 kilometre and Achilles will reach that point in .01 hour longer or .11 hour from the start. But in that last .01 hour the tortoise will then have run .001 kilometres and Achilles will reach that point in .001 hour more or .111 hour from the start. Thus the times at which Achilles will pass those previous places of the tortoise will be

.1 hour
.11 hour
.111 hour
.1111 hour
.11111 hour
etc.

from the start. But we have just seen that all those times are less than $\frac{1}{9}$ of an hour. So that in a ninth of an hour he will have run over them all. What is the wonder?

Do you say that the wonder is that the sum of an endless multitude of quantities all finite should not be infinite? Why should they be so, I ask? Because they must be at least infinitely greater than the smallest? No doubt, if there were any smallest; but since there is not, the only wonder I can see is that people should insist on reasoning inexactly when they can see for themselves that this way of reasoning leads them from true premisses to false conclusions.

B. LECTURE III (459)

Mathematics is the science which draws necessary conclusions. Such was the definition first given by my father, Benjamin Peirce, in 1870. At that day the new mathematics was in its early infancy and the novelty of this definition was disconcerting even to the most advanced mathematicians; but today no competent man would adopt a definition decidedly opposed to that. The only fault I should find with it is that if we conceive a science, not as a body of ascertained truth, but, as the living business which a group of investigators are engaged upon, which I think is the only sense which gives a natural classification of sciences, then we must include under mathematics everything that is an indispensable part of the mathematician's business; and therefore we must include the *formulation* of his hypotheses as well as the tracing out of their consequences. Certainly, into that work of formulation the mathematicians put an immense deal of intellectual power and energy.

Moreover, the hypotheses of the mathematician are of a peculiar nature. The mathematician does not in the least concern himself about their truth. They are often designed to represent *approximately* some state of things which he has some reason to believe is realized; but he does not regard it as his business to find out whether this be true or not; and he generally knows very well that his hypothesis only approximates to a representation of that state of things. The substance of the mathematician's hypothesis is therefore a creature of his imagination. Yet nothing can be more unlike a poet's creation. The reason is that the poet is interested in his images solely on account of their own beauty or interest as images, while the mathematician is interested in his hypotheses solely on account of the ways in which necessary inferences can be drawn from them. He consequently makes them perfectly definite in all those respects which could affect the ways in which parts of them could or [could] not be taken together so as to lead to necessary consequences. If he leaves the hypotheses determinate in any other respects, they are hypotheses of *applied* mathematics. The pure mathematician generalizes his hypotheses so as to make them applicable to all conceivable states of things in which precisely analogous conclusions could be drawn. In view of this I would define Pure Mathematics as the science of pure hypotheses perfectly definite in respects which can create or destroy forms of necessary consequences from them and entirely indeterminate in other respects.

I am confident that this definition will be accepted by mathematicians as, at least, substantially accurate.

As for the old definition that mathematics is the science of quantity, it first appears, I believe, in Boëthius, about A.D. 500 when mathematics was at its lowest ebb; and at that time three words that occur in it had entirely different meanings from those they now carry. Those three words are *Mathematics*, *Science*, and *Quantity*. First, under mathematics [were] then included only four sciences called Arithmetic, Astronomy, Geometry, and Music. But by arithmetic was not meant anything now called by that name; for our arithmetic was called *logistic* and was not included in mathematics. Secondly, by *science* was then meant *comprehension* through principles, which we now call *philosophy*, a thing which a modern mathematician would not touch with a nine foot pole. Thirdly, by *quantity* was meant simply things *measurable*. Therefore, the true meaning of that phrase Mathematics is the science of quantity was that four branches of learning, of which all but geometry are now utterly forgotten, constituted the philosophy of measurement. You see it is one of the many cases in which a phrase has quite survived its meaning. The relation of Quantity to Mathematics is, in fact this, that it is found that in one way or another the conception of quantity has an important bearing upon almost every branch of mathematics, as it has upon logic itself. But it is by no means the principal subject of all branches of mathematics. Thus the theory of Linear Perspective is a branch of Mathematics. Yet it is properly and primarily not concerned with quantity; and if it is made to appear so, it is badly taught.

Some very eminent and profound mathematicians go so far as to say that Mathematics is a branch of Logic. Dedekind is one of these, whose little book published by the Open Court Company under the title of Essays on Number I beg leave to recommend to your study. But Mathematics is not Logic for the reason that the mathematician deals exclusively with assumptions for whose truth he in no wise makes himself responsible, while logic deals with positive truth. The mathematician's interest in reasoning is to get at the conclusion in the speediest way consistent with certainty. The logician, on the contrary, does not care particularly what the conclusion is. His interest lies in picking the reasoning to pieces and discovering the principles upon which its leading to truth depends. As far as necessary inference is concerned, the mathematician and the logician meet upon a common highway. But they face in contrary directions.

Still, the mere fact that mathematicians of high rank consider mathematics as a branch of logic may serve as sufficient justification for my devoting a part of this course to the examination of mathematics.

There is no science more infested with a vermin of ignorant pretenders than logic; and there is one simple question by which they can commonly be detected. Ask your pretended logician whether there are any necessary reasonings of an essentially different character from mathematical reasonings. If he says no, you may hope he knows something about logic; but if he says "yes," he is contradicting a well-established truth universally admitted by sound logicians. If you ask for a sample, it will be found to be a very simple mathematical reasoning *blurred* by being confusedly apprehended. For a necessary reasoning is one which *would* follow under all circumstances, whether you are talking of the real world or the world of the Arabian Nights' or what. And that precisely defines mathematical reasoning. It is true that a *distinctively* mathematical reasoning is one that is so intricate that we need some kind of a diagram to follow it out. But something of the nature of a diagram, be it only an imaginary skeleton proposition, or even a mere noun with the ideas of its application and signification is needed in all necessary reasoning. Indeed one may say that something of this kind is needed in all reasoning whatsoever, although in induction it is the real experiences that serves as diagram.

One of the most striking characters of pure mathematics, — of course you will understand that I speak only of mathematics in its present condition, and only occasionally and with much diffidence speculate as to what the mathematics of the future may be, — but one of [the] characters of latter day pure mathematics is that all its departments are so intimately related that one cannot treat of any one as it should be treated without considering all the others. We see the same thing in several other advanced sciences. But so far as it is possible to break mathematics into departments, we observe that in each department there is a certain set of alternatives to which every question relates. Thus, in projective geometry, which is the whole geometry that is allied to perspective *without measurement*, namely, the geometry of planes, their intersections and envelopes, and the intersections and envelopes of intersections and envelopes, the question always is whether a figure lies in another figure or not, whether a point one way described lies on a point another way described, whether a point lies on a line or not, whether three lines coincide or not. Here there are two alternatives. In other departments, the alternatives are all the integer numbers; in still others, the alternatives are all the analytical numbers, etc. The set of alternatives to which a branch of mathematics constantly refers may be considered as a system of values; and in that sense, mathematics seems always to deal with quantity. It would seem that if any lines of demarcation are to be drawn between different mathematical theories they must be according to the number of alternatives in the set of alternatives to which it refers; but I am bound to say that this is a notion personal to myself — and that I have my doubts as to its worth as the basis of a complete classification of mathematics. We may, however, accept it in so far as it shows that the simplest possible kind of mathematics will be that all whose questions relate to which one of a single set of two alternatives is to be admitted. Now in Existential Graphs, all questions relate to whether a graph is true or false; and we may conceive that every proposition has one or other of two values, the *infinite* value of being true, and the *zero* value of being false. We have, therefore, in Existential Graphs an exposition of this simplest possible form of mathematics. It is Applied Mathematics, because we have given definite logical significations to the graphs. But if we were to define the graphs solely by means of the five fundamental rules of their transformation, allowing them to mean whatever they might mean while preserving those rules, we should then see in them the Pure Mathematics of two values, the simplest of all possible mathematics.

Were we to follow out the same principle, we should divide all mathematics according to the number of alternatives in the set of alternatives to which it constantly refers and also to the number of different sets of alternatives to which it refers. Perhaps that would give as natural a classification of pure mathematical inquiries as any that could at this time be proposed. At any rate, we may so far safely trust to it, to conclude that the very first thing to be inquired into in order to comprehend the nature of mathematics, is the matter of *number*.

Certainly, of all mathematical ideas, next after the idea of two alternatives, the most ubiquitous is the idea of whole numbers. Dr. Georg Cantor is justly recognized as the author of two important doctrines, that of *Cardinal Numbers* and that of *Ordinal Numbers*. But I protest against his use of the term *Cardinal Number*. What he calls cardinal number is not number at all. A cardinal number is one of the vocables used primarily in the experiment called counting a collection, and used secondarily as an appellative of that collection. But what Cantor means by a cardinal number is the *zeroness*, *oneness*, *twoness*, *threeness*, etc., — in short the *multitude* of the collection. I shall always use the word *multitude* to mean the *degree of maniness* of a collection. By *ordinal numbers* Cantor means certain symbols invented by him to denote the place of an object in a series in which each object has another next after it. The character of being in a definite place in such a series may be called the *posteriority* of the object.

Since I have alluded to Cantor, for whose work I have a profound admiration, I had better say that what I have to tell you about Multitude is not in any degree borrowed from him. My studies of the subject began before his, and were nearly completed before I was aware of his work, and it is my independent development substantially agreeing in its results with his, of which I intend to give a rough sketch. And since I have recommended Dedekind's work, I will say that it amounts to a very able and original development of ideas which I had published six years previously. Schröder in the third volume of his logic shows how Dedekind's development might be made to conform more closely to my conceptions. That is interesting; but Dedekind's development has its own independent value. I even incline to think that it follows a comparatively better way. For I am not so much in love with my own system as the late Professor Schröder was. I may add that quite recently Mr. Whitehead and the Hon. Bertrand Russell have treated of the subject; but they seem merely to have put truths already known into a uselessly technical and pedantic form.

The two doctrines of Cardinal and Ordinal Numbers, or of Multitude and Posteriority, though necessarily running parallel are curiously unlike one another. The doctrine of Ordinal Numbers is a theory of Pure Mathematics. For the relation of coming after is defined by this graph [Fig. 1]:

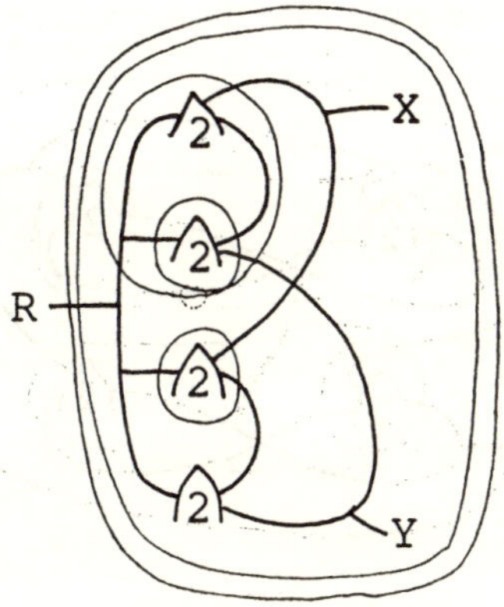

Fig. 1

That is to say, to say that X is posterior to Y is to say neither more nor less than that there is a relation R in which Y stands to whatever X stands in that relation to as well as to something to which X does not stand in that relation.

The relation R may be any relation whatever. The letter is attached to it and the two cuts simply to show that it must be a relation fixed in advance. To change from one relation, R, to another would generally be to change the order of sequence.

The relation of posteriority is composed of two parts which are the negatives of one another. It is that X is at least as late as Y while Y is not as late as X. The relation of being at least as late as is represented by this graph. I usually call it the relation of *inclusion of correlates*, for it implies that everything that U stands in any fixed relation to is included among the things to which V stands in that same relation [Fig. 2].

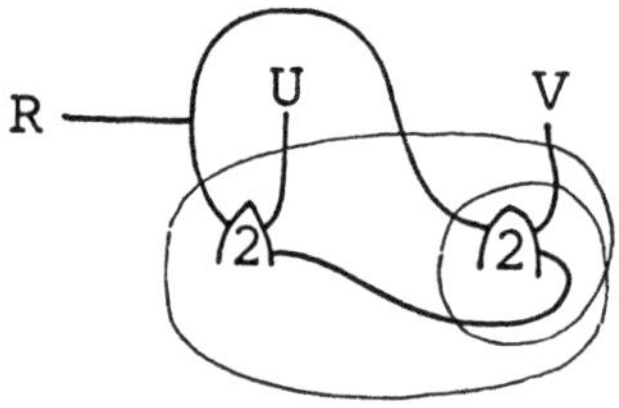

Fig. 2

This form of relation is of immense importance in mathematics. I will tell you presently what renders it so.

The relation of posteriority as defined by the graph is such sequence as there is among events in time but not such as there is among the milestones of a road. The difference is that two events in time may be contemporaneous, while two milestones cannot be reached simultaneously. We may express that in a given case the succession is serial by the graph

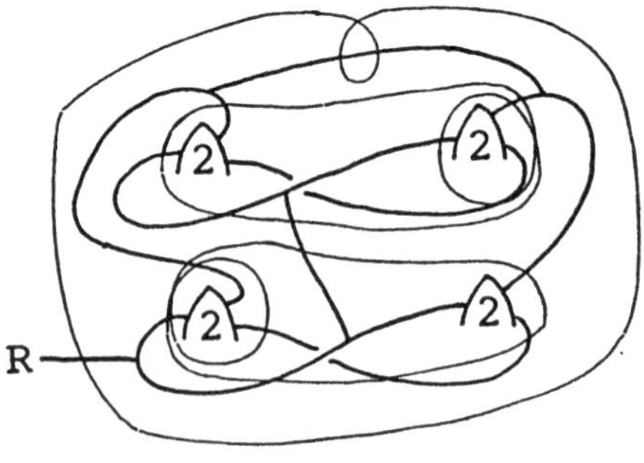

Fig. 3

[Fig. 3], that is, if U is as late as V and V is as late as U, then U and V are identical. We can express this more neatly by taking for the correlates certain qualities; namely, the quality of being the first in the assumed order in the series, the quality of being one of the first two, the quality of being one of the first three, and in general the quality of being as early in the series as some given one. Then taking p as an abbreviation

p = possesses one of the "as early" qualities we may write the graph thus [Fig. 4]:

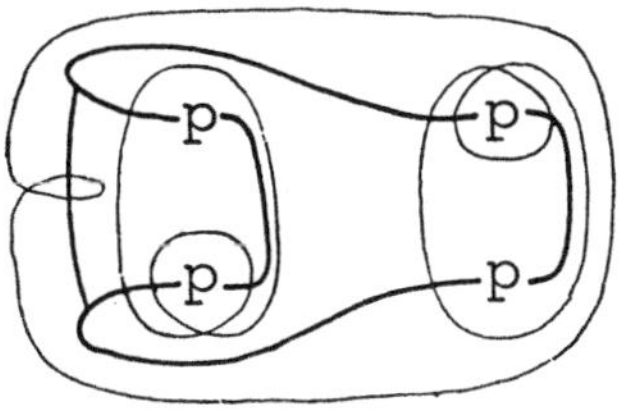

Fig. 4

I will now show you *why* the relation of inclusion is so important in mathematics. The system of existential graphs was invented for the purpose of representing the reasoning of mathematics in as analytical form as possible. It is not perfect. It is open to objections which I know well. For instance, the fact that a and  represent the same state of things, without either being more analytical than the other, is a grave fault which I shall have to leave future logicians to cure. The system is merely the best I could do. But it does express, in highly analytical form, all necessary reasoning. Now in constructing it, I was forced to introduce into its very forms four distinct kinds of signs of relation. Namely first, the relation expressed by writing anything on the sheet of assertion; 2nd, the relation expressed by writing two graphs together on the sheet; 3rd, the relation expressed by the heavy line; and 4th, the relation expressed by the scroll. Now since these four relations are the relations that thus enter into the very form and essence of necessary reasoning, it follows that every relation whose law has the same form as the laws of these must be of the highest importance for pure mathematics, since pure mathematics is precisely what results from abstracting from all the special meaning of necessary reasoning and considering only its forms. Of these four relations there is one which enters into the very definition of necessary reasoning. For necessary reasoning is that whose conclusion is true of whatever state of things there may be in which the premiss is true. Now this is expressed in a graph thus [Fig. 5]:

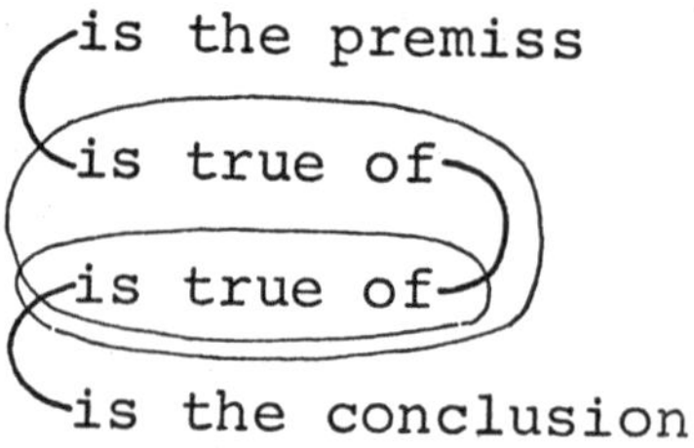

Fig. 5

The pure mathematician substitutes for these logical terms the indefinite symbols *x*, *y*, *z*, which are to mean whatever they may mean; and he thus gets this graph, [Fig. 6] which is precisely the graph of inclusion.

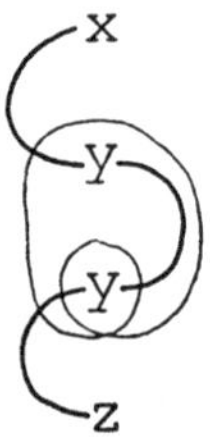

Fig. 6

I cannot stop to consider the other three relations but must hurry back to the subject of numbers. The doctrine of ordinal numbers, then, is a theory of pure mathematics and, as matters stand today, is the most fundamental of all branches of pure mathematics after the mathematics of the pair of values which existential graphs illustrate. The doctrine of multitude is not pure mathematics. Pure mathematics can see nothing in multitudes but a linear series of objects, having a first member, each one being followed by a next, with a few other such formal characters. A multitude, as such, Pure Mathematics knows nothing of. Multitudes are characters of collections; and the idea of a collection is essentially a logical conception. How would you define a collection, in general, without using the idea itself in your definition? It is not easy. In order to explain the matter, it is necessary to begin with the conception of a quality. There is an essential part of the doctrine of Existential Graphs, — essential to it, I mean, as a logical, not as a mathematical doctrine, — and of such importance as quite to overshadow all the rest which I have been forced to pass over for lack of time. It treats of the general properties of qualities and relations. Without it there are most important inferences that cannot be drawn. I call it the doctrine of *sub-*

stantive possibility, because qualities and relations are possibilities of a peculiar kind. In a secondary sense a quality may be said to exist when it has, as it were, a replica in an existing thing. But strictly speaking, a quality does not exist. For to *exist* is to be a subject of blind compulsion. A quality not only neither exerts nor suffers such force, but it cannot even be called an *idea* of the mind. For things possess their qualities just the same, whether anybody thinks so or not. The being of a quality consists in the fact that a thing *might be* such or such like. In saying this, I am not talking metaphysics nor epistemology. I am confining myself to logic. You can entertain whatever opinion seems good to you as to the real nature of qualities and as to the genesis of ideas in the mind. I have in this course quite nothing at all to say to all that. I simply say that you must use this form of thought, whether you regard it as corresponding to facts literally, metaphorically, symbolically, or however you may prefer. But you must use the form of thought or your threads will be inextricably entangled. For my part, when I think about logic, I dismiss irrelevant questions of metaphysics and psychology entirely from my mind. But that requires some training. Qualities, then, and Relations are pure possibilities; and as such they have no individual identity. Two qualities are more or less unlike. Identity belongs only to subjects of blind compulsion or force. There is no sense for example in asking how many shades of red there are, — unless you mean how many a man can distinguish, — which is a question of psychophysics, not logic. These substantive possibilities, — that is, qualities, relations, and the like, — are *prior* to existence, in the sense that non-existence is not a necessary proof of non-possibility, but non-possibility is a necessary proof of non-existence. For it is logically impossible that existence should exhaust pure possibilities of any kind. These truths are strictly deducible from the facts of phenomenology, or the analysis of the phenomenon; meaning by the *phenomenon* whatever is present to the mind in any kind of thought. From these truths there follows a most important rule of existential graphs, having two clauses. The first clause is that any graph which does not relate to what exists but only to pure substantive possibilities of the same order is true if the outermost parts of its innermost ligature [are] enclosed in an even number of cuts, but is false if that number be odd. The second clause is that in case existence is in question along with the substantive possibilities of one order, and if the truth or falsity of the proposition is contrary to what the first clause would be if it were not limited to cases in which no existence is spoken of, then the discrepancy may be regarded as due to insufficiency of the things that it says exist or to the excess of the things that it says do not exist. For example, although propositions about every quality are normally false yet it is true every quality is susceptible of varieties greater than existing things can embody. But this is because of the insufficiency of existing things.

What is meant by the order of substantive possibilities will be explained in the next lecture.

The doctrine of substantive possibility is an extensive one. You will understand that I only mention bits of it.

The variety of qualities, as is easily proved, literally exceeds not only all number, but all multitude finite or infinite; and anybody who should assume the contrary would be liable to great errors of reasoning. But the qualities with which we are familiar are a small number. Certainly a figure *one* with only twenty or thirty zeros after it would denote a greater number; and these are naturally regarded by us as composed of a *very* small number of qualities which we do not analyze. Therefore for all those purposes for which [we] regard the qualities themselves, they may be considered as comparatively few.

Qualities are general respects in which existing things *might* agree or differ. They are as I have said, mere possibilities. But qualities have themselves general respects in which they agree or differ. Thus, musical notes differ in respect to duration, intensity, pitch, timbre, stress, expression, and some other respects. These qualities of qualities differ very much from qualities of existing things. Considering the qualities of any one class of qualities, we find them to be innumerable indeed but not in excess of all multitude; and a set [of] three or four of some such small number of them that are independent of one another will fully suffice to describe the rest. These respects, or qualities of qualities, themselves again have general respects in which they agree and differ. Thus duration, pitch, and intensity are serial, that is each can only vary along one line of variation, while timbre, stress, and expression are multiform. These modes of variation of respects correspond to the possible whole numbers.

After these explanations, you will be able to understand this definition of a *quality*.

A quality is anything whose being consists in such logical possibility as there may be that a definite predicate should be true of a single subject. It is said to be actually *embodied in* or *possessed by* whatever there is of which that predicate is true.

But somebody may ask, Has a quality any being? I reply, Why of course it must have being because by the terms of its definition to say that it has being is at the very most, no more than to say that something is logically possible. Remember, we are not talking metaphysics; we are talking logic. A quality is an *ens rationis* of course. That is, it consists in a certain proposition having a meaning. The term *Essence* means being such as the subject of the essence necessarily is. Quality then has *essence*. But it has no *existence*, because it neither exercises nor suffers brute compulsion.

I am now prepared to give you the definition of a collection; and remember that by a collection I do not mean that whose members are in any sense actually brought together, nor even that whose members are actually thought together; but I mean that whose members *might*, in logical possibility, be thought together. But to think things together is to think that something might be true of them all that was true of nothing else. But to do this amounts to thinking that they have a common quality. Therefore the definition is plain:

A *Collection* is anything whose being consists in the existence of whatever there may exist that has any one quality; and if such thing or things exist, the collection is a single thing whose existence consists in the existence of all those very things.

According to this definition a *collection* is an *ens rationis*. If its members are actually brought together like the atoms which compose my body, it is more than a mere *collection*. As *collection*, it is an *ens rationis*, but that reason or *ratio* that creates it may be among the realities of the universe. A *collection* has *essence* and may have *existence*. It has *essence* from all eternity, in the logical possibility that it should be described. It has *existence* from the moment that all its members exist. Thus, all men constitute a collection; and not a very small one. But in the carboniferous period in a certain sense that collection had no existence. By saying it was so 'in a certain sense' I mean if by *men* be meant the men that live at the moment. In this same sense, the existence of this collection is constantly changing; the same collection in essence is becoming a different collection in existence. There is a collection of men with grass green hair; but having only essence and not existence, it has no individual identity. It is the collection that we call *Nothing*. It must be counted among collections; but it differs from all the rest in having no existence. Of course, for ordinary purposes, this is the emptiest nonsense. Nevertheless, it is a matter that has to be put straight for logical purposes. I may remark that nonsense often repays logical study and by that study enables us to avoid fallacious reasoning about serious questions. Another such little point is the following. According to the definition, there must be a collection of luminaries of the day. But there happens to be only one luminary of the day; namely, the Sun. Here then is a collection having but one member. Is not that collection the sun itself? I reply, Certainly not. For a collection is an *ens rationis*. Its being consists in the truth of something. But the Sun is not an *ens rationis* and its being does not consist in the truth of any proposition. It consists in the act of brute force in which it reacts with everything in its neighborhood. So then the Sun is one thing and exists, and this collection containing only the sun is something different and exists, and there would be a collection embracing as its sole member this collection, and this too exists and so on *ad infinitum*. This is true. Yet there is only one *existence*; for the existence of the collection *is* the existence of its sole member. Thus, that collection embracing the sun alone is different from the Sun but its existence is the same as the existence of the sun. In that sense, it is the same as the Sun.

In the next lecture I will show you what *multitude* is and what different grades of multitude there are; and then you will see how some of the hair-splitting of this lecture is, after all, very useful.

C. [FOR LECTURE III OR IV] (466)

Mathematics is the science which draws necessary conclusions. Such was the definition given first by my father as early as 1870. At that day, when the new mathematics was in its infancy, the novelty of this definition was disconcerting even to the profoundest mathematicians; but today nobody would propose a definition differing much from that. The only doubt that should entertain is whether we ought not to recognize as a part of mathematics, what is certainly a most important part of the mathematician's business, the formation of the assumptions on which his reasoning is to be based.

Some of the mathematicians who have the most deeply studied the fundamentals of their science have even gone so far as to pronounce mathematics to be a branch of logic. Dedekind is one of these whose two little books published in one volume in translation by the Open Court Company, which is doing so much for American culture, I should strongly recommend to your attention. The fact that so profound a mathematician can hold this opinion is a sufficient justification of my devoting several lectures of this short course to a study of the nature of mathematics.

I do not quite agree with Dedekind, myself; and I will tell you why presently. The question of whether mathematics was a branch of logic was the subject of careful discussion between my father and me at the time he had his definition under consideration. But first I had better notice an objection which will seem weighty to superficial minds. Namely, it will be said that much necessary reasoning is not at all mathematical. On that I take direct issue. Eminent jurists, moralists, and philosophers can be found whose powers of reasoning are famous, and who yet declare that they have no head for mathematics. This is, in part, a delusion owing to bad instruction which has given rise to such an aversion to everything that seems mathematical that as soon as one talks to them of x, y, z, they stop thinking. But what is also true of those persons is that they cannot hold clearly before their minds intricate relations between objects that are almost exactly alike except in respect to abstract relations. But when I ask one of those gentlemen to give me an example of a necessary reasoning that he considers not to be mathematical, it turns out to be one of those that are most readily amenable to mathematical representation, differing only from the reasoning he cannot grasp in its extreme simplicity. But that which conclusively stamps all necessary reasoning as mathematical is that in such reasoning, it makes not the slightest difference whether the premisses express observed facts (as strictly speaking they seldom do) or whether they describe wholly imaginary states of things. The conclusion follows as necessarily concerning the imaginary state of things as it would if that state of things had been observed. This, indeed, is precisely what the *necessity* of such reasoning consists in. For the purposes of the reasoning, therefore, the premisses are mere assumptions. If they happen to be more, it has nothing to do with the reasoning. Now the only science which deals with pure assumptions regardless of their real truth is mathematics. That is, on the whole, the best definition of mathematics. All necessary reasonings, therefore, are pieces of mathematics.

The only reason I do not agree with Dedekind in making mathematics a branch of logic is that logic is *not* a science of pure assumptions but is a study of positive truth. The mathematician seeks only to trace out the consequences of his assumptions in the readiest and speediest way. The logician does not care much what the conclusions from this or that system of assumptions may be. What he is interested in is in dissecting reasonings, in finding out what their elementary steps are, and in showing what positive facts about the real universe of things and of thoughts it is from which the necessity of the mathematician's reasonings and the validity of other kinds of reasonings depend, and exactly what the nature of that dependence is. I have already pointed out that the characters that make a system of symbols good for [mathematical] purposes are quite contrary to those which would make it good for logical purposes. The truth is that the mathematician and the logician meet in one department on a common highway. They *meet*; but one is facing one way while the other is facing just the other way. Each of them, it is true, finds it interesting to turn round occasionally and take a glance in the opposite direction.

The mathematician, however, has little or nothing to learn of the logician. Mathematics differs from all the special sciences whether of physics or of psychics in never encountering logical difficulties, which it does not lie entirely within his own competence to resolve. The logician on the other hand has everything to learn from the mathematician. Mathematics is, with the exception of Phenomenology and Ethics, the only science from which he can draw any real guidance. There is therefore every reason in the world why we should not delay examining the fundamental nature of mathematics.

The simplest possible kind of mathematics would be the mathematics of a system of only two values. Do we in experience meet with any case in which such mathematics could have any application? I reply that we do. There are just two values that any assertion can have. It is either true when it has all the value it can have, or it is false, and has no value at all. It follows that our system of Existential Graphs is precisely an application of the mathematics of two values. It will give you an idea of what Pure Mathematics is to imagine the Existential Graphs to be described without any allusion whatever to their interpretation, but to be defined as symbols subject to the fundamental rules of transformation. Namely,

1st, A graph is composed of spots with definite hooks from each of which one heavy line extends.

2nd, Any graph within an even number of cuts can be erased and any within an odd number of existing cuts can be inserted.

3rd, Any graph can be iterated or deiterated provided the iterated or deiterated replica is not outside of any cut that the other is inside of.

4th, Two cuts one inside the other with nothing between except heavy lines passing from within the inner to outside the outer, can either be made or destroyed anywhere.

5th, A heavy line of which both ends connect with lines inside a cut which lines do not connect spots connected inside the cut can anywhere be erased or inserted.

From those assumptions everything universally true of existential graphs could be deduced, and that would be the Pure Mathematical Treatment. I have defined Mathematics in general as the science which draws necessary conclusions and which formulates the assumptions from which such conclusions can be drawn. I now define *Pure* Mathematics as that Mathematics which leaves its assumptions entirely indeterminate in respects which have no bearing upon the manner in which they can be combined to produce conclusions.

The Pure Mathematics of the System of Two Values would leave us free to regard the Graphs as representing anything for which their fundamental transformations would hold good.

In like manner there would be a Mathematics of a System of Three

Values which would not be without utility and which has been in some measure developed. The theory of numbers furnishes partial developments of the mathematics of every system having a finite multitude of values.

I particularly call your attention to my use of the word *Multitude.* I never use this word to mean a *Collection* or *Plural.* I always use it as an abstract noun denoting that character of a Collection which makes it greater than some collections and less than others. Thus, twoness, tenness, three-hundred-and-sixty-five-ness, are individual Multitudes, as I use the word.

At the very outset of the study of the logic of mathematics our attention is forced to the subject of Multitude. Everybody knows that there is a Mathematics of Multitude; and I am going this evening to tell you some elementary things about it that you probably do not know.

It is desirable to remark at the outset that the Mathematics of Multitude is not pure mathematics. The reason is that the whole application of Numbers to Multitude is based on the four propositions. I will state these propositions although it is not necessary you should follow them exactly. It will be sufficient to apprehend their general character. The first is, that if you take any two numbers say 3 and 7 and there is a collection, [say of 5 dots], that 3 is used in counting: one, two, *three*, four, five, — and there is a collection that seven is used in counting, [say of 9 dots], then either 7 is used in counting the first or else 5 is used in counting the second. That is the first proposition represented by [Fig. 1] and the other

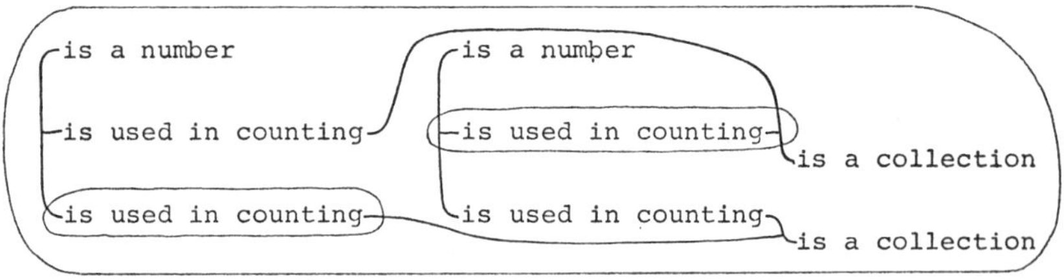

Fig. 1

proposition if two numbers *A* and *B* are not identical there is either some collection that *A* is used in counting and *B* is not, or there is some collection that *B* is used in counting and *A* not [Fig. 2].

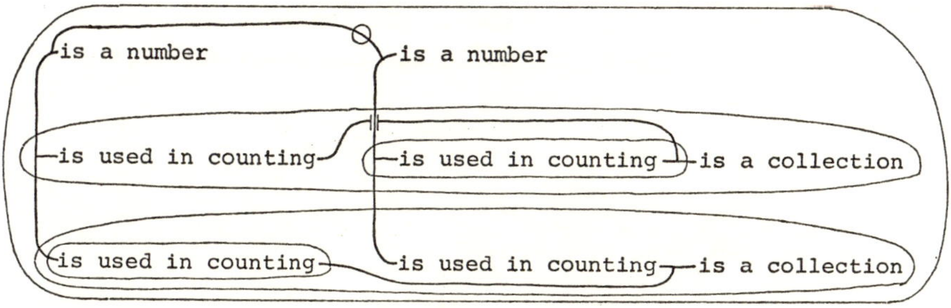

Fig. 2

The third proposition is that there is a number next greater than any given number or taking any number whatever, there is a number (namely, the next greater one) such that taking any number you please either this third number (being no greater than the first) is used in counting every collection that the first is used in counting or else (being as great as the next greater number) is not used in counting any collection unless that next greater is used in counting it [Fig. 3].

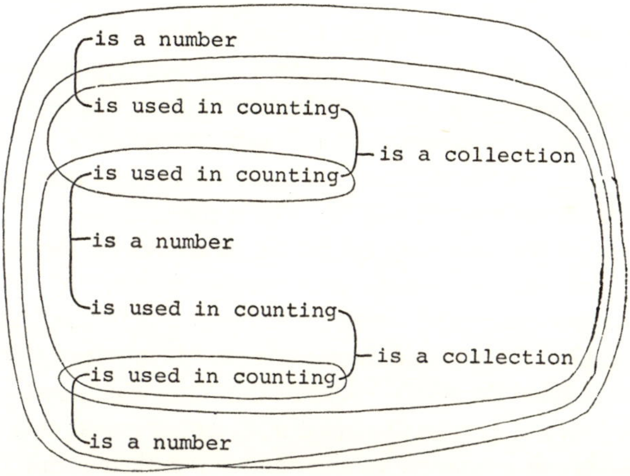

Fig. 3

The fourth proposition is that for any collection that some number is used in counting and that some number is not used in counting there is a number (namely the number of its count) used in counting it that is such that whatever number is used in counting this collection is used in counting every collection that this number is used in counting [Fig. 4].

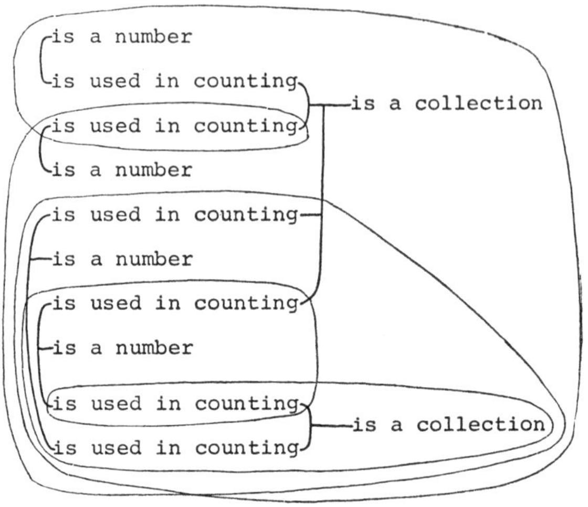

Fig. 4

(I might add a fifth proposition to express that there is an absolutely smallest number; but I do not think that of much importance.)

By manipulating those four graphs according to the rules of graphs we could reach every theorem of the mathematics of whole numbers. But the rules of the transformation of graphs do not depend upon what is written in the different spots. So long as they are preserved the same or different, as the case may be, letters or any other meaningless expressions might be substituted. In particular, in place of "collections" any other universe might be adopted, and in place of "is used in counting" a general symbol for any one dyadic relation. If this change were made in the four graphs, the result would be that they would define the relations of *ordinal* numbers to one another, except of course that they omit to say that there is a smallest number; and in fact there is not, if negative numbers are included.

Thus, the Arithmetic of Ordinal Numbers is Pure Mathematics; but the Arithmetic of Cardinal Numbers is Applied Mathematics. The conception of a Collection is, therefore, not a mathematical conception and the conception of Multitude is not a purely mathematical conception. But the conception of a collection is a highly important logical conception and for that reason, we will turn our attention to *Multitude*.

But why should the conception of Multitude be an important conception in logic? I will tell you. We have seen that there are three relations which subsist between the parts of graphs. The first is the relation expressed by the scroll [in Fig. 5]. This is the most important of all, since

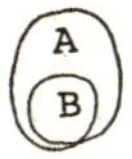

Fig. 5

this is the relation of premiss and conclusion; that is, if it be true that if *A* is true *B* is true, then should *A* occur as a premiss we have a right to conclude *B*. The second relation is that expressed by writing two graphs side by side *AB*, that is to say, the relation of coexistence, and the third is the relation of individual identity expressed by the heavy line. Now whatever relation is analogous to one of these three relations will be expressed by graphs into which the corresponding element of graphs enters and will therefore affect reasoning and hence will be of logical importance. Now the relation expressed by saying that one collection is at least as great as another is precisely analogous to the relation expressed by the scroll in graphs; and that is why *multitude* which is the character of a collection which constitues it as great as what it is as great as is so very important for logic.

Everything depends, in logic, upon formulating precise definitions. Now what precisely do we mean by saying that one collection is at least as small as another. The best definitions are those which express how the truth or falsity of the term defined will affect what we can do in the future. To say that a thing *is soft* is to say that it can easily be scratched. To say that is is hard is to say that it never will easily be scratched, unless it turns soft. The relation of greater and less was first defined in this fashion by the Catholic logician Bolzano in his posthumous work which appeared in 1851. He said a collection, say that of the *A*s is at least as small as another collection say that of the *B*s, if, and only if, every single *A* can be placed in a relation to *B* to which no other *A* stands in the same relation. But just as to say that a body *can* easily be scratched is the same as to say that it *is* soft, so to say that one thing *can be* placed in relation to another thing is to say that it already *is* in relation to that thing; and it is simpler to say that the collection of the *A*s is at least as small as the collection of the *B*s, if and only if there is a relation in which every *A* stands to a *B* to which no other *A* stands in the same relation.

There is a certain dyadic relation such that if there is an *A* it stands in that relation to a *B* and if there is any *A* and it stands in that same relation to that same *B* then it is identical with that first *A* [Fig. 6].

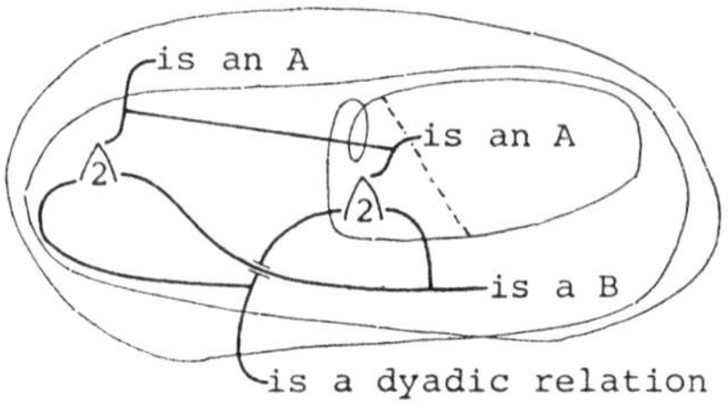

Fig. 6

To deny this is to say that there is no such relation, which we express by enclosing the whole in a cut. But every fact whatever consists in the existence of a relation. To say that there is no relation of one kind is to say that there is a relation of another kind. All things are related to one another in every combination. The brevity of this course forbids my going into the general theory. But if a relation not in itself absurd, and therefore having all the being that belongs to a relation in itself fails to exist in any case, it is simply because *existence* by its very nature must fall short of possibility. It must be because the things fail to exist that would bring the relation into existence in the particular case. If there is no one-to-one relation of whatever *A* there is to a separate existent *B*, it is simply because the *B*s which this would require do not exist. Add other things, say *C*s to the *B*s and every *A* would be provided with its own separate *B* or *C*. Take no more *C*s than are required, and conversely every *B* or *C* would have its own separate *A* and therefore every *B* has its own separate *A*.

Hence we get the proposition if the collection of *A*s is not as small as the collection of *B*s then the collection of *B*s must be as small as the collection of *A*s.

This proposition has been a great puzzle to the mathematicians. They have endeavored in vain to prove it. The truth is that it never can be mathematically proved because it depends upon the peculiar nature of relations which is a question of logic, not of mathematics. Hence, every proposed mathematical demonstration of it, of which there have been more than one, will necessarily be found upon examination to be fallacious.

Whatever things we may take it into our heads to regard as forming a collection, — say the milky way, this bit of chalk, and King Peter of

Servia, — have some character in common not possessed by anything else in the universe. For it is a fact that they might be regarded as forming a collection which should exclude everything else; and just as the fact that a piece of chalk *could* easily be scratched constitutes it eternally true that it *is* soft now, so the fact that any thing could be regarded as forming a collection containing nothing else constitutes a character which they all possess and they alone. *Qualities*, or characters, like *relations* are in themselves mere possibilities. Many possible things do not *exist* but whatever at any time exists or is even imagined to exist is *possible*. You must try to understand, what I know is difficult to understand without long explanations for which I have no time, that I do not at any time in this course touch upon metaphysics and that here at least I have nothing to do with epistemology. I must be understood in a purely logical sense, when I say that *qualities* and *relations* are not creatures of the mind, but are in themselves eternal possibilities, whether the mind thinks them to be so or not. They acquire existence only in the sense that they determine existent individual subjects. One of you may say to another "For all the lecturer says, I think qualities are creations of the mind." In saying that, you mean one or other of two things. You are either speaking as a metaphysician or as an epistemologist. I neither affirm nor deny what you say. I am simply talking logic. You are thinking either of the nature of the being of qualities or of the processes of human cognition. I have nothing to do with either. I simply say that in order to avoid getting wound up in your reasoning, you are to regard qualities in a certain *aspect*. You need not say what they really are nor what the genesis of the ideas is, but leaving that quite aside you are to regard *qualities* as eternal and immutable possibilities *prior* to existence in the sense that no fact of existence can affect in the least the qualities themselves. A *collection* is also a possibility, a mere possibility. But it differs from a quality in that it is a possibility *posterior* to existence [.·.·.]. Here are five dots and the existence of those five dots constitutes the *existence* of the collection. I might put in a sixth dot, and the possibility of that sixth constitutes the *possibility* of a collection of six dots. The existence of those five dots not only constitutes the existence of the one collection of five dots but also the existence of five collections of 4 dots each, as well as 10 collections of 3 dots each, 10 collections of 2 dots each, 5 collections of 1 dot each, and 1 collection *nothing*; and these 32 collections constitute one collection of collections. Or we may say there is a collection of six collections each of which latter collections is a collection of equal collections of dots.

The members of every collection possess some quality common to them and possessed by nothing else in the universe; and for that reason it is for some purposes much the same thing to say that an object belongs to a collection and to say that it possesses a quality. But for other purposes the distinction is important. Qualities possess no individual identity but only similarity while a collection is a single individual collection, though possessing it is true only a *derived* individuality.

A quality possesses, in itself, few positive qualities, which few are mostly of the particular kind called quantities. Thus, any particular kind of red, has its degree of light, its degree of fullness of color, and its peculiar quality of hue. This third is something more than a mere quantity. But it varies only in two opposite ways like a quantity. A collection, on the other hand, has all sorts of qualities discoverable by experience. Thus, a man's body is a collection of molecules; and its qualities are what is true of some of these molecules in relation to each other and to other things. *Multitude* is a quality of a collection or something like that; and accordingly it possesses only one quality which is its quantity. It consists in the collection's being more than some and less than others.

D. LECTURE V (469, 470)

I am going to speak to you tonight about *multitude* or *maniness* and shall, I hope, be able to give you a glimpse of its logical importance. Some kinds of *infinity* will come under our notice, but I fear I shall not have time to speak of other kinds. I shall say a few words only concerning *continuity*. All of these subjects become intensely interesting and of great importance if one can go into them in some detail. But the difficulty with a course of this kind is that it is not possible to go deeply enough into the subjects to reach the really valuable parts of them.

I must begin by a few words concerning gramma graphs; because it is by means of gamma graphs that I have been enabled to understand these subjects, and I do not believe it is possible really to understand them except by some method equivalent to that of gamma graphs.

In particular, it is absolutely necessary to representing the reasoning about these subjects that we should be able to reason with graphs about graphs and thus that we should have graphs of graphs.

It is essential to a graph or any other expression of a proposition that it should be represented by its interpretant sign to be *true*. But to say that it is true implies that it really is affected by its object; and in order that this object should have a real effect upon it, this object must be a subject of force, which is an individual. Consequently, an adequate interpretant of a graph must represent it as a sign of an *individual*. How, then, can there be a *graph* of a *graph*, considering that a graph is a *legisign*, or sign which has the mode of being of a *general type*, just as any word is a general type, and not a single individual object in a single definite place at a single instant. The answer is that a graph can only have a graph for its object indirectly. Directly, it can only refer to a graph replica. But it can assert what it asserts of *any graph-replica* you please so long as it be *equivalent* to a given graph replica.

The matter is unsuitable to presentation in a course like this, although it is indispensable that something should be said about it. If I can ever get the funds to print it, I shall be very happy to send copies to all who may care for them.

This matter of the graphs if I could present it would also illustrate the nature of Pure Mathematics, still better than it is illustrated on pages 20 to 22 of the Syllabus distributed last Thursday. Mathematics, in general, is the science of the logical possibility and impossibility of hypotheses. Upon that definition first put forth by my father Benjamin Peirce in 1870, mathematicians are now pretty unanimously agreed. *Pure* mathematics differs from mathematics in general in not admitting into its hypotheses any element that does affect their logical possibility or impossibility. Thus, the pure mathematics of graphs, as you see in those pages of the syllabus, says nothing at all about the logical interpretation of graphs but defines them exclusively by their logical relations to one another. So the pure mathematical presentation of graphs of graphs says nothing at all about the graphs considered being representations of graphs but merely defines the graphs representing the sheet of assertion, a cut, its area, a line of identity, and so forth, in terms of their logical relations to one another.

The theory of graphs of graphs rests upon a larger number of independent hypotheses than do most branches of pure mathematics. I divide these hypotheses into those which relate to physical possibilities and necessities and those which relate to permissions and prohibitions. The former class, about thirty in number, are all expressible in alpha and beta graphs. The latter class which do not much exceed half a dozen require, one and all of them, gamma graphs to express them. In particular the broken cut enters into every one, and into some of them with curious complications.

One is surprised to find how extremely complicated are the graphs that express some of these fundamentals of the nature of graphs; and this complication puts into a strong light two merits of this system of existential graphs. For, on the one hand, it shows the extraordinary analytical power of the system that analyzes conditions which seemed on our first acquaintance with them to be very very simple but which the system of graphs forces us to see are in reality very complex; while on the other hand, this same complication shows that the system of existential graphs possesses to an extraordinary degree the virtue that belongs more or less to all diagrams that of putting a matter really extremely complicated into a light under which it is fully and adequately represented and yet seems as easy and natural as slipping off a floating log.

But I must hasten to the subject of numbers.

Whole numbers can on the one hand be studied in two ways which are surprisingly different from one another throughout. They can be studied as qualities of collections, making the members of one collection many and those of another few, which is called by the Germans with their usual incapacity for language the doctrine of *Cardinal Numbers*; but which ought to be called the doctrine of *Multitude*. Or, on the other hand, numbers may be considered simply as objects in a sequence, as *ordinal numbers*. The latter study is a branch of pure mathematics, because it makes no difference what kind of objects they are that are in series, nor whether it is a series in time, in space, or in logic. The doctrine of multitude, on the other hand, is not pure mathematics. For the objects it studies, the multitudes are in a linear series exactly as the doctrine of ordinal numbers supposes; and since the doctrine of ordinal numbers permits the members of the series to be objects of any kind, it follows that it permits them to be multitudes. Thus the doctrine of multitude is nothing but a special application of the doctrine of ordinal numbers. But the special objects of its series have a special character which permits them to be studied from a special point of view; and that point of view is a logical point of view. It is not the pure mathematical forms that we study in the doctrine of multitude. It is on the contrary a branch of logic which, like all logic, is directly dependent upon mathematics. The first question we come upon in the study of multitude is very obviously a purely logical question; and there is nothing at all corresponding to it in the doctrine of ordinal numbers: it is the question what is multitude. Multitude is obviously a relative quality of *collections*, or plurals. Therefore the question becomes, What is a *collection*? That is obviously a most important question for logic; and it is about as difficult a one as could be found. In speaking of a collection, we do not mean that its members are physically or in any way existentially brought together. We mean by a collection merely a plural, whose objects are collected together by thought. The collection exists just as much as its members. Their existence is its existence. Yet in another point of view, it is a creation of thought. It is an *ens rationis*. An *ens rationis* may be defined as a subject whose being consists in a Secondness, or fact, concerning something else. Its being is thus of the nature of Thirdness, or thought. Any abstraction, such as Truth and Justice, is an *ens rationis*. That does not prevent Truth and Justice from being real powers in the world without any figure of speech. They are powers, just as much, and in the same way, as I am a power if I can open my window should the air seem to me stuffy. A collection is an abstraction, or is like an abstraction in being an *ens rationis.* But it is unlike an abstraction in that it *exists.* Truth and justice do not *exist,* although they are powers. I myself, properly speaking, do not exist. It is only a replica of me that exists, and I exist in that replica as the effect of my being as a law. A collection, however, exists, and this existence is derived from the existence of its members which may be pure Secondness. Our bodies are of course much more than so many collections of molecules; but as far as its *existence* is concerned, the existence of our bodies consists in the existence of the molecules. But the word *collection* and other words of the same general meaning have two different meanings with a very fine distinction between them. This makes a large part of the difficulty of defining a collection, and the non-recognition of this distinction makes a serious stumbling block in the doctrine of cardinal numbers. In accordance with my views of the Ethics of Terminology, I am going to make two new words to distinguish these two meanings. The one I shall call a gath which is simply the word 'gather' with the last syllable dropped. The other I call a sam which is the word 'same' with the last letter dropped. I also like this word because it is so much like the word *sum,* in the phrase *sum total.* It also recalls the German word *Sammlung.* A collection, in the sense of a *gath* is a subject which is a pure Secondness without Firstness, and whose only mode of being is whatever existence it may have; and this consists in the existence of certain other existents, or pure Seconds, called its *members.* Thus, the gath of human beings at this moment in Boston, consists in the existence of this man and that man. No matter how those men might be transformed, no matter if some of them were to leave Boston, that same gath exists, although it would cease to be the gath of the inhabitants of Boston. But were a single member of the collection to cease to exist, that particular gath would no longer exist. There would still be a gath of inhabitants of Boston but it would be a different gath. The description, 'All the inhabitants of Boston' describes a gath. But as time goes on it will describe a different gath. The description 'the inhabitants of Boston' is a proper name. It applies to but a single individual object, the whole of all the inhabitants of Boston. This whole is what I call a *sam* [and] is not exactly a gath; and it is important to get a distinct idea of the difference. Just as the molecules that compose a man's body are continually changing by the loss of some and the gain of others, while there remains the same man, so the population of Boston is every [day] changing, yet remains the same individual whole. I propose to say it is the same *sam.* But it does not remain the same gath. At each instant it is identical with a gath. Always there is a gath in the existence of which consists the existence of the sam of the inhabitants of Boston. Were the city to be devastated and not one inhabitant left, still, as long as it remained Boston, the 'sam, or sum total, of the inhabitants of Boston' would have a *being*, although it would under those circumstances have ceased to *exist*. It would continue to *be*, since the description would retain its meaning. The *essence* of the sam would remain, although its existence had departed. But as for the gath, since it has no other being than existence, and its existence consists in the existence of its members; and since under those circumstances no members would exist, the gath would altogether cease to be. It is important to have this distinction clearly in mind. I do not mean to say that [it] is usually important to hold this distinction clear in regard to any collection that we may happen to speak of; but I mean that for certain purposes it is indispensably necessary. Whatever sam there may be to whose members, and to them alone, any sign applies is called the *breadth* of the sign. This word *breadth*, originating with the Greek commentators of Aristotle, has passed into our vernacular. We speak of a man of *broad* culture. That means *culture* in many fields. *Breadth* of mind is the character of a mind that takes many things into account. If a man has *broad* and *deep* learning, the breadth consists in how many different subjects he is acquainted with, and the depth in how much he knows about whatever subject he is acquainted with. Now the *breadth* of a descriptive appellation has an *essence*, or Imputed Firstness; which is the signification, or *Depth*, of the appellation. Take the word *phenix*. No such thing exists. One naturally says that the name has *no breadth*. That, however, is not strictly correct. We should say *its breadth is nothing*. That breadth is precisely what I mean by a *sam*. Therefore I define a *sam* as an *ens rationis* having two grades of being, its *essence*, which is the being of a definite quality imputed to the sam, and its existence which is the existence of whatever subject may exist that possesses that quality. A *gath*, on the other hand, is a subject having only one mode of being which is the compound of the existence of subjects called the *members* of the *gath*.

You may remark that a *sam* is thus defined without any reference at all to a gath. I repeat the definition, so that you may observe this:

A *sam* is an *ens rationis* whose essence is the being of a definite quality (imputed to the *sam*) and whose existence is the existence of whatever subject there may be possessing that quality.

On the other hand, it is impossible to define a gath without reference to a sam. For when I say that a gath is a subject whose only mode of being is the compound existence of definite individuals called its members, what is the meaning of this *compounded* existence. It is plain that the idea of a compound is a triadic idea. It implies that there is some sign, or something like a sign, which picks out and unites those members. Now the fact that they are all united in that compound is a quality belonging to them all and to nothing else. There is thus here a reference to a possible *sam* which does this. Thus, we might as well at once define a *gath* as a subject which has but one mode of being which is the existence of a *sam*. From this fact, that a *gath* cannot be defined except in terms of a *sam*, it follows that if by a *collection* be meant, as ordinarily is meant, a *gath*; while a *gath* is not distinguished from a *sam*, it becomes utterly impossible to define what is meant by a collection.

This would not be true if the two clauses of the definition of the *sam* were two distinct ideas which have to be put together; but it is not so. Secondness involves Firstness, although it can be discriminated from it; and consequently the idea of the existence of that which has an essence, which is simple Secondness, is decidedly a simpler notion than that of existence without essence, or a Secondness discriminated from Firstness. For it is only by a rectification applied to the former notion that the latter can be attained.

No doubt the easiest way to conceive of the *sam* is to imagine that you have a common noun, without specifying what noun it is, and to think that that noun signifies some quality which is possessed by anything to which it does not apply. Now you are to imagine a single thing which is composed of parts. Nothing is done to these parts to put them into their places in the whole: their mere existence locates them in the whole. Now, think of this rule as describing the whole. If any individual object can properly have that common noun predicated of it, it is a part of the single object called a *sam*; if not, it is not. That gives you the idea of the *sam*. Now to get the idea of a *gath*, you are to consider that those individual objects might change their qualities without losing their individual identity; so that limiting ourselves to any instant, any individual object which at that instant forms a part of the *sam* forever forms a part of an object of which no object not at that instant a part of the sam is a part, and this individual composite whole which has nothing to do with the qualities of its members is a *gath*.

For every *gath* there must be a corresponding *sam*. This is what we should ordinarily express by saying that whatever exists is possible. Or, as DeMorgan puts it, the individuals of whatsoever collection have some quality common to them all that is peculiar to them i.e. possessed by nothing else. Kant, I dare say, would remark that this is a Regulative principle but that it cannot be proved to be a Constitutive principle. That is, it is proper to assume it, but you cannot prove it is so. But I reply that every principle of logic is a Regulative Principle and nothing more. Logic has nothing to do with Existence. And I should add: Herr Professor Dr. Hofrath Kant, permit me to say that in saying this is not a Constitutive principle you speak of *qualities* as if they were existent individuals. A quality has no other being in itself than possibility and to say that a quality is possible is to say that it has all the being that in the nature of things a quality could have. If as you say there *may* be a quality common and peculiar to all the members of a gath, then there certainly is such a quality; and you yourself have in this very same breath described one such quality, in saying that they are all members of the gath in question. So for every gath there is a corresponding *sam*. But it is not true that for every *sam* there is a corresponding *gath*. Since there is the *sam* of the phenix, although it happens not to exist up to date. But there is no such *gath* since there is no phenix. Another point which I observe puzzles the Hon. Bertrand Russell in his 'Principles of Mathematics' is whether a collection which has but a single individual member is identical with that individual or not. The proper answer is that if by a *collection* you mean a *sam*, the *sam* of the sun is not the sun, since it is an *ens rationis* having an essence, while the individual has no essence and is not an *ens rationis*. But if you mean the *gath*, the gath of the sun has no being at all except the existence of the sun which is all the being the individual existent sun has. Therefore, having precisely the same being they are identical and no distinction except a grammatical or linguistic one can be drawn between them. Mr. Russell's being puzzled by this is a good illustration of how impossible it is to treat of philosophy without making a special vocabulary such as all other sciences make. It is, however, far more needed in philosophy than in any other science, for the reason that the words of ordinary speech are needed by philosophy for its raw material.

What has been said of qualities is equally true of relations, which may be regarded as the qualities of sets of individuals. That is to say, if any form of relation is logically possible between the members of two given *gaths*, a relation of that form actually exists between them.

In order to illustrate this principle, I will consider a proposition which the ablest mathematicians have been endeavoring for half a century to demonstrate to be either true or false. The proposition assumes that there are two *gaths*, which we may call the *A*s and the *B*s. And now if we use the phrase a "one to one relation" to denote such a relation that no relate stands in that relation to two different correlates nor any two relates to the same correlate; like the relation of husband to wife in a monogamous country, or better of father to eldest child, then it is assumed that there is no one-to-one relation in which every *B* stands to an *A*. Then the question is whether there is necessarily any one-to-one relation in which every *A* stands to a *B*.

Since there is no one-to-one relation in which whatever *B* that may be stands to an *A*, it must be logically impossible that there should be any such relation. But since the form relation in itself is not absurd, and since in point of fact every *B* does stand in such a relation to some member of a gath; for every *B* stands in the relation of identity to some *B*, it follows that the logical impossibility lies in some existential limitation of the *A*s. However, dropping at the outset the assumption about the one-to-one relation and considering any two gaths, the *A*s and the *B*s, there is a relation in which no relate stands to two different correlates (though two relates may stand in it to the same correlate) in which every *B* stands to an *A*; for the *B*s may so stand to any one *A* [Fig. 1].

Fig. 1

Now setting out with such a relation, let us ask whether by a change of it so as to change each correlate into some other individual of the universe, it is possible so to change it as to make every *A* a correlate of it — Let us trace out the consequences of supposing this not to be possible. Let us under this hypothesis change our relation by a change of correlates so as to render it a one-to-one relation, and let us suppose this to be done in every possible way so as to give a vast multitude of different one to one relations. Among these different relations there must be some all of whose correlates are *A*s because every possible variation of our relation occurs, and they could not all agree in any respect unless it were logically compelled. But since we are going on the hypothesis that every one of these relations leaves some *A* that is not a correlate, it follows that there is nothing to compel any correlate to be non-*A*; for the only logical requirement is that no two correlates shall be identical and since even when we do not care whether they are so or not there are *A*s that are non-correlates, it follows that there is nothing to compel any correlate to be a non-*A*, since there are *A*s that it may be. But there cannot be any compulsion that does not affect any individual, and therefore there is nothing to compel correlates to be non-*A*s and consequently in some cases none will be non-*A*s. But now if we bring in the assumption that there is no one-to-one relation in which every *B* stands to an *A*, this result is rendered impossible, and with it disappears the hypothesis of which it is a necessary consequence; namely, that there is no relation in which no relate has two correlates and in which some *B* stands to every *A*. This hypothesis being overthrown appears then that there is such a relation in which no *B* stands to two *A*s yet of which every *A* is correlate to some *B*. But we have only to exclude from this all but one of the relations of different *B*s to the same *A* for every *A* to reduce its converse to a one-to-one relation of every *A* to a *B*. Thus we have proved that if there is no one-to-one relation of every *B* to some *A* there is a one-to-one relation of every *A* to some *B*.

Let us now look into another problem. Dr. Georg Cantor has a wonderful system of Ordinal Numbers which I must explain to you. It begins with the ordinary ordinal numbers

$$1,\ 2,\ 3,\ 4,\ 5,\ 6,\ 7,\ 8,\ 9,\ 10,\ 11,\ \ldots$$

That series is endless. That is to say, it has no last. But neither in logic nor in fact does that interfere with its being gone completely through provided the objects numbered are not in the nature of existences. When for instance Achilles runs after the Tortoise and we choose to divide his course into a first stage in which he goes to the place where the tortoise was at first, a second stage in which he goes to the place where the tortoise was at the end of the first stage, a third stage in which he goes to the place where the tortoise was at the end of the second stage, and so on, the series of stages is endless. But nothing in logic, or in fact, prevents Achilles from accomplishing them all, since they are only possibilities. So then there will be other ordinal numbers, after the first endless series is gone through. And the first of these Cantor denotes by ω then every ordinal number having one next after it, the series runs

$$\omega+1,\ \omega+2,\ \omega+3,\ \omega+4,\ \omega+5,\ \ldots$$

and after this endless series is accomplished there will come $\omega + \omega$ which he calls $\omega\cdot 2$. Then according to the rule must come

$$\omega\cdot 2,\ \omega\cdot 2+1,\ \omega\cdot 2+2,\ \omega\cdot 2+3$$
$$\omega\cdot 3,\ \omega\cdot 3+1$$
$$\omega\cdot 4$$
$$\omega\cdot 5 \qquad \text{etc.}$$

This series $\omega, \omega\cdot 2, \omega\cdot 3, \omega\cdot 4, \omega\cdot 5$ is endless and after it is accomplished we shall have $\omega\cdot\omega$ or ω^2. Then

$$\omega^2,\ \omega^2+1,\ \omega^2+2$$
$$\omega^2+\omega$$
$$\omega^2+2\omega \qquad \text{and so on}$$
$$\omega^2+\omega^2=\omega^2\cdot 2 \qquad \text{and so at length } \omega^3,\ \omega^4,\ \omega^5$$

An endless series, after which ω^ω. Then there will be

$$\omega,\quad \omega^\omega,\quad \omega^{\omega^\omega},\quad \omega^{\omega^{\omega^\omega}}$$

Cantor gives no other way of writing these. But suppose that instead of ω^2 we write $\omega(1,2)$. For ω^3, ω^4 we write $\omega(1,3)$, $\omega(1,4)$. For ω^ω, ω^{ω^ω} etc. we write $\omega(1,1,2)$, $\omega(1,1,3)$. So that $(\omega^{\omega^{\omega^\omega}})^5\cdot 6$ will be written $\omega(6.5.4)$. Then we shall at length get $\omega(1,1,\omega)$ [and] may write $\omega(1,1,1,2)$. So we shall get $\omega(1,1,1,1,2)$, $\omega(1,1,1,1,1,2)$, $\omega(1,1,1,1,1,1,2)$ and after all these we may put $\omega((1),2)$.

Then we shall have at length $\omega((1),(1),2)$ and so forth and then

$\omega(((1))2)$ and so $\omega((((1)))2)$
$\omega(((((1))))2)$

and after that we may put $\omega[2]$. In short, there will be no end to the need of new symbols. It all follows from two principles, 1st, every number has another number next after it. 2nd, every endless series of numbers accurately describable in any manner whatever has a number next after it.

Cantor calls such a series a *wohlgeordnet* series. But I propose, in admiration of the genius that has discovered it, to call it a *Cantorian succession.*

Cantor has conjectured that, given any collection whatever, any universe you please of independent quasi-individuals, there is a relation (and if one of course innumerable such) that in passing from relate to correlate and from correlate to correlates correlate, this relation arranges the whole universe in a Cantorian collection.

Indeed, Cantor puts forward this as more than a conjecture, — as a consequence of an unacknowledged law of thought. But the proposition has been received by mathematicians with the gravest doubt.

If I have time, I will say more about this very important question later; but at present I will only say this. The relation is *per se* possible. It only supposes that certain individuals shall have special one-sided connections. Now every two individuals have special one-sided connections.

Neither can any limitation of existence render this form of relation self-contradictory. Therefore, Cantor is right. This cannot be clear to you. It is the merest hint. But I might give several lectures elucidating this matter. I must hurry on to other things.

Having made it clear what a collection is, the next thing I have to do is to define *Multitude*. If we regard plural nouns, such as *men*, *horses*, *trolley-cars*, as names describing *sams men* meaning a sam of which any member is a man, etc. then the adjectives *two*, *ten*, *myriad*, *innumerable*, express qualities of collections of a certain class, and any quality of this class is a *multitude*. Of course, this is not a definition. It is only the framework for a definition. In order to define *multitude*, it is necessary to begin by analyzing our meaning when we say that one collection is greater than another. This analysis was first published in a posthumous work of Bernard Bolzano, which appeared in 1851 and has since been reprinted *Paradoxien des Unendlichen*. Bolzano was a catholic theologian of Hungary and the author of a logic in four volumes. He was far too clearheaded to escape persecution in his position, you may be sure. Bolzano's definition amounts to this. If one *gath*, say that of the *B*s, is so related to another *gath*, say the *A*s, that there can be no one-to-one relation in which every *B* stands to an *A*, then, and only then, the gath of the *B*s is *greater* than that of the *A*s, and the latter is *less* than the former. For example there is no possible one-to-one relation in which every one of five things stands to one of three things [Fig. 2]. 60 [ways] in all.

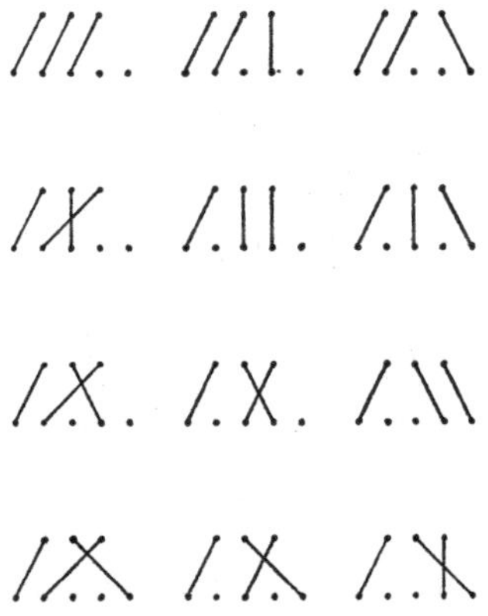

Fig. 2

On Bolzano's definition of greater and less can be based a definition of Multitude. That is the *multitude* of a collection will be defined as consisting in its being greater than those collections that [it] is greater than and being less than those it is less than. This definition is at least a practical one. It can be put to use in demonstrations. In that respect it is infinitely better than what Cantor puts forth as a definition in the style of Kant, who was a superb logical genius but who never studied logic and therefore sometimes commits grievous faults of logic. Namely Cantor tells us that if we think of a collection and then put out of view this and that element of our idea, what remains is multitude. But that is no definition at all. It is an experiment telling us, granting for the sake of argument that it obtains its aim, how to find and identify multitude. A definition ought to be an analysis of a conception useful in demonstration. The purely relative definition based on Bolzano's definition of greater and less does that. But it takes too narrow and formal a view of the matter. It fails to show the immense logical significance of *multitude*. It also makes it to be purely relative between one collection and others. That is not true. The multitude of a collection would be just what it is if there were no other collection in the world. Let me offer a definition. The multitude of a *gath*, — remember, it is a pure gath and the special Firstnesses or *flavors* of the characters of its units and sets of units is put out of the question ,— consists in the complexity of the relations existing between its units in virtue of its peculiar character purely as a gath. But what is meant by the complexity of relations? What does it consist in? In order to bring this home to you I compare a greater collection with a smaller. Among the units of any gath of multitude five or more there is the relation shown in this graph [Fig. 3].

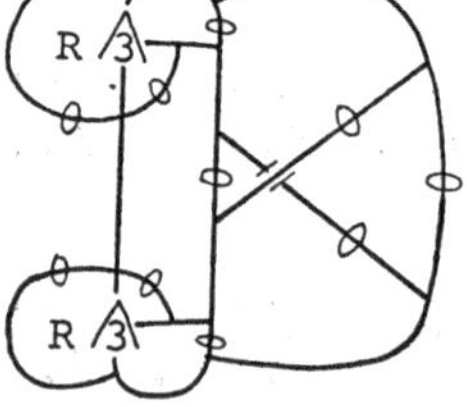

Fig. 3

In no smaller gath is such a scheme possible. It is in the possession of such schemes that multitude consists. But every such scheme holds for every gath, except for the cuts that ligatures traverse. That is to say every form of relationship exists between the members of all gaths alike, except in respect to the identity and diversity of the correlates. We may go further. Every scheme holds for every gath if there are no cuts. Therefore multitude consists in the complexity of indefinite relationships between the units of a gath as such. But this complexity consists solely in the diversities of correlates. Hence the multitude of a gath consists in the othernesses among its units that the multitude permits. This is so obvious that it is truly wonderful that nobody has ever said it before this minute, when I say it to you. On the other hand, the smallness of a collection expressed by enclosing such a scheme in a cut, consists in the necessitation of identities.

Now some logicians look upon every proposition as expressing a description of identity. That may be a contorted conception of a proposition, but it is not false. Adopting it for the moment, we see that the smallness of a collection compels descriptions of identities that in a larger collection are not compelled. But this is as much as to say that propositions follow as necessary consequences from premisses relating to members of smaller collections when in a larger collection no such conclusion would follow. If one collection is larger than another this will invariably be true. From this point of view you see the immense logical significance of multitude.

For instance, suppose you want to find out who committed certain murders, who some Jack the Ripper is. Plainly if you can make sure that he belongs to a certain small collection of persons your investigation may be greatly helped. You have ways of reasoning open to you that you had not before. Indeed if you can reduce the multitude of that collection to *unity*, your problem is solved.

Let us see then what all the different grades of *multitude* are. The lowest is *unity* for *Nullity* is not a multitude. *Unity* is that quality of a collection which consists in the absence of all diversity of its members. Add a single diverse individual to a collection of one, and you get a larger collection. Let me prove this by graphs. In each graph which I write, the universe of discourse will be the members of the *gath* I am dealing it. Then nullity will be expressed by [Fig. 4] which is of course

Fig. 4

the pseudograph. For a universe with nothing in it is absurd. I add one to this. That is I express that there is something such that if anything is different from this the graph of nullity holds for it. That is [Fig. 5] or [Fig. 6]. It is the graph of *Unity*. I add a unit to this. That

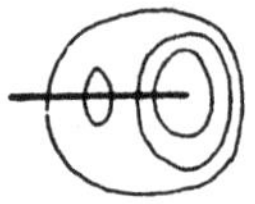

Fig. 5

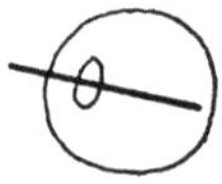

Fig. 6

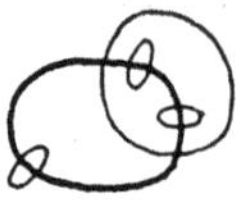

Fig. 7

is I express that there is something other than what the graph of unity asserts, for all that is not this that graph holds. This gives [Fig. 7]. It is the graph of Twoness. I again add a unit, asserting that there is something else but that apart from this new unit the graph of Twoness holds. This gives [Fig. 8] which is the graph of Threeness. In order to express that the gath taken as the universe is greater than a given number you must assert all the diversity that the graph of that number asserts

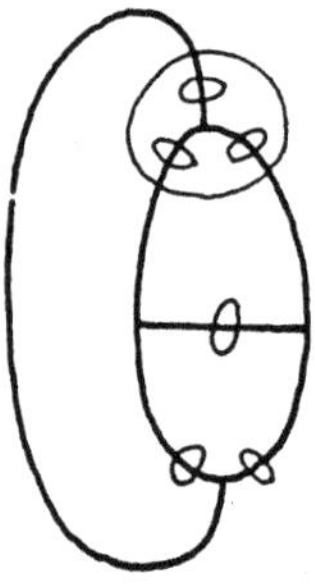

Fig. 8

Fig. 9

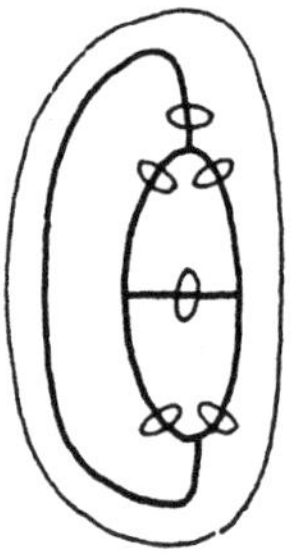

Fig. 10

and also the diversity it denies. That is, we must remove the large cut. Then in order to assert that our gath is not more than the number we must enclose the last graph in a cut. Thus to express that that *gath* in question is more than three we have the graph [Fig. 9]. To assert that it is not more than three we scribe [Fig. 10]. To assert that it is neither three nor more than three, that is that it is less than three, we scribe [as in Fig. 11 and Fig. 12]. We now iterate the right hand graph in the second cut of the left hand one thus [Fig. 13]. We extend the ligature by the rule of Iteration and join them inside 3 cuts by the Rule of Insertion. We thus get, after erasing the left hand part, [Fig. 14]. We now de-

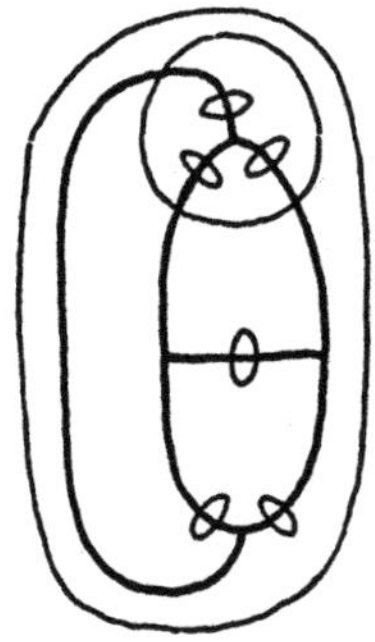

Fig. 11

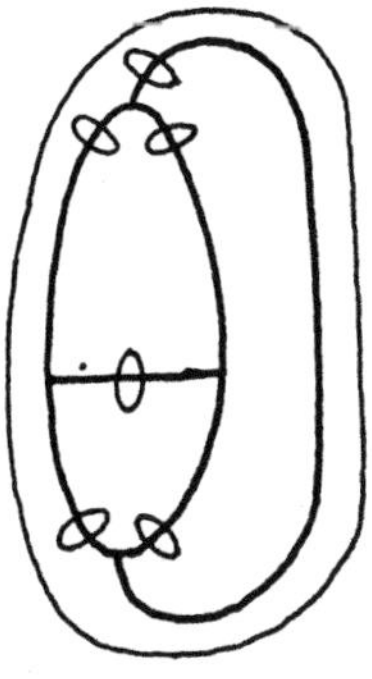

Fig. 12

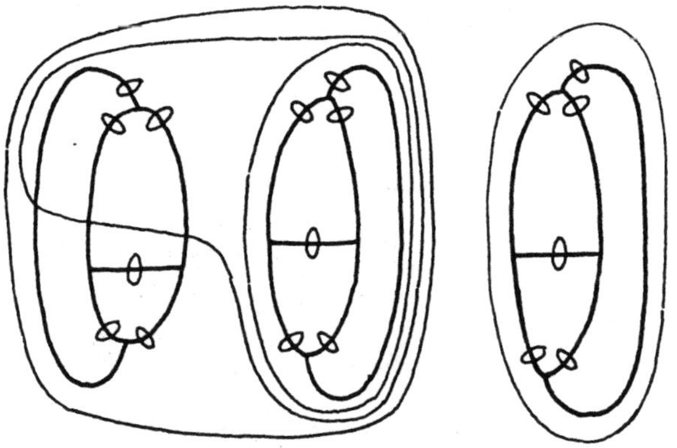

Fig. 13

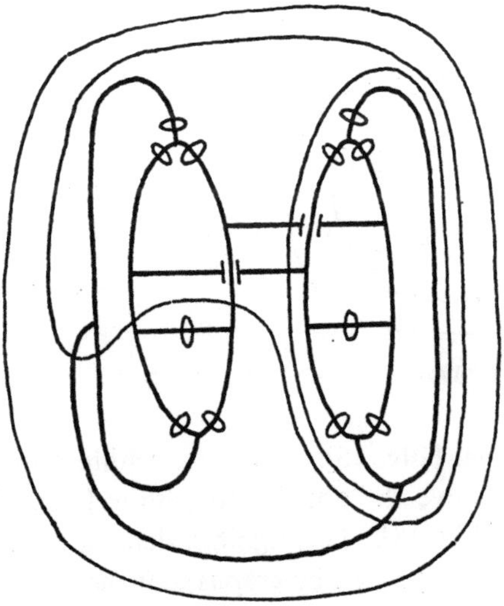

Fig. 14

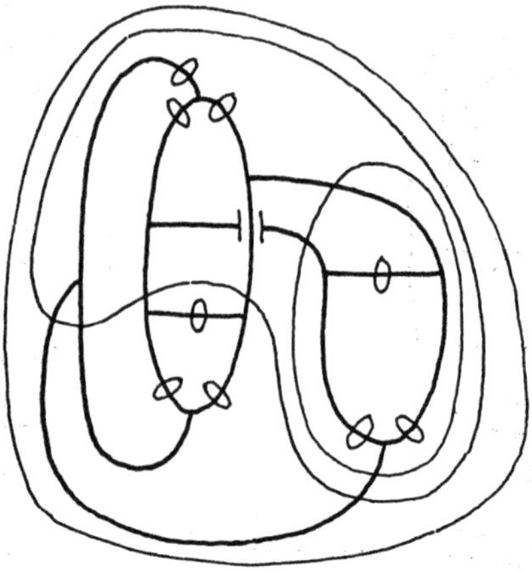

Fig. 15

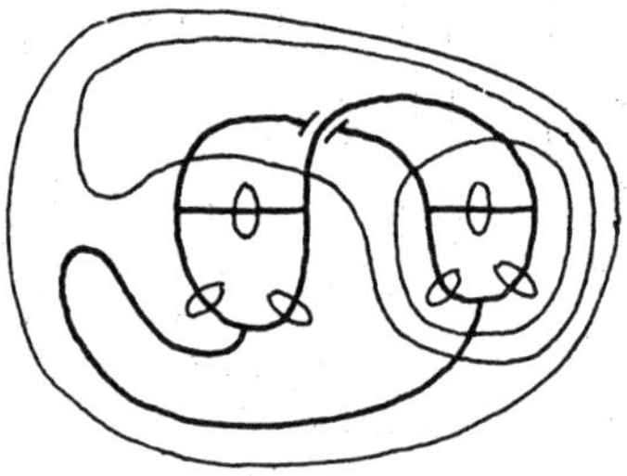

Fig. 16

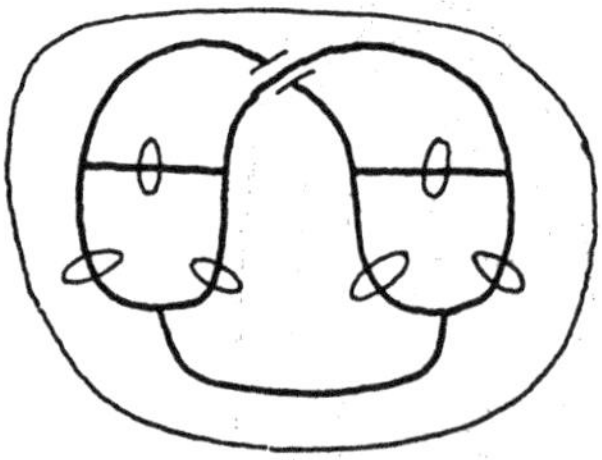

Fig. 17

Fig. 18

iterate so as to get [Fig. 15]. We erase within two enclosures giving [Fig. 16]. We remove the double cut getting [Fig. 17] and finally we deiterate getting [Fig. 18]. This signifies that the gath is not more than two. That is we have proved by graphs that there is no multitude intermediate between 3 and 2.

I am sorry there is no time to go further into the graphical treatment, because it becomes more and more interesting as we go on, and the results are very important. For instance, in order to prove by graphs that the Syllogism of Transposed Quantity holds good for all finite multitudes, since it is evidently impossible to scribe the graphs of all enumerable multitudes it is necessary to scribe a graph which shall describe what a graph of any enumerable multitude is like. The idea of a graph of a graph is in itself interesting. But it becomes far more so when we see what the character of that graph must be. In that way it can be proved what is, I confess, pretty plain, without the graphs that the smallest multitude that is greater than all finite multitudes is the multitude of a sam for which there is *some* relation, and some individual unit of the *sam*, such that whatever quality there may be that is possessed by that particular *unit*, that quality being of such a nature that if it be possessed by any *unit*, say M, it is necessarily possessed by a *unit* in that particular relation to the unit M, — any quality I say of which these two propositions hold is necessarily possessed by every member of the collection. This is the property of the denumeral multitude which is made use of in Fermation reasoning, a mode of reasoning that was invented by Pierre de Fermat, who was born in 1601 and died in [1665], and who was probably the greatest mathematical genius that ever lived. He also invented the mode of reasoning of the differential calculus. Fermatian reasoning has to be used in order to prove, rigidly *prove*, almost any proposition about whole numbers. For example, in order to prove $X+Y = Y+X$ whatever whole numbers X and Y may be, we first prove that it is so when $Y = 1$. That is we first prove $X+1 = 1+X$. To prove this we begin by remarking that it necessarily is so when $X = 1$. For then it is merely $1+1 = 1+1$. Now suppose it to be true of any whole number N that $N+1 = 1+N$. Then it is true also of $1+N$. That is $(1+N)+1 = 1+(1+N)$. For $(1+N)+1 = 1+(N+1)$. I suppose that to be proved already. And since $N+1 = 1+N$, $1+(N+1) = 1+(1+N)$. So then $X+1 = 1+X$ if true for any one value of X is true of the next greater value. But it is true when $X = 1$. Hence it is true for all values. Now then suppose $X+Y = Y+X$ is true when Y has a particular value say N. That is suppose $X+N = N+X$. Then it will also be true when $Y = 1+N$. That is $X + (1+N) = (1+N) + X$. For $X + (1+N) = (X+1) + N$ and $X+1 = 1+X$ as we have just proved so that

$$X+(1+N) = (X+1)+N = (1+X)+N = 1+(X+N)$$

But we are going on the supposition that $X + N = N + X$ so that $X+(1+N) = 1+(N+X) = (1+N)+X$. So the proposition $X+Y = Y+X$ if true when Y has any particular value is true when it has the next greater value. But it is true when $Y = 1$ whatever be the value of X. Hence it must be true for all values of X and Y.

We call a collection of the members of which the Syllogism of Transposed Quantity necessarily holds an *enumerable* collection and one of which this is not true, or where there is a *Fermatian* relation, a *denumeral* collection — in German *abzählbar*. It follows that it [i.e., the syllogism] is true of every *enumerable* collection.

All the whole numbers, properly so called, that is, all numbers of which our system of so called Arabic notation affords a definite symbol, — all the numbers up to any one form an enumerable collection. But the entire collection of whole numbers capable of representation in that system is a collection not enumerable, but innumerable. And the single multitude of all the whole numbers, or of any such endless series of which all the members up to any member form an enumerable collection is called the *denumeral* collection, "abzählbar" in German. The denumeral multitude is a single multitude of the class of innumerable multitudes.

I was saying that all collections less than a given collection can be reasoned about in ways that greater collections cannot. The mode of reasoning which is applicable to all enumerable collections, but to no others was first discovered, or first put into a logical treatise by Augustus DeMorgan. He called it the syllogism of transposed quantity. Here is an example:

Every dollar borrowed requires a borrowed dollar to be paid out
No borrowed dollar paid out covers more than one dollar borrowed
∴ Every dollar borrowed has to be paid out.

Now I need not tell you that sundry business schemes are based on the idea of evading this conclusion. The notion is that dollars borrowed can partly be spent and can all be paid out by means of new dollars borrowed. It was at the great fire of 1835 in New York, if I remember rightly that this scheme first emerged. All the insurance companies declared themselves bankrupt except only the Etna of Hartford which advertised that it would pay all losses. It had not the money to do so; but its advertisement brought so many new policies that it did pay, and laid the foundations of a great future. Paying out of the increase of one's business is, of course, fatal in the long run. And yet after all that is what we are all doing. A young man marries without having the money in the bank to support a dozen children. It is not in accordance with sound business; but the human race would progress little if they did not bank upon the future. But restricting myself to teaching logic, I limit myself to saying that the syllogism of transposed quantity is valid for every *enumerable* universe and for such alone. Here is a specimen of this reasoning applied to a question of philosophy:

No second of past time was immediately followed by two different seconds of past time.

But the last second of past time has no second of past time immediately following it.

Hence there must have been some second of past time that did not follow any second, that is time must have had an absolute beginning.

Whatever *definitely exists or has existed forms an enumerable collection, and the reasoning about time must be accepted* so long as you suppose the passage of seconds to be so many events that actually did take place. The past and the future stand upon an altogether different footing in this respect.

The entire collection of all whole numbers is not enumerable, since if we prefix *zero* to them, the resulting series has zero for its first, one for its second, two for its third member and so on, and has no capacity for diversity greater than that of the collection of whole numbers without the zero.

From a denumeral collection, such as that of the whole numbers, you may take away, not merely an enumerable part, but a denumeral part, equal to the whole, without thereby making the whole less.

Thus from all numbers, take away all the odd numbers, and the remaining collection of even numbers is equal to the whole, since corresponding to every number N there is a distinct and separate even number, its *double*. So likewise the multitude of powers of ten, or numbers written by a single figure 1, followed by zeros 1, 10, 100, 1000, 10000, etc. is just as great as that of all whole numbers.

The multitude of rational fractional values is denumeral, since any one may be reached in one way of writing them one by one. Namely, we begin with writing

$$\frac{0}{1} \qquad\qquad\qquad\qquad \frac{1}{0}$$

which are not properly rational fractional values, and now we proceed by the following rule. Having gone through the row, go through the whole row again, and insert between each successive pair of fractions a fraction having for its numerator and denominator respectively the sum of the numerators and the sum of the denominators of the two fractions

$\frac{0}{1}$ $\frac{1}{1}$ $\frac{1}{0}$

$\frac{0}{1}$ $\frac{1}{2}$ $\frac{1}{1}$ $\frac{2}{1}$ $\frac{1}{0}$

$\frac{0}{1}$ $\frac{1}{3}$ $\frac{1}{2}$ $\frac{2}{3}$ $\frac{1}{1}$ $\frac{3}{2}$ $\frac{2}{1}$ $\frac{3}{1}$ $\frac{1}{0}$

$\frac{\overline{1}}{1}$ $\frac{1}{4}$ $\frac{1}{3}$ $\frac{2}{5}$ $\frac{1}{2}$ $\frac{3}{5}$ $\frac{2}{3}$ $\frac{3}{4}$ $\frac{1}{1}$ $\frac{4}{3}$ $\frac{3}{2}$ $\frac{5}{3}$ $\frac{2}{1}$ $\frac{5}{2}$ $\frac{3}{2}$ $\frac{4}{1}$ $\frac{1}{0}$

$\frac{0}{1}$ $\frac{1}{5}$ $\frac{1}{4}$ $\frac{2}{7}$ $\frac{1}{3}$ $\frac{3}{8}$ $\frac{2}{5}$ $\frac{3}{7}$ $\frac{1}{2}$ $\frac{4}{7}$ $\frac{3}{5}$ $\frac{5}{8}$ $\frac{2}{3}$ $\frac{5}{7}$ $\frac{3}{4}$ $\frac{5}{5}$ $\frac{1}{1}$ $\frac{5}{4}$ $\frac{4}{3}$ $\frac{7}{5}$ $\frac{3}{2}$ $\frac{8}{5}$ $\frac{5}{3}$ $\frac{7}{4}$ $\frac{2}{1}$ $\frac{7}{3}$ $\frac{5}{2}$ $\frac{8}{3}$ $\frac{3}{1}$ $\frac{7}{2}$ $\frac{4}{1}$ $\frac{5}{1}$ $\frac{1}{0}$

[The next insertions between fractions from $\frac{0}{1}$ to $\frac{1}{1}$ are:]

$\frac{1}{6}$ $\frac{2}{9}$ $\frac{3}{11}$ $\frac{3}{10}$ $\frac{4}{11}$ $\frac{5}{13}$ $\frac{5}{12}$ $\frac{4}{9}$ $\frac{5}{9}$ $\frac{7}{12}$ $\frac{8}{13}$ $\frac{7}{11}$ $\frac{7}{10}$ $\frac{8}{11}$ $\frac{7}{9}$ $\frac{5}{6}$

[The insertions between these fractions from $\frac{0}{1}$ to $\frac{1}{1}$ are:]

$\frac{1}{7}$ $\frac{2}{11}$ $\frac{3}{14}$ $\frac{3}{13}$ $\frac{4}{15}$ $\frac{5}{18}$ $\frac{5}{17}$ $\frac{4}{13}$ $\frac{5}{14}$ $\frac{7}{19}$ $\frac{8}{21}$ $\frac{7}{18}$ $\frac{7}{17}$ $\frac{8}{19}$ $\frac{7}{16}$ $\frac{5}{11}$ $\frac{6}{11}$ $\frac{9}{16}$ $\frac{11}{19}$ $\frac{10}{17}$ $\frac{11}{18}$ $\frac{13}{21}$ $\frac{12}{19}$ $\frac{9}{14}$ $\frac{9}{13}$ $\frac{12}{17}$ $\frac{13}{18}$ $\frac{11}{15}$ $\frac{10}{13}$ $\frac{11}{14}$ $\frac{9}{11}$ $\frac{6}{7}$ [etc.]

between which it is inserted. Thus we get successively [the rows of the accompanying table].

What is true of all fractional values is equally true of all fractions. They can also be written down in the order of their values one by one, and that without even performing any addition.

The entire collection of all the objects of thought however minute to which men's attention has ever been attracted or ever will have been attracted is enumerable. It is a definite finite number. The entire collection of all the exact objects of thought that ever will interest men in the endless future is *denumeral*.

But we shall soon see how inexpressibly few are the objects of a mere denumeral collection compared with the infinitely overwhelming majority of innumerable multitudes.

Now in order to make out what the higher multitudes are, it will be necessary for me to prove to you a proposition which I was the first to enounce and to prove. The proposition is this.

Take any *gath* whatever, whether it be *enumerable* or *innumerable*, and call its units the *M*s. Now I am going to describe a sam which I call the *exponential* of the *M*s. Every member of this sam called the exponential of the *M*s is itself a sam composed exclusively of *M*s and this exponential sam comprises among its members *every* sam that is composed exclusively of *M*s.

For example suppose the *M*s are only four, say M_1, M_2, M_3, M_4. Then the exponential of the *M*s will be the sam whose single members are these sams

1	2	3	4	5
0	M_1	M_2	M_3	M_4

6	7	8	9	10	11
M_1M_2	M_1M_3	M_1M_4	M_2M_3	M_2M_4	M_3M_4

12	13	14	15	16
$M_1M_2M_3$	$M_1M_2M_4$	$M_1M_3M_4$	$M_2M_3M_4$	$M_1M_2M_3M_4$

Now my proposition is that, in every case, the multitude of the exponential is greater than that of the primitive gath. That is, no matter what the *M*s may be they are always fewer than all the possible sams of *M*s. That is it is impossible to find any one-to-one relation such that every sam of *M*s stands in that relation to an *M*. How am I going to prove this? I do it in this way: I say, imagine any one-to-one relation you please, and call this the relation r. Now I will exactly describe a sam of Ms which does *not* stand in that relation, r, to any M whatsoever. You might suppose it would be difficult to describe so exactly a sam of Ms for which this relation fails when we know nothing at all about this relation except that it is a one-to-one relation called r. Nevertheless I will describe a sam of Ms which does not stand in the relation r to any M at all. I will call it my *test sam*. This test sam shall be composed as follows. Whether or not it includes among its members any particular M, say M_x, depends upon what kind of a sam of Ms it is that is in the relation r to M_x. It may be that there is no sam of Ms in the relation r to M_x. In that case, I do not care whether M_x be included in my test sam or not. But if there be a sam of Ms that stands in the relation r to M_x then if this sam of Ms includes M_x, M_x shall be excluded from my test sam, while if the sam of Ms that stands in the relation r to M_x does not include M_x as a member of it, M_x shall be a member of my test sam. That describes precisely my test sam except as regards certain possible Ms which may be all included or all excluded or some included and some excluded as you please.

Anyway I say that this test-sam does not stand in the relation r to any M. For take any M you please, I care not what. Call this M_y. Now M_y belongs to one or other of three classes; namely 1st, the class of Ms which I positively require my test sam to include, or 2nd, the class of Ms which I require my test sam to exclude, or 3rd, the class of Ms as to which I do not care whether they be included in the test sam or not. If M_y belongs to the first class, it is one of the Ms that I require my test sam to contain, but those where the Ms that were not contained in the sams of M that were r to them. Plainly then if M_y belongs to this class my test sam is not r to it. For my test sam, in this case, contains M_y, while the only sam of Ms that is in the one-to-one relation r to M_y does not contain M_y. So this sole sam of Ms that is in the relation r to M_y is *not* my test sam. If M_y however belongs to the second class, it is one of those Ms which I expressly exclude from my test sam. It is however contained in the only sam of Ms that stands in the relation r to it. Plainly, then, this sam that alone is in the relation r to M_y if M_y belongs to the second class is not my test sam. The only remaining possibility is that M_y belongs to the third class. But in that case it is one of these Ms to which no sam of M stands in the relation r; so that my test sam cannot stand in that relation to it. Thus you perceive that to whatever class M_y belongs my test sam cannot stand in the relation r to it. That is, as I said, I have specified a sam of Ms that does not stand in the one-to-one relation r to any M, no matter what that one-to-one relation may be. But that is as much as to say that the exponential of the Ms or the sum of all sams of Ms is more multitudinous than the gath of Ms.

Therefore, the collection of all possible collections of whole numbers is greater in multitude than the whole collection of whole numbers. I will show you an instance of such a collection. You know the secundal system of arithmetical notation, I am sure. It runs like this

1	one
10	two
11	three
100	four
101	five
110	six etc.

So for the fractions

.1	is one half
.01	is one fourth
.11	three fourths
.001	one eighth
.011	three eighths
.101	five eighths
.111	seven eighths

and so on. Now consider all the values to which such fractions carried out indefinitely far could indefinitely approximate to.

The fractional places, being numbered correspond to the whole numbers. Those of these places that have *ones* in them are those which make up the whole collection of these places used to express the quantity. You see instantly, I am sure, that the entire collection of values to which such fractions can indefinitely approximate precisely corresponds to *all* possible collections of whole numbers. Such a collection I called in my paper *abnumeral.* But that is a poor word, and since it has not yet been taken up that I am aware of I will call it the *first ultranumerable* multitude. Cantor calls it [$\aleph_1$] aleph [1], though he does not define it in the same way. It is the multitude of all possible points that could be marked on a line without putting any two at an infinitesimal distance from one another. It is the multitude of all the quantities that the differential calculus and theory of functions consider.

But now consider the multitude of all possible collections of such objects. That will be the *second ultranumerable* multitude, and there will be an endless series of infinite multitudes each greater than the last.

This series of multitudes is endless. That is there is no *largest*. But that does not in the least prevent there being a multitude greater than them all, just as the multitude of all whole numbers is greater than each whole number though there is no largest whole number. What then would come next? Collections of each of these multitudes is possible, notwithstanding that the least of the ultranumerables far exceeds the multitude of all the atoms in the universe. Still a collection of any ultranumerable multitude not only is possible but once it *really IS*, in the sense in which any plural is apart from its having ever been thought of as such. Now, then, let collections of all these multitudes be aggregated. What sort of a collection will that be? It is easy to answer that question infallibly, though not so easy to comprehend the answer. The aggregate of those collections would falsify my proposition that the collection of collections of the *M*s is greater than the *M*s. It would not be so for such an aggregate. But that has been proved to be necessarily true for all collections. Then such an aggregate would not be a collection. It would be *too many* to be a collection. How so? I haven't time to show you; but I will tell you. It is because you have then so crowded the field of possibility, that the units of that aggregate lose their individual identity. It ceases to be a collection because it is now a continuum.

I regret exceedingly that I cannot give you a lecture on continuity because my discoveries in that field are of quite a different order of importance than those about multitude. I foresee that I shall die without getting my say said, although I strain every nerve in my work. The world will have to wait, I fear, a good while for the explanation of continuity if my work is not made public.

In the few minutes that remain I give you one or two scraps; but they are perfectly useless without the whole. One might as well, when a battle-ship is wanted, bring a quarter of a pound of boiler-iron.

I have in these lectures no concern with metaphysics. That is not saying that what I say has no importance for metaphysics. But for present purposes I care not what the real nature of time and space may be. But what I mean by a truly continuous line is a line upon which there is room for any multitude of points whatsoever. Then the multitude or what corresponds to multitude of possible points, — exceeds all multitude. These points are pure possibilities. There is no such gath. On a continuous line there are not really any points at all. Two lines which intersect, intersect in a point. That is true for the intersection breaks the continuity and makes a point where there was none before the intersection.

E. [LECTURE V] (471)

I have now said all I need say about rationals. It is plain that in order to consider Analytic continuity I must take up irrationals. I shall not need to consider any other ordinal. In order to consider true continuity I must consider the doctrine of multitude; and it will be more convenient to consider multitude first and irrationals afterward.

But at the very threshold of multitude, I am met by a great logical difficulty. For the whole doctrine of Multitude is founded upon a conception of the relation of one collection's being greater than another which is by no means the common-sense idea of that relation and which was first given to the world by Bernard Bolzano, who was at once a logician and a catholic theologian, a combination of specialties pretty sure to lead to grave personal inconvenience, as it did in his case and has done in most cases where the logician has attempted to advance his science. The common sense idea of being greater than is that if one collection contains a member representative of every member of another collection and more beside it is greater than that other; and consequently an infinite collection is greater than itself. But Bolzano's definition does not allow any collection to be greater than itself, if there are two collections the *A*s and the *B*s and if there is any possible relation, *r*, in which every *A* stands to a *B* to which no other *A* is *r*, then, says Bolzano, we will say that the collection of *A*s is *not greater* than that of the *B*s. If there be no possible relation of that sort the collection of *A*s is greater than that of the *B*s. But, thereupon, there at once arises the question, why may not two collections each be greater than the other? Now if that be possible, the idea of magnitude breaks down. There are logico-mathematicians who think it is possible.

But there are others, Cantor among them, who think not. To me the notion that every one-to-one relation between *A*s and *B*s should leave over some *A*s unrelated to any *B*s although there are *B*s to which no *A*s are related, and that there should be no relation whatever in which these unoccupied As could stand to those unoccupied Bs is contrary to the nature of *relation* and of *possibility.*

If by the *possible* you mean no more than that which you do not know to be false, then everything in mathematics not already demonstrated to be false is possible. No doubt, that logical situation is the basis of the conception of possibility. But so defining the possible a relation of the kind in question is possible until you have demonstrated that the collections of As and of the Bs have definite characters which conflict with the supposition that there is such a relation.

But I think it would be more convenient to say not that everything in mathematics about which you happen to be ignorant is possible but that only that is possible of whose falsity no demonstration ever will be given. The question then would be whether anything can render it demonstrable that the As cannot be put into a one-to-one relation to Bs except by showing that there are no Bs for them to be in one to one relation to. Now it is plain there cannot be such a demonstration.

It cannot be said that this definition of possibility is satisfactory.

Not to trouble you with logical subtleties where I can avoid it, let us say that an assertion is *general* if no amount of actual fact could certainly fulfill the truth of it. It says you *never will* find this falsified. The future never will have been exhausted. A general assertion, leaves it to the hearer to take any instance he can. To say "all men are mortal" is as much to say, you can take any man you please, and experience will prove he is a liar.

Of an opposite kind are indefinite assertions, where the speaker does not fully commit himself as to what is the instance he is speaking of. As the *general* is that which no amount of fact can constitute false; for its asserter only says the future will produce a case of such and such a kind.

Those two definitions will be found helpful in discussing the question whether or not two collections can be each greater than the other. But first we must answer the question, What is meant by a *collection*? I answer, It is a whole of ultimate parts which are *discrete* objects. What, then, is meant by discrete objects? I answer, objects which, whether it be in human power to distinguish them, are supposed to be in themselves of such a nature as to be susceptible of being so described that every one is distinguished from every other. In other words by *discrete* objects are meant individuals of a system of *definite* individuals — Definite each in its own nature although we may not be able to define them singly.

That being the case, I assert that given any definite collection whatsoever, there is some general character which belongs to all its members and belongs to nothing else, a character common and peculiar to the members of that collection. For if it were not so, the collection would not be a definite collection. If it were not so, you could instance such a collection. "Very well" you would say, "look there!" Look where, I should naturally ask; and you would have either to describe the collection unmistakably when your description would involve a common and peculiar character of its members, or you would have to *do* something bringing that collection and no other to my attention when you would *create* such a character. If you object that that character would not have existed before you had created it, I should reply that all general characters refer to what will take place in the future.

Now relations are characters of sets of objects; and therefore the proposition just enunciated includes the proposition that there is dyadic relation common and peculiar to all the pairs of objects of a collection.

[Notes]. The proof that two collections cannot be each greater than the other.

Then proof that $2^m > m$.

No maximum multitude

We now go to irrationals. These are the limits of converging series of rationals. It is easily proved that every positive real irrational is expressible to any desired degree of approximation though never exactly by a secundal fraction.

Expressions of Π and ϐ in secundals.

More zeros than units.

The multitude of irrationals nothing but the first abnumerable multitude.

Analytic continuity thus differs from true continuity most markedly.

F. LECTURE VI [PROBABILITY] (472, VOLUME 2)[1]

Now there are three characters which mark the universe of our experience in a way of their own. They are Variety, Uniformity, and the Passage of Variety into Uniformity. By the Passage of Variety into Uniformity, I mean that variety upon being multiplied almost in every department of experience shows a tendency to form *habits.* These habits produce statistical uniformities. When the number of instances entering into the statistics [is] small compared with the degree of their variation, the law will be extremely rough, but when the number runs up into the trillions, that is to say cubes of millions, or much higher, as in the case of molecules, there are no departures from the law that our senses can take cognizance of. You will find one such rough uniformity illustrated in two maps in "Studies in Logic by Members of the Johns Hopkins University." Here is another that I have dug out of our last Census on purpose for this lecture. The comparison is of the number of deaths per thousand of the population in the years 1890 and 1899 respectively, in over a dozen countries the comparison being made with numbers calculated from the law

$$r = R(1.000157954)^{(T-t)^2}$$

where

- t is the date of the year A.D.
- r is the number of deaths in that year t per thousand of the population in a given country
- T is the date when in that country the death rate will be lowest
- R is the lowest death rate, the same in all countries

[1] This is the second half of the lecture. The first half is found in the *Collected Papers* 6.88-6.97.

Country	Assumed Year when $r = R$	$R = 16.54$ Calculated r for 1890	Observed r for 1890	Calculated r for 1899	Observed r for 1899	
Sweden	1878	16.9	17.1	17.8	17.6	
Norway	1911	17.8	17.9	16.9	16.8	
Ireland	1916	18.5	18.2	17.3	17.6	
Denmark	1919	18.9	19.0	17.6	17.5	
Holland	1922	19.5	20.5	18.0	17.1	Bad
England	1923	19.7	19.5	18.1	18.3	
Scotland	1925	20.0	19.7	18.3	18.6	
Switzerland	1925	20.0	20.8	18.3	17.6	Bad
Belgium	1927	20.6	20.6	18.8	18.8	
France	1937	23.3	22.8	20.7	21.1	
Germany	1940	24.4	24.4	21.5	21.5	
Italy	1943	25.9	26.4	22.5	22.1	
Austria	1951	29.6	29.4	25.2	25.4	
Hungary	1955	32.2	32.4	27.1	27.0	
U.S. Reg.	[1922]	19.5	19.6	17.9	17.8	

The rates for the United States are separated from the others because they are separated by ten years instead of *nine*, like the others, and moreover they do not refer to any Calendar year but to years of 365 days ending May 30. I have taken no pains to get the best values of the constants. So it would be a good exercise in Least Squares to work these out. Of course, the uniformity is only rough. It was quite violated by Holland and Switzerland, countries that I mark as *bad* because they diminished their death rate in that novennium in a manner wholly unauthorized by any rule.

The best way of settling the meaning of a word is to take it in such a sense as shall render the word most useful. Now *chance* is the foundation of the great business of Insurance; and the doctrine of Chances has been without much exaggeration called the logic of the exact sciences. But when chance is said to consist in our *ignorance*, it certainly can be of no use except to those who desire to prey upon us. The Insurance business is not run on ignorance in any further sense than that if a man knew when he was going to die he would not insure his life. That which renders chance so important is that there is immense *diversity* throughout the universe. Diversity in many respects with uniformity in a few respects, and a great tendency among diversities to grow into uniformities are three real objective characteristics of the universe.

I will now drop metaphysics and consider the *doctrine of chances.* In order that this doctrine should have any useful application, it is necessary that we should *positively know* a number of propositions to be true. The matter by no means rests on mere ignorance. Of course, the doctrine of chances supposes a certain amount of ignorance, since it is a method of attaining knowledge; and every method of attaining knowledge must suppose that the information it teaches us how to obtain is not already in our possession. But there must be some large class of objects, say the *A*s, which have already presented themselves in experience and of which we have reason to believe that many other instances will present themselves in future experience. Now the *A*s of which we have had experience have some of them had a certain quality, while others of them have been without this quality. I shall say some *A*s are *Q* and some are not *Q*, using *Q* as an adjective expressive of their quality. Since we have experienced only a finite collection of *A*s, it must be that there has been some definite proportion of those *A*s we have experienced that have been *Q* and we must have some good reason for supposing that this same ratio or *some* ratio to which it approximates is going to hold in regard to our future experiences of *A*s. Or rather, what it is that we must know, — not with absolute certainty, for such knowledge is impossible as to future experience, but still what we must know, in such sense as the future can be known, — what we must know is *not* that our individual future experience will show the same ratio between the number of *A*s that are *Q* and the number of those that are not; — in case we have *that* kind of knowledge, which very rarely happens, the problem is specialized so as to become quite unlike the ordinary problem of chances. But what we must know is that, supposing the state of things does not undergo any change that makes any material difference in the result, that same *ratio would* hold good in the *long run.* A sufficiently clear understanding of the doctrine of chances to make it safe to apply it to any novel case requires an accurate conception of what a *long run* is. We shall soon have to give our closest attention to this matter. But it will be best to begin by taking note of all the elements of the problem, so that we may know what are the nice points that require analysis. We must not [only] have a sufficient knowledge that this ratio is going to continue substantially unchanged, but we must further know that there is not going to be any law of succession according to which *A*s that are *Q* and *A*s that are not Q are going to present themselves. If for example every other A is going to be Q and every other one not Q, the laws of chance will not hold. It is the same with any other law of succession. We must not only not know that [there] is such a law, but must have a sufficient assurance that there is *no* such law. The books on the subject are full of the word "independent." The instances must be independent. We must make sure that they are so. This "independence" that is so much insisted upon is nothing but the absence of any law of recurrence. All the good writers insist that we must assure ourselves of the *independence* of the "*events*," as they call the instances; so that they teach that it is not enough to be ignorant of law, but that it [is] requisite that we should make sure that there *is* no law of the kind that is pertinent to the question. Now information of this description does not involve any kind of knowledge of future individual instances. There may be in such knowledge ground for an inference as to individual instances; but the doctrine of chances institutes no *sound* inquiry into such inferences. Some writers on the subject endeavor to judge of such inferences by the principles of the calculus of probabilities; but what they say on the subject is, as I can clearly demonstrate, utterly worthless; and being so, insofar as it is apt to be accepted as sound, it is most mischievous. If a lady is afraid of going on board a steamer, it is a good argument to say to her that only one passage in thousands is accompanied by any serious harm to a passenger, and that she therefore ought not to hesitate to embark. But to say to her that the *doctrine of chances teaches* what it is wise to do in an individual case is a serious error of logic. Such inference is of a kind concerning which the doctrine of chances affords no direct knowledge. All that the doctrine of chances can do is to say what will happen in the *long run.* For example, suppose a pair of dice to be thrown. Call them die M, and die N. Now we know that die M will turn up an ace once in six times, or six times out of thirty-six, in the *long run*, and that die N will not only in the long run turn up in a deuce once in six times, but will do so in the long run once in six of *those* six times out [of] thirty-six in which die M turns up an ace. Therefore, once in thirty-six times in the long run M will turn up an ace and N a deuce, and by parity of reasoning, and then once in thirty-six times M will turn up a deuce and N an ace; so that deuce-ace will in the long run be thrown twice in thirty-six or once in eighteen times.

We do not know, however, that this will happen in any finite succession of throws. Therefore, by a *long run* we must understand an *endless* run. But there is no such thing as a half, or a third, or any other definite

finite proportion of the members of a *denumeral* collection. For instance, you might be tempted to say that one whole number out of every three is divisible by 3. But all whole numbers can be arranged in this way

$$\overline{1\ 3}\quad \overline{2\ 6}\quad \overline{(3)\ 9}\quad \overline{4\ 12}\quad \overline{5\ 15}\quad \overline{(6)\ 18}\quad 7\ 21\quad 8\ 24$$

You will evidently thus get all the whole numbers. Yet 3 out of every 5 are divisible by 3.

Or they may be arranged in two series so as to follow in different proportions in the two sequences

2 3 6 11 12 15 20 21 24 29 30 33 ...
1 4 5 7 8 9 10 13 14 16 17 18 19 22 24 26 27 28 29

A finite ratio of a denumeral collection only acquires a meaning when that collection has a fixed serial order. That amounts to saying that chance only relates to the *order of experience* or *order of existential succession*. What is called "geometrical probability" is not properly probability at all, unless it receives specifications not germane to the real problem, so as to put it into the guise of probability. Chance, then, in the sense in which the doctrine of chances studies it consists in a statistical law and no other law governing the succession of a species of events in the endless future.

Generally, in all its meanings, *chance* refers to variety, in contradistinction to uniformity. We call a fact *accidental*, if it is not governed by a specified or well-understood general formula or other general idea, such as a *general intention*. But it would be an excellent practice to restrict the expression "*happening by chance*" to meaning happening so in a series of experiences, which series must be distinctly specified to give the phrase any meaning, that the fact is not governed by any order of succession that holds in the *long run*, no matter whether it be intended, or otherwise necessitated, or not.

When we say that the frequency with which the *A*s are *Q*, or in other words the probability that an *A* will be *Q*, has a given value, *p*, what we mean is, that in any denumeral succession of occurrences of *A*s taken in the order in which they occur in the course of experience, if we take any finite range of conceivable values that includes *p* if, I say, whatever such range of values may be taken, there will come a time in the succession of the *A*s after which the proportion of *A*s that are *Q* among all the *A*s from the beginning never ceases to be included within that range, then, and only then *p* is the frequency with which the *A*s are *Q* in the long run, or in other words, is the probability that an *A* will be *Q*.

Of course, we cannot say at any time when that time will be or has been after which the value of the ratio in the evergrowing tally will never leave that range of values. For if we could do so, that would constitute a law in the succession and the occurrences would not be independent. But nevertheless the time will come, supposing the occurrences are independent. There must be some ultimate frequency; for otherwise there would be a law of alternation of some kind in the succession.

Suppose ten dice are thrown. Then there are $6^{10} = 602535996$ ways in which they may turn up, of which only one makes them all sixes. There are 50 ways in which all but one can be sixes. The other numbers are calculated as in the diagram [see p. 398].

Much of matter in the treatises on probabilities relates to mathematical methods of dealing with high numbers so as readily to get results of far greater accuracy than there can ever be any need of in practical questions about chance which are of their nature only rough.

When Pascal's triangle which forms the left hand part of our diagram is imagined carried out to many terms, the extreme parts of it are utterly insignificant. Now the central parts and all but the most extreme end are well represented by a curve called the probability curve and by a certain function called the theta of probabilities. The rule for using it is given in Studies in Logic.

The theory of probabilities is a very beautiful doctrine mathematically considered and is of the very extremest value for Logic.

No better illustration of chance distribution can be found than is afforded in Mendel's laws of heredity; and heredity is a subject to which probability is particularly applicable. Some important extensions of this branch of the subject are given in Pearson's Grammar of Science.

I now come to the most important part of this lecture. The calculus of probabilities proceeds entirely by mathematical reasoning, that is to say by *necessary* reasoning. Such inferences as it draws are *necessary inferences*. If it sounds incongruous to speak of necessary inferences of probability, this is perhaps merely an apparent incongruity not a real one. But if it is a real incongruity it is because the word probability ought not to be applied to ratios of frequency which are no more subjective or modal, in their nature than any other statistical averages. The use of the word probability for the average frequency in the long run of experience is a fault of a much graver nature than ill-chosen expressions usually are. It is a fault similar to that of measuring energy in "foot-pounds," which since a pound is a pound and a foot is a foot all the

										1										
									1											
								1		10	50									
							1		9											
						1		8		45	225	1125								
					1		7		36											
				1		6		28		120	600	3000	15000							
			1		5		21		84											
		1		4		15		56		210	1050	5250	26250	131250						
	1		3		10		35		126											
1		2		6		20		70		252	1260	6300	31500	157500	787500					
	1		3		10		35		126											
		1		4		15		56		210	1050	5250	26250	131250	656250	3281250				
			1		5		21		84											
				1		6		28		120	600	3000	15000	75000	375000	1875000	9375000			
					1		7		36											
						1		8		45	225	1125	5625	28125	140625	703125	3515625	17578125		
							1		9											
								1		10	50	250	1250	6250	31250	156250	781250	3906250	19531250	
									1											
										1	5	25	125	625	3125	15625	78125	390625	1953125	9765625

world over leads people to forget that a *footpound* in Edinburgh and a *footpound* say in the Philippines differ considerably; but the fault of the word probability is worse than that inasmuch as the errors it leads to cannot be considered as trifling, however rough a measurement may satisfy us. We speak of the probability of an *event*. It is not the *event* but the *fact* that ought to be spoken of, a *fact* being an element of an event resulting from an act of mental discrimination. However, that is, comparatively speaking, a trifle. When we so speak we only mention one term of the ratio. What would you think if I were to talk to [you] about the ratio of a house, as if a ratio were some absolute quality of the house? You would say that in the first place, it is not the house, but some dimension or measure of the house that can have a ratio. That is a fault similar to speaking of the probability of an *event* instead of the probability of a *fact*, or of the truth of a proposition. But this you would say is a small absurdity compared to that of speaking of the ratio of a thing without mentioning but one term of the ratio. If anything has a ratio, it must be a ratio to something. A proposition may have a probability without any second member, provided the term probability be used in another and perhaps more legitimate sense. But in the sense in which probabilities can be *evaluated* and made the subjects of mathematical calculation, a probability is a ratio of frequency in the long run of experience. This should never for one instant be lost sight of; and you should constantly ask yourself '*Is there going to be any such* experience in any way that can give the particular probability spoken of any real utility?' The majority of treatises on probability tells us that if a man who never heard of the tides were to wander to some shore where the tides are very noticeable and were to notice that the tide rose in each one any number of successive half days, say N for N half days then the probability that it would rise on the next half day would be $\frac{N+1}{N+2}$. But what is the long run of experience that this represents? The tide it is said would be found to rise the next half day in $N + 1$ cases out of every $N + 2$ taken at random. But what are these $N + 2$ cases, I ask. They will say, I suppose, that they are the cases in which I observe a new phenomenon on N successive occasions. But it is impossible to suppose any solid statistics with such vague conditions. It all depends on what sort of facts one would call a new phenomenon. I go out with a friend for a walk in the country. We successively meet three acquaintances and my companion remarks that all three were baptists. Has this any necessary bearing on what the communion of the next man we meet may be? Mind, I grant at once that some kind of argument might be founded upon it. But we are talking of arguments capable of being deduced mathematically; for the calculus of probabilities confines itself to mathematical reasoning. Two things are possible either that we meet successively four baptists or that we meet three baptists first and then a non-baptist. Now if these events are independent, which is the only case in which the calculus can be applied at all then representing by b the probability of a man met on that road being a baptist the probabilities of the two cases that are now the only remaining possibilities were at the beginning of our walk

$$b \times b \times b \times (1 - b) \text{ and } b \times b \times b \times b$$

and now that there are no other possibilities the probability of the last man being a baptist is simply $\frac{b^4}{b^3(1-b) + b^4} = \frac{b}{(1-b) + b} = b$. That is, the probability is just what it was in the first place. If the events are not independent still the probability of the fourth man being a baptist must be greater the greater the antecedent probability of it. Yet the *real argument* that he will be a baptist has no force except from the fact that it is a strange thing to meet three baptists successively. But of course, to say that the events are independent, precisely amounts to saying that you cannot argue from one to another.

If you carefully examine the line of thought of those writers, you will find that they assume throughout, that we must at all times be in a condition to know what the probability of a future occurrence is. Now if probability is to be anything but a delusion it must have some objective meaning. To know a probability must be to know something. To know that the probability of a coin turning up heads is $\frac{1}{2}$ must be to *know*, in a positive (I do not say in an infallible) sense that it will turn up heads half the time in the course of our future experience. Yet Laplace, the master of those writers, founds his whole theory on the express assumption that if we know nothing about a coin or even as he remarks if we know it is loaded one way or the other the probability that it will turn up heads is $\frac{1}{2}$, so long as we do not know which way it is loaded. This is the πρῶτον ψεῦδος of their whole argument. You note that in the formula $\frac{N+1}{N+2}$ if N is zero, that is if the new phenomenon has never been observed at all, its probability is $\frac{1}{2}$. Mr. F.Y. Edgworth, who is a thinker of no inconsiderable ability (notwithstanding his somewhat puerile delight in airing his contempt for my writings), regards it as an important basis for the probabilities, with the business of insurance and all that, that of just one half of all possible questions answerable by yes or no, "yes" is the true answer. To talk of extracting gold from sunbeams after that sounds rather flat.

The only security against utter nonsense in this subject is to give up talking of probabilities in connection with the doctrine of chances and to talk instead of ratios of frequency in the course of our future pertinent experience. I insert the word "pertinent" to indicate, for example, that an insurance company does not want to know what death-rate they may learn of from the returns of the board of health but what the death-rate is going to be among the people they insure. I do not care what proportion of possible questions can truly be answered by "yes;" but if I could know what proportion of my theories of logic were going to withstand criticism that information would be very welcome.

Eminent mathematicians have proposed to apply the calculus of probabilities to determining judicial decisions, and to express the "*veracity*" of a witness by a number equal to the proportion of questions that he will answer truly.

You remember the story in Rabelais of the judge who acquired such a great reputation for the eminent fairness and impartiality of his decisions that the King sent for him to explain his method. Whereupon he explained that his rule was first carefully to read the arguments on both sides and then take a pair of dice and give judgment for one or the other party according as the throw was odd or even.

There is one further feature of probability or unordered frequency to which I must draw your attention. It is quite obvious that a fact may happen independently with any frequency expressible by a rational fraction. It is less obvious that the ultimate frequency can have an irrational value without there being any law of succession; but there seems to be no doubt that this also is possible. For although it is impossible that the ratio of frequency should, at any stage of the endless succession of occurrences, be other than a vulgar fraction, yet given any two values, one greater and the other less than the irrational value, there is no reason why as the succession proceeds the ratio for all that part of it that has occurred should not at length forever cease to be so great as the greater of the two values while ceasing to be as small as the smaller one. Thus, that question seems to be satisfactorily answered.

A point which often puzzles those who are entirely ignorant of the doctrine of chances is that they think that if a coin has turned up heads say ten times in succession, — which will happen about once in a thousand times, — since it must turn up heads and tails in equal proportions, the long run of heads must be followed by throws among which there is an extra proportion of tails, to balance the extra heads. This of course is a great mistake. The extra heads though they were a million would not affect the relative frequency of heads in the long run, in the slightest degree. Everybody who has studied the doctrine of chances knows this, though perhaps not everyone understands it quite as clearly as he fancies he does, for not a few are puzzled by what is precisely the same thing, namely, the fact that it is not logically impossible that an event whose probability is zero should nevertheless occur on millions of occasions.

For example, three men, *A*, *B*, and *C*, agree to play a perfectly even game on the following conditions. Only two can play at a time, playing against each other; while the third stands by to take the place of the first one who goes out. At any one play one player wins a dollar from the other and the chances are perfectly even. The dollar is not actually paid over until one of the players retires temporarily from the table, and no player does retire until he has netted a gain during the sitting. Now the doctrine of chances shows that at a perfectly even game the probability that a player whose cash and credit amount to c dollars, will net a gain of n dollars before he is ruined will be $\frac{c}{c+n}$. At this game the credit of the players is unlimited, and thus the probability will be infinity divided by infinity *plus* one, which fraction is strictly equal to 1. That is the probability that a player *will* net a gain is 1, or certainty. Thus the three go on playing, each one as soon as he has made his absolutely certain net gain of a dollar, yielding his seat to the third; and so they keep on for all eternity, every man winning his dollar at every sitting, which dollar is paid in solid cash though the credit is unlimited while the sitting lasts.

I have known persons to find little conundrums similar to this quite puzzling.

Hume applies the doctrine of chances to disprove miracles. Now a miracle would not be a miracle if it were not entirely out of the ordinary course of nature, or what would be the course of nature to a materialist. As such the probability of a miracle is zero. But that does not logically conflict with its happening millions of times.

It is logically possible that a die should turn up a six; and therefore it is logically possible that it should turn up an endless succession of sixes and never any other side, although the probability of it is zero. But that a die should turn up a seven is logically impossible. That would be more than a miracle. It would be an absurdity.

7

EXISTENTIAL GRAPHS

A. LOWELL LECTURES, 1903. LECTURE II (455, 456)

Let us take up the subject of necessary reasoning, mathematical reasoning, with a view to making out what its elementary steps are and how they are put together.

In order to do this it is necessary to replace the confused syntax of ordinary language by a system in which the meaning of every form is exactly defined, which is free from forms that cast a tinge of passion or of any kind of subjective feeling on the facts, and which has no more forms than are requisite in order to express every kind of fact or truth in such a way as to enable us to carry the dissection of reasoning to its smallest steps.

Let us devote this evening's hour to forming such a system of expression.

Before beginning, let us distinctly recognize the purpose which this system of expression is designed to fulfil. It is intended to enable us to separate reasoning into its smallest steps so that each one may be examined by itself. Observe, then, that it is *not* the purpose of this system of expression to facilitate reasoning and to enable one to reach his conclusions in the speediest manner. Were that our object, we should seek a system of expression which should reduce many steps to one; while our object is to subdivide one step into as many as possible. Our system is intended to facilitate the *study* of reasoning but not to facilitate reasoning itself. Its character is quite contrary to that purpose.[1]

[1] MS. S-27 is a set of lecture notes on a subject related to that of this lecture. It reads as follows:

"Ladies and Gentlemen:

The first kind of reasoning to be studied is *Deduction*. Deduction is that kind of inference in which the fact expressed in the conclusion is inferred from the facts expressed in the premisses, regardless of the manner in which these facts have come to the reasoner's notice. Deduction is either necessary or probable. *Necessary deduction* is that sort of inference in which the fact concluded is conceived to be involved in the facts premised. It is the reasoning of mathematical demonstration.

It has taken two generations to work out the explanation of mathematical reasoning. This delay has been partly due to many writers entirely missing the point and directing their energies to ascertain the sequence of mental phenomena in reasoning instead of the logical sequence of the argument, which need not be closely related to the psychological sequence. The delay has also been due in part to the circumstance that some students attempted to divest thought of its garment of expression and to get at the naked thought itself, an attempt analogous to that to remove the peel from an onion so as to get at the naked onion itself. Reasoning is nothing but the discourse of the mind to itself. Divest thought of signs and it ceases to be thought, and becomes, at best, direct perception.

What is requisite is to take really typical mathematical demonstrations, and state each of them in full, with perfect accuracy, so as not to skip any step, and then to state the principle of each step so as perfectly to define it, yet making this principle as general as possible. For routine demonstrations there is no particular difficulty; but for the major theorems there is much. If we attempt to make the statement in ordinary language, success is practically impossible. Our syntax was not made with a view to such propositions; and it sometimes defies ingenuity to express them, I do not say clearly, but accurately with whatever intricacies of expression in words. At all times, the burden of language is felt severely, and leaves the mind no energy for its main work. Yet this is the least of the disadvantages of ordinary speech. After we have weeded out its ambiguities, it presents so many forms whose precise difference of meaning we are not prepared to define that we are very apt to pass over important steps of reasoning as mere grammatical transformations; and in many cases we see pretty clearly that an inference holds good but are provided with no sure way of stating its principle in general terms. Mathematicians have found themselves obliged to resort to algebraical arrays of letters in order to express themselves. But their algebras were devised for a purpose quite inconsistent with that of logical analysis, and are of no material help. It is necessary to devise a system of expression for the purpose which shall be competent to express any proposition whatever without being embarrassed by its complexity, which shall be absolutely free from ambiguity, perfectly regular in its syntax, free from all disturbing suggestions, and come as near to a clear skeleton diagram of that element of the fact which is pertinent to the reasoning as possible. I am going to devote this lecture to a brief description of a system of expression which, if it does not quite satisfy my ideal of what such a system ought to be, is at any rate the best I have been able to devise during forty years study of the problem. If you will learn this system and will then train yourselves to the use of it, I can promise that it will help you much to unravel tangles of thought.

Only, let not its aim be mistaken. I wish to declare distinctly and once for all that it is *not* intended to furnish a speedy or ready way by which to pass from premisses to conclusion. It aims in the diametrically opposite direction; namely, to break up reasoning into the greatest possible number of distinct steps, so that the constitution of reasonings may be studied. If we wished to obtain speedy passage from premisses to conclusion, we should, on the contrary, seek to make the steps as few and as large as we could. In short this system is meant not as an aid in reasoning but as an aid in the minute analysis of reasonings. Practice with it, however, will make thought clearer, and will so conduce indirectly to skill in reaching conclusions.

This system is a system of *diagrams*. A diagram has the advantage of appealing to the eye, and to that adds others due to the prominence it gives to *conventional signs*. Every conventional sign or other symbol is employed over and over again. The word *the* will occur several times on every page of English print; and it is everywhere one and the same word. Thus, it is the general type that constitutes the self-identity of the symbol; that is, it is its being formed in conformity to certain general precepts. But every one of those single embodiments of the word *the* which we find on a page; what are we to call them? A word is wanted for the purpose. I will call the single embodiments of a symbol, whether conventional or natural, its *replicas*.

The special system of diagrams that I am about to describe is called the *Method of Existential Graphs*. This system will repose upon 14 conventions or agreements into which you and I shall have to engage as to the significations of parts of our graphs. When I say there are 14 conventions, there is one that I do not count for the reason that it does not belong to this particular system any more than it does to any other system of conventions relating to no matter what subject. I will therefore number this Convention Number Zero. It reads as follows:

Convention No. 0

Whatever feature of a sign of this system is not the subject of any express convention can be varied indefinitely, without affecting the meaning of the sign.

For example, nothing being said about the size of certain parts of our diagrams, they can be made excessively small. Nothing being said about the length or shape of certain lines, we are free to give them any length and shape that may suit our purposes.

Convention No. 1 depends upon the consideration that an act of expression is a performance in which two persons take part, the one making the signs, the other interpreting them. It is essentially the same when a man deliberates with himself. He impulsively formulates an assertion, not necessarily in words, but in some kind of expression; and the self of a moment later criticizes and accepts or rejects, which requires him first to interpret the expression, that is, to reexpress it in his consciousness of a moment later. Vernacular language often embodies sound psychology, and the vulgar phrase "I says to myself," is endorsed by the most scrupulous science. Milton and Shakespeare speak of the "discourse of reason."

But if one person is to convey any information to another, it must be upon the basis of a common experience. They must not only have this common experience, but each must know the other has it; and not only that but each must know the other knows he has it, and must know the other knows that *he* knows the other has it; so that when one says "It is cold" the other may know that he does not mean that it is cold in Iceland or in Laputa, but right here. In short it must be thoroughly understood between them that they are talking about objects of a collection with which both have some familiarity. The collection of objects to which it is mutually understood that the propositions refer is called by exact logicians the *universe of discourse*. A certain amount of truth about this universe is taken for granted between the two. So far they have the same idea of the universe, upon that universe the attention of both is fixed; and when one makes any assertion to the other, and the other assents to it, what happens is simply that their common idea of the universe becomes more definite; for their whole discourse is about that and nothing else. Now the first convention, or agreement, which I shall ask you to join me in establishing is this:

Convention No. 1

The whole surface of this blackboard, excepting such parts as we may agree to sunder from the rest, together with any other surface which we may hereafter adjoin to it, shall be called our *sheet of assertion*, and every proposition the regular expression of which our imaginary graphist shall, at any time, render legible upon this surface, shall be understood to be thereby, asserted of the universe, thus rendering the blackboard a more definite representation of the universe.

For example, the graphist writes

Some orange has red pulp

We are thus informed that the universe is one in which there is an orange with red pulp.

Throughout this course, I shall avoid all questions of theoretical logic as far as I possibly can. I shall, therefore, not stop to inquire what assertion consists in. There will be the less need of doing so inasmuch as it will be quite foreign to our purpose to inquire into the truth of assertions placed on the board. The interpreter is supposed to assent to all that the graphist asserts. The word *assertion* means etymologically *braiding to*; and in this connection it is used [to] mean that that proposition which might be written elsewhere quite idly or to practice one's handwriting, or for any other purpose when put upon this board, is meant to be attached to the universe of discourse.

I ask attention to the following definition. Every expression of a proposition according to the conventions of this system is called an *existential graph.* But since we shall have no occasion to refer to other kinds of graphs, we may say *graph* simply when we mean an existential graph. The *graph* is the symbol, it is only a replica of it that can be placed on the sheet of assertion; but we may say that the graph is *scribed* on the sheet when a replica of it is placed there.

The sheet of assertion is itself a *graph*, and what is scribed upon it is a *graph*, and the two taken together form a *graph.*

I now hasten to the second convention which is, of all the conventions, the one that is the most arbitrary and the most characteristic of this special system of expression.

Convention No. 2

Two different graphs legibly scribed on the sheet of assertion in such a way that either might be removed without disturbing the other shall each have the same significance as if the other was not there.

For example, our graphist writes something further on the sheet

Some orange has red pulp
Naturalness is the last perfection of style.

Thus our idea of the universe becomes still more determinate.

We shall do well to conceive that the graph-replicas are movable over the sheet of assertion, and that the blackboard is not the whole sheet, but only the part of it to which attention is directed at the moment. Accordingly, when a replica on the board is rubbed out we need not understand that the graphist retracts his assertion, since he may merely have shoved the replica away to a distant part of the sheet.

I must here trouble you with three definitions. *Partial graphs* are graphs on the sheet of assertion that are not the only ones scribed on the sheet. The *entire graph* is all that is scribed on the sheet. The *total graph* is the entire graph together with the sheet of assertion itself.

Convention No. 3 will require a few introductory words. In ordinary language the conditional form of sentence, 'If one thing, then another thing' is employed for various purposes. Most commonly, in theoretical assertions, at least, it refers to a universe of possibilities, — a "range of possibility," as we say, — and means that, throughout this range of possibility, in whatsoever state of things the protasis, or antecedent, would be true, in that same state of things the apodosis, or consequent, would likewise be true. It is only rarely that in ordinary language the conditional form is used in a sense which is of the highest importance in logic, in which no range of possible states of things is considered, but merely the actual state of things. A conditional proposition which has this meaning is called a conditional *de inesse*. To say that there is a *connection* between one fact and another fact is to talk of *possibilities*. Since, therefore, the conditional *de inesse* does not refer to possibilities, but only to the actual state of things, it does not imply any connection between the facts expressed by antecedent and consequent. Take for example the proposition "If any orange has red pulp then naturalness is the last perfection of style," understanding this *de inesse*. Now there are four possible cases, as here shown:

[a]

Antecedent False / Consequent True	Antecedent False / Consequent False
Antecedent True / Consequent True	Antecedent True / Consequent False

[b]

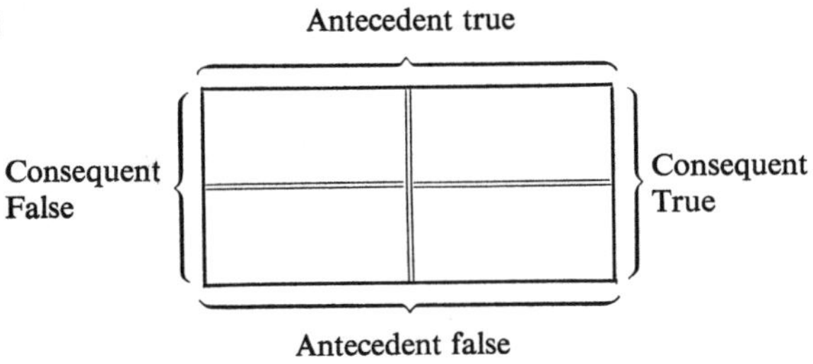

If the Antecedent is false, the proposition asserts nothing at all. It therefore contains no falsity. But every proposition is either true or false. Hence, the Conditional *de inesse* is in this case true, no matter how it may be with the consequent. If the Antecedent and Consequent are both true, then all that the Conditional asserts is true. Hence the Conditional *de inesse* is true if the consequent is true, regardless of the Antecedent. If however the Antecedent is true and the Consequent false, the Conditional *de inesse* is false. This therefore is the only one of the four possibilities which this form of proposition excludes.

It is absolutely indispensable that our system of expression should be provided with a way of expressing a conditional *de inesse*; and if it can express this kind of conditional, other conventions will enable it to express all other kinds of conditional propositions without difficulty. Now let the conditional *de inesse* be expressed in this way:

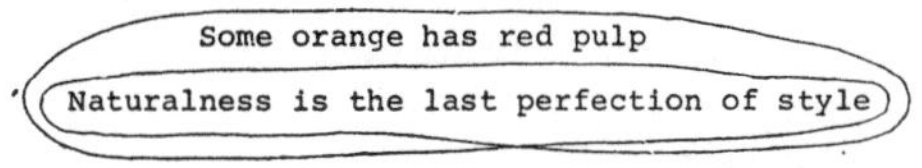

"If any orange has red pulp. then naturalness is the last perfection of style." The line consists of two ovals. The whole of it is called a *scroll*. Each oval is called a *sep*. The inner sep of a scroll is called its *loop*. The space enclosed by a sep is called its *close*. When one sep encloses another the space within the outer sep but not within the inner sep is called the *immediate* close of the outer *sep*, while this immediate sep together with the *close* of all the enclosed seps is the *entire* close of the outer sep. A sep together with all it contains is called an *enclosure*. It will be necessary to bear those terms in mind. A scroll is a curve consisting of one oval within another. A sep is a single oval. The inner sep of a scroll is its *loop*. The surface inside a sep is its *close*. Its *immediate close* is enclosed in no inner sep; its *entire* close embraces the whole space within it. An *enclosure* is a *sep* together with all its contains; or better, the *enclosure* is the sep considered as having its meaning as a sign determined by what it contains.

We now have two conventions, as follows:

Convention No. 3

1. A self-returning cut in the sheet of assertion or in any area enclosed within that sheet shall be considered to be itself, with every point of it, in that sheet or area.

2. It is not a graph and shall not affect the meaning of any graph-replica placed upon its line.

3. It shall cause the entire area within it to denote a world of ideas then and there created by the graphist.

4. The special quality of the cut shall assert a corresponding special relation between the world of ideas represented by the area it enclosed and the world represented by the area in which the cut is made.

Convention No. 4

A sep scribed on the sheet of assertion and containing another sep shall be understood to assert *de inesse* that if all that is expressed in its immediate close except the inner sep be true when interpreted as it would be if legibly scribed on the sheet of assertion, then all that is expressed in the close of that inner sep is true, when interpreted in the same manner.

You will probably be surprised that I should say that these conventions are less arbitrary than Convention No. 2. But that is because I have not time to develope the argument to show that the conditional *de inesse* can only be expressed in a manner essentially the same as that described.

The next seven conventions relate to the expression of individual identity.

Convention No. 5

The heavy marking of a point on the sheet of assertion causes that point to denote a single individual existing in the universe of discourse, and therefore asserts such existence without distinguishing that individual from others.

Thus • asserts that some individual object exists. But : does not assert that two individuals exist; for by Convention No. 2 each dot means just what it would if the other were not there; so that the pair merely asserts twice over that some individual object exists.

I am forced to inflict upon you one or two definitions more. A *rhema* is either a proposition or a blank form that would become a proposition if all its blanks were filled with proper names. Here are examples:

Rhemata

Cain killed Abel
      killed Abel
Cain killed
      killed

Let the blackboard represent the universe. As thus significant we will call it the sheet of assertion.

Let whatever I write upon the blackboard or sheet of assertion go toward making the representation of the universe more determinate. Thus, I write

A pear is ripe.

That represents that there is a pear and a ripe pear in the universe.

Necessary reasoning can never answer questions of fact. It has to assume its premisses to be true. Therefore, in order to avoid the possibility of questioning what is written on the board, let us say that it is not the real universe that is represented by the board, but a universe existing in my imagination, concerning which you have no source [of] information except my testimony. By the *universe* I mean the entire

---

A rhema which is made a part of a graph and is not itself compounded according to the rules of existential graphs is called a *spot* and the places where proper names might be inserted are called the *hooks* of the spot.

Convention No. 6

The placing of heavily marked points at all the hooks of a legible spot on the sheet of assertion shall be understood as asserting that there exist in the universe individuals related to one another as would be the individuals denoted by proper names filling the blanks of the rhema.

For example • kills • will assert something kills something.

• gives • to • in exchange for •

will assert that something gives something to something in exchange for something.

Convention No. 7

A heavily marked continuous line on the sheet of assertion shall assert that the individuals denoted by its two extremities are identical.

For example • is a bird / • is blue asserts that something is a bird and something is blue.

But 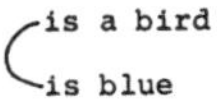

asserts that something is a bird and that that very same individual is blue. So 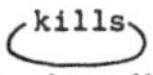

asserts that some individual kills itself.

Such a line is called a *line of identity*.

If three or more lines of identity have extremities at one point, then by Convention No. 5, those three lines denote the same individual. For example

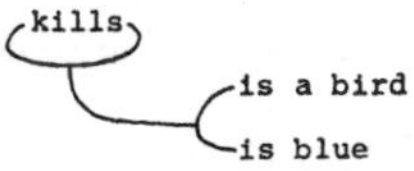

will mean that some blue bird kills itself.

Convention No. 8

If two different lines of identity cross one another, they must be interrupted and placed between parallels near the intersection, to show that this is not a point of branching." [Incomplete] collection of things, or subjects of force, to which all that is going to be written on the board will relate. Logicians call such a collection of things, or subjects of force, to which the whole of a discussion relates, the *universe of discourse*. This universe consists in the first place of certain mutually well-understood centres, or subjects of force well-understood to be different from one another; secondly, of certain subjects of force well-understood to exist, but not thoroughly understood to be known to be different from any of those of the first class; and thirdly of an indefinite supplement of subjects of force presumed to exist but of which there has been no definite recognition. Summing up the matter, we may say that the universe of discourse is the aggregate of subjects of the complexus of experience-forces well-understood between the graphist, or he who scribes the graph, and the interpreter of it.

The system of expression which I have thus begun to describe will be found to be a system of diagrams. The mathematicians call a diagram that is composed mainly of *spots* of different kinds and of *lines*, a *graph*. This system is called a system of *existential graphs*. In this system, the spots may be conveniently differentiated from one another by words written in them. The consequence is that a sign of this system is partly drawn and partly written. For brevity, I always say it is *scribed*. Any sign conforming to the rules of this system, which if it were placed on the board or sheet of assertion would assert some intelligible state of things to be true of the universe of discourse, is called a *graph*. Strictly, I ought to say an *existential graph*; but for brevity since existential graphs will be the only ones dealt with, I shall call it a *graph*, simply. For example, what I have just scribed is a *graph*. The board itself is a *graph*, since it represents the universe as consisting of single imaginary things. The board and what is written on it together make up another *graph*.[2]

[2] In MS. 479 Peirce writes as follows:

"ON LOGICAL GRAPHS

The word graph was introduced into algebra either by William Kingdon Clifford or by the great Sylvester, — I believe they attribute the invention to each other reciprocally, — to designate a diagram of dots and lines, similar to those by which the chemists represent the constitution of compounds, used as an icon of the relationships involved in invariants. Similar diagrams, though not called graphs, were employed by Kempe in his remarkable memoir on Mathematical Form to represent relationships of all kinds between individuals. I subsequently proposed (*Monist*, Vol. VII) a system capable of representing all facts of relation between classes as well as between individuals; but this was no sooner seen by me in type than I perceived that it was one of a pair of twin systems of which the other was to be preferred, and I wrote at once an elaborate paper on the subject, for which I vainly endeavored to find an asylum. At that time, I drew up an elaborate

It is quite important, however, to distinguish between a *graph* and a *graph-replica.* Suppose an editor writes to me and asks for an article of 4000 words. You know what he means by *words* in that case. In what I write the single *word* "the" may occur twenty times on every page. Every time it will count as a separate word. Yet in another sense, it is the same word. In this latter sense, the word *the* consists in the sum total of general conditions to which ink-marks or voice-sounds must conform in order to be understood in a certain way. In this sense the word *the* is, strictly speaking, never written; but what is written conforms to it, or, as we say, embodies it. In the other sense a written word is written once and only once, since every act of writing makes a new word. Now when I speak of a *graph,* I mean the general type of whatever means the same way, so far as the conventions of this system take cognizance of the ways; while that which is scribed once and only once and embodies the graph, I call a *graph-replica.* For brevity, however, I speak of "scribing a graph" just as we speak of writing the word *the.* The phrase may be defended as employing the word "scribe" in a special sense.

I will now put an additional replica upon the board, or sheet of assertion.

It rains

A pear is ripe

Let us agree to understand that each of those replicas has the same meaning as if it stood alone; and that it shall be the same with any other two replicas on the sheet. So that I now assert not only that there is a ripe pear, but also that it rains.

By calling this system a system of *existential graphs,* my meaning is that two graphs at different parts of the board, *whether far or near,* are both asserted, each, *just as much as if the other were not there.*

---

definition of a graph contemplating all sorts of possible generalizations; but I have since bestowed a great deal of study upon the matter both in its details and in its general aspects, and have been led to prefer a very much simpler definition which includes diagrams already in general use among logicians, — being one of the few things which all schools unite in finding valuable, and this catholic confession would seem to be an argument in favor of that intuitional theory of reasoning which was so forcibly defended by Friedrich Albert Lange.

I propose to use the term *logical graph* to designate any diagram which iconizes logical relations by means of geometrical relations."

In the *Preface* to *The Rules of Existential Graphs* in a small leatherbound notebook (MS. 1589) Peirce wrote that "The system of existential graphs is intended to afford a method for the analysis of all necessary reasonings into their ultimate elements .... The system was invented in January 1897. *The Monist* refused to publish an account of it."

The most immediately useful information is that which is conveyed in conditional propositions, "*If* you find that this is true, *then* you may know that that is true." Now in ordinary language the conditional form is employed to express a variety of relations between one possibility and another. Very frequently when we say "If *A* is true, then *B* is true," we have in mind a whole range of possibilities, and we assert that among all possible cases, every one of those in which *A* is true will turn out to be a case in which *B* is true also. But in order to obtain a way of expressing that sort of conditional proposition, we must begin by getting a way of expressing a simpler kind, which does not often occur in ordinary speech but which has great importance in logic. The sort of conditional proposition I mean is one in which no range of possibilities is contemplated, which speaks only of the actual state of things. "If *A* is true then *B* is true," in this sense is called a conditional proposition *de inesse*. In case *A* is not true, it makes no assertion at all and therefore involves no falsity. And since every proposition is either true or false, if the *antecedent*, *A*, is not true, the conditional *de inesse* is true, no matter how it may be with *B*. In case the *consequent*, *B*, is true, all that the conditional *de inesse* asserts is true, and therefore it is true, no matter how it may be with *A*. If however the antecedent, *A*, is true, while the consequent, *B*, is false, then, and then *only* is the conditional proposition *de inesse* false. This sort of conditional says nothing at all about any real connection between antecedent and consequent; but limits itself to saying "If you should find that *A* is true, then you may know that *B* is true," never mind the why or wherefore.

The question of the proper way of expressing a conditional proposition *de inesse* in a system of existential graphs has formed the subject of an elaborate investigation with the reasonings of which I will not trouble you. Suffice it to say that it is found that there is essentially but one proper mode of representing it. Namely, in order to assert of the universe of discourse that if it rains then a pear is ripe I must put on the blackboard this [Fig. 1]:

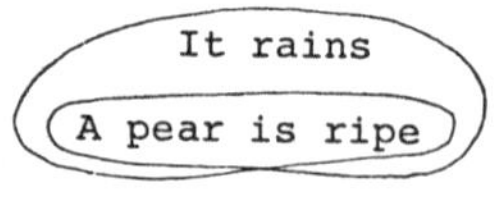

Fig. 1

I draw the two ovals which taken together I call a *scroll* in blue because I do not want you to regard them as ordinary lines. I want you to join me in making believe that they are cuts through the surface, and that inside the outer one the skin of the board has been stripped off disclosing another surface below. This I call the *bottom* or *area*. Therefore "It rains" is not scribed on the blackboard or, as I say, is not scribed on the sheet of assertion. For what is scribed on that sheet is asserted to be true of the universe of discourse; while the statement "It rains" is a mere supposition. Let us say that that *bottom* inside the outer cut represents another universe, a universe of supposition, and that it is only in that universe that it is said to rain. Besides this graph, "It rains" the *bottom* of the outer cut contains the inner cut which interrupts its surface; and inside the inner we will make believe that a *patch* is put on with a surface like that of the blackboard, although cut off from it. I use the word area for any part of the surface unbounded or bounded by cuts, never extending through a cut.

The outer cut is itself on the sheet of assertion although the whole of its interior is severed from that sheet. Now this outer cut, by being on the sheet of assertion, represents the conditional proposition *de inesse* to be true of the universe of discourse. A fixed terminology is a great comfort. Let us term the area on which a cut stands the *place* of the *cut*, while the *area* or *bottom* of the cut is the area within the cut. The cut itself is not a graph nor the replica of a graph. No more is the *scroll*. But the scroll with the two graphs scribed in its two *closes* or *areas* makes up a graph, or graph-replica; and this I call an *enclosure*. The term may be used indifferently to mean the graph or the replica.

In order to get an insight into *how* the scroll represents the conditional proposition *de inesse*, we must make a little experimental research.

Thus far, we have no means of expressing an absurdity. Let us invent a sign which shall assert that *everything is true*. Nothing could be more illogical than that statement inasmuch as it would render logic false as well as needless. Were every graph asserted to be true, there would be nothing that could be added to that assertion. Accordingly, our expression for it may very appropriately consist in completely filling up the area on which it is asserted. Such filling up of an area may be termed a *blot*.

Take the conditional proposition *de inesse*, "If it rains then everything is true" [Fig. 2]

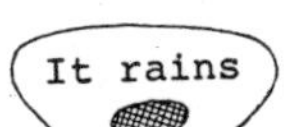

Fig. 2

That amounts to denying that it rains. But there is no need of making the inner cut so large. Let us write [Figs. 3a or even 3b].

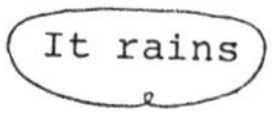

Fig. 3a

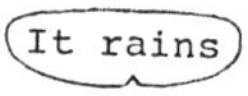

Fig. 3b

This suggests that the relation which the cut asserts between the universe of discourse and what is scribed within it is simply that what is scribed within is false of the universe of discourse.

Then we may interpret [Fig. 4]

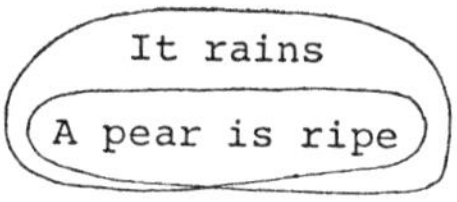

Fig. 4

as meaning "It is false that it rains and that a pear is not ripe." But we have already seen that this is precisely the whole meaning of the conditional *de inesse*; namely that it is false that the antecedent is true while the consequent is false. Thus, that which the cut asserts is precisely that that which is on its bottom is not, as a whole, true.

This agrees with the fact that if there is nothing on the bottom except the inner cut, its patch, and what is on that patch, the latter may be asserted. Thus [Fig. 5]

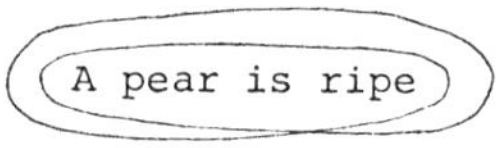

Fig. 5

since a blank merely asserts known truth means, "if the truth is true then a pear is ripe," or it is not false that a pear is ripe. So in the following [Fig. 6]

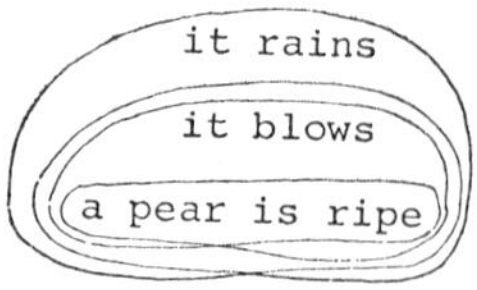

Fig. 6

or if it rains, then if it blows, a pear is ripe the two cuts with nothing between can be taken away without altering the fact expressed [Fig. 7] If it both rains and blows, a pear is ripe.

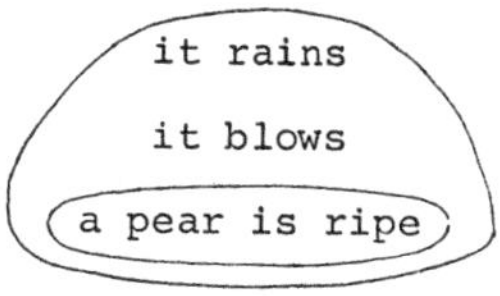

Fig. 7

So much of the system of existential graphs as I have thus far described I term the *alpha* part of the system. I shall presently describe a *beta* part; and in another lecture the system will be completed by a *gamma* part.

There are certain ways in which graphs that are scribed on the sheet of assertion can be modified without any danger of changing a true graph into a false one. Transformations according to any general method that can never change a true graph into a false one I term *permissible transformations*. In particular, those which the principles of the alpha part of the system render permissible shall be termed *alpha permissible transformations*, and in general I shall use "alpha" freely as a prefix to signify a reference to these principles. Thus, I shall say that the *alpha-signs*, that is to say the signs of this system heretofore described apart from the *graphs* themselves are two, and there are besides two peculiar graphs. Namely, the two peculiar graphs are the blank place which asserts only what is already well-understood between us to be true, and the *blot* which asserts something well understood to be false. The two signs which are not graphs are the putting of two graph-replicas upon the same area, which if we remember that a blank is a graph, is seen to include the scribing of a single graph as a special case. This idea that scribing a graph is a transformation of a graph already accepted is a very useful one. The other sign is the scroll.

I have just defined a *permissible transformation* as one which conforms to a general type which in no case can transform a true graph into a false one. But what do we mean by a *true* graph and a *false* one? I will not stop at present to analyze fully the meaning, because this system of existential graphs is not intended to inquire into the conformity of thoughts to experiences. The universe of our discourse is to be a universe of my imagination, and therefore any graph which I permit to be *scribed* will be true, unless it is absurd and so amounts to telling you that my universe has no being even in my imagination. Therefore, for our purposes, a true graph is nothing but a graph which will result from a special permission to transform the blank sheet of assertion into the sheet with that graph scribed upon it. As for the general alpha-permissions, they are nothing but the definitions of the four general alpha signs in terms of permissibility.

Let us draw up statements of them. You know that every definition consists of two propositions. Thus, if man is defined as a featherless biped, the one proposition, Every *man* is a featherless biped predicates the *definition* 'featherless biped' of the *definitum*, 'man'; while the other proposition 'Every featherless biped is a man' predicates the *definitum* of the *definition.*

Consequently, each sign of the system should furnish us with two permissions of transformation.

Beginning then with the sign which consists in scribing together, or as we may term it, *compounding* two graphs, the definition of it in terms of permission will be

1st, predicating the definition of the definitum, if it is permitted to scribe on the sheet of assertion a replica of a compound graph, then it is permitted to scribe on the sheet of assertion a replica of either component. Or, stating this in terms of transformations: Any replica of a compound graph may on the sheet of assertion be transformed into a replica of either component. That is to say, under a more [practical] aspect, any partial graph on the sheet of assertion may be erased or cancelled. This shall head our list of alpha permissions.

Permission No. 1. Any graph on the sheet of assertion can be erased.

2nd, predicating the definitum of the definition; if it be permitted to scribe on the sheet of assertion a replica of which we please of two graphs then it is permitted to scribe the replica of the compound graph of which those two are the sole components. Or in terms of transformation, if it be permitted to transform the blank sheet into either we please of two graphs, it is permissible to transform it into the compound of the two. That is to say, under a more practical aspect, whatever might be scribed on the sheet of assertion were this blank, can be scribed regardless of what is already scribed. This shall be our second alpha permission.

Permission No. 2. Whatever is permissibly scribable on the sheet of assertion is so regardless of what is already scribed.

Let us now treat the scroll in the same way. First, the predication of the definition with the definitum as subject is that when it is permitted to scribe upon the sheet of assertion a scroll with two graph-replicas, x and y, in its outer and in its inner close, or on its bottom and on its patch, respectively [Fig. 8], then whenever it is permitted to put the

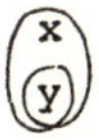

Fig. 8

graph x upon the sheet of assertion, it will likewise be permitted to put the graph y upon the sheet of assertion. Or in terms of transformation it will be permissible on the sheet of assertion to transform $\begin{matrix}x\\ x\end{matrix}$ by the insertion into it of $\begin{matrix}y\\ y\end{matrix}$ as a component of a compound graph $\begin{matrix}xy\\ xy\end{matrix}$.

In order to put this into a more explicit shape, I will first call your attention to a corollary from it. A *corollary* to a proposition of Euclid is a necessary consequence drawn from it by some editor of Euclid's Elements and inserted by him, originally, I suppose, marked with a little crown in the margin. These additions are, for the most part, propositions that Euclid thought too obvious for special notice. Hence, any easily drawn necessary consequence of a proposition is termed a corollary. Here I will tell you a secret about necessary consequences. It is a very useful thing to know, although most logicians are entirely ignorant of it. It is that not even the simplest necessary consequence can be drawn except by the aid of *Observation*, namely, the observation of some feature of something of the nature of a diagram, whether on paper or in the imagination. I draw a distinction between *Corollarial* consequences and *Theorematic* consequences. A corollarial consequence is one the truth of which will become evident simply upon attentive observation of a diagram constructed so as to represent the conditions stated in the conclusion. A *theorematic* consequence is one which only becomes evident after some experiment has been performed upon the diagram, such as the addition to it of parts not necessarily referred to in the statement of the conclusion. In the present case, I am going to draw a conclusion about a *double enclosure*, that is, two cuts one within the other and with the annular space between them blank like this [Fig. 9]. The observa-

Fig. 9

tion which I ask you to make is that in every such case there will be a graph of which one replica is in the outer close while another is on the sheet of assertion outside. Namely, that graph is the *blank*. And since the present principle permits us to transform [Fig. 10] *whatever x may be*, it allows this transformation when x is the blank; so that we can transform [as in Fig. 11]. We may count this as our third permission, so that we have

Permission No. 3. A graph within a double enclosure on the sheet of assertion may be scribed on the sheet of assertion, unenclosed.

Fig. 10

Fig. 11

The consideration of what further explicit permission is involved in the predication of the definition, the definitum being the subject, had better be postponed until we have considered the predication of the definitum the definition being the subject. This predication is that in case the permission to scribe on the sheet of assertion a replica of a graph, x, would carry with it in every case a permission to scribe on the sheet of assertion a replica of a graph, y, then it is permissible to scribe on the sheet of assertion a scroll containing in its outer close only a replica of x and in its inner close a replica of y. Or in terms of transformation, if it would be permissible to transform a graph, x, should it occur on the sheet of assertion, into a graph, y, then it is permissible to transform a blank on the sheet of assertion into a scroll having only x in its outer close, and having y in its inner close.

We may here draw a corollary analogous to, but much more obvious. Namely, to say that it is permissible to scribe any graph, y, on the sheet of assertion is to say that, it is permissible to transform a replica of the blank into a replica of y. But this, according to this part of the definition of the scroll, permits us to place on the sheet of assertion the scroll [Fig. 12]. Hence we have

Permission No. 4. If a graph could be permissively scribed on the sheet of assertion a double enclosure containing that graph may be placed on the sheet of assertion.

Fig. 12

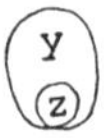

Fig. 13

Fig. 14

By combining the two parts of the definition of the scroll we get the highly useful graphical form of the *principle of contraposition.* Namely, suppose that a replica of the graph $\overset{x}{x}$, were it scribed on the sheet of assertion, would be permissively transformable into a replica of the graph, $\overset{y}{y}$, and suppose that the scroll [Fig. 13] were permissively placed on the sheet of assertion. Then, by the predication of the definition concerning the definitum, *y*, if scribed on the sheet of assertion, would be transformable into *z*. So *x* being transformable into *y* and *y* in its turn into *z*, it follows that *x* would on the sheet of assertion be transformable into *z*. Hence by the predication of the definitum concerning the definition, it would be permitted to place on the sheet of assertion the scroll [Fig. 14]. That is to say, the permissibility of the transformation on the sheet of assertion of *x* into *y*, carries with it the permissibility of the transformation on the bottom of a cut placed on the sheet of assertion, of the reverse transformation of *y* into *x*. Here is a principle which involves many permissions.

Permission No. 5. *The Principle of Contraposition.* Of whatever transformation is permissible on the sheet of assertion, the reverse transformation is permissible within a single cut [Fig. 15].

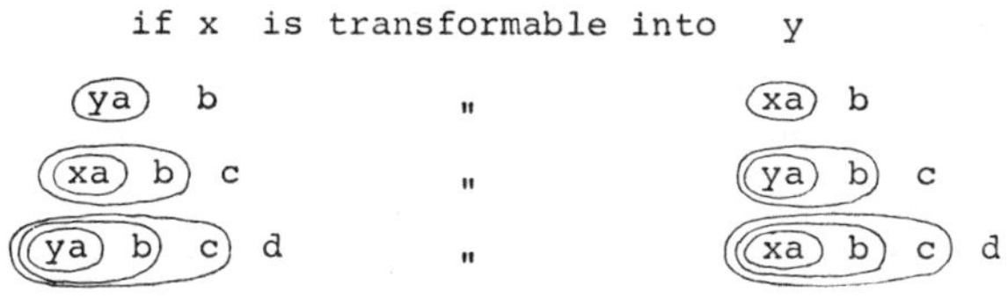

Fig. 15

And so on indefinitely. In short, whatever transformation is permissible on the sheet of assertion is permissible within any even number of cuts while the reverse transformation is permissible within any odd number of cuts. For example, since any graph can be erased on the sheet of assertion any graph can be erased within any even number of cuts while any graph can be inserted within any odd number of cuts. Since any graph already scribed on the sheet of assertion can by Permission No. 2 be iterated on the sheet of assertion, that is can have another replica of it placed on the sheet, it follows that if one replica of a graph is on the sheet of assertion and another replica of the same graph is oddly enclosed, the latter can be erased. Thus [Fig. 16] can be transformed into [Fig. 17] and thence by Permission No. 3 into xy, and thence by Permission No. 1 into y. This [is] what the logic books call the *modus ponens*. [Fig. 18] gives successively [Fig. 19]. The fact that our system breaks this up into three steps goes to show that our main purpose, that of dissecting reasoning into its simplest elements, has been, in some measure, at least, attained.

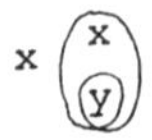

Fig. 16

Fig. 17

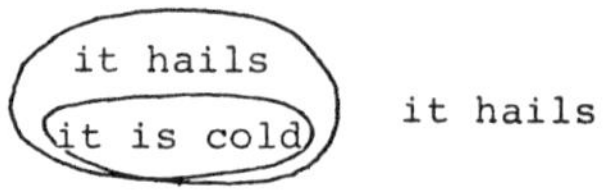

Fig. 18

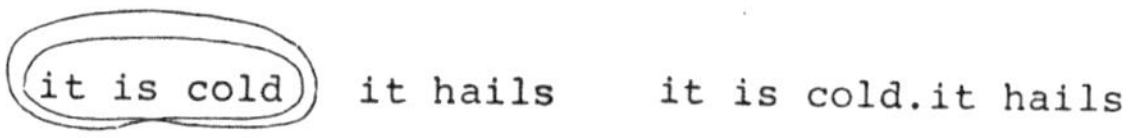

Fig. 19

It still remains to treat the blank and the blot in the same manner.

But when we undertake to predicate of the blank as definitum any definition we find that no specific permission follows from the existence of a blank, although the whole possibility of writing graphs depends upon it. That has therefore to be passed over. As to the predication of the definitum concerning the definition this renders it permissible for a blank to accompany any and every graph scribed on the sheet of assertion. It is truly fortunate that this is permitted, inasmuch as it would be physically impossible that a blank should not accompany every graph. The truth is that the system of existential graphs was intentionally contrived so that this matter should take care of itself. Here, again, therefore, no permission is called for.

Passing to the *blot*, or *pseudograph*, of which, you remember the meaning is that everything is true, the predication of the definition concerning the definitum is that within any even number of cuts where the blot is any graph we please may be inserted and within any odd number of cuts where the blot is any graph may be erased. The blot, it is true, fills its whole area, so as to leave no room for any other graph. But there is an equivalent of it of which this is not true. For since by

Permission No. 1 every graph on the sheet of assertion enclosed can be transformed into the blank, it follows, by the principle of contraposition that an enclosure containing nothing but a blank can when evenly enclosed be transformed into anything we please, and consequently into the pseudograph. The vacant enclosure is, therefore, a form of the pseudograph. For evenly enclosed it can be transformed [into] the blot, as the blot can be transformed into it. And since these two transformations are the reverse of one another, it follows, by the principle of contraposition that the same is true within any odd number of cuts. When the vacant enclosure is oddly enclosed as in this figure [Fig. 20], the enclosure

Fig. 20

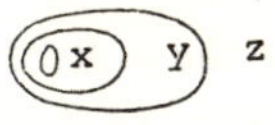

Fig. 21

on whose area it stands is evenly enclosed and can be erased by Permission No. 1. But when the vacant enclosure is evenly enclosed as in the next figure [Fig. 21] all other graphs, in the same enclosure (here represented by *x*), being evenly enclosed can be erased by Permission No. 1, and then Permission No. 4 which allows a double enclosure round any graph under even enclosures, permits the double enclosure to be removed under odd enclosures. I mean the double enclosure formed of the cut of the vacant enclosure together with the cut enclosing it. This done nothing but a blank remains; and so, within odd and even numbers of cuts alike, the whole enclosure containing the pseudograph may be suppressed. We thus get

Permission No. 6. Any enclosure containing a blot or other pseudograph may [be] suppressed whether evenly or oddly enclosed.

The predication of the definitum concerning the definition may be regarded as giving the last Alpha Fundamental Permission.

Permission No. 7. A vacant cut may be treated as a pseudograph.

These seven permissions being, however, somewhat confusing, I replace them by a compact little code which I call the three primary rules. It runs as follows:

Rule of Erasure and Insertion. Within even cuts (or none) any graph can be erased; and within odd cuts any graph can be inserted.

Rule of Iteration and Deiteration. Any graph of which a replica is already scribed may be iterated on the same area as the primitive replica or within any additional cuts already existing; and of two replicas of the same graph, one of which is enclosed by every cut that encloses the other, the former may be erased, this process being termed deiteration.

Rule of Insignificants and the Pseudograph. A double enclosure can be circumposed about any graph or be removed from any graph; and any enclosure containing a vacant cut or other form of pseudograph can be suppressed or inserted.

I now pass to the beta part of the system of existential graphs. It is far more interesting and important than the alpha part, but incomparably less so than the gamma part.

When one hears a proper name mentioned for the first time, one generally learns, of the individual person or thing denoted by that name that it exists. It may, of course, be identified with a subject of force already well-known; but that will be exceptional. It will frequently be apparent that it is a thing quite different from any hitherto mutually recognized. In this case, it will make an addition to the universe of discourse, brought about by means of the assertion a real relation of it to an object previously recognized. Sometimes, it will be doubtful whether it is one of the recognized subjects of force or not. But what I want to focus your attention upon is, that at the first mention of a proper name, apart from any special information about its subject that may then be conveyed, the name merely tells us that *something exists*, that is, is a factor of the entire complexus of forces that we partly have known by experience. But at any *subsequent* mention of the proper name, though this assertion of existence is reiterated, yet that being known already is of no importance. The importance of the name at all occurrences after the first is that it *identifies* what is mentioned with something we had heard of before.

If you bear in mind these characteristics of proper names, you will perceive that when lawyers and others use the letters *A*, *B*, *C* as a sort of improved relative pronouns, saying for example that if *A* owes *B* money and *C* owes *A* money, then *B* may "trustee" *C* for the debt (as you say in Massachusetts), these letters differ from new proper names only in the accidental circumstance that they are first introduced in the antecedent of a conditional proposition while proper names are first introduced in positive assertions. I call such improvised proper names *selectives*.

There is nothing to prevent our using the capital letters as such individual names, provided we distinguish the *first replica*, by scribing it heavily or otherwise. I cannot say that this is a bad way; it serves the purpose of putting out of view confusing triples. But I do say that it requires rather complicated rules, and from every other point of view except that of putting unimportant circumstances out of view and the convenience in printing, is usually inferior to another way of fulfilling the same purpose, which I proceed to describe.

Since the blackboard, or the *sheet of assertion*, represents the universe of discourse, and since this universe is a collection of individuals, it seems reasonable that any decidedly marked point of the sheet, should stand for a single individual; so that • should mean "something exists." We cannot make this : to mean that two things exist, since this would conflict with our convention that graphs on different parts of the sheet shall have each the same meaning as if each stood alone, so that consequently the second points merely *reiterate* that something exists.

You will ask me what use I propose to make of this sign that *something exists*, a fact that graphist and interpreter took for granted at the outset. I will show you that the sign will be useful as long as we agree that *although different points on the sheet may denote the same individual, yet different individuals cannot be denoted by the same point on the sheet.*

If we take any proposition, say

A sinner kills a saint

and if we erase portions of it, so as to leave it a *blank form* of proposition, the *blanks* being such that if everyone of them is filled with a proper name, a proposition will result such as

_______ kills a saint
A sinner kills _______
_______ kills _______

where *Cain* and *Abel* might for example fill the blanks, then such a blank form, as well as the complete proposition, is called a *rheme*, provided it be neither by logical necessity true of everything nor true of nothing. But this limitation may be disregarded. If it has one blank it is called a *monad* rheme, if two a *dyad*, if three a *triad*, if none a *medad* (from μηδέν).

Now such a *rheme* being neither logically necessary nor logically impossible, and represented as a part of a graph without being represented as a combination by any of the signs of the system is called a *lexis* and each replica of the *lexis* is called a *spot*. (*Lexis* is the Greek for a single word and a lexis in this system corresponds to a single verb in speech. A *lexis* is therefore an incomplex *contingent* graph. The plural of *lexis* is preferably *lexeis* rather than *lexises*.) Such a spot has a particular point on its periphery appropriated to each and every one of its blanks. Those points, which, you will observe, are mere places, and are not marked, are called the *hooks* of the spot. But if a *marked point*, which you remember we have agreed shall assert the existence of an individual, be put in that place which is a hook of a graph, it must assert that something is the corresponding individual whose name might fill the blank of the rheme. Thus

• gives • to • in exchange for •

will mean "something gives something to something in exchange for something."

Now let us further agree that a heavily marked line ———, all whose points are *ipso facto* heavily marked and therefore denote individuals, shall be a *graph* asserting the identity of all the individuals denoted by its points. Then [Fig. 22] will mean that there is a ripe pear, that is, something is a pear and that very same thing is ripe.

We call such a heavy line a *line of identity*. A point from which three lines of identity proceed has the force of the conjunction 'and' [Fig. 23].

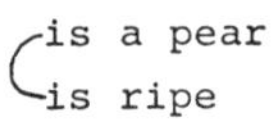

Fig. 22

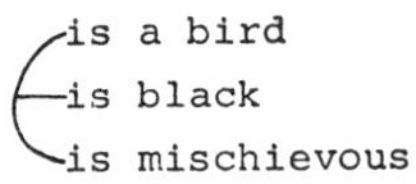

Fig. 23

There is no need of a point from which four lines of identity proceed; for two triple points answer the same purpose [Fig. 24]. Therefore a figure like this [Fig. 25] is to be understood as two distinct lines of identity crossing one another. Nevertheless, in order to avoid possible mistake a *bridge* may be represented thus [Fig. 26]. One line passes under the bridge, the other upon it.

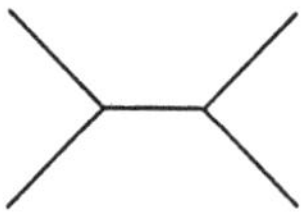

Fig. 24

Fig. 25

Fig. 26

The more you scribe on the bottom of a cut, the less you assert. Thus [Fig. 27] means: It is not true that somebody returns to earth nor is it true that somebody is translated. But [Fig. 28] merely says that *both* are not true. That is one or other is false. Either nobody returns to earth or else nobody is translated.

Fig. 27

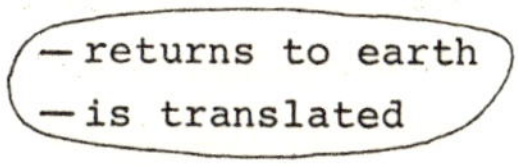

Fig. 28

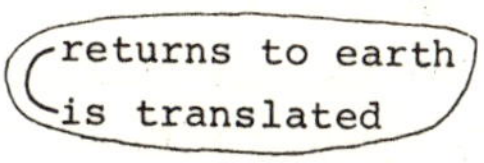

Fig. 29

Add to this a line of identity joining the two [Fig. 29], and still less is asserted. Either nobody is translated or if anybody is translated, that person does not return to earth.

Now take this [Fig. 30].

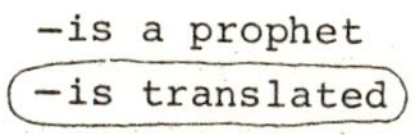

Fig. 30

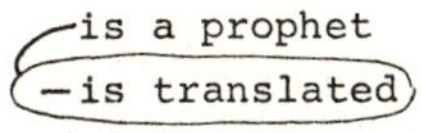

Fig. 31

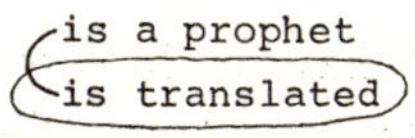

Fig. 32

That means somebody is a prophet but nobody is translated. If we continue the outer line to the cut, it will make no difference [Fig. 31]. For no significance attaches to the shape of the line. If, however, the inner line be extended to join the point on the cut, much less is asserted [Fig. 32]. This means: Somebody is a prophet and this person is not translated; that is, Some prophet is not translated.

Lines of identity bring only one new rule of illative transformation. (*Illative* transformation, by the way, is transformation of the nature of necessary inference.) But lines of identity require some slight changes to be made in the three primary rules already given.

Under the rule of omission and insertion, it is to be noted that a line of identity may be broken within an even number of cuts or on the sheet of assertion, while two lines may be joined within an odd number of cuts. We also have the curious fact that as far as this rule is concerned a point on a *cut* is to be treated as being within the cut, although in other respects it is to be treated as being outside the cut. The cause of this is that we take it for granted that something exists. How this cause produces this effect I shall leave it for you to make out.

The rule about the double enclosure receives an extension since not only are two cuts of no effect when no graph is between them, but they are equally so when nothing is between them except lines of identity that traverse the space between the two cuts.

The rule of iteration and deiteration takes a form which cannot easily be expressed without defining a new term, *ligature*. Namely, a line of identity is, as I have said, a graph and as such it cannot be part on one side of a cut and part on the other side. In the graph [Fig. 33] there are two lines of identity. But their having a common point on the cut identifies the individuals they denote. By a ligature is meant a line of identity together with all other series of identity that have points in common with it. For example [Fig. 34] means any man loves himself.

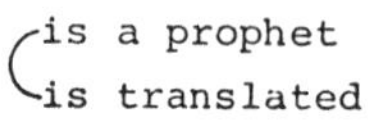

Fig. 33

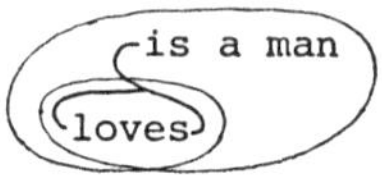

Fig. 34

It has four lines of identity, one attached to the monad spot "is a man," two attached to the dyad spot loves, and one joining the triple point to the inner cut. But all those make a single "ligature." Now the reformed rule of iteration and deiteration is, that any partial graph, detached or attached, may be iterated within the same or additional cuts provided every line or hook of the iterated graph be attached in the new replica to identically the same *ligatures* as in the primitive replica; and if a partial graph be already so iterated it can be deiterated by the erasure of one of the replicas which must be within every cut that the replica left standing is within. For example, suppose we have these premisses [Fig. 35].

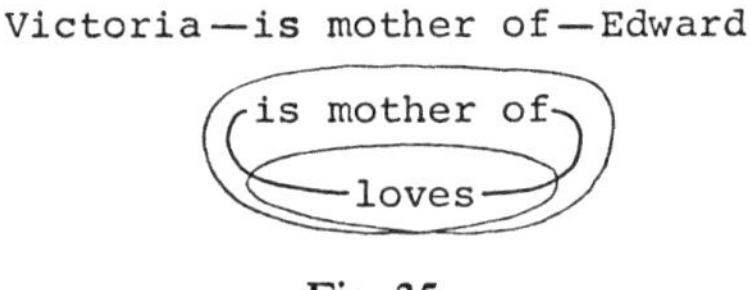

Fig. 35

We can iterate the two outside lines of identity within the outer cut, thus [Fig. 36]:

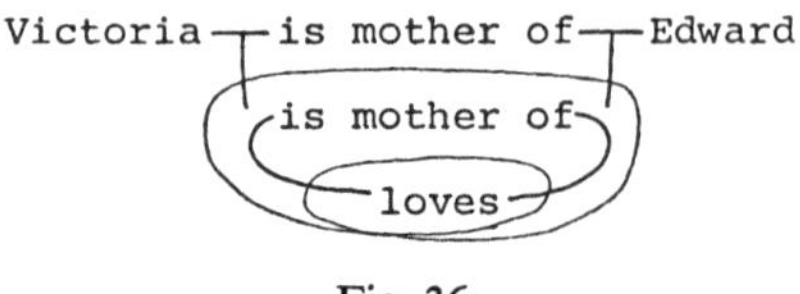

Fig. 36

Within one enclosure we can join the two lines on each side, thus [Fig. 37]:

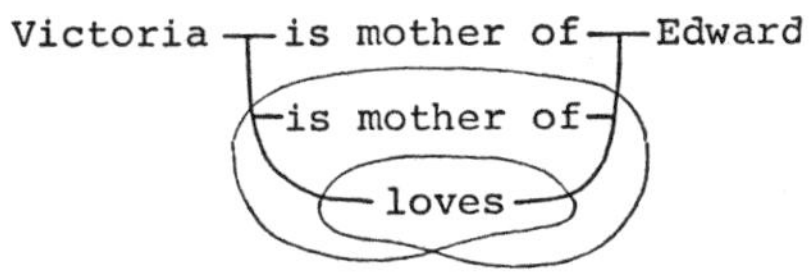

Fig. 37

We can now deiterate "mother of," thus [Fig. 38]:

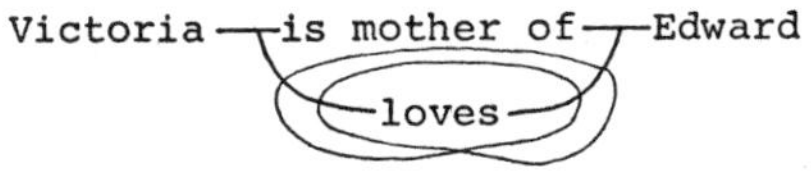

Fig. 38

We can now erase the two cuts which have nothing between them but lines of identity, thus [Fig. 39]:

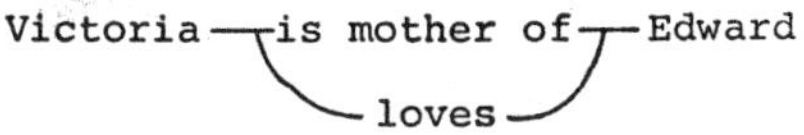

Fig. 39

We can now erase "mother of," thus [Fig. 40]:

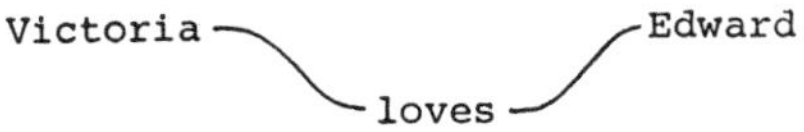

Fig. 40

I now proceed to the new fourth rule. It runs as follows:

> The innermost effective ligature between two spots lies within every cut that encloses both those spots.

In order to illustrate the meaning of this I take these [five] graphs [Fig. 41]:

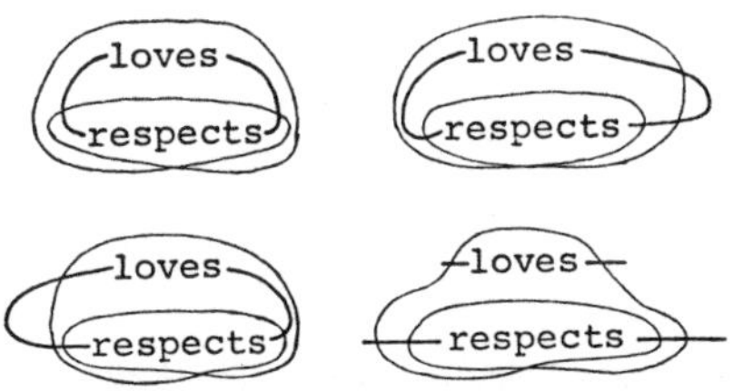

Fig. 41

The first three of these mean, respectively, "Nobody loves anybody whom he does not respect," "Somebody loves nobody whom he does not respect," "Somebody is loved by nobody who does not respect him." Those three propositions cannot be expressed, with the same degree of analysis without the ligatures, the innermost of which is within the cut that encloses both spots. But the fourth, which means "There is somebody whom somebody does not love unless he respects him" will not have its meaning changed by breaking both ligatures, as in the fifth graph, so as to make it read "Either there is somebody who non-loves somebody or else somebody respects somebody" or "If everybody loves everybody somebody respects somebody." The junctures protruding through two cuts could be cut without altering the meaning. By putting two cuts round the "loves" and retracting the junctures through two cuts we get the equivalent graph [Figs. 42a and 42b].

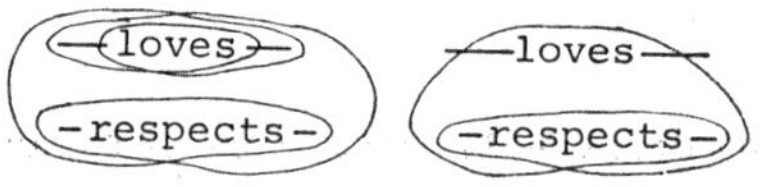

Fig. 42(a and b)

The third chapter of the exposition of existential graphs is by far the most important and interesting of the three. The whole gist of mathematical reasoning depends upon it. I shall have to remit it to another [lecture].[3]

[3] A good exposition of the gamma graphs is given in *The Existential Graphs of Charles S. Peirce* by Don D. Roberts (Mouton, The Hague, 1973).

B. REPLY TO MR. KEMPE (708)

I am glad to see the attention of logicians drawn to Mr. Kempe's *Memoir on Mathematical Form*,[1] which must endure as the classical work upon

[1] MS. 709 is a *Note on Kempe's Paper in Vol. XXI of the Proc. Lond. Math. Soc.* Its opening paragraph states that Kempe's paper, "which displays the author's mathematical power and still more his mathematical experience, also shows his native power in the direction of logic, and his sad want of training in this branch." A similar overtone is found in MS. 307 where Peirce writes:

"... — Throughout these ages, the chairs of Logic in the Universities have been turned over to a class of men, of whom we should be speaking far too euphemistically if we were to say that they have in no wise represented the Intellectual Level of their age. No, no; let us speak the plain truth. Modern logicians as a class have been distinctly puerile minds, the kinds of minds that never mature, and yet never have the *élan* and originality of youth. Just cast your eyes over the pages of a dozen average treatises, dismissing all preconceived estimates of their authors, and see if that is not the impression you derive from them. Why, in the majority of them, the greatest contribution to reasoning that has been generally applied during these centuries, — the Calculus of Probabilities is almost entirely ignored. If it were only the common run of logics that were affected by this state of things, it would not much matter; for if only one *per cent* of works on the subjects were what they should be, we should still be in possession of a splendid and extensive literature. But unfortunately the general standard has been so terribly lowered that even the treatises written by men of real ability have been but half thought out things. *Arnauld*, for example, was a thinker of considerable force, and yet *L'art de penser*, or the Port Royal Logic, is a shameful exhibit of what the two and a half centuries of man's greatest achievements could consider as a good account of how to think. (Lambert was so accurate a reasoner that he all but understood the outline of the Noneuclidean Geometry; and yet what a wretched disappointment his *Neues Organon* is!)

(Another effect has been that when great ideas have been born into the world they have not been properly cared for and developed. I should have no hesitation in saying that, excluding the mathematicians, Herbart was the greatest logical genius that nineteenth century Germany produced. But the state of science and of thought of his time forbade the proper development of his ideas, and look at what mountains of platitude his followers have inflicted upon the world!) You may retort that the past three centuries seem to have got on nicely without the aid of logic. Yes, I reply, they have, because there is one thing even more vital to science than intelligent methods; and that is, the sincere desire to find out the truth, whatever it may be; and *that* those centuries have been blessed with. But the subject to which it relates. When it first appeared, I feared that its logical importance might long escape recognition, since neither its title, nor the general character of the *Philosophical Transactions*, where it appeared, nor even the superficies of its substance would suggest its true significance. As to whether I have, or have not, misunderstood the system, what can the philosophical world care for that? Perhaps, I had better leave each student to decide for himself. Still, although for forty years I have engaged in no scientific inquiry, or business, or pleasure, without having in view its subserviency to my main task of penetrating into the secrets of logic, still I must confess I have not yet so depersonalized myself that I can with entire equanimity see myself accused of such careless study as to have overlooked the main feature of a logical system

---

according to such estimate, — not exactly mere guess-work, although rough enough, no doubt, — as I have been able to form, if logic during those centuries had been studied with half the *zeal* and *genius* that has been bestowed upon mathematics, the twentieth century might have opened with the special sciences generally, — particularly such vitally important sciences as molecular physics, chemistry, physiology, psychology, linguistics, and ancient historical criticism in a decidedly more advanced condition than there is much promise that they will have reached at the end of 1950. I shouldn't say that human lives were the most precious things in the world; but after all they have their value; and only think how many lives might thus have been saved. We can mention individuals who might probably have done more work; say Abel, Steiner, Galois, Sadi Carnot. Think of the labor of a generation of Germany being allowed to flow off into Hegelianism! Think of the extravagant admiration that half a generation of English, — decidedly the best average reasoners of any modern people, bestowed on that silly thing, Hamilton's New Analytic. Look through Vaihinger's commentary to see what an army of students have been entrapped by Kant's view of the relation between his Analytic and Synthetic Judgments, — a view that a study of the logic of relatives would at once have exploded.

Had logic not been sunk since the time of Copernicus into a condition of semi-idiocy, the Logic of Relatives would by this time have been pursued for three centuries by hundreds of students among whom there would have been no small number who in this direction or in that would have surpassed in ability any of the poor handful of students who have been at work upon it for the last generation or so. And let me tell you that this study would have completely revolutionized men's most general notions about logic, — the very ideas that are today current in the market-place and on the boulevards. One of the early results of such wide study of the logic of relatives must have been to cause the idea of *reaction* to be solidly fixed in the minds of all men as an irreducible category of *Thought*, — whatever place might have been accorded to it in metaphysics as a cosmical category. This I venture to say, notwithstanding that the lamented Schröder did not seem to see it so. Schröder followed Sigwart in his most fundamental ideas of logic. Now I entertain a high respect for Sigwart, — the kind of respect that I feel for Rollin as a historian, for Buffon as a zoölogist, for Priestly as a chemist, for Biot as a physicist, — a class of men whom οἱ πολλοί always place too high, and scientific specialists too low. He is one of the most critical and least inexact the fundamental importance of which I have taken pains to proclaim. I shall therefore beg leave to defend myself, hoping that I may be able to do this in such a way as to lend some general interest to my observations.

Each of Mr. Kempe's graphs consists of spots of various colors (using the word *color* to signify any qualities in respect to which any one spot can in itself differ from another) some of which spots are connected by simple lines of junction; and it is a point duly emphasized in the memoir

---

of the inexact logicians. Sigwart, like almost all the stronger logicians of today, present company excepted, makes the fundamental mistake of confounding the logical question with the psychological question. The psychological question is what processes the mind goes through. But the logical question is whether the conclusion that will be reached by applying this or that maxim will or will not accord with the *fact*. It may be that the mind is so constituted that that which our intellectual instinct approves, will be true to the extent to which that instinct approves of it. If so, that is an interesting fact about the human mind; but it has no relevancy for logic whatsoever. Sigwart says that the question of what is good logic and what bad must in the last resort come down to a question of how we feel; it is a matter of *Gefühl*, that is, a Quality of Feeling. And this he undertakes to demonstrate.

It is singular that as Schröder argues though without quite seeing what it was about against the admission of Category the Second even into the domain of logic so Category the Third finds an opponent in another man of high ability Mr. A. B. Kempe; and in this case I do not think there can be any possibility of attributing any naïve unconsciousness of hostility to the attacking party, although he does not explicitly say what the bearings of his remark are. It pretty clearly lies in the application thereof, like that of the illustrious pragmatist of David Copperfield. I certainly need not inform mathematicians who Mr. Kempe is. But I may say to the philosophical students that besides being a former President of the London Mathematical Society, he is interesting to us as the author of a very remarkable memoir on the Theory of Mathematical Form, which is from our point of view a splendid contribution to the Logic of Relations of almost Herculean grasp. Mr. Kempe in this memoir expresses different forms of relationship by means of so-called *graphs*. The word *graph* was invented by Sylvester as the designation of a kind of diagram, first employed, I think, with a purely formal signification by Dr. William Kingdon Clifford. These diagrams were suggested by those that are used by chemists to represent the constitution of chemical compounds. Kempe's graphs, like those of chemists, are composed of spots, variously marked, corresponding to the different atoms in the chemical formulae which spots are, some of them, connected by lines running from one to another, the lines being, as a general rule, of one kind only. Virtually, however, the lines are of two kinds, since disconnection is a second mode of connection. I am confident that I do not misinterpret Mr. Kempe's meaning, when I say that he is plainly of opinion that because he has no need of any third kind of element than those two, dots and lines, therefore two categories are all that enter into logic. At any rate, that argument against Category the Third cannot fail to suggest itself to any reader of his memoir; and I am bound either to confess its validity or to show wherein it fails. To this argument I have two distinct answers to make either of which completely refutes it.

that such simple junction amounts to a dichotomy between the pairs of spots and amounts to nothing more. This analysis of the idea of a "connexion," "mode of junction," or "dyadic relation" goes back at least as far as James Mill's *Analysis of the Human Mind.* It is fully accepted by me and, as I supposed, by Mr. Kempe, too. And yet his sole reason

---

In order to fix the ideas, I have drawn one of his graphs [Fig. 1].

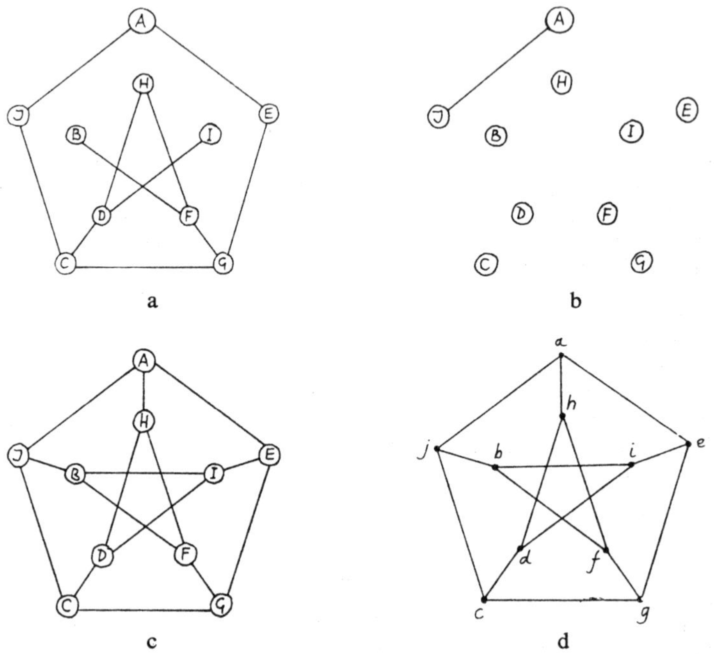

Fig. 1

Now it is quite obvious that the lines, each of which joins two spots and no more, represent Category the Second; and at first glance one naturally falls into the notion that the dots represent Category the First. But this is not so. Each dot reunites more than two lines namely three and reunites no less than six pairs of dots that are related to one another. The dots, therefore, represent Category the Third; and Category the First is represented by the black-board upon which the figure is immediately presented. So that after all Dr. Kempe has three kinds of elements not two as he supposes. That is one perfect reply to his objection. But in the second place, his argument rests on the assumption that all forms of relationship can be represented by his graphs. That, however, is manifestly not the case. For by such graphs he can in fact only represent forms of relationship between members of a finite collection of individuals. Consequently he cannot represent *general* systems of relationship at all. But I have invented a system of graphs which will represent all possible forms of relationship."

See *Collected Papers* 5.82-5.87.

for saying that I have misunderstood his system is that I said the lines of his graphs each connected two spots, or, in other words, represented a relation between the spots, while the truth of the matter is, he avers, *not that*, but that the lines mark out a class of pairs of spots. Has Mr. Kempe, then, all along failed to perceive that a "connexion," "mode of junction," or "dyadic relation," is nothing more nor less than a class of pairs? Or is it merely a case of *bonus dormitat Homerus*?

The presumption is that Mr. Kempe understands his own meaning, and that if he says I have misunderstood him, I have in reality done so. But this is not necessarily the case. For the discoverer of a logical idea does not acquire lordship over that idea, to give it any turn he likes. Much truer is it that it exercises rights over him, and that his duty is docilely to follow it whithersoever it may lead. Mr. Kempe understands his graphs as being composed of nothing but spots and connections (or specialized pairs) of spots. If that be true, the system is inadequate to represent reasoning; for reasoning involves the idea of an irreducible triad. Nor have any instances ever been published to demonstrate that Mr. Kempe's system is adequate to represent reasoning. It is true that my graphs seem at first sight to be nothing but a special variety of Mr. Kempe's. But mine, at any rate, do possess the third element in the *rules* which govern their transformations. There are traces of such rules even in Mr. Kempe's system. For, as he himself remarks, in place of any graph we may substitute another which has a bond between any two spots unbounded in the other, and *vice versa*. Nor is this the only rule of transformation in Mr. Kempe's system.

I do not know that Mr. Kempe actually asserts that there are only two elements in his graphs, the spots and the bonds. Perhaps he has never asked himself the question. But the question is of the first importance not merely for a Kantian application of logical conceptions to metaphysics, and in the eye of theoretical logic, but even for the purposes of practical reasoning. Setting aside rules for the moment, and looking only at a single graph, we see three elements, the bonds, and the spots, and the colors of the spots. In fact, spots are not paired merely by lines, and by the absence of connecting lines, but also by the likeness (and unlikeness) of the colors of the spots in themselves. (Hence, it is impossible to reverse *all* the specializations of pairs, — a remark which would lead to some interesting developments.) But why not docilely follow out the idea of representing connections by lines so far as to represent the connexion between a spot and its color by a line? That is, add one spot to represent each color, and instead of coloring the original spots, leave them uncolored, and draw a line between each and its color-spot. In that case, a formal distinction between spots at once emerges (even if it had not happened to appear in the original graph). Namely, some spots have a self-identity and distinction from all others, in themselves, while others are indeed distinct, but only so far as their connections with other spots are different. But continuing to docilely follow out the idea of representing all connections by lines, why not represent the connection between a spot and its individual character by a line, adding a special spot to represent that individual character? Each of the color-spots has, if the suggestion just made (but by no means as original with me) be followed out, an individual character. Consequently, when this last prescription of our governing idea is obeyed, the only kind of elements which have any distinctive characters in themselves, or say, which have any *monadic* characters, are spots each of which has but a single connexion with anything else, and each of which is the only thing which has that monadic character. As a matter of phraseology, since "color" suggests something general, or apt to inhere in many things, while letters are habitually used in geometry and algebra as proper names of single objects, we had better say that each of these *monads* bears a special letter. It will naturally be drawn as a pear-shaped outline with a letter at the broad end, 

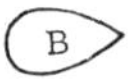

while at the pointed end it may be connected with something else. The second kind of element of the graph will consist of the dyadic elements, which may be represented by simple lines, or better by mere contacts. The third, and last, kind of element will consist of all the other spots, which, having now no distinctive colors nor identities (except by their connections), may be represented as mere nodes, or points of concourse of lines. These lines can be neither one, nor two, in number. If they are more than three, as in a four-rayed node,  , the node may [be] separated into a linking of three-rayed nodes, 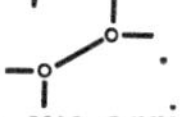. Accordingly, when we follow out the idea of Mr. Kempe's system, we arrive at the result that each graph consists of *monads*, *dyads*, and *triads*. So much it was necessary for me to say before I could explain the relation of my system to his.

I am now about to set forth one of the considerations which has led me to a different system. It is perhaps not the weightiest of my reasons, but it is particularly pertinent as showing one of the main distinctions, — the one which he thinks I do not understand, — between his graphs and mine. Its study will, therefore, lead to a further answer to his strictures. Each of Mr. Kempe's dyads, — I mean each of the lines, or bonds of his graphs, — is exhibited as having like terminals. I do not say the terminals *are* alike, because, when the spots they connect are different that *makes* the terminals different; and the spots almost always *are* different in *some* respect. But, as *exhibited* in the line itself, the two terminals of each dyad are alike. Mr. Kempe might equally well have exhibited them as always different, by giving, for instance, to one end an arrowhead. Had he done so, then in any special case in which it might be desired to represent a reciprocal connexion (or unordered pair) of spots, two oppositely place arrows could have been drawn between the two spots. Of these two systems, Mr. Kempe has chosen one. Was there, then, any good ground for preferring either to the other, and if so which system ought to have been preferred? I maintain that it is a matter, not of fancy nor of opinion, but of demonstration, that the system Mr. Kempe has chosen is the one which fulfils the less perfectly the purposes of logic. The distinction of prior and posterior, which makes the ordered pair, is the very soil out of which logic grows, the solid earth on which it is supported. One thing follows from another; one thing is true and primary, while its opposite is false and secondary. Else, logic cannot exist. What does it mean to say that *A* and *B* form an ordered pair, *AB*? Of course, any two things can be conceived as forming two ordered pairs, *AB* and *BA*. But there are certain modes of relation which at any moment we include in the universe of relations which we are thinking of, and to say that *AB* is an ordered pair means that *A* has some such relation to *B*. What does it mean to say that *A* and *B* form an unordered pair? It means that our universe of pairs is such that *AB* belongs to it and so does *BA*. *A* is connected with *B*, and *B* with *A*. That is to say, the ordered pair *AB* belongs to the class of pairs considered (and signified by a bond of the graph), and so does the ordered pair *BA*. Then, it is plain that to say that *AB* is an ordered pair of the class considered and marked by lines in the graph is to say *less*, then to say that *A* and *B* form an unordered pair of the kind considered and marked in the graph. For the latter assertion implies the former and further that *BA* is [a] pair of the class considered and marked. This is clear, because it is certain that if spot *A* is joined to spot *B* by a line then spot *B* is equally joined to spot *A* by the same line. Hence, the ordered collection has less implication, or is logically simpler, than the unordered pair. Hence, the direct representation of pairs as ordered gives a more analytical form of representation, which is what is needed for logical purposes. Even if it be contended that to say that *A* and *B* form an unordered pair means that either *A* is connected with *B* or else *B* is connected with *A* (and that this is *not* what is meant is shown by the fact that we can hardly make any sense of such a statement), even then the ordered pair is simpler. For we cannot form a single clear conception of such a statement. This is shown by the circumstance that we never say, John and Henry are father and son. We say, John is Henry's father, or else we say Henry is John's father. So, we do not say, Plaintiff or defendant will win the suit, but "the suit will be decided." When we say James and John are brothers, we mean James is John's brother *and* John is James's brother. When we say, "these two men are to fight and one will kill the other," in the first clause we make them an unordered pair, because each fights the other; but in the second clause this disjunction remains uncomprehended under a single conception. Consequently, even if the unordered connexion were understood in the disjunctive sense, the conception would be more complex that that of the ordered connexion. Again, unordered pairs immediately result from compounds of ordered pairings, but not the reverse. Thus, if *A* implies that *B* implies *C*, then *B* implies that *A* implies *C*, and there is a reciprocal relation between *A* and *B*. But now consider the device to which Mr. Kempe has to resort to represent an ordered relation between two black spots. He draws this figure [1]. Who does not see that the success of this (which is perfect) is due to the introduction of pairs that are necessarily ordered, since their terminals are unlike? For example, he might represent *A* to be father of *B* by this graph [Fig. 2].

Fig. 1

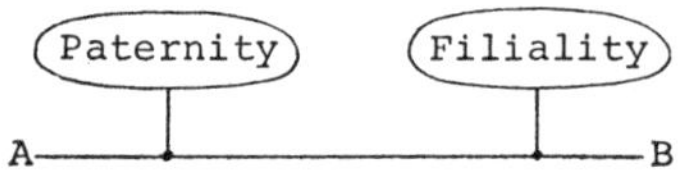

Fig. 2

Can we possibly overlook the non-reciprocal relation of Paternity to Filiality? Similar remarks apply to triads. Their three places of attachment should be different, just as those of dyads ought to be.

But the moment we once decide to accept this view, the whole idea of throwing the color, or distinguishing inward characters, entirely upon the monads has to be abandoned. It is true that any such character is of a monadic order, that is, it belongs to its subject within that subject itself. But as a pair or triad is composed of units, so likewise has the pair or triad its monadic character. Our graph to represent that *A* is father of *B* must, therefore, be something like this [Fig. 3]. The same thing is true of triads.

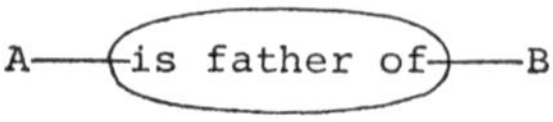

Fig. 3

This is a chief feature of the plan, upon which my system of graphs is constructed. Accordingly, I was quite right when I said that in Mr. Kempe's system the relations (that is, the dyads and triads) are represented by lines (or nodes of lines), *being exhibited as all alike except that some* (as being known) *are specialized* in one general way, while in my system the relations are spots, or, in other words, are positively colored.

Now as to the subjects, since in Mr. Kempe's system the lines represent the pairings, or rather the *asserted* pairings, the subjects of those pairings must be the spots which the pairs unite. Personally, Mr. Kempe regards mere nodes as spots, and mere nodes (I grant) ought not to be considered as subjects. But it seems to me that the spirit of his system is more docilely followed, if the nodes are regarded as junctions of sets of more than two objects to a set. In that case, all the spots ought to be regarded as subjects.

The spirit of my system, on the other hand, is to throw the color entirely upon the predicates, whether they be monads, dyads, or triads, and this leaves the subjects without color. That is to say, the subjects are either lines or topical singularities of lines, isolated points, extremities, or points of branching. Accordingly, I was equally right when I said that *in Mr. Kempe's system the subjects are represented by spots, while in mine they are represented by lines.*

I have never denied that there is another point of view from which my graphs are but a variety of Mr. Kempe's; but I do not think that point of view gives so true an idea of Mr. Kempe's work as the one I have here taken. For from the moment when he discards lines pointing one way, as he does in the main, in my opinion he specializes his system very much[2].

[2] MS. 714, dated January 15, 1889, contains the following statements:

"These ideas of Kempe simplified and combined with mine on the algebra of logic should give some general method in mathematics."

"I note that lines may be treated as monads and so a new graph made and the question is whether there would be any advantage in this.

"Also, what would be [the] inverse of this proceeding and what would be its advantages?"

"There will be certain general relations among the parts of the graph and these relations or rules may themselves be treated as monads."

"After the graph has been reduced to lines all alike and undirected and monads all alike, what is the proper algebra to express the general relations existing between its parts and to deduce others. Try this on the graph of the Boolian algebra."

"Kempe gets rid of triple relatives! But let us look closely into that! I have stated in my paper substantially that this will not do. But my idea from this point of view takes the form that no monad need have but 3 connections."

"I now go further than Kempe and say that every state of things or relationship can be represented by a graph or tree where all the joints are into branches and only the terminals have any material significations. And only one terminal to each signification. The joints having only formal signification. There are only two kinds of relation formally represented, the line and the non-line."

A second draft of MS. 708 has not been used here.

C. THE ALPHA PART OF EXISTENTIAL GRAPHS (s-26)

In all mathematics it is essential that there should be a multitude of objects which the student can perceive (usually see, but it would be sufficient if he could imagine them), and is able to modify at will (he usually creates and destroys them), but in doing this obeys certain general rules (usually conventional in part and physically compulsory in part). Then, his work consists in imagining problems which suppose the changeable objects to be at the beginning in a state defined in general terms and which require the discovery of an infallible general method by which they may be brought (without violation of the rules) to another state defined in general terms.

It would be, I suppose, psychologically impossible for a man to devote much energy to any branch of mathematics unless he supposed it likely that the rules of it were so analogous to the general laws of some relations of real objects, that the former might be considered as representing the latter; and thus we find that mathematicians usually speak of the mutable objects they study as signs, symbols, images, and the like. But their being such has no relation to any part of the mathematician's business except that of the formation and adoption of hypotheses.

The purpose and uses of the system of existential graphs cannot be explained until the system itself has been set forth. Certain features of the system have been adopted for trivial reasons; and it will be well that this should be understood at the outset.

The *replicas* (**1**), or things which the student sees and subjects to transformations, are represented as superficial objects, or films. This is convenient, because it is necessary that they should exhibit a great variety of easily recognized differences even in their *atoms* (**2**), that is, in those parts of them which are never cut in the transformations or even usually thought of as composite. The advantage, for this purpose, of making them visible surfaces is obvious; or at least a part of it is so. The different recognizable *types* of the atoms are called *elements* (**3**). Thus, the distinction between an *atom* and an *element* is no greater than that which we observe between two senses of the word "word." The word "the" usually occurs about a score of times upon a printed page. In one sense, it is always the same "word"; in another sense it is twenty different "words." The former sense is analogous to that of the term *element*, the latter to that of the term *atom*. The *atom* is the individual object that is seen and transmuted; the *element* is the perfectly definite *type* which is associated with atoms that are alike in their essential characters, though not in their existential relations of place, etc., this type being, in itself, such as it is, incapable of mutation, and only recognized by thought, not in itself a direct object of experience. The atoms and the objects that are composed of atoms are called *replicas* (**4**). The complex types, any one of which belongs only to replicas that are exactly alike in all essential respects and are different only in existential respects, are called *graphs* (**5**).

The replicas are imagined to be written or drawn, — *scribed* (**6**) is the term employed, — upon a sheet of paper or a blackboard. This is obviously convenient, since they are in great variety [and] must be recognized, made, unmade, and modified by the student. But there is hardly anything about the graphs which is analogous to any essential properties of a surface. In the Alpha Part all the atoms stand separate from one another; so that there is nothing like continuity. In the Beta Part, there is a relation sufficiently analogous to continuity to make its representation by means of continuity decidedly apt. But the restriction to a surface is severely felt as inconvenient and essentially unsuitable; but a third dimension, which is needed, would be still more inconvenient. No replica is scribed wholly or partly over another. But as to this certain explanations will be given presently.

Besides the replicas, we place upon the sheet certain fine self-returning lines which we do not regard as "scribed" and which we call *cuts* (**7**). No two cuts ever intersect; but there are frequently cuts within cuts. A series of cuts one within another is called a *nest* (**8**) of cuts. A nest of two cuts (generally drawn as one nodal line) is called a *scroll* (**9**). Two nests of cuts may have cuts in common; and one nest of cuts may be within another so as to constitute with it another nest. But of every pair of cuts of the same nest one is immediately or mediately within the other. One cut is said to be *immediately within* (**10**) another if it is within that other but is not within any third cut that is within that other.

The entire extent of that part of the sheet which is not within any cut, including the linear space occupied by the cut itself, is called the *area of assertion* (**11**), and relatively to any cut that does not lie within any other, it is called the *place* (**12**) of that cut. I intend very slightly to modify this statement and some others presently; but I think the easiest way to make the matter clear will be to state it as I am doing, in the first instance. The entire surface that is contained within any cut but is not contained within any cut within that cut is called the *area of the exterior cut* (**13**) and is called the *place* (**14**) of any cut immediately within that cut. The *area of a nest* (**15**) of cuts is the space within all the cuts of the nest; but in speaking of a scroll we often speak of the *outer* and *inner close* (**16**). By an *area* (**17**), simply, is meant either the area of assertion or the area of some cut. It is a superficial place which does not include its outer limit but does include any inner limit that it may have. An area either is, or may be, in part or in whole occupied by replicas; and the part occupied does not cease to be a part of the area.

Every replica is situated upon some one area which is its *place* (**18**), and is never traversed by a cut. Every area may have a replica scribed upon it. But now a fine distinction has to be noticed. Although a cut is not a replica, and surely is far from being a graph, yet every cut taken together with whatever replica is upon its area is regarded as constituting a replica (called an *enclosure* (**19**)) whose *place* (**20**) is the place of the cut. The area of the cut is called also the *area of the enclosure* (**21**). Since this is not an easy conception to grasp surely and steadily, I will suggest a way of gaining a sort of image of it. It is with reference to this that I said just now that some of my statements would be slightly modified. Instead of conceiving a *cut* (**22**) as separating the surface into two parts, it may be imagined to be a line of splitting of the surface such that inside the cut there are two sheets. Upon one of these sheets (say the under one, in order to "fix the ideas") are scribed whatever replicas are in the area of the cut; and this sheet is regarded as being no part of the main sheet outside. The other sheet that overlies that one is simply a part of the outer area, and is occupied with a replica the character of which depends upon the graphs that have replicas on the under sheet, but not in the same way as if they were scribed on this very same sheet. According to this, there would be two ways in which a graph may govern the character of a replica; namely, firstly, by a direct action of the graph upon a part of the sheet determining a replica of itself; and secondly, by the graph's acting directly in this way to produce a replica upon another sheet, which, being then brought under the main sheet, acts as a sort of photographic negative, to produce a second replica on the main sheet. Then, in case there is another cut within the first, this is to be conceived as a line of splitting of the lower sheet; so that there is first a direct replica of the graph upon the lower sheet of the split lower sheet, which is the area of the inner close; that influences a secondary replica upon the main sheet of the lower sheet, which is the outer close, and that again produces a tertiary replica upon the upper sheet, which is the place of the scroll. I do not know as every reader will find this fancy a help; but at any rate, he must in some way contrive to understand that while the area of a cut, together with what replicas there may be scribed upon it, are severed from the outer area which is the *place* (**23**) of the cut, yet the *enclosure* (**23**) is a replica which is upon that outer area as its place, and the peculiar graph of this enclosure is not the graph of the replica in the area of the cut but is determined to be the graph of the enclosure by the fact that the replica of the area of the cut is severed from the place of the cut. Moreover, when we come to the rules of these graphs, it will be found that the different areas, one within the other, of the cuts of a nest alternate in their natures in a way that harmonizes very well with the idea of the photographic negative. For this reason, it will be useful to distinguish replicas as *evenly enclosed* (**24**) and *oddly enclosed* (**24**) according as they are within even or odd numbers of cuts.

It will be necessary to correct what I have been saying in still another particular. For I have talked as if there might or might not be a replica upon the area of a cut. But it is far better to regard the parts of the sheet where nothing has been scribed and which remain in their original condition, nothing being there but *blank spaces* (**25**), as the phrase is, as being occupied with replicas of one graph, the *blank* (**25**). The view ought, at first blush, to recommend itself to mathematicians, to whom nothing is always a quantity when they are speaking of quantities. So when we are speaking of replicas filling a space, these spaces are filled with the absence of scribed replicas. In applying the system to logic, as I shall do later, this conception will become still more imperative. In this view, scribing and erasing are simply special cases of transformations of replicas; and this simplifies the rules. It will then be necessary to distinguish the *entire replica* (**26**) of an area from its *entire created* (**27**) (or *factitious* (**28**)) *replica*, the latter excluding the blank; while the *entire scribed replica* (**29**) will exclude enclosures. Graphs may be distinguished in the same way. *Partial replicas* and *partial graphs* (**30**), meaning replicas there are parts of the entire replica and their graphs, will also be distinguished, if necessary, by the adjectives *created* (**31**) (or *factitious*) and *scribed.*

I may mention here that it sometimes becomes necessary to speak of replicas as being *instantaneously created* (**32**). This means that although the graphist may be obliged to construct a replica part by part, and although in doing so he may perforce make something that seems to be a replica, yet it must be understood that all power of functioning as such is suspended until the construction is complete; as if it had been created in an instant.

There now remains, I believe, but a single group of definitions that need be added. It sometimes happens (and I leave the reader to guess whether the circumstance is unimportant or not) that there is upon an area a replica which renders it permissible to erase any and all other replicas upon that area and equally renders it permissible to scribe any and all other replicas upon that area. It is thus left to the option of the graphist to produce any effect that a replica in that area can bring about. How great, then, is likely to be the force of a replica in that area? An area so affected is said to be *opplete* (**33**) or to be *oppleted* (**34**) (from *opplēre*, to stuff up). Or we may prefer to say that it is the *annulus* (**35**), or annular space, comprising all that area except that occupied by the replica that effects the *oppletion* (**36**) that is *oppleted* (**37**). Or again, we may say that the enclosure in the area of which the *opplent* (**38**) replica occurs is *opplete* (**39**). Connected with this conception is that of a *vacant enclosure* (**40**), which is an enclosure whose area is entirely blank.

I have numbered [...] the definitions of this section, and I now add an alphabetical index to them.

Those of the rules, or rather the permissions, of the system of existential graphs, may, which are first principles, the essential definitions of the working of the system, be properly termed *postulates.* They relate, almost, if not quite, exclusively to what it is permitted to *scribe* or *insert*, and to what it is permitted to *erase* or *omit.*

Postulates of the Alpha Part

It is permitted

A. To scribe on the area of assertion whatever is needed to represent the data of the problem in hand;

B. To erase all there is on the area of assertion;

C. To make any transformation on the area of assertion, regardless of what else there may be there, that would be permissible if the subject of transformation were alone on that area;

D. To insert upon the area of assertion a replica of any graph of which it would be permissible to scribe a replica upon the inner close of a vacant scroll;

E. Instantaneously to create on the area of assertion, when entirely blank, a vacant scroll;

F. To perform on the area of any cut placed upon the area of assertion the reverse of any transformation that would be permissible on the area of assertion.

8

THE FOUR-COLOR PROBLEM

A. ON THE PROBLEM OF COLORING A MAP (153)

Some time in the early sixties Augustus DeMorgan mentioned in the Athenæum that experience showed that four colors would suffice to distinguish confine regions upon any map, and said that it was a reproach to logic and to mathematics that no proof had been found of a proposition so simple. It is plain from his interest in the logic of mathematics that he had himself attempted the demonstration in vain. In 1878, Arthur Cayley took up the question but abandoned it without discovering the desired proof. In the following year Mr. Kempe, since President of the London Mathematical Society, proposed a simple demonstration, which is unfortunately fallacious. Other strong mathematicians are known to have been equally unsuccessful.

The difficulty of the theorem becomes apparent when it is stated in analytical form. Let the regions be numbered, 1, 2, 3, etc. Let the four colors be A, B, C, D. Then let any of these four letters with any number of a region subscribed, as A_3, be a quantity which is equal to 1 if the region denoted by the subscript number has the color denoted by the letter, and which if this is not the case is equal to 0. Then for every region, say i, there are the following seven equations:

$$A_iB_i = 0 \quad A_iC_i = 0 \quad A_iD_i = 0 \quad B_iC_i = 0 \quad B_iD_i = 0 \quad C_iD_i = 0$$
$$A_i + B_i + C_i + D_i = 1.$$

These equations express the fact that every region has one of the four colors but not more than one. Moreover, every map furnishes a number of equations expressing the prohibition against coloring two confine regions alike. Thus, if regions i and j are confine, we have the equations:

$$A_iA_j = 0 \quad B_iB_j = 0 \quad C_iC_j = 0 \quad D_iD_j = 0.$$

Let such a set of four equations be called a map-condition. The theorem is that the equations can always be satisfied, no matter what map-conditions may be given, provided these are such as are possible on a spheroidal surface, and the definitions of such a spheroidal collection of map-conditions [are] as follows. A *mass* of regions may be defined as a collection of regions which cannot be separated into two subcollections such that no map-condition connects two regions of different subcollections. Two *masses* of regions may be said to be *confine* to one another if a map-condition connects two regions of the one and the other mass. Now, a collection of map-conditions is *spheroidal* if no five masses of regions are all confine to one another.

When we enunciate the proposition in this abstract form we perceive that it is very intricate. Its apparent simplicity is due to the ordinary statement dealing with ideas which experience and our mental constitution render very familiar without helping us much to analyze them.

It is doubtful whether the argument from experience is of the slightest weight.[1] Considering the intricacy of the problem, it may be that if the map satisfied some complicated condition not at all likely to be satisfied by any blindly constructed map, the coloring with four colors would be impossible. As long as we have nothing to assure us that if there were such a condition some map that we should be apt to draw would satisfy it, the logic of the argument is so dubious, that it is hard to say whether any reliance should be put upon it.

The problem may be somewhat simplified by some obvious considerations. Let a *boundary* be defined as a line at the common limit of two regions and extending as far in each direction as those two regions have a common limit. Let a *corner* be defined as a point at the common extremity of three or more regions. If more than three boundaries meet at any corner, we may slightly alter the map so as [to] separate this

[1] A fuller statement is found in another draft which reads:

"It is doubtful whether the slightest weight ought to be attached to the argument from experience. For we now perceive that the proposition is one of extreme complexity. It is, therefore, by no means clear that there might not be some intricately inter-related set of map-conditions which would render it impossible to satisfy all the conditions. If there were such sets of map-conditions they would probably have to fulfil conditions of inter-relation so complicated and peculiar, that it would be very unlikely that out of thousands of maps we might draw blindly, not knowing those conditions, there should be one that accidentally fulfilled them; the more so, since the maps we should be likely to draw are probably limited to certain kinds for psychological reasons. Now when the logical value of an argument is open to so much doubt, the utmost favor that ought to be accorded to a conclusion is that of a working hypothesis. A scientific induction is limited to such experiences as may occur under the same conditions as those of the experiment or sample from which the induction is drawn; and it is not absolute even under that limitation. But a mathematical proposition is not subject to any conditions of experience. Consequently, induction, as such, is of no value at all in mathematics, although a confused deductive reasoning may be mistaken for an induction."

corner into a series of corners at each of which three boundaries meet. For such an alteration cannot diminish the difficulty of coloring the map with four colors, since it will only add new map-conditions. I shall, therefore, assume that three boundaries meet at every corner. I shall further assume that there are no regions, and no masses of two regions, having any cyclosis. For if a map having such a region or mass of two regions requires five colors to color it, so also evidently does some map of fewer regions none of which nor any mass of two of them has any cyclosis. I shall further assume that there are no regions of fewer than four boundaries. For if there are, and each of these regions is destroyed by erasing one of its boundaries, and the map can then be colored with four colors, it will be easy afterwards to restore those regions and to color them without a fifth color. I shall further assume that there are no regions having four boundaries. For if there be such a region let two opposite boundaries of it be erased, thereby diminishing the number of regions by two. Continue this process until there is no longer any region of four boundaries. If the map can now be colored with four colors, after this is done, all the regions of four boundaries can be restored in the inverse order of their destruction. For example, the map of Fig. 1 has six regions of four boundaries. If these are destroyed by erasing the boundaries marked by cross-lines, the result will be a map of six regions all of four boundaries, — a quasi cube. These regions of this modified map are marked A_1, A_2, B_1, B_2, C_1, C_2. It is redrawn in Fig. 2. If the two opposite boundaries of A, marked by cross-lines are erased, we get a map which is redrawn in Fig. 3. It can be colored with colors A, B, C; and the original quadrilaterals can be colored with the fourth color D.

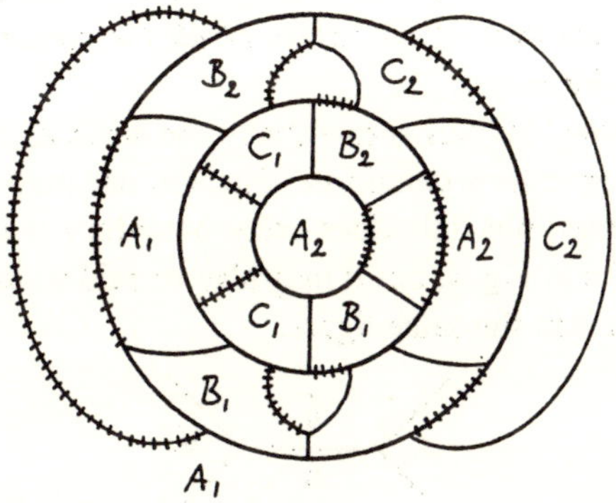

Fig. 1

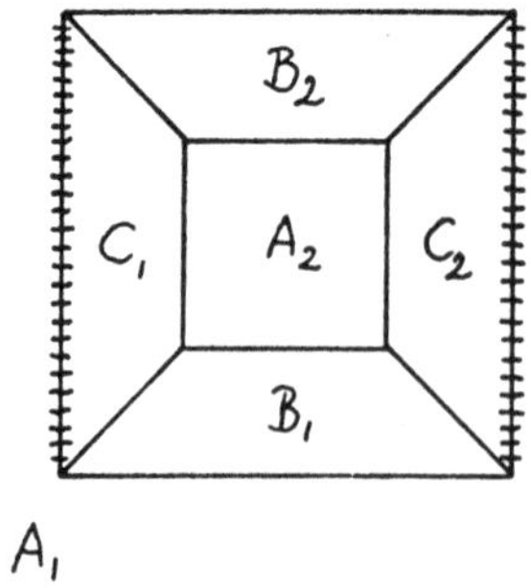

Fig. 2

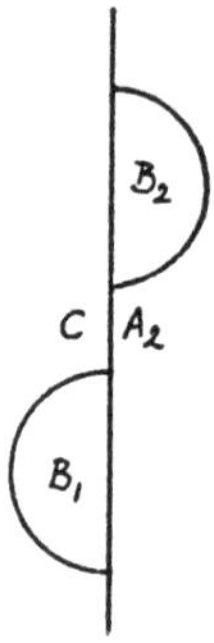

Fig. 3

When these modifications of the map have been made the Census Theorem of Euler* takes the form

$$N_5 = 12 + N_7 + 2N_8 + 3N_9 + 4N_{10} + \text{etc.}$$

where N_i is the number of regions of i boundaries. (*Legendre VII, 25. In a note Legendre refers to Euler's demonstration of 1758. Mr. Kempe wrongly attributes the theorem to Cauchy, a curious illustration of how Legendre's Geometry is unread in England.) It may be remarked that some sets of values of the Ns which this formula would permit are nevertheless impossible. For example, there cannot be a map of this sort with just one region of more than five boundaries. Moreover, many arrangements of the polygons are impossible.

It is to be expected that the problem would be simplified if we could reduce the number of classes, or colors, from four to two. In order to effect that result, I consider the corners instead of the regions. The problem then takes the following form:

It is required to prove that in every map of the kind considered, it is possible to assign to each corner one of the two values $+1$ *and* -1 *in such a way that the sum of the values of all the corners of every region shall be divisible by* 3.

In order to show that this is essentially the same as the original problem, let me first propose a definition. Since in our map three boundaries meet at each corner, not only does every boundary *separate* two *confine* regions, as *b* and *c* in Fig. 4, but it may also be said to *tie* two *peneconfine* regions, as *a* and *d* in the same figure.

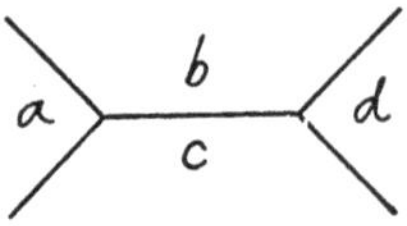

Fig. 4

Now suppose one of the values $+1$ and -1 has been assigned to each corner of such a map. Then, begin with any three mutually confine regions and color them A, B, C. Let us now proceed to color the others by the rule that two peneconfine regions tied by a boundary whose limiting corners have different values shall receive the same color, while if those values are the same the color to be assigned to the region to be colored shall not be that of the region that boundary ties to it nor that of either of the confine regions that boundary separates but shall be the fourth color. We have to prove that this rule will never lead to any contradictory results, by giving different colors to the same region according as it is reached by different paths, and also that it will never result in coloring confine regions with the same color. Let us form a notation for the regions, corners, and boundaries as follows: Beginning at any corner taken arbitrarily, denote that by A, and denote one of the three corners adjacent to it (that is at the other end of the same boundary) by B. Let the path along the boundary from A to B be denoted by $B-A$. Let the region on the right hand of this boundary be denoted by r, and a right-handed turn from one summit to the next round the region on the left-hand side of the same boundary be denoted by l. Let us for the moment denote that corner of r next after B in the sense that B is next after A, by C. We will then write

$$C - B = r(B - A)$$

and in the same way, if D be the next corner,

$$D - C = r(C - B).$$

We will extend this principle of notation to all bendings toward the right in going round any region. Let E be the third corner adjacent to B. Then upon that principle we must write

$$A - B = l(B - E).$$

It follows that we must denote the third region confine to r and l at B by $-\bar{l}\bar{r}$, where I put an obelus over a letter to denote its reciprocal. For, calling this region, for a moment, x, we have

$$E - B = x(B - C).$$

But we have just seen that

$$E - B = \bar{l}(B - A)$$

and that $$B - A = \bar{r}(C - B).$$

Consequently, $$E - B = -\bar{l}\bar{r}(B - C)$$

or $$x = -\bar{l}\bar{r}.$$

On the same principle the succession of regions confine to r is

$$
\begin{aligned}
&l \\
&-\bar{l}\bar{r} \\
&-(-\bar{l}\bar{r})^{-1}\,\bar{r} &&= rl\bar{r} \\
&-(rl\bar{r})^{-1}\,\bar{r} &&= -r\bar{l}\bar{r}^{2} \\
&-(-rl\bar{r}^{2})^{-1}\,\bar{r} &&= r^{2}l\bar{r}^{2} \\
&-(r^{2}l\bar{r}^{2})^{-1}\,\bar{r} &&= -r^{2}\bar{l}\bar{r}^{3} \\
&\text{etc.}
\end{aligned}
$$

So the succession of regions confine to $rl\bar{r}$, for example, will be

$$
\begin{aligned}
&r \\
&-\bar{r}(rl\bar{r})^{-1} &&= -\bar{l}\bar{r} \\
&-(-\bar{l}\bar{r})^{-1}\,(\bar{r}l\bar{r})^{-1} &&= rlr\bar{l}\bar{r} \\
&-(rlr\bar{l}\bar{r})^{-1}\,(rl\bar{r})^{-1} &&= -rl\bar{r}\bar{l}^{2}\bar{r} \\
&-(-rl\bar{r}\bar{l}^{2}\bar{r})^{-1}\,(rl\bar{r})^{-1} &&= rl^{2}r\bar{l}^{2}\bar{r} \\
&\text{etc.}
\end{aligned}
$$

In general, if u and v are any two regions confine to one another, the succession of regions confine to u going round it clockwise are

$$
\begin{array}{ll}
1 & v \\
2 & -\bar{v}\bar{u} \\
3 & uv\bar{u} \\
4 & -u\bar{v}\bar{u}^{2} \\
5 & u^{2}v\bar{u}^{2} \\
6 & -u^{2}\bar{v}\bar{u}^{3} \\
7 & u^{3}v\bar{u}^{3} \\
8 & -u^{3}\bar{v}\bar{u}^{4} \\
9 & u^{4}v\bar{u}^{4} \\
 & \text{etc.}
\end{array}
$$

If u is quinquilateral, we have

$$-u^{2}\bar{v}\bar{u}^{3} = v \quad \text{or} \quad u^{2}\bar{v}\bar{u}^{3}\bar{v} = -1,$$

where the factors may be cyclically transposed, or where the reciprocal, $vu^{3}v\bar{u}^{2} = -1$, may be written, where in place of v we may put any other region confine to u. Thus

$$+u^{2}(-\bar{v}\bar{u})^{-1}\bar{u}^{3}(-\bar{v}\bar{u})^{-1} = +u^{3}v\bar{u}^{2} = -1.$$

If u is septilateral, we have

$$u^{4}v\bar{u}^{3}v = -1,$$

and so forth for any odd number of boundaries.

If u is sexalateral, we have

$$u^3v\bar{u}^3 = v \quad \text{or} \quad u^3v = vu^3 \quad \text{or} \quad u^3v\bar{u}^3\bar{v} = 1.$$

If u is octilateral, we have $u^4v\bar{u}^4\bar{v} = 1$, etc.

Now it is obvious that the same region cannot as a result of the rule receive different colors except as a consequence of its having different symbols in this notation. For the only way in which the rule could give a region different colors would be by its being reached from the initial point by different paths. But insofar as the same region is reached from different paths it receives different symbols. But these different symbols are only due to the completion of circuits round single regions, and therefore if in going round each single region the rule brings the initial region to the same color it had at the outset, it will never lead to any inconsistency. Let us, then, see what successions of values of the corners lead from any color to that same color in going round one region. For this purpose, we may mark every corner whose value is -1 by drawing a little circle round it, leaving those corners whose value is $+1$ unmarked.

Suppose then we start as in Figure 5, with two adjacent summits whose values are $+1$ and -1. Their sum is 0 and the two regions tied by the boundary between them are alike. Suppose next, as in Fig. 6, that two

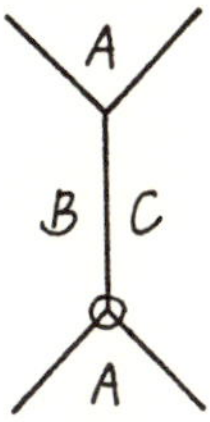

Fig. 5

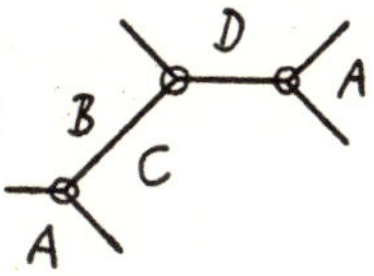

Fig. 6

successive corners have the same value. Then if the third corner going round the same region has the same value, their sum is congruent to zero for the modulus 3, and again the initial color is restored. It follows that any succession of corner values whose sum is congruent to zero for modulus 3 will restore the initial color. It follows that the rule cannot result in coloring any region two different ways.

Neither can it give confine regions the same color. For confine regions only have one corner between them. Now the value of this corner cannot be zero and consequently since the sum of the values of intermediate corners from one region to itself is congruent to zero for the modulus 3, that sum from one region to a confine region cannot be so congruent. Consequently, two confine regions cannot receive the same color.

This matter will appear more clearly from Fig. 7. It will be seen that if in going clockwise round any region colored D, in passing from A through a corner-value, -1, we come to B, then in passing from B through a corner-value, -1, or from A through a corner value $+1$, we come to C. And this will be so for every region colored D. Also in passing clockwise round any region colored C, every corner value -1 changes A to D, and D to B, and B to A. Also in passing clockwise round any region colored B a corner-value -1 changes A to C, C to D, and D to A. Also in passing clockwise round any region colored A a corner-value -1 changes B to D, D to C, and C to B. It is evident from this that confine regions cannot be colored alike by the rule unless the regions receive each two different colors.

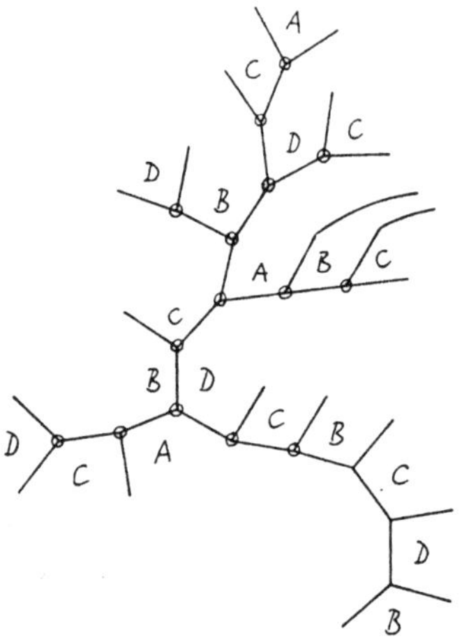

Fig. 7

Although the above reasoning is, in reality, perfectly cogent, yet an objection may suggest itself. It may be argued that (Fig. 8) if the sum of the corner-values round one region is $a+b+c \equiv 0 \pmod 3$ and round a confine region is $b+c+d \equiv 0 \pmod 3$, then the sum of the corner-values round the two combined, or $a+b+c+d$ will be congruent to zero only if $a \equiv d \equiv b+c \equiv 0 \pmod 3$. But the explanation of

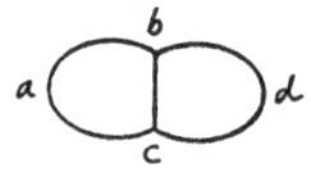

Fig. 8

this difficulty is shown by Fig. 9. Here all the corners are supposed to have the same value. Accordingly, between the first and second As there are 3 corners and between the first and third As there are 3 corners. But between the first and third As, there are 8 corners, since two corners occur along the edge of the second A. In the same way in Fig. 10, where again all the corners have the same value, there are 3 corners between all the successive As and Cs, and yet there are 10 corners all round the mass of two regions, without producing any conflict of colors. In forming the sum, all the corners from which branches pass inward are to have their signs changed. Thus reckoned, the sum is $8-2=6$. So in Fig. 11, the outward corners are $6-2=4$ and the inward corners are 4. Hence, the proper sum is $4-4=0$, according to the rule.

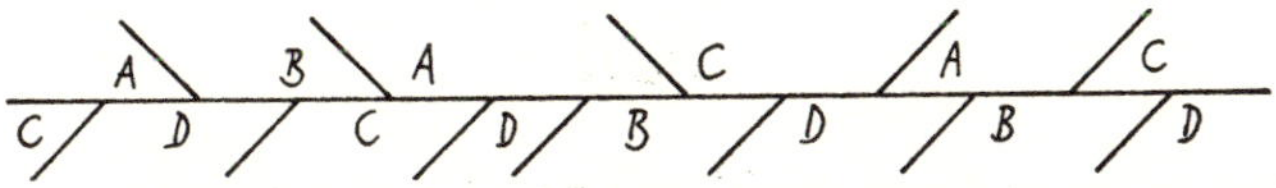

Fig. 9

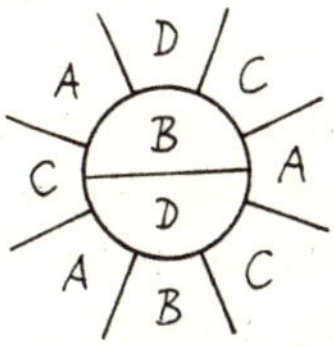

Fig. 10

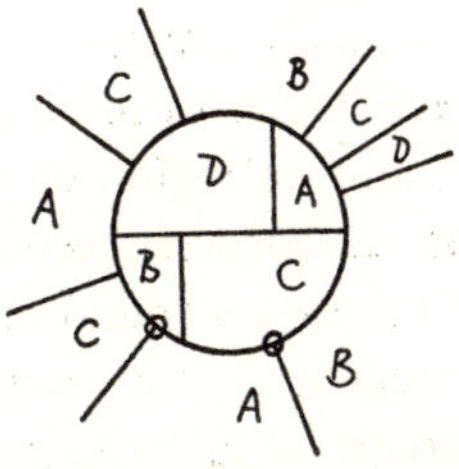

Fig. 11

I have now proved that the theorem relating to the corners is essentially the same as the original proposition about the four colors. It remains to prove that the values $+1$ and -1 can always be assigned to the corners so as to make the sum of the corner values for every region divisible by 3.

Before going on to this I will note by the way that the proposition accords with the census theorem. Namely, for regions of 5, 8, 11, 14, etc. boundaries the number of corners of value -1 is congruent to 1 for the modulus 3. For regions of 6, 9, 12, 15, etc. boundaries the number is congruent to 0. For regions of 7, 10, 13, 16, etc. boundaries the number is congruent to -1. Then, the total sum (since each corner is counted thrice) is congruent to

$$\tfrac{1}{3}(N_5 - N_7 + N_8 - N_{10} + N_{11} - N_{13} + N_{14} - \text{etc.})$$

But by the census theorem

$$\tfrac{1}{3}N_5 = 4 + \tfrac{1}{3}N_7 + \tfrac{2}{3}N_8 + N_9 + 1\tfrac{1}{3}N_{10} + 1\tfrac{2}{3}N_{11} + \text{etc.}$$

Hence the total is congruent to

$$4 + N_8 + N_9 + N_{10} + 2N_{11} + 2N_{12} + 2N_{13} + \text{etc.}$$

which is a whole number and is therefore possible. For instance, in Fig. 12, $N_5 = 14$, $N_6 = 4$, $N_8 = 1$, and the number of corners of value -1 is 5.

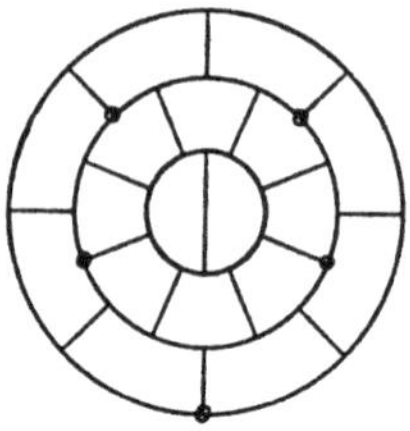

Fig. 12

In order to prove the proposition, I find it easiest to prove much more than that proposition. Namely, I shall prove that in any map of the kind considered, namely a map that can be drawn on a spheroidal surface, where three boundaries meet at every corner, no region of less than 5 boundaries, and no region or mass of two regions having any cyclosis, not only can the corners receive each the value $+1$ or -1 so as to make the sum of the values round each region divisible by three, but furthermore, any region of the map being arbitrarily selected and any two adjacent corners of that region, all the other corners of that region can receive alternately the two values $+1$ and -1.

I begin by proving that if this is true for all such maps of less than N regions, then, whatever number N may be, it is also true of any map of N regions. Namely, suppose we have a map of N regions, one of which is designated as having alternate corner-values except at two designated adjacent corners. We erase the boundary between these corners and so obtain a map of less than N regions. We select the new region as the one whose corner-values except two are to alternate and we select those two so as not to be among those of the region originally designated. We now by hypothesis can assign the values of the corners in the manner indicated. Having done so, we restore the erased boundary. We reverse the values of all the corner-values assigned to corners belonging to the region originally designated, and to each of the two new corners we assign the same value as the adjacent corner of the region originally designated. In order to prove that the sum of the corner-values of every region is now zero, we remark that it was so before the erased boundary was redrawn, and the only regions any of whose corner-values have been changed are the following:

1st The region originally designated.

2nd The region into which it was thrown by the erasure of the boundary.

3rd The regions confine to the region originally designated but not having either extremity of the erased boundary as a corner.

4th The two regions upon which the erased boundary abuts.

I will now show that the sum of the corner-values for every one of these regions is zero.

1st If the region originally designated has an even number of boundaries, the corners except the two of the boundary have half of them the value $+1$ and half of them the value -1, so that they sum up to zero. Of these corners one end one has the value $+1$ and one end one has the value -1, so that the corners at the ends of the erased boundary, receive one the value $+1$ and one the value -1. Thus the sum is zero.

If on the other hand the region in question has an odd number of boundaries, then of the corners not belonging to the erased boundary one more has one value than the other and the two end corners have that value. Consequently the two new corners take the same value and the sum is three times that value, so that it is divisible by 3.

2nd The region of the other side of the erased boundary has, let us say, a for the sum of its corner values except the terminations of that boundary. Then the sum of the corner values belonging to the originally designated region must before the restoration of the erased boundary have been $-a$. But these have all had their signs changed in restoring the [boundaries] so that they are now a, and since the two corners of the boundary reduce the one sum to zero it must reduce the other to zero.

3rd The regions confine to the region originally designated but not having for a corner either extremity of the boundary all had the sums of their corner values zero before the restoration of the boundary. But all that has been changed in them are two adjacent corner values changed from $+1$ and -1 to -1 and $+1$. Hence the sums of their corner values remain the same.

4th Each of the two regions upon which the erased boundary abuts has in the restoration of the boundary had the sign of a corner-value reversed and another corner value added. Suppose the reversed corner value originally was a. On reversal it becomes $-a$. Another corner-value $-a$ is added, making $-2a$. But $-2a \equiv a \pmod 3$. Therefore the sum of the corner values of each of these regions remains unchanged.

This completes the proof that if the proposition is true of every map of any given number of regions it is true of all maps of more regions. Now it is true of a map of two regions and hence, is true of all.

The following is an example. In each case an arrow within the designated region points to the boundary between the designated corners [Fig. 13].

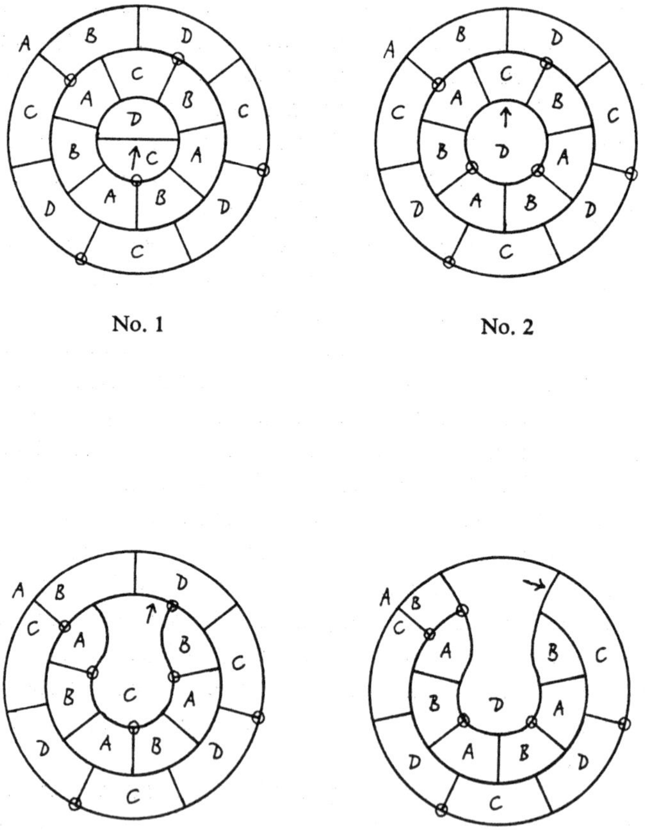

No. 1 No. 2

No. 3 No. 4

Fig. 13

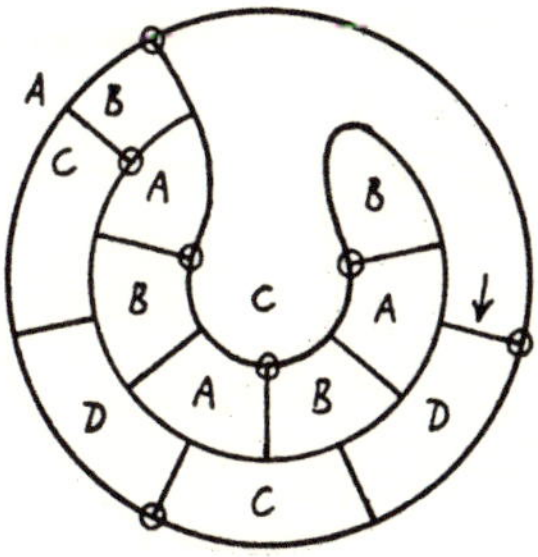

No. 5

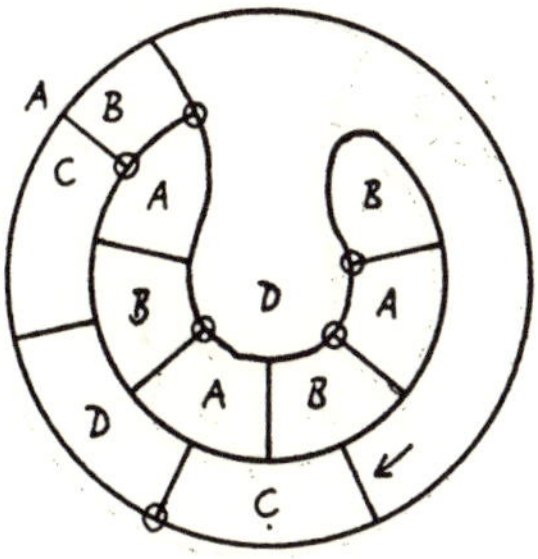

No. 6

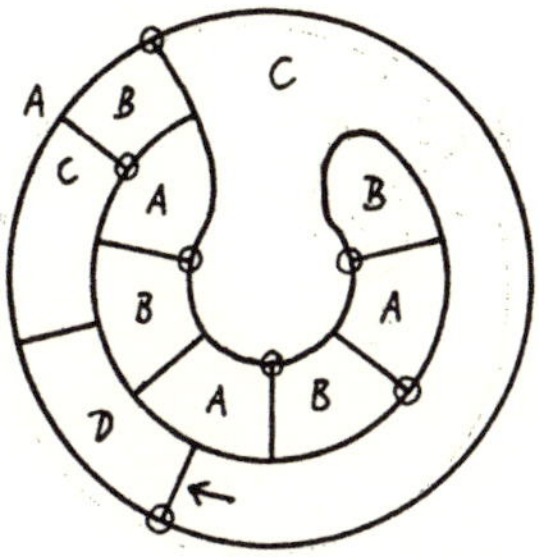

No. 7

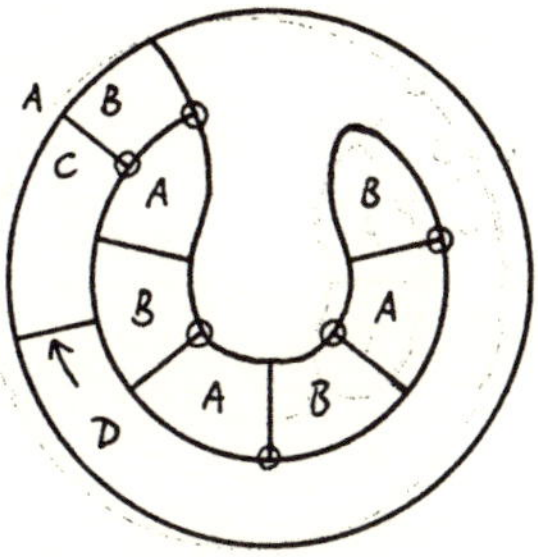

No. 8

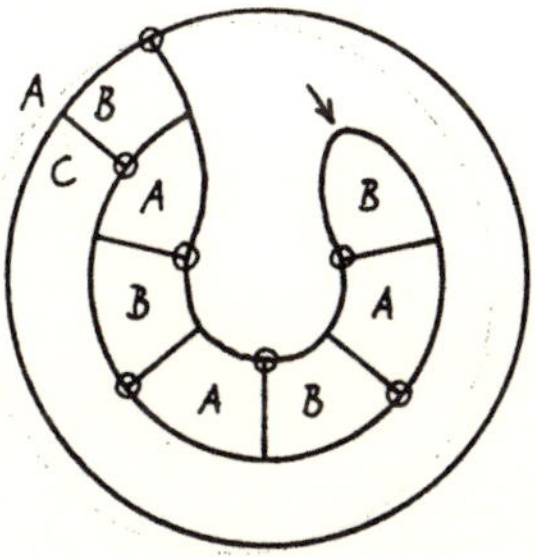

No. 9

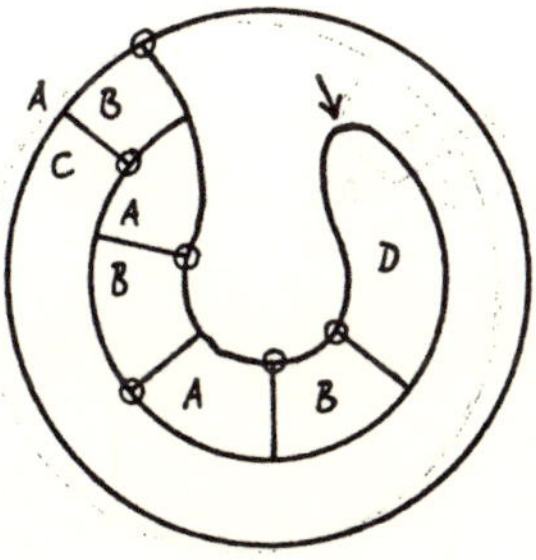

No. 10

Fig. 13 (continued)

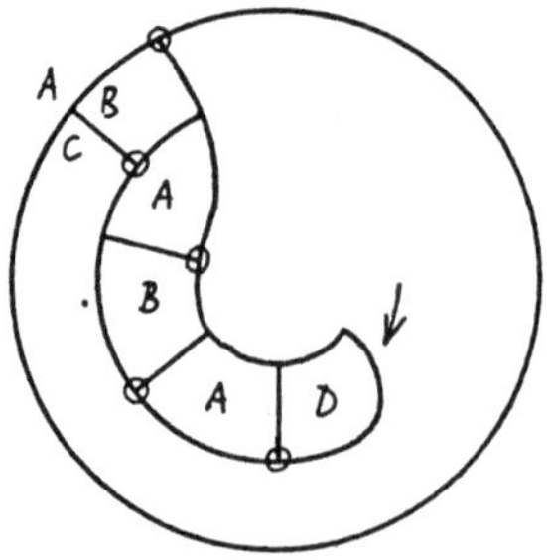

No. 11

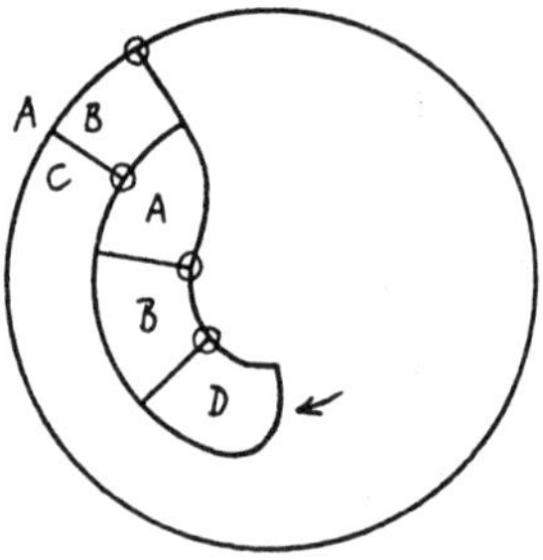

No. 12

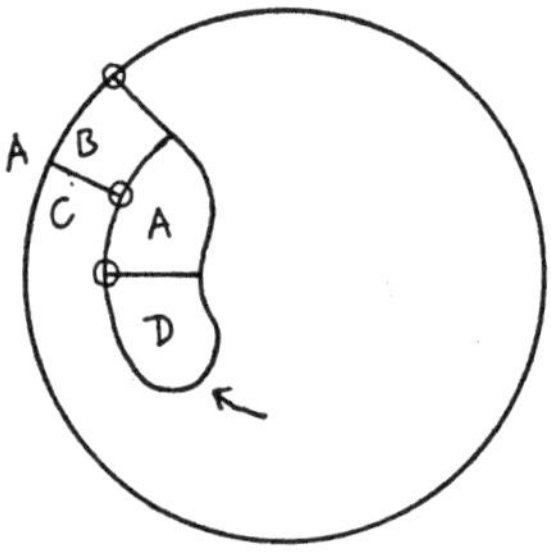

No. 13

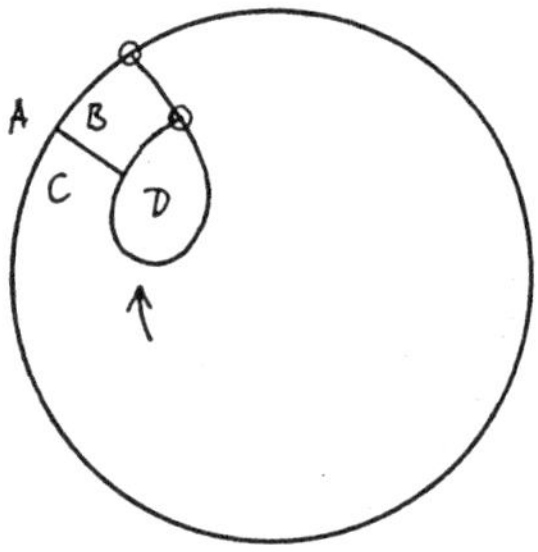

No. 14

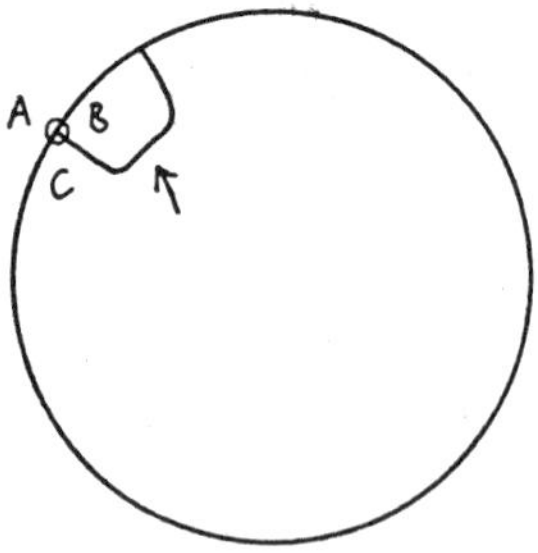

No. 15

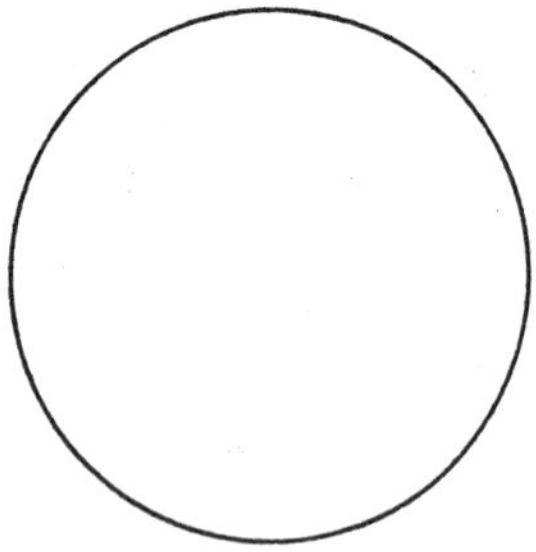

No. 16

Fig. 13 (concluded)

B. ON THE PROBLEM OF MAP-COLORING AND ON GEOMETRICAL TOPICS, IN GENERAL (154)

§1. The problem of map-coloring is [to] determine demonstratively the smallest number of colors that will suffice so to color any map whatever which can be drawn on a given surface, that no two confine regions (that is, two regions having a common boundary-line) shall have the same color.

It has long been known by experience that 4 is the required number for an ordinary surface; but De Morgan and Cayley failed to demonstrate the proposition. Mr. Kempe in 1879 proposed a proof in which a serious fallacy has been detected by [Heawood, 1890].

In undertaking to form a new proof, it has seemed to me proper to extend the problem to maps drawn upon any kind of surface whatsoever. The problem is plainly of a topical nature; yet Listing's doctrine of cyclosis and periphraxis (as well as Riemann's doctrine of connectivity) fails to make the necessary discriminations. For example, a spheroidal surface, which has periphraxis 1 and cyclosis 0, requires the same number of colors as a disk, which has periphraxis 0 and cyclosis 0. If a spheroidal surface has M caps upon it, in one of which N holes are made, its periphraxis is M and its cyclosis is $N-1$; but the number of colors is the same as if there were no such holes. Yet the cyclosis of a surface without any boundary or line of splitting suffices to determine the number of colors.

Listing's investigation is insufficient in that he takes no notice of topical singularities, or places where the ways of moving are greater or less than at ordinary points. Before we can handle with success the problem of map-coloring in its generality it is necessary to develope the doctrine of geometrical topics, or topology, in this respect.

§2. In order to proceed demonstratively we must set out from a pragmatic definition of continuity. I deny that the ideas of Cantor, as set forth in the paper in the second volume of the *Acta Mathematica* (the only papers of his that I have read), are adequate to this purpose.

He writes as if the points of an unlimited simple line could be placed in one-to-one correspondence in their order with the aggregate of all assignable real quantities in their order of magnitude. Such an aggregate consists of ultimate parts absolutely distinct from one another, which are the single exact quantities. Accordingly, we find Cantor occasionally supposing that a point or points are removed from a line, leaving it without those points, and therefore no longer continuous, nor perfectly a line. He thereby treats a line as if it were composed of points, as its ultimate parts, though he would, no doubt, grant that those points only constitute the matter of the line, whose form consists in their order. I propose to show that this conflicts with the indistinct, yet clear, notion of continuity which is familiar to us all.

Any definite whole of ultimate integrant parts may be called a *collection* of those parts. This statement tolerably expresses how I propose to use the word "collection", but cannot be said to afford any adequate analysis of the idea. For it leaves "definite", "ultimate", "integrant", and "part" as words expressive of confused notions. I know no concept of logic more difficult to analyze than that of a collection. It has to be remarked at the outset that a collection is a single object distinct from its parts. A nation is not a man; nor is it all the men it contains, since they are many while it is one. This remark, in its two clauses, that the collection is distinct from its parts, and that it is an individual object, will be found to gain in importance upon further consideration. It is to be remarked, in the second place, that a collection is a sort of abstraction, that is, it has a mode of being constituted by the possibility of something relatively concrete, — namely, the possibility that every existing object of a class should independently contribute to the construction of one effect, or should have a common relation to one correlate. The ultimate parts, or units, on the other hand, even if they have some abstractness, are not so abstract as the collection. They have, as already remarked, their existence distinct from the collection, and are, at least, so far independent of each other, that any one may enter into or be excluded from a collection independently of any other or others. From the circumstance that a collection has only a being *in posse*, and is devoid of *hecceity*, or thisness, except as derived from its units, it follows as a corollary that two collections in order to be distinct must differ in some generally describable respect, while the distinct units may be alike in all such respects and only be distinguishable in their reactions, like two drops of water. It is to be remarked, in the third place, that every respect in which two distinct collections differ is of this sort, that some unit of the one is not a unit of the other. For example, if every unit of a certain collection is one of the three individual objects, *A*, *B*, and *C*, that collection can only be one of eight, namely, one which includes all the objects, *A*, *B*, and *C*, excluding none, three pairs including two and excluding one each, three which include single units, and exclude each a pair, and one which includes none and excludes all.* (*Perhaps it will be objected that a collection must include, at least, two units. It is a trivial question, which I might pass by. It is not a question of the ordinary usage of language, but of mathematical and logical propriety. We have seen that a collection has quite another mode of being than that of its units, so that it cannot be identical with them. It is a mere *ens rationis*, which we create at will; and when we say that there is a collection we merely mean that its units may have a relation to one object, common and peculiar to them. In this sense, we may certainly say that there is a collection embracing but a single unit; for this unit may have a peculiar relation to another object. Nor is such a collection identical with its sole unit, since that unit has an existence, or some mode of being, independent of the collection, while the being of the collection is dependent on that of the unit. Similarly there may be a relation in which nothing stands to something, and thus there is a collection of nothing. This collection has the mode of existence of a single collection, while its units are null. There is, of course, but one collection of nothing.

For the benefit of the unmathematical, I will add a sort of *argumentum ad hominem*. For every cardinal number whatever, there are collections of that multitude. Now a cardinal number is the answer to the question, how many? But the answer to the question may be *one* or *none*. Hence, there are collections of one and of no units.) The precise respects in which these eight collections are distinguished from one another are determined by the units, while, given eight collections so related to one another, although the units are so determined yet the precise circumstances which distinguish them from one another are not thereby determined. It is to be remarked, in the fourth place, that the units of a collection are independent of one another as to their capabilities of forming collections. Let there be any general description, applicable only to units of a collection, and whatever units of the collection there may be of that description form by themselves a collection. Even if the description be self contradictory, it determined the null collection. Thus the idea of a collection carries with it that of a complete system of subcollections. It is to be remarked, in the fifth place, that although a collection has only a being *in posse*, and as such is only distinct from another so far as it has some point of difference from that other, yet that a collection is an individual object, not in the least indeterminate, and that further different collections may form the units of collections of collections. Thus, the eight collections formed by three objects themselves form 256 collections of which they are units, namely, one including all 8, 8 including 7 each, 28 including 6 each, 56 including 5 each, 70 including 4 each, etc. It is to be remarked, in the sixth place, that in order that a collection should have any definiteness, it is necessary that its units should conform to certain conditions. Thus, the population of the United States differs from minute to minute, and is definite only on the assumption that the date is fixed. Such conditions can never be fully described in purely general terms.

Reuniting all these considerations, we may analyze the concept of a collection into the indispensable concepts of logic (beyond which it would be aside from the present purpose to attempt to go); and this analysis is embodied in the following definition:

Given certain conditions, $\aleph$ (not necessarily altogether general), if any description, d, be such that, whatever description c may be, there is a logical possibility that there should be a single object, X, and a description of order pair, ϱ, such that if B be any individual object subject to the conditions, $\aleph$, and to which d and a both apply, then the description, ϱ, applies to the ordered pair $(B:X)$, while if B' be any object subject to the conditions $\aleph$, but to which d and c do not both apply, then the description, ϱ, does not apply to the ordered pair $(B':X)$, in that case, and only in that case, the object, — *ens rationis* (*terminus* for the Okhamist, *formalitas* for the Scotist) — whose being consists in this possibility conjoined to the applicability or non-applicability of d to definite objects conforming to $\aleph$, this object being considered as an individual identical with everything whose being consists in the same possibility and applicability, and also as capable of entering as a member into ordered pairs, is called *the collection of all the* ds *which conform to* $\aleph$. Any individual object to which the description d applies and which conforms to the conditions $\aleph$ is said to be a *unit* of the collection.

It may be objected to this definition that it is circular, inasmuch as it involves the idea of an ordered pair, which is a collection of two objects modified by distinguishing one of them, as first, from the other, as second. But I do not think that this objection is just, since we may adopt the following definitions. An *ordered pair* is an *ens rationis* whose being depends upon an individual, termed its *relate*, conforming to existential conditions while an individual, termed its *correlate*, conforms to existential conditions, possibly different from the others. A *general description* of an ordered pair is an *ens rationis* whose being depends upon the possibility of enunciating any definite proposition mentioning "a relate" and "a correlate". Such a description is *applicable to* a given ordered pair, provided the proposition upon which it depends is true when the relate and correlate of that pair are substituted respectively for "relate" and "correlate" in that proposition.

The main idea of the following group of definitions is adopted from Cantor:

If there is a possible relation in which every *A* stands to some *B* to which no other *A* so stands, then the collection of all the *A*s is said to be (at least) *as small as* the collection of all the *B*s, and the latter to be (at least) *as great as* the former. I omit the unnecessary words "at least".

If one collection is, at once, as small as and as great as another, it is said to be *equal to* that other.

If one collection is not as great as another, it is said to be *smaller than* that other, which is said to be *greater than* it.

In the *Monist* ... I showed that of any two collections whatever each is either as small as or as great as the other; and I have been told that Cantor, in a paper I have not read, had already proved the same proposition.

Evidently, the relation of being as small as is a transitive one. That is, if *A* is as small as *B*, and *B* is as small as *C*, then *A* is as small as *C*. The same is consequently true of the relations of being equal to and being smaller than. Further, if *A* is smaller than *B*, and *B* is equal to *C*, *A* is smaller than *C*; and if *A* is greater than *B*, and *B* is equal to *C*, *A* is equal to *C*.

The propositions in these last two paragraphs constitute a linear order among collections not equal to one another. The place of a collection in this order may be termed its *multitude*. That is, of two equal collections the *multitude* is *the same*, while of unequal collections the greater has the *greater multitude*.

In the number of the *Monist* just mentioned I proved that any multitude is smaller than some multitude. But the proof having been set forth in unmathematical language, and so that its cogency was not indubitable, I here repeat it. Given any multitude, by the definition of multitude, there is a possible collection of that multitude. By the definition of a collection, a description, *d*, is applicable to every unit of that collection. Further, by the same definition, if *c* is any description whatsoever, there is a possible single object, *X*, and a dyadic relation, ϱ, such that whatever object there may be to which the descriptions *d* and *c* both apply (but no other object) is in the relation ϱ to *X*. The combination of the descriptions *d* and *c* makes a description which we may call *d*′. Now if we take any description whatever, *c*′, any object to which *d*′ and *c*′ both apply (and no other) is an object to which both *d* and the combined description *c* and *c*′ apply, and therefore, by the same definition, there is a collection of all the *d*′s. Since every unit of this collection is a unit of the original collection of all the *d*s (for the description *d* applies to whatever *d*′ applies to), we may term the collection of all the *d*′s a *sub-collection* of the collection of all the *d*s, whatever may be the particular description, *c*, which conjoined to *d* makes *d*′.* (* If *c* is the same as *d*, the subcollection is the collection of all the *d*s, which is, therefore, a subcollection of itself.) Now the phrase "a sub-collection of the collection of all the *d*s" is a description which we may designate as *d*″. And since any sub-collection whatever is a collection, and as such, by the definition is an individual object capable of being the first member of an ordered pair to any individual object as second member, irrespective of all other such pairs, the same thing must be true of those subcollections of the collection of all the *d*s to which subcollections any other description whatever, *c*″, applies. Hence, by the definition, there is a collection of all the *d*″s, that is, the collection of all the subcollections of the collection of all the *d*s. In order to avoid confusing collocations of words, let us speak of "*a d*", meaning a unit of the collection of all the *d*s, and of "*a d*″", meaning a unit of the collection of all the *d*″s. It is to be borne in mind that a *d*″ is the same as a subcollection of the collection of all the *d*s. I now propose to prove that the collection of all the *d*″s is greater than the collection of all the *d*s. By the definition of "as small as," I shall have done this as soon as I shall have shown that if σ be such a relation as no two *d*″s bear to the same *d*, then it necessarily follows that there is some *d*″ that is not in the relation σ to any *d*. Every *d* is either, first, one to which no *d*″ is in the relation σ, or second, one to which no *d*″ which (as a subcollection of the collection of all the *d*s) includes it as a unit is in the relation σ, or third, one to which a *d*″ including it as a unit is in the relation σ. Unless there is some *d*, this is the case. If there is a *d*, but none to which any *d*″ including it as a unit is in the relation σ, then the collection of all the *d*s, considered as a subcollection of itself, is a *d*″ which is in the relation σ to no *d*, and the proposition is proved. If there is no *d* to which any *d*″ not including it as a unit is in the relation σ, then the null collection (which is a *d*″ even when there is no *d* at all) is a *d*″ that is in the relation σ to no *d*, and the proposition is proved. Let us define an *alpha d″* as a *d″* which, 1st, includes as a unit every *d* to which no *d″* including it as a unit stands in the relation σ, and which, 2nd, includes as a unit no *d* to which a *d″* including it as a unit stands in the relation σ. There must be some alpha *d″*; for the definition of an alpha *d″* augmented by any further limitation or not, is a description, which may be called *c*, and then it follows as before from the definition of a collection that there is a subcollection of the collection of all the *d*s which includes as a unit every *d* to which the description *c* implies and nothing else. But no alpha *d″* can stand in the relation σ to any *d* whatever. It obviously cannot do so to any *d* to which no *d″* stands in the relation σ. It cannot stand in the relation σ to any *d* to which no *d″* including it as a unit stands in the relation σ, since if it did it could not include this *d* as a unit, while by the definition of an alpha *d″* it does include as a unit every *d* to which no *d″* including it as a unit stands in the relation σ. Finally, it cannot stand in the relation σ to any *d* to which a *d″* including it as a unit stands in the relation σ, since in that case, there being but one *d″* that is σ to any one *d*, it must include that *d* as a unit, while by the definition of an alpha *d″* it includes as a unit no *d* to which a *d″* including it as a unit stands in the relation σ. Thus no alpha *d″* can be σ to any *d* at all, and since there is an alpha *d″*, some *d″* is not σ to any *d*. The collection of all the *d″*s is therefore not as small as, i.e. is greater than, the collection of all the *d*s; or every collection is smaller than a collection, and every multitude is smaller than a multitude.

ON THE PROBLEM OF COLORING A MAP

§1. THE ESSENTIAL CONCEPTIONS OF THE PROBLEM

A *map* is, for the purpose of this problem, nothing but an assemblage of *regions*, each of which is a definite surface consisting of a single piece, and no two of which have any superficial part in common.

If we ignore the distinction between different regions and consider them as making up one surface, this latter may be called the *subject surface* of the map.

The subject surface may consist of a multitude of detached pieces. Each of these pieces may have *topical singularities*, that is to say, isolated places from every part of which the possibilities of motion are different from those of motion from ordinary places of the same dimensionality within the same locus *in quo*. Thus, the topical singularities of lines are: 1st, *acnodes*, or outlying points from which no motion in the line is possible; 2nd, *extremities*, from which motion is possible only one way; and 3rd, *nodes*, or points of branching, from which motion is possible in more than two ways. In the plane of projective geometry there is but one way in which a line can move altogether away from a line that passes just once through infinity; yet the latter line is not a topical singularity, since it is not isolated from other lines presenting the same peculiarity and since, moreover, there are two ways in which a line can move away from any part of it. But the topical singularities of surfaces in general are either line-singularities or point-singularities. The line-singularities are, 1st, *pili*, or lines either entirely outlying or only connected with the rest of the surface at isolated points, so that the motion of a line away from any of them while remaining in the surface is impossible; 2nd, *edges*, from which lines while remaining in the surface can only move away on one side; and 3rd, *splits*, or lines from which several sheets diverge, so that lines can leave them on three or more sides. The point-singularities of surfaces are: 1st, *outlying points*, from which no motion in the surface is possible; 2nd, *extremities* of *pili*, from which a point can only move one way; 3rd, *nodes of pili*, from which a point can move in a multitude of ways but not expand into a line; 4th, 5th, 6th, *vincula*, that is, points where one or more pili meet an ordinary point of an edge, an ordinary point of the surface, or an ordinary point of a split; 7th, *split-stations*, that is, points on a slit where certain sheets cease to be connected with others; 8th, *split-edge-stations*, that is, points at each of which a split ends and an edge begins; 9th, *split-nodes*, that is, points from which radiate one or more splits and edges or splits only; 10th, *tacks*, or isolated points where different sheets are connected. But the problem of map-coloring assumes that no two parts of the subject surface are connected at isolated points; so that its only singularities are edges, splits, split-stations, split-edge-stations, and split-nodes.

Regions of maps may have all these singularities, and they may besides have *relative edges*, that is, lines where the subject surface has a split but where the region does not split into so many sheets as the subject surface does.

Independently of the singularities there is another respect in which surfaces differ upon which the number of colors required to color a map depends. I call this the *synĕsis* of the surface, and define it as the number of non-singular lines that can be drawn upon a single piece of surface while still leaving it possible that a point should move continuously on that piece from any position to any other, without crossing any of those lines. Or it may be defined as the number of self returning cuts that can be made in the surface without increasing the number of pieces. The synesis cannot be defined in terms of Riemann's "connectivity," which belongs only to surfaces with one or more edges but no split, and which is defined as one more than the number of cuts from edge to edge that can be made without increasing the number of pieces. For making holes in a surface never affects the synesis, while every hole in a surface having an edge increases the connectivity by one. On the other hand, an anchor-ring with one hole in it and a spheroidal surface with three holes in it have the same connectivity, 3, while the synesis of the former is 2 and that of the latter 0. Nor can the synesis be defined in terms of Listing's *cyclosis* and *periphraxis*, notwithstanding the value of those somewhat artificial conceptions. The periphraxis of a surface is the number of times it must be pierced to render it incapable of containing any edgeless surface, the cyclosis of a surface is the number of times, after its periphraxis has been destroyed by holes, that it must be cut from edge to edge to render it incapable of containing a non-singular line that cannot shrink to a point by a continuous deformation within the surface.

ON THE PROBLEM OF COLORING A MAP [I]

A *map* is, for the purpose of this problem, nothing but an assemblage of *regions*, each of which is a surface consisting of a single piece, and no two of which have any superficial part in common, these regions making up together a surface which may be called the *subject surface* of the map.

A point may be said to *belong to* a line, if it lies either upon or at an extremity of the latter. A point may be said to *belong to* a surface, if it either lies within the latter or belongs to its limiting edge. A line may be said to *belong to* a surface, if every point that belongs to the former belongs also to the latter.

By a *boundary*, I mean a line which either, 1st, lies wholly on a limiting edge of the subject surface and which stretches as far as the same region has a continuous line lying on that edge, or, 2nd, which belongs to at least two regions, every linear part of which belongs to the same set of regions and to no other, and which stretches as far as all those regions have a continuous line belonging to them and to no other region. Unless a boundary lies wholly on an edge or split (nodal line) of the subject surface, it belongs to two regions and to no more. Unless a boundary lies in part, at least, on a split of the subject surface, it lies on the edge of every region to which it belongs and can have no branching (node).

By a *corner*, I mean a point at the extremity of a boundary. There are cases in which a corner belongs to two boundaries only, or even to but a single one; but unless a corner is at a topically singular point of a topically singular line of the subject surface, it lies at the common extremity of three different boundaries. This feature of a map has an important bearing on our problem.

Two regions to both of which one boundary belongs are said to be *confine* to one another. Two corners which belong to one boundary may be said to be *next* to one another. Two boundaries belonging to one region and to both [of] which one corner belongs may be said to be *adjacent* to one another.

In coloring a map every region is colored in one color only and every two regions confine to one another must receive different colors.

This problem of map-coloring is this: Given a subject surface, it is required to demonstrate that some definite number of colors is the smallest that will suffice to color every possible map on that surface. This problem has been attacked by eminent mathematicians, but has, I believe, never been resolved for any description of surface, though it is known by experience that four colors suffice for an ordinary surface. The demonstration here offered is so far imperfect that it only applies to a map of a finite number of regions.

By a *premiss condition* I mean the condition that two given surfaces, being confine to one another, must be differently colored, or otherwise classed. Any distribution of the regions of a map into the proper number of mutually exclusive classes, if it satisfies all the premiss conditions, affords a permissible mode of coloring the map.

By an *ordinary surface*, I mean such a surface that any line within it without extremities separates it into two divisions such that a particle (or movable thing that at any one instant occupies a single point) cannot move continuously on the surface from a point within the one division to a point within the other division except by a path to which belongs a point belonging also to that line.

If I can prove that, whatever number N may be, every map of N regions on an ordinary surface can be colored according to the rule with four (or fewer) colors, provided every map of $N-1$ regions on that surface can be so colored, I shall have proved that every map of a finite number of regions on an ordinary surface can be colored according to rule with four (or fewer) colors. For a map of four regions can evidently be colored with four colors.

Let us suppose then that every map of $N-1$ regions on an ordinary surface can be legitimately colored with four colors. Taking, then, any map whatever of N regions on such a surface, if this map has a region of two or three boundaries only, let one of those boundaries be erased, as in Figs. 1-4 [which represent respectively region of two boundaries; after the erasure; region of three boundaries, after the erasure]. This will make a map of $N-1$ regions. (I do not undertake to prove proposi-

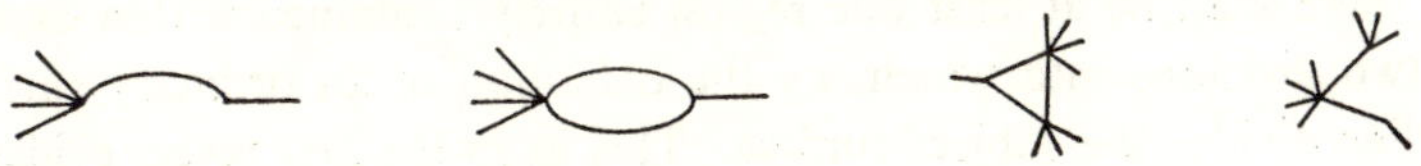

Figs. 1-4

tions intuitionally evident, since to do so would involve a logical analysis of continuity, which would not be germane to the present question.) This map can by hypothesis be colored according to rule with four, or fewer, colors. This having been done, the erased boundary can be restored and the region of two or three boundaries colored with a color different from any of its boundaries, so that the map of N regions is colorable with four colors.

If, however, the map of N regions has no region to which fewer than four boundaries belong but has a corner or corners belonging each to four or more boundaries, let the map be modified by erasing everything within a very small circle about each such corner and drawing an arc of this circle ending at two adjacent boundaries and passing through all the others, as shown in Figs. 5-8 [which represent respectively corner of

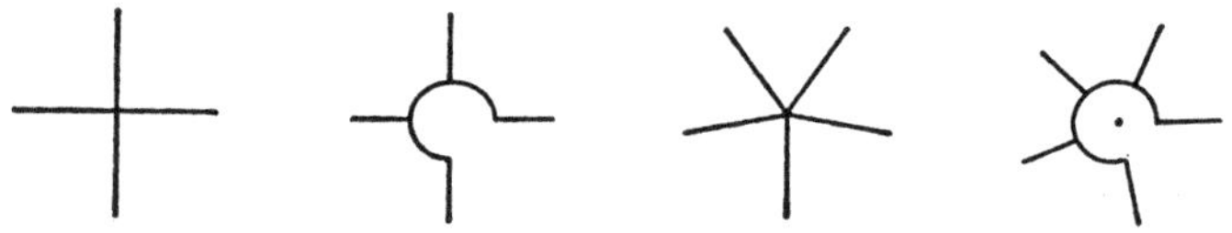

Figs. 5-8

four boundaries; the same modified; corner of five boundaries; the same modified]. This does not affect the number of regions. It adds to the premiss conditions to be satisfied, and annuls none. If therefore the map as so modified can be colored with four colors, so can the original map. In fact, the original boundaries can be restored without altering the coloring.

I shall assume that the subject surface is without a limiting edge; for an edge cannot enable any regions to be confine which otherwise could not be so. Besides, if the surface has edges, each hole may be filled over and a cap be fitted to the outer edge, and then one boundary at each edge may be erased. After the map, so modified, has been colored, the additions may be cut away again without disturbing the coloring.

I now propose to show that (except in a case in which it is quite evident that no more colors are required than if there were one region fewer) upon the map as now modified a self-returning line may be drawn which shall pass through no corner, and such that no region which it traverses shall be confine to three other regions that it traverses, and further such that there shall be at least one region entirely contained within each of the two divisions into which, by the definition of an ordinary surface, this line divides the subject surface. This is, in the first place, evidently so if any region has more than one edge. For a line drawn round near one of the edges is such a line, since the subject surface has no edge and the map covers the whole subject surface. The proposition is, in the second place, evidently true, if any two regions are confine to one another at two different boundaries. For then the line can pass through both boundaries, which would not be discontinuous, the one from the other unless there were other regions within both the divisions into which the line divides the subject surface. In the same way, in the third place, the proposition is evidently true if the obliteration of one boundary would cause two regions to be confine to one another at two distinct boundaries. But if none of these three states of things is the case, then consider any region, A. If this is confine to no region, the map without A would require as many colors as with A. But if A is confine to another region or regions, it must have an edge; for only upon a surface with a split can two regions be confine to one another elsewhere than at a common edge. Now in a surface with a split a self-returning line that crosses the slit need not divide the surface into two divisions; so that on any ordinary surface regions are only confine to one another at their mutually limiting edges. A has only one edge, it has at least four boundaries, and it is confine to no region at two boundaries. It is, therefore, confine to at least four different regions round its edge. We may denote these by B_1, B_2, B_3, B_4; etc. If each of these regions, A, B, is confine to only two others of them, then, since each is confine to at least four regions, there must be some other region, C. Then the line in question may be drawn just outside the edge of A through all the Bs. A will be the sole region within one of the divisions into which the line divides the subject surface. Hence, C will be within the other division, and the proposition will be true. But if one of the Bs, say B_i, besides being confine to B_{i-1} and to B_{i+1}, is also confine to a third B, say B_j, then the line in question may be drawn from A through B_i and B_j round to A, again; and B_{i-1} and B_{i+1} will be in the two different divisions into which this line divides the subject surface, so that the proposition is true in any case.

Let such a line, then, be drawn through a circular chain of regions each of which is confine to two others of the chain and to no more, while there is at least one other region in each of the two divisions into which this line divides the subject surface. Now let all that part of the map which lies in one of those divisions be obliterated and let the boundaries of the chain of regions be continued from the line into that division as follows: ...

ON THE PROBLEM OF COLORING A MAP [II]

§1. *Definitions.* A map is a figure covering a given surface and consisting of three kinds of parts, viz.: 1st, *regions*, or superficial parts, each in one piece, no two having a superficial part in common, and together including every superficial part of the surface covered; 2nd, *boundaries*, or linear parts, no two having a linear part in common, together including every linear part of every limit of the surface covered (if it has any) together with every linear part of every mutual limit between two regions, each boundary either belonging to a single pair of different regions, being at every part of it at the mutual limit of those regions, and continuing as far as those two regions have a continuous mutual limit, or else belonging to a single region that is limited everywhere along this line by the limit of the surface covered, the boundary continuing as far as the region continues to be so limited; and 3rd, *corners*, or punctual parts, each a single point where either three or more regions mutually limit one another, or where two regions are together limited by the limit of the surface covered.

Two regions are said to be *confine* to one another if they have a common boundary.

Two boundaries are said to be *adjacent* to one another if they at once limit one region and are limited by one corner.

An *ordinary* surface is a surface on which any two self-returning lines whatever that cross each other cross each other in a second point.

§2. In coloring a map every region is assigned to one of a number of mutually exclusive classes, distinguished by colors, and no two regions confine to each other are assigned to the same class.

It has long been known, as well as any mathematical proposition so intricate can be said to be known by experience, that four colors suffice to color any map on an ordinary surface. About 1860, De Morgan, in the *Athenæum*, called attention to the fact that this theorem had never been demonstrated; and I soon after offered to a mathematical society in Harvard University a proof of this proposition extending it to other surfaces for which the number of colors are greater. My proof was never printed, but Benjamin Peirce, J. E. Oliver, and Chauncey Wright, who were present, discovered no fallacy in it. In 1879, Mr. A. B. Kempe proposed a proof which I have been recently informed has been shown to be fallacious. I, therefore, here endeavor to recover my old proof, piecing out my recollections with new inventions, but restricting the demonstration to ordinary surfaces with finite numbers of boundaries and further restricting it by supposing the multitude of regions to be finite.

I shall begin by giving a rule for coloring any map of the kind in question in the required manner, interrupting the enunciation by remarks that will make it manifest to common apprehension that the method described will always lead to the desired result. But since such interest as the question has seems to be of a logical nature, I shall append a more formal proof. I shall then go on to show, though not demonstratively, what seem to be the required numbers of colors, for other surfaces.

§3. Let us begin (if necessary) by deforming the surface upon which the map is drawn, without rupturing it, so as to make it fit upon an ordinary closed surface, such as a sphere; and let its holes, or vacancies, be filled up. At each vacancy let one of the boundaries between the vacancy and one of the regions confine to the vacancy be erased, so that the map now covers the whole of an ordinary closed surface. (This can evidently not diminish the number of colors required, since it annuls no boundary between two regions, and consequently only increases the number of conditions to be fulfilled of unlikeness of color between confine regions.)

Now let the map be subjected to the following operations:

First Operation. Copy the map omitting one boundary of each region that is confine to fewer than four regions, thus throwing the former region into one of the latter. Repeat this operation as often as possible. (Since it always diminishes the finite number of regions, it cannot be repeated indefinitely; and therefore the final result must either be that every region is confine to more than three regions, or else that there is a region that has no boundary to be erased and that consequently covers the whole spheroidal surface and is the only region.)

Second Operation. Copy the map as last modified, substituting for each corner where more than three boundaries meet a short line at each end of which two adjacent boundaries are made to terminate, while all the others terminate at separate points of the line, without crossing one another, as shown in Fig. 9 [before and after 2nd operation]. (Since this annuls no condition that two regions must be differently colored, it cannot diminish the number of colors required.)

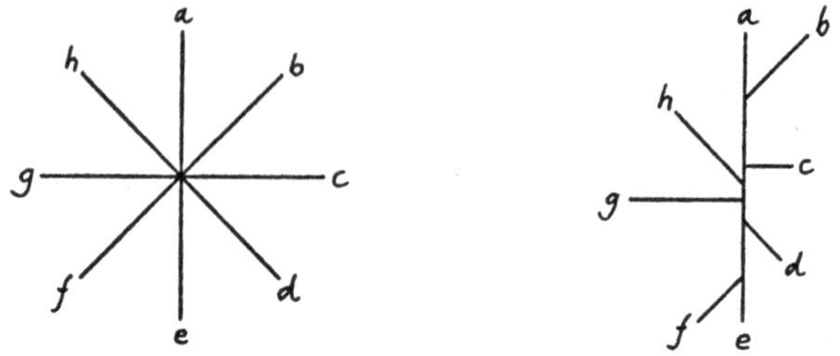

Fig. 9

Third Operation. If there are any two regions, A and B, which are confine to each other along several boundaries, at each of these places except one draw a line cutting off a portion of A so that the rest of A no longer is there confine to B, but so that no other region ceases to be confine to the rest of A, and then erase the boundary between this cut off portion and a third region, as shown in Fig. 10 [before and after 3rd operation].

Fig. 10

Fourth Operation. Unless the first operation has reduced the map to a single region covering the whole surface, there are now at least six regions. For by a well-known theorem of Euler (Legendre, *Géometrie*, VII, 25) the sum of the numbers of regions and corners exceeds the number of boundaries by 2. But if r_4, r_5, r_6, r_7, etc. are the numbers of regions of 4, 5, 6, 7, etc. boundaries respectively, since each boundary belongs to two regions, the number of boundaries is $\frac{1}{2}(4r_4 + 5r_5 + 6r_6 + 7r_7 + \text{etc.})$. And since three regions meet at each corner, the number of corners is $\frac{1}{3}(4r_4 + 5r_5 + 6r_6 + 7r_7 + \text{etc.})$, so that we have

$$r_4 + r_5 + r_6 + r_7 + \text{etc.} - \tfrac{1}{6}(4r_4 + 5r_5 + 6r_6 + 7r_7 + \text{etc.}) = 2,$$

or $$2r_4 + r_5 - r_7 - 2r_8 - 3r_9 - \text{etc.} = 12.$$

That is the number of regions of four and five boundaries each must by themselves be at least 6.

[ON THE PROBLEM OF COLORING A MAP (III)]

I assume that in topical geometry every line either comes to an end or ends at a point of the line itself. For example, if a spiral winds in and in towards a circle, which it approaches indefinitely, it must reach that circle at some definite point, although the law of the spiral may not determine what point of the circle that shall be. I conceive that this is involved in the idea of the continuity of the line. At any rate, I limit my considerations to lines of which this is true.

By the *topical singularity* of a line, I mean half the sum of the singularities of all its topically singular points. A topically singular point of a line is a point from which a particle can move along the line in more or fewer than two ways; and the singularity of such a point is the excess over two of those ways.

I define a *spheroidal surface* as one on which any line being drawn divides the surface into a number of regions equal to the number of separate pieces of the surface *plus* the number of separate pieces of the line *plus* the singularity of the line.

Let us suppose that upon a spheroidal surface in a single piece is drawn a line in a single piece and having no singular point whose singularity is other than $+1$. Then if we call any non-singular line forming a part of that line a *circuit*, every circuit divides the surface into two regions which we may call *circuit-regions*, in contradistinction to the regions into which the whole line divides the surface, which regions we may call *map-regions.* Now let us suppose the following state of things, the possibility of which we will consider later. Namely, suppose that all the singular points of the whole line are divided into two classes, which we may designate as class 1 and class 2, in such a way that taking any circuit whatever, there are on the circuit [a] number of singular points of the whole line of class 1 having branches extending into one circuit-region which added to the number of class 2 having branches extending into the other circuit region gives a number that is congruent for modulus 3 with the remaining singular points on the circuit. Let us now suppose that we make an *addition* to the original line by drawing a simple line between two ordinary points of the latter without crossing or touching it anywhere else and without having any other singularity.[1]

[1] In another draft of MS. 154 the following statement is found:

"The difficulty of the theorem becomes apparent as soon as it is thrown into algebraical form. Let the regions be numbered 1, 2, 3, etc. Let the four colors be *A, B, C, D*. Let any of these letters with a number subscribed, as A_3, be a quantity which is equal to one if the region bearing that number has the color signified by that letter, and which in the contrary case is equal to zero. Then whatever number of a region *i* may be, we have the equations,

$$A_i^2 - A_i = 0 \qquad B_i^2 - B_i = 0 \qquad C_i^2 - C_i = 0 \qquad D_i^2 - D_i = 0$$
$$A_iB_i = 0 \quad A_iC_i = 0 \quad A_iD_i = 0 \quad B_iC_i = 0 \quad B_iD_i = 0 \quad C_iD_i = 0$$
$$A_i + B_i + C_i + D_i = 1$$

Moreover, if *i* and *j* are the numbers of any two confine regions, we have

$$A_iA_j = 0 \qquad B_iB_j = 0 \qquad C_iC_j = 0 \qquad D_iD_j = 0$$

Let such a set of four equations be called a *map-condition*. Let a *mass* of regions be defined as a collection of regions which cannot be separated into two sub-collections such that no map-condition connects two regions of different sub-collections. Two masses of regions may be said to be *confine* to one another, if and only if a map-condition connects a region of the one with a region of the other. Then the proposition to be proved is, that so long as no five masses of regions are confine to one another, any map-conditions may be assumed at pleasure, and all the equations can be satisfied.

This statement shows that the simplicity of the proposition is only apparent, this apparence being due to the circumstance that it involves conceptions with which we are very familiar but which we do not commonly analyze."

C. [LINK COLORING] (157)

Question 1. In how many ways, with $c+1$ colors, can a simple chain of $l+1$ links be colored so that all adjacent links are colored differently?

Let p_l be the number of ways that give the first and last links the same color

Let q_l be the number of ways that give the first and last links different colors

(I write p for p_l, q for q_l, Ep for p_{l+1}, E^2p for p_{l+2}, etc.)

One easily sees that since if the last link is the same color as the first an additional link may have any one of c colors, and if first and last are different, any one of $c-1$ colors, therefore

$$Ep = q \qquad Eq = cp+(c-1)q$$
$$E^2p = Eq = cp+(c-1)q = cp+(c-1)Ep$$

The general solution is

$$p_l = C_1\, c^l + C_2(-1)^l \qquad q_l = C_1\, cc^l - C_2(-1)^l$$

But evidently if there is but one link, that is, if $l=0, p_0 = c+1, q_0 = p_1 = 0$

$$\therefore\ p_0 = C_1 + C_2 = c+1$$
$$q_0 = C_1 c - C_2 = 0$$
$$\therefore\ C_1(c+1) = c+1 \quad \text{or} \quad C_1 = 1 \quad C_2 = c$$

and thus the general solution becomes

$$p_l = c^l + c(-1)^l$$
$$q_l = c\cdot c^l - c(-1)^l$$

and the answer to the question is $p_l + q_l = (c+1)c^l$

If l is even $p_l = c^l + c = c(c^{l-1}+1)$ which is divisible by $(c+1)$

$q_l = c\cdot c^l - c = c(c^l - 1)$ which is divisible by $(c+1)$

If l is odd $p_l = c^l - c = c(c^{l-1}-1)$ which is divisible by $(c+1)$, $p_1 = 0$

$q_l = c\cdot c^l + c = c(c^l+1)$ which is divisible by $(c+1)$

If $c = 1$ and l is $\begin{Bmatrix} \text{even} \\ \text{odd} \end{Bmatrix}$ then $\begin{Bmatrix} q_l \\ p_l \end{Bmatrix} = 0.$

But if the chain begins with a determinate color and may end with that color in p_l ways with another determinate color in q_l ways and with some one of the remaining $c-1$ colors in r_l ways. Then

$$Ep_l = q_l + r_l \quad Eq_l = p_l + r_l \quad Er_l = (c-1)(p+q) + (c-2)r$$

$$E(p+q) = (p+q) + 2r \qquad p = \frac{c^l + c(-1)}{c+1}$$

$$E^2(p+q) = E(p+q) + 2Er \qquad q = \frac{c^l - (-1)^l}{c+1}$$

$$= p + q + 2r + 2Er$$

$$r = \frac{c-1}{c+1}(c^l - (-1)^l)$$

Question 2, In how many ways, with $c+1$ colors, can v different and separate given chains of l_1+1, l_2+1, l_3+1, ... l_v+1 links each be colored so as to give no two adjacent links the same color, and in how many ways in regard to the likeness or unlikeness of the $2v$ terminal links?

The solution of the last question gives us the answer to the first clause of this at once. Namely the total number is $(c+1)^v c^{\Sigma l}$.

Suppose that among the $2v$ terminal links different colors occur, and suppose the color of the last link of the μth chain is not one of these. Then there are a ways in which an additional link may be added to this chain so as to be of a color used before and $(c-a)$ ways in which it may be added so as to be of a color not used before. But if the previous link is of a color used before there are $a-1$ ways in which the additional link may take a color used before and $c-a+1$ ways in which it may take a color not used before.

But as for the first links of the different chains, there are $c+1$ ways in which the first can be colored, and no increase occurs to this number on account of other first links that are to be colored like it. For the second first link unlike a previous first link the number $(c+1)$ is to be multiplied by c. For the third by $c-1$ and so on. Thus if only one color is to be used in coloring the first links, the number of ways is $\frac{(c+1)!}{c!}$. If two $\frac{(c+1)!}{(c-1)!}$ if three $\frac{(c+1)!}{(c-2)!}$, if X colors are to be used in the first links, the number is $\frac{(c+1)!}{(c+1-X)!}$. These X colors, however, will require to be numbered, — or at least this is very convenient. And we may as well suppose the chains to be ordered according to these colors, which will only involve the trouble of finally multiplying our results by the possible ways of ordering the chains so as to conform to whatever conditions may be laid down as to their order. We will therefore suppose that all the chains beginning with the first color come first, those beginning with the second color coming next after these and so on. Thus the entire number of X initial colors are to be distributed among the v chains.

There being v_0	chains	beginning	with	color	0
v_1	,,	,,	,,	,,	1
$\vdots$					
v_{X-1}	,,	,,	,,	,,	$X-1$

$$\Sigma v_X = v$$

Each of these groups of chains will be composed of $c+1$ subgroups, of which the numbers of chains will be

$$v_{00}, v_{01} \cdots v_{0c}$$
$$v_{10}, v_{11} \cdots v_{1c}$$
$$\vdots \text{ etc.}$$
$$v_{X-1,0}, \cdots v_{X-1,c}$$

so that $\Sigma_c v_{ic} = v_i$ and any of those numbers may be 0. The individual chains may be ordinally numbered v_{001}, v_{002}, etc. The number of formations of any chain of the groups $v_{00}, v_{11}, v_{22} \ldots v_{X-1,X-1}$ will be the same as if it stood alone, that is it will be $\dfrac{c^l + c(-1)^l}{c+1}$ where $l+1$ is the number of its links. The number of ways in which it will end with any other *one* previously considered color (that is with 1 ... $X-1$, in the case of v_0) is $\dfrac{c^l - (-1)^l}{c+1}$.

D. THE BRANCHES OF GEOMETRY (97)

I begin with some rudimentary information about Geometry. This science has three parts, as unlike one another as possible. They are called Topics, Graphics, and Metrics. *Topics*, alone of the three, treats space itself. It studies the modes of connexion of different parts of places. It has only been known for a century and a half; and very little is known of it yet, for the reason that mathematicians have never been able to discover any good method of investigating it. For that reason, there are a number of propositions belonging to this branch of geometry which seem to be true but which nobody has ever yet succeeded in proving. Such for example is the proposition that four colors are sufficient for coloring any map of a globular surface; while six might be required for a map on a projective plane, and seven on a surface like that of an anchor-ring.

Of course, it has been known from ancient times that space has three dimensions, which is a proposition in Topics. But nothing further was added until Euler, a famous improver of mathematics, gave, in 1758, this proposition about bodies whose surfaces are composed of *faces* separated by *edges* which meet one another in *summits*, — like a cube for example. The proposition is that for every such body the sum of the numbers of faces and of summits exceeds by 2 the number of edges. But Euler did not succeed in proving this; and one reason, no doubt, of his failure to do so, was that it is not true in all cases; although it is true if the solid has no hole through it nor hollows within it and does not reach to infinity in any direction and in case it is true, it is true whether the faces are flat or bent, so long as none makes a band all round another. This proposition is now given as a theorem in every text book of geometry being the only proposition in *Topics* that such books contain, unless they mention that space has three dimensions. The first text book in which it found a place was Legendre's *Éléments de géometrie*. He gives an attempted proof of it which he himself admitted was fallacious. But finally Johann Benedict Listing, — a professor extraordinarius of Physics in Göttingen, — about 1848 put this proposition on a thoroughly satisfactory footing, extending it greatly and giving it the name of the *Census* theorem. I know no more instructive lesson in the art of thinking than that one may derive from the study of how it was that Listing, — who was far from being an intellectual giant — managed to succeed where such heroes of mathematics as Euler and Legendre had failed.

We shall do well to look into this, which is not only instructive logically, but may be useful as a matter of geometry; and at any rate, is both pretty and decidedly curious on account of the seemingly artificial nature of the *pattern*, so to speak, of the relations concerned.

But I have first to complete my statement of what the three categories of geometrical science are. The second is *graphics*.

Graphics is the doctrine of intersections and of tangencies; without any measurement. Linear perspective is the foundation and instrument of it. It depends on the fact that space contains a single plane through every three points of it, and that any three planes (according to the testimony of perspective) intersect in a single point or else in innumerable points. There are any number of such systems of surfaces. And the surfaces called planes are not distinguished by any purely geometrical peculiarities. They are simply surfaces in which light always travels if we disregard entirely diffraction and consider light only as it was known to the ancients. Graphics is the most fascinating, and perhaps the most useful department of geometry. But it makes no measures of lengths or of angles.

We recognize plane surfaces in two ways. First by looking along them; and this method is tolerably accurate. The other way is to grind two surfaces of rigid bodies *A* and *B* upon one another until they accurately fit in every position, and then take a third surface, *C*, and grind *B* and *C* together; then grind *C* and *A* together; then *A* and *B* together again, and so on *ad infinitum*. A third way, used in rougher work, depends on the fact that a taut thread is straight. A fourth way, very seldom actually applied, depends on the fact that a massive body moving with sufficient velocity and not deflected by any force, moves in a straight line, or *ray*.

The third department of Geometry is *Metrics*. What is called *Elementary* Geometry belongs entirely to this part, except the one topical theorem which Legendre introduced. It is called *elementary* because it is the subject of the geometrical part of Euclid's work entitled "Elements." Euclid apparently intended that it should be treated as rudimentary in teaching mathematics; and teachers almost always have done so from sheer stupidity. The rational way is the reverse: to begin with Topics, thence go on to Graphics, so far at least as to have a good understanding of perspective; and finally come to the difficult subject of Metrics, well prepared to comprehend a really profound view of it. Having sufficiently expatiated on the definitions of the three branches, I will, if I find the audience sufficiently interested, give specimens of the mode of thought of each.

For this purpose, in Topics, I must begin by explaining the four characters of places that are called Listing's Numbers, and whose names are

Chorisy,
Cyclosy,
Periphraxy,
Apeiry.

The *chorisy* of any place or figure, whether it be composed of points, lines, surfaces, or solids, or mixtures of these, is the number of simple interruptions of the same dimensionality as that of the place whose chorisy is spoken of, which are necessary in order to leave no point of that place vacant. In other words, it is the number of separate pieces of which the place in question is composed.

The *cyclosy* of any place is the number of simple interruptions each of dimensionality one less than that of the place whose cyclosy is spoken of, which are requisite in order not to leave room in the place for a non-singular (that is without terminations, or branching points) filament that cannot by a continuous motion (without breaks or weldings) within the place be reduced to indefinite approximation to a point or particle.

Thus the cyclosy of a globe is 0. That of the solidity of an anchor-ring is 1. That of the surface of an anchor-ring is 2, etc.

The *periphraxy* of a place is the number of interruptions of dimensionality 2 less than that of the place which are required in order to leave the place such that it can contain no non-singular surface that cannot by a continuous motion (without tearing or joining) within the place approach indefinitely toward the condition of a line.

Thus the periphraxy of one side of a disk is 0. That of the solidity of a globe is 0. That of the surface of a globe is 1. That of the solidity of a ring is 0. That of the surface of a ring is 1. That of the solidity of a box is 1. That of a solid with three hollows is 3. The periphraxy of any line and of any ordinary solid is 0.

The *apeiry* of a place is the number of interruptions of dimensionality

3 less than that of the place in question which are required in order to leave the place such that it can contain no non-singular solid that cannot by a continuous motion, without rupture or welding, within the place approach indefinitely foward being a surface.

The apeiry of all space is either zero or one according as we suppose its infinitely distant parts to be shaped. It is best regarded as having all its Listing Numbers equal to 1.

Every place has a "census-value" which consists of the Census number of its *point minus* that of its *lines plus* that of its *surfaces minus* that of its *solids*.

The Census number of any homogeneous space is equal to its Chorisy *minus* its Cyclosy *plus* its Periphraxy *minus* its Apeiry.

The "Census-theorem" is that the Census Value of any place is unaffected by cutting it up by boundaries of lower dimensionality.

I will give a few examples of the utility of this. Into how many compartments can the surface of an anchor ring be divided by simple lines meeting in summits three to a summit and so that every compartment abuts upon every other along a single boundary [Fig. 1].

Fig. 1

We calculate the census-value of the plain ring-surface.

Chorisy	+1	
Cyclosy		+2
Periphraxy	+1	
Apeiry	0	
Census-Value	(+2) − (+2) = 0	

Let this be cut up into x compartments. Each compartment will have $x-1$ bounding lines, and since each of these belong[s] to two compartments, the total number of these lines will be $\frac{x(x-1)}{2}$. Each compartment has $x-1$ summits; and since each belongs to three compartments their number will be $\frac{x(x-1)}{3}$. Hence the Census-Value after the division will be

Points	$\frac{x(x-1)}{3}$	
Lines		$-\frac{x(x-1)}{2}$
Surfaces	x	
Total	$\frac{7x}{6}-\frac{x^2}{6}$	

Equating this to 0, we get $x = 7$.

Into how many such compartments can a projective plane be cut up? Here we have

Chorisy	$+1$	
Cyclosy		$+1$
Periphraxy	$+1$	
Apeiry	0	
Census-Value	$(+2)-(+1)=+1$	

Hence $\frac{7x}{6}-\frac{x^2}{6}=1 \quad x=6$ [Fig. 2].

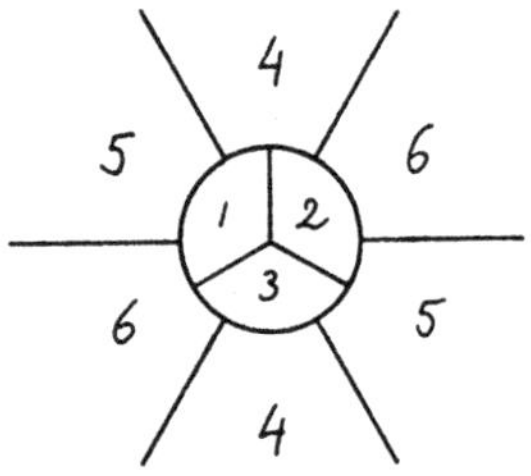

Fig. 2

Into how many can the surface of a sphere be cut up?

Chorisy	1	
Cyclosy		0
Periphraxy	1	
Apeiry		0
Census-Value	2	

$$\frac{7x}{6}-\frac{x^2}{6}=2 \quad x=4$$

E. [FROM PRAGMATISM] (318)

... I suppose, as a matter of fact, the majority of its votaries do, but to be quite negligible to a man whose earnest purpose is to do what in him lies toward a metamorphosis of philosophy into a genuine science. So I cannot turn aside into Schiller's attractive lane. No, if I ask what the interpreter's interest is in seeking to discover the logical interpretant, I do so not from Mr. Schiller's broad motive, — shall I venture to call it his fondness for strolling in paths where he may contemplate the varieties of his fellow-men? — but for the more definite reason that unless our hypothesis be rendered specific as to the interpreter's interest, we cannot trace out its consequences, since the way the interpreter will conduct his inquiry will depend upon the nature of that interest, very greatly. I will suppose, then, that the interpreter is not a man particularly interested in the theory of logic, which judging it by the logic-books he has looked into, he holds, I should hope, to be tiresome and profitless pedantry, calculated if early imbibed, and not shed as a duck sheds water, to convert a sincere intellect into a miserable caviller; but I will suppose that he has embarked a great part of the treasure of his life in the enterprise of perfecting a certain invention, for which purpose it is extremely desirable that he should have a demonstrative knowledge of the solution of a certain problem of reasoning which, being of an unusual kind, does not seem to offer anything to serve as a handle by which it might be firmly held and submitted to examination; so that it is continually slipping his mental grasp; and I will further suppose that it has occurred to him that if he could only get a perfectly distinct idea of the formal constitution of the predicate upon which this problem turns, he might be able to obtain, at last, that firm mental hold upon it that he has hitherto been unable to secure. There are many problems of this nature. Such, for example, is the question how many colors would be required to distinguish all possible abutting regions on a map of a curious planet which should have two holes through it, the surfaces of the holes being inhabited like the outer surface. It is well known that to color a map of the surface of an ordinary body with no hole in it no more than four colors can possibly be required, although I am not aware that any rigid proof even of this has ever appeared in print. It is not difficult to satisfy oneself that on a planet with one inhabited hole through it, so as to make it a ring, the number of colors would be seven. On a projective plane, which is a surface of the kind very misleadingly called "one-sided," the number of colors would be six, supposing the regions could extend without reference to distance, and even through infinity. For a projective plane with two holes through it, the number would be nine; with 3 holes, 10; with 7 holes, 13. But I, for my part, do not know what the number would be for a planet twice perforated. (Curiously enough, I can say that if the planet were six times perforated, the number would be 12.) It is a problem in topical geometry, the most fundamental, and no doubt, in its own nature, the easiest, of the three great divisions of geometry, of which the most studied is Metrics, the geometry of Euclid's *elements*, which is reducible to a particular application of the second division, Graphics, the geometry of perspective, which again is reducible to a particular application of this third division, Topics, which is a geometry that not only refuses to pay any attention to the relation between the greater and the smaller, but also equally refuses to recognize the distinction between flatness or straightness and curvature of any kind, but considers solely the continuous connexions between the different parts of figures. Of this, naturally so easy a theory, the awkwardness of geometers has prevented our knowing but a very little more of a general nature than a single proposition, the *census-theorem*. Our imaginary inquirer is probably right in his surmise that if we only had a clear, distinct, plastic, ductile, workable conception of what it is to require a given number of colors to distinguish abutting regions of any and every map upon a given topical form of surface, the difficulty of the problem that exercises him would disappear. It is certain that the most perfect definition of what he requires would be one that should point out to him how he ought to go to work, under what rule of procedure, in order actually to color any map on the given kind of surface with the minimum number of colors.

Under the high stimulus of his interest in the problem, and with that practical knack that we have supposed him to possess in coloring maps without too frequently being obliged to alter the colors that he had previously assigned to given regions, our inquirer will surely be put into a state of great activity in the world of fancies, in experiments of coloring maps and noting carefully just how it is that he manages to be as successful as he generally is. This activity is logically an energetic interpretant of the interrogatory he is putting to himself. If in that way he should be successful in working out a general rule of procedure, the demonstration of the theorem will be pretty sure to tread upon its heels; and it will certainly be the most welcome possible form that the solution of his problem could take. I will further say, — knowing that what I say will not be fully comprehended, — and not pretending to have myself any very definite meaning — that to some such form the solution of the problem must ultimately come. Suppose, however, what seems altogether the most probable near result, that the researcher comes, at length, to despair of finding, for his own part, any such solution. It will behoove him, then, to submit the problem to what I call a *theō'ric* transformation (because the demonstration of every difficult *theorem* involves such a transformation). This is an operation of necessary, or rather, of compulsive reasoning (granted the premisses, the interpreter of the reasoning is compelled to assent to the conclusion), which formal logic cannot possibly take into account, since from the point of view of formal logic there is no essential distinction between it and the most obvious *corollarial* transformation. Yet the key to mathematical methodeutic lies hidden here. It will consist in the transformation of the problem, — or of its statement, — due to viewing it from another point of view. I cannot here go into further explanations.

The first theoric transformation that our imaginary researcher is likely to try will consist in endeavoring to gain a distinct and manageable general conception of the conditions under which two regions not in immediate contact must be colored alike, as well as those under which such regions must be colored differently. This idea will produce an energetic interpretant quite similar to that his first idea produced. That is to say, he will be provoked to intense experimental activity in the inward world. But were the desired solution to be easily reached by any such highway as this, it would have been reached long ago. For it has baffled some of the greatest mathematicians of England. Mr. Kempe, who, outside the theories of multitude and of ordinal number, is perhaps the profoundest mathematical logician of our age, published some quarter of a century ago, a supposed demonstration that the number of colors necessary for a map of the earth is four, but I am informed that many years later a fatal fallacy was found to be involved in it. I confess I failed to detect it.

Supposing, then, that our researcher becomes discouraged in this second venture, it may occur to him that his problem, being a problem in *topics*, ought to be brought under the sole theorem that we know in that field, that is to say, the census-theorem. There is all the more reason for this in the rather obvious circumstance that the census theorem alone is not adequate to the demonstrative solution. For, such being the case, the study of this problem from that point of view is very likely to open up a new theorem in topics. Should our researcher be led to try that way, he will begin by reminding himself what that theorem, as limited to a map, is. It is that the number of bounding lines, each between two definite regions, but not an uninterrupted oval (such boundaries not counting), diminished by the number of regions that are bounded (an unbounded surface counting for two) and by the number of points in which several regions come together, is equal to the *Cyclosy* of the surface diminished by the sum of its *Chorisy* and its *Periphraxy*. These dreadful nouns, quite terrifying in the darkness, — turn out, when a light is thrown upon them to be rather ridiculously simple. They are the "Listing Numbers" of lowest "grade," — the chorisy of grade 0, the cyclosy of grade 1, the periphraxy of grade 2, the *Apeiry* of grade 3, etc. They may all be defined in one statement, thus: The Listing number of grade G of a place of dimensionality D is the number of interruptions by obstacles of dimensionality $D - G$ that would have to be established in that place in order that the uninterrupted part of it should be incapable of containing a continuously deformable thing of dimensionality G that, while remaining all the while in the uninterrupted part of the place *could not*, without rupture or welding of parts, shrink to an indefinite approximation toward losing a unit of dimensionality. Since a thing of dimensionality 0 can in no case lose a unit of dimensionality, the last clause becomes superfluous in defining the Chorisy, which is, in fact, nothing but the number of entirely separated parts of which the place it characterizes consists. What is called by Listing the *Census-Number* of a place is the sum of the Listing numbers of even grade belonging to singular parts of even dimensionality and of the Listing numbers of odd grade belonging to singular parts of odd dimensionality, diminished by the sum of the Listing numbers of even grade belonging to singular parts of odd dimensionality and those of odd grade belonging to singular parts of even dimensionality. The singular parts are 1st, the whole, if it be uninterrupted (by parts of lower dimensionality), and 2nd, all [uninterrupted] bounds of the whole or of parts. But no part that has a nodal point or nodal line is to be reckoned as a singular part in calculating the census-number. This represents Listing's original practice. It might be improved.

The census-theorem is that the census number of a place interrupted by a figure is the same as that of the uninterrupted place. This is an improvement upon Listing's enunciation of it. Of empty projective space all the Listing numbers are 1. But in dealing with unbounded rays, planes, and flat space, it must not be forgotten that, distance being entirely left out of account, deformable things may wholly or in part pass through the parts at an infinite distance.

This principle with the aid of some others, such as that, in order to necessitate so many colors, some one region must abut upon regions of all the other colors, every pair of these other regions being prevented, either by direct abuttal or otherwise, from being of the same color, will go a long way toward the solution of the problem, though it will leave some difficulties that can probably [be] overcome, and which may afford a hint of a new theorem of topics. The general form of our inquirer's activity will be as before experimental, though the experiments will be much more intelligent and purposive than they were before this pertinent idea had been suggested.

It is evident that when the problem is solved, the researcher will have acquired a new habit to which the various concepts, or general mental signs, that have arisen and been found valuable, are merely adjuvant.

Meantime, the psychological assumption originally made is in great measure eliminated by the consideration that habit [is] by no means exclusively mental. Some plants take habits; and so do some things purely inorganic. The observed laws of habit follow necessarily from a definition of habit which takes no notice of consciousness. Thus the facts that great numbers of individuals which die and are replaced by reproduction is favorable to a marked prominence of habit, and that highly complex organisms of which multitudes of parts exercise interchangeable functions are so, follow from such a definition.

Nevertheless, I am far from holding consciousness to be an "epiphenomenon," though the doctrine that it is so has aided the development of science. To my apprehension, the function of consciousness is to render self-control possible and efficient. For according to such analysis as I can make the true definition of consciousness is connection with an internal world; and the first impressions of sense are not conscious, but only their modified reproductions in the internal world.

I do not deny that a concept, or general mental sign, may be a logical interpretant; only, it cannot be the ultimate logical interpretant, precisely because, being a sign, it has itself a logical interpretant. It partakes somewhat of the nature of a verbal definition, and is very inferior to the living definition that grows up in the habit. Consequently, the most perfect account we can give of a concept will consist in a description of the habit that it will produce; and how otherwise can a habit be described than by a general statement of the kind of action it will give rise to under described circumstances?

This is the variety of pragmatism that I have urged. I do not deny the truth of the doctrine as defined by James; only I think it fails to go to the root of the matter. The most decided objection I have to the teachings of James and Schiller is that they seem to deny all infinity, including an infinite Being. No such denial follows from my form of pragmatism with which a critical acquiescence and assent to the utterances of common sense is especially in harmony even if it be not strictly implied therein.

Charles Santiago Sanders Peirce

9

MAP PROJECTION

A. THE UNITED STATES COAST SURVEY
APPENDIX No. 15

A QUINCUNCIAL PROJECTION OF THE SPHERE[1]

For meteorological, magnetological, and other purposes, it is convenient to have a projection of the sphere which shall show the connection of all parts of the surface. This is done by the one shown in the plate. It is an orthomorphic or conform projection formed by transforming the stereographic projection, with a pole at infinity, by means of an elliptic function. For that purpose, l being the latitude, and θ the longitude, we put —

$$\cos^2 \varphi = \frac{\sqrt{1 - \cos^2 l \cos^2 \theta} - \sin l}{1 + \sqrt{1 - \cos^2 l \cos^2 \theta}},$$

and then $\frac{1}{2}F\varphi$ is the value of one of the rectangular co-ordinates of the point on the new projection. This is the same as taking —

$$\cos am(x + y\sqrt{-1})(\text{angle of mod.} = 45°) = \tan \frac{p}{2}\left(\cos \theta + \sin \theta \sqrt{-1}\right),$$

where x and y are the co-ordinates on the new projection, p is the north polar distance. A table of these co-ordinates is subjoined.

Upon an orthomorphic projection the parallels represent equipotential or level lines for the logarithmic potential, while the meridians are the lines of force. Consequently we may draw these lines by the method used by Maxwell in his Electricity and Magnetism for drawing the corresponding lines for the Newtonian potential. That is to say, let two such projections be drawn upon the same sheet, so that upon both are shown the same meridians at equal angular distances, and the same parallels at such distances that the ratio of successive values of $\tan \frac{p}{2}$

[1] This material has been published in several places:
(a) *American Journal of Mathematics* 3 (1880)
(b) *Annual report of Coast and Geodetic Survey*, 1880, Appendix #15 (CS1877)
(c) *A Treatise on Projection* (1882) by Thomas Craig
Further details regarding it may be found in "Charles S. Peirce and the Problem of Map-Projection" by the editor in *Proceedings of the American Philosophical Society* 107:4 (August 1963).

is constant. Then number the meridians and also the parallels. Then draw curves through the intersections of meridians with meridians, the sums of numbers of the intersecting meridians being constant on any one curve. Also do the same thing for the parallels. Then these curves will represent the meridians and parallels of a new projection having north poles and south poles wherever the component projections had such poles.

Functions may, of course, be classified according to the pattern of the projection produced by such a transformation of the stereographic projection with a pole at the tangent points. Thus we shall have —

1. Functions with a finite number of zeroes and infinites (algebraic functions).

2. Striped functions (trigonometric functions). In these the stripes may be equal, or may vary progressively or periodically. The stripes may be simple, or themselves compounded of stripes. Thus, sin (a sin z) will be composed of stripes each consisting of a bundle of parallel stripes (infinite in number) folded over onto itself.

3. Chequered functions (elliptic functions).

4. Functions whose patterns are central or spiral.

I. *Table of rectangular co-ordinates for construction of the "quincuncial projection."*

x (for longitudes in upper line). y (for longitudes in lower line).

Lat.	0° 90	5° 85	10° 80	15° 75	20° 70	25° 65	30° 60	35° 55	40° 50	45° 45	50° 40	55° 35	60° 30	65° 25	70° 20	75° 15	80° 10	85° 5	Lat.
°																			°
85	.033	.033	.033	.032	.031	.030	.029	.027	.025	.024	.021	.019	.017	.014	.011	.009	.006	.003	85
80	.067	.066	.066	.064	.063	.061	.058	.055	.051	.047	.043	.038	.033	.028	.023	.017	.012	.006	80
75	.100	.100	.099	.097	.094	.091	.087	.082	.077	.071	.065	.058	.050	.042	.034	.026	.017	.009	75
70	.135	.134	.133	.130	.127	.122	.117	.110	.103	.095	.087	.077	.067	.057	.046	.035	.023	.012	70
65	.169	.169	.167	.163	.159	.154	.147	.139	.130	.120	.109	.097	.085	.072	.058	.044	.029	.015	65
60	.205	.204	.201	.198	.192	.185	.177	.168	.157	.145	.131	.117	.102	.086	.070	.053	.036	.018	60
55	.241	.240	.237	.232	.226	.218	.208	.197	.184	.170	.154	.138	.120	.102	.082	.062	.042	.021	55
50	.278	.277	.274	.269	.261	.251	.240	.227	.212	.196	.178	.159	.139	.117	.095	.072	.048	.024	50
45	.317	.316	.312	.306	.297	.286	.273	.258	.241	.223	.202	.181	.158	.134	.109	.083	.055	.028	45
40	.357	.356	.351	.344	.334	.321	.307	.290	.270	.250	.228	.204	.179	.151	.123	.094	.063	.032	40
35	.400	.398	.393	.384	.373	.358	.341	.322	.301	.279	.254	.228	.200	.170	.139	.106	.071	.036	35
30	.446	.443	.437	.427	.413	.396	.377	.356	.332	.308	.281	.253	.222	.190	.155	.119	.081	.041	30
25	.495	.492	.484	.471	.455	.435	.414	.391	.365	.338	.309	.279	.246	.211	.174	.134	.091	.046	25
20	.548	.545	.534	.518	.498	.476	.452	.426	.398	.369	.339	.307	.272	.235	.195	.151	.104	.053	20
15	.609	.604	.589	.568	.544	.517	.490	.461	.432	.401	.369	.336	.300	.262	.219	.173	.121	.062	15
10	.681	.672	.649	.620	.590	.559	.528	.497	.466	.434	.401	.367	.330	.291	.248	.200	.143	.076	10
5	.775	.752	.713	.673	.635	.600	.566	.532	.500	.467	.433	.399	.363	.324	.282	.234	.177	.102	5
0	1.000	.841	.774	.723	.679	.639	.602	.567	.533	.500	.467	.433	.398	.361	.321	.277	.226	.159	0

II. *Preceding table enlarged for the spaces surrounding infinite points.*[2]

x (for longitudes in upper line). y (for longitudes in lower line).

Lat.	0°	1°	2°	3°	4°	5°	6°	8°	10°	12½°	15°	75°	77½°	80°	82°	84°	85°	86°	87°	88°	89°	Lat.
	90	89	88	87	86	85	84	82	80	77½	75	15	12½	10	8	6	5	4	3	2	1	
°																						°
15	.609	.609	.608	.607	.606	.604	.602	.596	.589	.579	.568	.173	.147	.121	.098	.074	.062	.050	.038	.025	.013	15
12½	.643	.643	.642	.641	.639	.636	.634	.627	.618	.606	.594	.185	.159	.131	.107	.082	.069	.055	.042	.028	.014	12½
10	.681	.681	.680	.678	.675	.672	.668	.659	.649	.635	.620	.200	.173	.143	.118	.091	.076	.062	.047	.031	.016	10
8	.715	.714	.713	.710	.706	.702	.697	.686	.674	.658	.641	.213	.185	.155	.129	.100	.085	.069	.052	.035	.018	8
6	.753	.752	.750	.746	.741	.735	.728	.714	.700	.681	.662	.227	.199	.169	.142	.112	.095	.078	.060	.040	.020	6
5	.775	.774	.770	.765	.759	.752	.745	.729	.713	.692	.673	.234	.207	.177	.150	.119	.102	.084	.065	.044	.022	5
4	.798	.797	.793	.786	.779	.770	.761	.743	.725	.704	.683	.242	.215	.185	.158	.128	.110	.092	.071	.049	.025	4
3	.825	.823	.817	.808	.798	.788	.778	.757	.738	.715	.693	.250	.224	.194	.168	.137	.120	.101	.079	.053	.029	3
2	.857	.853	.843	.831	.819	.806	.794	.772	.750	.726	.703	.259	.233	.204	.178	.148	.131	.112	.090	.065	.035	2
1	.899	.889	.872	.854	.839	.824	.810	.785	.763	.737	.713	.268	.243	.215	.190	.161	.144	.126	.105	.079	.046	1
0	1.000	.929	.899	.877	.857	.841	.825	.798	.774	.747	.723	.277	.253	.226	.202	.175	.159	.143	.123	.101	.071	0

[2] The map is found as the frontispiece of Volume 3. Among so-called "fragments" in the Peirce Collection at the Houghton Library is a quincuncially-ruled chart on which Peirce plotted details of the celestial sphere. The constellation *Virgo* is sketched in quite distinctly, wings and all.

B. P.S. OF A LETTER TO J. M. PEIRCE (L 339: 5 APRIL 1894)

My chart is a skew Mercator, that is, a Mercator with the poles of an oblique great circle at infinity. Its value is great. The formulae are pretty. Start with a stereographic centred at the pole of that circle. Then the equation of every circle (meridian or parallel) is in polar coordinates

$$(r \cos \theta - M)^2 + r^2 \sin^2 \theta = N^2$$

or

$$2M \cos \theta = r + (M^2 - N^2)\, r^{-1}$$

But if x and y are the coördinates on the skew mercator $x = \theta$, $y = \log r$.

Hence

$$\cos x = \frac{\operatorname{Cosh}(y - A)}{B}$$

To take account of the compression *exactly* is a problem I have not quite solved. First, to make the angles preserved, for r a complicated function has to be substituted. But then the geodesic line does not return into itself and consequently is not a circle on the stereographic and it is necessary to take such a function of $r\mathfrak{G}^{\theta i}$ as to render it a circle. This is possible, and the approximate solution is easy. The exact solution would require some days of calm leisure. As it would evidently involve elliptic functions, the whole earth would be represented ∞^2 times but probably not in rectangular checkers. Nor is it quite clear to me that the scale of the map would be uniform over the whole of the geodesic taken as central and reduced to a straight line.

My notion about Father's geometry would be to make an introduction of three chapters. The first on the nature of mathematical reasoning and how to perform it. The second on projective geometry, especially perspective. The third on the metrical properties of space and the justification of the idea that an angle is a difference of directions.

I don't know that the body of the book would require any correction.

The thing would be to preserve its leading characteristic, which is to lay hold of and render definite our instinctive or natural ways of considering space, especially in reference to moving about in it.

I think, that done, the work would be a master-piece and would be much admired. Its sale would be good, too. Plimpton is anxious to have us do it. You and I should do it together, n'est-ce pas? What say you?

I don't know that I would not reform and enlarge the part relating to plane problems and append a chapter on Descriptive Geometry, and perhaps an appendix on Modern Geometry.

Rambling back to my chart again, the formulae are like this. For the meridians, I take Θ and H and η subsidiary quantities and have

$$\tan \Theta = \cos 51^\circ \tan \text{longitude}$$
$$H = \sin \Theta \tan 51^\circ$$
$$\sin (x - \Theta) = H \tan \eta = H \operatorname{Sinh} y$$
$$y = \frac{180^\circ}{M\pi} \log \tan \left(45^\circ + \frac{\eta}{2}\right)$$

For the parallels I have the subsidiary quantity ♈

$$♈ = -\frac{1}{M} \log (\cos 51^\circ + \sin \text{latitude})$$
$$\operatorname{Cosh} (y - ♈) = \sin 51^\circ \cos x = \sec \eta$$
$$y = \frac{180^\circ}{M\pi} \left\{\log (\cos 51^\circ + \sin \text{lat}) + \log \tan \left(45^\circ + \frac{\eta}{2}\right)\right\}$$

The approximate solution taking account of compression. In order to preserve the angles, all the books say that on the polar stereographic you must multiply r by

$$\left(\frac{1 + \varepsilon \sin \varphi}{1 - \varepsilon \sin \varphi}\right)^{\varepsilon/2}$$

where ε is the eccentricity of a meridian = 0.08227. If we make

$$\sin \varphi' - \varepsilon \sin \varphi$$

this is to say that in place of

$$r = \tan \left(45^\circ - \frac{\varphi}{2}\right)$$

we have

$$r = \tan^{\varepsilon} \left(45^\circ + \frac{\varphi'}{2}\right) \tan \left(45^\circ - \frac{\varphi}{2}\right)$$

or

$$\log r = y - \varepsilon y'$$

In order to make the geodetic line an approximate circle, let the difference of longitude of two successive intersections of a geodesic with the equator be $\frac{\pi}{1+k}$. Then raise $r\mathfrak{G}^{\theta i}$ to the $(1+k)^{\text{th}}$ power. Put $\tan^{\mathfrak{G}+1+k}\left(45^\circ + \frac{\varphi'}{2}\right)$ $\tan^{1+k}\left(45^\circ - \frac{\varphi}{2}\right) = \tan\left(45^\circ - \frac{\varphi''}{2}\right)$, $\theta'' = (1+k)\theta$.

Then for the latitude φ substitute in the formulae φ'' and for the long. $(1+k)$ long, and the central line of the chart is a geodesic. The meridians will then be broken at the poles, thus [Fig. 1].

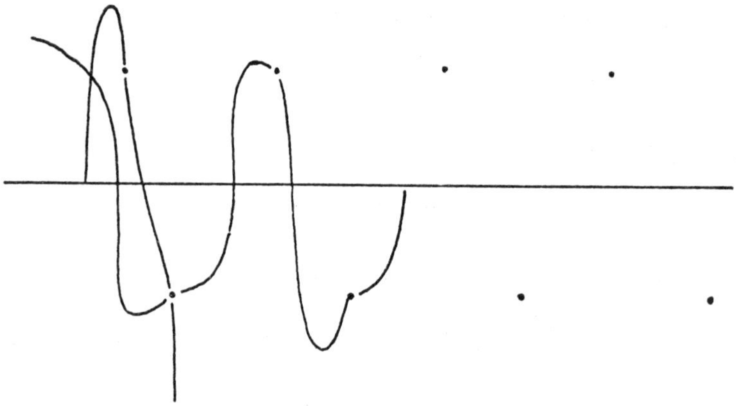

Fig. 1

The whole earth will be represented over and over again in stripes each differing from the last and so that every part of the earth between the extreme parallels (which I take at Latitude $\pm$ 51°) will somewhere appear on the central line without distortion.

C. A CAVEAT (1353)

"To the Commissioner of Patents:

"BE IT KNOWN, that I, CHARLES SANDERS PEIRCE, a citizen of the United States, residing on my wife's farm of ARISBE, in Westfall township, Pike County, Pennsylvania, have invented a new and useful improvement in MAP PROJECTIONS, and desiring further to mature the same, file this my Caveat therefor and pray protection of my right until I shall have matured my invention.

"The following is a description of my newly invented Map Projection, which is full, clear, and exact as I am able at this time to give.

"1. The object of this my invention is to represent one surface, say a part of the surface of the terraqueous globe, upon another, such as a sheet of paper or of glass, in such a systematic manner as shall facilitate the solution of certain practical questions, say how a given ship shall steer on a given day.

"2. The particular system of representation which constitutes my invention I term the Skew Mercator, and I would extend the same designation to any chart made according to its rules.

"3. While applicable to any two surfaces whatever, this system has been specially developed, by means of new and useful discoveries in mathematics, for the case of a part of an oblate ellipsoid of revolution having a compression of about 1/300 represented upon a plane surface.

"4. For the sake of clearness it will be well to remark that a geodetic line (which is a line drawn upon a given surface in such a manner that taking any two points upon it not too far apart, a portion of this line is the shortest that can be drawn upon that surface between those lines) does not generally return into itself at the end of any finite length of arc; and in particular on a oblate ellipsoid of revolution, excepting the geodetics which lie in plane through the axis of revolution or perpendicular to it, no geodetic returns into itself after traversing 360° of longitude; because it passes from an extreme latitude north (say) to an equal latitude south, not while it traverses 180° of longitude, but before it has quite got round so far, so that if the geodetic be indefinitely prolonged it will from a grating of its own cycles over all that part of the globe intermediate between its extreme latitudes, or more properly speaking a reticulation.

"5. In the Skew Mercator one oblique geodetic, which we may designate as *A*, is represented by a straight line, say *a*, on the plane, and is so represented that to equal distances on *A* correspond equal distances on *a*. Moreover, letting *M* represent any geodetic line on the ellipsoid (or other surface) which cuts *A* at right angles, this geodetic (and every one like it) is to be represented on the plane by a straight line, *m*, cutting *a* at right angles at the point that represents the point at which *M* cuts *A* at right angles. And the scale of the map is at every point the same in all directions."

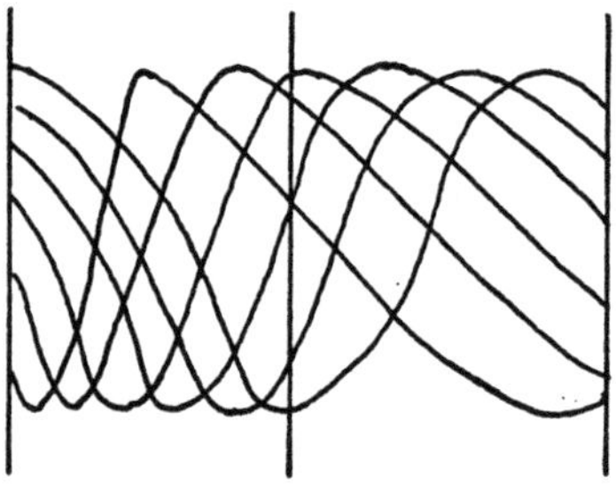

Fig. 1

D. FORMULAE AND TABLES FOR CONSTRUCTING TWO DIFFERENT CONFORMAL MAP-PROJECTIONS SUITABLE TO THE EXHIBITION OF ALL THE TERRITORY AND POSSESSIONS OF THE UNITED STATES OF AMERICA (1350)

EXPLANATION OF THE PROJECTIONS

A map is to be conceived as showing a part on the whole of a vertical projection of the Earth's surface as projected vertically (i.e. above the plane of gravity) upon the mean sea-level.

A "Conformal," or "Orthomorphic" map-projection is usually defined as one that shows all angles in their true magnitudes at their vertices; but perhaps a clearer definition is that one that has the same scale in all directions. This is not exactly true of any other kind of map nor even approximately so if the region represented covers as considerable part of the Earth's surface, as the possessions of the United States.

The form of the mean surface of the sea must, indisputably, always be regarded, by the cartographer, as ellipsoidal. For though departures from such a form may be made out by geodesists with more or less plausibility they will certainly be so minute, so irregular, and so liable to revision, that any correction of the distance between two points on such a basis will be far preferably made by means [of] an independent table of double entry rather than by a modification of the map-projection.

Everybody who has ever taken a course in Coördinatal Geometry (less precisely called *Analytical*, or *Algebraic Geometry*) may recall that the equation of an ellipsoid is

$$\left(\frac{x}{X}\right)^2 + \left(\frac{y}{Y}\right)^2 + \left(\frac{z}{Z}\right)^2 = 1,$$

where X, Y, Z are X the largest, Z smallest semidiameters, and Y is the semidiameter perpendicular to both the others; while x, y, z are the rectangular coordinates of any one point of the surface measured parallel respectively to the semi-axes X, Y, Z, from the centre of the ellipsoid.

In the case of the mean surface of the sea Z will be the distance from the centre of the earth to either pole (often called its polar *semi-axis*). As to X and Y, the only proper assumption in the present state of geodetical measurement is that they are equal. For the only attempt that I know of to determine their longitudes and relative length, that of Colonel Clarke, in 1878, being, I presume, deemed premature. Anyway he made their difference of length to be only 1524 feet, or say $\frac{2}{7}$ of a mile, each being about 4000 miles in length. If their difference can be neglected in geodetic calculations, it certainly ought to be neglected for any map on paper, which is liable, however great be the care taken of it, to shrink and swell unequally in different directions; while to take account of variations of the radius of the equator would multiply the number of oblique spherical triangles that have to be solved in making the table of coördinates for the projection, — would require this number of calculations, I say, to be multiplied by the number of meridians laid down in 45° of longitude. The labor (other complications being considered) would be from ten to twenty times as great, to save an error of one part in fourteen thousand. Now, who is so simple as to attempt to make measurement on paper accurate to one part in fourteen thousand?

E. [MATHEMATICAL NOTES ON THE SHAPE OF THE EARTH] (1071)

It is generally assumed that the earth's mean sea surface is an ellipsoid of revolution, and it has been generally regarded as proved that, neglecting tidal effects, such a surface might belong to a body in stable equilibrium. Poincaré, however, seems to have proved, or nearly proved, that other forms of stable equilibrium were possible. There is an interesting popular account of the matter in Sir George Howard Darwin's book entitled "*The Tides*," 1898 (Houghton, Mifflin, & Co.). In connection with this Sir G. H. Darwin has put forth the theory that our large moon, which is unparalleled in the solar system, originated by separation from the earth before the latter had acquired a stable equilibrium. This seems a reasonable conjecture, and if it be true, the earth probably must, after the separation, have gradually assumed a form approximately that of an ellipsoid with three unequal axes. Col. A. R. Clarke, an eminent geodesist, had undertaken (rather prematurely, perhaps, I think) to compute in 1878 what would be the ellipsoid, not necessarily one of revolution, that would best satisfy the geodetic observations. His result was that the difference between the greatest and least semi-diameters of the equator was only $\frac{1}{13721.2}$ part of the former being only about a quarter of a mile, and this supposed form did not satisfy the observations very much better than an ellipsoid of revolution would.

If the three semiaxes of an ellipsoid be denoted by X, Y, Z, and if x, y, z, be the sides of a rectangular parallelopipedon respectively parallel to those axes and having one vertex at the centre and the opposite vertex at any point on the ellipsoidal [at the sea-level] surface, these quantities will satisfy the equation

$$(1) \qquad \left(\frac{x}{X}\right)^2 + \left(\frac{y}{Y}\right)^2 + \left(\frac{z}{Z}\right)^2 = 1$$

which equation may be regarded as a definition of what is meant by an ellipsoid. If two of the semi-axes, say X and Y are equal, we may replace them by the single letter, R, when the equation will become

$$(2) \qquad \frac{x^2+y^2}{R^2}+\frac{z^2}{Z^2}=1.$$

If we are to regard this as representing the mean surface of the sea, the arduous and triumphant investigation of Mr. J. F. Hayford, which has rendered all other attempts to define the form of the mean sea-level obsolete (as far, at least, as the part of it covered by the United States is concerned), and which defines not the mean sea surface itself but a surface [with]

(3) R, the semidiameter of the equator

and

(4) (Z being the polar semiaxis) $R=\left(1+\frac{1}{294}\right)Z.$

But instead of z, the distance of any point of the surface from the equator, it will be more convenient to use a differently shaped zed, $\mathscr{Z}$, which is measured not from the plane of the equator, but from a parallel plane touching the ellipsoid at the North Pole, so that

$$(5) \qquad z=\mathscr{Z}-Z$$

while instead of x^2+y^2 I will write r, the distance of any point of the surface from the polar axis. With these substitutions, equation (2) becomes

$$(6) \qquad r^2=\left(1+\frac{1}{294}\right)^2(2Z\mathscr{Z}-\mathscr{Z}^2)$$

$$(7) \qquad r=\left(1+\frac{1}{294}\right)\sqrt{2Z\mathscr{Z}-\mathscr{Z}^2}.$$

Differentiating (7) we get

$$(8) \qquad dr=\left(1+\frac{1}{294}\right)\frac{Z-\mathscr{Z}}{\sqrt{2Z\mathscr{Z}-\mathscr{Z}^2}}(d\mathscr{Z})$$

whence

$$(9) \qquad (dr)^2=\left(1+\frac{1}{146\frac{2}{3}}\right)\left(\frac{Z^2}{2Z\mathscr{Z}-\mathscr{Z}^2}-1\right)(d\mathscr{Z})^2$$

and adding $(d\mathscr{Z})^2$

$$(10) \qquad (dr)^2+(d\mathscr{Z})^2=\frac{1}{146\frac{2}{3}}\left(147\tfrac{2}{3}\frac{Z^2}{2Z\mathscr{Z}-\mathscr{Z}^2}-1\right)(d\mathscr{Z})^2.$$

[Another version of this problem follows]

(1) $$\left(\frac{x}{X}\right)^2 + \left(\frac{y}{Y}\right)^2 + \left(\frac{z}{Z}\right)^2 = 1.$$

(2) If $X = Y$, $$\frac{(x^2 + y^2)}{X^2} + \frac{z^2}{Z^2} = 1.$$

If we denote $x^2 + y^2$ by r^2 and adopt Hayford's value of the compression of the earth

(2) becomes

(3) $$(0.99323 \pm V)\, r^2 + z^2 = Z^2$$

where $V = 10^{-5}$

Changing the origin to the N pole, by putting $z = \mathscr{Z} - Z$

(4) $$(0.99323 \pm V) r^2 = 2Z\mathscr{Z} - \mathscr{Z}^2$$

or

(5) $$(0.99661 \pm \tfrac{1}{2}V) r = \sqrt{2\mathscr{Z}Z - \mathscr{Z}^2}$$

or

(6) $$r = (1.003401 \pm .6V)\sqrt{2\mathscr{Z}Z - \mathscr{Z}^2}$$

Differentiating, we get

(7) $$dr = (1.003401 \pm .6V)\frac{\mathscr{Z}(Z - \mathscr{Z})}{\sqrt{2\mathscr{Z}Z - \mathscr{Z}^2}}\, d\mathscr{Z}$$

squaring

(8) $$(dr)^2 = (1.006814 \pm 1.1V)\left(\frac{Z^2}{2\mathscr{Z}Z - \mathscr{Z}^2} - 1\right)(d\mathscr{Z})^2$$

(9) $$(d\mathscr{Z})^2 + (dr)^2 = (1.006814 \pm 1.1V)\frac{Z^2}{2\mathscr{Z}Z - \mathscr{Z}^2}(d\mathscr{Z})^2 -$$

$$- (0.006814 \pm 1.1V)$$

$$= \left[\frac{Z^2}{2\mathscr{Z}Z - \mathscr{Z}^2} + (0.006814 \pm 1.1V)\frac{(Z - \mathscr{Z})^2}{2\mathscr{Z}Z - \mathscr{Z}^2}\right](d\mathscr{Z})^2$$

$$= \frac{Z^2}{2\mathscr{Z}Z - \mathscr{Z}^2}\left[1 + 0.006814 \pm 1.1V)\left(1 - \frac{\mathscr{Z}}{Z}\right)^2\right](d\mathscr{Z})^2$$

which will give nearly

(10) $$\sqrt{(d\mathscr{Z})^2 + (dr)^2} = \frac{Z}{\sqrt{2\mathscr{Z}Z - \mathscr{Z}^2}}\left[1 + .003407\left(1 - \frac{\mathscr{Z}}{Z}\right)^2\right] d\mathscr{Z}$$

F. NEW THEORY OF THE SKEW MERCATOR (1353)

1894 March 27

Suppose a geodesic line to be represented by a straight line and equal arcs by equal lengths, and that the projection is conform. The straight lines perpendicular to this central line will represent geodesics. For if from neighboring points on a geodesic, geodesic perpendiculars of equal length be erected (on the same side of the first) the junction of their extremities will by a known theorem be perpendicular to them. Therefore the angles will be preserved in making these perpendicular straight lines represent geodesics.

Therefore the poles will not lie on any of them except those that are meridians.[1]

THE MATHEMATICAL THEORY

§0. It is a comfort to me, who am reading mathematics constantly, to get a subject treated *ab ovo* provided the elementary part is abridged, even if I am quite familiar with it. Hence, I hope the navigators for whom this exposition is intended will not object to my going into elementary explanations.

§1. A map is a representation of a round surface upon a plane, by which we mean that,

1st. In a general way, *points* upon the round surface are represented by *points* on the map-plane, and *vice versa*,

2nd. *Continuous series* of points on the round surface are represented by *continuous series* of points on the map-plane.

Let us first imagine that

(1) Every point on the round surface is represented on the map-plane;

(2) Every point on the map-plane represents a point on the round surface;

(3) No point on the round surface is represented by more than a single point on the map-plane;

[1] In the worksheets on the geodesic Peirce makes use of Hoüel (Recueil, p. lxiii).

(4) No point on the map-plane represents more than one point on the round surface.

These things being assumed, imagine that round any point on the map-plane is drawn an oval, by which I mean a curve returning into itself without cutting itself. This oval will represent an oval on the round surface. Let the oval on the map-plane be itself enclosed in another oval. This second oval will represent an oval on the round surface not intersecting the first oval and not on the same [side] of the latter as the [origin point] represented by the point originally taken on the map-plane. Imagine an endless series [of] ovals to be drawn on the map-plane each enclosing the previous one and not approximating closer and closer to any finite oval. These will represent an endless series of ovals on the round surface enclosing or enclosed by one another not approximating to any oval. Then, these ovals on the round must either approximate to a point or a limited line. This point, or limited line, is represented by the infinitely distant parts of the plane. We may hold ourselves at liberty to assume anything we please about them, provided it involves no contradictory result. Hence we must either assume for our purpose that the parts of the plane infinitely distant are one point, or that they are a limited line. I shall not examine the latter assumption, but shall assume that for our purpose the infinitely distant parts of the plane are to be regarded as a single point.

[ADDITIONAL] MATHEMATICAL THEORY

§1. The position of a point may be expressed in rectangular coördinates, x and y, or polar coördinates r and θ having the same origin. But I reckon θ as on a clock from vertically up round to the right.

For the purposes of the conform projections, or those which preserve the angles, it is convenient to make one imaginary expression define the position of a point. Namely

$$z = re^{\theta i} = r\cos\theta + r\sin\theta . i \;= y + xi.$$

§2. If a projection have these two properties

1st, that for each place, there is some right angle in nature (I mean, the surface represented, earth, heavens or what) having its vertex there which is represented by a right angle on the map,

2nd, that along the two legs of such [a] right angle the scales are the same,

then the projection preserves all the angles. For let XOY be such a right angle in nature, represented by the right angle $X'O'Y'$ on the map. Imagine the portion seen to be infinitesimal, so that there shall be no curves. Then, completing the rectangle [to a fourth vertex Z and Z'], if $\frac{OY}{O'Y'} = \frac{OX}{O'X'}$ that is to say, if the scale of the map is the same along the two legs, it is plain that the angles $ZOY = Z'O'Y'$. Hence, obviously, all angles having their vertices at O are represented by equal angles.

THE SKEW MERCATOR PROJECTION

It is well known that the polar stereographic projection of the sphere is produced by making

$$r = R \tan\left(45° - \frac{\varphi}{2}\right)$$

where r is the radius vector on the map from the pole as origin, and φ is the latitude [λ is the longitude so reckoned].

Upon this projection, every circle is projected into a circle.

It is also known that in order to preserve the angles taking account of the compression of the earth we may put

$$r = \left(\frac{1 - \varepsilon \sin \varphi}{1 + \varepsilon \sin \varphi}\right)^{\frac{\varepsilon}{2}} \tan\left(45° - \frac{\varphi}{2}\right)$$

where ε is the eccentricity of a meridian $= 0.08227$.* (*For (Clarke's *Geodesy*, p. 104) the radius of curvature of a meridian is $\frac{a(1-\varepsilon^2)}{(1-\varepsilon^2 \sin^2 \varphi)^{3/2}}$ and that of a normal section perpendicular to the meridian is $\frac{a}{(1-\varepsilon^2 \sin^2 \varphi)^{\frac{1}{2}}}$.)

But upon this projection geodesic lines no longer appear as circles. The reason is that the difference of longitude between the intersection of a geodesic line with the equator and its northernmost point is not 90°, but is

Maximum Latitude	Diff Long
18°	90°-17′
36°	90°-15′
54°	90°-12′
72°	90°- 6′

A change $\delta\lambda$ in longitude, in passing along the outline of the last section makes a rotation of the normal equal to $\cos \varphi \cdot \delta\lambda$.[2] Hence, if

[2] On another fragment Peirce writes, "Hence we must have an angular motion along that section equal to $\cos \varphi . d\lambda$, represented on the map by a line whose ratio to $d\varphi$ is $\frac{1-\varepsilon^2}{1-\varepsilon^2 \sin^2 \varphi}$."

$\frac{\delta s}{\delta \lambda}$ is constant for all latitudes (as for a regular Mercator's chart)

$$\frac{\delta s}{\delta \varphi} = \frac{1}{\cos \varphi} \frac{1 - \varepsilon^2}{1 - \varepsilon^2 \sin^2 \varphi}$$

$$\frac{\delta s}{\delta \varphi} - \frac{1}{\cos \varphi} = \frac{1}{\cos \varphi}\left(\frac{1 - \varepsilon^2}{1 - \varepsilon^2 \sin^2 \varphi} - 1\right)$$

$$= \frac{1}{\cos \varphi}\left(\frac{-\varepsilon^2 \cos^2 \varphi}{1 - \varepsilon^2 \sin^2 \varphi}\right)$$

$$= \frac{-\varepsilon^2 \cos \varphi}{1 - \varepsilon^2 \sin^2 \varphi}$$

$$\frac{\delta s}{\delta \varphi} = \frac{1}{\cos \varphi} - \varepsilon^2 \frac{\cos \varphi}{1 - \varepsilon^2 \sin^2 \varphi}$$

Now $D\varphi \log \frac{1 - \varepsilon \sin \varphi}{1 + \varepsilon \sin \varphi} = \frac{-\varepsilon \cos \varphi}{1 - \varepsilon \sin \varphi} - \frac{\varepsilon \cos \varphi}{1 + \varepsilon \sin \varphi} =$

$$= -\varepsilon \cos \varphi \frac{2}{1 - \varepsilon^2 \sin^2 \varphi}$$

And $D\varphi \log \tan\left(45^\circ + \frac{\varphi}{2}\right)$

$$= \frac{1}{2 \tan\left(45^\circ + \frac{\varphi}{2}\right)} \frac{1}{\cos^2\left(45^\circ + \frac{\varphi}{2}\right)} = \frac{1}{\sin(90^\circ + \varphi)} = \frac{1}{\cos \varphi}$$

Hence

$$s = \log \tan\left(45^\circ + \frac{\varphi}{2}\right) + \frac{\varepsilon}{2} \log \frac{1 - \varepsilon \sin \varphi}{1 + \varepsilon \sin \varphi}$$

(This agrees with Craig's Projections, p. 65.)[3]

Hence we must have an angular motion along that section, equal to $\cos \varphi \cdot \delta\lambda$, represented in the map by a line whose ratio to $d\varphi$ is

$$\frac{1 - \varepsilon^2}{1 - \varepsilon^2 \sin^2 \varphi}.$$

[3] Craig was a member of the Mathematics Department at the Johns Hopkins University during Peirce's association with that institution and was later Editor-in-Chief of the *American Journal of Mathematics*. He had also served in the Coast and Geodetic Survey.

THE SKEW MERCATOR

On the Stereographic, the equation of a meridian is

$$(\xi - A)^2 + (\eta - B)^2 = C^2$$

When $\xi = 0$

$$\eta_1 = B + \sqrt{C^2 - A^2} = S \tan 25° \cdot 5$$
$$\eta_2 = B - \sqrt{C^2 - A^2} = S \cot 25° \cdot 5$$
$$(\xi - A)\,d\xi + (\eta - B)\,d\eta = 0$$

When $\xi = 0$ $\quad \frac{d\xi}{d\eta} = \frac{\sqrt{C^2 - A^2}}{A} = \tan \text{long}$

$$B = -S \cot 51°$$
$$\sqrt{C^2 - A^2} = S \operatorname{cosec} 51°$$
$$A = S \operatorname{cosec} 51° \cot \text{long} \qquad C = S \operatorname{cosec} 51° \operatorname{cosec} \text{long}$$

Putting $\xi = r \sin \theta \qquad \eta = r \cos \theta$
the equation is

$$(r \sin \theta - S \operatorname{cosec} 51° \cot \text{long})^2 + (r \cos \theta + S \cot 51°)^2 = S^2 \operatorname{cosec}^2 51° \operatorname{cosec}^2 \text{long}$$

$$S^{-1}r - Sr^{-1} = 2(\sin \theta \operatorname{cosec} 51° \cot \text{long} - \cos \theta \cot 51°)$$
$$= 2\,\frac{\sin \theta \cot \text{long} - \cos \theta \cos 51°}{\sin 51°}$$

Putting $\tan \Theta = \cos 51° \tan \text{long}$ and $E = \frac{\sqrt{\cot^2 \text{long} + \cos^2 51°}}{\sin 51°}$

$$S^{-1}r - Sr^{-1} = 2E \sin (\theta - \Theta)$$

Now $r = \mathfrak{G}^y \qquad \theta = x$

Put $\qquad S = \mathfrak{G}^Y$

$$\frac{r}{S} = \mathfrak{G}^{y-Y} = \operatorname{Cosh}(y - Y) + \operatorname{Sinh}(y - Y)$$
$$-\frac{S}{r} = \mathfrak{G}^{Y-y} = -\operatorname{Cosh}(y - Y) + \operatorname{Sinh}(y - Y)$$
$$\operatorname{Sinh}(y - Y) = E \sin (\theta - \Theta)$$

On the Stereographic the equation of a parallel is

$$\xi^2 + (\eta - \beta)^2 = \gamma^2$$

When $\xi = 0$

$$\eta_1 = \beta + \gamma = S \tan \frac{90° - \text{lat} + 51°}{2}$$
$$\eta_2 = \beta - \gamma = -S \tan \frac{90° - \text{lat} - 51°}{2} = S \tan \frac{51° - 90° + \text{lat}}{2}$$

$$\beta = \frac{S \sin 51^\circ}{\cos 51^\circ + \sin \text{lat}} \qquad \gamma = \frac{S \cos \text{lat}}{\cos 51^\circ + \sin \text{lat}}$$

$$r^2 \sin^2 \theta + \left(r \cos \theta - S \frac{\sin 51^\circ}{\cos 51^\circ + \sin \text{lat}}\right)^2 = S^2 \frac{\cos^2 \text{lat}}{(\cos 51^\circ + \sin \text{lat})^2}$$

[Incomplete]

VARYING SCALE OF SKEW MERCATOR

a Azimuth of geodesic
u Reduced lat
U ditto of northernmost points of geodesic
ε Eccentricity of meridian

$$\sin \alpha = \frac{\cos U}{\cos u}$$

$$\alpha' = 90^\circ - \alpha$$

$$\sin \alpha' = \sqrt{1 - \frac{\cos^2 U}{\cos^2 u}}$$

$$\cos U' = \sqrt{\cos^2 u - \cos^2 U}$$

$$\sin \alpha_2' = \frac{\sqrt{\cos^2 u - \cos^2 U}}{\cos u_2}$$

$$\alpha_2 = 90^\circ - \alpha_2'$$

$$\sin \alpha_2 = \frac{\sqrt{\cos^2 u_2 - \cos^2 u + \cos^2 U}}{\cos u_2}$$

$$\cos U'' = \sqrt{\cos^2 u_2 - \cos^2 u + \cos^2 U}$$

$$\therefore \sin^2 U'' = \sin^2 U - \sin^2 u + \sin^2 u_2 \quad \text{[Fig. 1]}$$

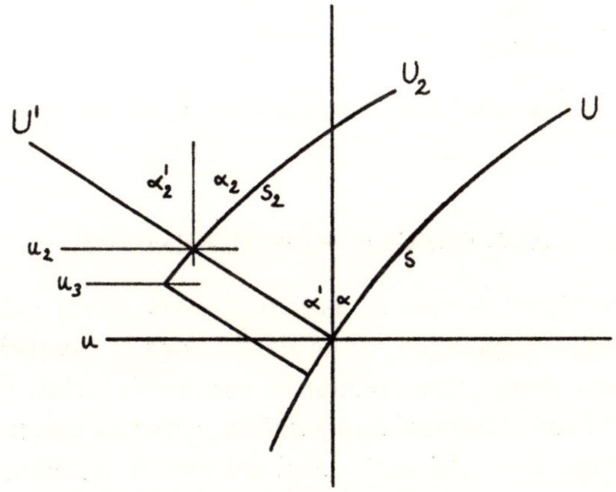

Fig. 1

Hence if rectangular geodesics are drawn between great circles whose northernmost points are constant, the difference of the squares of the sines of the reduced latitudes of their extremities is constant, or

$$\sin u \cdot d \sin u = \sin u_2 \, d \sin u_2$$

Let two such be drawn infinitely near, and compare the infinitestimal segments

$$ds = a\frac{\sqrt{1-\varepsilon^2\cos^2 u}}{\sqrt{1-\dfrac{\cos^2 U}{\cos^2 u}}}\,du = a\frac{\sqrt{1-\varepsilon^2\cos^2 u}}{\sqrt{\sin^2 U-\sin^2 u}}\,d\sin u$$

$$= a\frac{\sqrt{1-\varepsilon^2+\varepsilon^2\sin^2 u}}{\sqrt{\sin^2 U-\sin^2 u}}\,d\sin u$$

$$ds_2 = a\frac{\sqrt{1-\varepsilon^2+\varepsilon^2\sin^2 u_2}}{\sqrt{\sin^2 U_2-\sin^2 u_2}}\,d\sin u_2$$

$$= a\frac{\sqrt{1-\varepsilon^2+\varepsilon^2\sin^2 u_2}}{\sqrt{\sin^2 U-\sin^2 u}}\,d\sin u_2$$

$$= a\frac{\sqrt{1-\varepsilon^2+\varepsilon^2\sin^2 u_2}}{\sqrt{\sin^2 U-\sin^2 u}}\,\frac{\sin u}{\sin u_2}\,d\sin u$$

$$\frac{ds_2}{ds} = \frac{\sqrt{1-\varepsilon^2+\varepsilon^2\sin^2 u}}{\sin u_2}\cdot\frac{\sin u}{\sqrt{1-\varepsilon^2+\varepsilon^2\sin^2 u}}$$

$$= \frac{\sin \text{lat}}{\sin \text{lat}_2}$$

Though this is not what was wanted, it is a curious result. It is also puzzling.

[C. S. PEIRCE'S SKEW MERCATOR I]

C. S. Peirce's "Skew Mercator" is a projection *of the sphere* (i.e. not corrected for the compression of the Earth, since in measuring distances on a map of any considerable fraction of the earth's surface it will always be necessary to use numerical computation, either to assure oneself that the number read from the scale used sufficiently represents what one ought to mean for the special purpose in view, by the "distance" between two points on the earth's surface, or else to deduce this "distance" from the measure on the map) upon the plane of the great circle passing through latitude 60°, this projection being of the same kind as that of the ordinary Mercator's Projection, which ought to be a conform projection *of the sphere.*

But a map of a large fraction of the Earth's surface ought not to be regarded as intended to measure upon closer than to the first significant figure of our customary numerical notation, but rather to be viewed with the unaided eye. Now if any figure is drawn upon a plane surface held square before the eyes, and this figure is then turned in its own plane through a right angle, so that what was, before the turning, its height becomes its width right and left and *vice versa*, its *apparent* shape to a spectator not specially trained, will undergo a change ascribed to the greater effort required to turn our eyes up and down than to turn them right or left. But this change of apparent shape is not found to be a great inconvenience. Now that apparent change is just about double the exaggeration of [the] linear scale of Mercator's chart for a small area at 23° of latitude compared with a like object on the equator. Thus, to use the phrases of conscientious parlance, the accuracy of Mercator's chart within say 30° of the equator is truly astonishing, as compared with any visual aspect of it from without.

Now it happens that all parts and possessions of the United States of America, from Puerto Rico to Alaska and straight on to the Philippines lie within that distance of a Great Circle excepting Hawaii on one side and the Panama Canal on the other; and these are not very far outside of 30° from that central circle.

[C. S. PEIRCE'S SKEW MERCATOR II]

... If the sum of the orders of North poles is not equal to that of South poles, the point at infinity must make up the deficiency of the lesser sum. We, then, find a function which has these zeros and infinities; and finally we are free to multiply by a constant which serves to fix the latitude and longitude of any other selected point on the map. In this way, formulae can be obtained with great facility. For instance, suppose we wish the formulae for the stereographic projection when the plane of projection is tangent to a certain place. We calculate r and θ for that place, and denote them by r_0 and θ_0. Then we wish a function of $re^{\theta i}$ which shall vanish for $re^{\theta i} = r_0 e^{\theta_0 i}$ and shall become infinite for $re^{\theta i} = \frac{1}{r_0} e^{-\theta_0 i}$. This gives ($r_1$ and θ_1 being the coördinates on the new map)

$$r_1 e^{\theta_1 i} = C\frac{re^{\theta i} - r_0 e^{\theta_0 i}}{re^{\theta i} - \frac{1}{r_0}e^{-\theta_0 i}}$$

In the case of Mercator's projection the North pole is at $+\infty$ and the South pole at $-\infty$. Obviously, therefore,

$$x_1 + y_1 i = \log (\mathrm{re}^{\theta i})$$

or

$$x_1 = \log r$$
$$y_1 = \theta.$$

A map may in the first place have a finite number of North and South poles. In this case, it is connected with an algebraic function. If the number of poles is infinite, they cannot be specified singly and only the law of their arrangement can be described. The map may show the whole earth repeated over and over in stripes. These stripes may be all alike or they may be broader and broader indefinitely in one direction. Each stripe may show the whole earth but once, or several times, or even an infinite number of times. An example of the last case is the map resulting from the function sin · sin $a\,\varphi$, of which a sketch is given.[1] The map instead of being striped may be checkered. In this case, we have to do with a doubly periodic function. The pattern whether striped or checkered may be disturbed by irregularities, as in the case of the sum of a periodic function and an algebraic one. Finally, the map may have a pattern neither striped nor checkered. Thus, the whole earth might be represented in concentric rings, as by the function

$$\prod_{m=1}^{m=\infty} \frac{m-x}{m+x}$$

[1] No sketch accompanies this fragment. In another fragment Peirce writes: "The idea of the [Skew Mercator Map] is that since Mercator's Chart preserves the shapes and sizes of the globe within 40° of the equator sufficiently for comparison by the unaided and not excessively critical eye and since all the possessions of the United States (except Hawaii) lie within that angular distance of a certain great circle (namely one inclined 60° to the equator and cutting it in 35° West Longitude), it follows that a map which represents that great circle by a straight line, and preserves all angles unchanged, will give the best possible eye-view of those possessions."

G. [NOTES FOR] THE SKEW MERCATOR ON THE SPHERE (1584)

§1. Property of the Stereographic.

If a point in space X, Y, Z be projected from the origin upon the plane of $Z = 1$, its projection x, y will be

$$x = \frac{X}{Z} \qquad y = \frac{Y}{Z}$$

If we imagine the sphere through the origin

$$X^2 + Y^2 + (Z-1)^2 = 1$$

to be cut by the plane

$$\frac{X}{A} + \frac{Y}{B} + \frac{Z}{C} = 1$$

we have

$$\frac{x^2}{Z^2} + \frac{y^2}{Z^2} = -1 + \frac{2}{Z}$$

$$\frac{X}{AZ} + \frac{Y}{BZ} = \frac{1}{Z} - \frac{1}{C}$$

or eliminating Z

$$x^2 + y^2 + 1 = \frac{2}{A}x + \frac{2}{B}y + \frac{2}{C}$$

which is the equation of a circle.

If $C = \frac{1}{2}$, the plane passes through the centre of the sphere.

That the angles are preserved is evident from this: that the tangent plane to the sphere is as much inclined to the radius of projection as is the plane of the map.

In the case of the polar projection it is plain that

$$\frac{\sqrt{x^2 + y^2}}{Z} = \tan\left(45° - \frac{\varphi}{2}\right)$$

where φ = latitude.

§2. The Skew Mercator

We transpose to another pole on the sphere, distant we will say by an angle a from the a [Fig. 1]. The first pole we call B the second C.

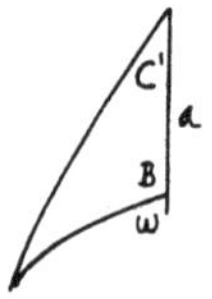

Fig. 1

Then the latitude = $90° - C$

Longitude = $180° - B$

reckoning longitude from the line of junction of the poles.

We might solve spherical triangle given $B = 180° - a$

a

$c = 90° - \varphi$

to find C and c.

We have two sides and included angle and could apply formulae of spherical trigonometry.

Namely,

$$\tan\tfrac{b}{2} = \sqrt{\frac{\dfrac{\tan\frac{a}{2}}{\tan\frac{c}{2}} + \dfrac{\tan\frac{c}{2}}{\tan\frac{a}{2}} + \frac{1}{2}\cos\omega}{\dfrac{1}{\tan\frac{a}{2}\tan\frac{c}{2}} + \tan\frac{a}{2}\tan\frac{c}{2} + \frac{1}{2}\cos\omega}}$$

$$\tan C = \frac{\sin\omega}{\tan\varphi\sin a + \cos a\cos\omega}$$

More simply

$$r'\,\mathfrak{S}^{\theta' i} = r\,\mathfrak{S}^{\theta i} + R = \sqrt{R^2+r^2}\left(\frac{r(\cos\theta+R)}{\sqrt{R^2+r}} + \frac{r\sin\theta}{\sqrt{R^2+r^2}}\,i\right)$$

$$\log r' + \theta' i = \log\sqrt{R^2+r^2} + \operatorname{arc\,tan}\frac{\tan\theta}{1+\dfrac{R}{r}\sec\theta}$$

Geodetic Line

ω Longitude

q Shortest distance station to axis ret.

s arc of geodesic

a azimuth

$q^2 d\omega = C ds$

$C \;= q \sin a$

φ Latitude

u Reduced latitude

a Semiaxis major

$$\sin\varphi = \frac{\sin u}{\sqrt{1-e^2\cos^2 u}}$$

$q \;= a\cos u$

e is eccentricity of meridian .08227

Compressed Earth

$$\frac{q^2}{a^2}+\frac{z^2}{b^2}=1$$

To preserve angles on polar projection

$$\begin{aligned} r &= q+(b-z)\tan\left(45^\circ+\frac{\varphi}{2}\right)\\ &= a\cos u + b(1-\sin u)\tan\left(45^\circ+\frac{\varphi}{2}\right)\\ &= \frac{a\cos\varphi}{\sqrt{1-e^2\sin^2\varphi}} + b\frac{\sqrt{1-e^2\sin^2\varphi}-\sqrt{1-e^2}\sin\varphi}{\sqrt{1-e^2\sin^2\varphi}} \end{aligned}$$

$$\tan\left[45^\circ+\frac{\varphi}{2}\right] = \frac{\cos\varphi}{1-\sin\varphi}$$

H. A NEW MAP OF THE UNITED STATES AND POSSESSIONS. EXPLANATIONS (1351)

The purpose of this map is to display to the eye with all practicable accuracy the shape of each considerable part of the territory of our republic, and the relations of the different parts to one another and to the adjacent areas beyond our borders. That principal purpose having been attained, I further provide a simple apparatus by which measurements of long distances in miles may be made with sufficient accuracy for the purposes of most persons.

Of course it would be impossible to *display* upon a flat map the true shape of the earth's surface on mountains with absolute precision. A map is conceived to show the "sea-level," or surface that the sea would have in the absence of tides, if it were allowed to flow through a tunnel to a point vertically below the point above it on the mountain-side. The distance between two such points would be altogether invisible to the eye on any map that should bring any large part of the whole earth within the range of distinct vision. Even the difference between the earth's polar and equatorial diameters, though over 26 miles, is many times too small to be perceptible to the eye on an accurately drawn outline of the globe.

10

LINEAR ALGEBRA AND MATRICES

A. NOTES ON B. PEIRCE'S LINEAR ASSOCIATIVE ALGEBRA (78)

§1. This definition of mathematics was entirely novel in 1870, when it was first put forth. Since then all important writers on the foundations of mathematics have been led to similar definitions, no doubt more or less influenced, directly or indirectly, by Peirce. After broad and deep study of the question, I am definitively convinced that the definition here given is better than any of the modifications of it. The chief points to be considered are as follows:

1st, Is not the definition too broad because mathematics is limited to deductions from a certain kind of hypotheses; namely, from such as are perfectly definite and somewhat complicated?

2nd, Is not the definition too broad, because mathematical reasoning is schematic while philosophical deductions deal with pure concepts?

3rd, Is not the definition too broad in that all sciences draw necessary conclusions? Should we not say that mathematics is the logic of necessary inference?

4th, Is not the definition too narrow in that some of the highest achievements of mathematics have consisted in the formation of mathematical hypotheses?

5th, Is not the definition too narrow in that induction, hypotheses, and every variety of reasoning is used in mathematics?

6th, Is it not, after all, true that when the conception of quantity is sufficiently generalized, all mathematics relates to quantity and is distinguished from other sciences by this?

The first question is suggested in the Article *Mathematics* in the 9th Edition of the *Encyclopaedia Britannica*, and elsewhere. But this question really is whether an improved definition is forthcoming; and it does not appear that anything more than a vague distinction between much and little is given, or can be given, to support an objection in this sense. No doubt, when hypotheses are simple, as vague hypotheses from which anything can be deduced are apt to be, people work out their consequences for themselves, while in cases where they are too complex for them to handle they consult mathematicians. But no more precise distinction has been offered; and the utility of it, if it should be offered, is very doubtful.

The second question is asked by nobody. But blustering intuitionalists, as ignorant of logic as they are arrogant, still maintain the opinion that there are conceptual deductions of a radically different nature from mathematical deductions. The only tenable distinction is that the conceptual deductions are loose and superficial reasonings that are inconclusive, while those of the mathematicians are accepted by every understanding that grasps them. My thirty years' study of the logic of relations would have brought to light any distinction that could have existed between mathematical reasonings and other deductions. But I only find distinctions of degree, especially the distinction that mathematicians seldom reason inconclusively and metaphysicians seldom conclusively.

The third question is amply justified by the fact that Dedekind holds mathematics to be a branch of logic. At the time my father was writing this book, I was writing my paper on the logic of relations that was published in the 9th volume of the *Memoirs of the American Academy.* There was no collaboration, but there were frequent conversations on the allied subjects, especially about the algebra. The only way in which I think that anything I said influenced anything in my father's book (except that it was partly on my urgent prayer that he undertook the research) was that when at one time he seemed inclined to the opinion which Dedekind long afterward embraced, I argued strenuously against it, and thus he came to take the middle ground of his definition. In truth, no two things could be more directly opposite than the cast of mind of the logician and that of the mathematician. It is almost inconceivable that a man should be great in both ways. Leibniz came the nearest to it. He was, indeed, a great logician, for all his nominalism, which clung to him like the coat of Nessus, but which he more and more surmounted as the idea of continuity gained strength in his mind. But his mathematical power, though far from being mean, was inferior to that of either of the two Bernouillis. The mathematician's interest in a reasoning is as a means of solving problems — both a problem that has come up and possible problems like it that have not yet come up. His whole endeavor is to find short cuts to the solution. The logician, on the other hand, is interested in picking a method to pieces and in finding what its essential ingredients are. He cares little how these may be put together to form an effective method, and still less for the solution of any particular problem. In short, logic is the theory of all reasoning, while mathematics is the practice of a particular kind of reasoning. Mathematics might be called an art instead of a science were it not that the last achievement that it has in view is an achievement of knowing.

The fourth question is one which has exercised me considerably. But I have now obtained a satisfactory solution of it. It is true that there is great exercise of intellect in framing a mathematical hypothesis like that of the theory of functions. But as a purely arbitrary hypothesis, there is no element of cognition in it, and consequently it has nothing to do with the nature of mathematics *as a science*.

The fifth question is raised by a remark of Sylvester. But the reply is that all those other kinds of reasoning are in mathematics merely ancillary and provisional. Neither Sylvester nor any mathematician is satisfied until they have been swept away and replaced by demonstrations.

The sixth question is suggested by modern studies of the nature of quantity. The reply to it is that it is so far from being possible to refute B. Peirce's definition by any reply to this question that no justification of the old definition can be based on such reply except so far as that reply justifies the new definition.

§3. This distinction is merely a distinction between a system of quantity in which there are only two values and a system of quantity in which are more than two values.

§12, 13. Whether or not mixed algebras ought to be excluded is a question of what one's aim may be. Certainly, some mixed algebras are very interesting. It will be found that, after all, no inconsiderable percentage of my father's algebras are mixed, according to the definition here given.

§28. The argument in favor of allowing the coefficients of the algebra to be imaginary is certainly inconclusive. Every logician must admit that imaginary quantity has two dimensions and therefore should be recognized as double. To say that where quantity is introduced "it should be received in its most general form and in all its variety," if sound would forbid our separating an imaginary into a modulus and argument. It is evident that every form of algebra which will result from making the coefficients imaginary would coincide with some algebra that would result from restricting them to being real; but it is not evident that the converse is true. Therefore, nothing can be gained but much may be lost by allowing the coefficients to be imaginary.

§41. It may be observed that no argument is introduced to prove the advantage of thus taking the question of whether there is an idempotent expression or not as the first question to be asked concerning an algebra; and the fact that there is not shows the difference between the mathematical and the scientifically logical type of mind.[1]

[1] The definition that Peirce refers to at the beginning of these remarks on his father's work is, of course: "Mathematics is the science which draws necessary conclusions."

B. NOTES ON ASSOCIATIVE MULTIPLE ALGEBRA (75)

I

A multiple algebra is one in which there are units of different qualities capable of being added and multiplied together. Thus, ordinary algebra is a double algebra the units being *unity* and a square root of negative unity. An associative algebra is one in which the multiplication of the units, though not generally commutative, is always associative. A linear algebra is one in which the number of linearly independent expressions is finite; but this restriction seems to be of little moment.

The theory of associative multiple algebra includes that of groups, a group being a special kind of associative algebra. My father in his work entitled *Linear Associative Algebra* has given a method for the discovery of all such systems of algebra, and has given

All the single algebras, 2 in number
All the double algebras, 3 in number
All the triple algebras, 5 in number
All the quadruple algebras, 11 in number
All the quintuple algebras, 70 in number
All the sextuple algebras containing any expression which is its own square, 65 in number. But he has reckoned the ordinary square root of negative unity as a scalar, so that each of his algebras will generally include several if the ordinary imaginary be reckoned as a separate unit. The tables given in this work may be applied to finding the group of the first five orders, and show at once that the list given by Professor Cayley (*Am. Jour. Math.* Vol. 1, p. 51) is extremely defective. Thus, the latter omits the following derived from Professor Peirce's b_i.

	1	i	j
1	1	i	j
i	i	i	j
j	j	i	j

II*

(* The main proposition of this note was presented to the American Academy of Arts and Sciences, May 11, 1875; and is published in the *Proceedings* of that body, p. 392.)

Multiplication expresses relation. Thus, iA expresses the quantity which is related to A in the manner signified by i. We may then for any algebra imagine another non-relative algebra, the expressions of which can only be multiplicands or products, never multipliers. Let i be any unit of a multiplying algebra, and let the general expression of the non-relative algebra be $aI + bJ + cK + dL +$ etc., where a, b, c, d, etc., are scalars. Let the product of this by i be written thus,

$$\begin{aligned} &(A_1 a + A_2 b + A_3 c + \text{etc.})I \\ &+ (B_1 a + B_2 b + B_3 c + \text{etc.})J \\ &+ (C_1 a + C_2 b + C_3 c + \text{etc.})K \\ &+ \text{etc.} \end{aligned}$$

Then, by an enlargement of the multiplying algebra, we may suppose i separated into a sum as follows

$$\begin{aligned} i = {} & A_1(I:I) + A_2(I:J) + A_3(I:K) + \text{etc.} \\ & + B_1(J:I) + B_2(J:J) + B_3(J:K) + \text{etc.} \\ & + C_1(K:I) + C_2(K:J) + C_3(K:K) + \text{etc.} \\ & + \text{etc.} \end{aligned}$$

where $(I:I)$, $(I:J)$, etc. are new units such that in general

$$\begin{aligned} &(I:J)J = I \qquad && (I:J)(J:K) = (I:K) \\ \text{and} \quad &(I:J)K = 0 && (I:J)(K:L) = 0 \end{aligned}$$

In this way it appears that every unit of every algebra may be regarded in at least one way as a linear polynomial in units following this law.

This proposition is equally true whether the ordinary imaginary be or be not regarded as a scalar.

In order to reduce ordinary double algebra to this form (not regarding the imaginary as a scalar) we put

$$\begin{aligned} 1 &= X:X + Y:Y \\ \sqrt{-1} &= X:Y - Y:X \end{aligned}$$

This representation is closely allied to the geometric representation of imaginaries, for it shows that the multiplication by $\sqrt{-1}$ converts the ordinate into an abscissa of equal value, and the abscissa into an ordinate of the negative of its value; in other words it produces a quadrantal rotation in a positive direction. It may be remarked, that strictness of logic in the development of the calculus, requires that $\sqrt{-1}$ should be regarded as a unit in a continuum limited to two dimensions. This is the way in which Cauchy treats the subject; and this is the only way in which his principles of integration can be established. Thus if we take the variable $z = e^{\theta\sqrt{-1}}$ and make it vary by a continuous change of its argument θ, it is essential to Cauchy's theory that while z changes from unity round to unity again $\sqrt{z}$ changes from unity to negative unity. Representing z by a line OP turning in a plane about O, and $\sqrt{z}$ by another line OQ turning about the same point in the same fixed plane, we easily see that this is so. But if P is free to move over the surface of a sphere, the position of Q which bisects the arc between the variable and the original positions of P, becomes indeterminate when OP is exactly in the inverse of its original position, that is when $z = 1$. If, then, in order to resolve the indeterminacy we let P make an infinitesimal detour around the "dead-point," we see that Q will make a detour infinitely near a great circle so that when P returns to its original position, that of Q is not reversed at all, but is also restored. Thus, Cauchy's whole theory of integration would fall to the ground if $\sqrt{-1}$ represented merely a square root of unity, or if the manifoldness of the variability exceeded 2.

The algebra of quaternions will appear in the new form as follows:

$$
\begin{aligned}
1 &= X{:}X + Y{:}Y + Z{:}Z + W{:}W \\
i &= X{:}Y - Y{:}X + Z{:}W - W{:}Z \\
j &= W{:}Y - Y{:}W + X{:}Z - Z{:}X \\
k &= Z{:}Y - Y{:}W + W{:}X - X{:}W
\end{aligned}
$$

In this point of view, quaternions appears as an algebra of the space of four dimensions, where the motion is so restricted that any rotation in any plane, say of XY, is accompanied by an equal rotation in the conjugate plane of ZW. As this restriction is no greater than requiring every particle to move in an elliptical or hyperbolic space of three dimensions, it would seem to follow that quaternions must be adequate to the representation of the non-Euclidean geometry. Inasmuch, indeed, as it has several times been shown that this geometry may be represented in our ordinary geometry, there is no reason to doubt the proposition.

The system of biquaternions of section IV of Professor Clifford's paper is represented as follows:* (*See Professor B. Peirce's remarks, Proceedings of Am. Acad. Arts & Sciences, May 1875.)

$$
\begin{aligned}
1 &= X{:}X + Y{:}Y + Z{:}Z + W{:}W + \Xi{:}\Xi + H{:}H + S{:}S + \Omega{:}\Omega \\
\omega &= X{:}X + Y{:}Y + Z{:}Z + W{:}W - \Xi{:}\Xi - H{:}H - S{:}S - \Omega{:}\Omega \\
i &= X{:}Y - Y{:}X + Z{:}W - W{:}Z + \Xi{:}H - H{:}\Xi + S{:}\Omega - \Omega{:}S \\
j &= W{:}Y - Y{:}W + X{:}Z - Z{:}X + \Omega{:}H - H{:}\Omega + \Xi{:}S - S{:}\Xi \\
k &= Z{:}Y - Y{:}W + W{:}X - X{:}W + S{:}H - H{:}S + \Omega{:}\Xi - \Xi{:}\Omega
\end{aligned}
$$

From Grassmann's calculus of extension we may deduce the following algebra:* (*Given by me, Proceedings of Am. Acad. of Arts and Sciences, Oct. 10, 1877.)

Three rectangular vectors

$$
\begin{aligned}
i &= M{:}A - B{:}Z + C{:}Y + X{:}N \\
j &= M{:}B - C{:}X + A{:}Z + Y{:}N \\
k &= M{:}C - A{:}Y + B{:}X + Z{:}N
\end{aligned}
$$

Three rectangular aspects

$$
\begin{aligned}
I &= M{:}X + A{:}N \\
J &= M{:}Y + B{:}N \\
K &= M{:}Z + C{:}N
\end{aligned}
$$

one solid

$$V = M{:}N$$

unity

$$1 = M{:}M + A{:}A + B{:}B + C{:}C + X{:}X + Y{:}Y + Z{:}Z + N{:}N$$

III

Proof that there are but three associative algebras in which division by finites always yields an unambiguous quotient.

1. If in any algebra division is unambiguous the product of two finites can never be zero. For if $pq = 0$, then $pr = p(r+q)$, and division by p may give r or $r + q$.

2. It is shown in the *Linear Associative Algebra*, §40, that every algebra that contains no expression whose square vanishes contains an idempotent expression, i. It is further shown that every other expression is resolvable into the sum of two such that

$$iA = A, \qquad iB = 0.$$

But in the case supposed, B must vanish, and every expression multiplied by or into i must give itself.

3. Although in the *Linear Associative Algebra*, the scalar coëfficients are permitted to be imaginary, while here they are restricted to being real, yet the reasoning holds by which it is shown, in §53, that A being any expression of the algebra, there is an equation

$$\sum_m (a_m A^m) + bi = 0.$$

If this equation has a real root, then, by the reasoning in the text, in the absence of expressions whose squares vanish there must be a second expression, besides i, which is idempotent, and further, §54, in that case there is a product of finites which vanishes. It follows that the above equation, in the class of algebras under consideration, can have only imaginary roots; and consequently two real scalars s and t may be found, such that

$$(A - s)^2 + t^2 = 0$$

or

$$\left(\frac{A - s}{t}\right)^2 = -1.$$

In other words, every expression in the algebra can, by a real linear transformation, be converted into a quantity whose square is negative unity, or say into a *versor vector.*

We have begun the inquiry into algebras in which the division of finites in unequivocal, that is in which

If $p \neq p' \quad q \neq q'$ then $pq \neq p'q \quad pq = pq'$

We have found that every expression in such an algebra is resolvable in one way and one only into a sum of two parts, the first of which is an ordinary number and the second such that its square is an ordinary negative number. Of course either of these parts may disappear. An ordinary number we call a scalar. A quantity whose square is negative we call a vector. A quantity whose square is -1 we call a unit vector. That scalar which subtracted from the quantity q yields a vector remainder we call the scalar of q. The vector remainder we call the vector of q. That positive quantity the negative of whose square is the square of the vector we call the tensor or modulus of v.

Our next step is to prove that the vector part of the product of two vectors is linearly independent of these vectors and of unity. That is, i and j being any two vectors, if

$$ij = s + v$$

s being a scalar and v a vector, we cannot so determine three scalars a, b, c so that

$$v = a + bi + cj$$

This is proved if we prove that ij is not of the form $a + bi + cj$, that is that there is no scalar which subtracted from ij leaves a remainder $bi + cj$. If this be true when i and j are any unit vectors whatever, it is true when these are multiplied by ordinary numbers, and so is true of every pair of vectors. We will, therefore, suppose i and j to be unit vectors.

Now $ij^2 = -i$. If therefore we had

$$\begin{aligned} ij &= a + bi + cj \\ -i = i(j^2) = (ij)j &= aj + bij - c \\ &= -c \qquad + aj \\ &\qquad + ba + b^2 i + bcj \end{aligned}$$

Then we should have

$$-i = b^2 i$$

or

$$b^2 = -1$$

Which is absurd, b being an ordinary number.

The next step is to prove that if the product of any two vectors i and j is $ij = s + v$, s being the scalar and v the vector, then $ji = r(s - v)$ where r is an ordinary number or scalar, positive or negative. If this is true when i and j are unit vectors it is true whatever vectors they are. We will therefore suppose them unit vectors.

Let us write

$$ji = s' + v'$$
$$vv' = s'' + v''$$

Then $\quad ij \cdot ji = (s + v)(s' + v') = ss' + sv' + s'v + v'' + s''$

But $\quad ij \cdot ji = ij^2 i = -i^2 = 1$

Hence $\quad v'' = 1 - ss' - s'' - sv' - s'v$

But since v'' is the vector of vv' this cannot be unless v'' vanishes.

Hence $\quad 0 = 1 - ss' - s'' - sv' - s'v$

Hence $\quad sv' + s'v = 0 \qquad$ or $\qquad v' = -\dfrac{s'}{s} v$

That is $\quad ji = s' - \dfrac{s'}{s} v$

or writing r for $\dfrac{s'}{s}$

$$ji = r(s - v)$$

Q.E.D.

The next step is to prove that this r is equal to unity, so that if $ij = s + v$ then $ji = s - v$. It is obviously sufficient to prove this when i and j are unit vectors. Now any quantity whatever upon having its scalar subtracted from it becomes a vector, unless it were a scalar at first. What is true then is that from any quantity such a scalar may be subtracted as to leave it either a scalar or a vector, that is to make the square of the remainder a scalar. We do not yet know whether the sum of two vectors is a vector or not. Let us, then, take any such sum as $ai + bj$ and suppose $-x$ to be the scalar which subtracted from it makes the square of the remainder a scalar. We have then $(x + ai + bj)^2$ a scalar.

But
$$\begin{aligned}(x + ai + bj)(x + ai + bj) = \; & x^2 + 2axi + 2bxj + abv \\ & -a^2 \qquad\qquad\qquad\quad - abrv \\ & -b^2 \\ & +abs \\ & +abrs\end{aligned}$$

Let c be the scalar which this equals. Then

$$ab(r-1)v = 2axi + 2bxj + x^2 - a^2 - b^2 + ab(r+1)s - c$$

But this is impossible unless the equation vanishes, because v is the vector of ij. Hence either

$$r - 1 = 0 \qquad \text{or} \qquad v = 0$$

But if $v = 0$ $ij = s$ and multiplying into j we should get

$$-i = sj$$

which is absurd. Hence $r = 1$ and $ji = s - v$. Q.E.D.

Next prove the sum of two vectors is a vector, because it cannot be a scalar and $(ai + bj)^2 = -a^2 - b^2 + 2abs$.

I now propose to prove that the number of independent vectors in any algebra fulfilling the required condition is either 0, 1, or 3. I have therefore to prove 1st that the number cannot be 2 and 2nd that it cannot be as great as 4.

That it cannot be 2 is obvious for the vector of ij is independent of i and j. The whole remaining difficulty is to show that it cannot be as great as 4.

Let us suppose, then, that we have two independent vectors i and j whose product is

$$ij = s_1 + v_1$$

Let us immediately substitute for j.

$$j' = s_1 i + j$$

Then we have

$$ij' = v_1 \qquad j'i = -v_1$$

This is a third vector linearly independent of i and j'. Let us suppose that we have a fourth vector k, linearly independent of all the others. Suppose we have

$$\begin{aligned} j'k &= s_2 + v_2 \\ ki &= s_3 + v_3 \end{aligned}$$

Let us then substitute for k

$$k' = s_3 i + s_2 j' + k$$

and we have

$$\begin{aligned} j'k' &= -s_3 v_1 + v_2 \qquad & k'j' &= s_3 v_1 - v_2 \\ k'i &= -s_2 v_1 + v_3 \qquad & ik' &= s_2 v_1 - v_3 \end{aligned}$$

Let us further suppose

$$(ij')k' = s_4 + v_4$$

Then because ij' is a vector

$$k'(ij') = s_4 - v_4$$

But $\quad k'j' = -j'k' \qquad k'i = -ik$

because both products are vectors. Hence

$$i \cdot j'k' = -i \cdot k'j' = -ik \cdot j' = +k'i \cdot j' = +k' \cdot ij'$$

Hence $\quad s_4 + v_4 = s_4 - v_4$

or $\quad v_4 = 0$

and the product of two vectors is a scalar. These vectors cannot then be independent or k cannot be independent of i and j. So that a fourth independent vector is impossible and the theorem is proved.

To ascertain of what associative algebras having a limited number of linearly independent units it is true that if $pq = p'q$ then $p = p'$ unless $q = 0$ and if $pq \equiv pq'$ then $q = q'$, unless $p = 0$.

I regard $\sqrt{-}$ as being a unit linearly independent of 1. My father in his work on linear associative algebra takes the opposite view but I shall be able to avail myself of certain propositions proved by him; and I commence by restating these propositions with the proofs.

The first is his §40 that in every linear associative algebra there is at least one idempotent or one nilpotent expression. If A be any expression in such an algebra there must be some equation

$$\sum_m a_m A^m = 0$$

where the a_ms are scalars. Collect all the terms except the last into one expression BA. Then the equation becomes

$$BA + a_1 A = (B + a_1)A = 0$$

Hence $(B + a_1)Aa = 0$

$$(B + a_1)B = 0$$

$$\left(-\frac{B}{a_1}\right)^2 = -\frac{B}{a_1}$$

so that $-\dfrac{B}{a_1}$ is idempotent unless $B^2 = 0$.

The next proposition I propose to borrow from my father is contained in his §41. If $i^2 = i$ and $iA = B$, then $iB = B$ $i(A - B) = 0$. So that every expression A may be separated into two parts B and $A - B$ such that one of these multiplied by i gives itself while the other gives zero. But either of these parts may vanish. So with multiplication into i.

Finally from my father's §53 I adapt the following. Given an expression $i^2 = i$ such that multiplied by or into any other A it gives A. Then

$$\sum_m (a_m A^m) + bi = 0.$$

I now pass to the new propositions. First the algebras we seek are precisely those for which $pq \neq 0$ so long [as] $p \neq 0$ $q \neq 0$. For if this be true if $pq = p'q$, $(p - p')q = 0$. Hence $p - p' = 0$ or $p = p'$ unless $q = 0$. Also if from $pq = p'q$ it follows that $p = p'$ if $q \neq 0$ then from $pq = 0 = 0q$ it follows that $p = 0$ unless $q = 0$.

Hence from the first proposition just proved it follows that in every algebra where this is true there is an idempotent expression.

And since we never have

$$iA = B$$
$$i(A - B) = 0$$

unless $A - B = 0$ it follows that

$$iA = A$$

invariably. The unit i is therefore arithmetical unity.

Let A be any other expression. We must have the equation

$$\sum_m (a_m A^m) + b = 0$$

This equation may by the ordinary theory of equations be resolved in the form

$$(A - x_1)(A - x_2)(A - x_3) \ldots \text{etc.} = 0$$

where $x_1 x_2 x_3$ etc. are ordinary real or imaginary quantities. If any one of these roots say x_1 be real, the equation is divisible by $A - x_1$ and denoting the quotient by B we have

$$(A - x_1) B = 0$$

so that in that case the algebra would not belong to the class we seek.

If all the roots be imaginary the whole equation may be resolved into a continued product of quadratics, and some one of these must be equal to zero. Hence there must be some quadratic equation of the form

$$A^2 + aA + b = 0$$

Hence

$$\sqrt{b^2 - \tfrac{1}{4}a^2}(A + \tfrac{1}{2}a)$$

is an expression whose square is negative unity.[1]

[1] This chapter is related to *C.P.* 3.289-3.305.

C. [QUATERNIONS] (90)

This manner of considering algebras will, I believe, prove useful. I will proceed to show some curious results of applying it to quaternions. Hamilton's form of the algebra of quaternions is resolved, thus.

$$\begin{aligned} l &= X{:}X + Y{:}Y + Z{:}Z + W{:}W \\ i &= X{:}Y - Y{:}X + Z{:}W - W{:}Z \\ j &= Y{:}W - W{:}Y + Z{:}X - X{:}Z \\ k &= Y{:}Z - Z{:}Y + X{:}W - W{:}X \end{aligned}$$

Now, an operation of the form $X{:}Y - Y{:}X$ is obviously a quadrantal rotation in the plane of X and Y. The quaternion algebra obviously therefore represents a space of four dimensions x, y, z, and w; a rotation in the plane of any two dimensions not being discriminated from a rotation in the plane of the other two. I have no doubt that the great secret of the power of quaternions consists in this implicit introduction of a fictitious fourth dimension.

In Professor Peirce's form quaternions appears thus:

$$i = A{:}A \qquad j = A{:}B \qquad k = B{:}A \qquad l = B{:}B$$

The geometrical interpretation of this is as follows. The four dimensions are to be divided into two pairs X, Y and Z, W. One dimension of each pair is to be regarded as the imaginary of the other. Thus A consists of x as its real part and y as its imaginary part and B of z as its real part and w as its imaginary part. Then the operations i and l are such that either reduces one pair of coordinates to zero and leaves those of the other unchanged; while the operations j and k are such that either substitutes for one pair of coordinates the values of the other pair and reduces this other pair to zero.

All these results were communicated some two years ago to the Philosophical Society of Washington but they have not all been printed.

D. TOPIC B. TRICHOTOMIC MATHEMATICS (from 431a)

Although the importance of the mathematics of three objects, or values, is vast, it has hardly been studied. Indeed, I know nothing relating to this subject in print, except some small contributions of my own.

In my memoir of Jan. 26, 1870 (*Mem. Am. Acad of Arts and Sci.* IX, p. 362, or p. 46 of the excerpt) I gave an algebra which Prof. W. K. Clifford wished to call nonions; but I prefer the designation *novenions*. Let there be three values of ingredients of quantities, u, v, w. Then, there will be nine ways in which an ordered pair of objects may be attached each to one of the three, u, v, w. Then, since an assertion concerning a pair may either exclude or admit each of these nine, there will be 512 possible assertions; or, excluding that one which is necessarily true and that one which is necessarily false, there will remain 510 forms of contingent assertions about the values of a pair of quantities. To treat these at the same length as I have the sixteen corresponding forms of dichotomic algebra would be most tedious. Considering them from the relative point of view there are nine forms which admit but a single pair of values. The corresponding operational quantities will be

$$\begin{array}{ccc} u{:}u & u{:}v & u{:}w \\ v{:}u & v{:}v & v{:}w \\ w{:}u & w{:}v & w{:}w \end{array}$$

The relative multiplication-table of these will be [Fig. 1]. Since my father had often discussed, and even lectured upon, the possibility of higher algebras analogous to quaternions, and since these novenions corresponded exactly to the algebra which he had shown to be closely connected with quaternions, it was inevitable that he and I should promptly inquire whether there were 9 polynomials in this nonions which were so many independent cube roots of

$$u{:}u + v{:}v + w{:}w,$$

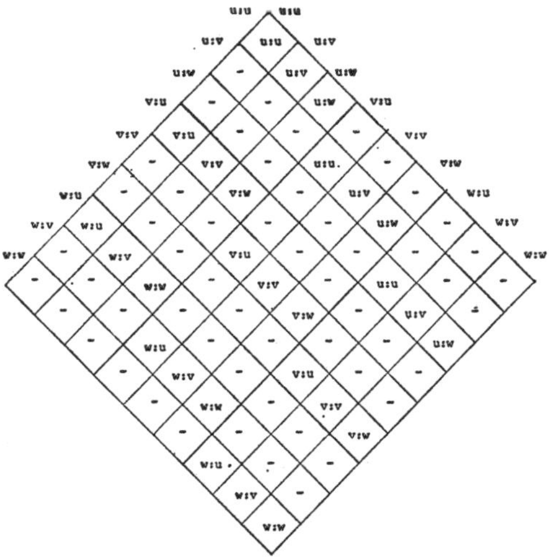

Fig. 1

and formed a group. I am sure I cannot remember which of us first did this; for as soon as either of us looked he must see that such a group of cube roots of unity were (putting g for an imaginary cube root of unity)

$$
\begin{aligned}
i &= u{:}u + v{:}v + w{:}w \\
j &= u{:}u + g \cdot v{:}v + g^2 \cdot w{:}w \\
k &= u{:}u + g^2 \cdot v{:}v + g \cdot w{:}w \\
l &= u{:}v + v{:}w + w{:}u \\
m &= u{:}v + g \cdot v{:}w + g^2 \cdot w{:}u \\
n &= u{:}v + g^2 \cdot v{:}w + g \cdot w{:}u \\
o &= u{:}w + v{:}u + w{:}v \\
p &= u{:}w + g \cdot v{:}u + g^2 \cdot w{:}v \\
q &= u{:}w + g^2 \cdot v{:}u + g \cdot w{:}v
\end{aligned}
$$

The relative multiplication table derived from the equations, $k = j^2$, $m = jl$, $n = j^2l$, $o = l^2$, $p = jl^2$, $q = j^2l^2$, $lj = \varrho jl$, is as here shown [Fig. 2].

That one or other of us, my father or myself, produced this in the early seventies, I testify positively. Probably it was I who actually effected the transformation, which was very easy after my father's corresponding transformation of quaternions. It was, as far as I was concerned, a performance of no remarkable merit, interesting as the algebra itself is. At any rate, while I was in Europe in 1875, my father wrote to me that

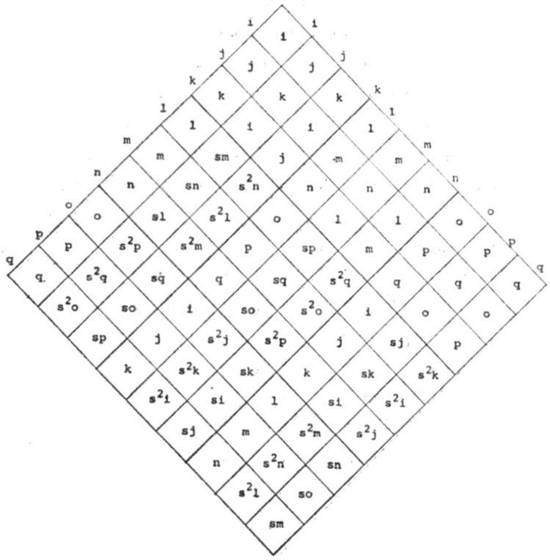

Fig. 2

he intended to make a supplementary communication to the American Academy of Arts and Sciences upon linear associative algebra, and that he wished me to write out a statement of what I had contributed to the subject for presentation along with his. I did so; but (I think) only saw the matter in print on my return. I was then astonished to find that various items had been omitted; among the rest this transformation. I thought at the time that part of my MS. must have been lost; but it now occurs to me as possible that I regarded this transformation as my father's (as the merit of it certainly was), while he thought it was mine. He perhaps did not understand that I would not wish to rest my claims to the attention of the scientific world, even in part, upon any such trifle as this. Nor should I now go into these details were it not that the imaginative Sylvester afterward practically accused me of trying to steal the discovery from him. Owing to the accident mentioned, and to my feeling that my part in the matter had been made too much of by various mathematicians who knew of it, it was only published in 1882 by Sylvester (*Johns Hopkins Circulars*, No. 15, p. 203, and No. 17, p. 242). In his first statement, Sylvester said "that in his recent researches in Multiple Algebra he had come upon a system of Nonions (a word borrowed evidently from Clifford; for otherwise such a stickler for linguistic analogies would have said *novenions*), the exact analogues of the Hamiltonian Quaternions ... *Mr. Charles S. Peirce, it should be stated, had to the certain knowledge of Mr. Sylvester arrived at the same result many years ago.*" Those are Sylvester's own words. Shortly after, Sylvester drew up a fuller statement of the transformation, and sent the proof-sheet to me with an oral message, requesting me, as he afterward said, "to supply such reference as he [I] might think called for." Let it be so; although the message delivered to me was a request to make the statement accurate. In consequence. I wrote on the margin without any indication to the printer that it was to be inserted, the following sentence: "These forms can be derived from an algebra given by Mr. Charles S. Peirce (*Logic of Relatives*, 1870)." The proof-sheet was returned by me to Mr. Sylvester, who must have drawn the crotchet and line to direct the insertion of the sentence before he sent it to the printer. The following year, in another frame of mind, Prof. Sylvester came out with a special note exclusively devoted to repudiating the sentence, which, he said had been inserted by me, leaving it to be supposed that I had surreptitiously interfered with his printing, and to declaring: "I know nothing whatever of the fact stated." I then published simply a narrative of the course of my cogitations upon the subject, without alluding either to Sylvester's previous interest in the matter or to an occasion in 1881 or early in 1882 when he and I sat at a table together with my memoir open before us, and I fully expounded the whole thing. On the contrary, I stated, what I never denied, or seemed to deny, that the priority of publication was his. He continued to insist, apparently, that my conduct had been of the blackest. I now restate the facts, because I find that that opinion is still maintained and propagated in some quarters. Lest it should be said that I make intangible insinuations, I will add that Sylvester was a man whose imagination and enthusiasm were incessantly running away with him; he was given to harboring the most ridiculous suspicions, and to making rasher assertions than became so great a man. His power of distinct recollection was most phenomenally weak, almost incredibly so; while his sub-conscious memory was not at all wanting in retentiveness. All this must be admitted by all who knew him. I suppose, as he said, that he "came across" the system of novenions in the form last given above, and remembered, or thought he remembered, that I had pointed out these forms. Subsequently, he got a suspicion that I was about to charge him with plagiarizing my "*Description of a Notation* etc.," and was anxious to declare that he had never read it, and knew nothing about it. He seems to have fancied that I had some deep-laid plot against him.

11

POLITICAL ECONOMY

A. [ON POLITICAL ECONOMY] (1569)

21 September 1874

It is true as said at the beginning of this cahier that Political Economy treats of the relations of Price, Demand, and Cost of Production.

Demand depends on price and not on cost; cost depends on demand and not directly on price; price is fixed according to sellers' idea of demand and cost.

Thus the facts of political economy are of three categories.

I. Dependence of demand on price

II. Dependence of cost on demand

III. Dependence of price on demand and cost, or other circumstances.

The dependence of demand on price arises from this fundamental proposition. The desire of a person for anything has a quantity of one dimension, and a person having a choice will take that alternative which will give him the greatest satisfaction. In other words if a person prefers A to B and B to C he also prefers A to C. This is the first Axiom of Political Economy.

The desirability of a thing depends partly on the possession of other things which are related to it in this respect either as alternative or as coefficient.

Alternatives. So far as things serve the same purpose more or less well the possession of one diminishes the desire for the others. So far as things are alternative the possession of the less valuable will leave a desire for the more valuable equal to the difference of desirability. This is strictly true. A person may desire superfluity but that is another desire and so far the things are not alternative.

Alternative things may be fit either of them to satisfy one of several desires but not any one thing more than one desire. In that case the first thing will be wanted for the greatest utility and consequently the second thing will be desired less. Under this *h* ... comes such a case as my desiring say, *t* ... I want one very much a second less and so on until I reach a limit beyond which my desire ceases.

Political Economy treats of the relations of these three quantities
x the Value per unit.
y the Demand.
z the Cost of production of the last unit.
We shall first assume the following propositions

1. The demand is a continuous function of the Value or Price and is independent of the Cost of production or

$$(1) \qquad y = Fx$$

2. The Cost of production is a continuous function of the Demand or amount produced and is independent of the price, or

$$(2) \qquad z = \varphi y$$

The seller is absolutely wise and sets the price so as to make his [total] profits or $xy - \int_y z = p$ a maximum. Then

$$(3) \qquad D_x p = y + D_x y(x-z) = 0$$

$$(4) \qquad D_x^2 p < 0$$

Then we have the following variations

$$(5) \qquad \frac{\delta x}{\delta y} = +\frac{D_x y D_y z - 1}{D_x^2 p}$$

$$(6) \qquad \frac{\delta x}{\delta (D_x y)} = -\frac{x - z}{D_x^2 p}$$

$$(7) \qquad \frac{\delta x}{\delta z} = \frac{D_x y}{D_x^2 p} = \frac{D_x y}{2D_x y + D_x^2 y(x-z)}$$

$$(8) \qquad \frac{\delta x}{\delta \int y^z} = 0$$

It is evident that x, y, and z, will always be positive and when (3) will hold $D_x y$ is negative if $x > z$. Consequently we have the following propositions.

From (5) An absolute increase of the demand will always increase the price unless the increased price would (by diminishing the sales again) increase the final cost of production more than by the increase of price. This supposes the sensitiveness of the market to be unchanged.

From (6) we have: An increased sensitiveness of the market will invariably lower the price.

From (7) An increase in the cost of production invariably raises the price. If $x = z$, that is if the demand equals the supply the additional cost would be equally divided between buyer and seller were it not that the buyers partly escape by leaving off buying which hurts the sellers' profits. When $x > z$ the increased cost falls more on the seller, except when the sensitiveness of the market increases with the price and then it only appears not to do so because it falls more on the *profits* of the seller and not so much on his *price.* But as $D_x^2 y$ is always small the price changes about half as much as the cost.

We will next consider the effect of exceptions to our three propositions. And first of points of discontinuity.

If there is a sudden change in the value of φy so that for values of y less than y_c, $z = \varphi_1 y$ and for values greater than y_c, $z = \varphi_2 y$ and if the first value of z gives y a value greater than y_c while the second gives a value less than y_c, y_c is the resulting value of y. In this case, ($D_x y$ being negative)

$$y_c + D_x y_c (x - \varphi_1 y) < 0$$
$$y_c + D_x y_c (x - \varphi_2 y) > 0$$

and $\varphi_2 y > \varphi_1 y$.

If there is a sudden change in the value of Fx so that for values of x less than x_c, $y = F_1 x$ and for $x > [x_c]$, $y = F_2 x$ and if the first equation would make $x > x_c$ and the other $x < x_c$ then x_c is the resulting value. In this case

$$F_1 x + D_x F_1 x (x - z) > 0$$
$$F_2 x + D_x F_2 x (x - z) < 0$$

This case cannot occur.

If a sudden change takes place in $F'x$ of the same sort, for $x = x_c$ we have

$$y + F_1' x (x - z) > 0$$
$$y + F_2' x (x - z) < 0$$

$F_1' x > F_2' x$ or the market must become suddenly more sensitive if the price is raised.

We will next suppose that z depends on x as well as on y and y on z as well as on x.

The dependence of y on z will make

$$\frac{\delta x}{\delta z} = \frac{D_x y - D_z y - D_x D_x y (x - z)}{D_x^2 p}$$

The dependence of z on $D_x y$ will make

$$D_{D_x y} D_x p = (x - z) - D_x y D_{D_x y} z, \quad \frac{\delta x}{\delta D_x y} = \frac{D_x y D_{D_x y} z - (x - z)}{D_x^2 p}$$

The dependence of z on x gives

$$D_x^2 p = D_x y (2 - D_x z) + D_x^2 y (x - z)$$

altering the value of the denominator of the variations.

B. CALCULUS OF WEALTH (s-86)

There are three unknown quantities

x, price

y, sales or demand

z, total cost or inconvenience.

z is nearly proportional to y and nearly independent (in its own nature) of x. If we put $z = \varphi y$ then $D_y z$ or $\varphi' y$ is always positive. When y is small $\varphi'' y$ is negative, but it must constantly increase, that is $\varphi''' y$ must be positive. And $\varphi'' y$ may become positive, and generally will for large values of y. In farming and mining, it obviously does so.

The value of y depends on that of x, by an equation not involving z. Put $y = Fx$. Then $F'x$ is necessarily negative. $F''x$ is doubtless for some articles positive, but in such case[s] it must increase.

As there are only these two equations $z = \varphi y$ and $y = Fx$ established between our three unknown quantities by the action of natural causes, it is left for the producer or seller to establish another by his wisdom. But the only element he can directly control is x. He therefore sets the value of x, so that

$$xy - z$$

shall be a maximum. Or using *d.* to denote infinitesimal changes such as he can produce

$$dx \cdot y + x \cdot dy - dx \frac{dy}{dx}\frac{dz}{dy} = 0$$

or $\quad dx \cdot y + x \cdot dy - dz = 0$

and $\quad 2dx \cdot dy + x \cdot d^2y - d^2z < 0$

This must have at least one solution giving a finite positive value of x. But the problem may be indeterminate in some special case.

Let us now consider the effect of variations in the functions φy and Fx.

A variation φy will affect x (and therefore y) only so far as it affects dz.

Therefore, a fixed tax irrespective of the sales falls wholly on the producer. An increase of $\varphi' y$ will since $D_x y$ is negative increase the first member of the equation of wisdom, so that it will be preserved only by increasing x which will for the same reason diminish the same equation. Therefore, *the price will increase with the cost.*

An increase of Fx will increase x; but an increase of $F'x$ will diminish x. But if to one demand $F_1 x$ a second demand $F_2 x$ be added independent of the first, $F'x$ will be diminished and therefore, almost universally, *the price will increase with the demand.* The increased demand may, however, so diminish the cost as to diminish the price.

Let us next consider the relations of different producers. These are either 1st between their *prices* or 2nd between their *sales.* The first is the case of competition. The second that of "concourse" or the case of their making things to be used together.

There is evidently a competition between different kinds of articles which can be put to the same use; but Cournot has overlooked this and has considered only the case in which all the prices must be the same. [...]

C. [LETTER TO BENJAMIN PEIRCE] (L 333)

U.S. Coast Survey Office
Washington, Dec. 19, 1871

My dear Father,

There is one point on which I get a different result fro m Cournot, and it makes me suspect the truth of the proposition that the seller puts his price so as to make his profits a maximum. Suppose two sellers only of the same sort of article which costs them nothing. Then

$$y_1 + D_{x_1}y_1 \cdot x_1 = 0$$
$$y_2 + D_{x_2}y_2 \cdot x_2 = 0$$

Now each seller must reflect that if he puts down his price the other will put down his just as much, so that if y is the total sales $\frac{1}{y}$ is a constant. Then the competition won't alter the price.

But suppose the seller does *not* make this reflection. Then if his price is lower than the other man's he gets all his customers besides attracting new ones. Then the price is forced down to zero, by the successive action of the two sellers for

$$D_{x_1}y_1 = -\infty.$$

Cournot's result is

$$y = 2D_xy \cdot x = 0.$$[1]

This is as though each seller in putting down his price expected to get all the new customers and yet didn't expect to get away any of the other

[1] A fragment states that "If the price *p* were nothing, the number of policies, *n*, that could be disposed of would be very great indeed. But it would not exceed the number of risks of that kind that exist. If you were to make *p* negative, that is pay people to take the policies, when the amount you paid was equal to the cost of creating such risks in the hope of loss, the number *n* would be infinite."

As regards Peirce's respect for the work of Cournot he mentions in 7.66 of the *Collected Papers* that "The scientific specialists — pendulum swingers and the like — are doing a great and useful work .... Cournot adapted to political economy the calculus of variations."

seller's customers, which is not consistent. He reasons in this way. The prices asked by the two sellers are the same. Put

$$x = f(y_1 + y_2)$$

Then to make xy_1 and xy_2 maxima we have

$$f(y_1 + y_2) + y_1 f'(y_1 + y_2) = 0$$
$$f(y_1 + y_2) + y_2 f'(y_1 + y_2) = 0$$

Summing etc. he gets his result. But his f' is indefinite. If it is a previous condition that the prices are the same

$$D_{y_1} f(y_1 + y_2) = 2D_y f(y_1 + y_2)$$

Am I not right? He seems to think there is some magic in considering $D_y x$ instead of $D_x y$. According to me the equation is

$$y + xD_x y - x(D_{x_1} y_2 + D_{x_2} y_1) = 0$$

if the sellers reckon on the others' price *not* changing and is $y + xD_x y = 0$ if they consider the change.

Your affectionate
C.S.P.

Best love to Mother, Helen, Jem, and Bertie
P.S. What puzzles me is this. The sellers must reckon (if they are not numerous) on all following any lowering of the price. Then, if you take into account the cost, competition would generally (in those cases) *raise* the price by raising the final cost. But is not this a *fact*?

12

AMAZING MAZES

A. THE THIRD CURIOSITY (199)

Since I had occasion, in describing the last curiosity, to introduce numeration with another base than ten, I will devote a few pages to a somewhat more general consideration of the subject. It is well known that the duodecimal notation has advantages in the measurement of timber in feet and inches; but a more generally convenient base would be *six*, since every prime number, except the factors of six, is either one more or one less than a multiple of six. Now it is particularly easy to ascertain whether a written number is divisible by a given number that is one more or one less than a multiple of the base, and if it be not so divisible, to ascertain what the remainder after division would be. Everybody knows, for example, that the sum of all the digits used in writing a number decimally diminished by the next less multiple of nine is the remainder after dividing the number by nine; and that the sum of the digits in the places for even powers of ten, diminished by the sum of those in the places for odd powers of ten, the remainder being further diminished by the next less multiple of eleven, gives the remainder after dividing the first number by 11. It is less generally known that an almost equally simple rule exists for ascertaining the remainder after dividing any given number by any given number of the form $10n \pm 1$. Namely, if the divisor is $10n-1$, you divide the first figure of the dividend by n, leaving the remainder attached to the next figure as is usual in "short division." You then add the quotient to the next figure, and repeat the process until you have added a quotient to the figure in the units' place. This will either be the remainder sought, or that remainder increased by a smaller multiple of the divisor than before. The rule for divisors of the form $10n+1$ is the same except that, instead of the quotients, the divisor *minus* each quotient, is to be added or the quotient itself subtracted. For example, required the remainder after division of 29160 by 29. I will first set down the figures of the work, and then go through the description of the proceeding.

3)2	$^{2}9$	$^{2}1$	$^{0}6$	$^{1}0$
	6	9	10	5
	3)29	3)30	3)16	15

Dividing the first figure, 2, of the dividend by 3 (which is the number of times ten that gives a product greater by one than the divisor, 29), I set down the quotient, 0, under the next figure, 9, of the dividend, and attach the remainder, 2, to the same next figure, making it 29. Adding to this the quotient, 0, I set down the sum, 29, below. This has now in turn to be divided by 3. The quotient, 9, is set down under the next figure, 1, of the dividend, to which the remainder 2 is attached, making it 21. Adding the 9 to the 21, the sum 30 is set down below it. This, in turn, must now be divided by 3. The quotient 10 is set down below the next figure, 6, of the dividend, which is preceded by the attached remainder, 0. Adding the 10 to the 6, and setting down the sum, 16, below, this has finally to be divided by 3, its quotient 5 to be set down under the next, and last, figure, 0, and its remainder, 1, to be prefixed to that 0, making it 10. Adding the 5 to the 10, we get 15, which is the remainder of 29160 after division by 29, as the long division shows it to be.

By the way, since long division is an operation in which peisons at all out of practice are particularly liable to make mistakes, instead of leaving them to the recourse of an old friend of mine who, in making up his accounts, used to add up each column five times and adopt the mean of his results (of which no two often agreed with each other), I will forthwith describe a method of conducting long division which affords almost absolute security against error, while it rather abbreviates than lengthens the time the division occupies, and greatly diminishes the strain and fatigue of mind involved. The whole secret of the thing lies, first, in the formation of a table of the first ten multiples of the divisor, secondly, in the expeditious and facile way making this, and thirdly, in the excellent check upon its accuracy.

In order to show how these multiples are formed, I will form, by this method the first ten multiples of the number 287165693. I begin with the units place and successively add three, *counting round a cycle of the ten digits*, to each multiple, beginning with 3, in order to get the next multiple. Observe carefully that there is no tens' place, since no number higher than 9 is contained in the cycle. This addition in counting round a cycle of the ten digits speedily becomes a remarkably facile operation in which an error is particularly unlikely. The result is shown in *Scheme* 1. The majority of the numbers in the column so formed are greater than the numbers just above them; but the fourth, seventh, and tenth are smaller. By a geological fancy, I denominate such exceptional numbers which differ from the majority of those in the same column in being greater instead of less or less instead of greater than the numbers immediately above them, the "faults" of the column. The second operation consists in putting in the tens' place a small dot against each fault of the units' column. The result is shown in *Scheme* 2. The number at the head of the tens' column is 9. I therefore pass round the cycle shown in Fig. 1,

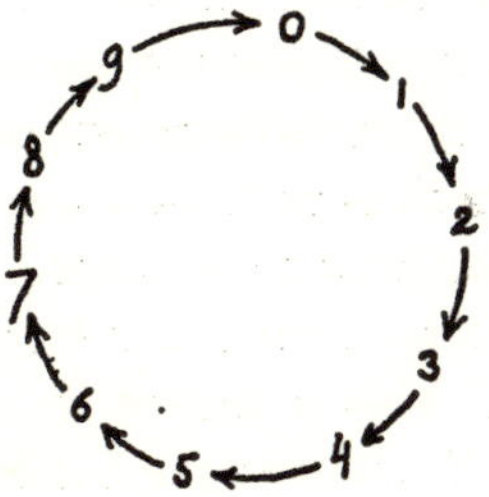

Fig. 1

nine places clockwise, or one place counterclockwise, in order to pass from any number in this column to the next, except that when I come to inserting a number in a dotted position I repeat the number just above it.

Scheme 1	*Scheme* 2
2 8 7 1 6 5 6 9 3	2 8 7 1 6 5 6 9 3
6	6
9	9
2	.2
5	5
8	8
1	.1
4	4
7	7
0	.0

Filling up the column according to this rule, I obtain *Scheme* 3. I now note that the bottom number of the tens' column is the same as the top number of the units' column, as it should be, since the bottom line is to contain the product of the top line multiplied by ten. Now this agreement is strong security that no error of computation has been committed in filling the column. For supposing that the computer puts, on an average, an erroneous figure out of every two hundred he sets down, which would be a degree of arithmetical skill below mediocrity, especially in such remarkably facile work, the chance of his making one error or more in a column would be about 1 in $22\frac{2}{3}$, that is, he would on the average fill one column erroneously out [of] every $22\frac{2}{3}$. But the bottom of a column in which an error had been committed could only come out right if a second error were committed, and there would be, on the average just four numbers among which that error could occur. This would happen once in $50\frac{3}{8}$ such sets of four numbers. The two things would happen together in one column out of every 1142. But then only out of every nine happenings would the second error just balance the first. Consequently, only in one column out of every 10278 would this check prove deceptive, which is perhaps a good many moıe times than it will ever be applied by so poor a computer.

There is no fault in the tens' column and therefore no dots to be made in the hundreds' column. This is headed by a 6; but the numbers 93 which follow it exceed half a unit in the hundreds' place, so that the 6 must be treated as a 7. That is, each number is 3 numbers removed, counterclockwise in the cycle of ten numbers, from the number above it. Applying this rule, we obtain *Scheme* 4; and we note that the check is

Scheme 3	Scheme 4
2 8 7 1 6 5 6 9 3	2 8 7 1 6 5 6 9 3
8 6	3 8 6
7 9	0 7 9
7 2	7 7 2
6 5	4 6 5
5 8	1 5 8
5 1	8 5 1
4 4	5 4 4
3 7	2 3 7
3 0	9 3 0

Scheme 3 *Scheme* 4

satisfied. This column has three faults, and setting dots against them in the thousands' place, we get *Scheme* 5. In the thousands' column, the 5 that heads it is followed in the column to the right by a number, 6, exceeding 5; and consequently the 5 must be treated as 6 except at the dots, and each number placed in an undotted position must be 4 distant in the cycle of ten, in the counterclockwise direction, from the number just above it in the thousands' column. This gives *Scheme* 6 (which I omit to figure). But in the dotted positions, of any of the pairs of numbers

```
2 8 7 1 6 5 6 9 3
            3 8 6
            0 7 9
          . 7 7 2
            4 6 5
            1 5 8
          . 8 5 1
            5 4 4
            2 3 7
          . 9 3 0
```

Scheme 5

```
2 8 7 1 6 5 6 9 3
          1 3 8 6
        . 7 0 7 9
          2 7 7 2
        . 8 4 6 5
          4 1 5 8
        . 9 8 5 1
          5 5 4 4
          1 2 3 7
        . 6 9 3 0
```

Scheme 7

```
2 8 7 1 6 5 6 9 3
        3 1 3 8 6
        9 7 0 9 7
        6 2 7 7 2
        2 8 4 6 5
        9 4 1 5 8
        5 9 8 5 1
        2 5 5 4 4
        9 1 2 3 7
        5 6 9 3 0
```

Scheme 8

```
2 8 7 1 6 5 6 9 3
      3 3 1 3 8 6
      4 9 7 0 7 9
      6 6 2 7 7 2
      8 2 8 4 6 5
      9 9 4 1 5 8
      1 5 9 8 5 1
      3 2 5 5 4 4
      4 9 1 2 3 7
      6 5 6 9 3 0
```

Scheme 10

05, 16, 27, 38, 49, if one member is just above the dotted position, the other must be in that position. Thus, 7 is followed by 2, 4 by 9, and 1 by 6. Since now 56, the numbers following the 6 at the head of the ten-thousands' column, exceed half a unit of that column, we dot in that column the positions opposite numbers in the thousands' column that exceed the numbers above them, giving *Scheme* 8; and treating the 6 that heads the ten-thousands' column as a 7, we go round the cycle of 10 counter-clockwise and put into every undotted position, the number removed by 3 in so going round that cycle, from the number next above it; but in the dotted positions the number must be 4, instead of 3, removed from the number just above it. This gives *Scheme* 7. Here the majority of the numbers in the column just completed are less than those above them, so that the faults are at the three 9s. I put dots opposite to them and so get unshown *Scheme* 9. The number 1, at the head of the next column toward the left, being followed by 6, must be treated as 2, except at the dotted positions, where it is treated as 1; and thus we obtain *Scheme* 10.

The column just filled has but one fault in the 7th place; and there only is the 7, that heads the next column to be filled, to be treated as 8. Accordingly, [only] in that one position am I to subtract 2 (that is recede counterclockwise to the second number of the cycle). In all other positions, I subtract 3. I now mark the faults, of which there are only two, in the next column to the left, and so get *Scheme* 13. The 8 that heads the next column to be filled is followed by a number greater than 5, and is, therefore, to be treated as a 9 at the undotted positions. That is, at all those positions, 1 is to be subtracted from the last number entered

```
2 8 7 1 6 5 6 9 3
    4 3 3 1 3 8 6
    1 4 9 7 0 7 9
  . 8 6 6 2 7 7 2
    5 8 2 8 4 6 5
    2 9 9 4 1 5 8
    0 1 5 9 8 5 1
  . 7 3 2 5 5 4 4
    4 4 9 1 2 3 7
    1 6 5 6 9 3 0
```

Scheme 13

```
  2 8 7 1 6 5 6 9 3
  7 4 3 3 1 3 8 6
  6 1 4 9 7 0 7 9
  4 8 6 6 2 7 7 2
  3 5 8 2 8 4 6 5
  2 2 9 9 4 1 5 8
  1 0 1 5 9 8 5 1
. 9 7 3 2 5 5 4 4
  8 4 4 9 1 2 3 7
  7 1 6 5 6 9 3 0
```

Scheme 15

in the column to get the number below. But where there are dots, 2 is to be subtracted. The column will have but one fault; namely at the 9 in the eighth line. Setting a dot against it, we get *Scheme* 15. The column headed by the 2 is now to be filled up by adding 3s (since the next figure exceeds 5) except at the one dotted place, where I add only 2. Reader, if you undertake to practice this rule, it will be indispensable to your getting the advantage of it, that you should acquire the faculty of adding and subtracting digits without the least thought of the tens' place or of the number that would go into it. In short, you must learn to proceed round the cycle of ten in either direction to the first, second, third, and fourth place, entirely without hesitation, almost without a thought. The column headed by the two being filled, I get *Scheme* 16 (not given). This has two faults, against which dots are to be placed. Then, a 0 being set to the left of the 2, as the head of another column, *Scheme* 17 results. The column headed by the 0 is then to be filled, in the usual way, giving *Scheme* 18 (not given). This column (or any headed by a 0) can never have a fault. It therefore only remains to number the nine first multiples as in *Scheme* 19. I have the habit of numbering them to the right; but probably it might be more convenient to put the multipliers to the left of

0 2 8 7 1 6 5 6 9 3
5 7 4 3 3 1 3 8 6
8 6 1 4 9 7 0 7 9
. 1 4 8 6 6 2 7 7 2
4 3 5 8 2 8 4 6 5
7 2 2 9 9 4 1 5 8
. 0 1 0 1 5 9 8 5 1
2 9 7 3 2 5 5 4 4
5 8 4 4 9 1 2 3 7
8 7 1 6 5 6 9 3 0

Scheme 17

0 2 8 7 1 6 5 6 9 3	1
0 5 7 4 3 3 1 3 8 6	2
0 8 6 1 4 9 7 0 7 9	3
1 1 4 8 6 6 2 7 7 2	4
1 4 3 5 8 2 8 4 6 5	5
1 7 2 2 9 9 4 1 5 8	6
2 0 1 0 1 5 9 8 5 1	7
2 2 9 7 3 2 5 5 4 4	8
2 5 8 4 4 9 1 2 3 7	9
2 8 7 1 6 5 6 9 3 0	

Scheme 19

the products. The advantage of the 0s in the first column is that they secure the computer against wrongly inserting or omitting 0s in the quotient, when the table is used in the aid of long division, and of not taking the proper account of 0s in the multiplier, when the table is used to aid long multiplication.

1	029
2	058
3	087
4	116
5	145
6	174
7	203
8	232
9	261
	290

Fig. 2

```
    1005
29)29160
   29
   ‾‾
    0160
     145
     ‾‾‾
      15
```

Fig. 3

1	071
2	142
3	213
4	284
5	355
6	426
7	497
8	568
9	639
	710

Fig. 4

This tedious explanation being finished at last, let us remember that we proposed to divide 29160 by 29 in order to verify the result we obtained that the remainder is 15. Here, in Fig. 2, is the table of multiples of 29 (such tables are always easier to make than to copy), and here, in Fig. 3, is the work of long division. The remainder is again found to be 15. Let us use the same rule, to find the remainder after dividing the same number by 71. I call it the same rule, the only variation being that the quotients have to be subtracted instead of being added, and that before performing the subtraction it is sometimes convenient to add the divisor to the minuend. Here is the work:

$$\begin{array}{lllll}
7)2 & {}^{2}9 & {}^{1}1 & {}^{0}6 & {}^{5}0 \\
 & \underline{0} & \underline{4} & \underline{1} & \underline{0} \\
 & 7)29 & 7)7 & 7)5 & 50
\end{array}$$

The remainder is 50. Here is the work of verification:

$$\begin{array}{r}
410 \\
71)29160 \\
\underline{284} \\
076 \\
\underline{071} \\
50
\end{array}$$

Every boy who has had decent instruction in arithmetic and who is fit for admission into a high-school knows that it is particularly easy to multiply and consequently to divide by numbers ending with 9 or 1. For example, what is 7 times 59. It $7 \times 60 - 7 = 420 - 7 = 413$. Now were 6 the base of numeration, instead of ten, these advantages would apply to all prime numbers except 2 and 3. But these being factors of the base, 6, would be the easiest numbers of all to divide by, every number ending with an even number and none ending with an odd number being divisible by 2; while every number whose expression, on the base 6, would end with a 0 or 3 being a multiple of 3, every one ending with 1 or 4 being one more, and every number ending with 2 or 5 being one less, than such a multiple. For my part, I seriously think it is a great pity that primitive man had not the gumption to use his fingers intelligently to count upon. It is true that it would require more figures to write a given number, with 6 as the base, than with base 10. But the numbers of figures would only be in the ratio of 9 figures with base 6 to every 7 with base 10. For let us express 9999999 and 1000000 in the notation with base 6. Here is the work of conversion:

$$\begin{array}{rc}
 & \textit{Remainders} \\
6)9^{3}9^{3}9^{3}9^{3}9^{3}9^{3}9 & 3 \\
6)1\ 6^{4}6^{4}6^{4}6^{4}6^{4}6 & 4 \\
6)2\ 7^{3}7^{1}7^{5}7^{3}7 & 1 \\
6)4\ 6^{4}2^{0}9^{3}6 & 0 \\
6)7^{1}7^{5}1^{3}6 & 0 \\
6)1\ 2^{0}8^{2}6 & 2 \\
6)2\ 1^{3}4 & 4 \\
6)3\ 5 & 5 \\
6)5 & 5 \\
0 &
\end{array}$$

Sextal equivalent $=$ 554200143

	Remainders
$6)1\ 0^4 0^4 0^4 0^4 0^4 0$	4
$6)1\ 6^4 6^4 6^4 6^4 6$	4
$6)2\ 7^3 7^1 7^5 7$	3
$6)4\ 6^4 2^0 9$	3
$6)7^1 7^5 1$	3
$6)1\ 2^0 8$	2
$6)2\ 1$	3
$6)3$	3
0	

Sextal equivalent = $3323334\underline{4}$

Thus the highest and lowest numbers — which in decimal notation are expressed in 7 figures — are in sextal notation expressed in 9 and in 8 figures. The ratio of the number of figures required for decimal expression to the number required for sextal expression is a bare trifle nearer equality than that of 7 to 9, on the average; since it is exactly the Briggsian logarithm of 6, which is 0.7781513. Now what, I pray, is this trifling disadvantage of the sextal system in comparison with the relative extent of the two multiplication tables (for it is multiplication that makes the vexation of ciphering), the decimal table containing $9^2 = 81$ products and the sextal only $5^2 = 25$, the usual one having $3\frac{1}{4}$ times the extent of the one I regret. If you object that this is not a fair statement, since every product is counted twice over and since multiplication of, or by, one is no multiplication, I fully admit the justice of your stricture. Making the amendment called for, the ordinary multiplication-table is not $3\frac{1}{4}$, but $4\frac{1}{4}$, times as difficult as the sextal. I admit, too, that multiplications by 2 and by 5 ought not to count as giving more than half the trouble of others. Making that allowance for factors of the base, the ordinary table is $5\frac{1}{3}$ times as vexatious as the one which the Creator of man's five fingers might (with the best *human* intelligence) have anticipated man's either using, or else (when it was too late for that), regretting that he had not used. It is a lesson not to wed oneself irrevocably to the first means of solving a difficulty that suggests itself.

The rule for translating the decimal or other expression of an integer number into the language of sextals, which has been illustrated above, may be formally stated as follows: Repeatedly divide the number by 6, throwing off, but preserving, the remainders. Then any remainder, say the Nth, is the figure in the place of 2^{N-1}, which ought to be called the Nth integer place. The rule for finding the sextal expression of a fraction is just the reverse of the above, as follows: Repeatedly multiply the fraction by 6, always throwing off, but preserving, the integer part of the product. Then any product, say the Mth, is the figure in the Mth fractional place of sextals. This pair of rules is entirely general, and gives a ready means of expressing any number in any system of the form Σ_i (single figure) $_i$ (Base)^i.* (*All such systems ought to be denominated *Aryan Systems*, since the evidence is quite as good as could be expected that they originated in Arya or thereabout; possibly in Bactria, not likely Chorasmia, but more likely in Arya.) For example, Required to express $\frac{1}{16}$ in decimals and in sextals. Here is the work:

$$10 \times \tfrac{1}{16} = 0\tfrac{10}{16},\ 10 \times \tfrac{10}{16} = \tfrac{100}{16} = 6\tfrac{4}{16},\ 10 \times \tfrac{4}{16} = \tfrac{40}{16} = 2\tfrac{8}{16},$$
$$10 \times \tfrac{8}{16} = \tfrac{80}{16} = 5\tfrac{0}{16},\ 10 \times \tfrac{0}{16} = 0\tfrac{0}{16},\ \text{etc.}$$

$\therefore$ $\frac{1}{16}$ decimally expressed, is $.06250\dot{0}\dot{0}$

$$6 \times \tfrac{1}{16} = 0\tfrac{6}{16},\ 6 \times \tfrac{6}{16} = \tfrac{60}{16} = 3\tfrac{12}{16},\ 6 \times \tfrac{12}{16} = \tfrac{72}{16} = 4\tfrac{8}{16},$$
$$6 \times \tfrac{8}{16} = \tfrac{48}{16} = 3\tfrac{0}{16},\ 6 \times \tfrac{0}{16} = 0\tfrac{0}{16}.$$

$\therefore$ $\frac{1}{16}$ sextally expressed is $.0343\dot{0}\dot{0}$

The following is an example involving a circulating fraction in each of the two systems (but I have something to communicate about circulating fractions, to which I think I will devote another chapter). Required to express $\frac{1}{7}$ decimally and sextally.

$$10 \times \tfrac{1}{7} = 1\tfrac{3}{7},\ 10 \times \tfrac{3}{7} = 4\tfrac{2}{7},\ 10 \times \tfrac{2}{7} = 2\tfrac{6}{7},\ 10 \times \tfrac{6}{7} = 8\tfrac{4}{7},$$
$$10 \times \tfrac{4}{7} = 5\tfrac{5}{7},\ 10 \times \tfrac{5}{7} = 7\tfrac{1}{7}.\ \text{Now } \tfrac{1}{7} \text{ is the original fraction.}$$

Hence $\frac{1}{7}$ decimally expressed is $.\dot{1}4278\dot{5}$

$$6 \times \tfrac{1}{7} = 0\tfrac{6}{7},\ 6 \times \tfrac{6}{7} = 5\tfrac{1}{7}.$$

Hence $\frac{1}{7}$ sextally expressed is $.\dot{0}\dot{5}$

Much the prettiest of the Aryan systems, is the *secundal* or that one having two for its base, which was invented and very finely applied to the theory of simultaneous linear equations by Leibniz, making the embryon of Determinants and Matrices, being precisely the conception that Sylvester called "*umbrae*," and greatly plumed himself, though quite erroneously, with having invented. It is also the foundation of the logic of necessary reasoning; or rather, of the mechanical part thereof. It is, to say the least, extremely convenient in logic, especially in the logic of denumeral and abnumeral linear series.

The four rules of arithmetic, and therefore all its processes, are performed in this system with no other addition-table than that exhibited in Fig. 5, and with no other multiplication-table than once-one is one, shown in Fig. 6. In other words, secundal arithmetic does not call upon the computer for any recollection whatever of any particular sum or product of plural numbers.

	0	1
0	0	1
1	1	10

Addition-table of Secundal Arithmetic.

Fig. 5

1

Multiplication-Table of Secundal Arithmetic.

Fig. 6

The rule of addition of secundal arithmetic is a follows: Part I: when no more than one circulating fraction is involved in the summands. Having written the summands so that all ones that stand for the same power of two shall be in the same vertical column, perform the following process. Draw a fine line that, starting in that column containing more than one ones that is further to the right hand than any other such column that there may be, crosses two previously uncrossed ones in that column, and then passing to the left crosses a single uncrossed one in each column that it passes through, comes to an end in a large dot (which for all purposes shall be regarded as an uncrossed one) in a column that previously contained no one. Repeat this process until it can no longer be performed, owing to there being no column that contains more than one uncrossed one. Then and finally, copy below and in their columns all the remaining uncrossed ones, placing zeros in the vacant places; and the result will be the sum required.

Part II: used when there are two or more circulating secundals among the summands. (N.B. That part of an Aryan number which, as it happens to be written, extends, *inclusively*, from one of the two dots that indicate the limits of a period of a circulating fraction (though the right hand dot, by the way, is rather an idle member of the expression), may be called the *circulant*, while the part of the expression to the left of both dots may be called, by opposition, the *stagnum*, or *stagnine* part.) By repeating, in their cyclical order, the figures of the circulants and moving the dots, as may be necessary, bring all the left hand dots of the circulants to the same secundal place, namely, to the left-hand-most place that is to the right of all the stagnine places of all the summands, and make the number of places of the different circulants, thereafter to be known as the reformed circulants, the same; namely, the same as the number that is the least common multiple of the numbers of places in the periods of the different circulating secundals, however they may be *written*. Then add the circulants by themselves, using Part I of this rule. But whenever the fine lines drawn in adding carry ones any number of places beyond the left hand place of the reformed circulants, not only must corresponding ones be put as additional stagnine summands, at the same ordinal place of the stagna counting the extreme right hand place of any of them as the first, but also corresponding ones must be taken as additional summands of the reformed circulants, counting their right-hand-most place as the first. The stagna and the reformed circulants having been thus separately added, and the additional summands having been added in with both, the resulting sums will be respectively the stagnum and the circulant of the required sum, subject to possible simplifications of expression. *Example*: Add these three numbers,

```
  1 0 1 1 1 0 0 0 0 0 : 1 0 0 1 0
1 1 1 1 1 1 0 0 0 0 0 1 0
1 0 0 0 1 1 1 0 0 0 0
```

We first reconcile their circulants, thus

```
  1 0 1 1 1 0 0 0 0 0 0 1 1 0 0 1 0 0 1 1 1 0 0 0 0 0 0 1 1 0 0 1 0 0 1
1 1 1 1 1 1 0 0 0 0 0 1 0 1 1 1 0 0 0 0 0 1 0 1 1 1 0 0 0 0 0 1 0 1
1 0 0 0 1 1 1 0 0 0 0 1 1 0 0 0 0 1 1 0 0 0 0 1 1 0 0 0 0 1 1 0 0 0 0
```

The sum of their stagna is taken thus:

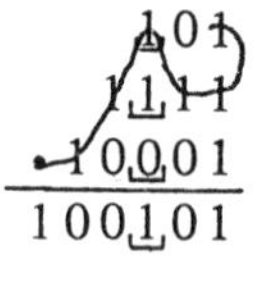

The sum of the reformed circulants is found as follows:

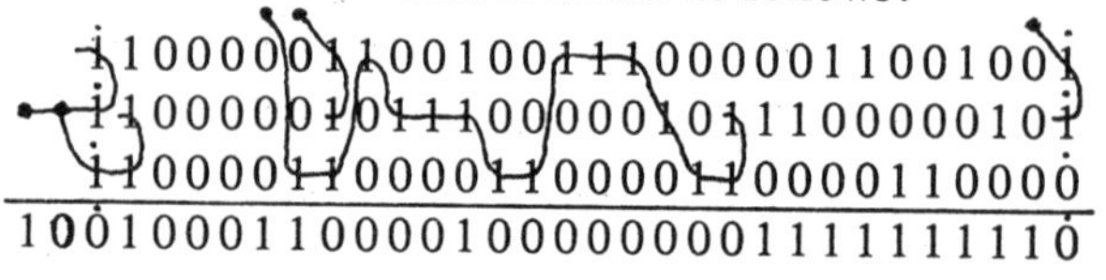

Next, the 1 0 which projects beyond the circulants to the left must be added both to the sum of the stagna and to that of the circulants. The addition to the sum of the stagna is performed as follows:

$$\begin{array}{r} 1\,0\,0\,\underline{1}\,0\,1 \\ 1\,0 \\ \hline 1\,0\,0\,\underline{1}\,1\,1 \end{array}$$

The addition to the sum of the circulants is performed as follows:

$$\begin{array}{r} \dot{0}\,1\,0\,0\,0\,1\,1\,0\,0\,0\,0\,1\,0\,0\,0\,0\,0\,0\,0\,0\,\not{1}\,\not{1}\,\not{1}\,\not{1}\,\not{1}\,\not{1}\,\not{1}\,\not{1}\,\not{1}\,\dot{0} \\ \not{1}\,0 \\ \hline \dot{0}\,1\,0\,0\,0\,1\,1\,0\,0\,0\,0\,1\,0\,0\,0\,0\,0\,0\,0\,1\,0\,0\,0\,0\,0\,0\,0\,0\,0\,\dot{0} \end{array}$$

Writing this result after the previous one, we get for the final sum,

$$1\,0\,0\,\underline{1}\,\dot{0}\,1\,0\,1\,0\,0\,0\,1\,1\,0\,0\,0\,0\,1\,0\,0\,0\,0\,0\,0\,0\,1\,0\,0\,0\,0\,0\,0\,0\,0\,0\,\dot{0}.$$

Subtraction can be variously performed. I will give two rules. *First Rule for Subtraction.* Write the subtrahend, place by place, vertically under the minuend. (That is, whatever real integer n may be, 2^n would appear in the same vertical column in both.) Cancel together the extreme uncancelled left hand 1 of the subtrahend and the right hand most of those uncancelled 1s of the minuend that are not further to the right than that 1 of the subtrahend, and change into a 1 every 0 of the minuend that is further to the right than the cancelled 1 of the minuend but not than the cancelled 1 of the subtrahend. Repeat this until no. 1 remains uncancelled in the subtrahend, when the uncancelled 1s of the minuend form the expression of the remainder.

The second rule for subtraction may be prefaced with a leash of small remarks. The first is that, just as 1 0 1 $\underline{0}$ 1 (decimally: 10.5) is just the same as 0 0 0 1 0 1 $\underline{0}$ 1 0 0 0 0, so any Aryan expression remains unchanged in value, though a 0 or any plural of 0s be prefixed to it or be suffixed to it, or both; provided, however, that the places of "the significant figures"* remain determinate and unchanged. (* "The significant figures" is an expression frequently used and never misunderstood, which is probably why it has never (so far as I know) been defined. Some of my friends, the nominalistic pragmatists, may think that since nobody misunderstands the phrase there would be no good in defining it. Well, I do not mean to waste space in refuting that opinion; so let it go, as my private opinion, publicly expressed, that there may be, and in this case is, something to be gained by defining an expression that nobody misunderstands. It means those figures which would exactly express a rational number, if only that power of the base a multiple of which any one of them would then express were stated. The number [multitude] of significant figures is one more than the excess of the exponent of highest place in which stands a figure not 0 over the exponent of the lowest place in which stands such a figure.) The second remark is that if we use a horizontal bar, — a 1 turned over on its nose, — as a numeral figure to express *minus* 1, then, just as in decimal augrim 999900 is equal to 1000$\dashv$00, so in the secundal augrim an uninterrupted row of 1s in any places denotes a number otherwise expressed by a pair of figures; to wit by a 1 in the right-hand-most of the places at the left of the row of 1s, while the other figure of the pair is the horizontal bar, — the $\dashv$, — in the right-hand-most place of those occupied by the row of 1s. To this brace of remarks may be added a third, still more obvious, if such a thing be possible; namely, that if to any number secundally expressed we add what may be called its *arithmetical supplement*, meaning the number which would result from changing every figure of a secundal expression into its contrary, — every 0 into a 1 and every 1 into a 0, — then the sum of any number and of its arithmetical supplement will be expressed by an uninterrupted row of 1s. Upon the substance of these remarks is based the following

Rule for Algebraic Summation. When you have given a collection of numbers, one at least of them being positive, and one, at least, negative, and you desire their algebraic sum, you must *first* ascertain, by inspection or otherwise, which of the two subcollections of that collection of numbers, each subcollection being composed of all the numbers of one and the same algebraic sign, has the larger sum; and this subcollection may be called *the majority*, and the other *the minority*, regardless of how many members the one and the other may consist of. This, in somewhat unusual cases, will require a preliminary summation to be made of all the figures in or above a certain secundal place which we may call the μth, in all the positive numbers* to be added, and again in all the negative ones. (* In order to avoid using the word "number" in puzzling multiple senses, I shall avail myself of the well-established algebraical application of the word 'term,' to denote an algebraically additive part of a complex expression; and shall speak of the numbers to be algebraically added as 'terms.' It was, I guess, Aristotle who first deflected the word 'term' (or rather, ὅρος) from its natural line of signification, so that it has come to mean independent part.) In performing this operation, we may assume, unless information or observation (which latter should never go quite unmade) should lead to a decidedly different conclusion, that the average value of all the neglected figures in places whose numbers are less than μ is a 1 in the $(\mu - 2)$ place. But the decision in regard to which of the two subcollections of numbers is the majority, and which is the minority, should not turn upon an assumption as to the average values of the sums of certain parts of the numbers, unless a person with a good practical training in the application of the doctrine of chances pronounces that it is prudent so to do. Otherwise, another secundal place must be brought into the computation. All this, however, is rather idle, considering the kind of work for which secundal augrim is likely to be employed.

Having ascertained which subcollection of numbers to be algebraically added is the Majority, and which is the Minority, you are to proceed, *Secondly*, to write down all the numbers, (1) secundally, (2) equispatially, and (3) one below another so that all the figures that are in one secundal place shall (to whatever numbers they may belong) appear in one vertical column. Moreover, (4) the numbers of the majority are to be written in ink, or otherwise distinctively, while the numbers of the minority are to be written lightly in pencil-lines, or otherwise so as to be at once instantly distinguishable, figure by figure, from the numbers of the majority, and, at the same time, so as to be capable of being changed into their contraries — the 0s into 1s, and the 1s into 0s.

Thirdly, you are to perform upon the entire body of terms to be algebraically added an operation of *atonement* by which they are all brought to the algebraic sign of the "majority." This atonement is effected by taking up all the minority-terms, one after another and subjecting each to an operation of *conversion* by which it gets replaced by a majority-term. The operation of conversion is not only itself a composite one, consisting of a *transmutation* of the minority-term itself, followed by two *pulses* by each of which a single majority-figure gets changed, but moreover has to be preceded by a *preparation* for pulsation, which again consists of two parts, an *initial* and a *final preparation.* The whole rule for atonement is as follows:

Treat all the minority terms, in single succession as follows

First, make the initial preparation of the minority-term under treatment by either prefixing pencil-mark 0s to it, or a single 0, or else striking off one or more 0s forming the left-hand part of the expression of the term, or else finally leaving the expression quite unchanged, the sole requirement in the initial preparation, and it must be watched for, is that as the term remains after the preparation the last secundal place of its *field* must contain in some horizontal line a majority-1.

As soon as the initial preparation is complete, the final preparation of the same number must at once be made. This consists in adding or removing one or more zeros at the right hand end of the secundal expression of the minority-number, the sole condition that must imperatively be observed being that in case the number after preparation ends in a zero, then there must be a majority-zero somewhere in that place. In case the required condition for either preparation cannot be fulfilled as the quantities stand, it may usually be rendered capable of fulfilment by replacing certain majority-1s by their sum.

As soon as both preparations of any minority-term have been made, it should be subjected to the operation of conversion, which consists in the simultaneous performance of two changes in it. Namely, its arithmetical supplement should be substituted for the number itself, or, in other words, every 0 in its expression should be changed to a 1, and every 1 to a zero; and moreover, the changed figures should be written in ink, or otherwise be given the character of majority-figures. This done the two *pulses* which are parts of the conversion of the same minority-term are to be executed. The initial pulse consists in cancelling, or otherwise changing to 0, a majority-1 in the righthandmost, or last, place in the *field* of the expression number just transmuted as the preparation left it, while the final pulse consists in changing a majority-0 to a 1 in the last place of the same number in its prepared state.

This concludes the conversion of this minority-term. With that of all the minority terms, the atonement is accomplished, and it only remains to add the terms by the rule of addition.

Although, in this analytic statement, this rule seems highly complicated, yet in practice it is the simplest thing in the world, *exceptis exciptendis*; being nothing but the rule of the arithmetical complement, so stated as to exhibit its symmetries.

The occurrence of circulants has the same effect as in addition, except that there is no final pulse, on account of the expression representing a series that has no end. The whole consideration of circulants naturally should come after that of division.

The rule of multiplication depends upon whether the exact product is required (which can only be for purposes of mathematical theory), or whether it is only required that a certain number of *significant figures* is wanted to be correct. The number of significant figures is one more than the excess of the number of the place of the extreme left-hand over that of the extreme right-hand 1 in any expression in secundal notation. For almost all purposes ten significant figures are wanted; for very few more than twenty; for almost none more than thirty. If there are no circulants the matter is simple enough. I proceed to state the rule in connexion with figures showing the stages of multiplying 1 0 0 0 $\underline{1}$ 1 0 0 1 1 1 (decimally written, 17 · 609375 or $17\frac{39}{64}$) by 1 0 1 $\underline{0}$ 1 0 1 1 0 0 1 (decimally, 10 · 6953125 or $10\frac{89}{128}$).

First Clause of the Rule. Write the multiplicand (i.e. either facient) giving precisely equal spaces to all the secundal places, and marking the zero, or unit, place by placing a square bracket in a horizontal position beneath the 0 or 1 in that place. Draw a long horizontal line under the multiplicand.

Second Clause of the Rule. Upon the lower edge of a card or slip of stiffish paper, write the figures of the multiplier (i.e. the other facient) in reversed order, giving spaces to the secundal places accurately equal to those of the multiplicand; and mark the zero-place as in the multiplier.

Third Clause of the Rule. Place the reversed multiplier horizontally over the multiplicand, so that its highest secundal place (which is that of the 1 which now, in the reversed order of the figures, is at the extreme right hand of the row) in the same vertical column as the highest, or left-hand-most 1 of the multiplicand; and set down a 1 as an additive part of the product beneath the line under the multiplicand, in that vertical column in which the bracket shows the zero-place of the multiplier to be. N.B. When, in the course of this rule you are directed to write a number in a certain column below the multiplicand, what is intended is that the zero-place of that number should be in that column with the other places in columns situated relatively to that column exactly (in distance as well as in direction), as the same places of the multiplicand are situated relatively to the zero-place of the expression of that quantity as it has been written in accordance with the first clause of this rule. Fig. 7 shows the work just after the directions of this clause have been fulfilled.

|1 0 0 1 1 0 1 $\underline{0}$ 1 0 1|

1 0 0 0 $\underline{1}$ 1 0 0 1 1 1

1

Fig. 7

Fourth Clause of the Rule. As soon as any additive part of the product has been duly set down in accordance with this rule, the card or slip bearing the reversed figures of the multiplier is to be moved horizontally to the right through this distance between the middles of two adjacent secundal places of the multiplicand. Then passing along the multiplicand from right to left, count those 1s of the multiplicand of which each has a 1 of the multiplier vertically over it, on the card or slip, and set down the number of the count in the secundal place then occupied by the zero-place of the multiplier, and in any convenient horizontal line below the line drawn under the multiplicand, as an additive part of the product. To follow out the working of our example, as soon as Fig. 7 has been formed, the card carrying the reversed multiplier is to be shifted into the position shown in Fig. 8, and then a 0 is to be set down, as there

```
|10011010101|
       ---------------
          10001100111
       ---------------
       10
```

Fig. 8

shown, under the zero-place of the multiplier, because no 1 of the multiplicand is now directly under a 1 of the reversed multiplier. As soon as this 0 is set down, the card or slip bearing the figures of the reversed multiplier is to be shifted one place further to the right, as shown in Fig. 9; and thereupon a 1 is to be written under the zero-place of the

```
|10011010101|
      ----------------
        10001100111
      ----------------
      101
```

Fig. 9

multiplier, because one 1 of the multiplicand is now directly under a 1 of the reversed multiplier. As soon as that is done, the card is again to be shifted in the same manner, as shown in Fig. 10, and a 0 is to be set

```
|10011010101|
     ----------------
       10001100111
     ----------------
     1010
```

Fig. 10

down under the zero-place of the reversed multiplier, because no 1 of the multiplicand is now directly under a 1 of the reversed multiplier. Figs. 11-25 explain themselves. The inspection of Fig. 25 evidently shows that there will follow two 1s and nothing else. In Fig. 26 these are set down.

```
|10011010101|
      10001100111
 ____________________
  1010
      10
```

Fig. 11

```
|10011010101|
      10001100111
 ____________________
  1010  1
      10
```

Fig. 12

```
|10011010101|
      10001100111
 ____________________
  1010  1
      10
         10
```

Fig. 13

```
|10011010101|
      10001100111
 ____________________
  1010  1
      10  10
         10
```

Fig. 14

```
|10011010101|
      10001100111
 ____________________
1010  1  10
    10  10
       10
```

Fig. 15

```
  |10011010101|
      10001100111
 ____________________
1010  1  10
    10  10
       10  10
```

Fig. 16

```
       |10011010101|
        -----------
        10001100111
  -----------------
  1010 1 10
      10 10
         10 10
            100
```

Fig. 17

```
       |10011010101|
        -----------
        10001100111
  -----------------
  1010 1 10 11
      10 10
         10 10
            100
```

Fig. 18

```
       |10011010101|
        -----------
        10001100111
  -----------------
  1010 1 10 11
      10 10   11
         10 10
            100
```

Fig. 19

```
       |10011010101|
        -----------
        10001100111
  -----------------
  1010 1 10 11 1
      10 10   11
         10 10
            100
```

Fig. 20

```
       |10011010101|
        -----------
        10001100111
  -----------------
  1010 1 10 11 1
      10 10   11
         10 10   11
            100
```

Fig. 21

```
          |10011010101|
   10001100111
--------------------------
1010 1 10 1111
   10 10    11 11
     10 10
        100
```

Fig. 22

```
          |10011010101|
   10001100111
------------------------------
1010 1 10 11 1 10
   10 10    11 11
     10 10    11
        100
```

Fig. 23

```
          |10011010101|
   10001100111
------------------------------
1010 1 10 11 1 10
   10 10    11 11 1
     10 10    11
        100
```

Fig. 24

```
          |10011010101|
   10001100111
------------------------------
1010 1 10 11 1 10 1
   10 10    11 11 1
     10 10    11
        100
```

Fig. 25

```
   10001100111
------------------------------
1010 1 10 11 1 10 1
   10 10    11 11 1 1
     10 10    11     1
        100
------------------------------
1011110001010110011111
```

Fig. 26

Fifth Clause of the Rule. The process described in the fourth clause is to be continued, in case the exact product is required, until the left-hand-most 1 of the reversed figures of the multiplier has reached the place of the right-hand-most 1 of the secundal expression of the multiplicand. A horizontal line will then be drawn under all the additive parts of the product obtained by the processes prescribed in Clause the Third and the Fourth of the Rule; and these parts are to be added; taking account of their secundal place. The sum will be the product desired.

This sum is exhibited, in the case of our example, in Fig. 26. Its decimal expression is 188 · 3377685546875.

I shall postpone until after treating division the rather delicate question of the best way of performing the multiplication in case the facients contain circulants, — a subject, by the way that I have never seen treated either in a text-book of arithmetic or elsewhere.

Long division becomes exceeding simply in secundal computations. Everybody knows that the idea of this algorithm is successively to subtract from the dividend readily formed multiples of the divisor, while setting down the multipliers of the divisor in those multiples as additive parts of the quotient. What remains of the dividend still to be divided at any stage of the algorithmic procedure may be called the *re' manent*; so that the dividend itself is the initial remanent. In secundal augrim (or logistic, or the art of computing, if you prefer either of these terms to the good old Chaucerian 'English undefiled'), the only multiples of the divisor that will be used in long division, the only ones that can be formed without the aid of long multiplication or some such process, consist in the dividend itself removed, it may be, to another series of secundal places.

If in Long Division you desire to perform the successive subtractions by the rule of the arithmetical supplement then having formed this supplement, having prefixed the 1 lying 'noselings,' as the old writers say (which for short may be called the 'supine') — the character denoting negative unity, — and having added 1 according to the rule, and having thus obtained what may be called the *divider*, you may be puzzled to know under what place of the remanent, or dividend remainder, the supine ought to go; and if you stop to reason it out, you will lose all the advantages both of accuracy and of dispatch of this rule of subtractions. What you need, therefore, is a rule easy to remember, by the aid of which you may dispense with thinking. The following is such a rule:

Compare the figures of the divider, one by one, with those of the remanent or dividend remainder, beginning with the figure of the former next following the supine, which you compare with the first significant figure of the remanent, or dividend-remainder, and proceeding figure by figure toward the right in both numbers, until you come to a figure of the supine which agrees (in being 1 or in being 0) with the figure of the remanent, or dividend-remainder, with which it is compared. The number denoted by this figure will be the number of places (either 1 or 0) by which the supine is to be placed to the left of the first significant figure of the remanent, or dividend-remainder.

Whenever it happens that any remanent, or dividend-remainder, consists of the same figures as a remanent previously obtained in the same division, this shows that the figure of the quotient just obtained is the last figure of a circulant of which the first figure was the first figure of the quotient obtained after the previous occurrence of a remanent composed of the same figures. As an example, divide 1 1̲ 0 1 (decimally expressed, 3 · 25) by 1 0 1̲ 1 (decimally expressed, 5 · 5)

```
          0̲10̇010111010̇001
         ________________
10 1̲1    11̲01
         1011      11
         ____
           10      101̲1
                   ____
           101̲1       1
          _____
            101       101̲1
                     _____
             101̲1     101
             _____
             1001
              101̲1
              ____
               111
              101̲1
              ____
                11
```

Fig. 27

When the entire circulant of the quotient is wanted, the Rule of 'Direct Division' is the easiest. It is as follows:

First, free dividend and divisor of circulants, by these three principles: 1st, that a number expressed by a given series of figures is twice the number expressed by the same series of figures but with the mark of the zero-place removed one place to the left among the figures of the series; 2nd, that each unit of a circulant beginning at the -1-place and ending at the $-n$-place is equal to 1 divided by $2^n - 1$; and 3rd, by the algorithms for the greatest common divisor, and that for the series of approximations of continued fractions. In this way, a dividend or divisor involving a circulant can be brought to the form of a vulgar fraction expressed in its lowest terms. For example, given $1\,0\,\underline{\dot{0}}\,0\,1\,\dot{1}$ is twice $1\,\underline{0}\,\dot{0}\,0\,1\,\dot{1}$, or twice $10\frac{11}{1111}$. Now $(10 \times 1111) + 11 = 11110 + 11 = 100001$, so that $10\frac{11}{1111} = \frac{100001}{1111}$. The Greatest Common Divisor is 11; and dividing numerator and denominator by this, we find $\frac{100001}{1111} = \frac{1011}{101}$; so that $1\,0\,\underline{\dot{0}}\,0\,1\,\dot{1} = \frac{10110}{101}$. That done, the numerator of the dividend must be multiplied by the denominator of the divisor to make a new dividend, and *vice versa*.

Secondly, note the secundal place of the right-hand-most 1 of the dividend, the exponent of which we will here call m; and note likewise that of the divisor, whose exponent we will call n. Then, for the present, treat the places of those last 1s as zero-places.

Thirdly, strike off the last 1 from the divisor and add it to the truncate remnant as if the latter ended in the zero-place. Call the sum the current addend. Write the dividend in the upper right-hand corner of your paper, draw a vertical line to the left of its last 1, continue this line one line of writing lower, write the current addend, place by place, beneath the truncate remnant of the dividend, draw a horizontal line, write the sum of the truncate remnant and the current addend beneath the horizontal line, and treat this, in every respect, as the dividend has just been treated. Repeat this process of cutting off the last 1 with any 0s that may follow it, and of adding the current addend to the truncate remnant, to make a new quasi-dividend, until you finally get a quasi-dividend, which is just the same (except for its secundal place) as one you have already had. Then stop. Bring down in their relative secundal places (as shown by the columns they are in) the figures that have been cut off to the right of the vertical lines from the quasi-dividends; or it will be sufficient to bring down a few of the last figures of the circulant together with all those that were written before (and which therefore belong to the right of) any of the figures of the circulant. Subtract these from the figures of the circulant following in order those of the circulant already written, repeating as many figures of the circulant as have figures not belonging to it from which subtraction has to be made. Then the last figure of the subtrahend is to be regarded as being in the zero-place and all that follow it as belonging to the circulant.

Thirdly, move the supine bracket that marks the zero place $(m - n)$ places algebraically toward the right, and the result is the quotient.

Example. Divide $1\,0\,0\,0\,0\,\underline{0}\,0\,0\,0\,0\,0\,1$ by $1\,0\,\underline{1}\,1\,1$. Here $m = -1\,1\,0$ and $n = -1\,0$. Cutting the last 1 off from the divisor and adding it to the truncate remnant, thus

```
1011|1
   1|
----
1100
```

, we find the current addend to be 1 1 0 0. We then proceed as shown (in Fig. 28).

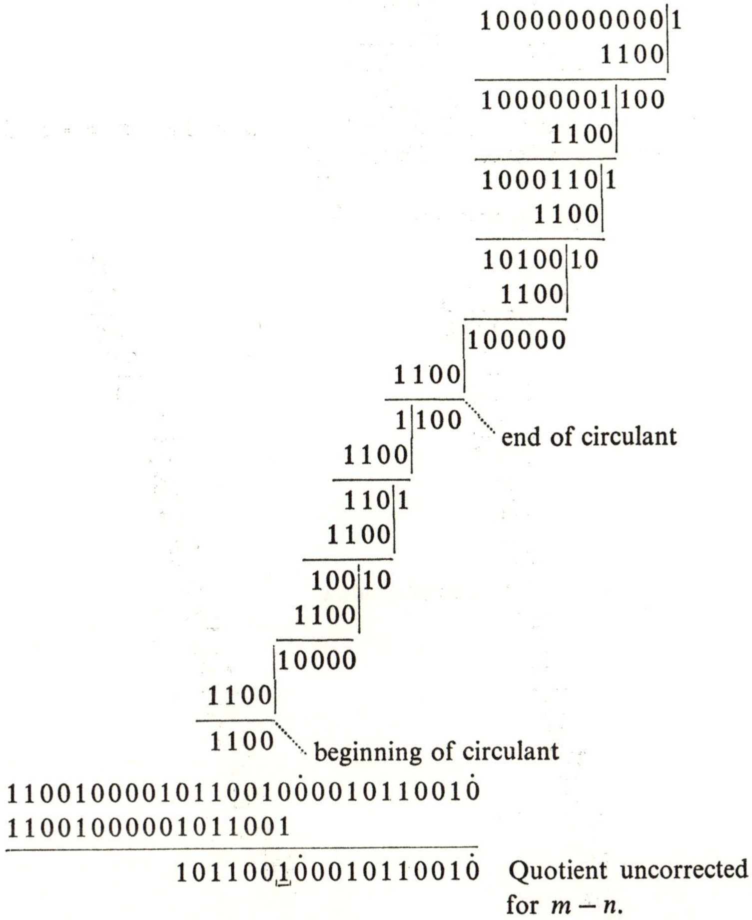

Fig. 28

But since $m - n = -100$ the supine bracket must be moved *minus four* places to the right; that is, four places to the left and the true quotient is $10\lfloor 1 \rfloor 1001\dot{0}0010110010\dot{0}$.

In division, the secundal system has an unqualified advantage over the decimal system. In order to illustrate this, I will exhibit the work of this same division performed decimally. The rule of "Direct Division," as it applies to decimal augrim, will be found briefly stated in the Century Dictionary under the word *division.*

The dividend is 32·015625, and the divisor is 5·75. According to the rule, both must first be multiplied by 12 thus

```
 5·75         32·015625
 12           12
 ----         ---------
 11·50        64·031250
 57·5        320·15625
 -----       ----------
 69·00       384·187500
```

Now $m = 0$ and $n = -4$. We proceed as follows:

```
                    Divisor  6|9
                             1|
                             --
Current Multiplier:          7

        384187|5                                 6|6
            35|                                 42|
        ------                                  --
        38422|2                                  4|8
           14|                                  56|
        -----                                  ---
        3843|6                                   |60
          42|                                 42|
        ----                                  --
        388|5                                  4|2
         35|                                  14|
        ---                                   --
        42|3                                   1|8
        21|                                   56|
        --                                    --
         6|3  end of circulant                 5|7
        21|                                   49|
        --                                    --
         2|7                                   5|4
        49|                                   28|
        --                                    --
         5|1                                   3|3
         7|                                   21|
        --                                    --
         1|2                                  24
        14|                                  28|
        --                                   --
         1|5                                   |30
        35|                                 21|
        --                                  --
         3|6                                 2|1
        42|                                  7|
        --                                   --
         4|5                                  |9
        35|                                 63|
        --                                  --
         3|9                                63  beginning of circulant
        63|
        --

1739130433̇478260869565217391304̇
17335625
---------------------------------
    5·56793̇478260869565217391304̇      Correct Quotient.
```

Fig. 29

It is a mere accident that the secundal quotient has not as long a circulant as the decimal one. On the average they would be about equal in length. But that the decimal operation has a current *multiplier* in place of a current *addend* is no accident. Had I compared the work of 'Long Division' in the two systems, the advantage of the secundal over the decimal would have appeared much greater yet.

For the extraction of roots in secundal augrim, I am inclined to think that, barring exceptional cases, the most convenient rule is one which I invented myself, and thought myself the *first* inventor of it, until some friend informed me that it had been given by Euler, and since I make no pretension to being (in the phrase of them who 'chaw') any great shakes of a mathematician, I was no little elated to learn that the one-eyed Leonard had deemed my invention worth his *preplagiarizing*. This rule has the advantage over several others of being extremely easy to remember, which is quite an important merit in a rule so seldom wanted, and wanted so very badly when it is wanted. It also has the advantage of furnishing any desired root, real or imaginary, of any algebraic equation with numerical coefficients, by successive approximations, of course. It has the disadvantage, in decimal augrim a serious one, of giving the answer in the form of a vulgar fraction. But in secundal arithmetic division is so facile as to diminish this objection to the point where the infrequency of the problem, and the consequent unimportance of a slight lengthening of the operation, makes the combined consideration of the generality of a rule applicable to all numerical equations together with the almost absolute unforgetableness, the decisive one, I think. As to the rapidity of the convergence, in the case of quadratic equations, the rule coincides with that of the solution by continued fractions, which is well-known to be the most rapidly convergent possible. But generally speaking, one can only say that it gives an indefinitely close approximation, at last. It cannot even be said that in all cases the last approximation is closer than the one which preceded it. It is usually very rapid; yet in some cases is most provokingly slow, so that it had better then be abandoned in favor of the differential method. The rule is as follows: The question to be solved being

$$\sum_0^n {}_i a_i x^i = 0,$$

where n is positive and a_i is a known quantity, of course, no process of computation can give more than one root at a time, though if all the coefficients are real, any one imaginary root immediately points out another. It is necessary therefore to introduce some condition which shall distinguish the root desired from every other. For that purpose, you eliminate x by means of the equation $x = Y+y$, where Y is a known quantity such as to render that value of y (which is unknown) that corresponds to the desired value of x, either greater or less than every other value that satisfies the equation. If it be greater, by making $1/y$ the unknown instead of y, you still have the rule that you seek the smallest root; and by the "smallest," is to be understood that root whose modulus is less than that of any other root. Let the equation so transformed be

$$\sum_0^n {}_i b_i y^i = 0.$$

Then, dividing by b_0, denoting $-\dfrac{b_i}{b_0}$ by β_i and transposing you will put it into the form

$$\sum_1^n {}_i \beta_i y^i = 1.$$

Now write down in a column any n numbers you please, except that all must not be 0s. I should recommend that all but the last be made 0s, and the last 1. Imagine these numbers to be in their order *substitutes*, for the purpose of calculating a 'substitute' for 1, of $y^n, y^{n-1}, \ldots y^2, y$. Now calculate the substitute for 1 by the equation

$$\sum_1^n {}_i \beta_i y^i = 1,$$

where the substitutes for the different powers of y are to be substituted for those powers and the result, instead of being 1, or y^0, the "substitute" for 1, which set down under the others. As soon as that is done, change the substitution by making the last n numbers of the column, as it now stands, the substitutes in their order, for the n powers of y, every number that has been substitute, say for y^j, now becoming the substitute for y^{j+1}. The calculated substitute for y^0 becomes the substitute for y^1. With these new substitutes calculate, by the same equation, a new substitute for 1. Repeat this process as often as you see fit. Generally speaking, the oftener the better. Then, the quotient of the division of the substitute for y^1 by the substitute for y^0 will be an approximate value of the smallest root of the equation in y.

Example 1. Required the square root of 10. The equation $x^{10} = 10$ has, of course, no root having a modulus less than that of the other. Put

$x = y+1$. Then $x^{10} = y^{10}+10y+1 = 10$ or $y^{10}+10y = 1$. We may now proceed as follows:

$$
\begin{array}{r}
0\\
1\\
10\\
101\\
1100\\
11101\\
1000110
\end{array}
$$

Then, the successive approximations to $\sqrt{10}$ will be

$$1,\ \underline{1}1,\ 111/101 = \underline{1}\dot{0}11\dot{0};\ 10001/1100 = \underline{1}01\dot{1}\dot{0};$$
$$101001/11101 = \underline{1}\dot{0}11010011110111001011000010\dot{0};$$
$$1100011/1000110 = \underline{1}0110101000001110\ldots$$

Example 2. Required both roots of the equation $x^{10}-101x+101 = 0$. Since the discriminant $(101)^{10}-101\cdot 100 = 11001-10100$ is positive, the roots are real. Since the last term is positive, they have the same sign; and since the second term is negative that sign is that of *plus*. The sum of the roots is five. Comparing the different values of the product $\xi(10\underline{1}-\xi)$ in the following table, we see that the ratio of the greater root to the lesser one is not much more than that of $11\underline{1}$ to $1\underline{1}$; — say, about $100\underline{0}$ to $1\underline{1}$. Now this is not large enough to make the approximation comfortably rapid; for though theoretically it is only requisite that the moduli of the two roots should not be exactly equal, yet if the least root has not a modulus *much* less than that of any of the others, the process of approximation would wear out the patience of Job. Here is the table:

ξ	$10\underline{1}-\xi$	$\xi(10\underline{1}-\xi)$
$\underline{1}$	$10\underline{0}$	$10\underline{0}$
$\underline{1}1$	$1\underline{1}1$	$10\underline{1}01$
$1\underline{0}$	$1\underline{1}$	$11\underline{0}$
$1\underline{0}1$	$1\underline{0}1$	$11\underline{0}01$

Since, then, the lesser root is pretty nearly $1\underline{1}$, let us substitute $x = -y+\underline{1}1$, since the table shows that the lesser root is somewhat less than $\underline{1}1$. Then we shall transform the equation $x^{10}-101x+101 = 0$, by substituting $x = -y+\underline{1}1$, as follows:

$$
\begin{array}{rcl}
x^{10} & = & y^{10} - 1\underline{1}y + 1\underline{0}01 \\
-10\underline{1}x & = & +10\underline{1}y - 11\underline{1}1 \\
+10\underline{1} & = & +10\underline{1} \\
\hline
0 & = & y^{10} + 1\underline{0}y - \underline{0}01
\end{array}
$$

or
$$\underline{1} = (y^{10} + 1\underline{0}y) \times 10\underline{0}.$$

Now in regard to the two arbitrary substitutes for y^{10} and y, although there is a slight, — a very slight, — advantage in almost every case, in making those numbers all 0s, except the last, which is best made 1, yet for the sake of imparting a little stronger *chiaroscuro* to the exhibition of a certain relation to which the reader's attention will shortly be drawn, I will make both my arbitrary substitutes equal to 1. And, by the way, for readers who are little skilled in the trade of ciphering, — a condition in which the wretched schooling we get in it leaves most of us, — it will not be superfluous to suggest that in order to avoid much copying of figures, and yet to have figures that are to be added in the same vertical column, the computation of the substitutes may conveniently be arranged as in the accompanying Fig. 30, in explanation of which I will say that the

$$
\begin{array}{lrl}
S_0 = & \underline{1} & \\
S_1 = & \underline{1} & \\
S_2 = & 110\underline{0} & \\
S_3 = & 110010\underline{0} & \quad \text{etc.}
\end{array}
$$

Fig. 30

sum of the figures of the last two substitutes obtained as they lie in their vertical columns, is to be set down four places to the left of the column added, and two 0s are to be affixed to this sum, the last of the two being marked as being in the zero place. And in these additions carryings are to be made from one place to the place next to the left of it, in the usual way. You will perceive that the effect of this arrangement of the work is that with great facility you make the number in each horizontal line equal to four times the sum of twice the number in the line next higher *plus* the number in the line higher than that. In short, you thus make

$$S_r = 4(S_{r+2} + 2S_{r+1})$$

where, r being any ordinal number positive or negative, S_{r+2} is the substitute that comes the second before S_r. Our arbitrary substitutes are, first, S_0, and next after that S_{-1}. From these we calculate the next which I call S_{-2}. You will see more fully, before long, why I thus number them backwards; but you can already discern a certain appropriateness in the numbering, since if S_r be the substitute for y^2, S_{r-1} is the substitute for y^1, and S_{r-2} is the substitute for y^0. Or if S_r be regarded as the substitute for y^r, then S_{r-1} is the substitute for y^{r-1}, and S_{r-2} is the substitute for y^{r-2}. But of all this you shall gain a much clearer vision before you get to the end of this prolusion.

I here set down a few of the first substitutes and the resulting approximations to the values of y.

		Approximations to y.
S_0 =	$\underline{1}$	$\underline{1}$
S_{-1} =	$\underline{1}$	$\underline{0}$0001
S_{-2} =	110$\underline{0}$	$\underline{0}$000111101
S_{-3} =	110010$\underline{0}$	$\underline{0}$00011110001100
S_{-4} =	110101000$\underline{0}$	$\underline{0}$0001111000110111111
S_{-5} =	11100000100 0$\underline{0}$	The last two figures should be 00

The last approximation is nearer correct than it could be expressed in 5 decimal places.

I have now something to say which applies equally to the numerical solutions of all equations, and is not limited to the present example. It is that not only will this method give the root of smallest modulus, if the equation in hand has any root whose modulus is smaller than that of any other root, but it will likewise give the root of greatest multitude if any root has a greater multitude than that of any other root; and if either or both these conditions be unfulfilled, an easy substitution will overcome the difficulty; and though it is sometimes a pretty heavy handicap to deal with the equation under the same form in the evaluation of both roots, yet it can be done, and in many cases can be done very easily. All that is necessary is to prolong the column of substitutes in the upward direction. I will express my meaning more explicitly. The equation of whatever degree, say the nth is of the form

$$a_0 x^n + a_1 x^{n-1} + a_2 x^{n-2} + \dots + a_{n-2} x^2 + a_{n-1} x + a_n = 0$$

where the as are the coefficients. It may be written more compactly, and in some respects more conveniently thus:

$$\sum_{0}^{n} {}_{i} a_i x^{r+i} = 0,$$

for to multiply the whole by x^r is only to add r new roots all of which are zero. Now if the series of substitutes are denoted by S_i where i takes a consecutive series of integral values, then the equation for calculating the substitutes is

$$\sum_0^n {}_i a_i S_{r+1} = 0.$$

The arbitrarily taken substitutes run from S_1 up to S_n inclusive. To calculate the substitutes with negative indexes, we put the equation, as I have said, into the forms

$$-\frac{1}{a_0}\sum_1^n {}_i a_i S_i = S_0$$

$$-\frac{1}{a_0}\sum_0^{n-1} {}_i a_i S_i = S_{-1}$$

$$-\frac{1}{a_0}\sum_{-1}^{n-2} a_i S_i = S_{-2} \quad \text{etc.}$$

Whereupon, the successive approximations to the root of smallest modulus, if there be such a root, will be $\frac{S_0}{S_{-1}}, \frac{S_{-1}}{S_{-2}}, \frac{S_{-2}}{S_{-3}}$, etc. In case there should be no root having a modulus less than any other, this series of values will not converge toward any limit. In order to get the root of largest modulus, we throw the very same equation

$$\sum_0^n {}_i a_i S_{r+i} = 0$$

into the form

$$S_{r+n} = -\frac{1}{a_n}\sum_0^{n-1} {}_i a_i S_{r+i};$$

and then giving r successively the values 0, +1, +2, +3, etc., calculate successively S_n, S_{n+1}, S_{n+2}, etc. Whereupon, exactly as before $\frac{S_{n+1}}{S_n}$, $\frac{S_{n+2}}{S_{n+1}}$, etc. will approximate, — but this time to the root of largest modulus. Here in [the accompanying figure 31] are the substitutes and the approximations for our Example No. 2. (The printer has been obliged to put in two sections on two opposite pages.) [In this edition they are on successive pages. – Ed.]

$$S_n = 100(S_{n+10} + 10\,S_{n+1}) = -10\,S_{n-1} + {}^{1}/_{100}\,S_{n-10}$$

n	
11	+ 1011110110 1 1101100111
10	− 1011001100 0 1010000111
9	+ 10101001 0 01011001
8	− 1001111 1 10111111
7	+ 100101 1 011011
6	− 10001 1 100111
5	+ 1000 0 1101
4	− 11 1 1111
3	+ 1 1 11
2	− 1 11
1	+ 1
0	+ 1
−1	+ 1100
−2	+ 1100100
−3	+ 1101010000
−4	+ 1110000010000
−5	+ 1110110111000000
−6	+ 1111101111001000000
−7	+ 1000010101010010000000000
−8	+ 10001101001100000100000000
−9	+ 1001010110000101011000000000000

Fig. 31 First Section

Approximations $= S_{n+1}/S_n$

n		
11		1 8 16 24 32 40
10	−	10000111100011011101111001101110100011111
9	−	10000111100011011101111001101110010101001
8	−	10000111100011011101111001101111001010010
7	−	10000111100011011101111001100000001101001
6	−	10000111100011011101111101101100100111111
4	−	1000011110001101110011 . . .
4	−	100001111000$\dot{1}$11100$\dot{0}$
3	−	10$\dot{0}$00111$\dot{0}$11$\dot{1}$
2	−	0001$\dot{0}$0$\dot{1}$
+1	−	111
0	+	1
−1	+	00001$\dot{0}$$\dot{1}$
2	+	00001111010111 . . . .
3	+	0000111100011000001
4	+	000011110001101111110000000
5	+	00001111000110111011101
6	+	0000111100011011101111010000
7	+	000011110001101110111100100110
8	+	00001111000110111011110011011 1001*
9	+	0000111100011011101111001101110010*1111
		5 10 15 20 25 30

Fig. 31 Second Section

Now, of course, Reader, I don't know how good a mathematician you may be, further than this, that no very brilliant and active-minded mathematician would have been likely to wade so far through obviosities: but I guess it may not be clear to you why this method should give the root of greatest, and that of least, modulus of every algebraic equation having such roots. The explanation, however, is simplicity's self. It rests on three principles that can hardly be unknown to you. The first is, that an equation of any degree, say the *n*th, has *n* roots, of which, however, some

* This figure is a 0 followed by eleven 1s.

may be equal to one another. Now we begin, according to the rule, by adopting arbitrarily n successive substitutes, $S_0, S_1, \dots S_{n-1}$. First, suppose all the roots to be unequal. Denote them by $\alpha, \beta, \gamma, \dots \nu$. Then it would always be possible to satisfy the following n equations, by giving appropriate values to the n unknown quantities, $A, B, \Gamma, \Delta, \dots N$; namely, the equations may be written thus:

$$\begin{array}{llllll} A & +B & +\Gamma & +\Delta & +\dots N & = S_0 \\ A\alpha & +B\beta & +\Gamma\gamma & +\Delta\delta & +\dots N\nu & = S_1 \\ A\alpha^2 & +B\beta^2 & +\Gamma\gamma^2 & +\Delta\delta^2 & +\dots N\nu^2 & = S_2 \\ \dots & & & & & \\ A\alpha^{n-1} & +B\beta^{n-1} & +\Gamma\gamma^{n-1} & +\Delta\delta^{n-1} & +\dots N\nu^{n-1} & = S_{n-1} \end{array}$$

Now according to the rule, r being any number, from n consecutive values of the Ss, we calculate one more, either S_r or S_{r+n}, from the equation

$$a_0 S_{r+n} + a_1 S_{r+n-1} + a_2 S_{r+n-2} + \dots a_n S_r = 0,$$

where the coefficients $a_0, a_1, \dots a_n$, are those of the equation to be solved,

$$a_0 x^n + a_1 x^{n-1} + a_2 x^{n-2} + \dots a_n x^0 = 0.$$

The second principle is that every root of this equation satisfies every equation of the form

$$a_0 x^{r+n} + a_1 x^{r+n-1} + \dots a_n x^r = 0,$$

whether r be a positive or negative integer.

It follows that, — waiving for a moment the question of whether the equations of the following array of equations be correct or not, — if they are correct then the sum of them will be zero, because the sum of corresponding terms, that is, the sum of all terms in one vertical column, will be zero.

$$\begin{array}{lllll} a_0 A\alpha^{r+n} & +a_0 B\beta^{r+n} & +a_0 \Gamma\gamma^{r+n} & +\dots a_0 N\nu^{r+n} & = a_0 S_{r+n} \\ a_1 A\alpha^{r+n-1} & +a_1 B\beta^{r+n-1} & +a_1 \Gamma\gamma^{r+n-1} & +\dots a_1 N\nu^{r+n-1} & = a_1 S_{r+n-1} \\ a_2 A\alpha^{r+n-2} & +a_2 B\beta^{r+n-2} & +a_2 \Gamma\gamma^{r+n-2} & +\dots a_2 N\nu^{r+n-2} & = a_2 S_{r+n-2} \\ \dots & & & & \\ \dots & & & & \\ a_n A\alpha^r & +a_n B\beta^r & +a_n \Gamma\gamma^r & +\dots a_n N\nu^r & = a_n S_r \end{array}$$

and, therefore, not only will the first n consecutive Ss be so related to α, β, γ etc., but if the Ss are calculated according to the above equation, all subsequently calculated Ss will have the same constitution. That is, if n of these equations hold good, all hold good.

The third principle, being complex, might fairly be reckoned as three distinct principles, but it is more conveniently stated without that dissection. It is that, in the first place, the modulus of any power of a quantity is the same power of the modulus of the quantity that is raised to that power; that, in the second place a modulus greater than unity raised to a sufficiently high positive power becomes indefinitely great, and raised to a sufficiently high negative power becomes indefinitely small, while for a modulus less than unity it is just the other way; and that, in the third place, the addition to the finite numerator and denominator of sufficiently small addends (that is addends whose moduli are sufficiently small as compared with the augends to which they are added) alters the value of the fraction indefinitely little, so that as the addend successively approximates toward zero, the value of the fraction approximates indefinitely toward the quotient of the fixed augend numerator divided by the fixed augend divisor.

Consequently, supposing the root of greatest modulus to be α and that of smallest modulus to be ν, and employing p and $-m$ to denote integers both of considerable absolute values, but p having the *plus* and $-m$ the *minus* signs, we shall have

$$\frac{S_{p+1}}{S_p} = \frac{A\alpha^{p+1} + B\beta^{p+1} + \Gamma\gamma^{p+1} + \ldots + M\mu^{p+1} + N\nu^{p+1}}{A\alpha^{p} + B\beta^{p} + \Gamma\gamma^{p} + \ldots + M\mu^{p} + N\nu^{p}} =$$

$$\alpha\frac{1+(B/A)(\beta/\alpha)^{p+1}+(\Gamma/A)(\gamma/\alpha)^{p+1}+\ldots+(M/A)(\mu/\alpha)^{p+1}+(N/A)(\nu/\alpha)^{p+1}}{1+(B/A)(\beta/\alpha)^{p}+(\Gamma/A)(\gamma/\alpha)^{p}+\ldots+(M/A)(\mu/\alpha)^{p}+(N/A)(\nu/\alpha)^{p}}$$

and by taking p large enough all the terms of the denominator after the first and *a fortiori* those of the numerator will become inconsiderable and the approximation to α will be indefinitely close; while

$$\frac{S_m}{S_{m-1}} = \frac{A\alpha^{-m} + B\beta^{-m} + \Gamma\gamma^{-m} + \ldots + M\mu^{-m} + N\nu^{-m}}{A\alpha^{-m-1} + B\beta^{-m-1} + \Gamma\gamma^{-m-1} + \ldots + M\mu^{-m-1} + N\nu^{-m-1}} \text{ [becomes]}$$

$$\frac{S_m}{S_{m-1}} = \nu\frac{(A/N)(\nu/\alpha)^m + (B/N)(\nu/\beta)^m + \ldots + (M/N)(\nu/\mu)^m + 1}{(A/N)(\nu/\alpha)^{m+1} + (B/N)(\nu/\beta)^{m+1} + \ldots + (M/N)(\nu/\mu)^{m+1} + 1}$$

which when m is sufficiently great will differ inconsiderably from ν.

B. SOME AMAZEMENTS OF MATHEMATICS (202)

I desire to make two contributions to your "Mazes of mathematics." The first of them consists in some commonplace, yet pertinent, remarks concerning a curiosity which has been noticed in your pages. The second contribution communicates a fact which has surprised every person, including some very eminent mathematicians, to whom I have shown it, and will, I am sure, interest every intelligent reader who shall actually go through the operations I shall describe. But having described the surprising fact, I shall go on to explain it upon well-known and evident principles.

The first of my contributions is suggested by your revival of interest in the number 142857, whose peculiarity is that it is the only number, written in the usual way, of which every multiple, $M \times N$, where N is this number and M is an integer multiplier, is written with the same figures in the same cyclical order, so long as M does not exceed the number of figures with which the number is written. Thus,

$$\begin{array}{rclcrcl}
142857 &=& N & \qquad & N &=& 142857 \\
714285 &=& 5N & & 2N &=& 285714 \\
571428 &=& 4N & & 3N &=& 428571 \\
857142 &=& 6N & & 4N &=& 571428 \\
285714 &=& 2N & & 5N &=& 714285 \\
428571 &=& 3N & & 6N &=& 857142
\end{array}$$

I propose to explain why this is. In the first place, it is obvious that the same thing could not be true of any number, N, which is written with more than nine figures, since in that case one of the multiples, $M \times N$, would be $10N$. Now this is written with more figures than N, unless we allowed so far to depart from the usual way of writing N as to place one zero or more to the left of the significant figures. If this be allowed, there are indefinitely many numbers having the same property. Thus, take this number of 18 figures:

052631578947368421 = N
105263157894736842 = $2N$
210526315789473684 = $4N$
421052631578947368 = $8N$
842105263157894736 = $16N$
684210526315789473 = $13N$
368421052631578947 = $7N$
736832105263157894 = $14N$
473684210526315789 = $9N$
947368421052631578 = $18N$
894736842105263157 = $17N$
789473684210526315 = $15N$
578947368421052631 = $11N$
157894736842105263 = $3N$
315789473684210526 = $6N$
631578947368421052 = $12N$
263157894736842105 = $5N$
526315789473684210 = $10N$

Let us ask what sort of numbers they are which have this property. For the benefit of any reader who may wish to pursue this inquiry retroductively, I will set down two good large numbers of this kind. One is (in so-called "English" numeration)

002724	sexagintillions,	795640	undesexagintillions,
326975	duodesexagintillions,	476839	quinquagintaseptillions,
237057	quinquagintasextillions,	220708	quinquagintaquintillions,
446866	quinquagintaquadrillions,	485013	quinquagintatrillions,
623978	quinquagintabillions,	201634	quinquagintamillions,
877384	quinquagintillions,	196185	undequinquagintillions,
286103	duodequinquagintillions,	542234	quadragintaseptillions,
332425	quadragintasextillions,	068119	quadragintaquintillions,
891008	quadragintaquadrillions,	174386	quadragintatrillions,
920980	quadragintabillions,	926430	quadragintamillions,
517711	quadragintillions,	171662	undequadragintillions,
125340	duodequadragintillions,	599455	trigintaseptillions,
040871	trigintasextillions,	934604	trigintaquintillions,
904632	trigintaquadrillions,	152588	trigintatrillions,
555858	trigintabillions,	310626	trigintamillions,
702997	trigintillions,	275204	undetrigintillions,
359673	duodetrigintillions,	024523	vigintaseptillions,

160762	vigintasextillions,	942779	vigintaquintillions,
291553	vigintaquadrillions,	133514	vigintatrillions,
986376	vigintabillions,	021798	vigintamillions,
365122	vigintillions,	615803	undevigintillions,
814713	duodevigintillions,	896457	septendecillions,
765667	sedecillions,	574931	quindecillions,
880108	quattuordecillions,	991825	tredecillions,
613079	duodecillions,	019073	undecillions,
569482	decillions,	288828	nonillions,
337874	octillions,	659400	septillions,
544959	septillions,	128065	quintillions,
395095	quadrillions,	367847	trillions,
411444	billions,	141689	millions,
373297.			

The accuracy of the above has been checked by doubling it and comparing the figures. No error was found except one due to miscopying the originally obtained result. The mode of calculation is explained below.

The following, which is in the sexagintatrillions, has not been checked in any way:

002638,	522427,	440623,	245382,	585751,	978891,
820580,	474934,	036939,	313984,	168865,	435356,
200527,	704485,	488126,	649076,	517150,	395778,
364116,	094986,	807387,	862796,	833773,	087071,
240105,	540897,	097625,	329815,	303430,	079155,
672823,	218997,	361477,	572559,	376754,	617414,
248021,	108179,	419525,	065963,	060686,	015831,
134564,	643799,	472295,	514511,	873350,	923482,
849604,	221635,	883905,	013192,	612137,	203166,
226912,	928759,	894459,	102902,	374670,	184696,
569920,	844327,	176781.			

I now return to the task of showing what sort of numbers they are which possess this property; and in doing so, I will not confine myself to decimal arithmetic but will suppose that the base of numeration, B, has for its prime factors, the prime numbers, α, β, γ, etc. raised, respectively, to the powers a, b, c, etc.; so that the base, $B = \alpha^a \cdot \beta^b \cdot \gamma^c \cdot$ etc. Then no fraction, $\frac{N}{D}$ reduced to its lowest terms whose denominator has any prime factor, μ, different from α, β, γ etc.

can be exactly expressed in that system of numerical notation. For if it could, this fraction multiplied by a suitable power of B (which multiplication would be effected by simply shifting the point that separates the units' place from the fractional places) would give a whole number as the product; and thus a number not containing μ as a factor would be divisible by μ, contrary to the elementary properties of a prime number. Moreover, if D_1 and D_2, the denominators of two fractions reduced to their lowest terms, and neither of them contain any of the prime factors of B, their sum and difference will be $\frac{N_1}{D_1} \pm \frac{N_2}{D_2} = \frac{N_1 \cdot D_2 \pm N_2 \cdot D_1}{D_1 \cdot D_2}$, which will be incapable of exact expression in the system of notation whose base is B.

It is evident, — far more so than is the meaning of the words and phrases that must be used to express the fact in its generality, — that, given a system of notation whose base is B, by using only one place to the right of the point that separates units from fractions we can express any quantity whatever within a maximum possible error of $\frac{1}{B}$; that by using two places, we can express it with a maximum possible error of $\frac{1}{B^2}$, and so on. Granted that one could use an endless series of places, as one virtually can when one is able to form an exact *general description* of the succession of figures, and the limit of possible error is $\frac{1}{B^\infty}$, which is the "infinitesimal" of Leibniz. Until the revelations of Dr. Georg Cantor's genius, mathematicians regarded infinity as absolutely inconceivable, just as Euclid regarded an unlimited line; and even at this day there linger obsolete mathematicians who cling to their incapacity to conceive the infinite, just as that traditional old lady, who was certainly remarkably well grounded in orthodox and catholic theology, clung to her total depravity, saying (with absolute theological accuracy), "Take away my total depravity, and you take away my salvation, and I revert to being 'a little lower than the angels,' instead of being higher, as I am." But they are wrong: endlessness, ∞, is perfectly conceivable; and $+\infty$ and $-\infty$ differ endlessly. So the endlessly little, $\frac{1}{+\infty}$, differs from $\frac{1}{-\infty}$. But when any considerable quantity, B, is raised to a negatively endless power, so as to give $B^{-\infty}$, the result is something far less than the mere endlessly little, and is the true infinitesimal.

I acknowledge that I am straying from my subject when I drag in this consideration; but I do not see any alternative except that of conveying ideas which I know to be inaccurate, which, when one professes to be setting forth theoretical truth, comes infinitesimally near to lying.

Let B be the base of numeration (or of numerical notation, rather, if one cares to draw the distinction); and let b be the number one less than B. Thus in vulgar arithmetic, B is 10; so that b will be 9. Then the fogey mathematicians, pressing their darling agnosticism to their breasts, declare that

$$0 \cdot bbbbbb \textit{ et cetera, ad infinitum} = 1.$$

It is not so, although the difference, being infinitesimal, is less than any number can express. The difference exists all the same, and sometimes takes a quite easily intelligible form. Thus, in the ordinary development of the doctrine of chances, 1 signifies absolute certainty, or the probability of an event which, according to the hypothesis of the problem, logically must and will occur, under all logically conceivable circumstances. Now suppose a man tosses up a coin with the privilege in case it turns up 'tails' of trying again, no matter how often he may have tossed it up already. What is the probability that he will eventually turn up a 'head'? We may conveniently employ the secundal system of numerical notation, in which 10 denotes two, 0.1 one half, .01 one quarter, .001 one eighth, and so on indefinitely. Then, the probability that his first throw will be 'heads' will be 0.1 or one half. If not, he will throw again; and the probability that his first throw will be 'tails' and his second 'heads' will be 0.01 or one fourth. This added to the other result will give 0.11 as the probability that he will throw 'heads' in one or other (or both) of his first two throws. By similar reasoning it would be shown that the probability that he would throw at least one 'head' in his first three throws is 0.111; that he would do so in his first four throws, would be 0.1111; and so on indefinitely. Consequently, the probability that he would some time or other throw a 'head' will be 0.1111111 *et cetera ad infinitum*. But this is not the certainty that 1 expresses; for the conditions of the problem leave open the possibility that the turning up of 'heads,' which may occur at any throw, independently of what occurs at other throws, should occur at all throws The endless series of figures is not, therefore, strictly equal to 1, but is only so close an approximation that no number can measure its error We may say that it is *equivalent* (or more explicitly, *finitely equivalent*) to one, and may use the sign $\doteq$ to signify this 'equivalence,' writing

$$0.11111111 \text{ et cetera ad infinitum} \doteq 1 \quad (\text{Base} = \text{two})$$

In general, if though a quantity be inexpressible in a system with a given base, yet if we can find an *exact general description* of the succession of figures, we can, by that description, express the finite equivalent of the quantity in the given system

Let us seek that description for the different fractions less than 1 that have for their common denominator, P, a prime number that is not a divisor of B. Let the equivalent of $\frac{1}{P}$ in the given system be

$$\frac{1}{P} \doteq 0 \cdot abcde \text{ etc.},$$

where, a, b, c, d, e, etc. are numbers less than B. Multiplying this by B, we get $\frac{B}{P} \doteq a \cdot bcde$ etc. and $\frac{B - aP}{P} = 0 \cdot bcde$ etc. Let us save words by calling this operation of shifting the fractional point (the dot between the units place and the $\left(\frac{1}{B}\right)$s place) one place to the right and then dropping the figure to the left of that point one *diabasis*. It is evident that a diabasis will always alter the value of an expression, unless $a = b = c = d = e =$ etc. So will two diabases, unless $a = c = e =$ etc. and $b = d =$ etc. So will three, unless $a = d =$ etc., $b = e =$ etc. etc. And so on. Moreover, from the above expression, $\frac{B - aP}{P}$, or the result of the diabasis, it is plain that P being a prime that does not divide B, the diabasis will not alter the denominator, but only the numerator of the fraction represented. Thus, taking $B = 10$ and $P = 13$, the first diabasis will change $\frac{1}{13}$ to $\frac{10}{13}$; and since the first nine multiples of 13 are those shown in the accompanying little table, the second diabasis will change $\frac{10}{13}$ to $\frac{100-91}{13} = \frac{9}{13}$;

Multiples of 13

1	13
2	26
3	39
4	52
5	65
6	78
7	91
8	104
9	117

the third will change this to $\frac{12}{13}$; the fourth will give $\frac{3}{13}$; the fifth $\frac{4}{13}$; and the sixth will restore $\frac{1}{13}$. It follows that the decimal equivalent of $\frac{1}{13}$ is a circulating decimal, and that, in general, the B-expression for $\frac{1}{P}$ is a circulating expression.

Let that diabasis which first restores the expression for $\frac{1}{P}$ be the qth diabasis. It may be that none of the diabases of $\frac{1}{P}$ will give the equivalent of $\frac{2}{P}$. In that case this latter must be obtained by simply doubling the equivalent of $\frac{1}{P}$, so that the circulating expressions must contain the same number, q, of figures. An easy modification of this reasoning will show that the equivalents of all the fractions whose denominator is P and whose numerators are less than P must consist of circulating expressions of the same number of figures. All these must be different, and there are $P - 1$ of them in all. Consequently, the value of q must always be some aliquot part of $P - 1$. In other words $P - 1$ diabases must restore the value of every fraction in P upon which they successively operate. In particular, whatever the value of B, $B^{P-1}\frac{1}{P} = W + \frac{1}{P}$, where W is some whole number. Multiplying this equation by P, we get

$$B^{P-1} = WP + 1,$$

which is the important proposition, named after its great discoverer, *Fermat's theorem*. It would be easy (were it of any present concern) to prove from this that if D be any number indivisible by every divisor (except 1) of B, and if TD denote the 'totient' of D, that is to say, the number of numbers less than D of which none is divisible by any divisor (except 1) of D, then

$$B^{TD} = W \cdot D + 1,$$

which is called "the extended Fermat's theorem."

My demonstration has been constructed so that its concepts should be as germane as possible to the subject under discussion. Two excellent proofs are found in the common books. For one of them the reader is referred to Cayley's article "Numbers, Theory of" in the 9th Edition of the *Encyclopedia Britannica*. The other is given in Dedekind's redaction of Lejeune Dirichlet's "Vorlesungen über Zahlentheorie" §20, p. 41. I will restate this.

The polynominal theorem will be most intelligibly stated by first expressing it in symbols and then explaining these symbols. It is that

$$(\Sigma_b b)^n = \Sigma_b b^n + n!\,\Sigma_{(\Sigma\beta=n)}\Pi_b \frac{b^\beta}{\beta!}$$

Any number followed by a note of admiration, as $m!$, denotes the continued product of all the integer numbers from 1 up to and including the number written before the note of admiration; and $0! = 1$. In the formula $\Sigma_b b$ denotes the sum of all the bs, the bs being any collection of numbers, equal or unequal, or some equal and others of different values. $(\Sigma_b b)^n$ denotes the nth power of this sum. $\Sigma_b b^n$ denotes the sum of the nth powers of all these numbers. $\Sigma_{(\Sigma\beta=n)}$ means that a sum is to be formed *in each term of which* certain numbers called the βs are to be attached, one to each b, no β being negative and none being as great as n; and the different terms are to exhaust all possible ways of assigning a β to each b, number the condition that the sum of the βs in each term must be equal to n. How each term is composed is shown by $\Pi_b \frac{b^\beta}{\beta!}$. That is, each term is the product of the different bs each raised to a power whose exponent is β and this power is to be divided by $\beta!$ An example will render the statement clearer. $(a+b+c+d)^5$ is equal to $a^5+b^5+c^5+d^5+120$ times the sum of 52 terms of which the following are specimens: $\frac{cd^4}{1\cdot 24}, \frac{c^2d^3}{2\cdot 6}, \frac{bcd^3}{1\cdot 1\cdot 6}, \frac{b^2cd^2}{2\cdot 1\cdot 2}, \frac{a^2bcd}{2\cdot 1\cdot 1\cdot 1}$. Since no β is as great as n, it follows that if n is a prime number, every term excepting those that contain a single b will be divisible by n. Make every b equal to 1, and let B be the number of them; and let p be a prime number. Then the theorem will be $B^p = B +$ terms divisible by p. If B is not divisible by p, we can divide by B, and get $B^{p-1} = 1 + pW$.

In order to illustrate this, I will give the 11th power of each of the first ten numbers, and subtracting from this the number itself will show show that the sum of the figures in the odd places differs from the sum of those in the even places, if at all, by a multiple of 11, which is the usual test of a number being divisible by 11.

	Odd	*Even*	
	0		
$10^{11} = 100000000000$	9	9	$11^{11}-11 = 11\times(11^{10}-1)$.
	9	9	
$10^{11}-10 = 99999999990$	9	9	
	9	9	
	9	9	
	45	− 45 = 0	

It has thus been amply proved that, p being prime, and B a number not divisible by p, $B^{p-1}-1 = Wp$, where W is some whole number. Thus, taking $B = 10$ and $p = 7$, we have $10^6 = 1000000$ and $10^6-1 = 999999$, which must be divisible by 7. Now since, as nearly as numbers can express quantities, $1 \doteq 0.999999999999$ etc. *ad infinitum*, it follows that since $\frac{1}{7} \doteq 0.142857142857$ etc. *ad infinitum*, the whole number 142857 must divide 999999 without a remainder. To show that it not only *must be* so, but *is* so, let us perform the division, thus:

```
7)999999
  7 ... 1
  --
  29
  28 ... 4
  --
   19
   14 ... 2
   --
    59
    56 ... 8
    --
     39
     35 ... 5
     --
      49
      49 ... 7
      --
       0
```

The reason of the property of this number is thus apparent. For since

$$\tfrac{2}{7} \doteq 0{\cdot}\dot{2}8571\dot{4}$$

$$\tfrac{3}{7} \doteq 0{\cdot}\dot{4}2857\dot{1}$$

etc.

it follows that the whole number

$$285714 = \tfrac{2}{7} \times 999999$$
$$428571 = \tfrac{3}{7} \times 999999$$

etc.

We now naturally ask this question: Will that whole number which is expressed by the first $p-1$ figures of the decimal value of $1/p$ have these same properties *whatever prime number p may be*? This must be answered in the negative, for although B^{p-1} will in every case exceed by 1 some multiple of p, yet the same may be true of some lower power of B, whose exponent is some aliquot part of $p-1$. Thus, taking $B = 10$ and $p = 3$, not only is $10^{3-1}-1$, or 99, divisible by 3 but so is $10^{(1/2)(3-1)}-1$, or 9, equally so divisible. In the case of $p = 37$, not only does 10^{36} exceed by 1 a multiple of 37, but so does $10^{(1/12)(37-1)} = 10^3$. For $27 \times 37 = 999$. Indeed, the majority of primes fail to possess the properties of 7 for the same reason; and the smallest numbers, after 142857, which have the peculiarities of this number are

0588235294117647
052631578947368421
0434782608695652173913
0344827586206896551724137931

Such numbers are simply periods of circulating decimals, and are to be calculated by division. If the period of the decimal expression of $1/p$ turns out to consist of $p-1$ figures, and not of an aliquot part of this number, it has the property in question. But when the entire circulating expression of the quotient is desired, the following rule for division is far more expeditious, facile, and free from danger of error, than any other with which I am acquainted, although it is entirely unsuited to ordinary calculations. I shall give the rule so that it will apply to any base, B, of numerical notation.

Rule of Direct Division. To divide one number, N, by another, D, *and obtain the entire circulating expression for the quotient*, first find the greatest common divisor, G, of N and D, divide each of these numbers by G, and use the quotients, $N' = N/G$, as a dividend instead of N, and $D' = D/G$, as a divisor instead of D. Next find the greatest common divisor, G', of D' and B (the base of numeration), and divide both D' and N' by this, using the quotient $D'' = D'/G'$ as the divisor in place of D', and the quotient $N'' = N'/G'$ (which cannot involve a circulating fraction) as the divident in place of N'.

The next step depends upon the facility with which you can perform multiplication. If you cannot instantly multiply two figures by two with accuracy, you are to inspect the multiplication table and find what number less than B, multiplying the last figure of D'', will give a product whose last figure is $B-1$, and you are to multiply D'' and N'' by this to obtain a new divisor and a new dividend. But if you are dealing with ordinary decimal numbers, and have at hand Dr. A. L. Crelle's "Rechentafeln," which is a multiplication-table of three figures by three, find what number multiplying D'' will give a product ending with the figures 999, and multiply by it both D'' and N'' to obtain a new divisor and dividend. Strike off the $B-1$ or $(B-1)$s from the end of the expression for the new divisor, and treating the last place that remains as the units' place, add 1 to this fragment, and call the sum the "current multiplier." Write the new dividend in the upper right hand corner of a sheet of paper, and separate from the end of it, by a "*line* of separation," as many figures as you have struck off from the end of the new divisor and this figure or these figures, so separated from the dividend, will be the last figures of the circulating expression. Multiply the figure or figures of the quotient at the right of the line of separation by the current multiplier and set down the products so that its last figure comes under the last figure of the part of the dividend to the left of the line of separation, to the whole of which this product is to be added. Cut off from the end of the sum, by a new line of separation, as many figures as before. They will be the figures of the quotient next preceding the figures already obtained. Multiply these by the current multiplier and set down the product so that its last figure shall come under the last of the figures at the left of the last line of separation. Proceed in this way until you perceive that if you were to go further you would only reproduce the circulating expression already obtained. You then have the whole quotient.

An abridgment may be effected when the last figure or figures of a sum is a zero, or are zeros. You may then leave on the right of your line of separation an additional figure for each zero, and in the multiplication are to neglect the zeros. I will give two examples.

Example 1. To find in decimals the value of $\frac{4}{13}$. Multiplying numerator and denominator by 3, the fraction becomes $\frac{12}{39}$. The current multiplier is 4. The work is shown in Fig. 1. But if you are ready at multiplying two figures by two, you may multiply numerator and denominator by 23, giving $\frac{4}{13} = \frac{92}{299}$. Then the current multiplier will be 3 and the work

Current
Multiplier = 4

```
        1|2
         8
       ---
     3|6|9
    24|
   ----
    2|7
   28|
 12|30
```

Answer = $0.\dot{3}0769\dot{2}$

Fig. 1

Current
Multiplier = 3

```
      2|76|92
     228|
    ----
   69|230
```

Answer = $0.\dot{3}0769\dot{2}$

Fig. 2

will proceed as in Fig. 2. In the upper right hand corner you set down 92. Separate the whole. Multiplying by the current multiplier 3 gives 276 which you set down. Separate the last two figures, multiply them by the current multiplier 3, giving 228. Set down this so that the 8 comes next to the left of the line of separation. Adding 228 to the 2 you get 230. Owing to the zero, you separate the whole of this. Multiplying the 23 by the current multiplier you get 69 which you set down. But you have evidently been running into a second period, and the answer is as before.

Example 2. Express 1/706 in decimals. The greatest common divisor of the base 10 and the divisor 706 is two. Dividing numerator and denominator by this, you have $1/706 = \frac{0.5}{353}$. Calculate $\frac{5}{353}$ and then shift the decimal point. Crelle tells you that 353 × 983 = 346999. Multiplying the numerator, 5, by 983, you find the product to be 4915. Fig. 3 shows the work.

```
Current Multiplier = 347                                   4|915
                                                     317 505|
                                                 17 6623|509
                                             240 818|6940
                                         289 745|835
                                     341 795|985
                                           1|084
                                     2 9148|
                                 329 303|9490
                             105 835|305
                                   1|164
                              56 908|
                                   1|014
                               4 858
                                   |915
```

Answer: $0.0\dot{0}141643059490084985835694050991\dot{5}$

Fig. 3

SECOND COMMUNICATION

If, in my anxiety to make my first communication readily intelligible, I have made it pretty dry, I am sure the reader will be interested in my second, *provided he does not content himself with simply reading it, but actually goes through the operations described.*

A pack of 52 cards is required, — a good smoothly running pack of American cards. I prefer the little Fauntleroy cards made by the "U.S. Playing Card Co." You pick out all the cards of a red suit and all the cards but the king of a black suit. Using the abbreviations *X* for ten, *J* for knave, *Q* for queen, and *K* for king, you arrange the cards of the red suit in a pack in the following order from the back to the face of the pack:

Back. 1 2 3 4 5 6 7 8 9 *X J Q K Face*.

You arrange the cards of the black suit in a pack in the same order, except that there will be no king.

I shall repeatedly speak of the cards of a little pack being dealt out into a given number of piles. Such dealing is to be performed as follows. You hold the pack in your hand, back up, and deal them out one by one, turning each one face up. The first card will form the bottom and back of the first pile; the second of the second pile, which is to be placed at the right of the first. When you have placed a card to form the bottom of the last pile, you place the next on the first pile; and so you proceed to deal them upon the piles in rotation.

You place the pack of red cards on the table, face down. You then deal the black cards into two piles, except that when you come to the last card, instead of placing it on the top of the second pile, you put it by itself on the table, face up, to form the bottom card of a separate pile, which will have nothing to do with the others; and where that card would naturally be put, at the top of the second pile, you substitute the top card of the pack of red cards, which you turn face up in putting it on that second pile. You now clutch up the first pile in your hand and put it on the top of the second pile; and then you take up the whole of the pile thus formed of the first and second piles, and turning it over, back up you proceed to deal it again into two piles just as before,except that, as before, the last card, instead of being put at the top of the second pile is to be put, face up, on that separate pile, of which you have already laid down the first card. On the second of the two piles resulting from the dealing, you put, as before, the top card of the pack of red cards. It will be the 2. You take up the piles just as before, and continue the same operations until, when you have performed twelve dealings you will have only red cards in your hand in place of black ones. You will then place the red king on the face of this pack of red cards, and facing the same way as the rest of them.

You will now take up the separate pile, consisting of all the twelve black cards, and turning it over in your hand back up, you will range these cards, one by one from left to right, in a row on the table (which should be a pretty large one), all with their backs up. You may take a peek at the seventh card to see that it is the ace, as it should be; *and in all the subsequent operations you must bear well in mind the place of the ace counting from the left-hand end of the row.* You now inform anybody who is looking on that you will have no right to do anything more with those cards of the row except to shift a certain number, one by one, from one end of the row to the other.

You now take up the red cards, backs up, and request a person to cut the pack. As usual you place the part he leaves on the part he removes.

You then ask him into how many piles he will have these red cards dealt, and you will deal them out accordingly. If any other persons would like to cut the pack and to say how many piles it should be dealt into, you may allow it. Only each time that the cards have been dealt out, you must gather the piles together in the following way, without error. You first clutch up the pile on which the last card was placed. You will then note how many piles there are to the right of that pile. If there is only one you place the pile in your hand on the first pile to the left and clutch up the two together; and so on always taking the first pile to the left of the one last taken, and taking the right-hand-most pile last, so that it comes at the back. But if when you have taken up the pile on which the last card was dealt, you find there are two to the right of its place, you must put the pile taken on the pile in the second place to the left of it, and always skip a place. For you must count the original places, and not the piles, that are left. This will happen when there are five piles, a number often desired. They will, according to the rule be taken up in this order

24135

If the last card falls on the first, or left hand-most, pile, which most frequently happens, you simply take up the piles in regular succession from left to right. If you have dealt the cards into more than 8 packs, an easier rule for gathering them up will be to note how many piles are to the *left* of the pile on which the last card was placed, and then, going round toward the *right*, *skip* that number of piles. In all cases, the pile at the extreme right is to be gathered last.

When the dealing into piles is done, and the cards are again gathered up, you will explain that the black cards are to form the key to the places of the red ones; but that since there is no king among the black cards, it will be necessary to bring the king to the face of the red pack. You will, therefore, cut the red pack, so as to bring about this. To do so you will necessarily turn the red cards face up, which will give you an opportunity slyly to see what card is next to the king toward the face of the pack, if indeed you have not already noticed what card, in the last deal, was laid directly upon the king. If the king should be at the face of th peack, you will have to look at the card at the back of the pack. The number of this card will be the number from the left hand of the row of black cards where the ace should be. You are to make it so by carrying the requisite number of cards from one end of the row to the other. Thus, if the card is a nine, remembering that the ace was originally the seventh card from the left, you carry two cards from the right hand end to the left hand end.

You now ask one of the company what red card he would like to find. Whatever number he mentions, you count along the row from the left aloud, putting your finger on each card as you count until you come to the number desired. You turn up that card, and its number will be the number of the card desired from the back of the red pack. You hold this pack back up and pass card after card from the top from one hand to the other (without deranging their order) until you come to the number indicated when you will turn up the card and show that it is the one sought. The company will probably desire to see the thing done a second time; and if you have not deranged the order of the cards you can go on and do it as many times as you like. Only you must always remember where the black ace is in the row.

The same thing can be done with two complete packs of cards, one of them being provided with a joker. But such long dealings are wearisome. Instead of putting fifty-two cards in a row you can make four partial packs of 13 each; and in shifting the same number of cards must be shifted from each pack to the next. Very likely, you may yourself be puzzled to know just why the operation comes out as it does. In the first place if you take any prime number of cards in consecutive order, and deal them out and gather them up according to the above rule, let us see what the effect will be.

The number of piles plainly cannot be greater than the number of cards in the pack; and the first, or bottom, cards of the successive piles will be the cards which before the dealing occupied the places whose ordinal numbers from the back of the pack were the successive numbers from 1 up to the number of piles, which number we will call H (the initial of 'heap,' since P is generally used to denote a prime number). If H is less than the total number of cards, one at least of the piles will receive a second card, and this second card will be the one whose previous ordinal place was greater by H than that of the first card of the same pile. And if the whole number of cards is greater than a given multiple of H, say mH, then one or more piles will contain more than m cards; and the original place-number of each will be greater by H than that of the card that precedes it in the same pile, that of the mth in the pile being $(m-1)H+$ the ordinal number of the pile. So things will be in any one pile. It cannot be so after the cards are gathered, since then next after the card at the top of one pile must come the bottom card of another pile, whose ordinal place before the deal was evidently a less number than that of the card at the top of any pile. Under these circumstances, the natural, if not the only, way of securing regularity is to count round and round, the number in the cycle being the total number of cards. Thus, if this number is 13, we count round as on this cycle of numbers. [See Fig. 4.] The total number of cards is supposed to be prime; let them

Fig. 4

be dealt into any number, H, of piles. Let the total number, P,of cards be equal to mH-g, where g must be prime to H, that is, must have no integer divisor greater than 1 that is a divisor of H. For otherwise, if, for example $g = sd$ and $H = td$, where s, t, and d are integers, and the last greater than 1, then $P = mH - g = mtd - sd = (mt - s)d$; so that P, instead of being prime, would be a multiple of d. The first $H - g$ piles will contain m cards each; the remaining g piles will contain $m - 1$ cards each. After the piles are gathered up, the last pile, pile number H, will be the first from the back of the pack; and its first card will be the one which before the deal was the Hth from the back of the pack. Since the ordinal number (from the back) before the deal, of each card is H more than that of the card that, after the gathering, immediately precedes it (as one runs them over from back to face), except that in passing from the last of one pile to the first of the next, P must be subtracted to get the ordinal number before the deal, it follows that the card that after the gathering is in the m place, being the bottom card of the pile which now comes second, is the card that before the deal stood number $mH - p = mH - (mH - g) = g$; and this is the ordinal number of that pile, after the deal, but before the gathering. Now g may be either less or greater than $\frac{1}{2}H$. We have seen that it cannot be in any rational proportion to it. If $g < \frac{1}{2}H$, the pile so numbered contains m cards, and the first card of another pile that follows after the gathering, will be m places further toward the face, so that its ordinal number before the deal will be $g+mH-(mH-g) = 2g$; and this will be the ordinal number of that pile while the cards are on the table. But if $g>\frac{1}{2}H$, then the pile that after the gathering comes second, contains only $m-1$ cards, and the first card of the third pile will be the one that before the deal stood as number $g+(m-1)H-(mH-g) = 2g-H$; and this will be the ordinal number of the pile while the cards are on the table. Obviously, it could not be $2g$ since this would be greater than H, the total number of piles. It begins to look as if each successive pile, passing from back to face, after the gathering was the one that while the cards were on the table was g piles further to the right, counting the piles round and round, in a cycle of H units.

Let us now return to the case where $g<\frac{1}{2}H$. Then $2g$ may be either less or greater (it cannot exactly equal) than $H-g$. If $2g<H-g$, or $g<\frac{1}{3}H$, the $2g$ pile contains m cards, and the first card of the next pile, after the gathering, will be m places further toward the face than the first card of the pile that then stands third. Consequently it will be the card that before the deal stood $2g+mH-(mH-g)=3g$. But if g while $<\frac{1}{2}H$ is $>\frac{1}{3}H$, the $2g$ pile will only contain $m-1$ cards, and then the pile that after the gathering follows the $2g$ pile will be $2g+(m-1)H-(mH-g)= 3g-H$. Of course, it could not be $3g$, since $3g>H$.

Now taking again the case in which $g>\frac{1}{2}H$, so that the first card of the pile that stands third after the gathering is $2g-H$, this pile may contain m cards, which will be when $2g-H<H-g$, so that $g<\frac{2}{3}H$, or it may contain only $m-1$ cards, which will be when $g>\frac{2}{3}H$. In the former case, the first card of the pile that comes fourth after the gathering will be m cards in advance of the first card of the third pile in the same order, and it will be the card that before the deal was number $2g-H+mH-(mH-g) = 3g-H$, which will be the number of this pile while the cards are on the table. In the other case, where $g>\frac{2}{3}H$, the first card of that fourth pile will be only $m-1$ cards in advance of the first card of the third pile, and will be the card which before the deal was number $2g-H+(m-1)H-(mH-g) = 3g-2H$ which will be the number of the pile while the cards are on the table. It is now quite plain that each successive pile as the piles stand after the gathering is just g piles to the right of the last while the cards are on the table, counting the H piles round and round. This, of course, could readily be proved in general terms; but I really do not think it worth the space the proof would occupy, though that would not be much. What then would be the place before the deal of the card that after the gathering stands last? The place before the deal of the card that after the gathering stands at any place x is Hx *minus* $(mH-g)$ times the number of times one passes from one pile to another. For the last card $x = mH-g$, and the number of times one passes from one pile to another is $H-1$. Its place before the deal is therefore $(mH-g)H-(mH-g)(H-1) = mH-g = P$. That is to say the card that is last after the gathering is also the very same card that was last before the deal. Now this succession of piles is the same succession that results from the rule first given for picking up the piles. For we pick them up in the reverse order of that of the places which they will occupy, counting from back to face, after they are picked up, since each pile picked up is placed at the *back*, and not at the *face*, of the piles already clutched. The first clause of the rule is that the pile on which the last card was dealt is to be clutched first. But the effect of that injunction is that the last card will come last after the gathering, and its pile will be the last of the piles, going from back to face. The second clause of the rule is that we are to note how many piles there are to the right of the place of the pile topped by the last card. This number is obviously g, and we are told to count round and round to the left and take always the pile in the gth place after the pile we last took. But since we take them up from face to back, this amounts to saying that the order of the gathered piles from back to face is that every one as gathered is on the pile g piles further to the *right* than the preceding one in gathered order from back to face, as we now find it must be, in order to produce regularity. But we have found that in the gathered order the backmost pile is the pile that on the table is the last, that is, the furthest to the right. Does our rule, then, necessarily lead to the result that the last pile taken is the first, from right to left? Of course, it must do so, since we have found that beginning with right-hand-most pile and proceeding in the reverse order makes the last pile the one containing the card that was last before the deal. But *how* is the result necessitated by our rule. Let W denote *some* whole number, but not necessarily always the *same* whole number, but rather the reverse. It may be zero. Counting the piles from right to left, we first take the pile numbered $g+1$. The second pile we take is the one whose ordinal number from right to left is $2g+1-WH$. The third is $3g+1-WH$; and so on. The last is number H in the order of taking, and it must be the pile whose ordinal number from right to left is $Hg+1-WH$. But g is itself a whole number; so that this comes to $1-WH$. Now counting H piles carries you just through the whole series back to the pile from which the counting began. Hence $WH = 0$, its only effect being to prevent the formulae from indicating impossible numbers. Consequently $1-WH = 1$. That is the way the last pile taken must be the right-hand-most.

We have found, then, that if b denotes the ordinal place of any card before dealing into H piles, while a denotes the ordinal place of the same card after gathering up the H piles, then $b = Ha - WP$. Transposing, $b + WP = Ha$. Dividing by H, $a = \frac{b+WP}{H}$. But what operation of counting the cards is signified by the division? The answer is that by adding to any number a suitable multiple of any prime number, you will get a sum that is exactly divisible by any predesignated number H, that you may have named, provided H is neither 0 nor a multiple of P. This must be so, if it be so when 1 is substituted for b. For if $\frac{1+WP}{H} = A$, then $\frac{b+WP}{H} = bA$. Therefore the question is whether, whatever prime P may be, and whatever number H may be, provided it be neither 0 nor WP, there is some multiple AH of H which is 1 more than some multiple WP of P. You can convince yourself that there always is by imagining the circumference of a circle to be divided into P equal parts, the dividing points being numbered in regular succession from 0 up to $P-1$. If H is greater than P, its remainder, R, after division by P, the quotient being Q, may be used instead of $H = PQ + R$. For evidently, if some multiple of R exceeds some multiple of P by 1, the same must be true of the same multiple of H. Imagine, then, that lines are drawn in the circle, each beginning on the circumference where the last ends, each stretching over precisely R (or H, if $H < P$) divisions, and the first beginning at the zero division point. Let this broken line continue until it crosses the zero point. If it then ends at the 1-point, that proves the proposition; but if not, it must end at some point short of the R-point (or the H-point), since all the stretches are of equal length, and the last one stretches over the zero-point. Call the point at which it ends R'. Then $R' < R$ or H; and if the proposition is true of R' it is true of H, since R' is simply a multiple of R or of H diminished by P. Imagine, then, a broken line drawn from 0 to R' and continued exactly like the broken line from 0 to R (or H). If the stretch of this that passes the zero-point ends at the 1-point, it proves the proposition. If it does not, it must end at some point between 0 and R', for the same reason that the former line had to end at some point between 0 and R (or H). Let the point at which it ends be called R''. Then $R'' < R'$. If this process be continued, a length of stretch, $R^{[x]}$, must ultimately be reached such that its broken [line] will end at 1.

For this is no more than to assert that the number of positive whole numbers less than any given whole number is finite, and that no collection that is finite in one order of counting is infinite in another order of counting. But the fact that one of this series of broken lines ends at the 1-point proves that all the others would do so, if continued round and round a sufficient number of times, and thus proves the proposition.

In fact, counting round a cycle of 13 numbers, we shall have
$1/2 = 7$ since $2 \times 7 = 13 + 1$
$1/3 = 9$ since $3 \times 9 = (2 \times 13) + 1$
$1/4 = X$ since $4 \times X = (3 \times 13) + 1$
$1/5 = 8$ since $5 \times 8 = (3 \times 13) + 1$
$1/6 = J$ since $6 \times J = (5 \times 13) + 1$
$1/7 = 2$ as above
$1/8 = 5$ as above
$1/9 = 3$ as above
$1/X = 4$ as above
$1/J = 6$ as above
$1/Q = Q$ since $Q \times Q = (J \times 13) + 1$

It follows that the effect of dealing a pack of a prime number, P, of cards, and then gathering them up according to the rule, is to multiply the ordinal number of each card by a whole number; namely by the whole number $\frac{WP+1}{H}$.

We attain the same result with a pack of $P-1$ cards, by the simple device of adding an *imaginary card* at the face of the pack, always treating this just as if it were real. For example, the first pile to be clutched must be the one that has the imaginary card at the top of it. The reason that this works is that the Pth card (when the total number is the prime, P) will always be the last card (the one at the face of the pack), since, in counting round and round, $P = 0$, and 0 times any [number] is always 0 ...

C. SPECIMENS OF MATHEMATICAL AMAZES (201)

Mazes intricate,
Eccentric, intervolv'd, yet regular
Then most, when most irregular they seem.

Paradise Lost, V. 623.

About 1860, I cooked up a combination of effects of elementary principles of cyclic arithmetic, and ever since, at the end of some sitting at card-play, have occasionally exhibited it in the form of a "trick" (though there is really no trick about the phenomenon), with the uniform result of interesting and surprising all the company, though their mathematical powers might range from the minimum requisite to an altruistic tolerance of cards all the way up to those of some of the greatest mathematicians of the age, who assuredly with a little reflection could have unraveled the rationale of the thing ...

I now return to the subject of cyclical systems. My definition of such a system may very well be thought queer. I venture, on the contrary, to think that it is the simplest and most straightforward of all possible definitions, — that really are definitions, — of this system. I can imagine somebody asking why I do not say simply that it is a system that returns into itself. My answer would be that that conception is a confused, unanalyzed, and indefinite one. I might be asked why I do not say that the numbers succeed one another regularly until they reach or exceed the value of the modulus, when the remainder after subtracting from each the highest possible multiple of the modulus is substituted for the number. I answer that that is a highly artificial, not to say false, way of conceiving the system, which has no break in its law of succession, and has, in its pith and marrow, no connection with any system of endless numeration. My definition introduces the relation of *A*-hood without any definition of it. I only say how it enters into every cyclical system;

and since these properties are introduced into the *definition* of a cyclical system, it follows that *A*-hood, or *A-age*, is only defined as a certain relation that enters, in a described way into some possible collections. But the truth is that, further than that, *A-age* may be any relations. The "numbers" of a cyclical system are not necessarily those vocables which we are accustomed to call numbers. The days of the week, Sunday, Monday, Tuesday, Wednesday, Thursday, Friday, Saturday, are cyclical numbers; and if we choose to take them in their Ptolemaic planetary, or horary order, Monday, Wednesday, Friday, Sunday, Tuesday, Thursday, Saturday, then Monday becomes *A* to Wednesday instead of to Tuesday, and so on. In short *A*-age, or immediate antecedence in a cyclical course, belongs essentially to a given cyclical system.

... not thoroughly grounded in cyclic arithmetic, and that you have before you the keen pleasure of reading the first three *Abschnitte* (indeed, the first hundred pages will be sufficient to unlock the mysteries of my little invention) of Dedekind's lucid and elegant redaction of the unerring Lejeune Dirichlet's "*Vorlesungen über Zahlentheorie.*" Meantime, I can give you a few hints which will enable you, if you like, to work out the explanation for yourself. In the first place, the dealing out into any number, *P*, of piles and regathering as the rule directs a prime number, *C*, of cards will, as you will easily make out, have the effect of making the ordinal number of any card before the dealing equal to the cyclic product of its ordinal number after the dealing multiplied by *P*, the number of piles. ...

In one of my classes of prize stupids, to which I gave less than a dozen lessons, two of my pupils afterward led all the rest of the school in mathematics. I once sent De Morgan a copy of a geometry that was required to be used in the New York public schools, — including, as I was informed, the Free Academy; but of that I am not sure, — in the hopes of getting some of his wit in return; but he was so saddened, shocked, and sickened by the exhibit that I got only lamentations. I myself wrote a book called the Elements of Mathematics to teach algebra and geometry scientifically and together. I still think it was a beautiful book. With much difficulty I persuaded my brother, Prof. J. M. Peirce, who was regarded as a very lucid teacher, to allow me to read him a portion of it. Though he had evidently dreaded the infliction he soon became much interested in it. But the publishers would not take it:

it did not accord with their preconceived ideas. School-book publishers are apt to be infected with the highly erroneous notion that they understand something of the logic of geometry. They think it is a simple matter, quite within their scope. It is not so.

There is, I believe, no other thing taught at school which contributes so much to the happiness and success of life as ciphering. It ought to be well taught in every grammar school. It never is well taught. The principles of the art can hardly be said to be taught at all, unless perhaps in the commercial colleges, which, however narrow and somewhat sordid their spirit may be, seem to be about the only schools which really do teach what they pretend to teach. The others often teach some things that are highly useful, but which they do not profess and hardly mean to teach. The utter ignorance of mathematics manifested in the many pedagogical works I have read, including about sixty of the most esteemed school arithmetics and dozens of works devoted to the treatment of the pedagogy of mathematics, — I mean of the real spirit of mathematics, as well as of the psychological etiology of backwardness in arithmetic and geometry, and of the true logic of the subject, — has struck me as most deplorable. I have more than once induced the master of a school of excellent repute to make up for me a class composed entirely of pupils reported to be utterly incapable of learning any mathematics and who simply detested the subject. ...

This pretty much exhausts the merely arithmetical interest of the secundal notation. From the point of view in which I have led the reader to place himself, it is a curiosity indeed; but is about as idle a curiosity as I know of. Yet as Leibniz viewed it, there is no better illustration of the genius of that very great philosophical mind. To learn something of it, one may consult the admirable work of Muir in which that writer developes the elements of determinants on the historical plan. I note that the more a science is made up of exact thought, the more profitable it becomes to study its historical development.

I think I ought to say a few words more concerning the advantages of the sextal system, which I seriously hold to be so superior to the decimal or any other that were I a radical, by which I understand one of those minds who hold that it is every man's duty to endeavor to improve the condition of humanity in every way, and bestir himself to bring about every condition of society which he is quite sure would be the best, I would certainly agitate in favor of revolutionizing numeration. The argument in favor of radicalism I take to be that since any reform will cost less the earlier it is undertaken, and since the expenditure will cease as soon as the reform is accomplished, while an abuse or bad state of things involves a fixed disadvantage *per* head of the population and *per annum*, it follows that its cost must in the indefinite future indefinitely exceed the cost of the reform. This is a specious argument, but it does not convince me by any means. I grant, at once, that it would not be right to calculate upon the dying out of the human race or otherwise upon the ultimate cessation of the cost of any abuse, in the absence of any definite and sufficient reason to believe that it will cease within a time capable of being estimated. But my objection is that every man finds many more good works any one of which he might probably bring to a successful issue, if he devoted himself to it, than the wildest dreamer could think of sensibly furthering all at once. Man must economize his life. He must select his first task, and not undertake a second until he sees his way to achieving it without unduly interfering with the first. He should begin with the most solemn and mature deliberation upon the present value to him of the different ends, to each of which he should attach a numerical estimate. For everything, eternal felicity included, bears a finite present value. Next he should consider the probabilities of each of these different good ends coming about without any particular effort on his part, the probable cost to him to each one, and the present subtractive value to him of the uncertainties of his various estimations. It will generally be found that those things which he has peculiar facilities for doing are the best worth doing; while undertakings that are unlikely to succeed are least worth doing. Most tasks are of such a nature that the achievement may be followed out further and further almost indefinitely. These will be worth prosecuting as long as the man has great facilities for prosecuting them, and ought to be dropped as soon as another man's increased facilities, or some other circumstances have rendered some other task better worth the labour. For example, in the present case, I can perhaps more clearly present the advantages of sextal numeration than anybody else, my attention having been drawn to it, and also because I am practised in sanely estimating the values of intangible things. But when I have done so much, I shall neither myself engage further in the conceivable reform, nor counsel anybody else with whom I am at present acquainted to undertake anything so quixotical.

Beyond the considerations already adduced, the chief advantages of one base of numeration over another consist in the simplicity with which it expresses multiples, powers, and especially reciprocals of powers of the prime numbers that in human affairs naturally occur most frequently as divisors. Now the study of wide ranges of different classes of facts has led me to the provisional opinion that, but for the influence of the decimal system which fosters division by 2 and by 5, nearly 5/7, or more precisely 0.709 of all divisions by primes would, in the run of practice, [be] divisions by 2; nearly a quarter, or 0.236 would be divisions by 3; about one twenty-first, or 0.047 would be divisions by 5; about a one hundred and fortieth, or 0.007 would be divisions by 7; and about one thousandth would be divisions by higher primes. Let us then compare the expressions of the multiples and powers of these numbers in the sextal and the decimal systems. The multiples are as follows:

Multiples of 2

Decimal		Sextal	
Multiplier	Multiple	Multiplier	Multiple
1	2	1	2
2	4	2	4
3	6	3	10
4	8	4	12
5	10	5	14
6	12	10	20
7	14	11	22
8	16	12	24
9	18	13	30

Multiples of 3

Decimal		Sextal	
Multiplier	Multiple	Multiplier	Multiple
1	3	1	3
2	6	2	10
3	9	3	13
4	12	4	20
5	15	5	23
6	18	10	30
7	21	11	33
8	24	12	40
9	27	13	[43]

Multiples of 5

Decimal		Sextal	
Multiplier	Multiple	Multiplier	Multiple
1	5	1	5
2	10	2	14
3	15	3	23
4	20	4	32
5	25	5	41
6	30	10	50
7	35	11	55
8	40	12	104
9	45	13	113

It will be seen that the decimal expressions of the multiples of 3 follow no other law than that the sum of the digits is either 3, 6, or 0; while the sextal expressions of the analogous multiples of 5 (which is the lowest prime not a factor of the base 6, as 3 is the lowest prime not a factor of the base 10) follow the simpler and more perfect rule that the sum of the digits is 5, that the units decrease cyclically by 1, and that the sixes increase cyclically by 1 except when the sixes of the multiplier increase. ...

One of my inventions is a curious rule for solving numerical algebraical equations. There are several objections to employing it as the regular method of treating such equations. Namely, when the coefficients are large, it involves considerable calculation, and it does not give the answer in decimals, but as a vulgar fraction. But its advantages are, first, that it is very easily remembered, and secondly, that the successive approximations converge toward the truth much more rapidly than in any other way (except in the solution of a quadratic by continued fractions); so that in many cases it is by far the swiftest way.

... The usage of mathematicians, however, since Gauss, is to use what I think a detestably involved way of conceiving of cyclic arithmetic which carries with it an irregular use of the word *modulus*.[1] Legendre and

[1] A manuscript fragment reads as follows:

all the earlier writers on cyclic arithmetic conceived of its numbers as ordinal numbers, and spoke of 18 being *equal* to 1 on a cycle of 17, just as we say that the 1st, 15th, 22nd, and 29th days of August fall on *the same* day of the week; and just as we say that Longitude 270° West is *the very same* longitude as Longitude 90° East. Gauss, however, seems to have entertained the now generally condemned opinion that the pure abstract numbers are not ordinal but cardinal, and as such essentially extend indefinitely. He therefore did not regard 18 and 1 as *the same* number in a cycle of 17, but as two numbers of the same "class"; and all the books since his time are incommoded and burdened, in purely cyclic arithmetic, with the perfectly irrelevant reference to the extension of the number-system beyond the modulus of the cycle. To be consistent, Gauss ought not to have said that places of which the Longitude of one was by the count 90° East while the other was 270° West have *the very same* longitude: but he ought to have said (and for aught I know, he did) that their longitudes are "of the same class." But this is not greater exactitude of expression: it is simply confusion of ideas. The pure abstract numbers are ordinal, and there is no necessity for their increasing *ad infinitum*; they may do so, or they may come round in a finite cycle in which case, letting k (for κύκλος, as is usual) denote the modulus of the cycle $(k-1)+2 = 1$, — *equals*, I say, i.e. *is* the very same place along the line or series of places. Mind, Reader, if you please, that I am by no means denying that there are regions of higher arithmetic in which it is proper to regard any two numbers which are separated by a

---

"CHAPTER I

The arithmetic of modulus two

In higher arithmetic, we speak of numbers being "congruent," and "congruences." A congruence is nothing but an equation relating to a system of values, few in number, and returning into itself in a cycle. For example, two numbers a and b are said to be "congruent for the modulus 10," and we write

$$a \equiv b \pmod{10}$$

if, when divided by 10, they give the same remainder. The two numbers are then equal, in respect to the last figure in the usual way of writing them. Such congruences are just like ordinary equations, as far as addition, subtraction, and multiplication go; but in regard to division, they have some peculiarities. I mean that if

$$a \equiv b \pmod{10}$$
$$c \equiv d \pmod{10},$$

then

$$a+c \equiv b+d \pmod{10}$$
$$a-c \equiv b-d \pmod{10}$$
$$ac \equiv bd \pmod{10} \ldots.\text{"}$$

certain constant interval as "congruent" rather than as *equal.* I know that there are. I only say that in pure cyclic arithmetic, which is from a human point of view (at least) the most important part of higher arithmetic, that mode of conception is a confusing complication. But Gauss was too apt to think that complication of thought involved accuracy of thought; which is far from true. It sometimes means confusion of thought. Whether this or what else was the reason why he did not call *cyclic logarithms* by that eminently correct designation, I cannot tell. In fact, I called them *indices*, a general and colorless word, but no doubt correct so far as it carries any signification. While I am thus sharply dissenting from the notions of a very great man in a manner which mathematicians are silly enough to think next door to blasphemy, I will say that while I hold the mathematical mind in high respect, because mathematicians are the only class of men who habitually reason correctly, and for sundry other reasons, yet this does not prevent me either from recognizing their general tendency to decide difficult practical questions on the most absurdly slight grounds (as all the world knows many of them are apt to do), nor does it prevent my holding that the minds that are mathematically strong are by no means equally strong in analyzing the first principles even of their own subjects. Thus, with all my admiration of Dedekind. ...

... I enunciate the governing principle of succession of numbers in a finite cycle as follows: *A cyclical system of numbers is such a system that every number belonging to it stands to some number of it or other in a certain relative capacity,* A, *such that, taking any definite predicate whatever, if this predicate is not true of any number of the system without likewise being true of at least one number of the system that is* A*'d by that number, then this predicate is true either of all or of none of the numbers of the system.* From this definition of a cyclical system (for so I shall regard it), all the properties of the system necessarily follow, without exception. It is particularly remarkable that it follows from this that the multitude of numbers composing the system is enumerable, or finite; for this, one would say, could only be by a logical mode of inference hitherto unrecognized. The proofs will be indicated below.

The governing principle of succession of the system of *all positive integers* is as follows: *In this system there is a certain relative capacity,* A, *in which every number stands to some number, and there is a certain number,* 1, *to which no number is* A; *and whatever definite predicate is not true of any number of the system without likewise being true of at least one number to which that number is* A, *is either true of all numbers of the system or is false of* 1.

The governing principle of succession of the system of *all real integers* is as follows: *In this system there is a certain relative capacity,* A, *in which every number of the system stands to at least one number of the system, while in the converse respect all numbers of the system are alike; and if any definite predicate is not true of any number of the system without at once being true of some number to which it is* A, *and being true besides of some number that is* A *to it, then from both these conditions conjointly, but not from either of them alone, it follows that that predicate is true either of all or of none of the numbers of the system.* It is remarkable that from this principle, which differs so little from that of a cyclical system, far from following that the multitude of numbers of the system is enumerable, it follows that it is denumeral.

The governing principle of succession of the *Cantorian system of ordinals,* if I rightly understand that system (which I am not quite sure has been logically described by Cantor), is as follows: In this system there is an ordinal, 1, to which none of the ordinals of the system is A_1, where I use A_1 as a relative term, that is, followed by the construction with *to,* to signify a relative character to which that which is denoted by the word or symbol following *to* is correlative; and the character of A_1 is such that every ordinal of the system is A_1 to another. The character subjacent to the A is an ordinal of the [system] here used to distinguish the relative character signified by A_1 from other more or less analogous relative characters which will be signified by A with other ordinals subjacent to it. Moreover, there is a relative term, B_1 (the ordinal subjacent to which is used like that subjacent to A_1), such that if any ordinal, l, is A_1 to an ordinal, m, then l is also B_1 to m; and if m is B_1 to a third ordinal, n, then l is B_1 to n; but no ordinal is B_1 to any other unless this rule determines it to be so. But no ordinal is A_1 to any ordinal that is B_1 to it. The rule of A_1-ness, as I shall call it, is that every ordinal is A_1 to some ordinal that is not B_1 to it. There is, moreover, a relative term, A_2, such that if, by the C_1s, we denote the entire collection of all the ordinals of the system whose existence in it is necessitated by the existence of 1, according to the rule of A_1-ness, then the collection of the C_1s is A_2 to an ordinal that is not a C_1. There is, moreover, a relative term, B_2, such that if one ordinal is B_1 or A_2 to another it is always B_2 to that other, which, however, is not B_2 to it; and further any ordinal that is B_2 to another that is B_2 to a third is always B_2 to that third, while that third is not B_2 to it. And calling the C_ns the entire collection of the ordinals whose existence is necessitated by that of 1, according to any part of the principle of succession this collection is always A_2 to an ordinal that is not one of the C_ns; and every C_n in B_2 to this new ordinal, which is not B_2 to any of the C_ns.

It is by no means clear that Cantor's ordinals fulfill the purpose for which they were designed. [Schönfliess] positively denies their doing so; and I much doubt it. I have, however, devised a system of ordinals which seems to leave no doubt of the truth of the propositions that Cantor thought he had proved by his own ...

The most important part of a demonstration is a step in reasoning of a kind which I call the *theoretic inference*. Although I have for a quarter of a century had in mind the importance of systematizing theoric inference, it is only of late that I have begun to do some real work upon it, and what I have done so far is but little. Nevertheless, I am sure that such comments as I can make will be useful to all who have not a natural genius for mathematics, of which I have myself but a small share.

Everybody knows that mathematics, which covers all necessary reasoning, is as far as possible from being purely mechanical work; that it calls for powers of generalization in comparison with which all others are puny, that it requires an imagination which would be poetical were it not so vividly detailed, and above all that it demands invention of the profoundest. There is, therefore, no room to doubt that there is *some* theoric reasoning, something unmechanical, in the business of mathematics. I hope that, before I cease to be useful in this world, I may be able to define better than I now can what the distinctive essence of theoric thought is. I can at present say this much with some confidence. It is the directing of the attention to a sort of object not explicitly referred to in the enunciation of the problem in hand ...

13

LOGIC MACHINES

LOGICAL MACHINES[1]

In the "Voyage to Laputa" there is a description of a machine for evolving science automatically. "By this contrivance, the most ignorant person, at a reasonable charge, and with little bodily labor, might write books in philosophy, poetry, politics, laws, mathematics, and theology, without the least assistance from genius or study." The intention is to ridicule the Organon of Aristotle and the Organon of Bacon, by showing the absurdity of supposing that any "instrument" can do the work of the mind. Yet the logical machines of Jevons and Marquand are mills into which the premises are fed and which turn out the conclusions by the revolution of a crank. The numerous mathematical engines that have been found practically useful, from Webb's adder up to Babbage's analytical engine (which was designed though never constructed) are also machines that perform reasoning of no simple kind. Precisely how much of the business of thinking a machine could possibly be made to perform, and what part of it must be left for the living mind, is a question not without conceivable practical importance; the study of it can at any rate not fail to throw needed light on the nature of the reasoning process. Though the instruments of Jevons and of Marquand were designed chiefly to illustrate more elementary points, their utility lies mainly, as it seems to me, in the evidence they afford concerning this problem.

The machine of Jevons receives the premises in the form of logical equations, or identities. Only a limited number of different letters can enter into these equations — indeed, any attempt to extend the machine beyond four letters would complicate it intolerably. The machine has a keyboard, with two keys for the affirmative and the negative form of each letter to be used for the first side of the equation, and two others for the second side of the equation, making four times as many keys as

[1] From the *American Journal of Psychology* (November 1887).

letters. There is also a key for the sign of logical addition or aggregation for each side of the equation, a key for the sign of equality, and two full stop keys, the function of which need not here be explained.* (* *Phil. Trans.* for 1870.) The keys are touched successively, in the order in which the letters and signs occur in the equation. It is a curious anomaly, by the way, that an equation such as $A = B$, which in the system of the transitive copula would appear as two propositions, as All *A* is *B* and All *B* is *A*, must not be entered as a single equation. But although the premises outwardly appear to be put into the machine in equations, the conclusion presents no such appearance, but is given in the form adopted by Mr. Mitchell in his remarkable paper on the algebra of logic. That is to say, the conclusion appears as a description of the universe of possible objects. In fact, all that is exhibited at the end is a list of all the possible products of the four letters. For example, if we enter the two premises All *D* is *C*, or $D = CD$, and All *C* is *B*, or $C = BC$, we get the conclusion in the following shape, where letters in the same vertical column are supposed to be logically multiplied, while the different columns are added or aggregated:

A	*A*	*A*	*A*	*a*	*a*	*a*	*a*
B	*B*	*B*	*b*	*B*	*B*	*B*	*b*
C	*C*	*c*	*c*	*C*	*C*	*c*	*c*
D	*d*	*d*	*d*	*D*	*d*	*d*	*d*.

The capital letters are affirmatives, the small letters negatives. It will be found that every column containing *D* contains *B*, so that we have the conclusion that All *D* is *B*, but to make this out by the study of the columns exhibited seems to be much more difficult than to draw the syllogistic conclusion without the aid of the machine.

Mr. Marquand's machine is a vastly more clear-headed contrivance than that of Jevons. The nature of the problem has been grasped in a more masterly manner, and the directest possible means are chosen for the solution of it. In the machines actually constructed only four letters have been used, though there would have been no inconvenience in embracing six. Instead of using the cumbrous equations of Jevons, Mr. Marquand uses Professor Mitchell's method throughout.* (* It would be equally true to say that the machine is based upon Mrs. Franklin's system. The face of the machine always shows every possible combination; putting down the keys and pulling the cord only alters the appearance of some of them. For example, the following figure 1 re-

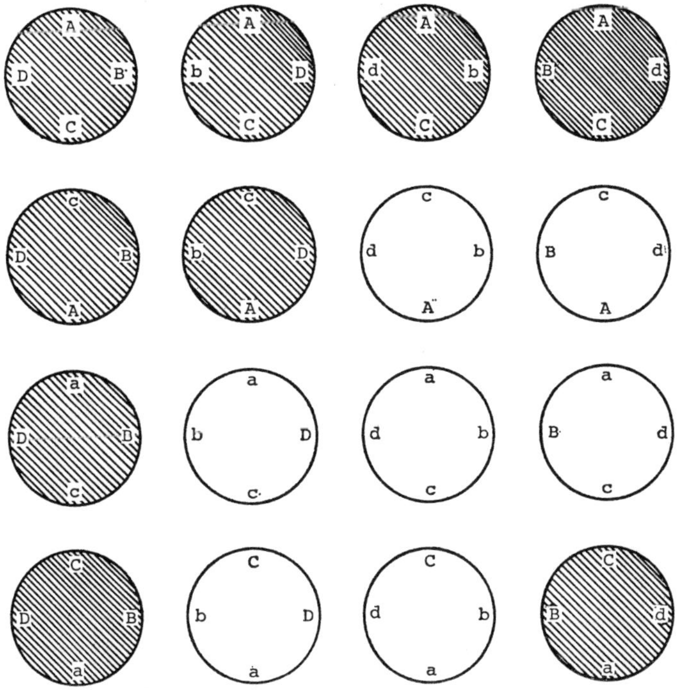

Fig. 1

presents, diagrammatically, the face of such a machine with certain combinations modified.

This face may be interpreted in several different ways. First, as showing in the shaded portions —

$$(A+B+C+D)\,(A+b+C+D)\,(A+b+C+d)\,(A+B+C+d)$$
$$(A+B+c+D)\,(A+b+c+D)$$
$$(a+B+c+D)$$
$$(a+B+C+D) \qquad\qquad (a+B+C+d),$$

which is the same as what is seen on the unshaded portions if we regard the small letters as affirmative and the capitals as negative, and interchange addition and multiplication, that is, as —

$$aBCD + abCD$$
$$+ ABCd + ABCD + AbCD$$
$$+ ABcd + ABcD.$$

Or, looking at the unshaded portion, we may regard it as the negative of the above, or —

$$(A+b+c+d)\,(A+B+c+d)$$
$$(a+b+c+D)\ (a+b+c+d)\ (a+B+c+d)$$
$$(a+b+C+D)\,(a+b+C+d),$$

or, what is the same thing, as —

$$\begin{aligned}&abcd+aBcd+aBcD+abcD\\&+abCd+aBCd\\&+AbCd\\&+Abcd \qquad\qquad\qquad +AbcD.\end{aligned}$$

There are two other obvious interpretations. We see, then, that the machine always shows two states of the universe, one the negative of the other, and each in two conjugate forms of development. In one interpretation simultaneously impressed terms are multiplied and successively impressed combinations added, and in the other interpretation the reverse is the case.)

There are virtually no keys except the eight for the letters and their negatives, for two keys used in the process of erasing, etc., should not count. Any number of keys may be put down together, in which case the corresponding letters are added, or they may be put down successively, in which case the corresponding combinations are multiplied. There is a sort of diagram face, showing the combinations or logical products as in Jevon's machine, but with the very important difference that the two dimensions of the plane are taken advantage of to arrange the combinations in such a way that the substance of the result is instantly seen. To work a simple syllogism, two pressures of the keys only are necessary, two keys being pressed each time. A cord has also to be pulled each time so as to actualize the statement which the pressure of the keys only formulates. This is good logic; philosophers are too apt to forget this cord to be pulled, this element of brute force in existence, and thus to regard the *solvet ambulando* as illogical. To work the syllogism with Mr. Jevon's machine requires ten successive movements, owing to the relatively clumsy manner in which the problem has been conceived.

One peculiarity of both these machines is that while they perform the inference from $(A+B)C$ to $AC+BC$, they will not perform the converse inference from $AC+BC$ to $(A+BC)$. This is curious, because the inference they refuse to perform seems to be merely syllogistic, while the one they do perform, and in fact continually insist on performing, whether it is wanted or not, is dilemmatic, and therefore essentially more complicated. But in point of fact neither of the machines really gives the conclusion of a pair of syllogistic premises; it merely presents a list of all the possible species in the universe, and leaves us to pick out the syllogistic conclusions for ourselves. Thus, with Marquand's machine, we enter the premise All *A* is *B* in the form $a + B$, and the premise All *B* is *C* in the form $b + C$; but instead of finding the conclusion in the form $a + C$, it appears as —

$$\begin{aligned}&ABCD + ABCd\\&+ aBCD + aBCd + abCd + abCD\\&\qquad\qquad + abcd + abcD.\end{aligned}$$

As we only want a description of *A*, we multiply by that letter, and so reduce the conclusion to $ABCD + ABCd$, but there is no elimination of the *B* nor of the *D*. We do not even get the full conclusion in the form $ab + BC$, although it is one of the advantages of Marquand's machine that it does give the conclusion, not only in the form just cited, but also, simultaneously, as

$$\begin{aligned}&(a + B + c + d)\ (a + B + c + D)\\&(a + B + C + d)\,(a + B + C + D)\,(a + b + C + D)\ (a + b + C + d)\\&\qquad\qquad\qquad\qquad (A + b + C + D)\,(A + b + C + d).\end{aligned}$$

The secret of all reasoning machines is after all very simple. It is that whatever relation among the objects reasoned about is destined to be the hinge of a ratiocination, that same general relation must be capable of being introduced between certain parts of the machine. For example, if we want to make a machine which shall be capable of reasoning in the syllogism

If *A* then *B*,
If *B* then *C*,
Therefore, if *A* then *C*,

we have only to have a connection which can be introduced at will, such that when one event *A* occurs in the machine, another event *B* must also occur. This connection being introduced between *A* and *B*, and also between *B* and *C*, it is necessarily virtually introduced between *A* and *C*. This is the same principle which lies at the foundation of every logical algebra; only in the algebra, instead of depending directly on the laws of nature, we establish conventional rules for the relations used. When we perform a reasoning in our unaided minds we do substantially the same thing, that is to say, we construct an image in our fancy under certain general conditions, and observe the result. In this point of view, too, every machine is a reasoning machine, in so much as there are certain relations between its parts, which relations involve other relations that were not expressly intended. A piece of apparatus for performing a physical or chemical experiment is also a reasoning machine, with this difference, that it does not depend on the laws of the human mind, but on the objective reason embodied in the laws of nature. Accordingly, it is no figure of speech to say that the alembics and cucurbits of the chemist are instruments of thought, or logical machines.

Every reasoning machine, that is to say, every machine, has two inherent impotencies. In the first place, it is destitute of all originality, or all initiative. It cannot find its own problems; it cannot feed itself. It cannot direct itself between different possible procedures. For example, the simplest proposition of projective geometry, about the ten straight lines in a plane, is proved by von Staudt from a few premises and by reasoning of extreme simplicity, but so complicated is the mode of compounding these premises and forms of inference, that there are no less than 70 or 80 steps in the demonstration. How could we make a machine which would automatically thread its way through such a labyrinth as that? And even if we did succeed in doing so, it would still remain true that the machine would be utterly devoid of original initiative, and would only do the special kind of thing it had been calculated to do. This, however, is no defect in a machine; we do not want it to do its own business, but ours. The difficulty with the balloon, for instance, is that it has too much initiative, that it is not mechanical enough. We no more want an original machine, than a house-builder would want an original journeyman, or an American board of college trustees would hire an original professor. If, however, we will not surrender to the machine, the whole business of initiative is still thrown upon the mind; and this is the principal labor.

In the second place, the capacity of a machine has absolute limitations; it has been contrived to do a certain thing, and it can do nothing else. For instance, the logical machines that have thus far been devised can deal with but a limited number of different letters. The unaided mind is also limited in this as in other respects; but the mind working with a pencil and plenty of paper has no such limitation. It presses on and on, and whatever limits can be assigned to its capacity to-day, may be overstepped to-morrow. This is what makes algebra the best of all instruments of thought; nothing is too complicated for it. And this great power it owes, above all, to one kind of symbol, the importance of which is frequently entirely overlooked — I mean the parenthesis. We can, of course, dispense with parentheses as such. Instead of $(a+b)c = d$, we can write $a+b+t$ and $tc+d$. The latter t is here a transmogrified parenthesis. We see that the power of adding proposition to proposition is in some sort equivalent to the use of a parenthesis.

Mr. Marquand's machines, even with only four letters, facilitate the treatment of problems in more letters, while still leaving considerable for the mind to do unaided. It is very desirable a machine on the same principle should be constructed with six letters. It would be a little more elegant, perhaps, instead of two keys to each letter, to have a handle which should stand up when the letter was not used, and be turned to the right or left, according as the letter was to be used, positively or negatively. An obvious extension of the principle of the machine would also render it possible to perform elimination. Thus, if six letters, A, B, C, D, E, F, were used, there could be an additional face which should simply take no notice of F, a third which should take no notice of F or E, a fourth which should take no notice of F, E or D; and these would suffice. With such a machine to represent $AB+CD$, we should proceed as follows: Put down handle E to the left. (The left hand would naturally signify the negative.) Leaving it down, put down handle A to the right and then bring it back after pulling the cord. Put down handle B to the right and pull the cord, and then restore handles B and E to the vertical. Next, put down handle F to the left and successively put down the handles C and D to the right, as before. After restoring these to the vertical, put down handles E and F to the right, and pull the cord. Then we should see on the third face

$$(A+B+C+D)\,(A+b+C+D)\,(A+b+C+d)\,(A+B+C+d)$$
$$(A+B+c+D)\;\;(A+b+c+D)$$
$$(a+B+c+D)$$
$$(a+B+C+D) \qquad\qquad (a+B+C+d)$$

or, what comes to the same thing,

$$aBCD+abCD$$
$$ABCd+ABCD+AbCD$$
$$ABcd+ABcD$$

I do not think there would be any great difficulty in constructing a machine which should work the logic of relations with a large number of terms. But owing [to] the great variety of ways in which the same premises can be combined to produce different conclusions in that branch of logic, the machine, in its first state of development, would be no more mechanical than a hand-loom for weaving in many colors with many shuttles. The study of how to pass from such a machine as that to one corresponding to a Jacquard loom, would be likely to do very much for the improvement of logic.

C. S. Peirce[2]

[2] An exhibition showing "the origins and first lines of development of the computer" has been set up at the IBM center in New York City. *A Computer Perspective* (Harvard University Press, 1973) with introduction by I. Bernard Cohen describes the exhibition in detail. In the 1890-1900 section one finds Marquand's logic machine and, under a description of it and a picture of Marquand, a picture of Peirce and part of a letter from him to Marquand dated 30 December 1886. The letter, discovered by Max Fisch, reads in part as follows:

"You spoke, when I saw you, as if disappointed with the reception your machine met with. I wish I could see it. My impression is that it has two defects; first, I believe it only extends to four simple terms instead of to six as it should; and second, I believe it does not reduce the solution to its simplest expression. It ought to perform 4 operations, or 3 at least. First it should develope any expression as a into $abcdef + abcde\bar{f} + abcd\bar{e}f +$ etc. Second it should reduce expressions; for instance $abcdef + abcde\bar{f} + abcd\bar{e}f$ into

$abcde + abcdf.$

Third, it should multiply two developed polynomials, if not any two. Fourth, though not absolutely required, it would be well to have it capable of adding. I think you ought to return to the problem, especially as it is by no means hopeless to expect to make a machine for really very difficult mathematical problems. But you would have to proceed step by step. I think electricity would be the best thing to rely on.

Let A, B, C be three keys or other points where the circuit may be open or closed. As in Fig. 1 there is a circuit only if *all* are closed; in Fig. 2 there is a circuit if *any one* is closed. This is like multiplication and addition in logic.

Yours faithfully,
C. S. Peirce

P.S. If you will send me a copy of your last paper on your machine, I will act as Devil's Advocate, by attacking it."

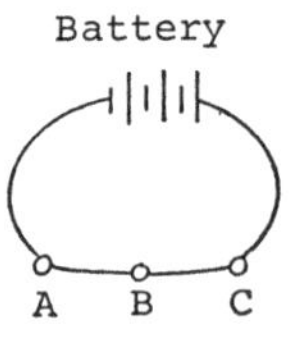

Fig. 1

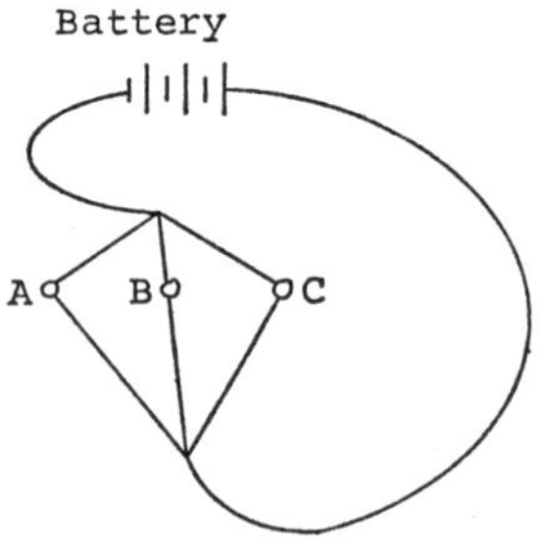

Fig. 2

14

MEASUREMENT

A. [FROM NATURAL METRIC SYSTEM] (from 427)

Now let us consider that branch of physics which is concerned with the precise measurement of constants. There are only two physical constants which are not restricted to particular kinds of matter. They are the absolute modulus of gravitation and the velocity of light. They are just enough to establish relations between the units of time, length, and mass. The siderial revolution of the earth is the inevitable unit of time. We already know the velocity of light almost well enough to use the distance traversed by light in a siderial day as the unit of length.[1] The absolute modulus of gravitation is still very roughly ascertained. But if we could take, as our unit of mass, that mass whose gravitation would in a siderial day impart to another mass at the unit of distance a component velocity equal to that of light, we should have a metric system of which the connections would be the only fundamental laws of nature. This, of course, is supposing that the ether is not merely a certain chemical body or mixture of chemical bodies that happens to be widely diffused in space, and also that gravitation is not the effect of some analogous accidental circumstance.

All the rest of this business of determining constants may be called chemical physics. Its first part is the determination of atomic weights, which is usually reckoned as a branch of chemistry because only a chemist has the requisite experience and skill to undertake it. This is another illustration of how the domains of the different sciences are determined by the different training of different classes of men. A man makes such studies as he is equipped to make; and studies that nobody is properly equipped to make are necessarily either postponed or botched, — like psychical research.

1 In 1879 Peirce determined the length of the meter from the wavelength of light. See "Charles S. Peirce, Nineteenth Century Man of Science", *Scripta Mathematica,* 24 (1959), 316, by the editor.

B. [DRAFT OF REVIEW OF ALEXANDER ZIWET'S

AN ELEMENTARY TREATISE ON THEORETICAL MECHANICS.

PART I. KINEMATICS.
PART II. INTRODUCTION TO DYNAMICS (1893)] (1381)

We do not await the third volume treating "Kinetics," as a modern jargon names dynamics, to notice this work. The volumes are thin, and very [openly printed]; so that the three will make but a small treatise on Analytic Mechanics; and we confess to a prejudice against small treatises on analytical mechanics. Why use the analytical method at all, unless the student can pursue it far enough to attain greater expertness in the consideration of practical problems than he would have attained in the same time devoted to a more synthetic treatment of Mechanics? The student of a small treatise is presumably not destined for any mathematical calling. Mechanical problems will not, therefore, present themselves to him in after life oftener than a dozen a year; and every mathematician knows that that method of attacking a problem which is the most economical when your intellectual plant, so to speak, is used often enough to keep it in good running order is not by any means always the most economical if it is to be used only semi-occasionally. For ordinary men an "elementary" treatise, not excluding analysis but chiefly synthetical is we believe the most suitable.

We admit that the collegian who takes up Mechanics has other purposes than that of solving practical problems in after life. For that stage of philosophical reflection which every reflecting man aims to attain, nothing is more important than a clear apprehension of the conceptions of science. And these are at bottom mechanical for the most part. The history of European man for the past three centuries may be pronounced with little exaggeration unintelligible without a clear apprehension of Force. Still, all such concessions made, it remains that the only sure way to get this comprehension without mistake is to learn to apply the idea of force in actual problems.

The plan of this book being somewhat novel, it may be well to run through it, to see, what the two parts so far published contain. The first chapter treats the geometry of motion; the second only is headed "Kinematics," by which the author means apparently the general analysis of rotation and acceleration. This is certainly an improvement on Thomson and Tait, who include under Kinematics everything in solid geometry that strikes their fancy.

Professor Ziwet defines motion as change of position. Change implies time; but it does not necessarily imply continuity. Motion would be better defined as continuous variation in the position of an object or objects relatively to the time. But as variation in time presents no peculiarities that are considered in the discussion, it would have been better to head the chapter, "Excerpts from geometry relating to certain general modes of variation of positions." The truth is there is no special geometry of motion. Nothing characterizes this branch of geometry except a slightly different phraseology, and a strictly metrical point of view. A translation is defined as a *displacement*. This fails to draw the usual distinction between a displacement and a motion; but it makes no difference certainly in a chapter in which motion is not considered under an aspect in which it differs from any other variation of position. Every treatise on geometry, and much more a work on mechanics, ought to explain at the very outset the mechanical significance of the straight line and circle. And perhaps it would be as well to inquire at once into the reality of absolute space and time. The truth is that inasmuch as geometry cannot be philosophically comprehended without mechanical considerations, the subject of dynamics ought to be broached at the very outset, — apart from the rule of rhetoric that an author ought at once to outline what he has to say. This might be done in an introductory chapter and would not prevent the subject being afterwards dissected according to the scheme used by Professor Ziwet. We are happy to see the C.G.S. system introduced at once, but we note that both the metre and the yard are wrongly defined.

"The original standard metre is a platinum bar preserved in the *Palais des Archives* in Paris, a legalized copy of which has been deposited at Washington, D.C. The metre can be defined as the distance between two marks on the standard metre when at a temperature of 0°C."

The *mètre des archives* has no lines upon it. It is an end-measure. To legalize a copy of it would be to change the metre. A copy of it has, it is true, been legalized; namely the new international metre preserved in the *Pavillon de Breteuil* in the Park of St. Cloud at Sèvres. And this new metre, — for the *mètre des archives* is deposed, by the legalization of the copy, — is, it is true, a line-measure. There are of course copies of this at Washington of high authority.

"The original British standard yard is a bronze bar preserved in London. For the United States the yard is defined as the distance between the twenty-seventh and sixty-third divisions of the brass standard yard kept in the Bureau of Weights and Measures at Washington, when the bar is at a temperature of $16\frac{2}{3}$°C. or 62°F."

Of course, there were early standard yards; but at the time of our Declaration of Independence, no authoritative yard existed which was so made as to be susceptible of accurate comparisons. It was, therefore, the practice of scientific men to define lengths by reference to Mr. Bird's scale, Mr. Bird being an instrument maker. Subsequently, a scale sent to Washington (it is not a standard yard), which had been very carefully (for those days) compared with scales in use in London, was used as the standard by our Office of Weights and Measures. For Congress has never created a yard; and has neglected, from a silly motive, to adopt the British Imperial yard which was created after our Independence. Practically the British yard is used, as is right; for standards of one name ought not to be multiplied. The chief moral virtue connected with metrology is veracity. The manner in which our government has treated the subject is not conducive to that virtue. An office of weights and measures where a mendacious system prevails is in danger of being distrusted, and even of becoming materially mendacious.

C. ON THE THEORY OF ERRORS OF OBSERVATIONS, BY ASSISTANT C. S. PEIRCE

(FROM THE UNITED STATES COAST SURVEY REPORT FOR 1870) APPENDIX NO. 21

The object of this paper is to give a general account of the theory of errors of observations, with the design of showing what the limitations to the applicability of the method of least squares are, and what course is to be pursued when that method fails. We shall begin with an elementary account of the general principles of the subject, in order to state them with a little more accuracy than is commonly done.

The notation employed is one which has been suggested by the study of the logic of relations. Small Roman letters will denote objects partly indeterminate. Thus m may denote a man, without saying what man. Small Italics will be used for relative terms; thus l may denote a lover. The correlates of such relative terms will be written after them on the same line; thus t m or l m may denote a tooth of a man or a lover of a man, if m denotes man, t a tooth of and l a lover of something undetermined. Then, l w will denote a lover of a woman, it being indeterminate what lover and what woman. $t\,l$ w will denote a tooth of a lover of a woman. If we wish to denote that which is a lover of all women, we must have a symbol to denote all women. As [x] is commonly used in the method of least squares to denote the sum of all the quantities x, so we may write [w] to denote all women, and then l[w] will denote something which is a lover of all women, or we may write the same thing thus, l^{w}. A relative term has a double indeterminacy, being indeterminate in reference to the relate and also in reference to the correlate. A lover may be this lover, that lover, or the other lover, and each of these may be lover of this, or that, or the other. Corresponding, therefore, to the [l], which denotes all lovers, we may write $\{l\}$ to denote the lover to whomever he is a lover. Thus, $\{l\}^{\mathrm{w}}$ will denote a lover of nothing but women, or we may write the same thing thus, ${}^{l}\mathrm{w}$. We may denote "loved by" by K^{l}.

Corresponding to any absolute term, as man, there is a relative term, "man that is," as in the expression "a man that is rich." I shall denote a relative of this sort by the symbol for the absolute term, with an inverted comma after it, as $\mathrm{m}_{‘}$. Thus, if b denotes anything black, $\mathrm{m}_{‘}\,\mathrm{b}$ will denote a man that is black.

Let V be the relative "a general name which is applicable to." Thus, $V\mathrm{m}$ will denote a general term which is applicable to some man. V^{m} will denote a general term which is applicable to every man. $V^{\mathrm{m}}{}_{‘}\,V^{\mathrm{w}}$ will be a general term which is applicable to every man and to every woman. $K\,V[V^{\mathrm{m}}{}_{‘}\,V^{\mathrm{w}}]$ will denote that to which every general term is applicable which is applicable to every man and to every woman; in other words, this denotes either a man or a woman. I shall write this for short m w.

Zero is defined by the general equation $x \mathbin{+\!\!,} 0 = x$, whatever x may be. Then, *zero* generally denotes nothing.

Unity is defined by the general equation $\mathrm{x}_{‘}\,1 = \mathrm{x}$, whatever x may be, and then 1 generally will denote anything.

But 1 and 0 have sometimes to be interpreted as relative terms. Now, it can be proved by the principles of the logic of relatives that so considered $0^{\mathrm{x}} = 0$, unless $\mathrm{x} = 0$, when $0^0 = 1$; and that $1\,\mathrm{x} = 1$, unless $\mathrm{x} = 0$, when $10 = 0$. It follows that 0^{x} is such a logical function* (* A mathematical function of x, such as ϕx, is something whose value is obtained by mathematical processes when x is given. A logical function of x, of which we have, as common examples, letters with a subscript x, as P_x, is something whose signification is logically deducible when the signification of x is known.) of x that it signifies "the case of the non-existence of," while $1x$ is such a logical function of x that it signifies "the case of the existence of."

Since [m] denotes all men, we may naturally write $\frac{[\mathrm{m}]}{\mathrm{m}}$ to denote what all men become when that factor is removed which makes [m] refer to *men* rather than to anything else; that is to say, to denote the *number* of men. We may write this for short **[m]** with heavy brackets. Then t being a relative term, ("a tooth of,") by **[t1]** will be denoted the total number of teeth in the universe. But **[t]** will be used as equivalent to $\frac{[t1]}{[1]}$, or the average number of teeth that anything has. But "anything" is not to be taken here in an absolute sense. We always limit our considerations entirely to a certain class. As De Morgan expresses it, we always have a limited universe. When we reckon up the number of all things to find the average number of teeth *per* thing, it would be absurd to count among things the virtues, shades of color, days or seconds of time. Anything which belongs to the limited universe under consideration is called, in the theory of probabilities, an *event*. An expression like [*t*], where *t* is a relative term, is termed a relative number, average number, or probable number. If the relative term to which the average number refers is one of those relatives [which are] formed by adding a comma to the symbol for an absolute term, as m, then the relative number is called a *probability*. For example, [m,] is the average number of men that anything is, but it is usually called the probability that anything is a man.

The importance of average numbers arises from the fact that all our knowledge really consists of nothing but average numbers; for all our knowledge is derived from induction, and its analogue, hypothesis. Now, the scientific conduct of this kind of reasoning is highly complex, because all sorts of precautions have to be attended to, and it has to be accompanied by a great deal of deduction. But the general nature of induction is everywhere the same, and is completely typified in the following example. From a bag of mixed black and white beans I take out a handful, and count the number of black and the number of white beans, and I assume that the black and white are nearly in the same ratio throughout the bag. If I am in error in this conclusion, it is an error which a repetition of the same process must tend to rectify. It is, therefore, a valid inference. But it clearly teaches me nothing in reference to the color of any particular bean. Of that I am as ignorant as before. The case is not in the least altered if I find all the beans of my handful to be black, or all to be white. I can still infer only the approximate general ratio, and it is only this I express when I say the observation makes the probability of any one bean being white or being black very great; for a probability is itself only an average number. At first thought it is hard to admit this; but the difficulty will be in great measure removed if we consider how it is that the knowledge of average numbers becomes useful in particular cases. Suppose we know the relative number of black beans in a certain bag; then, if we draw a large number of beans out of it, we know that the total number of black beans we shall draw will be equal to the number of drawings multiplied by the average number of beans in the bag. Suppose we know the relative numbers of black beans in a large number of bags, containing different proportions; then, if the beans are well mixed up in each, we may only draw a single bean from each, and yet we can predict nearly the total number of black beans which would be drawn by simply taking the sum of the relative numbers. If the black beans had a value while the white ones were worthless, then the total number of black beans which would be drawn would be the important thing to know. But as knowledge derives its practical importance from its influence upon our conduct, let us suppose that at every drawing we have our choice between two bags to draw from; then the man who knew the relative number of black beans in every bag would act in every case as though the bean he would draw from the bag which contained the larger proportion of black ones were known to be black, and as though the bean he would draw from the other were certainly white. Strictly speaking, he would know nothing about the beans that would be drawn in the particular case, but he would have a knowledge which would be so far equivalent to that that it would influence his conduct in the particular case. This is the only knowledge we ever have, a knowledge of what assumption to make in the particular case in order to do the best in the long run. Whenever, then, we have to do with a *value*, the sum-total of which in the long run is the only thing which concerns us, the average amount of it is important to be known; but in all other cases the average numbers are of no consequence.

It is evident that in the example just given it would be a valuable increase of knowledge to know, for instance, what the difference in the relative number of black beans at the top and bottom of a bag was, and any limitation of the "universe" used which should separate a relative number into two different ones would be advantageous.

There are many problems in probabilities, which, being solved, give a relative number composed of two terms, one known and the other unknown. Such an indeterminate result shows that a wider "universe" must be adopted for one of the terms of the relative number.

The fundamental arithmetical formulae relating to relative numbers are as follows:

We have seen that the relative number of things that are men, or the probability that a thing is a man, is equal to $\frac{[m_{\varsigma} 1]}{[1]}$. By "thing" here is meant any object of our limited universe, as, for example, an animal. But we may wish to consider the relative number of animals that are men when our limited universe is a wider class than animal. In this case, *a* denoting an animal, we write this probability $\frac{[m_{\varsigma} a]}{[a]}$. Let us suppose that, our universe being "day," we wish to know the probability that if it thunders on any day it will rain on that day. To say that if it thunders it will rain on the same day is the same as to say that every day on which it thunders is a day on which it rains. Then let t be a day on which it thunders and r a day on which it rains, and the probability in question is $\frac{[t_{\text{c}} r]}{[t]}$. In general, the probability that if one event happens another will happen is equal to the probability that both will happen, divided by the probability that the first will happen.

Let us now see how to express the probability that a certain quantity will have a certain value. It is clearly implied that the quantity is defined in some other way than by its value. It might be, for instance, the length of time a bird will be on wing. Let x be this quantity, and let n be any definite value. Then $\frac{[x_{\text{c}} n]}{[x]}$, or the number of cases in which the time a bird is flying has that value, divided by the whole number of cases of a bird's flying, is the probability that a bird will fly for that length of time. But since time is continuous, the length of time a bird may be up may have an infinite number of different values. Consequently, the probability of any one is *zero*. We, therefore, seek the probability that the time lies between n and n + $\triangle$ n; and if $\triangle$ n is infinitesimal, (say d n,) then the probability is proportional to d n. We may, therefore, write this probability thus, $[n_x]$ d n. Here n_x denotes the case of the value of x being about n.

The probability that if one quantity, x, has a value lying between m and $m + \mathrm{d}\, m$, then another quantity, y, has a value lying between n and $n + \mathrm{d}\, n$, will, according to what has been said, be equal to the probability that both x and y will have the supposed values, divided by the probability that x will have the supposed value, or will be $\frac{[m_{x\text{c}}\, n_y]\, \mathrm{d}m \cdot \mathrm{d}n}{[m_x]\, \mathrm{d}m}$, or, since the d$m$ disappeared by cancellation, $\frac{[m_{x\text{c}}\, n_y]}{[m_x]}\mathrm{d}n$.

Given, the probability of A, the probability of B, and the probability that if A happens, B happens. Required, the probability that if B happens, A happens.

The probability of A is [A].
The probability of B is [B].

The probability that if A then B is $\frac{[A_{\text{c}} B]}{[A]}$.

The probability that if B then A is $\frac{[A_{\text{c}} B]}{[B]}$.

Then we have—

$$\frac{[A\,B]}{[B]} = \frac{[A_{\text{c}}\,B]}{[A]} \times [A] \div [B]$$

or the probability, if B happens that A happens, is equal to the probability if A happens that B happens, multiplied by the probability of A and divided by the probability of B.

We now pass to the theory of observations. An observation gives us the value of a certain quantity which is connected with an unknown quantity in such a way as to be partly dependent on the latter value, and partly on accidental circumstances, not capable of being separately taken account of.

These accidental variations are, however, in all cases subject to a statistical law, so that (observations of a certain kind forming the limited universe, X being the unknown quantity, Ξ the quantity observed) the quantity,

$$\frac{[\xi_{\Xi\,\text{c}}\,x_{x}]}{[x_{x}]}d\xi$$

or the probability that if the unknown quantity X has a certain value x, then the observed quantity Ξ will have a value between ξ and $\xi + d\xi$, is a certain arithmetical function of the values ξ and x. If we write ε for $\xi - x$, or the error of observation, then we may put φ for such a function that —

$$\frac{[\xi_{\Xi\,\text{c}}\,x_{x}]}{[x_{x}]} = \varphi(\varepsilon, x)$$

The special form of the function φ is called the law of the facility of the errors. Except so far as this law is known, an observation can afford us no information whatever. The following conditions are invariably fulfilled by this function. (It must be understood that only *real* quantities are considered.)

1. It has but one value for each set of values of its variables.
2. Its value is always positive and less than unity.
3. It vanishes when ε is infinite.
4. Its integral relatively to ε from $-\infty$ to $+\infty$ is always unity.

Beyond this the form of the function is determined by the peculiarities of the kind of observations.

The probability that if the observed quantity Ξ has the value ξ, then the unknown quantity X has the value x is —

$$\frac{[x_{x\,\text{c}}\,\xi_{\Xi}]}{[\xi_{\Xi}]}dx$$

The value of this probability is for any particular kind of observations an arithmetical function of ε and ξ, which we may write $\psi(\varepsilon, \xi)\, d\varepsilon$.

The probability that the unknown quantity X has the value x without reference to observation is $[x_x]dx$. This is in any case a function of x, which may be written $\Psi x \cdot dx$.

The probability that the quantity given by observation Ξ has the value ξ, without reference to the value of the unknown quantity, is $[\xi_\Xi]\, d\xi$. This an arithmetical function of ξ, which may be $\Phi\xi \cdot d\xi$.

If $\Phi\xi$, Ψx, and $\varphi(\varepsilon, x)$ are given, then we can obtain $\psi(\varepsilon, \xi)$ thus:

$$\psi(\varepsilon, \xi) = \frac{\varphi(\varepsilon, x)}{\Phi\xi} \Psi x$$

Suppose a number of independent observations to be made. Then we shall have a series of functions —

$$\begin{array}{ll} \varphi_1(\varepsilon_1, x) & \Phi_1\xi_1 \\ \varphi_2(\varepsilon_2, x) & \Phi_2\xi_2 \\ \varphi_3(\varepsilon_3, x) & \Phi_3\xi_3 \\ \&c. & \&c. \end{array}$$

then the probability that if the quantities observed have the values ξ_1, ξ_2, ξ_3, &c., the unknown quantity X has the value x will be —

$$\Psi x \cdot \frac{\varphi_1(\varepsilon_1, x)}{\Phi_1\xi_1} \cdot \frac{\varphi_2(\varepsilon_2, x)}{\Phi_2\xi_2} \cdot \frac{\varphi_3(\varepsilon_3, x)}{\Phi_3\xi_3} \cdot \&c.$$

or—

$$\Psi^x \cdot \prod_1^n {}_i \frac{\varphi_i(\varepsilon_i, x)}{\Phi_i\xi_i}$$

The probability $\Psi x.\, dx$ antecedent to *all* observations will be simply dx, and, therefore, the factor Ψx may be omitted in the above expression.

It will be perceived that observation never gives us to know a *number* expressing the value of the unknown quantity, but only a *function* expressing the probability of each value. It happens, however, in a very comprehensive case, that this function assumes a form which involves but two constants, so that in this case observation may be said to give us two numbers, a value for the unknown quantity, and the probable error of that value.

Mr. Crofton's method of considering this case seems to me to make it more comprehensible than any other. Suppose that the unknown quantity X has been observed twice, the values given by observation being ξ_1 and ξ_2. Put $[\xi]$ for $\xi_1 + \xi_2$. What then is the probability that if x is the value of X, the sum of the values given by the two observations will be $[\xi]$. It is clearly —

$$\int_{-\infty}^{+\infty} \varphi_1(\varepsilon_1, x) \cdot \varphi_2([\varepsilon] - \varepsilon_1, x) \cdot d\varepsilon_1$$

Developing $\varphi_2([\varepsilon) - \varepsilon_1, x)$ by powers of ε_1, this integral becomes—

$$\varphi_2([\varepsilon], x) - \varphi_2'([\varepsilon], x) \int_{-\infty}^{+\infty} \varepsilon_1 \varphi_1(\varepsilon_1, x) \cdot d\varepsilon_1$$

$$+ \tfrac{1}{2}\varphi_2''([\varepsilon], x) \int_{-\infty}^{+\infty} \varepsilon_1^2 \varphi_1(\varepsilon_1, x) \cdot d\varepsilon_1$$

$$+ \tfrac{1}{6}\varphi_2'''([\varepsilon], x) \int_{-\infty}^{+\infty} \varepsilon_1^3 \varphi_1(\varepsilon_1, x) \cdot d\varepsilon_1 - \&c.$$

If in place of two we have n observations, the probability that the sum of all will be $[\xi]$ is—

$$\prod_{1}^{n-1}{}_{i}\left(1 - \int_{-\infty}^{+\infty} \varepsilon_i \varphi_i(\varepsilon_i, x) \cdot \mathrm{D}_i + \tfrac{1}{2} \int_{-\infty}^{+\infty} \varepsilon_i^2 \varphi_i(\varepsilon_i, x) \cdot \mathrm{D}^2 \varepsilon_i - \&c.\right) \varphi \mathrm{n}([\varepsilon], x)$$

Of these co-efficients—

$$\int_{-\infty}^{+\infty} \varepsilon_i \varphi_i(\varepsilon_i, x)$$

is the probable or mean value of the error of the observed quantity—

$$\tfrac{1}{2} \int_{-\infty}^{+\infty} \varepsilon_i^2 \varphi_i(\varepsilon_i, x)$$

is half the probable value of the square of this error, and so on. The probable value of the error is often less than the probable value of its square, owing to positive and negative errors balancing. But the co-efficients which involve the cube and higher powers of the error may often become insignificant. This, for example, will be the case if n is very great; for then, in comparison with the sum of all the errors, the value of any one will be very small. In fact, in this case we may neglect a quantity a certain number of times, say M, larger than what we could neglect before, and may take a unit of measurement M times larger. Then if —

$$[\varepsilon], \quad [\varepsilon^2], \quad [\varepsilon^3], \quad \&c.,$$

be the probable value of the error, the error square, the error cube, &c., on the old scale of measures, their numerical values on the new scale will be —

$$\frac{[\varepsilon]}{\mathrm{M}}, \quad \frac{[\varepsilon^2]}{\mathrm{M}^2}, \quad \frac{[\varepsilon^3]}{\mathrm{M}^3}, \quad \&c.$$

Consequently if M is sufficiently large, the higher co-efficients may be neglected. It also frequently happens that the error of each observation is due to the combined effect of a great number of independent or nearly independent causes, each one of which alone would produce but an insignificant effect. In this case, by the same reasoning the higher co-efficients will disappear.

The manner in which sensation and volition are propagated through the nerves is unknown, but it must be by some very complicated process, because it is very slow compared with the rate of propagation of sound. It is, therefore, probable that there may be variations of the rate of passage, and that the velocity through each small portion of the whole length of the nerve is to some extent independent of the velocity through the other parts. If this is so, the whole time of propagation would be subject to a variation, the probable values of whose cube and higher powers would be insignificant. However this may be, it appears to be a fact that in all carefully-made observations, the error of which is due to the inevitable inaccuracy of the action of the human nerves, the probable values of the cube, &c., of the errors are very small.

Whenever these quantities disappear it can be proved by an analytical process which need not be reproduced here that —

$$\varphi(\varepsilon, x) = \frac{h}{\sqrt{\pi}}\, e^{-h^2(\varepsilon - E)^2}$$

where h and E are quantities which depend upon x. We thus reach the fundamental formula of the method of least squares.

It is not the object of this paper to explain that method itself. Only it may be remarked that in the deduction of it which is usually given, it is assumed that what we wish to obtain from observations in any case (whether the method of least squares is applicable or not) is the most probable value of the unknown quantity. This is not the case, for there is but an infinitesimal probability in favor of any one value. Suppose that the cause of error in observing the place of a star were a nearly simple oscillation of the image about its mean point. Then the most probable errors would be the extreme ones. But we should much prefer to assume as the value the probable or mean value than the most probable value, which would be removed the furthest possible from that value. The conception of the matter is this. What observation has to teach us is a *function*, $\psi(\varepsilon, \xi)$, not a mere number. But in cases to which the method of least squares is applicable, this function is completely determined by two numbers, which are the value of the unknown quantity derived from the application of that method and the value of its probable error.* (* The term probable error is here used in its usual but unanalogical sense, and not for the probable value of the error, which is always assumed to be zero in least squares, or else determined by some special research.)

It is assumed in treatises on least squares that $\Phi\xi \cdot d\xi$, or the probability (without regard to the value of the unknown quantity) that the quantity observed will have a value between ξ and $\xi + d\xi$, is equal to $d\xi$. When this is not the case it is only necessary to weight each observation by dividing by $\Phi\xi$.

If the probability of error does not follow that law which the method of least squares supposes, that is to say, if the probability of the error x in the mean of a large number of observations is not equal to $\frac{h}{\sqrt{\partial}} 6^{-h^2x^2}$, where h is a constant independent of x, then it must at least be of this form if h be considered to be a function of x. Now, h^2 is the weight which has to be assigned to an observation in the application of the method of least squares; and therefore when this method does not apply in its unmodified form, it is only necessary to find what function of x, h must be in order to give the required law of facility of errors, and then proceed according to the method of least squares, after having weighted each observation by h^2. Let the equation which represents the facility of error be plotted, the error being taken as abcissæ, and the probability of that error as ordinate; then plot on the same diagram the curve $y = \frac{1}{\sqrt{\partial}} 6^{-x^2}$. Let us suppose, then, that a certain value of x, y, is A times as great for the actual curve of errors as it is for the normal least-squares curve. Now, if h be increased in the ratio A, the ordinates will be increased in this same ratio, and the abcissæ will be diminished in the inverse ratio so that the area of the curve is preserved. But if the ordinates at any one point are to be increased in the ratio A, then the abcissæ at that point must be contracted in the inverse ratio, so as to preserve the area; so that if we had a function fx such that $D_x fx = \frac{1}{a}$, then the probability of this function of the error would follow the law $y = \frac{1}{\sqrt{\partial}} 6^{-(fx)^2}$, and consequently $\frac{1}{(D_x fx)^2}$ is the weight which has to be assigned to any such observation in applying the method of least squares. It will be observed that since the weight depends on the value of the error, it will be necessary first to make an approximate solution of the problem in order to get an approximate value of the error from which to determine the weight, so that when the method of least squares is not applicable in its unmodified form, an approximate method must necessarily be resorted to.

Let us now proceed to consider the method of ascertaining the law of facility of error. In any case, if the error is compounded of an infinite number of infinitely small errors, or approximates to being so, then the law of facility of errors is of that general form which the method of least squares prescribes, and nothing remains indeterminate excepting the value of h. Observation has sufficiently shown that in transit-observations the law of error is of this sort. I copy, for example, from Chauvenet's Astronomy the following table, taken from Bessel's "Fundamenta Astronomiæ," showing the errors made by Bradley in observing Sirius and Procyon:

Between—	No. of errors by the theory	No. of errors by experience
0.0″ and 0.1″	95	94
0.1 and 0.2	89	88
0.2 and 0.3	78	78
0.3 and 0.4	64	58
0.4 and 0.5	50	51
0.5 and 0.6	36	36
0.6 and 0.7	24	26
0.7 and 0.8	15	14
0.8 and 0.9	9	10
0.9 and 1.0	5	7
Over 1.0	5	8

Fechner, in his "Elemente der Psychophysik," has, in connection with his psychophysical law, discussed the applicability of the method to cases of sensation generally such as the estimate of the relative weights of two masses, and he finds that if v be the energy of the force which produces the excitation of any nerve, then if log v be considered as the observed quantity, the errors of the observed quantity will follow the law of least squares. Strictly speaking, the law of least squares recognizes the possibility of any error positive or negative, however great, although the probability of indefinitely great errors will be indefinitely small. It occurred to me that in the case of the emersion of a star from an occultation, since it was impossible to strike the chronograph-key too early, while it might be struck indefinitely too late, the law of least squares could hardly apply, and I have made some experiments upon this point, which I will narrate in detail at the end of this paper, merely remarking in this place that the divergence from the theoretical law proved to be insignificant. On the other hand, there are many cases in which we have no reason to expect that the errors will vary according to the least-squares curve. Let us consider, for example, a chemical analysis. Here the error is generally not due to the combined action of a very great number of very small causes, but, on the contrary, there are generally two or three leading causes of error, depending upon some defect in the theory of the analysis or on some error in the manipulation, which is likely to result from a single one, such as cannot be regarded as in any way compounded of a large number of independent parts. Thus, a drop may be spilled or a portion thrown out of the crucible too small to be detected, but the whole drop is spilled at once, or the whole portion goes at once. In very exact analysis, in which such causes are almost altogether eliminated, the remaining and chief cause of error will be an error of weighing, due to a want of delicacy of the balance, and will be of the same nature as an error of computation, due to the fact that the number of decimal places used has been limited. The method of considering such errors is treated by Dr. Bremiker, in the preface to his "Tabula Logarithmorum Sex Decimalium," a work to which the attention of chemists ought to be drawn. When the law of facility of errors cannot be deduced *a priori* from the consideration of the causes to which it is due, a large number of experimental observations must be made upon a known quantity in order to find in what manner the errors vary, or the same series of observations may be used both to determine what the value of the unknown quantity is and also what the law of the variation of errors is. Thus, in the method of least squares, h is to be determined empirically, and the common way of doing it is to use the actual observations by which the unknown quantity is determined to determine also the value of h. In doing this, it should be remembered that a precise and trustworthy determination can only be obtained from a large number of observations. This procedure amounts, in fact, to adding an additional unknown quantity, a very obvious fact, and yet one which is habitually overlooked. Encke, in his "Astronomisches Jahrbuch" for 1834, has given the most complete formulae that I have anywhere found for determining the value of h, as well as its uncertainty. These formulae require correction on account of the circumstance just mentioned.

When it is necessary to combine, by least squares, observations of different orders of precision, they are weighted proportionately to h^2.

If we have two series of observations, one of which is as accurate as you please, and the other as inaccurate as you please, a better result than that which the most accurate series of measures gives can always be got by combining with it the least accurate series, provided the proper weights are given to the two series. This proposition seems paradoxical, and is not admitted by many very competent heads, but I cannot see how the conclusion can possibly be evaded. It does not depend at all upon any of the peculiar principles used by the method of last squares, but rests on the fundamental axioms of probabilities. Indeed, it may conveniently be based directly upon the principles of logic itself. The least accurate series of measures offers certain facts, which may be used as premises, and it cannot be that if these facts are properly used they leave us in greater ignorance than we were before. Additional facts must increase our knowledge, if the proper inferences are made from them, and, therefore, an additional series of observations, if it have any weight at all, must, if its proper weight be assigned to it, improve the value of the unknown quantity. On the other hand, when it is considered that there is an uncertainty in the value of h, it may be that if the two series of observations differ greatly in accuracy, and the value of h is not determined with much precision, it may be better at once to take the result of the best series of observations than to combine the two series with the best weights that we are able to give.

Let —

x_i be the value from any set of observations;
ε_i the mean error of this set;
$g_i\,\varepsilon_i$ the mean error of ε_i;
w_i the true weight;
E the mean error of weight *one*; and
x the truly weighted mean.

$$x = \frac{\Sigma_i(w_i x_i)}{\Sigma_i w_i}$$

The best approximation we can get to x will be subject to two kinds of error: first, those arising from errors of x_i; and, secondly, those arising from errors in our assumed values of ε_i, from which we derive w_i by the formula,

$$w_i = \frac{E^2}{\varepsilon_i^2}$$

The mean error of w_i will be $w_i^2\ g_i^2$

$$D_{x_2}x = \frac{w_2}{\Sigma_i w_i}$$

$$D_{w_i}x = \frac{x_i \Sigma_i w_i - \Sigma_i w_i x}{(\Sigma_i w_i)^2} = \frac{x_i - x}{\Sigma_i w_i}$$

Then we have—

$$\varepsilon^2 = \Sigma_i\left(\frac{x_i - x}{\Sigma_i w_i}\right)^2 w_i^2\ g_i^2 + \Sigma_i\left(\frac{w_i}{\Sigma_i w_i}\right)^2 \varepsilon_i^2$$
$$= \frac{\Sigma_i (x_i - x)^2 w_i g_i^2 + E^2 \Sigma_i w_i}{(\Sigma_i w_i)^2}$$

These are the two common rules by which ε may be calculated. Call their results ε' and ε'', so that—

$$\varepsilon' = \frac{E}{\sqrt{\Sigma_i w_i}}$$

$$\varepsilon'' = \sqrt{\frac{\Sigma_i w_i (x_i - x)^2}{(m-1)\Sigma_i w_i}}$$

where m is the number of determinations.

The first gives—

$$\varepsilon^2 = \varepsilon'^2 + \frac{\Sigma_i (x_i - x)^2 w_i^2\ g_i^2}{(\Sigma_i w_i)^2}$$

If $m = 2$,

$$(x_1 - x) = x_i - \frac{w_1 x_1 + w_2 x_2}{w_i + w_2} = \frac{w_2}{w_1 + w_2}(x_1 - x_2)$$

$$x_2 - x = -\frac{w_1}{w_1 + w_2}(x_1 - x_2)$$

$$\varepsilon^2 = \varepsilon'^2 + \frac{w_1^2\ w_2^2}{(w_1 + w_2)^4}(g_1^2 + g_2^2)(x_1 - x_2)^2$$

Put $\dfrac{w_2}{w_1} = r$,

$$\varepsilon^2 = \varepsilon'^2 + \frac{r^2}{(1+r)^4}(g_1^2 + g_2^2)(x_1 - x_2)^2$$

$$\varepsilon'' = \sqrt{\frac{w_1 w_2^2 + w_1^2 w_2}{(w_1 + w_2)^3}(x_1 - x_2)^2} = \sqrt{\frac{w_1 w_2}{(w_1 + w_2)^2}(x_1 - x_2)^2}$$

$$\varepsilon''^2 = \frac{w_1 w_2}{(w_1 + w_2)^2}(x_1 - x_2)^2 = \frac{r}{(1+r)^2}(x_1 - x_2)^2$$

$$\varepsilon^2 = \varepsilon'^2 + \frac{r}{(1+r)^2}(g_1^2 + g_2^2)\varepsilon''^2$$

Now, suppose $\varepsilon^2 < \varepsilon_1^2 < \varepsilon_2^2$; then $r < 1$; then—

$$\varepsilon_1^2 > \varepsilon'^2 + \frac{r}{(1+r)^2}(g_1^2 + g_2^2)\varepsilon''^2$$

But—

$$\varepsilon' = \varepsilon_1 \sqrt{\frac{w_1}{w_1 + w_2}} = \frac{\varepsilon_1}{\sqrt{1+r}} = \varepsilon_2 \sqrt{\frac{r}{1+r}}$$

$$\varepsilon'^2 = \varepsilon_1^2 \frac{1}{1+r} = \varepsilon_2^2 \frac{r}{1+r}$$

$$\frac{\varepsilon_2^1}{\varepsilon_2^2} = \frac{w_2}{w_1} = r$$

$$\varepsilon_1^2 = r\varepsilon_2^2$$

$$\varepsilon_1^2 - \varepsilon'^2 = \varepsilon_1^2 \frac{r}{1+r}$$

$$\varepsilon_1^2 \frac{r}{1+r} > \frac{r}{1+r^2}(g_1^2 + g_2^2)\varepsilon''^2$$

$$\frac{\varepsilon_1^2}{\varepsilon''^2}(1+r) > (g_1^2 + g_2^2)$$

$$\varepsilon_2^2 r = \varepsilon_1^2$$

$$r\frac{\varepsilon_2^2}{\varepsilon''^2}(1+r) > (g_1^2 + g_2^2)$$

$$r\frac{\varepsilon_1^2 + \varepsilon_2^2}{\varepsilon''^2} > [g_2^1 + g_2^2]$$

Unless this condition is satisfied, the combination is worse than the best determination.

It is generally admitted that m_i being the number of observations from which ε_i has been determined,

$$g_i = \frac{1}{\sqrt{2m_i}}$$

Then the condition is—

$$2r\frac{\varepsilon_1^2+\varepsilon_2^2}{\varepsilon''^2} > \frac{1}{m_1}+\frac{1}{m_2}$$

or we may write—

$$\frac{\varepsilon_1^2+\varepsilon_2^2}{(x_1-x_2^2)} > \frac{g_1^2+g_2^2}{(1+r)^2}$$

or—

$$\frac{\varepsilon_1^2+\varepsilon_2^2}{(x_1-x_2)^2}(1+r)^2 > \frac{\frac{1}{m_1}+\frac{1}{m_2}}{2}$$

or we may write—

$$\frac{\frac{(x_1-x)^2}{\varepsilon_1^2}+\frac{(x_2-x)^2}{\varepsilon_1^2}}{\frac{1}{\varepsilon^2}+\frac{1}{\varepsilon_2^2}} < \frac{-4\varepsilon_1^2}{\frac{1}{m_1}+\frac{1}{m_2}}$$

Some writers upon the subject have wished to assign a smaller weight to those observations which differ largely from the mean than to those which come close to it. They have reasoned as if h, or the precision of an observation, were something which belonged to a single observation; whereas, in fact, it is a statistical quantity altogether, and belongs only to an observation as a member of a certain series. We may have a large series of observations for which h has a certain value, and those observations may perhaps be separated into two series on some principle or other for which h shall have two different values, and if this can be done there is an advantage in doing it. It is, in fact, limiting our universe. In probabilities generally h is a mean or probable quantity for a series of observations, and if we can divide our universe into two parts, getting different values of h, it will be an increase of knowledge to do so. For example, suppose that some of the observations were taken under one set of circumstances and the rest under another set of circumstances. That would afford a principle upon which the observations could be distinguished, and if the value of h for the two sets turned out different, it would be an advantage to separate them and to give them different weights. Now, the degree of discordance of observations from the most probable value of the unknown quantity may be taken as a means of estimating the relative degrees of care, &c., used in making them, and so to discriminate between them. But it would certainly be very absurd to allow no weight to the fact that we have endeavored to make them all with equal care. It must never be forgotten that h is a statistical quantity; not one which belongs to a single observation, but one which belongs to an infinite series of observations.

It is entirely in accordance with the method of least squares to reject discordant observations, and this has always been done, even by those who object to an exact criterion for determining what observations should be rejected. For example, Mr. Glaisher says that no observation should be rejected excepting obvious mistakes, thereby admitting that it is proper sometimes to reject observations, and nobody is more opposed to the rejection of observations than he. But no line of demarkation can be drawn between mistakes which are obvious and mistakes which are not obvious. In some cases it may be obvious that 53 has been written by the recorder instead of 35. In other cases it may be doubtful whether it ought to be called an obvious mistake or whether there may be some doubt hanging over it, and there is every grade of probability, from the greatest to the least; and when we examine into the facts of observation, and do not attempt to make our way through a vacuous space of pure theory, it will be found that the occasional rejection of observations is justified from every point of view; and if observations are to be rejected, exact criteria are necessary to determine upon principles of probabilities in what cases they should be rejected. The criterion of Professor Peirce is the only one which conforms rigidly to those principles, and, indeed, I am not aware that it has been attacked upon the ground of not conforming to the principles of probabilities, although it has been attacked on the ground that no such criterion should be used, and that no observation should be rejected except upon principles of guesswork, for that is what it amounts to to say that none but obvious mistakes should be rejected. Experience has shown that the errors which this criterion rejects are almost precisely those which a person of sound judgment would pronounce to be obvious mistakes, but some other criteria have been proposed, which are confessedly inexact, but which have the advantage of involving less calculation, but these are no better than the unaided judgment of an experienced person, and in some cases not so good.

ACCOUNT OF THE EXPERIMENTS

These experiments were made in order to study the distribution of errors in the observation of a phenomenon not seen coming on, as in the case of a transit, but sudden, as in the case of the emersion of a star from behind the moon. The time was noted upon a Hipp chronoscope,which is a modification of an invention of Wheatstone's. The train of clock-work moved by weight is regulated by the vibration of a little spring or reed striking against a toothed wheel a thousand times a second. There are two hands, one of which marks tenths of a second, and the other thousandths of a second. These hands are thrown into gear when the first event occurs, and out of gear when the second event occurs, so that the amount that they have moved measures the interval. The manner in which they are thrown in and out of gear is this: The axis of one of the wheels of the main train is hung, and the axis of one of the wheels of the hand-gearing passes completely through it and comes out behind, where it rests upon a spring, which spring is influenced by an electro-magnet. There are two crown-wheels, one upon the hollow axis belonging to the main train already mentioned, the other facing it at a very small distance from it, and fixed in position and upon the axis of the wheel belonging to the hand-gearing, which moves backward and forward inside, and the other axis as described. There is a little arm, which will catch in the teeth of one or other of these crown-wheels. Before the first event it is in the teeth of the fixed crown-wheel, which thus prevents the hands from turning round. When the first event occurs this arm is thrown forward into the teeth of the rapidly rotating crown-wheel, and thus the hands begin to turn round. When the second event occurs the arm is thrown into the teeth of the first crown-wheel, and so the hands are suddenly stopped.

It will be observed that although the instrument only registers to thousandths of a second, yet if an event can be repeated many times with a variation of time much smaller than that, the instrument ought, theoretically, in the mean of a large number of observations, to give a much closer result than 0.001 second; for when the first event occurs, and the arm is thrown into the moving crown-wheel, it probably will not strike exactly in any catch, but will strike on the inclined side of a tooth. If it strikes on the forward side of the tooth, the hands will be carried forward by the fraction of a thousandth of a second more as the arm glides down this side to the bottom of the catch. But if it strikes on the back side of the tooth, the hands will be carried relatively back the fraction of the thousandth part of a second as the arm glides back to the bottom of the catch. Now, if the top of the tooth is midway between the bottom of two catches, it is equally likely to be carried forward or back. The same thing occurs when the second event happens and the arm strikes upon the fixed crown-wheel. An error in the marked interval will thus result, which error may amount to -0.001 second in the extreme, or to $+0.001$ second, and any one error between these limits is as likely as any other; consequently, these errors, being entirely incidental and independent of one another, they will balance one another in the mean of a large number of observations, and thus a much higher degree of accuracy may be reached. This, however, is a matter which has no influence on the experiments which I have made, inasmuch as the interval measured by me was a variable one, and in point of fact I have never been able to make the instrument work with such nicety as to measure much closer than 0.001 second. In the descriptions of this instrument which I have seen, only two instrumental corrections have been mentioned; one is owing to the rate with which the instrument goes, and the other is with reference to the time occupied by the arm in passing from one crown-wheel to the other. To determine these two constants, a little apparatus accompanies the instrument, by which the time of the fall of a ball from different heights may be registered, and by registering the time from two different heights these two corrections, one of which is proportional to the time and the other is a constant, may be determined. The ball is held in a pair of jaws; when these jaws separate, the contact is broken, the hands begin to move, and the ball begins to fall. Care should be taken that the ball is so small that the jaws cannot be separated for any appreciable time before the ball is free to fall, but if the spring with which they open is sufficiently strong the ball may fall freely from the very first. At the bottom the ball strikes upon a platform made of wood and covered with green cloth, and it throws this platform down upon two metallic springs below it, through which contact is made again, so that the hand stops, and then the platform is held down by a catch.

As the instrument came from the makers it was found that when the ball struck upon the platform it threw one of the springs down, so that the contact was made, and then immediately interrupted before there was time for the hand to stop, so that a slight error of about 0.001 second arose in this way. This was usually corrected by putting little wooden wedges beneath these springs, so that they could not be thrown down in this way. In order further to test the instrument, I made use of a break-circuit chronometer, and measured the interval of two seconds upon the instrument for this purpose. It was necessary to employ two telegraph-repeaters. There are two ways in which this can be arranged, so as to correct a break-circuit chronometer with a Hipp chronoscope. It is sufficient to describe one of them. The arrangement is shown in the accompanying figure [no diagram is given]:

Bat. is the battery; Ch., the chronoscope; Chr., the chronometer; R., a resistance-coil; and I and II, two telegraph-repeaters. I is a common telegraph-repeater, with the nonconducting screw so far raised that when the armature once flies up the magnet cannot bring it down again. F is the magnet end of this repeater; B, the end at the second circuit. When the first circuit is broken the armature flies up and instantly breaks the second circuit. II is arranged differently from common repeaters. As long as there is a current through the first circuit and the armature is held down, there is no connection in the second circuit. When the first circuit is broken, the armature, under the influence of a very strong spring, flies up for a distance of a tenth of a millimeter, and there makes the connection in the second circuit.

This repeater can be extemporized out of a common relay. The resistance-coil should always be used in connection with the chronoscope in such a way that when the circuit is broken in the first place the current shall be so weak as just to be able to hold the hands still, while when it is made again the current shall be so strong as to make the circuit as quickly as possible.

The rate as given by the break-circuit chronometer did not agree with that found by the fall-apparatus, and indeed there was a slight discrepancy in the rate given by the latter for different heights of fall. Professor J. E. Oliver suggested to me that this discrepancy was due to a retardation of the instrument when the hands were geared in, which took place somewhat gradually, and I found that this was the case, and that the ear could detect that when the hands were geared in, during the space of three-fourths of a second, the note produced by the vibration of the reed was lowered about the sixth of a tone. The supposition that this took place uniformly sufficiently accounted for all the discrepancy, and this gave me two more instrumental constants, viz: the amount of retardation on gearing in the hands, and the time during which that retardation was brought about. With this instrument as well as with the other chronometer I made a large number of experiments upon the time occupied in answering signals of various kinds, such as the emission of points upon paper from behind a screen, the appearance of induction-sparks from a Ruhmkorff coil, flashes of light thrown upon a screen, sudden changes from one magic-lantern figure to another, &c., the general result of which was to confirm the facts already in our possession, and which are due to the researches of Hirsch, Daumbusch, and others. But there was one series of experiments which deserves particular description. I employed a young man about eighteen years of age, who had

had no previous experience whatever in observations, to answer a signal consisting of a sharp sound like a rap, the answer being made upon a telegraph-operator's key nicely adjusted. Five hundred observations were made on every week-day during a month, twenty-four days' observations in all. The result are given in the accompanying table, and are also shown upon [the following diagrams]. On this plate the abscissæ represent the interval of time between the signal and the answer as indicated on the Hipp chronoscope, the ordinates measure the number of observations, which were subject to a large amount of error. The curve has, however, not been plotted directly from the observations, but after they have been smoothed off by the addition of adjacent numbers in the table eight times over, so as to diminish the irregularities of the curve. The smoother curve on the figures is a mean curve for every day drawn by eye so as to eliminate the irregularities entirely. It was found that after the first two or three days the curve differed very little from that derived from the theory of least squares. It will be noticed that on the first day, when the observer was entirely inexperienced, the observations scattered to such an extent that I have been obliged to draw the curve upon a different scale from that adopted for the other days. It will also be seen that the personal equation from the mean amount by which the answer came too late rapidly decreased for the first five days, until it was about one seventh of a second, and that it then gradually increased until the twelfth day, when it amounted to about 0.22 seconds. But while the personal equation was thus first diminishing and afterward increasing, the probable error or range of errors was constantly decreasing after the twelfth day. There was some variation in the personal equation, but not much, but the range of errors continually decreased as long as the observations lasted, and so remarkably that for the twenty-fourth day the probable error does not exceed one-eightieth of a second. I think that this clearly demonstrates the value of such practice in training the nerves for observation, for it can hardly be supposed that the best observer has so small a range of error as this, and I would therefore recommend that transit-observers be kept in constant training by means of some observations of an artificial event which can be repeated with rapidity, so that several hundred can be taken daily without great labor, and I do not think that it is essential that these observations should very closely imitate the transit of a star over wires, inasmuch as it is the general condition of the nerves which it is important to keep in training more than anything peculiar to this or that kind of observation.

Details of the experiments

FIRST DAY, JULY 1, 1872

Thousandths of a second	Number of observations	Thousandths of a second	Number of observations	Thousandths of a second	Number of observations	Thousandths of a second	Number of observations	Thousandths of a second	Number of observations	Thousandths of a second	Number of observations	Thousandths of a second	Number of observations	Thousandths of a second	Number of observations	Thousandths of a second	Number of observations
158	1	348	0	389	1	430	0	471	1	512	1	553	0	593	1	633	2
240	1	9	1	390	3	1	2	2	4	3	2	4	1	4	1	4	1
261	1	350	0	1	2	2	0	3	4	4	2	5	1	5	0	5	1
277	1	1	0	2	3	3	3	4	2	5	3	6	2	6	0	6	3
296	1	2	0	3	4	4	2	5	1	6	2	7	2	7	1	7	1
312	1	3	1	4	2	5	1	6	2	7	1	8	0	8	0	8	1
3	0	4	2	5	1	6	5	7	1	8	0	9	1	9	1	9	0
4	0	5	2	6	4	7	1	8	3	9	2	560	3	600	1	640	0
5	1	6	0	7	2	8	2	9	1	520	1	1	1	1	0	1	0
6	1	7	1	8	0	9	0	480	5	1	0	2	1	2	2	2	0
7	0	8	2	9	3	440	2	1	2	2	0	3	0	3	0	3	0
8	0	9	1	400	1	1	3	2	1	3	1	4	0	4	0	4	0
9	0	360	0	1	3	2	1	3	0	4	0	5	2	5	0	5	0
320	0	1	3	2	0	3	1	4	4	5	3	6	0	6	1	6	1
1	2	2	3	3	0	4	1	5	2	6	1	7	2	7	0	7	0
2	0	3	2	4	3	5	3	6	2	7	1	8	0	8	1	8	0
3	1	4	1	5	2	6	3	7	0	8	2	9	1	9	1	9	0
4	1	5	0	6	2	7	5	8	1	9	1	570	1	610	0	650	0
5	2	6	2	7	2	8	2	9	6	530	2	1	1	1	1	1	0
6	0	7	3	8	3	9	1	490	2	1	2	2	2	2	0	2	2
7	0	8	1	9	3	450	1	1	0	2	1	3	0	3	0	3	0
8	0	9	1	410	2	1	1	2	6	3	2	4	1	4	0	4	0
9	0	370	1	1	3	2	1	3	2	4	0	5	0	5	0	5	0
330	0	1	1	2	3	3	2	4	0	5	3	6	1	6	1	6	0
1	0	2	1	3	1	4	2	5	1	6	0	7	1	7	0	7	0
2	1	3	0	4	4	5	3	6	2	7	1	8	1	8	0	8	0
3	0	4	1	5	3	6	3	7	3	8	1	9	2	9	0	9	1
4	0	5	1	6	6	7	2	8	2	9	2	580	1	620	1	660	1
5	0	6	0	7	2	8	2	9	2	540	3	1	2	1	1	1	0
6	0	7	2	8	2	9	1	500	0	1	3	2	0	2	1	2	0
7	2	8	1	9	2	460	3	1	3	2	1	3	3	3	2	3	0
8	0	9	1	420	2	1	6	2	3	3	1	4	2	4	0	4	1
9	1	380	1	1	1	2	1	3	5	4	1	5	0	5	1	682	1
340	0	1	1	2	0	3	3	4	0	5	0	6	2	6	1	692	1
1	0	2	0	3	3	4	1	5	2	6	1	7	1	7	0	732	1
2	1	3	0	4	2	5	1	6	4	7	0	8	0	8	2	748	1
3	1	4	1	5	4	6	3	7	6	8	2	9	1	9	0	757	1
4	2	5	0	6	3	7	6	8	4	9	2	590	1	630	0	900	1
5	2	6	4	7	0	8	4	9	1	550	1	1	2	1	1	919	1
6	0	7	2	8	4	9	4	510	2	1	1	592	2	632	2	978	1
347	3	388	1	429	2	470	4	511	0	552	2						

Details of the experiments — Continued

SECOND DAY, JULY 5,1872

Thousandths of a second	Number of observations	Thousandths of a second	Number of observations	Thousandths of a second	Number of observations	Thousandths of a second	Number of observations	Thousandths of a second	Number of observations	Thousandths of a second	Number of observations	Thousandths of a second	Number of observations	Thousandths of a second	Number of observations	Thousandths of a second	Number of observations
97	1	176	3	203	3	230	8	257	3	283	1	309	1	335	0	361	1
106	1	7	1	4	4	1	10	8	7	4	5	310	0	6	1	2	0
119	1	8	1	5	3	2	6	9	2	5	0	1	0	7	0	3	1
124	1	9	2	6	1	3	5	260	3	6	1	2	0	8	0	4	0
127	1	180	2	7	4	4	1	1	2	7	0	3	2	9	1	5	0
129	1	1	2	8	4	5	5	2	3	8	4	4	2	340	0	6	2
134	1	2	1	9	3	6	3	3	5	9	0	5	0	1	0	7	0
136	1	3	1	210	3	7	7	4	5	290	1	6	2	2	0	8	0
139	1	4	0	1	6	8	7	5	1	1	2	7	1	3	0	9	1
140	1	5	2	2	4	9	7	6	3	2	1	8	3	4	1	370	0
144	1	6	0	3	5	240	6	7	2	3	1	9	1	5	0	1	0
153	1	7	1	4	4	1	1	8	8	4	1	320	0	6	2	2	0
154	1	8	2	5	4	2	2	9	1	5	0	1	0	7	0	3	0
155	1	9	0	6	7	3	3	270	3	6	4	2	1	8	0	4	0
159	1	190	1	7	4	4	6	1	3	7	4	3	0	9	0	5	0
164	1	1	2	8	8	5	4	2	3	8	1	4	0	350	1	6	0
5	1	2	4	9	2	6	7	3	1	9	3	5	0	1	0	7	0
6	0	3	2	220	7	7	3	4	1	300	5	6	1	2	3	8	0
7	1	4	1	1	7	8	4	5	0	1	1	7	1	3	0	9	1
8	1	5	4	2	3	9	2	6	2	2	3	8	0	4	0	380	0
9	0	6	1	3	7	250	5	7	2	3	0	9	0	5	1	1	1
170	1	7	4	4	6	1	1	8	2	4	1	330	0	6	0	2	0
1	0	8	3	5	3	2	7	9	5	5	2	1	0	7	0	3	1
2	2	9	3	6	4	3	5	280	3	6	1	2	0	8	0	4	1
3	0	200	2	7	2	4	5	1	1	7	2	3	0	9	1	5	1
4	0	1	6	8	2	5	6	282	1	308	3	334	0	360	0	386	1
175	6	202	7	229	9	256	3										

Details of the experiments — Continued

THIRD DAY, JULY 6, 1872

Thousandths of a second	Number of observations	Thousandths of a second	Number of observations	Thousandths of a second	Number of observations	Thousandths of a second	Number of observations	Thousandths of a second	Number of observations	Thousandths of a second	Number of observations	Thousandths of a second	Number of observations	Thousandths of a second	Number of observations	Thousandths of a second	Number of observations
90	**2**	142	1	164	2	186	4	208	5	229	3	250	0	271	1	292	1
106	1	3	1	5	3	7	4	9	4	230	2	1	1	2	2	3	1
111	1	4	3	6	3	8	2	210	3	1	6	2	2	3	0	4	0
113	1	5	1	7	5	9	2	1	4	2	4	3	5	4	2	5	1
124	2	6	3	8	2	190	2	2	3	3	2	4	1	5	1	6	0
5	2	7	5	9	5	1	6	3	1	4	3	5	2	6	1	7	0
6	2	8	5	170	4	2	4	4	9	5	2	6	2	7	0	8	0
7	2	9	2	1	8	3	5	5	9	6	2	7	2	8	1	9	0
8	2	150	4	2	1	4	10	6	5	7	1	8	2	9	0	300	0
9	0	1	7	3	3	5	10	7	1	8	4	9	4	280	1	1	0
130	1	2	2	4	4	6	1	8	2	9	3	260	1	1	1	2	0
1	0	3	0	5	6	7	4	9	7	240	2	1	2	2	3	3	1
2	2	4	3	6	2	8	5	220	3	1	1	2	1	3	1	318	1
3	0	5	1	7	3	9	10	1	4	2	3	3	2	4	0	323	1
4	1	6	1	8	6	200	8	2	7	3	1	4	1	5	1	347	1
5	0	7	5	9	2	1	3	3	5	4	0	5	3	6	0	356	1
6	2	8	1	180	6	2	4	4	3	5	3	6	0	7	1	364	1
7	0	9	3	1	5	3	4	5	2	6	4	7	0	8	0	365	1
8	1	160	4	2	2	4	4	6	5	7	2	8	1	9	1	381	1
9	1	1	6	3	2	5	8	7	8	8	1	9	1	290	0	445	1
140	3	2	1	4	8	6	4	228	3	249	1	270	0	291	1	452	1
141	1	163	3	185	2	207	5										

FOURTH DAY, JULY 8, 1872

65	1	148	1	169	2	190	6	211	10	231	3	251	1	271	2	291	0
97	1	9	1	170	4	1	9	2	6	2	5	2	2	2	1	2	0
111	2	150	0	1	7	2	9	3	9	3	3	3	2	3	2	3	0
126	1	1	1	2	5	3	4	4	8	4	3	4	1	4	0	4	1
131	1	2	2	3	4	4	7	5	6	5	3	5	1	5	0	5	0
2	0	3	2	4	5	5	9	6	1	6	3	6	2	6	0	6	0
3	0	4	1	5	3	6	5	7	6	7	2	7	0	7	1	7	2
4	1	5	2	6	6	7	6	8	1	8	3	8	0	8	0	8	0
5	1	6	0	7	10	8	5	9	5	9	2	9	2	9	1	9	2
6	0	7	3	8	4	9	5	220	5	240	3	260	1	280	0	314	1
7	0	8	0	9	4	200	10	1	5	1	1	1	2	1	0	319	1
8	1	9	5	180	2	1	12	2	5	2	1	2	0	2	0	323	1
9	0	160	2	1	1	2	8	3	4	3	1	3	4	3	1	327	1
140	0	1	3	2	8	3	6	4	8	4	1	4	3	4	0	343	1
1	1	2	2	3	13	4	10	5	3	5	0	5	0	5	0	347	1
2	0	3	1	4	10	5	6	6	4	6	1	6	0	6	0	352	1
3	1	4	3	5	5	6	6	7	3	7	3	7	1	7	1	355	1
4	1	5	2	6	4	7	3	8	3	8	0	8	1	8	0	365	1
5	0	6	3	7	3	8	6	9	5	9	1	9	0	9	0	381	1
6	1	7	3	8	6	9	6	230	4	250	0	270	0	290	0	521	1
147	1	168	2	189	6	210	8										

Details of the experiments — Continued

FIFTH DAY, JULY 9, 1872

Thousandths of a second	Number of observations	Thousandths of a second	Number of observations	Thousandths of a second	Number of observations	Thousandths of a second	Number of observations	Thousandths of a second	Number of observations	Thousandths of a second	Number of observations	Thousandths of a second	Number of observations	Thousandths of a second	Number of observations	Thousandths of a second	Number of observations
62	1	82	0	101	7	120	3	139	8	158	2	177	6	196	1	215	0
3	0	3	1	2	5	1	3	140	7	9	8	8	3	7	0	6	1
4	0	4	0	3	2	2	4	1	5	160	6	9	6	8	0	7	0
5	2	5	0	4	3	3	5	2	8	1	9	180	5	9	0	8	1
6	0	6	2	5	4	4	1	3	8	2	7	1	3	200	1	9	0
7	1	7	0	6	0	5	4	4	11	3	3	2	5	1	0	220	1
8	1	8	0	7	1	6	2	5	4	4	4	3	3	2	1	1	0
9	0	9	0	8	3	7	4	6	8	5	3	4	5	3	2	2	0
70	0	90	2	9	2	8	4	7	6	6	3	5	3	4	2	3	0
1	0	1	1	110	3	9	1	8	4	7	3	6	4	5	0	4	0
2	1	2	3	1	2	130	6	9	3	8	4	7	4	6	0	5	1
3	1	3	3	2	2	1	6	150	5	9	2	8	3	7	1	237	1
4	1	4	0	3	0	2	5	1	4	170	7	9	2	8	2	247	1
5	0	5	3	4	3	3	4	2	5	1	6	190	2	9	2	249	1
6	1	6	2	5	4	4	7	3	8	2	5	1	1	210	2	256	1
7	2	7	2	6	6	5	7	4	4	3	8	2	4	1	2	259	1
8	2	8	0	7	4	6	8	5	7	4	6	3	3	2	3	270	1
9	1	9	0	8	4	7	6	6	10	5	3	4	1	3	0	285	1
80	2	100	3	119	1	138	9	157	5	176	3	195	1	214	0	310	1
81	0																

SIXTH DAY, JULY 10, 1872

Thousandths of a second	Number of observations	Thousandths of a second	Number of observations	Thousandths of a second	Number of observations	Thousandths of a second	Number of observations	Thousandths of a second	Number of observations	Thousandths of a second	Number of observations	Thousandths of a second	Number of observations	Thousandths of a second	Number of observations	Thousandths of a second	Number of observations
66	1	117	0	137	2	157	5	177	4	197	3	217	1	237	1	257	0
72	1	8	1	8	0	8	6	8	3	8	3	8	3	8	1	8	0
75	1	9	1	9	5	9	7	9	7	9	1	9	2	9	2	9	1
87	2	120	1	140	5	160	7	180	3	200	5	220	3	240	0	260	1
88	1	1	1	1	3	1	7	1	4	1	1	1	1	1	0	1	0
101	2	2	3	2	6	2	3	2	11	2	8	2	1	2	1	2	0
2	0	3	2	3	3	3	10	3	9	3	2	3	1	3	0	3	0
3	0	4	2	4	4	4	6	4	7	4	4	4	1	4	3	4	1
4	1	5	1	5	1	5	12	5	6	5	0	5	2	5	1	272	1
5	1	6	0	6	6	6	2	6	8	6	2	6	1	6	1	277	1
6	1	7	0	7	8	7	4	7	9	7	1	7	0	7	0	280	1
7	1	8	1	8	3	8	5	8	2	8	2	8	1	8	0	285	1
8	1	9	2	9	4	9	6	9	7	9	1	9	3	9	0	287	2
9	2	130	1	150	5	170	9	190	7	210	4	230	1	250	0	290	1
110	0	1	4	1	4	1	5	1	6	1	3	1	0	1	0	316	1
1	1	2	2	2	7	2	9	2	7	2	3	2	0	2	0	327	1
2	2	3	0	3	4	3	5	3	5	3	1	3	1	3	0	367	1
3	2	4	5	4	7	4	5	4	6	4	4	4	0	4	0	376	1
4	0	5	4	5	4	5	5	5	2	5	3	5	0	5	0	392	1
5	1	136	1	156	5	176	7	196	7	216	3	236	0	256	0	411	1
116	3																

Details of the experiments. — Continued

SEVENTH DAY, JULY 15, 1872

Thousandths of a second	Number of observations	Thousandths of a second	Number of observations	Thousandths of a second	Number of observations	Thousandths of a second	Number of observations	Thousandths of a second	Number of observations	Thousandths of a second	Number of observations	Thousandths of a second	Number of observations	Thousandths of a second	Number of observations	Thousandths of a second	Number of observations
65	1	107	0	129	2	151	5	173	6	195	4	217	3	238	1	259	0
67	1	8	1	130	6	2	4	4	3	6	4	8	5	9	0	260	0
76	1	9	2	1	1	3	6	5	0	7	7	9	4	240	1	1	1
82	1	110	1	2	2	4	5	6	5	8	4	220	1	1	1	2	0
89	1	1	0	3	2	5	2	7	4	9	6	1	4	2	0	3	1
90	1	2	1	4	1	6	5	8	2	200	2	2	2	3	2	4	0
1	0	3	0	5	0	7	4	9	6	1	4	3	4	4	1	5	1
2	0	4	4	6	1	8	5	180	6	2	1	4	3	5	2	6	0
3	0	5	3	7	1	9	2	1	9	3	5	5	1	6	3	7	0
4	0	6	2	8	2	160	3	2	13	4	5	6	5	7	2	8	1
5	0	7	3	9	2	1	4	3	6	5	2	7	3	8	2	278	1
6	1	8	0	140	1	2	5	4	8	6	4	8	4	9	0	282	1
7	0	9	0	1	1	3	4	5	7	7	7	9	4	250	2	301	2
8	0	120	0	2	2	4	2	6	6	8	4	230	1	1	1	314	1
9	0	1	1	3	3	5	7	7	7	9	6	1	5	2	1	336	1
100	1	2	0	4	2	6	4	8	4	210	4	2	1	3	0	357	1
1	0	3	0	5	3	7	14	9	5	1	5	3	3	4	0	373	1
2	2	4	0	6	2	8	4	190	4	2	4	4	2	5	1	376	1
3	0	5	0	7	7	9	2	1	7	3	8	5	2	6	0	386	1
4	1	6	1	8	4	170	7	2	1	4	2	6	1	7	1	459	1
5	0	7	2	9	2	1	3	3	7	5	5	237	3	258	0	687	1
106	1	128	1	150	1	172	5	194	5	216	0						

EIGHTH DAY, JULY 16, 1872

61	1	133	0	151	2	169	5	187	9	204	9	221	5	238	1	255	1
85	1	4	0	2	2	170	7	8	10	5	10	2	7	9	1	6	0
105	1	5	0	3	1	1	3	9	3	6	3	3	2	240	0	7	1
111	1	6	0	4	4	2	3	190	11	7	8	4	1	1	2	8	0
119	1	7	2	5	5	3	5	1	6	8	5	5	4	2	1	9	0
120	0	8	2	6	3	4	7	2	8	9	5	6	1	3	1	260	1
1	0	9	1	7	2	5	4	3	11	210	4	7	2	4	1	1	0
2	1	140	0	8	4	6	4	4	8	1	3	8	3	5	0	2	1
3	1	1	0	9	1	7	8	5	8	2	10	9	5	6	0	3	0
4	0	2	0	160	3	8	3	6	8	3	7	230	4	7	1	4	1
5	0	3	3	1	2	9	5	7	7	4	9	1	2	8	0	5	1
6	1	4	1	2	2	180	6	8	8	5	3	2	3	9	0	274	1
7	0	5	0	3	4	1	9	9	8	6	3	3	1	250	0	288	2
8	0	6	4	4	0	2	5	200	6	7	6	4	1	1	1	312	1
9	0	7	2	5	3	3	6	1	4	8	5	5	3	2	0	318	1
130	0	8	2	6	4	4	10	2	13	9	2	6	0	3	2	335	1
1	1	9	2	7	4	5	4	203	7	220	10	237	1	254	0	360	1
132	1	150	1	168	3	186	11										

Details of the experiments — Continued

NINTH DAY, JULY 17, 1872

Thousandths of a second	Number of observations	Thousandths of a second	Number of observations	Thousandths of a second	Number of observations	Thousandths of a second	Number of observations	Thousandths of a second	Number of observations	Thousandths of a second	Number of observations	Thousandths of a second	Number of observations	Thousandths of a second	Number of observations	Thousandths of a second	Number of observations
71	1	136	1	154	4	172	2	189	6	206	9	223	2	240	1	257	0
77	1	7	0	5	3	3	6	190	9	7	3	4	4	1	0	8	1
88	1	8	1	6	1	4	5	1	6	8	6	5	5	2	0	9	0
106	1	9	2	7	3	5	8	2	5	9	8	6	0	3	0	260	0
114	1	140	1	8	3	6	6	3	8	210	5	7	5	4	2	1	1
115	1	1	1	9	2	7	6	4	6	1	8	8	1	5	0	2	0
124	1	2	1	160	1	8	4	5	10	2	6	9	4	6	4	3	0
5	0	3	3	1	2	9	7	6	7	3	7	230	2	7	1	4	1
6	0	4	0	2	2	180	2	7	6	4	5	1	0	8	0	276	1
7	1	5	1	3	1	1	8	8	14	5	6	2	3	9	1	281	2
8	0	6	3	4	5	2	7	9	7	6	6	3	4	250	2	286	1
9	2	7	2	5	4	3	5	200	8	7	4	4	2	1	0	301	2
130	0	8	0	6	5	4	8	1	10	8	7	5	1	2	0	302	1
1	0	9	2	7	2	5	5	2	6	9	3	6	2	3	1	307	1
2	2	150	2	8	2	6	8	3	3	220	8	7	2	4	1	347	1
3	0	1	2	9	5	7	6	4	8	1	3	8	3	5	1	368	1
4	2	2	1	170	8	188	11	205	6	222	5	239	4	256	0	505	1
135	1	153	2	171	2												

TENTH DAY, JULY 18, 1872

81	1	165	0	183	5	200	5	217	7	234	4	251	6	268	0	285	1
122	1	6	2	4	1	1	4	8	10	5	2	2	2	9	0	6	0
130	1	7	2	5	6	2	5	9	9	6	9	3	4	270	0	7	0
150	1	8	0	6	4	3	6	220	10	7	6	4	2	1	0	8	1
1	1	9	1	7	5	4	5	1	9	8	4	5	5	2	2	9	0
2	0	170	2	8	4	5	7	2	5	9	9	6	2	3	0	290	0
3	0	1	3	9	5	6	7	3	4	240	4	7	1	4	1	1	0
4	0	2	3	190	3	7	10	4	11	1	4	8	2	5	1	2	0
5	0	3	1	1	4	8	13	5	4	2	1	9	1	6	0	3	0
6	1	4	1	2	3	9	6	6	9	3	6	260	2	7	0	4	1
7	1	5	2	3	7	210	11	7	8	4	7	1	2	8	0	5	0
8	0	6	1	4	1	1	8	8	10	5	0	2	2	9	0	6	0
9	1	7	1	5	9	2	11	9	5	6	2	3	0	280	0	7	0
160	0	8	5	6	7	3	5	230	8	7	3	4	1	1	1	8	0
1	1	9	2	7	6	4	8	1	9	8	2	5	3	2	0	9	1
2	2	180	2	8	2	5	10	2	7	9	2	6	0	3	0	302	1
3	1	1	4	199	8	216	4	233	1	250	3	267	1	284	0	416	1
164	3	182	5														

Details of the experiments — Continued

ELEVENTH DAY, JULY 19, 1872

Thousandths of a second	Number of observations	Thousandths of a second	Number of observations	Thousandths of a second	Number of observations	Thousandths of a second	Number of observations	Thousandths of a second	Number of observations	Thousandths of a second	Number of observations	Thousandths of a second	Number of observations	Thousandths of a second	Number of observations	Thousandths of a second	Number of observations
68	1	158	1	178	1	198	7	218	11	238	8	258	1	278	2	298	0
114	1	9	3	9	0	9	8	9	5	9	5	9	1	9	0	9	0
135	1	160	1	180	3	200	7	220	10	240	2	260	3	280	0	300	1
141	2	1	1	1	1	1	9	1	7	1	7	1	1	1	1	1	0
2	2	2	3	2	4	2	5	2	3	2	5	2	2	2	0	2	0
3	0	3	1	3	2	3	4	3	3	3	3	3	0	3	0	3	1
4	1	4	0	4	2	4	7	4	9	4	1	4	1	4	0	4	0
5	1	5	1	5	2	5	4	5	11	5	6	5	1	5	0	5	1
6	1	6	3	6	3	6	5	6	2	6	2	6	2	6	0	6	0
7	0	7	0	7	6	7	9	7	8	7	3	7	0	7	1	7	1
8	0	8	2	8	3	8	6	8	4	8	3	8	1	8	0	8	0
9	1	9	3	9	5	9	7	9	2	9	6	9	0	9	0	9	0
150	0	170	0	190	3	210	12	230	5	250	4	270	2	290	0	310	0
1	1	1	2	1	5	1	12	1	3	1	5	1	1	1	0	1	0
2	0	2	0	2	12	2	5	2	7	2	2	2	0	2	3	2	3
3	0	3	0	3	9	3	5	3	4	3	2	3	1	3	1	360	1
4	2	4	2	4	10	4	7	4	5	4	2	4	3	4	0	379	1
5	2	5	2	5	10	5	5	5	4	5	2	5	1	5	0	400	1
6	2	6	3	6	5	6	7	6	4	6	0	6	0	6	0	437	1
157	0	177	3	197	4	217	5	237	3	257	4	277	1	297	0	491	1

TWELFTH DAY, JULY 20, 1872

95	1	175	1	193	2	211	1	229	6	247	3	265	5	282	0	299	0
109	1	6	0	4	2	2	2	230	8	8	1	6	3	3	2	300	1
126	1	7	0	5	0	3	6	1	4	9	2	7	5	4	3	1	0
141	1	8	0	6	5	4	5	2	3	250	2	8	1	5	0	2	0
157	1	9	1	7	0	5	5	3	9	1	7	9	4	6	1	3	0
162	1	180	0	8	4	6	7	4	6	2	3	270	4	7	0	4	1
3	0	1	1	9	1	7	5	5	6	3	3	1	2	8	0	5	1
4	1	2	1	200	3	8	8	6	5	4	3	2	1	9	2	314	1
5	0	3	0	1	2	9	6	7	4	5	5	3	1	290	0	316	1
6	0	4	0	2	2	220	12	8	4	6	6	4	3	1	0	318	1
7	0	5	0	3	1	1	8	9	6	7	5	5	1	2	1	322	1
8	0	6	0	4	1	2	11	240	11	8	1	6	1	3	0	329	1
9	0	7	1	5	2	3	4	1	4	9	3	7	1	4	0	354	1
170	1	8	0	6	4	4	4	2	7	260	2	8	0	5	0	365	1
1	0	9	0	7	5	5	6	3	6	1	4	9	1	6	1	374	1
2	2	190	0	8	3	6	5	4	5	2	3	280	0	7	1	390	1
3	1	1	3	9	4	7	5	5	5	3	1	281	3	298	1	391	1
174	0	192	0	210	2	228	13	246	6	264	1						

Details of the experiments — Continued

THIRTEENTH DAY, JULY 22, 1872

Thousandths of a second	Number of observations	Thousandths of a second	Number of observations	Thousandths of a second	Number of observations	Thousandths of a second	Number of observations	Thousandths of a second	Number of observations	Thousandths of a second	Number of observations	Thousandths of a second	Number of observations	Thousandths of a second	Number of observations	Thousandths of a second	Number of observations
133	1	166	0	185	0	204	0	223	8	242	8	261	14	280	3	299	0
140	1	7	0	6	0	5	1	4	5	3	11	2	8	1	0	300	0
148	1	8	0	7	0	6	3	5	9	4	8	3	4	2	3	1	0
9	1	9	0	8	1	7	3	6	5	5	9	4	7	3	6	2	0
150	0	170	0	9	0	8	4	7	5	6	7	5	2	4	2	3	0
1	0	1	0	190	0	9	2	8	9	7	9	6	6	5	1	4	1
2	0	2	0	1	0	210	2	9	4	8	7	7	4	6	0	5	0
3	0	3	0	2	0	1	6	230	8	9	9	8	5	7	4	6	1
4	0	4	0	3	0	2	4	1	13	250	5	9	6	8	0	7	0
5	0	5	0	4	1	3	1	2	7	1	9	270	5	9	0	8	0
6	0	6	0	5	2	4	3	3	6	2	8	1	6	290	1	9	0
7	1	7	1	6	0	5	2	4	13	3	8	2	3	1	1	310	1
8	1	8	0	7	0	6	3	5	8	4	5	3	2	2	0	1	1
9	0	9	0	8	2	7	2	6	6	5	13	4	5	3	1	2	1
160	0	180	0	9	0	8	4	7	3	6	2	5	5	4	1	3	0
1	0	1	0	200	1	9	4	8	10	7	7	6	3	5	2	4	0
2	0	2	0	1	0	220	5	9	5	8	4	7	2	6	3	5	0
3	0	3	0	2	0	1	10	240	5	9	7	8	2	7	1	6	1
4	1	184	1	203	2	222	5	241	7	260	5	279	3	298	2	373	1
165	0																

FOURTEENTH DAY, JULY 23, 1872

63	1	185	0	203	4	221	6	239	13	256	12	273	2	290	2	307	0
105	1	6	0	4	2	2	9	240	6	7	1	4	0	1	0	8	
119	1	7	4	5	3	3	5	1	10	8	6	5	3	2	0	9	
132	1	8	2	6	5	4	10	2	9	9	2	6	0	3	0	310	0
157	1	9	2	7	2	5	9	3	11	260	6	7	0	4	0	1	
158	1	190	0	8	1	6	6	4	8	1	6	8	2	5	0	2	
166	1	1	3	9	8	7	6	5	9	2	2	9	1	6	0	3	
174	1	2	1	210	5	8	11	6	11	3	5	280	1	7	1	4	0
5	0	3	1	1	8	9	14	7	5	4	3	1	1	8	0	5	0
6	0	4	0	2	1	230	10	8	8	5	1	2	1	9	0	6	0
7	0	5	2	3	5	1	11	9	3	6	4	3	1	300	0	7	2
8	0	6	4	4	4	2	5	250	5	7	2	4	0	1	0	8	0
9	1	7	3	5	3	3	8	1	7	8	1	5	0	2	0	9	1
180	1	8	2	6	2	4	6	2	9	9	0	6	1	3	0	359	1
1	1	9	1	7	4	5	9	3	9	270	1	7	0	4	1	470	1
2	0	200	4	8	7	6	8	4	8	1	3	8	0	5	1	693	1
3	0	1	4	9	3	7	11	255	7	272	4	289	1	306	2	705	1
184	3	202	1	220	4	238	12										

Details of the experiments — Continued

FIFTEENTH DAY, JULY 24, 1872

Thousandths of a second	Number of observations	Thousandths of a second	Number of observations	Thousandths of a second	Number of observations	Thousandths of a second	Number of observations	Thousandths of a second	Number of observations	Thousandths of a second	Number of observations	Thousandths of a second	Number of observations	Thousandths of a second	Number of observations	Thousandths of a second	Number of observations
73	1	177	0	195	2	213	4	231	5	249	4	267	1	285	1	302	1
105	1	8	0	6	1	4	9	2	11	250	7	8	7	6	0	3	0
110	1	9	1	7	1	5	8	3	8	1	4	9	1	7	1	4	0
112	1	180	0	8	3	6	4	4	8	2	6	270	4	8	2	5	0
140	1	1	0	9	1	7	8	5	9	3	3	1	3	9	1	6	0
148	1	2	0	200	0	8	7	6	15	4	4	2	4	290	1	7	0
158	1	3	0	1	1	9	9	7	7	5	7	3	1	1	0	8	0
166	1	4	1	2	4	220	5	8	12	6	8	4	1	2	1	9	2
7	0	5	0	3	1	1	5	9	10	7	6	5	2	3	1	310	1
8	1	6	1	4	2	2	10	240	15	8	3	6	1	4	0	1	1
9	1	7	3	5	2	3	12	1	9	9	1	7	1	5	2	2	0
170	0	8	0	6	1	4	7	2	8	260	2	8	1	6	1	3	0
1	0	9	0	7	0	5	6	3	14	1	3	9	3	7	0	4	0
2	0	190	0	8	5	6	7	4	4	2	8	280	1	8	0	5	1
3	0	1	0	9	4	7	8	5	4	3	3	1	0	9	0	406	1
4	2	2	2	210	6	8	11	6	7	4	1	2	1	300	0	443	1
5	1	3	4	1	5	9	8	7	7	5	1	3	0	301	0	467	2
176	0	194	1	212	5	230	12	248	8	266	2	284	1				

SIXTEENTH DAY, JULY 25, 1872

67	1	170	2	186	0	202	2	218	6	234	8	250	6	266	4	282	2
86	1	1	0	7	3	3	4	9	8	5	10	1	10	7	5	3	0
103	1	2	0	8	3	4	1	220	8	6	8	2	6	8	6	4	2
114	1	3	0	9	0	5	10	1	5	7	5	3	3	9	3	5	0
157	1	4	0	190	1	6	3	2	3	8	7	4	5	270	4	6	0
8	0	5	0	1	0	7	6	3	3	9	7	5	4	1	4	7	1
9	0	6	3	2	2	8	4	4	6	240	6	6	10	2	1	8	0
160	0	7	0	3	4	9	3	5	10	1	11	7	7	3	3	9	0
1	1	8	1	4	1	210	1	6	7	2	11	8	4	4	2	290	0
2	1	9	1	5	0	1	7	7	8	3	12	9	3	5	0	1	1
3	0	180	2	6	2	2	5	8	9	4	9	260	0	6	0	303	1
4	0	1	1	7	1	3	3	9	12	5	7	1	2	7	0	312	1
5	0	2	2	8	3	4	3	230	15	6	6	2	3	8	1	313	1
6	0	3	2	9	4	5	2	1	13	7	7	3	2	9	1	333	1
7	0	4	1	200	5	6	4	2	6	8	7	4	0	280	4	357	1
8	0	185	3	201	6	217	5	233	8	249	8	265	2	281	2	784	1
169	0																

Details of the experiments — Continued

SEVENTEENTH DAY, JULY 26, 1872

Thousandths of a second	Number of observations	Thousandths of a second	Number of observations	Thousandths of a second	Number of observations	Thousandths of a second	Number of observations	Thousandths of a second	Number of observations	Thousandths of a second	Number of observations	Thousandths of a second	Number of observations	Thousandths of a second	Number of observations	Thousandths of a second	Number of observations
76	1	195	0	216	1	237	5	258	9	279	10	300	4	321	1	342	1
104	1	6	2	7	0	8	2	9	7	280	4	1	0	2	0	3	0
128	1	7	0	8	1	9	5	260	11	1	7	2	1	3	1	4	1
147	1	8	0	9	2	240	7	1	9	2	6	3	1	4	1	5	3
162	1	9	0	220	3	1	2	2	8	3	7	4	1	5	1	6	0
166	1	200	1	1	3	2	4	3	5	4	7	5	1	6	1	7	1
180	1	1	1	2	0	3	5	4	7	5	3	6	2	7	0	8	2
1	0	2	0	3	4	4	4	5	8	6	6	7	0	8	0	9	0
2	0	3	1	4	3	5	9	6	5	7	7	8	2	9	0	350	0
3	0	4	0	5	2	6	2	7	3	8	3	9	5	330	0	1	0
4	0	5	0	6	0	7	5	8	7	9	5	310	6	1	0	2	0
5	0	6	1	7	2	8	4	9	2	290	5	1	1	2	1	3	1
6	3	7	2	8	1	9	8	270	6	1	6	2	4	3	0	366	1
7	0	8	5	9	3	250	5	1	5	2	7	3	0	4	0	372	1
8	0	9	4	230	6	1	7	2	6	3	5	4	1	5	0	380	1
9	0	210	2	1	2	2	4	3	7	4	5	5	4	6	0	390	1
190	1	1	2	2	4	3	5	4	5	5	3	6	1	7	1	392	1
1	0	2	2	3	3	4	7	5	6	6	5	7	2	8	1	394	1
2	0	3	2	4	3	5	7	6	11	7	1	8	0	9	1	448	1
3	1	4	2	5	4	6	5	7	2	8	3	9	0	340	0	467	1
194	0	215	2	236	3	257	7	278	5	299	3	320	2	341	1		

EIGHTEENTH DAY, JULY 27, 1872

184	2	201	3	218	0	235	4	252	16	269	5	286	6	302	2	318	0
5	0	2	1	9	3	6	10	3	6	270	11	7	4	3	1	9	0
6	0	3	2	220	2	7	5	4	10	1	10	8	1	4	0	320	0
7	0	4	2	1	3	8	2	5	11	2	6	9	1	5	1	1	0
8	1	5	0	2	0	9	5	6	8	3	6	290	1	6	1	2	1
9	0	6	6	3	4	240	6	7	7	4	5	1	3	7	0	3	0
190	0	7	1	4	6	1	5	8	7	5	7	2	1	8	0	4	0
1	0	8	1	5	2	2	3	9	8	6	5	3	1	9	0	5	0
2	1	9	3	6	0	3	8	260	13	7	5	4	2	310	0	6	0
3	0	210	0	7	5	4	6	1	9	8	3	5	0	1	0	7	0
4	0	1	0	8	3	5	5	2	5	9	6	6	0	2	0	8	0
5	0	2	3	9	8	6	7	3	16	280	3	7	0	3	0	9	0
6	0	3	3	230	7	7	14	4	11	1	1	8	0	4	0	330	1
7	1	4	1	1	2	8	8	5	4	2	4	9	2	5	1	341	1
8	0	5	3	2	10	9	11	6	6	3	5	300	0	6	0	366	1
9	1	6	4	3	3	250	12	7	8	4	5	301	0	317	0	367	1
200	0	217	4	234	6	251	3	268	5	285	4						

Details of the experiments — Continued

NINETEENTH DAY, JULY 29, 1872

Thousandths of a second	Number of observations	Thousandths of a second	Number of observations	Thousandths of a second	Number of observations	Thousandths of a second	Number of observations	Thousandths of a second	Number of observations	Thousandths of a second	Number of observations	Thousandths of a second	Number of observations	Thousandths of a second	Number of observations	Thousandths of a second	Number of observations
119	1	200	0	217	0	234	5	251	11	267	8	283	2	299	1	315	0
148	1	1	0	8	1	5	2	2	10	8	6	4	1	300	1	6	0
152	1	2	0	9	2	6	7	3	13	9	11	5	3	1	2	7	0
157	1	3	1	220	3	7	7	4	15	270	6	6	5	2	1	8	1
187	1	4	0	1	3	8	9	5	11	1	10	7	4	3	0	9	0
8	0	5	1	2	3	9	4	6	12	2	12	8	4	4	0	320	0
9	0	6	0	3	0	240	11	7	7	3	9	9	5	5	0	1	1
190	1	7	2	4	1	1	8	8	11	4	10	290	1	6	0	2	0
1	1	8	0	5	0	2	7	9	6	5	4	1	0	7	0	3	0
2	0	9	0	6	2	3	8	260	11	6	5	2	1	8	2	4	1
3	0	210	2	7	7	4	7	1	8	7	5	3	2	9	1	334	1
4	0	1	0	8	0	5	10	2	11	8	4	4	1	310	1	359	1
5	0	2	1	9	3	6	11	3	4	9	2	5	1	1	1	366	1
3	0	3	0	230	3	7	8	4	4	280	7	6	1	2	0	591	1
7	0	4	1	1	4	8	11	5	4	1	5	7	0	3	0	779	1
8	1	5	1	2	3	9	13	266	8	282	4	298	1	314	0	851	1
199	1	216	1	233	6	250	7										

TWENTIETH DAY, JULY 30, 1872

130	1	195	0	214	2	233	10	252	12	271	1	290	1	309	1	328	0
159	1	6	0	5	3	4	4	3	9	2	8	1	1	310	0	9	1
177	0	7	0	6	1	5	8	4	4	3	7	2	3	1	0	330	1
8	0	8	1	7	5	6	6	5	7	4	5	3	1	2	1	1	0
9	0	9	2	8	2	7	9	6	9	5	3	4	0	3	0	2	0
180	0	200	0	9	2	8	5	7	8	6	8	5	1	4	0	3	0
1	0	1	0	220	9	9	6	8	8	7	6	6	2	5	0	4	0
2	0	2	1	1	6	240	13	9	6	8	2	7	3	6	0	5	0
3	1	3	2	2	2	1	12	260	8	9	4	8	0	7	0	6	0
4	0	4	2	3	2	2	9	1	9	280	5	9	1	8	0	7	1
5	0	5	0	4	4	3	4	2	8	1	2	300	0	9	0	8	0
6	1	6	1	5	3	4	6	3	14	2	7	1	2	320	1	9	0
7	1	7	0	6	4	5	7	4	7	3	2	2	0	1	0	340	0
8	0	8	2	7	4	6	8	5	11	4	0	3	1	2	0	1	0
9	0	9	0	8	3	7	7	6	6	5	1	4	0	3	0	2	1
190	0	210	2	9	2	8	9	7	8	6	1	5	0	4	1	360	1
1	0	1	2	230	4	9	3	8	7	7	3	6	1	5	0	486	1
3	0	2	1	1	4	250	12	9	6	8	2	7	0	6	0	754	1
2	0	213	0	232	4	251	9	270	12	289	3	308	1	327	1	817	1
194	0																

Details of the experiments — Continued

TWENTY-FIRST DAY, JULY 31, 1872

Thousandths of a second	Number of observations	Thousandths of a second	Number of observations	Thousandths of a second	Number of observations	Thousandths of a second	Number of observations	Thousandths of a second	Number of observations	Thousandths of a second	Number of observations	Thousandths of a second	Number of observations	Thousandths of a second	Number of observations	Thousandths of a second	Number of observations
75	1	193	0	209	0	225	7	241	10	257	8	273	1	288	0	303	0
137	1	4	0	210	1	6	7	2	7	8	7	4	4	9	3	4	0
160	1	5	0	1	3	7	7	3	10	9	10	5	2	290	1	5	0
161	2	6	1	2	1	8	6	4	8	260	5	6	3	1	2	6	0
181	1	7	1	3	3	9	9	5	3	1	11	7	1	2	2	7	0
2	0	8	2	4	0	230	8	6	10	2	6	8	3	3	2	8	0
3	1	9	0	5	3	1	10	7	11	3	8	9	4	4	0	9	0
4	0	200	1	6	5	2	6	8	11	4	6	280	4	5	0	310	0
5	1	1	3	7	5	3	10	9	9	5	7	1	1	6	0	1	0
6	0	2	2	8	2	4	4	250	7	6	3	2	2	2	1	2	1
7	0	3	0	9	2	5	8	1	8	7	7	3	1	8	1	3	1
8	0	4	2	220	9	6	8	2	8	8	5	4	0	9	1	331	1
9	2	5	2	1	4	7	11	3	9	9	3	5	4	300	0	348	1
190	0	6	3	2	6	8	9	4	7	270	4	6	1	1	0	349	1
1	2	7	0	3	5	9	10	5	10	1	1	287	5	302	1	438	1
192	1	208	2	224	3	240	10	256	10	272	3						

TWENTY-SECOND DAY, AUGUST 1, 1872

139	1	204	0	219	2	233	2	247	7	261	14	275	12	289	6	303	1
157	1	5	2	220	0	4	7	8	10	2	8	6	3	290	0	4	0
171	1	6	1	1	1	5	6	9	3	3	12	7	4	1	1	5	0
192	1	7	2	2	6	6	6	250	8	4	7	8	8	2	1	6	0
3	1	8	0	3	1	7	6	1	13	5	8	9	4	3	0	7	1
4	1	9	1	4	4	8	12	2	8	6	17	280	2	4	2	8	0
5	0	210	1	5	4	9	10	3	14	7	8	1	4	5	0	9	1
6	1	1	2	6	4	240	4	4	8	8	6	2	7	6	2	325	1
7	1	2	2	7	4	1	7	5	6	9	5	3	3	7	1	339	1
8	0	3	2	8	1	2	10	6	8	270	5	4	1	8	2	342	1
9	0	4	1	9	6	3	7	7	6	1	9	5	4	9	1	382	1
200	0	5	1	230	5	4	3	8	8	2	7	6	5	300	1	406	1
1	0	6	3	1	10	5	6	9	10	3	9	7	3	1	0	410	1
2	0	7	0	232	6	246	10	260	9	274	2	288	1	302	0	689	1
203	0	218	0														

Details of the experiments — Continued

TWENTY-THIRD DAY, AUGUST 2, 1872

Thousandths of a second	Number of observations	Thousandths of a second	Number of observations	Thousandths of a second	Number of observations	Thousandths of a second	Number of observations	Thousandths of a second	Number of observations	Thousandths of a second	Number of observations	Thousandths of a second	Number of observations	Thousandths of a second	Number of observations	Thousandths of a second	Number of observations
142	1	184	0	202	1	220	1	238	5	256	10	274	7	292	0	310	1
166	1	5	1	3	0	1	6	9	9	7	10	5	3	3	4	1	0
7	2	6	0	4	1	2	5	240	8	8	11	6	1	4	1	2	0
8	0	7	0	5	0	3	6	1	8	9	9	7	3	5	2	3	0
9	0	8	0	6	1	4	5	2	7	260	8	8	4	6	0	4	1
170	0	9	1	7	1	5	6	3	12	1	9	9	7	7	0	5	0
1	0	190	0	8	0	6	4	4	9	2	13	280	3	8	2	6	1
2	2	1	0	9	0	7	3	5	7	3	6	1	2	9	0	7	0
3	0	2	0	210	1	8	4	6	7	4	11	2	3	300	0	8	0
4	0	3	1	1	0	9	5	7	6	5	7	3	4	1	0	9	0
5	1	4	0	2	3	230	4	8	6	6	6	4	1	2	1	320	0
6	0	5	0	3	3	1	9	9	12	7	8	5	0	3	0	1	0
7	0	6	3	4	2	2	8	250	9	8	8	6	3	4	0	2	1
8	0	7	0	5	4	3	6	1	8	9	6	7	1	5	0	3	1
9	1	8	0	6	1	4	6	2	10	270	14	8	1	6	1	340	1
180	1	9	0	7	0	5	6	3	12	1	3	9	1	7	0	371	1
1	0	200	0	8	5	6	4	4	9	2	7	290	2	8	0	406	1
2	0	201	0	219	2	237	2	255	5	273	4	291	1	309	1	661	1
183	1																

TWENTY-FOURTH DAY, AUGUST 3, 1872

119	1	198	2	213	3	228	9	243	12	257	12	271	3	285	1	299	0
134	1	9	1	4	4	9	4	4	10	8	5	2	1	6	4	300	0
156	1	200	2	5	1	230	7	5	16	9	8	3	4	7	0	1	0
172	1	1	1	6	1	1	11	6	9	260	6	4	1	8	1	2	1
175	1	2	0	7	4	2	12	7	7	1	9	5	5	9	0	3	0
188	2	3	0	8	7	3	18	8	11	2	2	6	0	290	2	4	0
9	0	4	1	9	5	4	5	9	13	3	6	7	2	1	0	5	0
190	1	5	1	220	3	5	7	250	13	4	4	8	0	2	4	6	0
1	0	6	1	1	6	6	11	1	14	5	1	9	0	3	0	7	2
2	1	7	1	2	3	7	10	2	14	6	1	280	3	4	3	8	0
3	0	8	2	3	1	8	8	3	12	7	2	1	2	5	0	9	1
4	0	9	6	4	4	9	8	4	12	8	2	2	1	6	0	334	1
5	0	210	3	5	8	240	6	5	6	9	3	3	1	7	1	340	1
6	0	1	3	6	6	1	9	256	8	270	3	284	0	298	0	374	1
197	0	212	3	227	5	242	8										

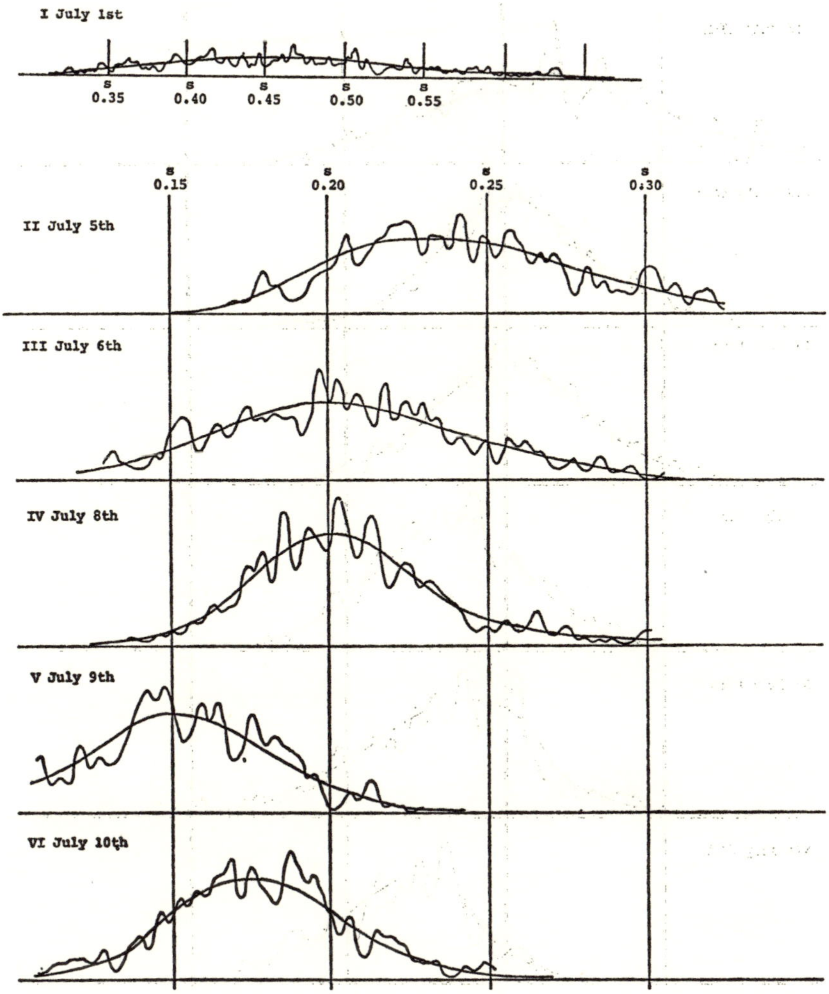
I July 1st
s 0.35
s 0.40
s 0.45
s 0.50
s 0.55
s 0.15
s 0.20
s 0.25
s 0.30
II July 5th
III July 6th
IV July 8th
V July 9th
VI July 10th

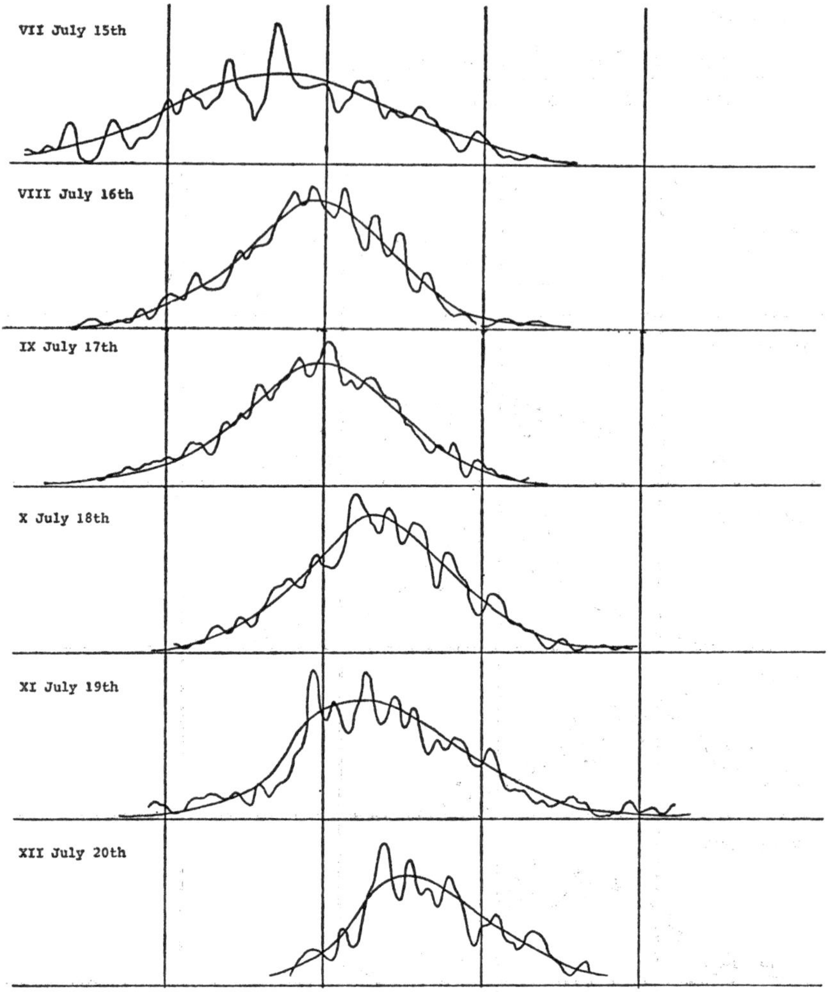
VII July 15th
VIII July 16th
IX July 17th
X July 18th
XI July 19th
XII July 20th

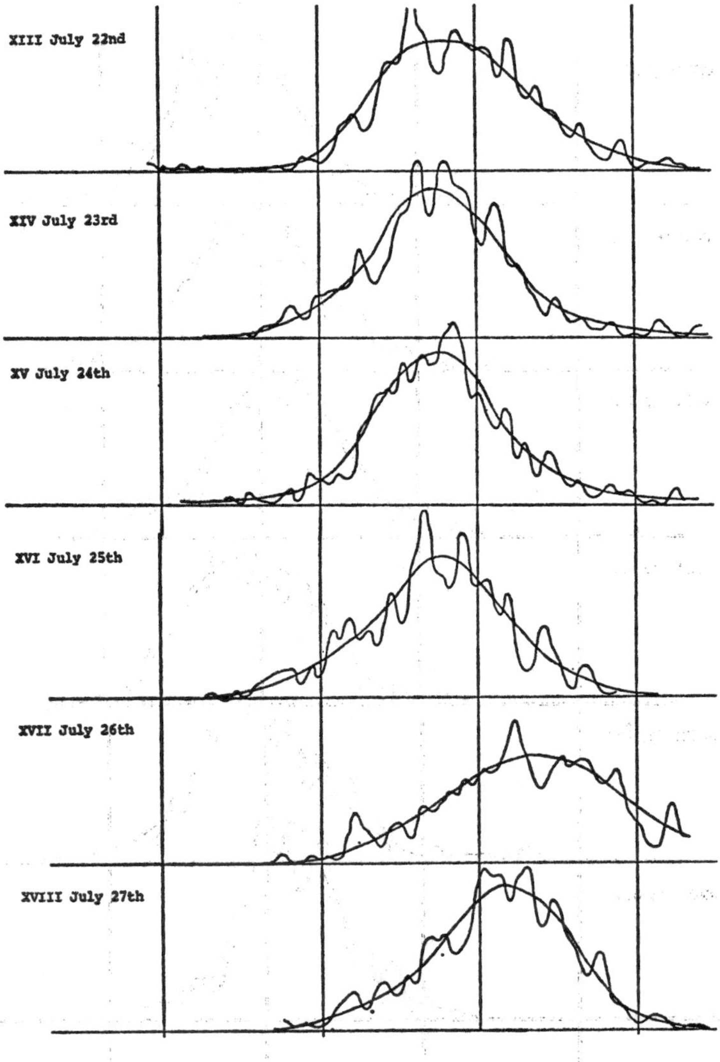
XIII July 22nd
XIV July 23rd
XV July 24th
XVI July 25th
XVII July 26th
XVIII July 27th

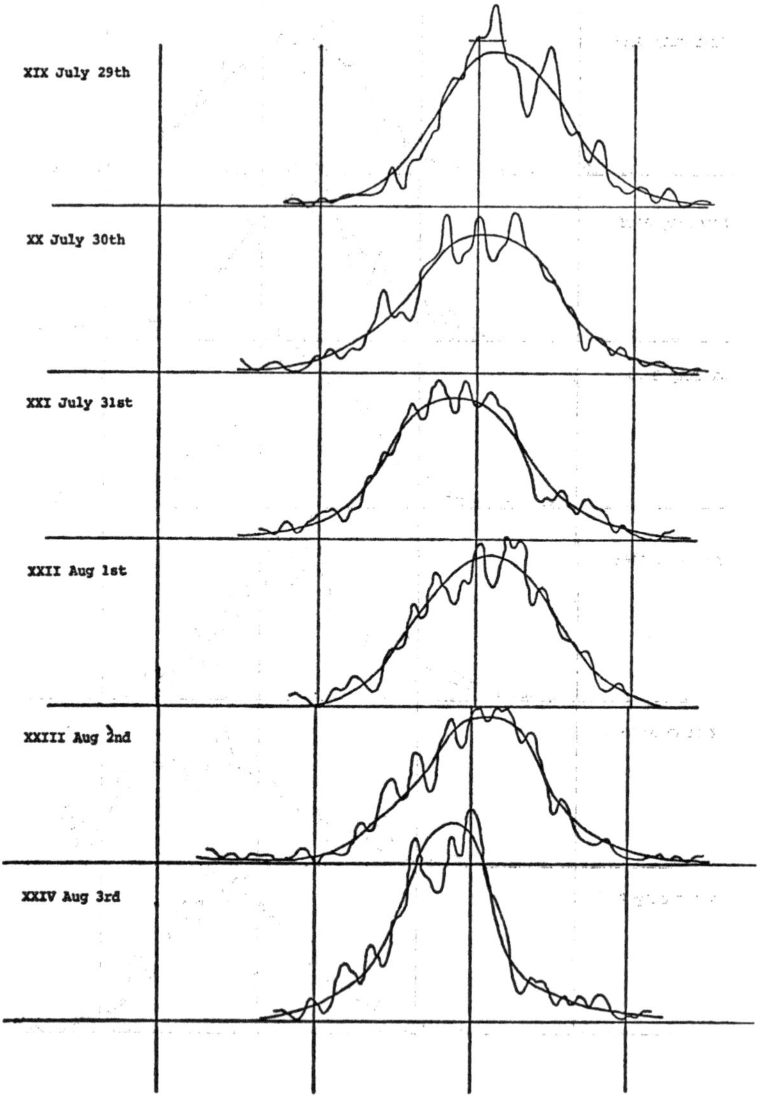
XIX July 29th
XX July 30th
XXI July 31st
XXII Aug 1st
XXIII Aug 2nd
XXIV Aug 3rd

D. THE METRIC SYSTEM OF WEIGHTS AND MEASURES

Published by the Hartford Steam Boiler Inspection and Insurance Co. (1898)

[REVIEW] (1410)

Of the many collections of tables for the comparison of the Metric and English weights and measures this is decidedly the best. We know no other manual which makes the inch 2.5400 centimetres which is the most probable value ... The book is unusually pretty and convenient, though it has the fault of not carrying its name on its back. The name of Mr. A. D. Risteen as compiler will be assurance to all who know the world of calculators that the work has been performed in every respect with superior care and judgment. The tables are remarkably clear, owing to a new mode of spacing. Another new method is used in the table of data. These are interesting features to those who make tables and to those who use them. The scales of centigrade and Fahrenheit thermometers are shown as they appear on a thermometer carrying both graduations, and modern presswork has permitted the insertion of an accurate scale of millimeters.

Mr. Risteen prefaces the tables with a full account of the Metric system, including an interesting history of its institution. He gives credit to Talleyrand for putting life into the movement by his great speech of May 1790. This speech bears the marks of long research, and we are inclined to suspect that the idea of unifying the weights and measures originated with Talleyrand when he was agent of the clergy. The idea of a unification of measures is not mentioned by Pancton who[se] large *Métrologie* was published in 1780, and who was certainly ignorant of many facts collected by Talleyrand. It were to be wished that Mr. Risteen had been as explicit in his explanations concerning our usual weights and measures as he is about the French ones, for the generality of well-informed persons know much more about the latter than about the former. For example, we find in the Table of Data that 1 Kilogram has as its "U.S.

Equivalent," "2.2046212 avoirdupois pounds." But in fact this figure is simply the number which Prof. Miller found for the ratio of the *kilogramme des archives*, not at all to an "*avoirdupois pound*," but to the British *imperial pound*, which has never been legalized in this country. The only prototype avoirdupois weights are those of Queen Elizabeth, dated 1588. They weigh less than the imperial pound by one 3500th part; and they must be held to define the only legal avoirdupois unit that there is today. In England, there is no legal avoirdupois pound. It was always considered as a rough sort of weight; and when the British commissioners appointed in 1818 endeavored to ascertain how many grains it contained, they found the best specimens they could collect differed so much, that it was impossible to say to a grain what the standard weight was; but since it approached 7000, that round number was taken as suitable for the British imperial pound. The latter, however, is defined by a certain particular piece of platinum; and the one 7000th part of that mass is a British grain. Originally the grain was the one 5760th part of the troy pound; for though there is (we believe) no legal troy pound in England, the old troy pound of 1758 would be legal here, had not Congress about 1830 adopted another standard troy pound, on the recommendation of John Quincy Adams. That remains the only standard weight or measure this government has ever ordained, though it has *permitted* the metric system. That U.S. troy pound is an unsatisfactory standard because it is of brass and hollow; so that, as nobody dares put it into water, its specific gravity and consequent buoyancy in air are unknown. Although the U.S. office of weights and measures has, without authority, quietly substituted British weights for those originally distributed to the States, and will not acknowledge that there is any difference, yet there is little doubt there really is a difference. We must not be too much shocked by this procedure; for the offices of weights and measures of all countries down, at least, to the time when the influence of the International Commission became potent, habitually practised such reticences as to make one think they had sat rather at the feet of Talleyrand than at those of any single-minded man of science. But Mr. Risteen is quite right in adopting the figure he does; for the person who consults the table wants to know what a pound in practice is, and not what, by law, it ought to be.

In law, a U.S. or old wine gallon is 231 cubic inches. But it is impossible to measure a cubic inch with any accuracy, without labors occupying a first-rate physicist long months, if not years. The definition is, therefore, perfectly impractical. For that reason, our Treasury

Department in 1832, without any law, adopted the following definition: "The gallon is a vessel containing 58372.2 grains of the standard pound of distilled water, at the temperature of maximum density of water, the vessel being weighed in air in which the barometer is 30 inches at 62° Fahrenheit." The English statute is more scientific in making the imperial gallon 10 pounds of water. Ours is $8\frac{1}{3}$ pounds. If a gallon were 231 cubic inches, then (supposing a litre to be a decimeter cube) it would be 3.78542 litres. But, according to the formula given by Miller (*Phil. Trans.* for 1856, p. 785), one litre of common air under the specified conditions weighs 1.21199 grammes or $\frac{1}{825.913}$ of a kilogram. It follows that in order to correct for buoyancy, the weight given in the Treasury regulations must be increased by $\frac{1}{824.913}$th part of itself or 70.8 grains, making the mass 58443.0 grains of water at maximum density, which is 3.787045 litres. The figure Mr. Risteen gives is 3.78679 litres; but the discrepancy of one fifteen thousandth part is insignificant for a measure so essentially rough as is any volumetric standard. The same remark applies to the bushel which we make 35.253252 litres against Mr. Risteen's 35.23911. He may very likely be in possession of some information of which we are unaware which explains the small difference.

During a tolerably close examination, we have met with but one positive error in the volume; namely, 16.387083 for the number of centimetre-squares in a square inch, which should be $(39.37006)^{-3}$ or 16.387087. This error is due merely to the computer having used seven place logarithms, and indicates so high a degree of accuracy that we know of no similar work which could sustain comparison with this in that respect.

While we are on the subject of illegal units, we notice two curious facts.

First, the assize of bread was only abolished by law in 1815, so that it would seem that the legal pound for bread in this country must still be the ancient "commercial pound" of England of about 7600 grains.

Second, the unit of domestic postage is by law "an ounce." We note, by the way, that one might expect such valuable matter to be weighed by troy weight; but the law being derived from a statute of England, where there is but one pound, then "ounce" is understood to be an ounce avoirdupois. But for international postage the Post Office Department recognizes 15 grammes as a half-ounce, though this is 6 per cent in excess of an avoirdupois half-ounce and near 4 per cent short of a troy half-ounce. The lesson of all this is that for weights and measures the convenience of business is stronger than statute law, in a country like ours.

Mr. J. M. Allen, who writes a preface to Mr. Risteen's little book, seems to think it possible to cause the metric system to be "adopted" in this country. Doubtless, it would be very easy to train children to more accurate acquaintance with the French long measures and weights than one person in a hundred has with the English ones. But between this and the *displacement* of the English measures there is a great gulf. Were it a mere question of making the word "quart" mean a liter, and the word "pound" mean a half kilo, there would be no insuperable difficulty. But when it comes to inducing people to abandon units of long measure that have been used in objects that will last for many years, and the measurements of which must be again referred to in making repairs, etc., there is the pinch. For a semi-civilized country, again, to "adopt" the metric system is a small, — hardly more than a nominal, — matter; but in a civilized community the effective prohibition of familiar units, — an ingrown part of vernacular speech, — has always required despotic laws to be rigorously enforced by a government deaf to the protests of the people. In Prussia, for example, for some years it was made a misdemeanor, punishable by a considerable fine, to have a foot-rule in one's possession. Yet, such was the force of life-long habit, that notwithstanding all the rigors of the law, when, during that period, a carpenter was called into the sacred precincts of the Berlin *Eichungsamt* (or Office of Weights and Measures) by an American to whom a room had been assigned there for certain operations, this artisan insisted upon receiving his instructions and making his measurements in Rhein feet and inches. In France any butcher will today give you prices in *sous per livre;* but should he put out a placard so expressed, instead of in *centimes per kilo*, he would be promptly arrested. How would our legislators maintain such laws? In France, the metric system altogether antedated modern machinery. Even in Germany there was comparatively very little machinery at the epoch of the introduction of the centimetre.

The only way to bring this unit into general use would be to prohibit the selling or manufacturing anything in terms of the inch; and that would almost amount to duplicating every piece of machinery in the country during the first year or two. No wonder great machinists rather favor the change.[1]

[1] This last paragraph has been taken from another draft on the back of p. 1 of the 6 pages in the manuscript.

For some recent comment on the adoption of the metric system in the U.S.A. see *Science*, 29 November 1963, "The Metric Question" and "Exobiology," and 9 July 1965, editorial on "Adoption of the Metric System."

E. [NOTE ON METRIC SYSTEM]

(From a report of the National Academy of Sciences Meeting — *The Nation*, 24 April 1902)

Mr. William Sellers read a paper on the compulsory introduction of the metric system into the United States. This referred to a bill which the doctrinaires of the metric system, with their usual utter neglect to ascertain the state of facts, have introduced into Congress requiring every bureau of the Government (including the Bureau of Weights and Measures, the Mint, the Bureau of Construction of the Navy, etc.), from and after a given date, to use no other than the units of the metrical system for any purpose whatsoever. That this would render every plan in the Navy Department worthless, that it would be impossible to repair the engines of any ship, are among the smallest inconveniences which would result from carrying out the purposes of this fantastic measure, which is, however, urgently pushed by Gen. Comstock. At present the American screw system is in use generally upon the Continent of Europe. There has been, of late years, some attempt to revolt against it; but if America only maintains her position, those countries must ultimately come to the inch for mechanical purposes, because it and its modes of subdivision are more convenient and advantageous for those purposes. America is now, said Mr. Sellers, fifty years in advance of the rest of the world in mechanics. Really, to discard the inch would be to surrender our preëminence, which could not, under those circumstances, continue, such advantage should we be at once putting into the hands of England.

F. THE NUMERICAL MEASURE OF THE SUCCESS OF PREDICTIONS

(*Science* 4, 14 November 1884)

Suppose we have a method by which questions of a certain kind, presenting two alternatives, can in every case be answered, though not always rightly. Suppose, further, that a large number of such answers have been tabulated in comparison with the events, so that we have given the following four numbers: —

(*aa*), the number of questions for which the answers were the first way and the events the first way;
(*ab*), the number of questions for which the answers were the first way and the events the second way;
(*ba*), the number of questions for which the answers were the second way and the events the first way;
(*bb*), the number of questions for which the answers were the second way and the events the second way.

Then the problem is, from these data to assign a numerical measure to the success or science of the method by which the answers have been produced. Mr. G. K. Gilbert (*Amer. Meteorological Journal*, September 1884) has recently proposed a formula for this purpose; and I desire to offer another.

I make use of two principles. The first is, that any two methods are to be regarded as equal approximations to complete knowledge, which, in the long-run, would give the same values for (*aa*), (*ab*), (*ba*), and (*bb*). The second principle is, that if the answers had been obtained by selecting a determinate proportion of the questions by chance, to be answered by an infallible witness, while the rest were answered by an utterly ignorant person at random (using *yes* and *no* with determinate relative frequencies), then the approximation to knowledge in the answers so obtained would be measured by the fraction expressing the proportion of questions put to the infallible witness. The second witness may know *how often* he ought to answer 'yes'; but I give him no credit for that, because he is ignorant *when* he ought to answer 'yes.'

Let i be the proportion of questions put to the infallible witness, and let j be the proportion of questions which the ignorant witness answers in the first way. Then we have the following simple equations: —

$$\begin{aligned}(aa) &= i\{(aa)+(ba)\} + (1-i)j\{(aa)+(ba)\},\\ (ab) &= (1-i)j\{(ab)+(bb)\},\\ (ba) &= (1-i)(1-j)\{(aa)+(ba)\},\\ (bb) &= i\{(ab)+(bb)\} + (1-i)(1-j)\{(ab)+(bb)\}.\end{aligned}$$

Now, whatever the method of predicting, these equations can always be satisfied by possible values of i and j, unless the answers are worse than if they had been taken at random. Consequently, in virtue of the two principles just enunciated, the value of i obtained by solving these equations is the measure of the science of the method. This value is,

$$\begin{aligned}i &= \frac{(aa)}{(aa)+(ba)} - \frac{(ab)}{(ab)+(bb)},\\ &= \frac{(aa)}{(aa)+(ba)} + \frac{(bb)}{(ab)+(bb)} - 1,\\ &= \frac{(aa)(bb)-(ab)(ba)}{\{(aa)+(ba)\}\{(ab)+(bb)\}}.\end{aligned}$$

Mr. Gilbert's formula has the same numerator, but a different denominator. It is, in the present notation

$$i = 2\frac{(aa)(bb)-(ab)(ba)}{\{(aa)+(ab)+(ba)+(bb)\}^2-(aa)^2+(ab)^2+(ba)^2-(bb)^2}$$

For Sergeant Finley's tornado-predictions, $(aa) = 28$, $(ab) = 72$, $(ba) = 23$, $(bb) = 2{,}680$. From these data, Mr. Gilbert finds $i = 0.216$, while my formula gives $i = 0.523$.

If the questions should present more than two alternatives, it would be necessary to assign relative values or measures to the different kinds of mistakes that might be made. I have a solution for this case.

Another problem is to measure the utility of the method of prediction. For this purpose, let p be the profit, or saving, from predicting a tornado, and let l be the loss from every unfulfilled prediction of a tornado (outlay in preparing for it, etc.); then the average profit per prediction would be,

$$\frac{p\cdot(aa)-l(ab)}{(aa)+(ab)+(ba)+(bb)}.$$

15

NON-EUCLIDEAN GEOMETRY

We have an *apriori* or natural idea of space, which by some kind of evolution has come to be very closely in accord with observations. But we find in regard to our natural ideas, in general, that while they do accord in some measure with fact, they by no means do so to such a point that we can dispense with correcting them by comparison with observations.

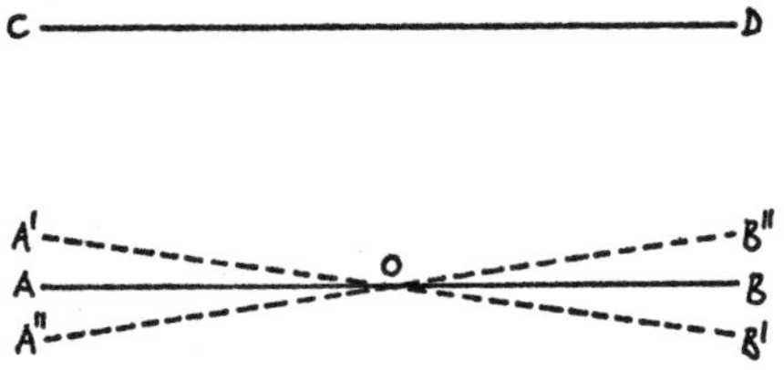

Fig. 1

Given a line CD and a point O [Fig. 1]. Our natural (Euclidean) notion is that

1st there is a line AB through O in the plane OCD which will not meet CD at any finite distance from O.

2nd that if any line $A'B'$ or $A''B''$ through O in the plane OCD be inclined by any finite angle, however small, to AB, it will meet CD at some finite distance from O.

Is this natural notion *exactly* true?

A. This is not certain.

B. We have no probable reason to believe it so.

C. We *never* can have positive evidence lending it any degree of likelihood. It *may* be disproved in the future.

D. It may be true, perhaps. But since the chance of this is as $1:\infty$ or $\frac{0}{1}$, the logical presumption is, and must ever remain, that it is not true.

E. If there is some influence in evolution tending to adapt the mind to nature, it would probably not be completed yet. And we find other natural ideas require correction. Why not this, too? Thus, there is some reason to think this natural idea is *not* exact.

F. I have a theory which fits all the facts as far as I can compare them, which would explain how the natural notion came to be so closely approximate as it is, and how space came to have the properties we find it has. According to this theory, this natural notion would not be exact.

To give room for the non-Euclidean geometry, it is sufficient to admit the first of these propositions.

Either the first or the second of the two natural propositions on page 1 of this article may be denied, giving two corresponding kinds of non-Euclidean geometry. Though neither of these is quite so easy as ordinary geometry, they can be made intelligible. For this purpose, it will suffice to consider plane geometry. The plane in which the figures lie must be regarded in perspective.

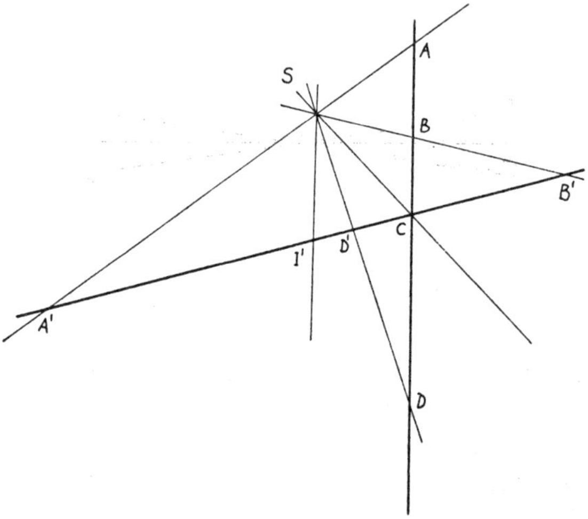

Fig. 2

Let *ABCD* be this plane, which I call the natural plane,* (* merely because so-called by writers on Perspective. Nothing to do with the "natural assumptions" of page 1) seen edgewise [Fig. 2]. Let *S* be the eye, or point of view, or centre of projection. From every point of the natural plane, rays, or straight lines proceed to the point of view, and are continued beyond it if necessary. If three points in the natural plane lie in one straight line, the rays from them through the point of view will lie in one plane. Let $A'B'$ bc the plane of the delineation or picture seen edgewise. It cuts all the rays through S in points, and so many of these rays as lie in one plane it cuts in a straight line; for the intersection of two planes is a straight line. The points in which this plane cuts those rays are the perspective delineations of the natural points, i.e. the corresponding points in the natural plane. We extend this to cases in which the point of view is between the natural and the delineated points.

It is readily seen that the delineation of a *point* is a *point*, and that to every point in the picture corresponds a point in the natural plane. And to a straight line in the natural plane corresponds a straight line in the picture. For the first straight line and the point of view lie in one plane, and the intersection of this plane by the plane of the picturc is a straight line.

All this is just as true for the non-Euclidean as for the Euclidean geometry.

But now let us consider the parts of the natural plane which lie at an infinite distance and see how they look in perspective.

First suppose the natural, Euclidean, or parabolic geometry to be true. Then all the rays through S from infinitely distant parts of the plane, themselves lie in one plane. For let SI' be such a ray, then if SI' be turned about S the least bit out of the plane parallel to the natural plane it will cut the latter at a finite distance.

These rays through S from the infinitely distant parts of the natural plane, since they lie in one plane, will cut the plane of the picture in a straight line, called the vanishing line of the natural plane.

The delineations of any two parallel lines will cross one another on this vanishing line.

Next, suppose that the 2nd natural proposition on p. 1 [of this article] is false and that SI' may be tilted through a finite angle without cutting the natural plane at a finite distance.

Then, I will *state*, what there is no difficulty in proving, that the delineation of the infinitely distant parts of the natural plane will occupy a space on the picture bounded by a conic section.

The picture looks exactly as before, *only* that the *real distances* of certain parts which on the first assumption were finite, now become infinite.

The following two figures [Fig. 3 and Fig. 4] show the straight line which on the first assumption alone represents parts *really* infinitely distant, as well as the conic which on the second assumption bounds

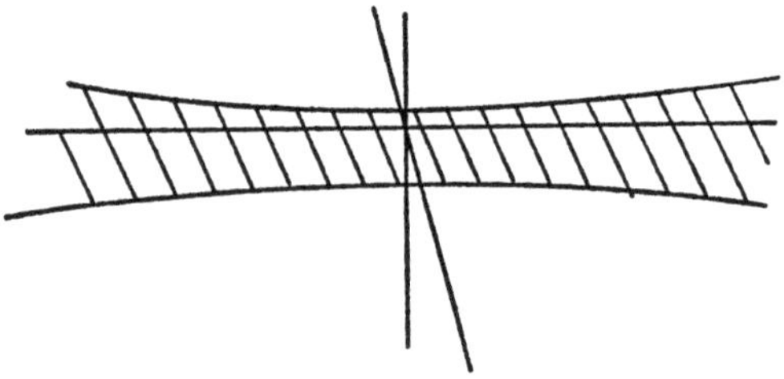

Fig. 3

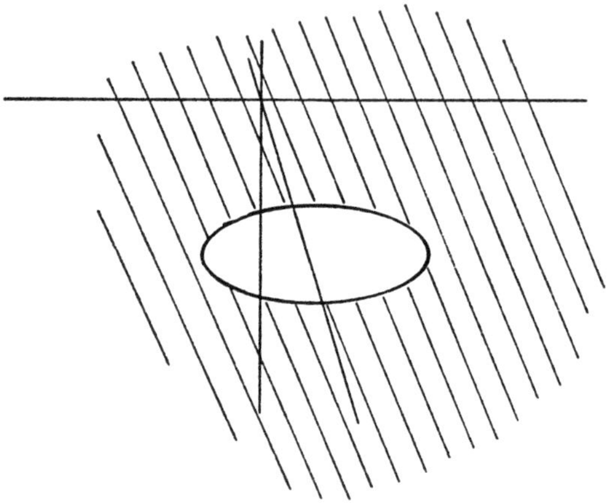

Fig. 4

the delineation of the really infinitely distant parts. With two lines parallel on the first assumption.

In this case, space is *limited.* But though *limited,* it is *immeasurable.* Imagine, for instance, that every kind of unit of linear measure, should shrink up as it was removed from a fixed centre, and perhaps differently according as it was radially or peripherally placed, so that it never could in any finite number of repetitions, get beyond a certain spherical surface. Then that surface would be at an infinite distance and no moving body could ever traverse it, for the distance moved over in a unit of time would be a unit of distance. But unless this linear unit were to shrink according to a peculiar law, the result would be that different parts of space would be unlike. That is to say, for example, if we were to draw the plane figure *ABCD*, by means of given lengths *AB*, *AC*, *AD*, *BC*, we should find the resulting length of *CD* to be different in different parts of space [see Fig. 5]. Now, geometers assume, perhaps with little reason, that the fact that rigid bodies move about readily in space without change of proportions, shows that all the parts of space are alike.

With this condition, I will *state*, what is again readily proved that the law of the representation of equal distances and equal angles is in this second or hyperbolic geometry for the distances [see Fig. 6]. Let the conic be the "absolute," or the perspective representation boundary of infinitely distant regions. Let AB be a unit of distance. Through AB draw the line IJ, let $I'J'$ be any other line. Draw straight lines through their intersections II' and JJ' with the absolute. Let O be the point where these lines intersect; from O draw rays to A and B. Then the distance $A'B'$ cut off by these rays on the line $I'J'$ is equal to AB.

The rule for angles is this. Let the conic be the absolute [Fig. 7]. Let ACB be a given angle. From C draw CI and CJ tangent to the absolute. Take any other vertex C', and draw tangents $C'I$ and $C'J$ to the absolute. Through I the intersection of CI and $C'I$ and J the intersection of CJ and $C'J$ draw the straight line IJ cutting AC and BC in A and B. Then $AC'B = ACB$.

The third kind of geometry, which denies the 1st natural assumption supposes space to be *unlimited* but *finite*, that is, measurable. It is as if the linear unit so expanded in departing from a centre as to enable us to pass through what is naturally supposed to be at an infinite distance.

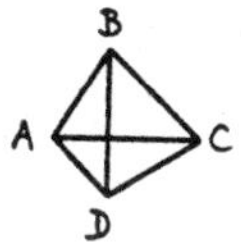

Fig. 5

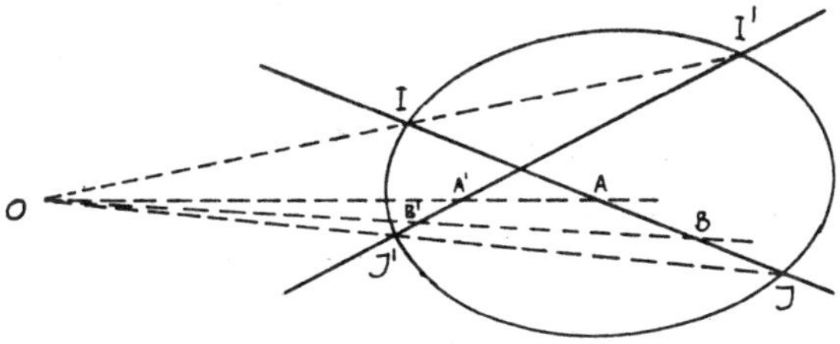

Fig. 6

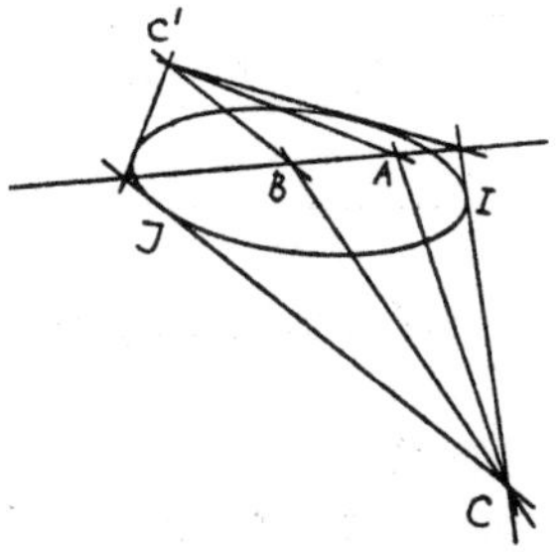

Fig. 7

B. REFLECTIONS ON NON-EUCLIDEAN GEOMETRY (118)

Although the ideas of imaginaries and of the absolute are the creation of algebra, yet I think it quite in the spirit of modern algebra, the ideas once gained, to treat them as synthetically as possible.

The non-Euclidean geometry treats exclusively of the projective relations of a certain individual surface, the absolute.

This surface is a quadric, so that every line either lies wholly within this surface, all its points and planes being elements of the surface, or else it has but two points and two planes in common with the surface. If either pair coalesces, both coalesce. Taking a plane and point incident the one on the other, in that plane through that point there are but two lines tangent to the surface, unless all such lines are so, and in that case, there are but two which lie wholly within the surface. Either of the pairs of lines spoken of may coalesce.

No definition of a plane is possible except that it is a linear surface. All transportation of a rigid body is representable by the result of a series of projections. These are subject only to the condition that they must not move the absolute away from itself or bring any other surface into tangency with it.

A pair of points or planes may have 9 singular positions in reference to the absolute.

1. It may coalesce on the absolute.
2. It may be distinct, but on a generator of the absolute.
3. It may be on the absolute but not on one generator.

4. One terminal only may lie on the absolute, and the other on a tangent line having that same point or plane of tangency.
5. One terminal may lie on the absolute, but the pair not on a tangent line.
6. Neither terminal lying on the absolute, the line may be a tangent line.
7. The pair may be harmonically situated with reference to the absolute.
8. One terminal and one common point of the line of the pair and the absolute may be harmonically situated with reference to the other two.
9. The pair may be equianharmonic relatively to the absolute.

In reference to the principles of measurement.

1. A spread which can be transported so as more than to cover a former place is infinite.
2. A spread which can be transported so as partly to fall short or partly extend beyond a former place but cannot more than cover a former place is finite.

Commensurable ratios of magnitude are determined by counting successive superpositions.

Incommensurable ratios depend on approximations which require us to take notice that of magnitudes not coincident one is more nearly so than another.

The section of the absolute by a plane, as well as any segment of it, is infinite. A sector is also infinite. But the area of a triangle inscribed in the absolute is finite. (In fact, a triangle cannot be infinite.) For it may partly fall short of and partly overlap a former position but cannot more than cover it. Moreover, all such triangles are plainly equal; since any one may be superposed on any other [Fig. 1]. All the angles of such triangles are zero. If such a triangle is separated into two by a

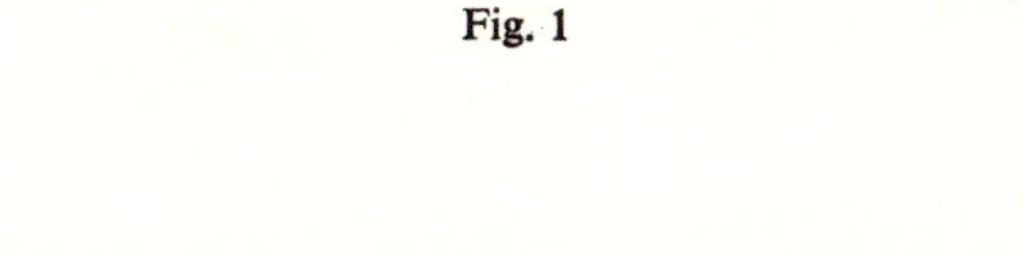

Fig. 1

line from the vertex to an opposite side, their angles are supplements. The areas of such triangles are determined as follows. Let a and β be respectively the angles of two triangles whose other vertices are on the absolute. Place the triangles side by side with their finite angles at [a] coincident vertex. From the quadrilateral so formed we may subtract a triangle having three vertices on the absolute or else we may subtract the quadrilateral from the triangle. The result will be a triangle with two vertices on the absolute and its finite angle equal to $a+\beta$, in the first case or to $360° - (a+\beta)$ in the second. It follows that the areas of triangles having two vertices on the absolute are proportionate to the difference of the sum of the angles from 180°. The difference between one such triangular area and another is equal to the area of a triangle having one vertex on the absolute, and the difference of the areas of two such triangles is equal to the area of a triangle with all its vertices.

So far we have assumed the triangle inside the absolute, the angles measuring round and round through 360°. In like manner, we might have set out from a triangle circumscribing the absolute having all its angles infinite, and all its sides zero [Fig. 2]. We easily prove that area is the sum of the three sides subtracted from the total length of a line.

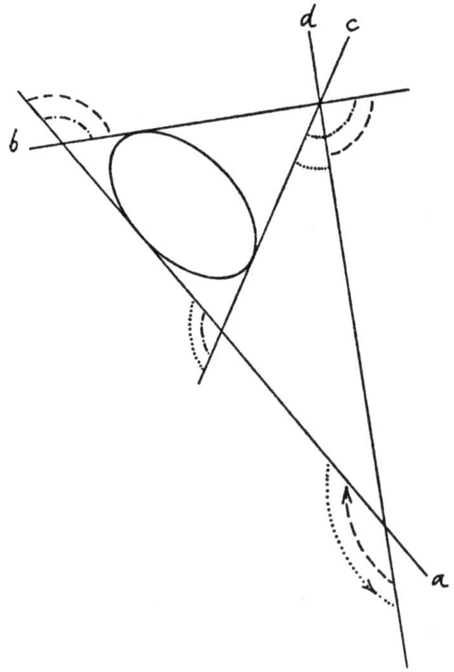

Fig. 2

C. NON-EUCLIDEAN GEOMETRY (122)

Suppose the parts of a line are displaced along that line in any way so long as the continuity of points (including those that are imaginary) is not broken and so that the new position of each point shall coincide with the old position of just one point, and *vice versa*. Then, let x and y be the homogeneous coordinates of a point on the line, these quantities being connected by a linear equation, but otherwise distributed over the line in any continuous and [ordinal] manner. If then (x',y') is a point which in its new position coincides with the old position of the point (x, y), we have a lineolinear equation between the two pairs of coordinates which may be written

$$axx' + exy' + iyx' + oyy' = 0.$$

Putting $x = x'$ and $y = y'$ we get the quadratic

$$ax^2 + (e + i)xy + oy^2 = 0,$$

which defines a pair of points on the line which remain unmoved by the displacement.

The manifoldness of displacements is evidently ∞^3, because there are 3 independent constants in the equation. There are ∞^2 positions of the unmoved points and ∞ displacements for each. But considering imaginaries, all the exponents must be doubled. All the displacements having any one position of the unmoved points are iterations of one infinitesimal displacement, and may collectively be called an area of displacement.

Let the equation of a displacement be written

$$xx' - \tfrac{1}{2}(\alpha + \beta + \tau)x - \tfrac{1}{2}(\alpha + \beta - \tau)x' + \alpha\beta = 0$$

or, writing Fx for x',

$$Fx = \frac{\tfrac{1}{2}(\alpha + \beta + \tau)x - \alpha\beta}{x - \tfrac{1}{2}(\alpha + \beta - \tau)}$$

This gives the equation of finite differences

$$F^{n-1}\cdot F^n - \tfrac{1}{2}(\alpha+\beta+\tau)F^{n-1} - \tfrac{1}{2}(\alpha+\beta-\tau)F^n + \alpha\beta = 0,$$

the general solution of which (that in Boole's Finite Differences is wrong) is (α and β being unequal)

$$F^n = \frac{C_1\alpha(\tau+\alpha-\beta)^n + C_2\beta(\tau-\alpha+\beta)^n}{C_1(\tau+\alpha-\beta)^n + C_2(\tau-\alpha+\beta)^n}$$

To determine the ratio of the arbitrary constants, we have

$$F^{\circ}x = x = \frac{C_1\alpha + C_2\beta}{C_1 + C_2}$$

and so finally

$$F^n = \frac{\alpha(x-\beta)(\tau+\alpha-\beta)^n + \beta(\alpha-x)(\tau-\alpha+\beta)^n}{(x-\beta)(\tau+\alpha-\beta)^n + (\alpha-x)(\tau-\alpha+\beta)^n}$$

If the moduli of $\tau+\alpha-\beta$ and $\tau-\alpha+\beta$ are not equal assume the former to be the larger. Then, $F^{\infty}x = \alpha$, $F^{-\infty}x = \beta$.

If the moduli in question are equal, then for some real value of ω, and in any case for some imaginary value

$$F^{\omega}x = x.$$

Considering the plane of imaginary quantity, every displacement of a given area of displacements may be conceived of composed of two different kinds. The displacements of the first kind are those for which $\tau+\alpha-\beta$ and $\tau-\alpha+\beta$ have the same argument, so that $x-\beta$ and $\alpha-x$ retain the same argument, and these displace all the points along circumferences through the unmoved points. The displacements of the second kind are those which displace points along circumferences orthogonal to the last, $\tau+\alpha-\beta$ and $\tau-\alpha+\beta$ having the same modulus.

The only general mathematical condition to which the rigid displacement of a line along itself is subject is, that a positive or negative, integral or fractional, but real number of repetitions of any such displacement will carry any real point to some real point and not to any imaginary point. Such a path of displacement must therefore appear as a straight line upon the plane of imaginary quantity. But there are only two straight lines among the system of circles just considered. Either therefore α and β are conjugate imaginaries, and a finite number of rigid displacements brings all the points of the line back to their original positions, or α and β are real and then they are absolute limits to rigid displacement.

SKETCH OF A SYNTHETIC TREATMENT

Chapter I. Geometry of points on a line

1. Taking any three points *A*, *B*, *C*, any fourth point *X* is either 1st on the way from *A* to *B* not passing *C*, 2nd on the way from *B* to *C* not passing *A*, or 3rd on the way from *C* to *A* not passing *B*.

2. In each of these divisions the multitudes of points is such that they cannot in any way be placed in one-to-one correspondence with the integral numbers.

This does not say that there is a continuous succession; but I do not see that the idea of continuity is possible without a second dimension.

Chapter II. Correspondences between 2 lines of points

Only a few topics of this vast subject to be noticed.

A *consecutive* correspondence is one in which every four points on one line have the same order of succession as the corresponding four points on the other.

A two-to-one consecutive correspondence is either pursuing or meeting. The pursuing correspondence has nothing particular, except that going once round one line corresponds to going twice round the other. But in the meeting correspondence, as a point moves part way round one line the corresponding point moves all the way round the other and then as the first point finishes the circuit of the first line the corresponding point returns and makes its circuit in the opposite direction; so that these ought to be self-evident, and indubitable apart from experience; and then many attempts were made to replace Euclid's assumption by one which should seem undeniable. General Perronet Thomson enumerated thirty of these, which the reader will find in an appendix to this treatise in an article by DeMorgan from the *English Cyclopædia.* Perhaps the two most successful of these attempts are those of Playfair. The first adopted in his edition of Euclid assumes that of two intersecting straight lines one or other must cut every third straight line in the same plane. The second is that a figure turned completely round in a plane, although not always turning round the same point must be rotated through four right angles; and this is substantially the assumption in my father's geometry. But all such premises were felt, after all, to be assumptions which might or might not be exactly true; and finally Lobachevski produced a geometry in which they were assumed to be false in some unknown measure. The falsity might, as a special case of his general admission, be nothing.

But all this time another demonstration by Euclid which is completely fallacious, had been admitted as sound reasoning. It still appears in the best treatises [Euclid's *Elements*. Book I. Prop. XVI]. ...

ON THE NON-EUCLIDEAN GEOMETRY

In this geometry there is a fixed quadric surface (it might be called a sphere, or rather the exemplar of a sphere) called the absolute. Every superposition or rigid displacement is by projection and section which does not move the absolute, unless in its own surface.

Thus the distance *AB* becomes in this way superposed upon *AC* (Fig. 1). In like manner the angles *ab* and *ac* are superposed (Fig. 2).

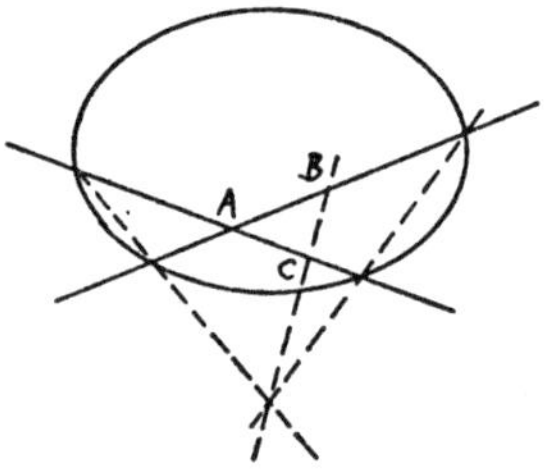

Fig. 1

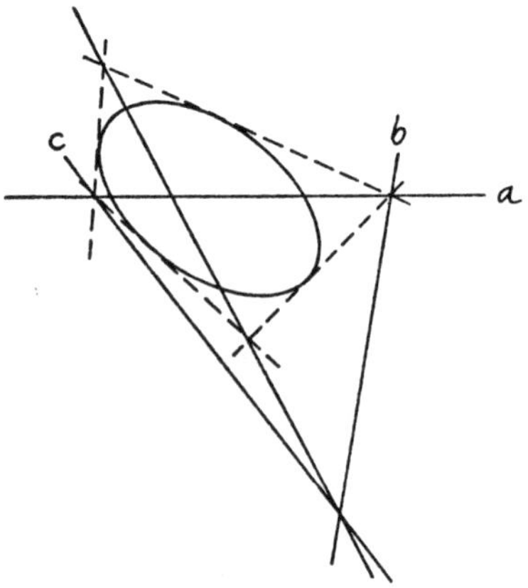

Fig. 2

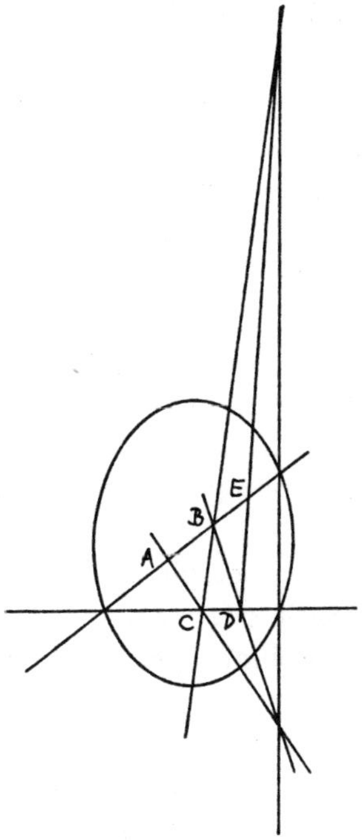

Fig. 3

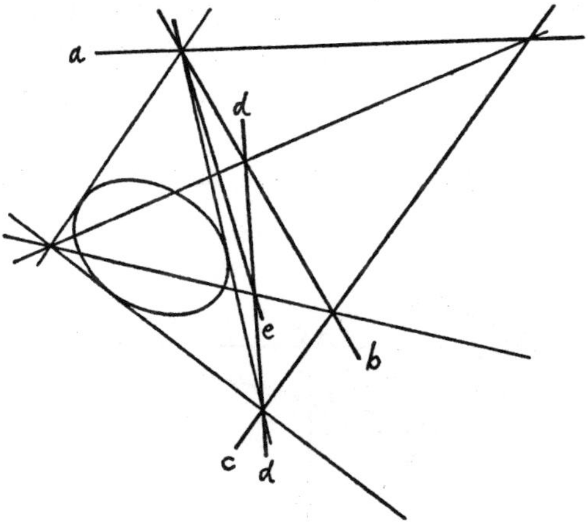

Fig. 4

So also (Fig. 3) *AB* is superposed first on *CD* and then on *BE*. And in like manner the angle *ab* is superposed first on *cd* and then on *be*.

THE PRINCIPLES OF NON-EUCLIDEAN GEOMETRY

This geometry treats of the projective relations of a certain individual quadric surface called the absolute.

This quadric surface, like every other, has in common with every line two points and two tangent planes, and has in every plane through every point in it two tangent lines. Only, in every tangent plane, intersecting at the point of tangency are two lines lying wholly in the surface.

Every transformation of this geometry is composed of displacements each of which carries any two points into new positions such that all four lie in one plane and carries any two planes into new positions so that all four pass through one point; and no point, line, or plane is carried to or from the absolute. In such displacement, all points at first in one plane continue in one plane, and all planes at first through one point continue to meet in one point.

There are six singular positions which a pair of points or a pair of planes may have, as follows:

1st, they may be in coincidence.

2nd, their line may be a tangent line. In this case a displacement may bring them into coincidence. They are then called "parallel."

3rd, one of the pair may belong to the absolute.

4th, they may be harmonically situated each on the polar or pole of the other.

5th, one of them may be harmonically situated towards one of the absolute points or planes of their line with reference to the other and the other absolute point or plane of their line. [In Fig. 1], *ABC* are harmonically situated.

6th, they may be equianharmonically situated with reference to the absolute.

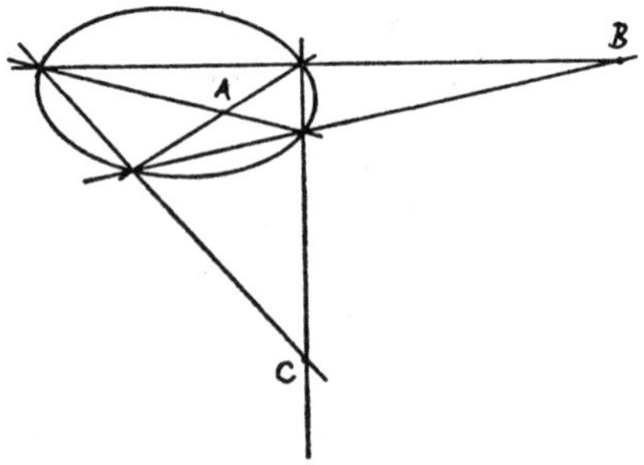

Fig. 1

The mode of superposition is as follows: Given the point *O*. Draw two lines through it, and through the intersections two lines meeting at two points *P*, *Q*, harmonic to *O*. Then by projection from *Q*, *OA* is superposed upon *OB*, by projection from *P*, *OB* is superposed on *OC*, and by projection again *OA* or *OC* is superposed upon *OD* [see Fig. 2].

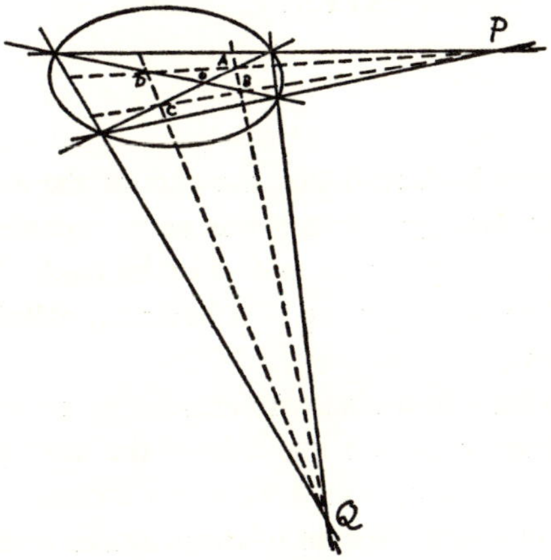

Fig. 2

D. THE ELEMENTS OF NON-EUCLIDEAN GEOMETRY. PREFACE (120)

The attempted proof by Euclid that the sum of the angles of a triangle cannot be less than two right angles was never regarded as satisfactory, not even, it is safe to say, by himself. For he baldly begs the question in his fifth postulate (often, owing to a bad text, called the 11th axiom) which is as follows:

"If two straight lines in a plane are crossed by a third, so as to make the sum of the inner angles on one side of the last less than two right angles, then those two lines will meet, if sufficiently produced, on that side of the third on which the sum of those angles is less than two right angles."

But this, I say, is begging the question; for if anybody thinks that the sum of the angles of a triangle can be less than two right angles, he will naturally hold this to be case for all triangles, and in fact it can be proved that the proposition must be true of all or of none. He will therefore hold that on a given base there will be a maximum value which the sum of two angles cannot exceed.

Euclid calls his premise a postulate, that is, an assumption; and he seems purposely to have stated it in a form in which the importance of it could be recognized. He did not seek to entrap his pupils. But the idea arose that the principles of geometry should be self evident, and indubitable apart from experience; and accordingly many attempts were made to replace Euclid's assumption by one which should seem undeniable. We reprint in an appendix a valuable article by DeMorgan from the *English Cyclopædia* enumerating some 30 of these. The most successful was that of Playfair that two straight lines which intersect must one or other cut every third straight line in the same plane.

E. [FROM A PAPER FOR THOMAS S. FISKE] (121)

... had told him had made it clear to him that the *strict geometrical truth* is that the sum of the three angles of a triangle is 180° *minus* a hypothetical correction.[1] (He should have said a constant proportion of the area.) He adds that the Euclidean geometry is practically true, at least for terrestrial figures. The limitation is significant, considering that the business of Bessel's life was with non-terrestrial figures. Gauss, after fourteen months interval, expresses his delight at learning that Bessel had come to that conclusion. The certainty of arithmetic and algebra, he adds, is absolute; but then they do not relate to anything but our own creations (unseres Geistes Product). But space has a reality without us (the Newtonian, anti-Leibnizian, anti-Kantian doctrine), and its laws cannot be known a priori. In other words, Gauss divides Geometry into Physical Geometry, a branch of physics which inquires what the properties of real space are, and all whose conclusions are affected with a probable error, and Mathematical Geometry, which in consequence of the suggestions of physical geometry, develops certain ideas of space. I do not merely say to trace out the consequences of hypotheses; for certainly Gauss would have held that it was part of the mathematician's function to invent the *n*-dimensional and other varieties of the space-idea.

It is singular that two fallacies in the reasoning of the first book of Euclid should have remained unnoticed to our day. One of these, in the 16th proposition, is that in the figure, where $AE\Gamma$ are coradial (a word I prefer to collinear), BEZ are coradial, $B\Gamma\Delta$ are coradial, $AE = E\Gamma$, and $BE = EZ$. Euclid assumes

$$\angle E\Gamma\Delta > \angle E\Gamma Z \quad \text{[Fig. 1].}$$

1 Professor Fiske's name is written in C.S.P.'s hand on the back of the last page of manuscript, which had been folded as if once sent to him. Peirce had read a paper entitled "Rough Notes on Geometry, Constitution of Real Space" at a meeting of the American Mathematical Society on 24 November 1894. It is highly probable that this paper was the basis of his remarks at that time.

The missing first page apparently referred to a letter to Gauss from Bessel.

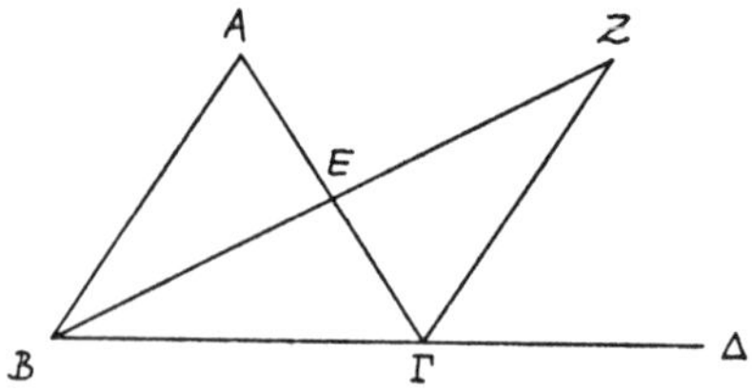

Fig. 1

That this does not follow was overlooked by Saccheri, and is overlooked by the Society for the Improvement of Geometrical Teaching, and by many other authors; but it is pointed out by Stäckel. The other, in the 7th proposition, where the construction is $A\Gamma = A\Delta$, $B\Gamma = B\Delta$, consists in inferring that $\angle A\Delta\Gamma > \angle \Delta\Gamma B$, because $\angle A\Gamma\Delta = \angle A\Delta\Gamma$

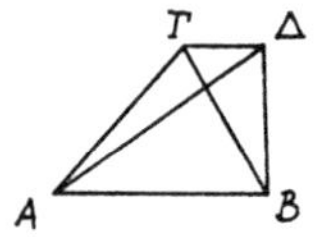

Fig. 2

[Fig. 2]. This is overlooked even by Stäckel. Yet it is not true, even in Euclidean geometry, when Γ and Δ are at infinity; and it is not generally true in that variety of non-Euclidean geometry in which the intersection of the plane with the absolute is a point. It is true, that simple observations refute that geometry; but they are not put forward by Euclid.

Both of these fallacies arise from Euclid's confidence in his 8th Axiom, that the whole is greater than its part. These are the only places in the first book in which he makes use of that axiom, except in 18, 20, 24, and 26, which, properly understood, are equally fallacious. It has been abundantly shown by Cauchy, G. Cantor, and others, that the whole is only necessarily greater than its part when that part is finite and positive. Were the logic of relatives, either with or without any of the algebras I have proposed to facilitate its study, made use of, such fallacious reasoning would be impossible.

I maintain that Euclid was himself a non-Euclidean geometer. I do not mean, in the complete, Gaussian and Besselian sense, but more so than Saccheri and Lambert. I proceed to state the evidence of this. In the first place, the effort of all writers on the theory parallels, all defenders of the Euclidean geometry, was to produce some proposition in place of the 5th Postulate which should be more readily acceptable. This was very easy. It was so easy that Euclid could not but have known it.

For instance, he must manifestly have known that it was sufficient to say that if of three rays in one plane one was perpendicular to a second while the third was not perpendicular to that second, then the first and third must meet on the side of the acute interior angle. But Euclid disdains every effort to gloss over his assumption, and prefers to state it in the form in which its hypothetical character is most apparent. The only reasonable explanation is that he was fully alive to the hypothetical character of the proposition, and thought truth demanded its being presented in that light. Besides, consider the order of propositions in the first book. The first six are valid for all varieties of non-Euclidean geometry. The next eight (that is, to I.14, inclusive) are false in case the intersection of the plane with the absolute is a point, and in no other case. Proposition 15, that vertical angles are equal seems to the modern geometer no theorem at all; but Euclid's treatment of it is as if he conceived it to be of the same class as the preceding eight. All the next twelve propositions, in the sense in which Euclid, no doubt, would understand them, are false of the elliptic geometry alone. Owing to the confusion between internal and external angles, some of these would now be considered true in all cases. Finally, all the remaining propositions of the book are false both of elliptical and of hyperbolical geometry. It seems, therefore, that the propositions of the first book, which is, of course, the book of Fundamenta, are arranged in reference to the different kinds of non-Euclidean geometry. It seems to me that the most probable, and I might almost say the only rational, explanation of this arrangement is to suppose that Euclid had studied the non-Euclidean geometries.

In this connection let me ask: is it possible that Euclid, the author of the Data and Porisms, was unaware that metrical geometry is not the real basis of geometry?

The following count of the different kinds of propositions of the elements may be useful.

Book I, on Fundamentals, contains 23 definitions, 5 postulates, 9 (5 genuine) axioms, and 48 demonstrated propositions, of which 1, 2, 3, 9, 10, 11, 12, 22, 23, 31, 40 (11 in all) are problems.

Book II, on Areas, contains 2 definitions, and 14 demonstrated propositions, of which the 14th is a problem.

Book III, on Circles, contains 11 definitions, and 37 demonstrated propositions, of which 1, 17, 30, 33, 34 (5 in all) are problems, and 1, 16, 31 have corollaries.

Book IV, on Polygons, has 7 definitions, and 16 demonstrated propositions, all problems, of which 5 and 15 have corollaries.

Book V, on Proportions, has 18 definitions and 25 demonstrated propositions, all theorems, of which the 7th has a corollary.

Book VI, has 5 definitions and 33 demonstrated propositions, of which 9, 10, 11, 12, 13, 18, 25, 28, 29, 30, 31 (11 in all) are problems, and the 8th, 19th, 20th have corollaries. There is a lemma before the 23rd.

Book VII, on Arithmetic, has 22 definitions and 39 demonstrated propositions, all theorems, of which the 2nd has a corollary.

Book VIII, on Arithmetic, has 27 demonstrated propositions, all theorems, of which the 2nd has a corollary.

Book IX, has 36 demonstrated propositions, all theorems, of which the 11th has a corollary.

Book X, has 4 definitions before the first proposition, 6 more before the 48th, and 6 more before the 85th, 115 regular propositions all theorems, of which the 3rd, 4th, 6th, 9th, 23rd, 111th, 114th have corollaries. There are lemmas before 10, 14, 17, 19, two before 29, before 23, 42, 54, 60. There is a scholium, or remark, before the 72nd.

Book XI, on Stereometry, has 28 definitions, 39 regular propositions, of which 11, 12, 23, 26, 27 (5 in all) are problems, and 33 and 35 have corollaries. There is a lemmatic problem before the 24th.

Book XI has 18 regular propositions of which 16 and 17 (two in all) are problems, and 7, 8, 17 have corollaries. There are lemmas before the 3rd and 5th.

Book XIII has 18 regular propositions, all theorems, of which 16 and 17 have corollaries. There are so-called Lemmas before the 3ıd and 14th and at the end. There is also a scholium, or remark, at the end.

Total: 132 definitions, 465 regular propositions, 27 corollaries, 17 "lemmas," 2 remarks, 5 postulates, and 9 axioms.

It is worth notice that Ptolemy's theorem (which lies at the foundation of the Brocard geometry) holds good for every form of non-Euclidean geometry.

The properties of Space may be stated (more fully, I believe, than has hitherto been done) as follows:

I. *Intrinsic Properties of Space.*

1. Space is continuous.

2. Space is tridimensional.
3. Space has no topical singularities, that is, no surfaces, lines, or points from which it extends in more or less multiple ways than from neighboring surfaces, lines, or points.
4. Space is perissid; that is, there are innumerable multitudes of unbounded surfaces of which any three have an odd number of points in common.
5. The chorisis (number of separate point-regions), cyclosis, periphraxis, and immensity of space are as small as the fourth property allows, namely, each 1.

II. *Property of the Image of Space, or the Optical Property of Space.*

6. Through any three points there is just one plane surface; and the place common to two planes is a ray, or free path of vision; and any plane and ray not lying in that plane have just one point in common.* (* The point of intersection of two rays in one plane is, in one sense, a real point, even though such a system of measurement be adopted that its distances from other real points of those rays be imaginary. We conceive of vision as making the complete circuit of the ray, irrespective of the length.)

III. *Property of the Displacement of Rigid Bodies in Euclidean Space, or the Metrical Property of Euclidean Space.*

7. When Space is measured by a rigid body, there is a plane firmament, or surface which no particle can either leave or arrive at.

It follows from what Bessel implies, in the letter above referred to, that the physical geometry of celestial triangles needs examination, in order to ascertain whether the constant of space may not have a sensible magnitude. I have undertaken such an examination. I began by forming a list of all possible methods of determining this quantity by means of the following observations: 1st, the parallaxes of stars; 2nd, the numbers of stars of each parallax; 3rd, the proper motions of stars; 4th, the numbers of stars of different proper motions; 5th, the spectroscopic determinations of the motions of stars in the line of sight; 6th, the magnitudes of stars; 7th, the numbers of stars of each magnitude. My list of possible methods was long. All of them, it is true, involved some hypothetical element; but that is true of any research, whatever, into the value of a physical quantity; and it is possible so to modify the methods that the hypotheses that appear the most dangerous may probably be eliminated. I applied several methods: they seemed to indicate a hyperbolic space with a constant far from insignificant. The best method seemed to be by proper motions. We know that our ordinary dynamics gives a definite value to absolute, or nonrelative velocity of rotation. Thus, Foucault's pendulum experiment shows that the earth has an absolute motion of rotation, and not merely a rotation relatively to the sphere of the heavens. All German attempts to escape this conclusion are metaphysical word-spinning. In the Euclidean geometry, translation is an infinitesimal rotation about an axis lying in the firmament. In the non-Euclidean geometry there is no plane firmament and, consequently, no translation. In every rotation, every particle lying in the ray which is the polar conjugate with respect to the absolute of the axis of rotation moves along that ray. All other particles describe circles. Consequently, there is an absolute velocity in all motion. The same apparent effect cannot be produced by a motion of the stars and an opposite motion of the solar system. The proper motion due to the absolute motion of a star varies inversely as the sine of the product of the distance by the square root of a constant. The proper motion due to the absolute motion of the sun varies inversely as the tangent of the product of the distance by the square root of the same constant. If, therefore, we had all the proper motions plotted upon a globe, and if we were first to throw out all those which are excessively large, as probably belonging to stars so near the sun that they may be supposed sensibly to partake of the system of motion to which the sun's motion belongs, — whether they move in the same direction or with the same velocity or not, — we should expect to find the remaining proper motions presenting one or other of two appearances. Namely, if space is elliptical (that is, finite), and by consequence the square root by which the distance is multiplied be real, the proper motions due to the sun's motion would for very distant stars become insensible (because divided by indefinitely great tangents) and therefore those stars whose proper motions were very small would be moving about in every direction quite irrespective (or, at least, relatively irrespective) of the motion of the sun. If, however, space is hyperbolical, and that square root aforesaid imaginary, we should expect to find that those stars whose proper motions were very small would (except for the effect of errors of observation) move more directly away from the apex of the sun's motion than those whose proper motions were relatively large. I made the necessary computations for a selection of stars from Newcomb's catalogue. Only, instead of judging which were the most distant stars by the proper motions alone, I formed my judgment by combining the evidence from proper motion and from magnitude.

The result was most markedly in favor of the hyperbolical geometry. A rough calculation of probabilities showed that it could not be with probability attributed to chance. I was in hopes sufficient interest would be excited by this result to enable me to obtain the funds to employ a computer for a few months to extend the calculations to all the proper motions of Auwers. But as the question has no immediate relation either to business or comfort, this has not hitherto been the case. Its decision must strongly influence the scientific conception of the universe; and might thus lead to practical results sooner than one might, at first sight, suppose possible. I have now laid out a new plan of the work; so that for $150, a rough graphical result may be obtained; and if this should be of a convincing character, $500 more may be expended in computations, without the determination of any connected constants. It might ultimately be thought worth while to work in a more fundamental way. It will be remarked, that if a constant value results from such calculations, as the work I have already done gives me no room to doubt, then that constant must mean *something*. Even if I am wrong in thinking that it can only be attributed to a property of space, it still must have some significance for the visible universe. I shall be happy to receive subscriptions payable when enough have been received to cover the expense of the graphical solution.[1]

C. S. Peirce

[1] At this point the reader is referred to pp. 421-424 of "The Charles S. Peirce–Simon Newcomb Correspondence" by the editor in *The Proceedings of the American Philosophical Society*, 101:5 (October 1957), 409-433.

F. ON TWO MAP-PROJECTIONS OF THE LOBATSCHEWSKIAN PLANE[1]

The proper motions of the stars show very strong indications that our space is really hyperbolic, or, what comes to the same thing, that the law of dynamics, or kinematics, is such that if two stars move in the same plane with the same uniform velocity, without being acted upon by any forces, and each appears at one moment as seen from the other to be abreast with it, then as time goes on each will fall behind the other. Other indirect arguments tend to confirm me in this opinion; and thus, entertaining for this species of geometry something closely approaching belief, I have found it convenient to give it intuitional shapes by map-projections.

I use two, both very simple. To avoid confusion, I will write all formulae applying to the noneuclidean space in *red* (in printing, Old English can be used), all formulae applying to the representation of that space in Euclidean space in *blue* (in printing, Roman type), all formulae relating to the connection between the two in *brown* (in printing, heavy-faced type; on the blackboard, yellow chalk), and all formulae applicable everywhere since they merely relate to algebra in *black* (in printing,

[1] This manuscript is to be found in the archives of the Smithsonian Institution. For details on Peirce's association with that Institution when Samuel P. Langley was the Secretary see "The Scientist-Philosopher C. S. Peirce at the Smithsonian" by the editor in the *Journal of the History of Ideas* XVIII:4 (October 1957). Filed with the manuscript at the Smithsonian are two letters. Peirce's covering letter to Langley dated 18 April 1901 reads: "If the Academy is in session when this reaches you I should like to have it presented. I fear it is too trifling to be printed as a Memoir, though I know not where else to print it ...." It was received at the Smithsonian on April 20.

Langley's reply was written on 22 April 1901 and reads: "Your paper 'On two Map-Projections of the Lobatschewskian Plane' only reached me the day after the Academy had completed its session. What would you like to have me do with it?"

MS. 114 at the Houghton Library is a partial draft of this paper. It is entitled "On Hyperbolic Geometry."

Italics; on the blackboard, white chalk.) I also use the sign $\backsim$ to mean "represents," or "corresponds to."

The formulae of Lobatchewskian trigonometry can be obtained by the following system of correspondences.

$$(1)\quad \begin{cases} \mathbf{A} \text{ (an angle in spherical trig.)} \backsim \mathbf{A} \text{ (in Lobat. plane trig.)} \\ \sin \mathbf{a} \text{ (a being a side in sph. trig.)} \backsim -\mathbf{i}\, \text{Sinh}\, \mathbf{a} \text{ (in Lob. pl. tr.)} \\ \cos \mathbf{a} \text{ (a being a side in sph. trig.)} \backsim \text{Cosh}\, \mathbf{a} \text{ (in Lob. plane trig.)} \end{cases}$$

Although only a few of the following formulae will be needed, the others are given for convenience.

Right Triangles

$$\text{Cosh}\, \mathfrak{h} = \text{Cosh}\, \mathfrak{a}.\ \text{Cosh}\, \mathfrak{b}$$

$$(2)\quad \text{Cosh}\, \mathfrak{h} = \cot \mathfrak{A}.\ \cot \mathfrak{B}$$

$$(3)\quad \cos \mathfrak{A} = \text{Cosh}\, \mathfrak{a}.\ \sin \mathfrak{B}$$

$$\text{Sinh}\, \mathfrak{h} = \frac{\text{Sinh}\, \mathfrak{a}}{\sin \mathfrak{A}}$$

$$(4)\quad \text{Tanh}\, \mathfrak{h} = \frac{\text{Tanh}\, \mathfrak{a}}{\cos \mathfrak{B}}$$

$$\text{Sinh}\, \mathfrak{b} = \frac{\text{Tanh}\, \mathfrak{a}}{\tan \mathfrak{A}}$$

Oblique Triangles

$$\frac{\text{Sinh}\, \mathfrak{a}}{\sin \mathfrak{A}} = \frac{\text{Sinh}\, \mathfrak{b}}{\sin \mathfrak{B}} = \frac{\text{Sinh}\, \mathfrak{c}}{\sin \mathfrak{C}}$$

$$(5)\quad \text{Cosh}\, \mathfrak{c} = \text{Cosh}\, \mathfrak{a}.\ \text{Cosh}\, \mathfrak{b} - \text{Sinh}\, \mathfrak{a}.\ \text{Sinh}\, \mathfrak{b}.\ \cos \mathfrak{C}$$

$$\cos \mathfrak{C} = -\cos \mathfrak{A} \cos \mathfrak{B} + \sin \mathfrak{A} \sin \mathfrak{B}\, \text{Cosh}\, \mathfrak{c}$$

$$\text{Sinh}\, \mathfrak{a}.\ \text{Coth}\, \mathfrak{b} = \cot \mathfrak{B}.\ \sin \mathfrak{C} + \text{Cosh}\, \mathfrak{a}.\ \cos \mathfrak{C}$$

A few formulae of hyperbolic functions may also be added, as follows.

$$(6)\quad \text{Tanh}\, x = \frac{\text{Sinh}\, x}{\text{Cosh}\, x} \qquad \text{Coth}\, x = \frac{\text{Cosh}\, x}{\text{Sinh}\, x}$$

$$(7)\quad \text{Sech}\, x = \frac{1}{\text{Cosh}\, x} \qquad \text{Cosech}\, x = \frac{1}{\text{Sinh}\, x}$$

$$(8)\quad \begin{cases} \text{Sinh}\, 0 = 0 & \text{Cosh}\, 0 = 1 & \text{Tanh}\, 0 = 0 \\ \text{Sinh}\, \infty = \infty & \text{Cosh}\, \infty = \infty & \text{Tanh}\, \infty = \infty \end{cases}$$

$$\text{Sinh}\, x = \frac{\mathfrak{S}^x - \mathfrak{S}^{-x}}{2} \qquad \text{Cosh}\, x = \frac{\mathfrak{S}^x + \mathfrak{S}^{-x}}{2}$$

$$\text{Cosh}^2 x = 1 + \text{Sinh}^2 x$$

$$(9)\quad \text{Tanh}^2 x + \text{Sech}^2 x = 1$$

(10) $\operatorname{Coth}^2 x = \operatorname{Cosech}^2 x + 1$

$\operatorname{Sinh}(x+y) = \operatorname{Sinh} x.\ \operatorname{Cosh} y + \operatorname{Cosh} x.\ \operatorname{Sinh} y$

$\operatorname{Cosh}(x+y) = \operatorname{Cosh} x.\ \operatorname{Cosh} y + \operatorname{Sinh} x.\ \operatorname{Sinh} y$

(11) $\operatorname{Coth} x + \operatorname{Cosech} x = \operatorname{Coth}\frac{x}{2}$

(12) $\operatorname{Coth} x - \operatorname{Cosech} x = \operatorname{Tanh}\frac{x}{2}$

(13) $$\operatorname{Tanh} x = \frac{2\operatorname{Tanh}\frac{x}{2}}{1+\operatorname{Tanh}^2\frac{x}{2}} = \frac{2\operatorname{Coth}\frac{x}{2}}{1+\operatorname{Coth}^2\frac{x}{2}}$$

(14) $\operatorname{Coth} x.\ \operatorname{Coth}\frac{x}{2} = \frac{1}{2}(1+\operatorname{Coth}^2\frac{x}{2})$

(15) $\operatorname{Coth} x.\ \operatorname{Tanh}\frac{x}{2} = \frac{1}{2}(1+\operatorname{Tanh}^2\frac{x}{2})$

(16) $$\operatorname{Cosech} x.\ \operatorname{Coth}\frac{x}{2} = \frac{1}{2\operatorname{Sinh}^2\frac{x}{2}}$$

(17) $$\operatorname{Cosech} x.\ \operatorname{Tanh}\frac{x}{2} = \frac{1}{2\operatorname{Cosh}^2\frac{x}{2}}$$

A CONFORMED PROJECTION.

Let us now proceed to construct a projection, fulfilling the following conditions:

1st, all angles to be preserved unchanged;

2nd, there is to be a centre of projection all straight lines through which are represented by straight lines [Fig. 1].

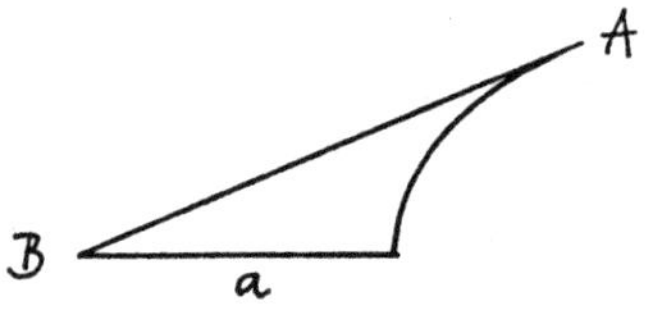

Fig. 1

Let B be at the centre of projection, and denote the angle between two straight lines by the same letter. Upon one of these lines, at a distance a from the centre, erect a perpendicular, intersecting the other line in an internal angle, A. Then (3) applies.

Upon the map, let the point corresponding to B be taken as an origin of polar coordinates. Let the other extremity of a be represented by a point whose coordinates are:

(18) $\boldsymbol{\theta} = \mathbf{0} \qquad \mathbf{r} = \mathbf{Fa} = \mathbf{l}$

Then, by (3),

(19) $\cos A = \operatorname{Cosh} a \cdot \sin \theta$

Upon the map, we shall have a plane triangle having, at the origin, an angle $d\theta$, included between two sides whose lengths are r and $r+dr$, while one of the other angles is A. Then, by plane trigonometry, we shall have

$$\tan A = \frac{r \cdot \sin d\theta}{r + dr - r \cdot \cos d\theta} = r\frac{d\theta}{dr}$$

whence

$$\text{(20)} \qquad \frac{d \log r}{d\theta} = \cot A = \frac{\cos A}{\sqrt{1-\cos^2 A}} = \frac{\operatorname{Cosh} \mathrm{a} \cdot \sin\theta}{\sqrt{1-\operatorname{Cosh}^2 \mathrm{a} \cdot \sin^2\theta}} = \frac{\sin\theta}{\sqrt{\cos^2\theta - \operatorname{Tanh}^2 \mathrm{a}}}$$

[see (9)]. Multiply both sides by $\dfrac{d\theta}{d\cos\theta} = -\dfrac{1}{\sin\theta}$; and we have

$$\text{(21)} \qquad \frac{d \log r}{d \cos\theta} = -\frac{1}{\sqrt{\cos^2\theta - \operatorname{Tanh}^2 \mathrm{a}}}$$

The integral of this is

$$\text{(22)} \qquad \frac{\mathrm{C}}{r} = \frac{\cos\theta}{\operatorname{Tanh} \mathrm{a}} + \sqrt{\frac{\cos^2\theta}{\operatorname{Tanh}^2 \mathrm{a}} - 1}$$

It is obvious that the two values resulting from the ambiguous sign of the square root are reciprocals of one another; for

$$(x+\sqrt{x^2-1})(x-\sqrt{x^2-1}) = x^2-(x^2-1) = 1.$$

Multiplying (22) by r, and putting

$$\text{(23)} \qquad r\cos\theta = x \quad r^2 = x^2+y^2,$$

we get

$$\text{(24)} \qquad \left(x - \frac{\mathrm{C}}{\operatorname{Tanh} \mathrm{a}}\right)^2 + y^2 = (\mathrm{C} \operatorname{Cosech} \mathrm{a})^2;$$

so that the equation represents a circle.

In order to determine the value of C, we use (18) and find from (22)

$$\text{(25)} \qquad \frac{\mathrm{C}}{l} = \operatorname{Coth} \mathrm{a} \pm \sqrt{\operatorname{Coth}^2 \mathrm{a} - 1} = \begin{cases} \operatorname{Coth} \mathrm{a} + \operatorname{Cosec} \mathrm{a} = \operatorname{Coth}\frac{a}{2} \\ \operatorname{Coth} \mathrm{a} - \operatorname{Cosec} \mathrm{a} = \operatorname{Tanh}\frac{a}{2} \end{cases}$$

[see (10), (11). (12)]. Substituting this in (24), we get, by (14), (15), (16), (17),

$$(26) \qquad [x - \tfrac{1}{2}l(1 + \mathrm{Coth}^2 \tfrac{a}{2})]^2 + y^2 = \left(\frac{l}{2\,\mathrm{Sinh}^2 \frac{a}{2}}\right)^2$$

$$[x - \tfrac{1}{2}l(1 + \mathrm{Tanh}^2 \tfrac{a}{2})]^2 + y^2 = \left(\frac{l}{2\,\mathrm{Cosh}^2 \frac{a}{2}}\right)^2$$

The difficulty of conceiving a straight line to be represented by a circle which is topically untransformable into a straight line will be considered below, and shown not to be serious.

Since the leg b of the triangle is represented by a circle, it follows that, as B increases, the angle A must ultimately become zero. In fact, by (2), when $A = 0$, Cosh $h = \infty$, $h = \infty$. But the circle representing the leg b is a finite one, and thus the locus of points infinitely distant from the centre must be represented by a finite circle. Let the equation of the circle representing the absolute be

$$(28) \qquad r = L$$

Then, every circle orthogonal to this circle on the map will represent a straight line in the Lobatchewskian plane. The general equation of such a circle is

$$(29) \qquad \left(x - \frac{L^2 + l^2}{2l}\right)^2 + y^2 = \left(\frac{L^2 - l^2}{2l}\right)^2$$

The same equation precisely would hold if l, instead of representing the minimum distance of the circle representing leg b from the origin, represented the maximum distance. If (29) be compared with (26) and with (27), each of these last two equations affords two ways of determining the value of l. From (26), we have

$$(30) \qquad \frac{L^2}{l^2} = \mathrm{Coth}^2 \tfrac{a}{2}, l = L\,\mathrm{Tanh} \tfrac{a}{2}; \text{and } \frac{L^2}{l^2} - 1 = \mathrm{Cosech}^2 \tfrac{a}{2}, l = L\,\mathrm{Tanh} \tfrac{a}{2}$$

while from (27) we have

$$(31) \qquad \frac{L^2}{l^2} = \mathrm{Tanh}^2 \tfrac{a}{2}, 1 = L\,\mathrm{Coth} \tfrac{a}{2}; \text{ and} \frac{L^2}{l^2} - 1 = -\mathrm{Sech}^2 \tfrac{a}{2}, l = L\,\mathrm{Coth} \tfrac{a}{2}.$$

In order to make the last right, since a hyperbolic tangent of a real quantity is essentially less than 1, we have to assume that l is the maximum distance of the circle representing the leg b, from the origin. The circle on the map is cut twice by the radius vector, although the straight line it represents is cut but once by the hypotheneuse represented by the radius vector. In short, one point on the Lobatchewskian plane is represented twice over on the map; and generally, all that part of the map outside the circle representing the absolute only represents again what is represented inside. Equations (26) and (30) refer to the inside map; equations (27) and (31) to the outside map. On comparing (25) with (30) and (31), we get

(32) $\quad C = L.$

It is now proved that (30) gives the rule for the projection, if there is any such projection; but it is perhaps not proved that that rule does not lead to contradictory results. At any rate, it will be worthwhile to test it by applying (4). Substituting (13) in this, we get

(33) $$\operatorname{Tanh}\frac{\mathfrak{h}}{2} + \operatorname{Coth}\frac{\mathfrak{h}}{2} = \left(\operatorname{Tanh}\frac{\mathfrak{a}}{2} + \operatorname{Coth}\frac{\mathfrak{a}}{2}\right)\cos\mathfrak{B}$$

In the figure [2], the distance between 0, the origin, and C, the centre of the circle representing the leg b is $\dfrac{L^2+l^2}{2l}$ and the radius of that circle is $\dfrac{L^2-l^2}{2l}$. Hence, by plane trigonometry,

(34) $$\left(\frac{L^2-l^2}{2l}\right)^2 = \left(\frac{L^2+l^2}{2l}\right)^2 + r^2 - 2\frac{L^2+l^2}{2l}r\cos\theta$$

which reduces at once to

(35) $$\frac{L}{r} + \frac{r}{L} = \left(\frac{L}{l} + \frac{l}{L}\right)\cos B$$

Now since we know, by (30), that

(30) $$\frac{\mathbf{l}}{\mathbf{L}} = \operatorname{Tanh}\frac{\mathbf{a}}{2}$$

it follows from the comparison of (33) and (35) that

(36) $$\frac{\mathbf{r}}{\mathbf{L}} = \operatorname{Tanh}\frac{\mathbf{h}}{2}$$

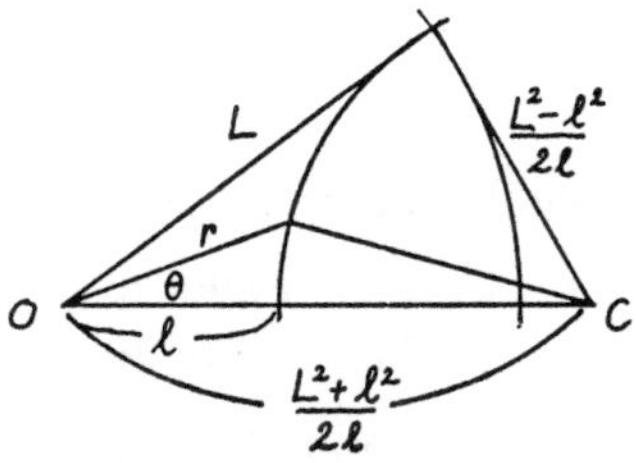

Fig. 2

The projection is thus established. It may be very appropriately called the hyperbolic stereographic projection, since its rule of construction only differs from that of the stereographic projection of the elliptic plane in that the radii vectores are proportional to Tanh $\frac{h}{2}$ instead of to tanh $\frac{h}{2}$, where h is the distance of the point to be represented from the centre. It also agrees with the elliptic stereographic projection in preserving the angles. The other important property of the latter is that every circle represents a circle (or straight line), and conversely. In order to ascertain whether the hyperbolic stereographic projection has this property or not, we assume the equation of a circle, which by plane trigonometry is

$$(37) \qquad R^2 = r^2 + Q^2 - 2rQ\cos\theta$$

Now, of course, the circle cannot represent a real circle unless

$$(38) \qquad L > Q + R$$

so that $\sqrt{(L+Q+R)(L-Q-R)(Q-R-L)(R-L-Q)}$ is real. Call this square root V for the sake of brevity, $\dfrac{L+Q-R}{V}$ and $\dfrac{L-Q+R}{V}$ will, in case of reality, be positive quantities greater than 1 and, therefore, of the nature of hyperbolic cosines. We may assume, then,

$$(39) \qquad \text{Cosh}\,\mathbf{b} = \frac{\mathbf{L}+\mathbf{Q}-\mathbf{R}}{V}$$

$$(40) \qquad \text{Cosh}\,\mathbf{c} = \frac{\mathbf{L}-\mathbf{Q}+\mathbf{R}}{V}$$

Solving these equations for Q and R, we find

$$\mathbf{Q} = \frac{\mathbf{L}\,\text{Sinh}\,\mathbf{b}}{\text{Cosh}\,\mathbf{b} + \text{Cosh}\,\mathbf{c}}$$

$$\mathbf{R} = \frac{\mathbf{L}\,\text{Sinh}\,\mathbf{c}}{\text{Cosh}\,\mathbf{b} + \text{Cosh}\,\mathbf{c}}$$

Putting also

$$\boldsymbol{\theta} = \mathbf{C}$$
$$\mathbf{r} = \mathbf{L}\,\text{Tanh}\,\tfrac{\mathbf{a}}{2}$$

the equation (37) of the circle becomes

$$\text{Sinh}^2\,\mathfrak{c} = \text{Tanh}^2\tfrac{\mathfrak{a}}{2}\,(\text{Cosh}\,\mathfrak{b} + \text{Cosh}\,\mathfrak{c})^2 + \text{Sinh}^2\mathfrak{b} - 2\text{Tanh}\tfrac{\mathfrak{a}}{2}\cdot\text{Sinh}\,\mathfrak{b}\,(\text{Cosh}\,\mathfrak{b} + \text{Cosh}\,\mathfrak{c})\cos\mathfrak{C}$$

or, dividing by Cosh b + Cosh c, and observing that

$$\text{Sinh}^2 c - \text{Sinh}^2 b = (\text{Cosh}\,c - \text{Cosh}\,b)(\text{Cosh}\,c + \text{Cosh}\,b)$$

we get

$$\text{Cosh}\,\mathfrak{c} - \text{Cosh}\,\mathfrak{b} = \text{Tanh}^2\tfrac{\mathfrak{a}}{2}(\text{Cosh}\,\mathfrak{b} + \text{Cosh}\,\mathfrak{c}) - 2\text{Tanh}\tfrac{\mathfrak{a}}{2}\cdot\text{Sinh}\,\mathfrak{b}\cdot\cos\mathfrak{C}$$

Dividing by $1-\text{Tanh}^2\frac{a}{2}$, and observing that

$$\text{Cosh}\,a = \frac{1+\text{Tanh}^2\frac{a}{2}}{1-\text{Tanh}^2\frac{a}{2}}$$

$$\text{Sinh}\,a = \frac{2\,\text{Tanh}\frac{a}{2}}{1-\text{Tanh}^2\frac{a}{2}}$$

we get

$$\text{Cosh}\,\mathfrak{c} = \text{Cosh}\,\mathfrak{a}\cdot\text{Cosh}\,\mathfrak{b} - \text{Sin}\,\mathfrak{a}\cdot\text{Sin}\,\mathfrak{b}\cdot\cos\mathfrak{C}$$

which, when a and C are variable is, by (5), the equation of a circle. Hence, this projection represents circles (and straight lines) by circles, and conversely. Summing up its properties:

1st. All angles are preserved.

2nd. Representing any point on the hyperbolic plane by the origin on the map, all straight lines through that centre are represented by straight lines.

3rd. A point distant by h from the centre is represented by a point distant by L Tanh $\frac{h}{2}$ from the original.

4th. Every straight line in the hyperbolic plane is represented by a circle orthogonal to the circle of radius L with its centre at the origin.

5th. Every circle lying wholly within that absolute circle on the map represents a real circle of the hyperbolic plane with a real radius, and conversely. Straight lines not through the origin and circles cutting the absolute circle represent real circles of imaginary radius, a sort of locus peculiar to hyperbolic geometry.

Practically, this projection is distinctly more convenient even than the ordinary stereographic; but theoretically it presents greater inconveniences. As the elliptic stereographic represents every point twice over, so also does the hyperbolic stereographic projection. This arises from the fact that it is conformal. For conformal projections are exclusively between artiad surfaces, that is, surfaces of even cyclosis, while the projective plane, or plane upon which unbounded lines can intersect in an odd number of points, cannot by any distortion be brought into coincidence with an artiad surface. In the hyperbolic plane, we have to distinguish between imaginary points and points at imaginary distances from the origin. That every pair of real rays in a plane intersect in a real point is a proposition of projective geometry, which no mode of measurement, that is, no system of designating the points, can falsify. The elliptic stereographic divides the plane along the whole extent of a ray and there joins it to a similarly cut copy of itself turned upside down;

so that every part of the elliptic plane is twice represented in relatively perverted images. The hyperbolic stereographic cuts a disk out of the plane, along the absolute conic, and joins that to a perverted image of itself, leaving all real points at imaginary distances entirely unrepresented. It obviously cannot represent those points conformedly since the angles are there imaginary. This is a grave inconvenience for many purposes.

The projection is subject to another grave inconvenience which manifests itself, for example, if we attempt to apply the projective definition of a circle as a conic having double contact with the absolute. Since the real part of the absolute circle is shown in the projection, we might naturally expect to find some traces of the double contact of circles with the absolute, although the plane of a conformal projection has no imaginary parts, being, on the contrary, itself the representation of a single imaginary variable. But Cayley's theory of distance, beautiful as it is, takes but a narrow view of the subject. The system of distances is the expression of the law which defines the possibilities of displacement of a rigid body. In our universe, a part of that law is that rigid bodies can receive no imaginary displacements. Nor need imaginaries be called in to define the law in general terms, as I intend to show in another paper.

A CENTRAL PROJECTION OF THE LOBATCHEWSKIAN PLANE

The Lobatchewskian plane can also be so represented, that every ray, α, is represented by a ray cutting at right angles that radius vector from the origin which represents the ray through a fixed centre that the ray α cuts at right angles; and so that all angles about that centre are represented by equal angles about the origin [Fig. 3. △B, red; △O blue].

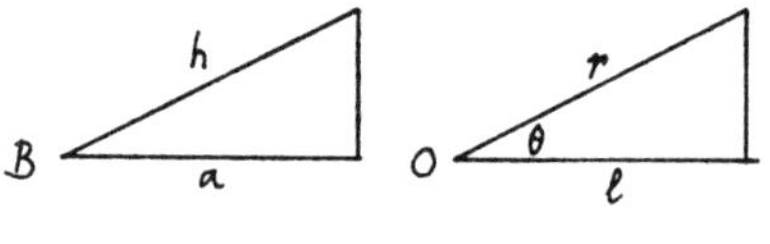

Fig. 3

A right triangle having one of its angles, B, at the centre will be represented by a right triangle having the corresponding angle at the origin. Then we shall have

$$\mathbf{l} \backsim \mathbf{a}$$
$$\mathbf{r} \backsim \mathbf{h}$$
$$\boldsymbol{\theta} = \mathbf{B}$$

$$\begin{aligned} \mathbf{l} &= \mathbf{Fa} \\ \mathbf{r} &= \mathbf{Fh} \\ \mathbf{r} &= \mathbf{l}\sec\boldsymbol{\theta} \\ \mathrm{Tanh}\,\mathfrak{h} &= \mathrm{Tanh}\,\mathfrak{a}\cdot\sec\mathfrak{B} \end{aligned}$$

Hence the projection of any point will be determined by the equations

$$\begin{aligned} \boldsymbol{\theta} &= \mathbf{B} \\ \mathbf{r} &= \mathrm{L}\,\mathrm{Tanh}\,\mathbf{h} \end{aligned}$$

where L is the radius of the circle which represents the circle at infinity.

Distances are constructed with the greatest ease. Suppose, for example, that having given two points, C and H, on the map, we wish to know where the ellipse representing a circle about C as a centre and having the radius represented by CH will cut another ray through C. The figure [4] almost explains itself. The line CH is produced to PR, its

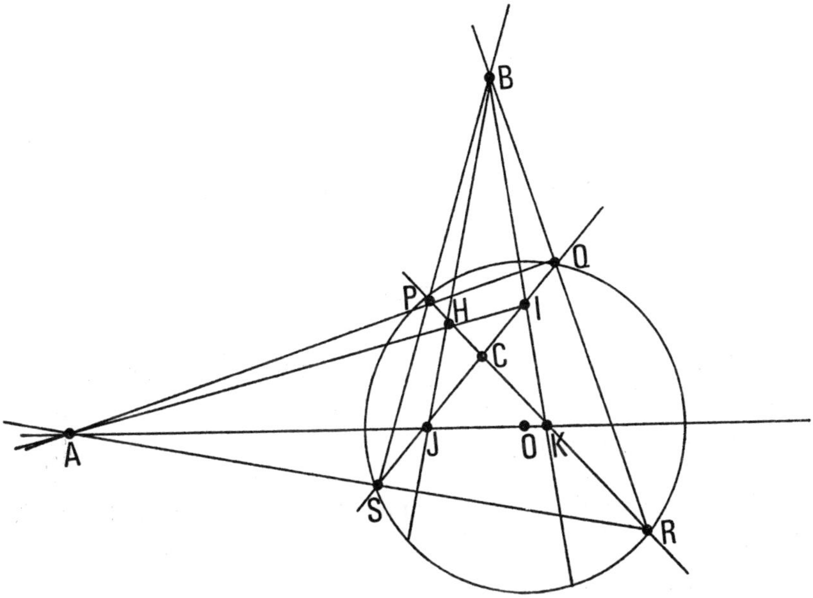

Fig. 4

intersection with the absolute, and the other line is produced to QS, its intersections with the same circle. The quadrangle $PQRS$ is completed so that A is collinear with P and Q and with R and S, while B is collinear with P and S and with Q and R. Then construct I at the intersection of AH and [QS], J at the intersection of BH and [QS], and K at the intersection of AJ and BI. [Then] CH, CI, CJ, CK represent equal distances. By drawing a diameter through C and O, the centre of the circle representing the absolute and laying down four points situated symmetrically to H, I, J, K, on the other side of this diameter, we have eight points on the ellipse representing the circle. There are other obvious modes of constructing equal distances. [Fig. 4 follows Peirse's instructions.]

I do not know any simple geometrical construction for angles, or any which is purely projective. There probably exists such a real projection. Triangles can, however, readily be solved by means of the following

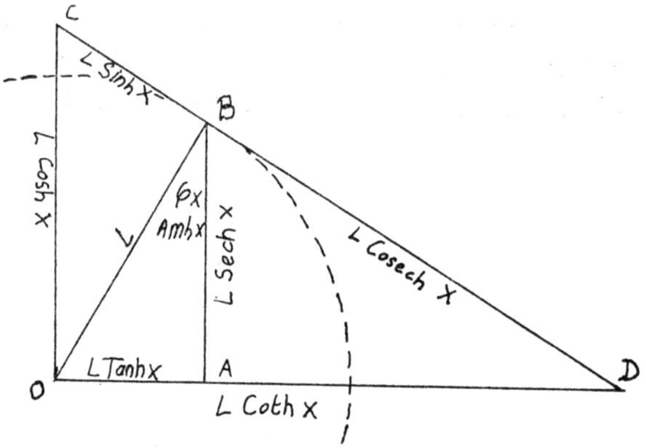

Fig. 5

figure [5]. Let O be the centre of the circle of radius L representing the absolute. Let OA represent the distance x, so that

$$OA = L \operatorname{Tanh} x$$

At A erect a perpendicular to OA cutting the circle of the absolute in B, so that $OB = L$. Then,

$$AB = L \operatorname{Sech} x$$

$$\angle OBA = 2 \arctan \mathrm{e}^x - \frac{\pi}{2}$$

Through B draw a perpendicular to OB cutting OA produced in D. Then

$$OD = L \operatorname{Coth} x$$
$$BD = L \operatorname{Cosech} x$$

At O raise a perpendicular to OA cutting BD produced in C. Then

$$OC = L \operatorname{Cosh} x$$
$$BC = L \operatorname{Sinh} x.$$

On these principles, angles can be brought to the centre; but with some trouble [see Fig. 6].

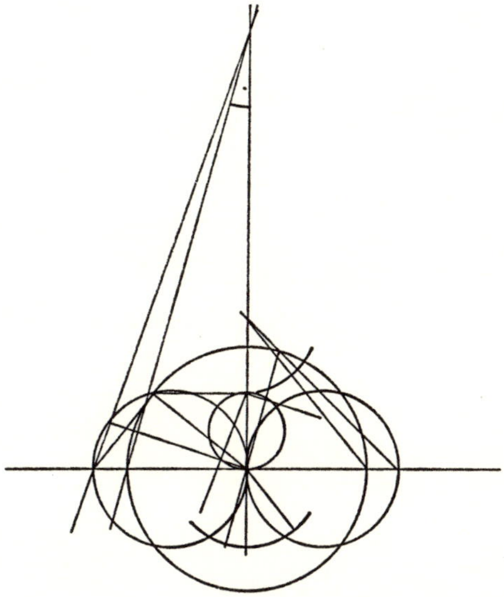

Fig. 6

16

APPLICATION OF MATHEMATICS

A. [FRAGMENT OF REPORT ON MORISON'S BRIDGE PROJECT] (1360)[1]

The entire theory of the general catenary, whether festoon or arch, is contained in the above two theorems, namely:

I. The horizontal component of the stress is constant throughout the cord.

II. The vertical component of the stress at any point equals the weight between that point and the horizontal stretch.

Yet there are a few relations, which might be deduced from these, which are worth mention. It may also do no harm to point out that there are two other very good ways of reasoning about the catenary besides the method of the parallelogram of forces used above.

One of these is the method of virtual velocities. To prove the above Theorem I, that taking any two points, say the *m*th and the *n*th the horizontal stress is the same at the two, sever the cord at those points, but hold it in place. Now increase the value of x_n by an infinitesimal amount, and the change of $\eta(m, n)$ will measure the force. Next increase the value of x_n by the same amount, when $\eta(m, n)$ will be restored. For the whole curve has only received a horizontal displacement. It follows the two horizontal forces measured are equal and opposite. Q.E.D. To prove Theorem II, cut out the portion of the cord between the middle of the horizontal stretch and the middle of any other. Now move the whole vertically by an infinitesimal amount. Since the only force on the horizontal stretch is a horizontal pull (or thrust in the case of the arch), the change of the centre of gravity measures the upward force at the other point, and shows the whole intervening weight equals that force. Q.E.D.

[1] During the 1890s Peirce served, whenever he could, as mathematical consultant on projects req''iring the service of a mathematician. For a time he was a consultant to the civil engineer George S. Morison whose offices were at 49 Wall Street, N.Y.C. Peirce received $1000 from him, 13 August 1895, for his services in investigating the theory of the stiffened suspension bridge.

The other method consists in imagining any portion of the cord to become rigid, not thereby varying the forces upon it. Of such forces there are three, the pulls on the end stretches (the bounds of the solidification not being at weighted points) and the weight applied to the centre of gravity. These being in equilibrium, their lines are concurrent. So,

Two tangents to any catenary meet in the vertical of the centre of gravity of the intervening weights.

Theorems I and II also follow at once.

Another way of thinking about a stretched cord, whether pliant or elastic, due to Kirchhof, if not very serviceable may truly be called poetical, for it excites the esthetic emotion in consequence of assimilating things widely remote, just as poetry does. The essence of this method is to think of the dynamical problem whose differential equation has the same form but with *time* substituted for the *arc* of the cord. The only use of this is that it may suggest something. For example, keeping the meaning of $d\omega/ds$ that it is the (vertical) force at the point, if s means the time, ω will be vertical component of the velocity *plus* a constant. Supposing the horizontal component is constant, as it must be if the force is vertical, the problem of the catenary is that of finding by what path a particle will move from one tower to the other in a given time supposing the force to vary with the time in a fixed way. Such similes may in an extensive treatise save solving the same equations twice over; and they are rather pleasing. They are toys. It seems to me there is no better image to think with than an analytical expression, provided it be clear of irrelevant coördinates, etc. But no doubt to different imaginations different kinds of images come the easiest.

The weight may be resolved into a tangential and a normal component. The former is $\dfrac{d\omega}{ds}\sin\tau$ at each point. Just as in a fluid the tendency to motion is proportional to the space-variation of the pressure. That is,

$$\frac{dq}{ds} = \frac{d\omega}{ds}\sin\tau.$$

This agrees with (1) and (2); for $\dfrac{dq}{ds} = Q\sin\tau\sec^2\tau\cdot\dfrac{1}{\varrho}$ and $\dfrac{d\omega}{ds} = Q\sec^2\tau\dfrac{1}{\varrho}$.

The second member may be integrated by parts in two ways. One gives

$$q - Q = \frac{d\omega}{ds}y' - \int y\frac{d^2\omega}{ds^2}\,ds'.$$

Now the last term vanishes in the common catenary; and any other may practically be considered as made up of pieces of ordinary catenaries joined together. The other way of integrating gives

$$q - Q = \omega' \sin \tau - \int \frac{\omega'}{\varrho} \cos \tau \cdot ds.$$

The normal component of the weight $\frac{d\omega}{ds} \cos \tau$ is balanced by the resultant of q_n and q_{n-1} resolved in the direction bisecting the angle between them, which is $\frac{1}{2}(q_n + q_{n-1})(\tau_n - \tau_{n-1})$.

If there is no horizontal stretch we imagine an infinitely short one interposed at the lowest point. If this is not at either end, but at $n = L$, we make the stretch horizontal by separating the weight $\omega_L - \omega_{L-1}$ into two parts $\omega_L - \omega_H$ and $\omega_H - \omega_{L-1}$ such that

$$\omega_H = \omega_L - Q \tan \tau_L = \omega_{L-1} - Q \tan \tau_{L-1}$$

which (4) shows to be possible. If the lowest point of the cord is at one tower, there is an upward pull on that tower. We balance it by an equal weight at an infinitesimal distance on the cord, making

either $\quad \omega_H = Q \tan \tau_0$

or $\quad \omega_H = -Q \tan \tau_N;$

for τ_N will be negative if N is the lowest point.

We may apply the principle of virtual velocities.[2] This at once shows the horizontal component of the stress is the same at all points. For consider the portion of the cord between any two points. Move one horizontally by an infinitesimal distance. The centre of gravity will be raised or lowered. Carry the other extremity horizontally in the same direction by the same amount. The two motions will merely shift the whole horizontally. Hence there are equal and opposite horizontal forces on the two points due to the intervening weight. We also show that the vertical stress equals the weight between the point and the horizontal stretch. For consider the portion of the cord between any point and the horizontal part. Raise both ends vertically. The form of the intervening portion is unchanged, and its centre of gravity is raised

[2] In his writings on mechanics the problem of virtual velocities is raised frequently. He was responsible for the *Century Dictionary* definition of the term. It is mentioned in his work on Mach and is further discussed by the editor in "C. S. Peirce and the Scientific Philosophy of Ernst Mach," a forthcoming publication in the *Actes* of the *XIIth International Congress for the History of Science*, August 1968 (Paris). Also see "The Influence of Galileo on the Thought of C. S. Peirce," by the editor in the *Atti del Simposio su "Galileo Galilei nella Storia e nella filosofia della scienza"* (Firenze-Pisa, Settembre 1964).

by the same amount. But there is evidently no vertical stress on the horizontal part. Hence, this measures the vertical stress at the other end.

The Common Catenary. Let

$$Fs = ks$$

where k is constant. Then

$$\tan \tau = \frac{ks'}{Q}$$

is the "intrinsic" equation of the curve. To transform to rectangular coordinates,

$$\sec^2 \tau \cdot d\tau = \frac{k}{Q}\, ds$$

and remembering k is constant the integrals expressing x' and y' can be resolved. Namely, attending to the lower limit,

$$\frac{kx'}{Q} = \frac{k}{Q}\int \cos \tau \cdot ds = \int \sec \tau \cdot d\tau = \log \tan\left(45^\circ + \frac{\tau}{2}\right)$$

$$\frac{ky'}{Q} = \frac{k}{Q}\int \sin \tau \cdot ds = \int \sin \tau \cdot \sec^2 \tau \cdot d\tau = \sec \tau - 1.$$

Using the notation of Gudermannians,

$$\tan \tau = \operatorname{Sinh} \frac{kx'}{Q} = \sqrt{\left(\frac{ky'}{Q} + 1\right)^2 - 1} = \frac{ks'}{Q},$$

$$\sec \tau = \operatorname{Cosh} \frac{kx'}{Q} = \frac{ky'}{Q} + 1 = \sqrt{1 + \left(\frac{ks'}{Q}\right)^2} = \frac{q}{Q},$$

$$\sec^2 \tau = \operatorname{Cosh}^2 \frac{kx'}{Q} = \left(\frac{ky'}{Q} + 1\right)^2 = 1 + \left(\frac{ks'}{Q}\right)^2 = \frac{k\varrho}{Q},$$

$$\tan \frac{\tau}{2} = \operatorname{Tanh} \frac{kx'}{2Q} = \frac{y'}{s'}.$$

Finally,

$$\frac{k\xi'}{Q} = \frac{kx'}{Q} - \frac{y'}{s'}$$

$$\frac{k\eta'}{Q} = \tfrac{1}{2}\left(\frac{x'}{s'} + \frac{ky'}{Q} - 1\right).$$

If the computer has a good table of Gudermannians he is properly equipped. If he has Legendre's *Fonctions elliptiques*, he will find Vol. II, p. 256 a table of log tan $\left(45° + \frac{\varphi}{2}\right)$. Hülsse's *Sammlung* 1st Ed., p. 468, has the best table of e^x with its Briggsian logarithm that I know. Hoüel's *Recueil de formules et de tables numériques* has a good 4-place table of Gudermannians and trigonometric functions combined, natural and logarithmic face to face, the argument being the "amplitude." But the centesimal division of the quadrant is used. I have no other table of Gudermannians, nor have I Legendre's *Fonctions elliptiques*. Hoüel's table I did not consider sufficiently precise. I have therefore calculated log tan $\left(45° + \frac{\tau}{2}\right)$ by 7-place logarithms, occasionally resorting to Namur's 12-place table and to Hoüel's 15-place table, by which I have calculated some of the values by means of the series

$$\log\tan\left(45° + \frac{\tau}{2}\right) = \frac{\pi}{180}\frac{\tau}{1°} + [93.9474808524]\left(\frac{\tau}{1°}\right)^3 +$$
$$[89.82917560]\left(\frac{\tau}{1°}\right)^5 + [87.7760409]\left(\frac{\tau}{1°}\right)^7 +$$
$$[81.758583]\left(\frac{\tau}{1°}\right)^9 + [77.7629]\left(\frac{\tau}{1°}\right)^{11} + [73.7615]\left(\frac{\tau}{1°}\right)^{13}$$

There are several approximations which are valuable. One function may be said to be an *approximation* to another when the first two terms of their developments are identical. If the development of the quotient of one divided by the other is 1 plus a very small fraction of the fourth power of the variable the approximation may be called *very close.* Adhering to these definitions we must not call the parabola

$$+\frac{ky'}{Q} = \tfrac{1}{2}\left(\frac{kx'}{Q}\right)^2$$

an approximation to the catenary. But we may call it a *rough approximation*, the quotient being

$$\frac{y' \text{ for Catenary}}{y' \text{ for Parabola}} = 1 + \tfrac{1}{12}\left(\frac{kx'}{Q}\right)^2 + \tfrac{1}{360}\left(\frac{kx'}{Q}\right)^4 + \text{etc.}$$

	$\tau = 10°$	20°	30°
y for Catenary	.0154266	.0641778	.1547005
y for Parabola	.0153868	.0635024	.1508686

We have

$$\frac{x'}{s'} = \text{Very approximately } \sqrt[3]{\frac{1}{\frac{ky'}{Q}+1}}$$

$$\text{In fact } \frac{x'}{s'}\sqrt[3]{\frac{ky'}{Q}+1} = 1 + \tfrac{1}{45}\left(\frac{kx'}{Q}\right)^4 + \tfrac{8}{3267}\left(\frac{kx'}{Q}\right)^6 + \text{etc.}$$

The small value of the third term is remarkable. If we carry this approximation one step further we have

$$\log\tan\left(45^\circ + \frac{\tau}{2}\right) = \text{To a second approximation}$$

$$\tan\tau\left(\sqrt[3]{\sec\tau} + \tfrac{8}{90}(\sec\tau - 1)^2\right)$$

The following are three values

τ	10°	20°	30°
$\log\tan(45^\circ + \frac{1}{2}\tau)$	.1754258	.3563782	.5493061
Second Approximation	.1754258	.3563735	.5492077
First Approximation	.1754295	.3565013	.5503214

Again, we have

$$\frac{k\xi'}{Q} = \text{Very approximately } \frac{y'}{x'}.$$

The quotient is

$$\frac{Qy'}{kx'\xi'} = 1 + \tfrac{1}{90}\left(\frac{kx'}{Q}\right)^4 - \tfrac{1373}{1300320}\left(\frac{kx'}{Q}\right)^6 + \text{etc.}$$

The following comparison shows the accuracy

τ	10°	20°	30°
$\frac{k\xi'}{Q}$	.0879371	.1800512	.2813569
$\frac{y'}{x'}$	.0879380	.1800835	.2816289

But the most remarkable property of this approximation is that it (or rather, its analogue) holds, not only for the common catenary, *but for every curve which is likely to be assumed by the cable of a suspension bridge.*

B. HINTS TOWARD THE INVENTION OF A SCALE-TABLE
(221)

A system of logarithms is a set of numbers in a table one for each natural number, such that different pairs of natural numbers having the same ratio correspond to pairs of logarithms having the same difference. Thus, since 14:21 = 22:33, log 14 – log 21 = log 22 – log 33.

A table of antilogarithms is a table in which the numbers are entered for constant intervals of the logarithms. In other words, it is a geometrical progression.

A logarithmic scale is one having numbers set down at distances from the origin measured by their logarithms, so that the operation of addition is replaced by measurement along the scale.

Logarithms were invented in 1614 by Napier, and the logarithmic scale in 1624 by Edmund Gunter. Things patentable are combinations of ideas. What is the difference between a table of antilogarithms and a logarithmic scale?

In a table of antilogarithms, what use are the logarithms anyway?

When you make a scale of anything is it not simpler to have the divisions equal even if the numbers attached are irregular (so that the measurement is speedy and precise), rather than have regular numbers at irregular distances the measurement of which requires infinite refinement?

Here make invention No. 1.

Now make a model of the thing you have invented, whether you call it scale or table. To do so you want a geometrical progression from 1 to 10. For your convenience I set down several such.

1 3.1623 10

1 2.1544 4.6416 10

1 1.7783 3.1623 5.6234 10

1 1.2589 1.5849 1.9953 2.5119 3.1623 3.9811 5.0119 6.3096 7.9433 10

Rules of logic are unfortunately not patentable and therefore the following will have to be used without protection. *Rule for Invention.* Take any useful thing. Make an accurate logical analysis of the good it does, and of the means whereby that is effected. Consider each element by itself and generally, and compare it with analogous processes; tabulate all possible methods of reaching the given result; consider the precise advantages and essential disadvantages of each, and select the combination which answers best in the case considered.

We will apply this rule to Invention No. 1, bearing in mind the aphorism: "Anything is applicable to anything."

Your invention, what does it do? It works the rule of three (of which multiplication and division are special cases). Good, but take a broader view of it so as to make it as useful as possible to the persons for whom it is intended. Say, it computes a number from 3 given numbers (including computation from 2 numbers as a special case).

How does it do this? By the transference of one part of a scale to another. The essence of this is the transference. The thing transferred only has to have its parts immovable or capable of being moved into a certain right position. No reason why it should be stretched out into a line. If 100 points can be measurably distinguished on a line 10000 can be so on a sheet, and 1000000 in a book of a hundred sheets or pages.

Here make general invention No. 2.

Now as to transference. Who constantly practise accurate transference? Draughtsmen. What means do they find the most convenient and speedy?

Here make patentable application No. 3.

It is to be remarked in passing that photography is the most marvellous *record* of transfer. Note side-patent No. 4. Also when a record is not required a *camera-lucida* is very accurate. Note side patent No. 5. These we won't use at present.

Old things may be patentable very properly as new inventions. For seven hundred years all the world, even the most ignorant, have been in daily use of logarithms, — nay, of a logarithmic scale, — nay, of *the very system of measuring down to a unit of measurement and then using computation,* which is precisely what we propose to patent in our scale table. I mean the decimal places of the Arabic system of numeration.

In crowding our numbers into the scale-table of course we shall use the decimal places of these numbers and the spaces between them as units of the scale $..\underset{\cdot}{5}.\underset{\cdot}{6}.\underset{\cdot}{8}.\underset{\cdot}{2}$. This will subdivide the interval into ten parts. *Patentable feature No. 6.*

In this way there is no difficulty in making numerical and trigonometrical scale-tables of a million divisions. It is perfectly practicable. Such a thing, equivalent to a 6 place table of logarithms would be used with extreme facility. Anybody who wanted to measure on it (merely adding and subtracting the numbers of the pages) could do so. Anybody who preferred to set down the numbers and use arithmetic (though this would be folly) could do so. *Patent this No. 7.*

We have now to consider the management of the interpolation. Suppose we wish to calculate $\dfrac{(A+a)\cdot(B+b)}{C+c}$ where A, B, C, are the tabular numbers nearest to $A+a$, $B+b$, and $C+c$, respectively. We have

$$\frac{(A+a)\cdot(B+b)}{C+c} = \frac{AB}{C} + \frac{Ab}{C} + \frac{aB}{C} - \frac{AB}{C'}$$
$$+ \frac{ab}{C} - \frac{Ab}{C'} - \frac{aB}{C'} + \frac{AB}{C''}$$

where $C' = C\dfrac{C}{c}$ and $C'' = C'\dfrac{C}{c}$.

In the case of extensive tables where the terms on the last line must be exceedingly small, this formula may be useful The expressions $\dfrac{Ab}{C}, \dfrac{aB}{C}$ and $\dfrac{AB}{C'}$ can be calculated by a scale-table.

In other cases, this formula is inconvenient. We may then use

$$\frac{(A+a)(B+b)}{C+c} = \frac{AB}{C} \times \frac{(1+\frac{a}{A})(1+\frac{b}{B})}{1+\frac{c}{C}}$$

In this case, we may insert into the table opposite or under each number, say n, the value of $\log(l+\frac{n}{100})$ or $\log(l+\frac{n}{1000})$ or whatever may be convenient. In other cases the fractions $\frac{a}{A}$ etc. may be calculated in the head, when it will be convenient to use such a system of logarithms that $\log(1+\frac{a}{A}) = \frac{a}{A}$ nearly. This system must be *nearly* the natural system or some system where the logs are multiples of the natural system. That is, the logarithm of 10 must be nearly 2.3026 or 23.026 or some such number. But to get the advantage of the ordinary system the log of ten must be a whole number.

Here invent the most convenient system for the scale-table.

Invention No. 8.

I will now put a simple scale table on the length of this paper.[1]

This scale is carried two significant figures too far. A scale table for 4 significant figures should have 230 divisions. One for 6 figures should have 2300 divisions, occupying 10 pages. The advantage of the scale-table will only be decidedly great when 6 places is reached. For the present table, it is quite as good as a 3 place table and only slightly inferior.

[FURTHER] HINTS TOWARD THE INVENTION OF A SCALE TABLE

§1. A system of logarithms is a system of numbers corresponding, one-to-one, to natural numbers in such a way that pairs of natural numbers which are in the same ratio to one another have logarithms which differ from one another by the same amount. Thus, since 10:15 = 14:21 it follows that

$$\log 10 - \log 15 = \log 14 - \log 21.$$

Logarithms were invented by Napier, 1614.

§2. A logarithmic scale is a scale on which natural numbers are set down at distances from the origin measured by their logarithms. If we apply a piece of paper to such a scale and mark off the distance of 15 from 10 and measure this on from 14 we shall find 21; thus solving the proportion.

The logarithmic scale was invented by Edmund Gunter, 1624.

§3. All linear logarithmic scales are similar. Consequently, different systems of logarithms are only different scales of measurement along a logarithmic scale.

[1] The numbers run the vertical length of the table. It is necessary here to break the line into several sections.

994	1098	1213	1340	1480	1634	1805	1993	2201
1000	1105	1222	1350	1492	1650	1823	2015	2228

2430	2682	2960	3267	3605	2977	4386	4837
2462	2721	3008	3325	3648	4062	4489	4962

5333	5819	6478	7136	7859	8651
5484	6062	6700	7406	8185	9047

The lower scale is repeated on the vertical edge of a separate sheet of paper.

§4. Suppose I convert the edge of this sheet into a rude logarithmic scale, using the spaces between the lines as units of measurement.[2] If I have no means at hand of subdividing them, except that of writing the numbers regularly, the proper subdivision of the scale may be treated in the use of it as a separate problem.

Practice with this scale will suggest several patentable inventions. You will see the scale is at the same time a table of antilogarithms, that is of numbers corresponding to given logarithms.

§5. How long must the slip of paper be for use with this scale?

§6. With how many places of figures can this scale be usefully inscribed? If too many are used the subdivisions of one space will not be in equal proportion throughout the space. Thus, if the last numbers were instead of 670, 741, 819, 905

6700190
7405684
8185466
9047342

The differences would be

705494
779782
861876
952658

and the second differences would be

74288
82094
90782

The intervals evidently could not be subdivided proportionally.* (* Now the rule is that when the second differences of a table are greater than 2, the intervals cannot be subdivided proportionally without danger of error. The close approximation of the second differences to the original numbers and of the third differences to the first will not escape you. Nor that the first differences are halfway between the numbers, nearly.)

But if not enough places were inscribed, the table would lose *very much* of its utility.

[2] Peirce repeats the second line in the table in footnote 1, using three significant figures only, i.e. 100, 111, 122, 135, etc.

§7. You will notice that the first differences are very nearly (though not exactly) tenths of the means between successive pairs of numbers. This can evidently be put to use in subdividing the intervals.

Then what should be number of spaces on the scale-table?

§8. Suppose you have the problem: As 25 is to 55 so is 28 to the answer. How do you proceed? The roughest use of the scale gives 61; but how to get the next figure. We have

$$\text{Answer} = \frac{55 \times 28}{25} = \frac{1}{10} \cdot \frac{(548+2)(272+8)}{246+4} =$$

$$\frac{1}{10} \times \frac{548 \cdot 272}{246} \times \frac{(1+\frac{2}{548})(1+\frac{8}{272})}{1+\frac{4}{246}}$$

This is very nearly $\frac{1}{10} \times \frac{548 \cdot 272}{246} \times \left(1 + \frac{2}{548} + \frac{8}{272} - \frac{4}{246}\right)$

$$= (\tfrac{1}{10} \times 606) + \tfrac{1}{10} \times (2\tfrac{606}{548} + 8\tfrac{606}{272} - \tfrac{606}{246})$$
$$= \tfrac{1}{10}(606 + 2 + 18 - 10) = 61.6$$

This is precisely right.

Required to multiply 23 by .0434. This is

$$(223+7)(449-15) = 223.449 \times (1 + \tfrac{7}{223})(1 - \tfrac{15}{449})$$
$$= 1000 \times (1 + 7 \times .0434 - 15 \times 23)$$
$$= 1000 \times (1 + 0.314 - .335) = 1000(1 - 0.021) = 0.9979$$

The true answer is 0.9982.

Had we inscribed the scale with an additional figure, we should have had for 226 222.8

449 448.9

and the answer would have been .9981.

17

N-VALUED LOGIC

A. ON THE SIMPLEST POSSIBLE BRANCH OF MATHEMATICS (1 and 1250)

Each branch of mathematics makes certain essential suppositions, in addition to leaving room for other suppositions in special problems. The development of the branch of mathematics will consist in tracing out the consequences of the essential suppositions as well as the consequences of their being besides whatever further suppositions there may be. The deepest-cutting difference between different branches of mathematics lies in the different degrees of complexity of the essential suppositions; and this degree of complexity is measured by the multitude of distinct objects which are essentially supposed. The essential suppositions may be considerably modified in essential respects, so as to remain essentially the same. By such modification, if necessary, these distinct objects that are essentially the same may be made to take the character of *values*; and thus, in this sense, we may say that the deepest-cutting difference between different branches of mathematics lies in the different multitudes of values which these different branches suppose.

It is possible to develope, and in fact I have developed, a system of mathematics in which only one value is positively supposed, provided another value or other values are not positively denied. Since every problem asks what value some quantity has, there could be no system of mathematics which should suppose that there was but a single value which any quantity could have; but there may be, and is, a branch of mathematics which positively supposes one value and never broaches the question whether there be any other or not. Such a mathematics would be fit to represent the possible thoughts of a baby who has some ideas but who has never thought of there being any such thing as *falsity*, and who consequently can have no idea of *truth*. The kind of logic which might govern such thought, I call *paradisaical logic*; but of course there is very little to be said about it.

The branch of mathematics to which I desire to direct your attention is not that, but is the simplest kind of mathematics which makes any positive supposition as to what essentially supposed objects, or values, there are. This is the system which supposes only two values. If these two values are represented by v and f, then we might say that, in this branch of mathematics, any quantity whatever, say x, is virtually subject to the quadratic equation

$$(x-v)(x-f) = 0.$$

Boole's original algebra of logic was nothing but the special form which this branch of mathematics takes when we make $v=1$ and $f=0$, and retain the ordinary operations of addition, subtraction, multiplication, and division. But this special form is altogether artificial. We ought not to introduce conceptions simply because they are needed for the mathematics of number, in general, when we have nothing to do with number in general. Moreover, it was shown, first, I believe, by me, that Boole's system was inadequate to the representation of the ordinary syllogisms, inasmuch as it furnished no means of consistently expressing a particular proposition like

Some *S* is a *P*.

Nor did it afford any means of developing the *Logic of Relations*, a branch of logic recognized in all ages as existing, and into which Augustus De Morgan had made a highly scientific research sufficient to show that it was a matter of great importance, although it was not sufficient to show how vast that importance was.

Boole's algebra appeared in 1846, De Morgan's memoir on the Logic of Relations in 1860. In 1870, I supplied the two defects I have just mentioned in Boole's algebra by a method which, for present purposes, may be described as the introduction into the algebra of characters which do not in themselves denote quantities but which are such that various combinations of them build up a symbol of a quantity. There was an algebra for this building up, and therefore there was a system of values, or of essentially supposed objects. But what I wish to say is that in my algebra of relatives it remains true that there are only two values which any quantity can have.

Now in order to explain a matter of great importance for the logic of all kinds of mathematics, I shall be obliged to introduce a small item of personal history. All mathematics either virtually or actually employs multiplication and addition, operations which are subject to what is called the distributive formula

$$(x+y)z = xz + yz.$$

The present rigid style of dealing with mathematics requires that this formula like every other should be traced to the original supposition or suppositions from which it is a necessary consequence. But this logical sequence itself consists in the truth of a proposition in this dyadic mathematics; and upon careful analysis it will be found that that proposition is of the very same form

$$(x + y)z = xz + yz$$

so that it appears that the truth of the distributive principle in any other mathematics depends upon its truth in dyadic mathematics. It thus becomes a matter of importance for all mathematics to know upon what the distributive principle of dyadic mathematics depends.

Now in the month of April, 1880, while I was lecturing in logic in the Johns Hopkins University, we had a spell of bleak weather, and I got a severe influenza and was ordered by my doctor to remain in my chamber. I thus got the leisure to write a memoir on the fundamentals of logical algebra, that is to say, of dyadic mathematics, and in that memoir, having expressly mentioned the premises of which that formula was the consequence, I said that I omitted the proof itself as being too tedious. By this I meant that while it was long it was also quite obvious. But soon Prof. Schröder wrote and said that he could not make out how that proposition followed from my premisses. By the time I got his letter I was off swinging my pendulum, and had not by me my original notes on which my memoir was based; and my own ingenuity failed as much as that of Professor Schröder had done to reproduce the demonstration; and before I could get back to Baltimore I heard from Schröder that he had succeeded in proving that there could be no such demonstration. I thought, therefore, that he must be right about it; and concluded that my statement must be set down to the clouding of my mind under the influenza when I drew up my memoir. It was not until many years later, that, [...]

B. MATHEMATICAL LOGIC (1147)

The logical analysis of mathematics. It will be sufficient to consider only those forms of mathematics whose problems relate to the "values" of "quantities"; for any mathematical system can readily be thrown into this form. Each "quantity" has a single "value." Then the simplest mathematical logic will be one in which there are but two values. This accords with ordinary deductive logic in which every proposition has one or other of two values, that of being true (i.e. of involving no falsity) or that of being false. For each finite number of values of which the system of values may consist, there will be a special logic, differing from all those logics of smaller systems of value only in not admitting certain forms of inference which are valid for them, and differing from all logics of greater systems in that for it certain forms of inference are valid which are not valid for them. Thus, if there are only two values, the inference holds

A differs in value from *B*,
B differs in value from *C*;
∴ *C* has the same value as *A*.

For a system of three values this inference does not hold but the following is valid, though not for any system of values greater than three:

A differs in value from *B*;
A differs in value from *C*;
A differs in value from *D*;
B differs in value from *C*;
B differs in value from *D*;
∴ *C* has the same value as *D*.

Of these logics of finite systems, only that of the system of three values seems likely to repay study. As long as the system of values is finite, the following inference will hold good:

An operation, F, is such that being performed upon any two quantities of different values, x and y, it gives two quantities, Fx and Fy, of different values;

∴Every equation of the form $Fx = y$ has a solution in the assumed system of values.

If the system of values is not finite but is not more multitudinous than the cardinal numbers, the above inference obviously will not hold good (as may be seen by supposing $Fx = x+1$; for the equation $x+1 = 0$ has no cardinal number as its solution), but there will always be *some* relation, r, such that the following Fermation inference will hold good:

Every value which stands in the relation r to, but not in that of r to anything itself r to, a value v, has the property P;

But whatever value there may be which stands in the relation r to, but not in that of v to anything itself r to, a value which at once stands in the relation r to v and which at the same time has the property, P, itself has the property, P;

∴Every value that stands in the relation r to v has the property, P.

It has been substantially shown by Dr. G. Cantor (*Acta Mathematica*, Vol. II) that no system of values each of which can be distinguished from all the others by means of a finite collection of cardinal numbers is more multitudinous than the system of cardinal numbers themselves. If a system of values is more multitudinous than that, the values cannot be distinguished, every one from every other, by any method whatsoever. Nevertheless, if the system is not more multitudinous than that of all possible collections of different cardinal numbers, an indefinite approximation may be made toward distinguishing its values. For, in that case, we may find a method of assigning to each value such a collection of numbers, and then, we may give as many of the numbers in each collection as we like, in the order of their magnitudes, beginning with the smallest; and thus we shall narrow indefinitely the class of values to which the value in question belongs. Such is the system of quantities contemplated by the theory of equations and the differential calculus, to each of which we can indefinitely approximate toward distinguishing by using an indefinite number of places of decimals. This is made instantly clear by reflecting that we might equally well use a binary system of numerical notation, in which case each numeral place would either contain a 0 or a 1; and we might simply give the numbers of the places which are filled by 1s. Thus,

.10011101 ... would be 1, 4, 5, 6, 8, ...

There are numberless other ways of connecting the system of values of the calculus with endless series of whole numbers; as, for example, by endlessly continued fractions. There is no special virtue in the binary notation except that it makes certain things evident. If the quantity is imaginary, we could number the numerical places of the real part with even numbers, and those of the imaginary part with odd numbers; and a similar device would enable us to take account of the integer parts of the quantities. The same considerations would apply to a quantity of any finite number of dimensions. For such a system of values the Fermatian inference ceases to hold good; but we may still use the method of limits. This method virtually assumes, and assumes no more than, that all the values of the system may be put into one-to-one correspondence with different collections of different values of another system to which latter the Fermatian inference applies. This will become apparent on due reflection upon two points; first, that the method of limits shows that though something is not exactly true of any finite part of an endless series of quantities that are rational or, at any rate, each of them calculable or determinable, yet that thing would be exactly true of the whole series, were every member that the rule of its formation provides for, present, and consequently is exactly true of the value which that series represents; and secondly, that a proposition true of an endless series but not of any finite part of it can only have been proved by the Fermatian inference. This inference holds good of every endless series of which each member can be individually specified in the following form.

First Premiss. All the members of the series up to the Mth inclusive, stand, in their order, in the relation, R, to one another;

Second Premiss. If all the members up to the Nth exclusive stand, in their order in the relation, R, to one another, and if $N > M$, then all the members up to the Nth *inclusive* stand, in their order, in the relation, R, to one another;

Conclusion. All the members of the series stand, in their order, in the relation, R, to one another.

It might be thought that to suppose an endless series to be completed involves a contradiction; and so it would to suppose it completed by a last completing member. But that there is no contradiction in supposing *every* member the rule calls for to be present is susceptible of rigid proof, of which only the main points can here be indicated. If to suppose, for example, all finite numbers exist involves contradiction, there must be a contradiction in the rule that every number has a number next greater than it. For mere existence is not a predicate that can contribute to a contradiction. But if the rule is self-contradictory there is some particular greatest finite multitude, so that no conceivable collection of that multitude could conceivably receive an additional unit. But if so the very idea of a collection would involve a contradiction, since a collection consists of members present *independently*, so that the presence of any does not prevent the presence of any others that could be present were the former absent. But if no collection were possible, no class would be possible, and consequently no class of inferences. But every reasoning, good or bad, refers to a possible class of inferences. Now the court of logic will take cognizance of the fact that there may be a reasoning. If there is an endless series of convergent approximations, we may very conveniently say that the exact value is that of the "infinity-eth," or *infinitesima*, approximation. It is true that this involves a contradiction, inasmuch as the series consists exclusively of approximations that can be calculated, which no "infinity-eth" can be. For that reason, it has long been the fashion to explain the word "infinitesimal" as a figure of speech. But this rejection is not logical nor in the spirit of mathematics. Before the representation of imaginary quantities by points on a plane was invented, a number whose square should be negative was self-contradictory; but that did not prevent imaginaries from being used with great convenience. The contradiction involved in supposing an "infinitieth" approximation merely lies in calling it one of the series of calculable approximations. For there is manifestly no contradiction in the supposition of a perspective picture of a straight line having one extremity and extending to infinity such that the parallel to it through the viewpoint cuts the picture. Nor is there any contradiction in supposing that picture to show a point upon that line at each whole mile, from an initial point. The vanishing point of that line will be the representation of the infinitieth mile-point. That is a conception perfectly familiar to mathematicians to which no objection is raised. But if the infinitieth member (which of course ceases to be a member of that series) is actually capable of being accurately pictured, there can be no contradiction in supposing it to exist. Only, we must be careful to note that it just surpasses the series. To suppose it to exist is to accept *infinitesimals* in the Leibnizian sense.

The multitude of values to which decimals or continued fractions can indefinitely approximate has been called the *first abnumeral multitude*. Let the *A*s be a collection of that, or any other, multitude. Then, the multitude of the different possible collections of different *A*s is greater than that of the *A*s themselves. That is to say, there is no relation, *r*, in which every collection of *A*s stands to an *A* to which no other collection of *A*s stands in this relation, *r*. For whatever relation *r* may be, no collection of *A*s which includes every *A* to which no collection including it is the only one standing in the relation *r*, and which includes no *A* to which a collection containing it is the only one in the relation *r*, — no such collection, I say, is *r* to any *A*. For, on the one hand, it is not *r* to any *A* to which some collection containing that *A* is the only one in the relation, *r*. For if it were it would contain that *A*. But by hypothesis it excludes all such *A*s. Nor is it, on the other hand, an *r* to any *A* to which no collection containing that *A* is the only one that stands in the relation *r*. For, by hypothesis, it contains all such *A*s; and to no one of these *A*s is a collection containing it in the relation, *r*.

It follows that the method of limits is not applicable to a system of values of such *second abnumeral multitude*; and consequently, it is impossible always, or generally, to approximate indefinitely to a single value. But this will not prevent all reasoning about it; although nobody has attempted to investigate such a system. Yet it would be possible to make use of the principle that each value of the system could be represented by a first-abnumeral collection of values of a first-abnumeral system. An endless succession of such abnumeral systems are possible, each being obliged to surrender some principle of reasoning which all those which precede it could use. Nevertheless, every one of those systems can make use of the principle that it consists of independent individual values, independent, that is to say, for the purpose of entering into collections of values; and reasoning can be performed by the aid of that principle.

We now come to systems of values for which this principle fails. Namely, we will suppose that we have to do with such a system of values that any collection of independent quantities can all have different values in this system. It follows that the system cannot consist of independent values; for if it did, a collection of quantities, one for every collection of those values, could not all have different values in that system. In order to avoid contradiction it is necessary to suppose that the system contains one or more (up to any multitude) of unit-systems each correspondentially equal to the whole system; and even each unit-system may be separated (by a difference in any respect) in to two parts, as long as one part is correspondentially less than the whole; but it cannot be, properly speaking, separated into two parts each correspondentially equal to the whole, these parts being distinguished by the fact that the values of certain quantities absolutely belong in the one and are absolutely outside of the other; but there will always be a third part, correspondentially less, of which the proposition in question may equally well be regarded as true or as false. In short, the principle of excluded middle, or that of contradiction, ought to be regarded as violated here; or there is a limitation here to the applicability of the relation of negation which those principles define. The definitions of otherness and of identity proper (identity ἀριθμῷ) presuppose a universe of individuals; in a universe not consisting of individuals, where every part consists of parts of the same kind, they are only applicable so far as that universe admits of individuals. Suppose, for example, a ball rolling on a billiard table to come to rest. The system of dates in time (not of *assignable* dates, since these can only form, at most, a first-abnumeral system, but of instants such as our conception of flowing time supposes) is a system of values of the kind considered. That ball is moving down to a certain instant, and [at] all subsequent instants is at rest. It cannot, however, absolutely move through *all* the time previous to that instant and be at absolute rest during all subsequent time, without at that instant being in a state which is neither one of absolute motion nor one of absolute rest. The method of limits would explain this, were it applicable; but it is only applicable to a collection of assignable instants, which is a very limited kind of multitude. If time flows, no instant has an absolutely independent identity. It is so far independent that an instantaneous state of things may be supposed to exist absolutely at that instant alone. But a duration which begins or ends at that instant cannot properly be said absolutely to contain or absolutely to exclude that instant. No contradiction is involved in this hypothesis: it is simply the hypothesis of a system which, since it is capable of affording separate values for every multitude of independent quantities, and since there is no maximum multitude of independent quantities, does not itself consist of independent individual quantities; so that the relations of identity and negation have but a limited applicability to it. Without this conception it is impossible to set forth the logic of topical geometry; and the philosopher will fall into insoluble difficulties about the flow of time, and our consciousness of it. Suppose a filament to occupy the whole of a limited line and to be restricted to that line. Let the terminal particle which occupies an absolute point, be thrown off. Then, according to our ordinary idea of space, will that filament remain without any terminal particle? No. But if it now had *another* particle as its termination, it would consist of particles next to one another, and so not be continuous. Thus, we are forced to say that it is *the same* particle which remains after being thrown off, — in short, we must admit that, according to our ordinary idea of space, points and particles occupying points have no absolute identity. Kant defines a continuum as that of which every part consists of parts, and though he confuses this definition with infinite divisibility, it is really a different hypothesis, and may (in the present writer's opinion) be accepted as an approximate definition of a continuum. For it is the same as to say that it is not a collection of individuals. Applying this to the circumference of a circle, there are no points on that circumference. There is only room to put upon it any multitude of points whatever. Each point put upon it, if it be regarded as a part of the figure, breaks the continuity at that point, because it is a part which does not consist of parts. The difficulty of the conception must excuse the length of this explanation. Reflection upon it will render the matter perfectly clear.

The smallest kind of system (correspondentially speaking) of continuous systems of values is the line-figure. Every line is correspondentially equal to every other, since it may be conceived as being stretched and bent so that that other shall coincide with a part of it. A line-figure may consist of any multitude of separate lines. It may have any multitude of outlying points. Each piece may have any multitude of furcations, and may have a multitude of terminal points dependent on the multitude of furcations. A piece without furcations may have no terminal, or one, or two at most. The number of separate parts of a line figure is a property which it shares with discrete collections. But the *cyclosis*, or return into itself, is a property, properly speaking, peculiar to continua. Surfaces have an additional property, the *periphraxis*, or bagging; solids another corresponding property; and each additional dimension confers an additional property of this kind. These properties are called *Listing's Numbers*. Each additional dimension restricts the modes of reasoning which are applicable to continua. General logic applies to them all, provided we take care not to assume that objects have independent identity, after it has already been assumed that they have not. To a continuum of an infinite multitude of dimensions it seems that no reasoning can be applied, except that of general logic; so that we, thus, return to that very logical system which belongs, in a peculiar way, to the simplest of all possible systems of value, that of two values.

In regard to the peculiarities of mathematical reasoning about all the above systems of values alike, which as we have seen, have each its peculiar logic, it is first to be noted that the mathematical reasoning, proper, is confined to deducing the properties of hypotheses with the truth or falsity of which, for the actual universe, the mathematician has no concern. The framing of the hypothesis is a work of thought demanding the greatest genius and good judgment; and this is a part of the business of every mathematician. Nevertheless, this sort of thought is not called mathematical reasoning. The hypothesis of the mathematician must have such a degree of definiteness as to permit formal deductions. For that purpose, it must be definitely supposed that certain forms of relation subsist between objects. In other respects, the less definite the hypothesis is, the better. Thus, it would be a hindrance rather than a help to suppose a geometrical figure to have any particular color. Finally, the hypothesis of the mathematician is always of an intricate kind, so that all the relations involved cannot be seen at a glance. Mathematical power is the power of dealing with intricate hypotheses and of finding a point of view from which they will appear simpler. It is, therefore, a sort of generalizing power; and all strong mathematicians show great ability in generalization. Dedekind considers the theory of numbers and algebra to be parts of logic; and geometry considered as tracing out the consequences of hypotheses seems to be as much so. But while the logician is occupied with analyzing reasoning and describing the essential character of its different steps, the mathematician's interest is in the practice of reasoning, — so that the only purely ideal science is the only one which entirely consists in practice.

In order to analyze mathematical reasonings three, if not four, distinct branches of logic are required, ordinary logic, the logic of relatives, the logic of abstraction, and the logic of possibility. But the last two branches have been so little investigated as yet, that it is impossible to be sure that they are distinct.

The procedure of the mathematician is, first, to state his hypothesis in general terms; second, to construct a diagram, whether an array of letters and symbols with which conventional "rules," or permissions to transform, are associated, or a geometrical figure, which not only secures him against any confusion of *all* and *some*, but puts before him an icon by the observation of which he detects relations between the parts of the diagram other than those which were used in its construction. This observation is the third step. The fourth step is to assure himself that the relation observed would be found in every iconic representation of the hypothesis. The fifth, and final, step is to state the matter in general terms.

Among the general peculiarities of mathematical reasoning may be mentioned the great use it makes of extreme cases, which are here often the key of the problem. Thus, almost the whole theory of functions lies in ascertaining where they become *zero* or infinite. Another peculiarity is the extraordinary generalizations it forms. Of still greater importance is the practice of making operations and relations of all kinds objects to be operated upon.

C. [THE MODUS PONENS] (748)

The relation of truth and falsity as formal logic conceives it is defined in two clauses, as follows: 1st, No proposition is both true and false. Every proposition is either true or false. The first clause is called the principle of contradiction, the second the principle of excluded middle.

A different conception is quite possible. We might consider that a proposition is never, or hardly ever, perfectly true, and might obtain some mode of measuring the proportion of truth that a proposition contains. Geometry, we know, has its descriptive and metrical portions, the former considering whether points coincide or not, etc., the latter measuring how far distant from one another they are. We can import this distinction into logic, and say that the conception of truth and falsity as it is conceived in formal logic, is descriptive not metrical.

We cannot reason or think at all without making a distinction between truth and falsity; while we can perform some elementary kinds of reasoning (in point of fact, all that are considered in the traditional logic) without absolutely excluding the possibility of an intermediate state between the two, say the state of being sometimes true and sometimes false. The principle of contradiction is more elementary than that of excluded middle, so that we may begin by considering the consequences of the former while leaving the latter out of account.

Let us call the State of a proposition as being true or false (or something else) its *value*. That will be a convenient phrase. More than that, it will suggest a convenient application of arithmetic. For we may choose any two numbers at pleasure to represent the values of truth and falsity. Denote these by v and f respectively: then the value of every true proposition is v and that of every false proposition is f, and if there be any propositions neither true nor false, their values are represented by other numbers. Using the letters of the alphabet to denote the values of different propositions distinguished by those letters, we write $x = v$ to mean that the proposition x is true, or just the same as x written alone, and $x = f$ to mean that the proposition x is false. This last we may also write $v + f - x$, for this is something the value of which becomes f if $x = v$ and becomes v if $x = f$, or in other words it is true if the proposition x is false and false if the proposition x is true, but if the proposition x is neither true nor false, so neither is this. In short, $v + f - x$ represents the negative of the proposition x.

Let us write $f(x, y)$ to signify some algebraical expression which when $x = v$ never itself equals v unless $y = v$. Further than that, the value of $f(x, y)$ is not defined; no matter what values it takes when x is not equal to v, and no matter if it never becomes equal to v itself, it will be a quantity denoted by $f(x, y)$ provided it does not become equal to v with $x = v$ and y unequal to v. So defined, the expression $f(x, y)$ signifies precisely "If x is true y is true" for this hypothetical statement is not true if x is true unless y is true also, but whether it be true if x is not true or if both x and y are true depends upon circumstances.

This definition is the formula of the simplest and most rudimentary of all inferences, the *modus ponens* of hypothetical inference.

If x is true, y is true,
But x is true,
Therefore y is true.

Suppose that being at the theatre, I were to hear a rumor of a house being on fire near where I lived. I should say to myself "If this be true I had better be going." I make inquiries and ascertain that it is true, and I hurry home accordingly. This is an example of an inference of this kind.

THE MODUS PONENS

The relation between truth and falsity, as we must begin by conceiving them, is thus defined. 1st, Nothing is both true and false: 2nd, Every proposition is either true or false. The former clause of the definition is called the principle of contradiction, the latter the principle of excluded middle.

A somewhat different view is familiar to physicists.[1] Dealing as they do with matters of measurement, they hardly conceive it possible that

[1] By working in the logic of substantive possibilities, Peirce was led to the concept of an infinite class of infinite multitudes. To him these multitudes were apparently the alephs of Cantor, but Peirce occasionally expressed uncertainty. However, his investigations made Peirce believe in room on a line segment and in an interval of time

the absolute truth should ever be reached, and therefore instead of asking whether a proposition is true or false, they ask how great its error is. Just as geometry has its descriptive and its metrical portions, the former considering whether points coincide or not, the latter measuring how far distant from one another they are, just as chemical analysis has its qualitative and quantitative divisions, so logic has first to decide whether a proposition or reasoning be true or false, and, secondly in the latter case, to measure the amount of its falsity.

The two principles of contradiction and excluded middle do not stand at all upon the same plane. All motive for the criticism of our own thoughts or those of others springs from the conviction of a distinction between truth and falsity; and no philosopher has yet been found to maintain that any proposition is in precisely the same sense absolutely true and false at once. Hegel, it is true, professes to do this; but that was because he mistook the relation actually existing between his own thought and that of ordinary men. On the other hand, it has often been held, even when not professed, that concerning some matters there is no truth nor falsity. All philosophical sceptics and all those who believe in limits to human cognition, since they hold that certain propositions are never to be either accepted nor rejected, even if they say those propositions are either true or false can really mean nothing by it. But putting aside, for the present, these topics of higher logic, what concerns us now is that certain rudimentary forms of reasoning, embracing all those that the traditional logic has handed down to us, depend only upon the impossibility of a fact's being both true and false, and remain equally sound arguments, if we suppose that some things are neither true nor false.

It will be advantageous, from the very outset, to introduce certain arithmetical conceptions into logic. These notions are really somewhat extraneous to the subject; but nevertheless they will be found very valuable in giving our thoughts a definite and concrete form. Let us call the state of a proposition as being true or false (or whatever else it may be) its value. We may choose any two numbers at pleasure, to represent the values of truth and falsity, and denote these by the letters v and f respectively, so that the value of every true proposition is said to be v,

---

for more than "aleph-one" elements of the Cantorian analysis. In the consideration of boundary problems there arose the possibility of triadic and, indeed, of n-value systems in which the Principles of Excluded Middle and of Contradiction do not necessarily hold. Mention has already been made of Peirce's treasured *Logic Notebook* (MS. 339) in which Professor Fisch discovered that Peirce had formalized in his own mind by 23 February 1909 the basic conception of a three-valued logic.

and that of every false proposition is said to be f, while propositions neither true nor false, if there be any, take other values. Thus two propositions will have the same value if they are either both true or both false, notwithstanding the difference in their meaning, just as two triangles may have the same area and yet be very differently shaped. Accordingly, using the letters of the alphabet to designate different propositions, if we write $x = y$ it will not at all signify that the propositions x and y are equivalent in meaning, but only that both are either true or false at once. The equation $x = v$ or $x - v = 0$ will mean that the proposition x is true, — precisely what x written alone would mean. The equation $x = f$ or $x - f = 0$ will signify that the proposition x is false; and the same thing would be signified by writing any algebraical expression which becomes equal to v when $x = f$ and equal to f when $x = v$. The simplest such expression is $v + f - x$. This then represents the negative of the proposition x.

The simplest of all possible inferences is what is commonly called the modus ponens of hypothetical syllogism, though some logicians very properly object to its being called a syllogism at all. It is this:

> If x is true, y is true,
> But x is true;
> Hence, y is true.

Suppose, for example, that being at the theatre, I were to hear a rumour of a house being on fire near where I lived. I should say "If this be true, I had better be going." Suppose then, that on making inquiries, I were to find that the rumor was true: I should hurry home, in accordance with the inference.

Let us represent this in algebraical notation. The hypothetical proposition "If x is true, y is true," is such that if it be true and if x be true, then y is true, but if x be not true or if both x and y be true, then the truth of the hypothetical proposition depends on other circumstances. All this is perfectly represented by the equation,

$$(x - v)A = (y - v)B.$$

where A and B are quantities whose values are not determined by those of x and y alone, and which may vanish but cannot become infinite. For if $x = v$ the equation cannot hold unless $y = v$. It may not hold even then, if $B = 0$. If x does not equal v, the equation may or may not be true, according to the values of y, A, and B. From this equation together with $x = v$ we at once deduce $y = v$, thus performing the inference of the modus ponens by an arithmetical process. ...

D. [A SEARCH FOR A METHOD: FRAGMENTS] (594)

1. *Of deductive Arguments. The Simple Consequence.* By a *simple consequence*, I mean an argument from a single premise.* (* The scholastic doctors generally, and Scotus especially, made great use of this form of reasoning, to which the name *consequentia* was sometimes applied. Thus, Ralph Stroud, in one of the ablest expositions of the doctrine, says, "Consequentia est illatio consequentis ex antecedente," and even Scotus sometimes falls into that way of speaking. But more properly, the *consequentia* was the leading principle. Thus, Tartaretus [In Petr. Hisp. tr. 6] says, "Consequentia est habitudo consecutionis antecedentis ad consequens." The classical definition is that of Scotus [Ad 1. Anal. Pr. qu. *x*]: "Consequentia est propositio hypothetica, composita ex antecedente et consequente, mediante conjunctione conditionali vel rationali, quae denotat quod impossibile est ipsis, silicet antecedente et consequente, simul formatis, quod antecedens sit verum et consequens falsum."

The doctrine of the *consequentia* is decidedly important in medieval logic. We find in the treatises of Albertus Magnus on the first book of the Prior Analytics, some traces of this doctrine. In Scotus it is developed at some length. He divides *consequentiae* into the *formal*, "quae tenet in omnibus terminis, stante consimili disopositione et forma terminorum," and the *material*. Of the latter, the consequence simply true is one which can be reduced to a formal consequence by the assumption of a necessary proposition. This is distinguished from the *consequentia bona ut nunc*, which to be converted into a formal consequence requires the assumption of a true but contingent proposition. Ockham [Logica aurea. Part 3 of part 3, cap *i*] makes the distinction between the simple consequence and the consequence *ut nunc*, the fundamental one. His definition is, "Consequentia simplex est quando pro nullo tempore poterit antecedens esse verum sine consequente." He draws a new distinction between consequences which hold "per medium intrinsecum" and those which hold "per medium extrinsecum." (The words *intrinsecum* and *extrinsecum* are here adjectives.) The former are such as hold in consequence of some fact relative to the objects spoken of, as "Sortes non currit; ergo, homo non currit." The latter are such as hold good by virtue of a general principle of language, as "Tantum homo est asinus; ergo, omnis asinus est homo." Ockham draws many other distinctions; but the rest are not worth notice. His rules for consequences turn on the consideration of the distribution.)

2. *Illogical Consequences.* Let us follow Scotus and the scholastics succeeding him in using the word consequence (L. *consequentia*) to denote either a necessary argument from a single premise, which may, however, be a copulative proposition* (* Among that collection of errors which still makes up the current notion of medieval scholasticism is this, that the sententiary doctors generally reasoned in syllogisms. I shall endeavor in another note to characterize their method of thought: I now speak of the form of its expression.) or, more properly, the leading principle of such an argument. In the latter case, instead of premise and conclusion, we speak of *antecedent* and *consequent*.

By an illogical consequence, I mean an incomplete argument which cannot be rendered complete, because the thinker is not prepared to insist upon the leading principle as a premise. He may believe it; but to adopt it as a premise would be to confess that it was a bald and unwarranted assumption. There are many such arguments used among ordinary people, but the most remarkable instance of the deliberate use of an illogical consequence by a great philosopher is the celebrated *Cogito, ergo sum* of Descartes. It is true that in his systematic treatise he introduces it as a syllogism; but he had avoided this in the first presentation of it, and at a later date expressly denies that it is syllogistic. To complete the argument would be repugnant to his philosophy. He has just urged the reader to doubt everything except that he doubts, upon which immediate consciousness he proposes to erect the whole fabric of philosophy. In this state of doubt, the truth of a *general* proposition cannot be admitted. It cannot, therefore, be immediately admitted that *whatever thinks exists*; or, if this can, and must be immediately admitted, the whole theory of cognition of Descartes falls to the ground. The proposition *whatever thinks exists* appears, as is, irrefragable to a person who has no distinct idea of what he means by existence; because he does not know how to go about to dispute the statement. But, in fact, it is open to much doubt. It is fully as likely that we ought to say, whatever thinks, in so far as he thinks, has not attained the entelechy of existence. But be that as it may, according to the Cartesian view, the *cogito ergo sum* is altogether prelogical. Logic has not yet any standing at the time when, seeing that I doubt, I am irresistibly led to believe I exist. I am sensible that this belief arises from my recognition of my doubting; but precisely how or why it does so, that is, on what principle, I cannot say before I have as yet admitted any principles. Besides, the conclusion that I exist is regarded by Descartes as more certain than any principle, — more certain than logic itself. Its certainty, therefore, does not rest on that of any principle. It is above all logic, — a pretension uncommon only in the mouth of a philosopher. Logic, on the other hand, can lend no countenance to such an argument. If you find the argument irresistible, you must accept the conclusion: you can no more help it than you can arrest the beating of your heart. But it is no more a rational procedure than your pulses are rational. Such mental operations are on the border between reasoning and the unconscious suggestions of associated ideas.

3. *That Deductive Reasoning is the first kind to be studied.* A deductive argument governed by a leading principle according to which if the facts stated in the premise are true, then the fact stated in the conclusion is true. This being a conditional assertion, to make meaning distinct it is requisite to define the meaning of a conditional assertion, in general. To say, then, that if A is true, B is true, means that every possible state of things in which A should be true would be a state of things in which B would be true. Thus, the necessity of the conditional proposition is merely an aspect or species of the universality of a categorical proposition.* (* Note the analogy, *necessity*: *hypothetical*: *universality*: *categorical*, contrary to the implication of Kant's table.) We refer to a universe, or range of possibility. The possible, in its fundamental sense, is that which we do not know not to be true; but in special uses of the word, we are supposed to be in special states of information. This will be considered in another essay.

Most of our reasoning is not deductive. It does not result from the consideration of a world of possibilities, or states of things we do not know not to be true, but results from the observation of the course of events in experience. In other words, it results from the estimation of probabilities. Probability is closely allied to possibility: it has been called the negative measure of our ignorance. But possibility is much the easier conception; and therefore necessary reasoning must be considered first. Besides, it will be found that every non-deductive argument refers to a necessary consequence, upon which its validity depends.

Deductive reasoning is not conterminous with necessary reasoning. On the contrary, we may reason deductively about probabilities. But necessary reasoning is all deductive, and is the simplest kind of deductive reasoning.

We pass to the consideration of arguments properly so called. And, first, we consider deductive arguments. These are arguments governed each by a leading principle according to which if the facts stated in the premise are true, then the fact stated in the conclusion is true. (This excludes arguments from the mode in which facts present themselves, inductions and arguments from signs.) To say that if the premise is true, the conclusion is true, means that every possible state of things in which a proposition in certain respects similar to the premise should be true would be a state of things in which a proposition corresponding to the conclusion would be true. Thus, a general range, or "universe," of possibility is assumed in the very idea of a consequence. This range of possibility is different in different cases.

We begin with consequences *ut nunc*. A consequence *ut nunc* is one in which the range of possibility is limited to the actual state of things. To speak of *the* actual state of things implies a great assumption, namely that there is a perfectly definite body of propositions which, if we could only find them out, are the truth, and that everything is really either true or in positive conflict with the truth. This assumption, called the principle of excluded middle, I consider utterly unwarranted, and do not believe it. Still, I hold that there is reason for thinking it to be very nearly true. But, for our present purpose, it is no matter whether it is true or not. A consequence *ut nunc* refers to some single possibility, whether this be realized or not. To say, then, that *ut nunc* if P is true, C is true, is the same as to say that either P is false or C is true. Thus, the consequence holds *ut nunc* if P is false, no matter what C may be. It also holds *ut nunc* if C is true, no matter what P may be. It fails *ut nunc* only if P is true while C is false.

Supposing it to be taken for granted that some consequences do not hold, so that there are some true propositions and some false propositions, we may define a true proposition as one such that every consequence holds *ut nunc* of which it is the conclusion, and a false proposition as such a one that every consequence holds *ut nunc* of which it is the premise.

---

4. *Consequences ut nunc.* A consequence is said to hold *ut nunc*, when the range of possibility is limited to the actual state of things, so

that whatever we do not know not to be true is true. It may be objected that this supposes we are in a state of omniscience; and that omniscience precludes reasoning. It may further be objected that the definition of a consequence *ut nunc* is self-contradictory; for a consequence has been defined as an argument, and an argument as referring to a genus of arguments, that is, to an extended field of possibility. The answer is that a consequence which was in every respect *ut nunc* would cease to be an argument, but it would be a proposition that was in a mathematical sense the limiting case of an argument; and on account of its simplicity it should be considered before the study of arguments proper. Besides, the range of possibility is often, if not usually, of a multiple character; it has, in the language of geometry, several dimensions; and a consequence may be *ut nunc* in some important respect, although general in other respects. There may be a range of possibility as to *time*, as to *place*, as to the *objects* spoken of, as to the *knowledge* of the speaker. In some one of these respects, the consequence may be limited to a single object. Every categorical proposition, that is every proposition which affirms or denies a definite predicate of a subject more or less definitely indicated, may be considered as a consequence (or more accurately as the assertion of the validity of a consequence). To say that every man is mortal is the same as to say that *if* you take a man, you will find he is mortal; or *if* you limit yourself to the attributes of a man and discover all of them, you will discover mortality. It is not necessary to inquire here whether or not the metaphysical relation of substance and attribute presents any peculiarity. It suffices that, for the purposes of ordinary reasoning, it may be treated in this way. In this point of view, every singular proposition, or assertion about a definite individual object, is a consequence *ut nunc*. Nay, we may go further, and say that even a universal proposition is a statement of the validity of a consequence which holds *ut nunc* after a certain individual or individuals have been selected. Thus, suppose I say, every man is born of a woman. This is as much as to say that, taking any man you please, after he has been chosen a woman can be found who was that man's mother. After the selections are made, the proposition is singular, and therefore is a consequence *ut nunc*. The principal object of treating consequences *ut nunc* first is that in this way we separate the study of the logical processes concerned about the selection of instances from the logical processes concerned about what is asserted of the individuals after they have been selected.

No doubt there is an assumption involved in speaking of *the* actual state of things as in the definition of a consequence *ut nunc*, namely, the assumption that reality is so determinate as to verify or falsify every possible proposition. This is called the *principle of excluded middle.* For reasons which will appear in another essay, I do not believe it is strictly true.* (* Epicurus denied it. Cicero, de Natura Deorum, I. 25. The Eleatics, on the other hand, naturally affirmed it. The great importance of the proposition was not generally recognized till very late. The Greeks confounded it with the principle of contradiction; and I do not know what authority Goelenius has for saying they distinguished it as the *affirmative* principle of contradiction. The great scholastics treated it only as a canon of contradictory opposition. But Franciscus Mayronis, in the XIVth century, erected it into the "first complex principle" of knowledge. Thenceforth, it was known as the "axioma de quolibet," from the formula *De quolibet dicitur affirmatio vel negatio,* or as the "axioma quodlibet," from *Quodlibet est vel non est.* The modern designation, "the principle of excluded middle," must have been invented by some late Leibnitzian ...) But that is nothing: logic does not inquire into the truth of premises. It is convenient, not only in a practical but in a philosophical sense, to commence with the study of arguments which assume such an absolutely determinate state of things, without ourselves asserting that such a state is quite realized.

As a *nota consequentiae* in a proposition, no other symbol, so far as I know, had been already proposed in 1870, except De Morgan's spicula, $A)B$, or $A))B$. As a matter of scientific *savoir vivre,* I should have felt myself bound to adhere to that master's precedent, had it not been requisite for my purpose to use a symbol of algebra, or unless some other very substantial and weighty reason had forbidden. I did not, I admit, occur to me to raise this objection to Mrs. Franklin's $\bar{V}$ and V, in place of De Morgan's)·(and ( ); but I am to blame for it. And when my friend Dr. Schröder adheres to $\notin$ in place of $\leqq$ and my $\prec$, between which he had a choice, he behaves just as a naturalist does who creates a new name for an old genus. Mac Coll's colon embodies a distinct idea (though one that is not in perfect harmony with the rest of the algebra of logic); yet he felt bound to apologize when he found that another symbol was already applied to that use. But the case is somewhat different when, having for good reasons already adopted the symbol $\prec$, I find in this particular section on the "algebra of the copula" that those reasons lost their force. Still, perspicuity and a regard for the reader's convenience prescribe my confining myself to a single symbol to express one and the same meaning. I accordingly adopt $\prec$, throughout.

After all the principal symbols of algebra (mostly of English invention) had long been in use, no less than 81 years after the introduction of =, Descartes, with an ugly spirit of nonconformity, chose to substitute for it the character ꝏ or ∞, as the copula of algebra. This dissymmetry of this figure, which disqualified it for that use, would well fit it to serve as the copula of logical inclusion. Perhaps I might have suggested reversing it, and writing "*A* is as small as *B*" in the form $A \propto B$, if the happy thought had not occurred to me twenty years and more too late.

From a purely algebraical point of view, such reasons might not have much weight; and perhaps even some of my own old students may think them rather trifling. But I never have treated logic from a purely algebraical point of view; and deeply significant as the relations between syllogistic and algebra are, they seem to me to throw more light on the latter than on the former. In logic, the use of algebraic symbols is all but indispensable, but after all its use is but a convenience. However, when once it has been decided to make use of an algebraic symbol in the exposition of a system of logic, it is absolutely requisite that the symbol should not be so compounded as to imply a denial of the doctrines of that system. That is my apology for not using the symbol $\leqq$.

As for the three dots $\therefore$ I first find them used as a sign of illation about a century ago in connection with the theory of proportions, where two dots and four dots are also used as symbols; but my inquiry into this little question of history has been quite hasty.

Let us in the present section confine ourselves to the study of the copula of inclusion, $\prec$, considered as the connecting sign of a consequence *ut nunc*. We shall find when we come to consider other consequences, that these virtually assert merely every one of a class of consequences *ut nunc*; so that, from that point of view, the copula does not occur except in a consequence *ut nunc*.

I proceed to define the copula, first, verbally, then analytically.

Verbal definition of the copula of inclusion. If $A \prec B$, then, and then only, from *A* follows *B*, *ut nunc*.

Analytical definition of the copula of inclusion. $A \prec B$ is true if *A* be false or if *B* be true, but is false if *A* be true while *B* is false.

I now give certain purely algebraical rules for the use of the copula, which are deducible from the analytical definition and can be supplied without reference to either definition. But if the verbal definition is to be admitted, then each of these rules can be expressed as a general consequence-formula ...

E. [ON POSSIBILITY] (145)

We certainly have the notion of objective possibility, whether there be such a thing or not. It may be defined as that mode of being which is not subject to the principle of contradiction since if it be *merely* possible that *A* is *B*, it is possible that *A* is not *B*. Necessity, on the other hand, is that mode of being which is not subject to the principle of excluded middle, since it may neither be that *A* is necessarily *B*, nor that *A* is necessarily not *B*. How can the principle of contradiction fail to apply to anything? By something being held in reverse and not expressed. A man is rich; and yet a man is poor. That may be because you do not say what man. Possibility is that mode of being in which something is held in reserve, so that actuality is not attained. How can the principle of excluded middle fail to apply? By the function of determination being given over to the person addressed. It is not true that *any* man you please is rich nor is it true that any man you please is poor. That is because you have surrendered to another person your right to say what you are talking about. Necessity is that mode of being the determination of which does not lie with the subject of being.

Another thing which the Gamma part of existential graphs, or some subsequent part, must enable us to express is *hypostatic abstraction. Abstraction* names two wholly different operations. One of them consists in supposing some feature of the fact to be absent, or at least leaving it out of account. I call that *prescissive abstraction.* The other changes 'This man is shy' to 'This man is affected with shyness.' It may be called clipping the wings of words provided we call those words in a sentence which show us upon what our attention is to rest because something is about to said of that, the ἔπεα ἀπτερόεντα and those words which say something of those subjects the ἔπεα πτερόεντα. In more prosaic language it changes a predicate into a subject (extending the term subject beyond the subject nominative to the subject accusative and subject dative, — in short, to what are called the direct and indirect *objects* of the verb). "The rose smells very sweetly" is by hypostatic abstraction converted into "The rose possesses a delightful perfume." So "Cain killed Abel" is changed to "Cain caused the death of Abel." Perfume and death are *hypostatical abstractions*. They denote *entia rationis*, whatever that may mean. They are predicates; namely, qualities, dyadic relations, triadic relations, etc. Logically simple predicates are easily shown to be necessarily of one of those three kinds. For a tetradic relation between *A*, *B*, *C*, and *D*, may be regarded as a triadic relation between *A*, *B*, and a combination of these, which combination is in a triadic relation to *C* and *D*. It is impossible to analyze a triadic relation in this way, because *combination* is essentially a triadic relation or combination of such relations. The difficulty of representing a hypostatic abstraction in existential graphs (which I trust may be conquered eventually) is that what suggests itself is to distinguish individuals regarded as determinate in every respect, so that the principle of excluded middle applies to them, by (for example) using a different colored ink say red from that say blue used in scribing predicates such as 'is wise.' But then the dot which denotes 'something,' should be red, while the continuous line which has a dot at every part of it should be blue. Perhaps the remedy would be to make this line purple.

Solomon ——— is wise

But when the operation of hypostatic abstraction is performed, the proposition takes the form 'Solomon *possesses wisdom*' or 'Solomon is possessor of wisdom.' I must interpose a special dyadic relative between two parts of the line, as well as changing the color of is wise.